12542

~~D. Eisenhut~~
~~544-68-7279~~

SANDY SMITH
929-4862

Genetics, Evolution, and Man

Genetics, Evolution, and Man

W. F. BODMER
OXFORD UNIVERSITY

L. L. CAVALLI-SFORZA
STANFORD UNIVERSITY

W. H. FREEMAN AND COMPANY
SAN FRANCISCO

Cover: A computer-generated map showing the approximate distribution of the blood-group gene *A* in the aboriginal populations of the world. See page 565 for key. (Map courtesy of D. E. Schreiber, IBM Research Laboratory, San Jose, California.)

Library of Congress Cataloging in Publication Data

Bodmer, Walter F 1936-

Genetics, evolution, and man.
Includes bibliographical references and index.
1. Genetics. 2. Human genetics—Social aspects.
3. Human evolution. I. Cavalli-Sforza, Luigi Luca, 1922- joint author. II. Title. [DNLM: 1. Genetics, Population. 2. Evolution. QH455 B668g]
QH430.B64 575.1 75-33990
ISBN 0-7167-0573-7

Printed in the United States of America

1 2 3 4 5 6 7 8 9

To Julia and Alba

Contents

Preface

Like every other organism, man owes his peculiarities to the inherited blueprint contained in the unit determinants of biological inheritance, the genes. Yet social factors may also go a long way toward shaping individuals, and especially their behavior. Two opposite streams of thought compete for the attention of students of human activities. One stream views man as entirely controlled by his genes. Because individuals differ in their genetic backgrounds, some hereditarians are led to the fatalistic view that your genes determine your destiny. The other stream of thought holds that differences in genetic constitution have trivial effects and that the environment molds the individual. Both of these extreme views are wrong and lead inevitably to serious errors. The correct viewpoint lies somewhere between the two extreme views. Genes do predetermine our reaction to the environment to a large extent, but the final outcome of individual development necessarily reflects both the initial genetic endowment and the conditions in which genes have acted—that is, the environment of growth and everyday life. Many aspects of the organization of our lives depend on an understanding of this principle and of the means of ascertaining which traits are sensitive to which influences. It is important to know which traits of which individuals can be affected by stimulation of education of a given sort, and to know the nature and extent of the effects that can be expected.

Physical adaptation is a part of the subject of physiology, while abnormal conditions are a part of the study of medicine. Patterns of normal and abnormal mental behavior are studied by psychologists, psychopathologists, and psychiatrists. Yet individuals and perhaps groups may be unique in their physical or psychological responses. The basic knowledge needed to explain such differences between individuals or groups is found in the study of genetics. Human genetics, a discipline younger than most others, is the key to our understanding of individual and group differences.

This book is divided into four parts. The first covers the essentials of general genetics, using examples drawn almost entirely from man. The topics covered include the nature of genes and chromosomes, present knowledge about gene action, and the basic methods of genetic analysis. This part should be readable by students with no biological or chemical background, but even within this limitation it attempts to give a view of some of the most modern developments of research on mechanisms of biological inheritance. It may seem surprising, in fact, how much general genetics can now be taught using only human material for illustration. This approach should, we hope, heighten students' interest in genetics through its relevance to human problems.

In animal and plant genetics, much progress was made through crossbreeding experiments. These techniques, of course, cannot be applied to human genetics, where comparable information must be acquired through study of whatever matings happen to have occurred in the population. This study is the subject matter of population genetics and of the second part of this book. Population genetics is essential to a full understanding of human inheritance; it also forms the background for evolutionary theory.

In the third part, we consider the common and important class of traits that have a complex mode of inheritance. We discuss many behavioral traits (including IQ) and some significant mental diseases. These traits have been the focus of much research and public attention recently. As in the discussion of population genetics, our approach here is less quantitative than that used in most textbooks. We have tried to avoid mathematical formulas almost entirely in order to make the book accessible to the student with little interest or ability in mathematics. We have collected in the appendix some mathematical proofs and results of more general use for those who are interested. We hope that students who prefer a more deductive approach will avoid frustration through use of the appendix.

The fourth part of the book concentrates on those aspects of human genetics and evolution that are most relevant to social problems. Human biological evolution, the formation of races, and some major contributions of cultural change to the present state of man are delineated. The discussion focuses on recent human evolution with very little attention paid to primate evolution or to the earliest human evolution; these topics are well covered in several good short books designated for courses in physical anthropology—books that are quite complementary to this book. In the final part we also consider various aspects of medical genetics. We have concentrated particularly on those aspects that have come into prominent attention because of recent discoveries and applications in the field of prevention and treatment of genetic disease. This field of research will undoubtedly move quickly during the coming decades, for it responds to widely felt needs of individuals and of societies. On the other hand, its full social impact will be achieved only if the relevant concepts are widely disseminated; they should become known to the general public, extending well beyond the limits of the medical profession. There is an urgent need for education in this field. In addition, much public discussion has recently centered on social aspects of genetics, eugenics, and related problems. Ethical and legal problems have emerged, and await solution. We have tried to give the student a general background for the understanding of all these problems and issues.

Today, the emphasis in higher education is increasingly toward the human sciences. "Know thyself" begins with the biology of man, and human genetics is a critical part of this study.

31 October 1975

W. F. Bodmer
L. L. Cavalli-Sforza

Acknowledgements

We are indebted to many persons who have read parts of the manuscript or have been helpful in other ways: Albert Ammerman, Martin Bobrow, Howard Cann, Ian Craig, Marc Feldman, Glynn Isaac, Joshua Lederberg, Mark Noble, Marilyn Parsons, David Roberts, Glenys Thompson, Diane Wagener. Mark Noble in particular has given substantial help in adding illustrations to Part I of the book. Many colleagues have supplied original photographs, and their generous contribution is acknowledged in the figure legends. We are also indebted to other persons, and are especially grateful to the manuscript editor, Larry McCombs, and to Margaret Muller, who directed the art work, for their untiring and valuable collaboration.

Gene frequency maps—made by computer—that appear in the text were prepared thanks to the collaboration of D. E. Schreiber (of IBM Research Laboratory at San Jose, California) and also of R. Matessi, A. Piazza, and L. Sgaramella-Zonta. The procedure for map construction is still in an experimental stage; it was thought appropriate to include the maps even if the final product may differ in details from those available at present.

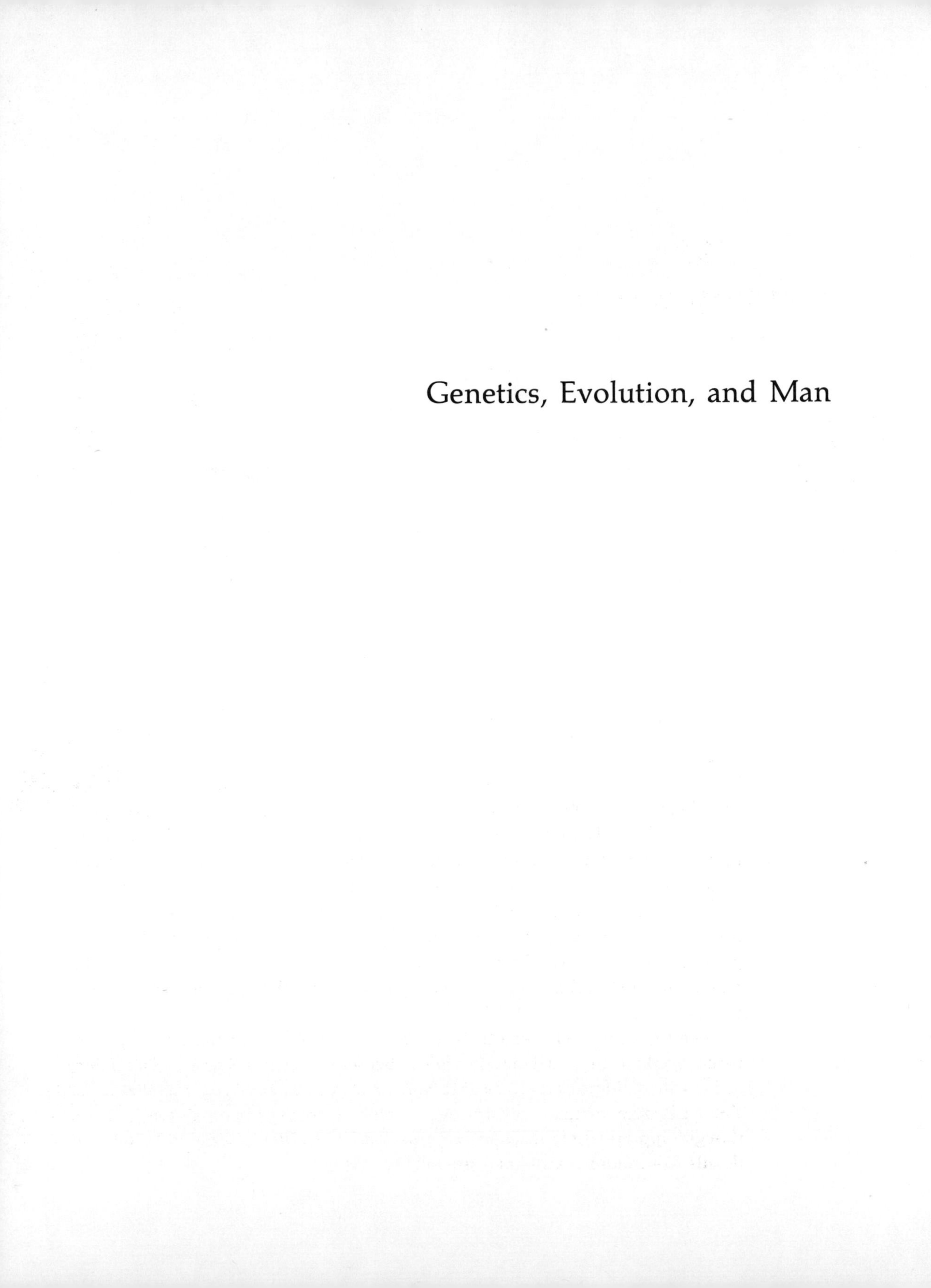

Genetics, Evolution, and Man

PART I

MECHANISMS OF INHERITANCE

The first chapter serves as a general introduction to the topics of the book and the remainder of this part reviews basic genetic principles. The basic laws of inheritance were worked out by Gregor Mendel through careful study of the characteristics of parents and offspring. Microscopic study of cell division led to the theory that inheritance is controlled by genes located on the chromosomes. Recent developments have made it possible to construct chromosome maps, sometimes with considerable detail. Also in recent years, studies by geneticists and molecular biologists have revealed many details of the chemical mechanisms underlying heredity. Genetic information is stored in the nucleotide sequences of the DNA molecules that make up the core of the chromosomes; RNA plays a major role in the synthesis of proteins whose nature is determined by the DNA sequences; the proteins, acting as enzymes and antibodies and structural components, in turn determine the phenotypic traits of the individual.

Abnormal cell division can lead to chromosomal aberrations (chromosome breakage, loss of chromosomes, presence of extra chromosomes). Chemical events can produce changes in DNA sequences (mutations). These disturbances of the normal genetic process can produce abnormal phenotypes. They are the sources of genetic disease and of evolutionary change. We conclude this part with a look at some of the genetic information obtained through studies in cell biology.

1

Genetics and Man

This chapter summarizes in simple terms the problems considered in this book and the contents of its four parts.

Unless they are identical twins, any two individuals have some difference in appearance. But how much of this difference is genetic? The outward features—the characters by which we usually tell people apart—generally are not inherited in simple patterns. But there do exist a number of simply inherited traits, such as the blood groups, for which common variants are found. These simple genetically determined variations do seem to parallel the apparent differences between people.

One of the most striking results of genetic studies over the last decade is the discovery of just how much genetic variability exists in natural populations, including the human population.

These genetic differences have many implications for people and for society. They range from rare genetic diseases to common differences, such as the ABO blood types, where no one type is more "normal" than another. Genetic factors influence behavioral attributes, susceptibility to disease—in fact, virtually all human characteristics, however they are defined. The main aims of this book are (1) to give the basic genetic background needed for understanding the inheritance of these genetic differences, (2) to analyze the principles of population genetics

1865 analyzed segregation patterns of inherited trait – independent assortment

GREGOR MENDEL

Mendel

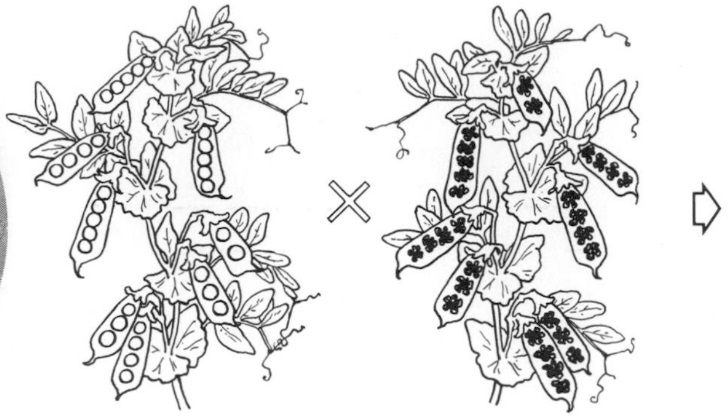

(1890's ?) hereditary units are on chromosomes + gametes have ½ #

Weismann

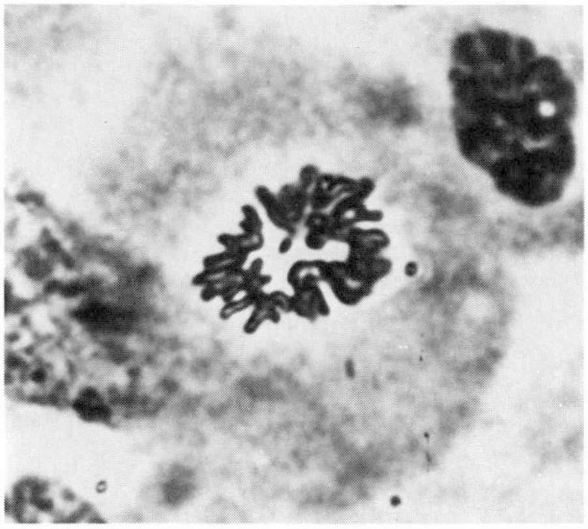

1869 isolated nucleic acids from pus cells

Miescher

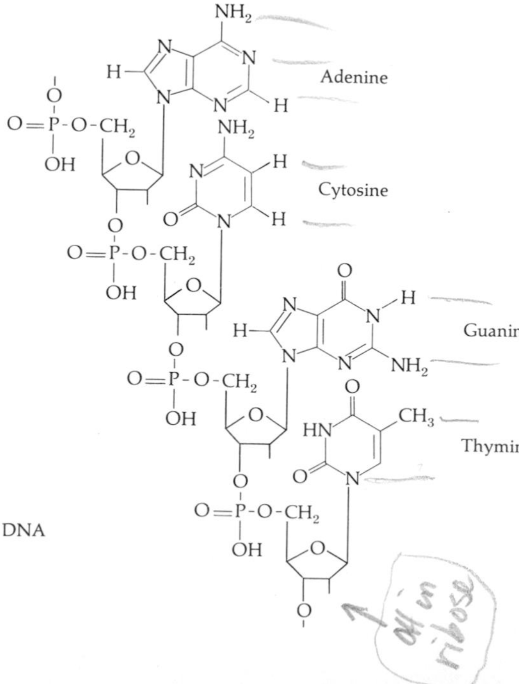

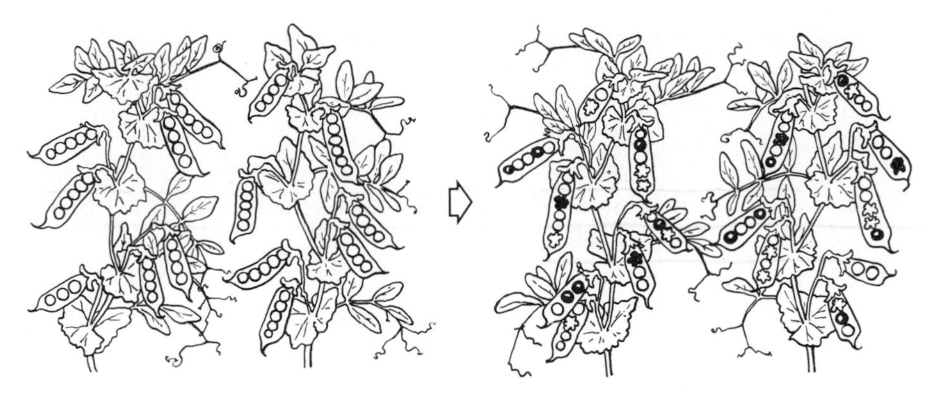

Figure 1.1
Historical figures in genetics. Gregor Mendel (*top*) was the first to analyze segregation patterns of an inherited trait. Using seed shape and albumen color as his chief experimental characteristics, Mendel observed that some traits behave as though under the control of discrete, independently assorting units. August Weismann (*center*) adopted the idea of Wilhelm Roux that hereditary units are linearly arranged along the chromosomal thread, and then went on to propose that the original number of hereditary units is halved in the germ cells of sexually reproducing organisms. Weismann put forth his theory while Mendel's work was still unknown. The nucleic acids were identified within four years after Mendel's (then unknown) formulation of the gene concept. Friedrich Miescher (*bottom*) began biochemical studies of the nucleus by examining cells (such as pus cells and salmon sperm) where the nucleus represents a large fraction of the total cell mass. Miescher found a hitherto-unknown, phosphorus-rich, acid substance that he named "nuclein," soon rechristened "nucleic acid." (Portrait of Weismann courtesy of G. Montalenti.)

that explain the origin, maintenance, and evolution of genetic differences within and between populations, and (3) to assess some of the social impacts of modern knowledge of genetics, especially in relation to medical problems. The aim of this chapter is to provide a brief survey of these topics as a guide to the rest of our book.

The first step in a discussion of genetic differences must deal with the mechanisms of inheritance and their underlying molecular basis.

This topic, therefore, forms the main theme for the first part of our book. There are three main avenues in the development of genetics, each of which has progressed somewhat independently of the other two (Fig. 1.1). The first avenue originates with Gregor Mendel and his breeding experiments with peas, described in his famous paper of 1865. Using simple character differences, he worked out the statistical rules that determine the patterns of inheritance in most organisms. These rules laid the foundations for much of our present knowledge of genetics.

The second avenue began just before the turn of the century with the description of the behavior of chromosomes in cell division. The chromosomes (literally "colored bodies") can be made visible by appropriate stains only at certain times of the cell's life cycle, shortly before it divides. August Weismann and others recognized that the regularity of chromosome behavior in cell division made them likely candidates as carriers of the hereditary material. However, it was not until 1903 that W. S. Sutton and Theodor Boveri pointed out the parallels between Mendel's statistical laws and chromosome behavior, so uniting these first two avenues in the development of genetics.

The third major avenue in the development of genetics was biochemical. It started with Friedrich Miescher's description of nucleic acid in 1869, only four years after the publication of Mendel's paper. Yet it was not until 1953, 84 years later, that James Watson and Francis Crick elucidated the structure of DNA (deoxyribonucleic acid), one of the components of Miescher's original nucleic acid, and that this material was finally accepted as the chemical substance of heredity. In the resulting era of molecular biology, we have gained a growing chemical understanding of the nature of heredity and, more generally, of almost all biological processes.

The study of genetics can start along any one of these three avenues. We shall begin with chromosomes and their behavior in cell division, proceeding from there to Mendel's laws and their extensions, and then to the chemical basis of heredity.

The simply inherited differences that form the basis for most studies of genetic variation in human populations are identified chiefly by laboratory tests on blood cells (Fig. 1.2) or serum.

Blood is used for these studies because it can be obtained readily from almost any healthy person. Perhaps the best known of the differences are the ABO blood types, which are important in blood transfusion for matching potential blood donors and their recipients. More than 30 sets of genetic differences, similar to the ABO blood types, are now known; these alone give rise to something like 1,000,000 different genetic types. Yet this is only a small proportion of the total of genetic differences that must exist between people. Studies along these lines really do seem to confirm that each of us differs *genetically* from everyone else. The population distribution of genetic variants can be effectively described by simple models. These models form the basis for our understanding of population genetics, and they are the main theme for the second part of our book.

Some 50 years ago American physician J. B. Herrick, working in Chicago, described a disease of the blood that he found to occur only in the American black population.

Most people, especially in the United States, are now quite familiar with sickle-cell anemia and its associated sickle-cell trait. The anemia has become a notorious

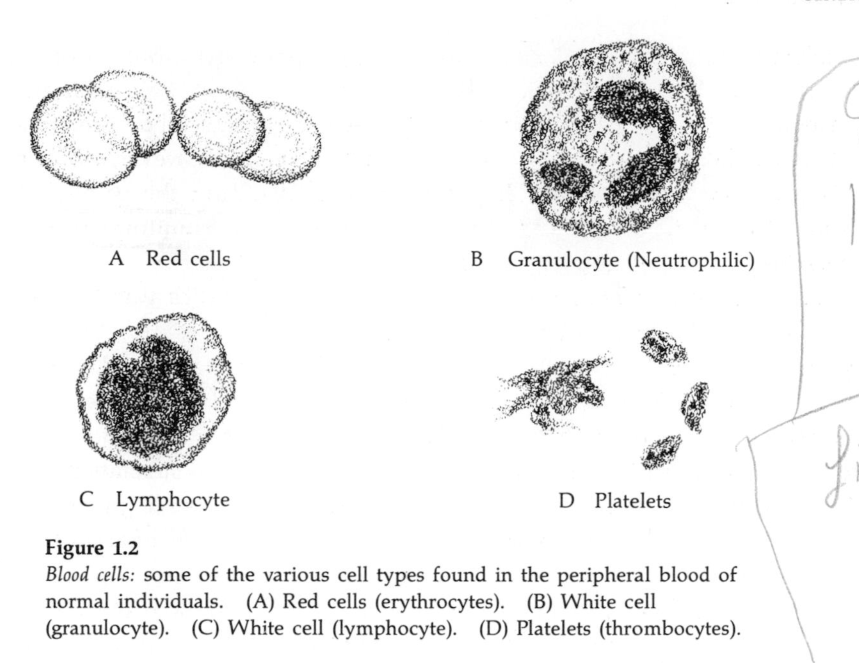

Figure 1.2
Blood cells: some of the various cell types found in the peripheral blood of normal individuals. (A) Red cells (erythrocytes). (B) White cell (granulocyte). (C) White cell (lymphocyte). (D) Platelets (thrombocytes).

example of a genetic disease. The fact that the anemia cases seemed to cluster in certain families suggested that it is genetically determined, but this was not established until many years later, in 1949, by J. V. Neel. In that same year, Linus Pauling and his colleagues showed that the disease is associated with the presence of an abnormality of the blood protein hemoglobin. Hemoglobin is the major constituent of red blood cells and is the substance responsible for their color. It is also the main vehicle for transporting oxygen in the blood.

The disease is called "sickle-cell" anemia because the red cells of affected individuals are easily deformed to a sickle shape, in contrast to the neat disk of normal red blood cells (Fig. 1.3). Parents of individuals with the anemia are usually normal from a doctor's point of view, but they have the "sickle-cell trait." Their red cells sickle to some extent and part of their hemoglobin is abnormal. If two individuals with the sickle-cell trait should mate, then the laws of inheritance predict that, on average, one-quarter of their children will have the severe anemia.

The occurrence of the sickle-cell anemia and trait in black but hardly ever in white Americans suggests that the disease and trait were brought to America by the African ancestors of present-day blacks. African populations do have particularly high frequencies, while the trait is virtually unknown in most other parts of the world. The frequency of the sickle-cell trait is as high as 30 percent in some parts of Africa, while that of the severe anemia can range over 2 percent at birth. (At least until recently, not many individuals survived for very long if they had the severe anemia.) A single genetic disease of this gravity occurring in one out of every 50 newborns certainly poses a severe public health problem. Though the

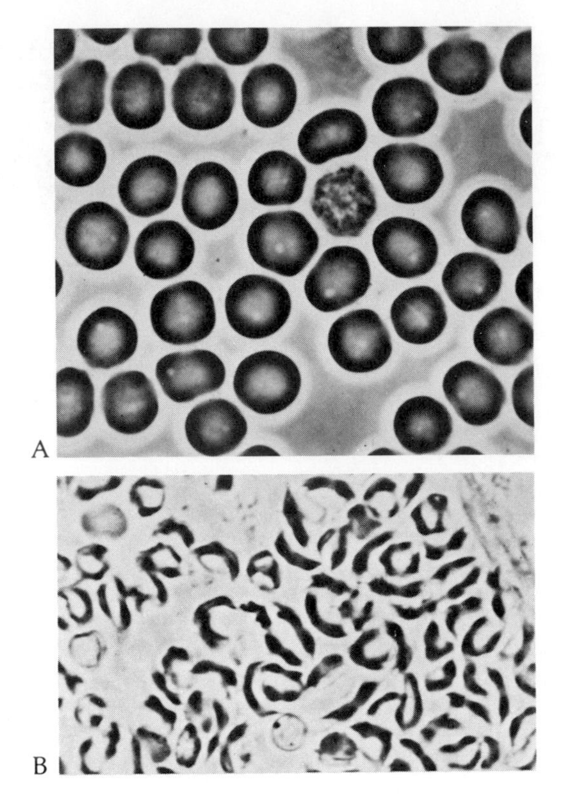

Figure 1.3
Normal and sickle red blood cells: phase-contrast photomicrographs of deoxygenated blood (A) from normal individuals and (B) from persons with sickle-cell trait or sickle-cell anemia. Persons with sickle-cell trait normally do not experience any untoward effects, and even have some advantage in malarial environments. Persons with sickle-cell anemia, however, have a low chance of survival. (Courtesy of A. C. Allison.)

frequency of the disease in American blacks is lower, it still poses an important health problem. The racial connotations of the disease have, unfortunately, tended to make the political problems of the disease almost worse than the medical ones.

Why should such a severe genetic disease have such a specific and restricted distribution?

Why doesn't natural selection cause the disease to be eliminated from the population? After all, until quite recently, people with the anemia hardly survived at all, let alone lived to have children. The clue to the answers to these questions was given by J. B. S. Haldane in 1949, the same year in which the inheritance and chemistry of the trait were first described. He noted that the distribution of some anemias of which sickle-cell anemia is one, seemed to correspond closely to the distribution of malaria, a severe disease that is particularly prevalent in many parts of Africa. Haldane therefore suggested that the anemia might occur with a high frequency because individuals with the associated sickle-cell trait were more resistant to malaria, and so in a malarial environment had a survival advantage over normal nonsicklers. This hypothesis has now been substantially confirmed for the sickle-cell trait, providing a most satisfying explanation for its geographic and racial distribution. The story of sickle-cell anemia and its associated trait is a model of the way we can explain the frequency and geographical distribution of a genetic trait.

Many other inherited abnormalities of hemoglobin are now known, but nearly all of them are extremely rare, much rarer than the sickle-cell hemoglobin.

Some of these rare variants have been found in just one family, and only some of them are associated with diseases. Many genetically determined rare diseases, analogous to those associated with abnormal hemoglobins, are now known. The first clearcut example of an inherited disease was described by the English physician Archibald Garrod in 1901, just one year after the rediscovery of Mendel's work. Garrod described 11 cases of a disease called alkaptonuria, which makes the urine turn black on exposure to air, and produces symptoms of arthritis. He noted that three of the cases had parents who were related, a fact that gives a most important clue to the mode of inheritance of the disease. This connection depends on an understanding of the consequences of inbreeding, one of the topics discussed in the population genetics section. Most significantly, Garrod realized that alkaptonuria is probably due to a simple biochemical malfunctioning of a metabolic process, and so he related inherited differences to biochemical defects. Many years later, in 1958, George Beadle and E. L. Tatum won the Nobel prize for working out in the mold *Neurospora* the details of this relationship which, unknown to them, had been predicted so much earlier by Garrod on the basis of his observations of human patients. Garrod called these biochemical genetic diseases "inborn errors of metabolism."

Phenylketonuria, often better known simply as PKU, is another example of an inborn error of metabolism. Its special significance lies in the fact that the mental retardation associated with PKU can be effectively prevented by an appropriate diet starting from birth. For this treatment to be useful, however, PKU individuals must be identified at birth. This can now be done using simple biochemical tests. Nowadays in the United States and the United Kingdom, for example, essentially all newborn babies are screened for PKU. Not many of the inborn errors of metabolism can yet be treated in such a comparatively simple way, but fortunately, these errors are rare diseases. Why are they rare? What is their overall incidence? What forces determine their frequencies in the population, and will

Figure 1.4
Differences between populations. A Caucasian anthropologist (Colin Turnbull) and an African Pygmy (Makubasu, from Epulu, Ituri forest, Zaire). There are considerable differences in average height between populations, but there are also considerable differences among individuals of the same population (see Fig. 1.5). Height is in part genetically determined, but also is strongly affected by environmental factors. Attributes determined strictly by heredity alone also vary between populations. For example, ABO blood types show some differences between Caucasians and Pygmies. Among Pygmies, individuals of type B are somewhat more frequent and those of type A somewhat less frequent than among Caucasians. Other genetic attributes may show greater or smaller differences than those found in ABO blood types.

these frequencies change as a consequence of medical treatment? These are other questions that population genetics tries to answer.

The differences we have been talking about so far are those that are found between individuals. Are there also genetic differences between population groups or races? Can race be properly defined in biologically meaningful terms? The clue to the answer to these questions lies in the observation that genetic attributes that are simply determined, like the ABO blood types, vary in frequency from one population to another (Fig. 1.4). Thus, a population or race can, to some extent, be characterized by the frequency with which different genetic attributes are found in it. In this average statistical sense, populations do differ genetically, though the range of genetic variation within any human population is generally far greater than the average difference between any two populations.

What are the factors that give rise to the high levels of genetic variability found in man, and in nearly all species? Why are there differences between populations, and what factors determine the extent of such differences? These are evolutionary questions that take us back to the sickle-cell trait and its advantage in a malarial environment. This is one small example of the kind of change known to form part of the evolutionary process that has produced the variety of types (or species) of biological organisms in the world today. Biochemical studies based on genetic principles have allowed us to pose evolutionary questions with much more precision than was previously possible. We can now estimate the average number of genetic differences separating two species, and hence can estimate rates of evolution at the genetic level. These estimated rates can be compared with the predictions made from population genetic models.

To what extent are there genetic differences for behavioral traits such as intelligence, or for apparently simple quantitative characters such as height or weight?

These are characters that must be measured on a continuous scale in which there are no obvious gaps. Though the inheritance of some rare extremes, such as dwarfism, may be simply determined, there are in general no obvious ways of subdividing people into different categories with respect to height, and then studying the patterns of inheritance associated with these categories (Fig. 1.5). In any case, all sorts of environmental factors such as diet and home background must surely have some effect on height, weight, and intelligence. How, then, can we say anything about the contribution of heredity to such characters? The answer comes from a statistical approach to studying the similarities between relatives as a function of how closely they are related; this is the main subject for the third part of our book.

The comparison between identical and nonidentical twins is one of the simplest and most widely used approaches for studying the inheritance of quantitative characters. Identical twins (Fig. 1.6) are derived from a single fertilized egg, and so are genetically identical. Any differences between them must, therefore, be due to the environment. Nonidentical twins are just like any brothers and sisters, but

Figure 1.5
Differences in height in the same population: heights of conscripts over 60 years ago. (From A. Blakeslee, *Journal of Heredity,* vol. 5, 1914.)

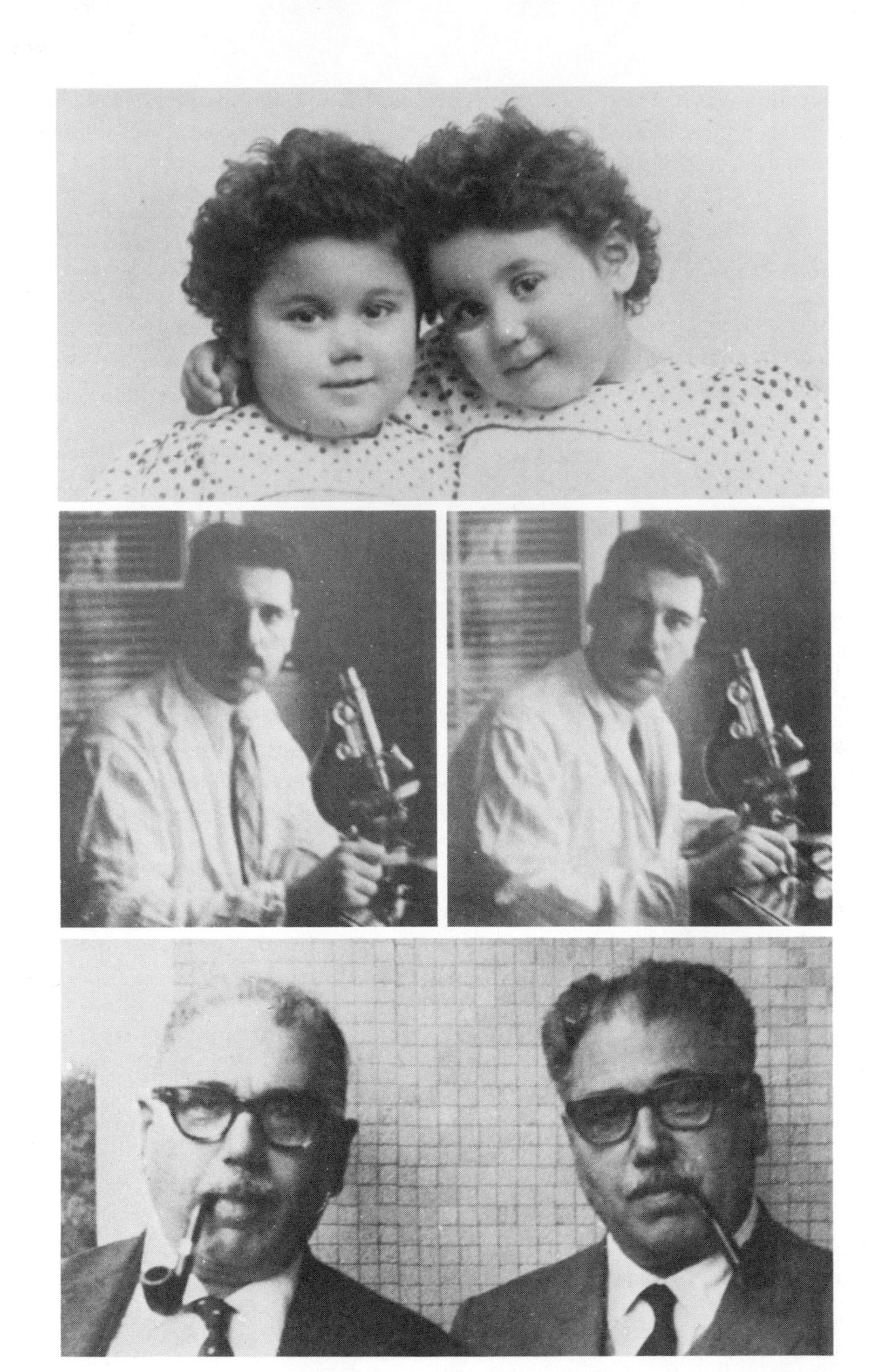

happen to be born at the same time. The extent to which nonidentical twins differ from one another more than identical twins is thus one relatively simple way of looking for genetic factors influencing characters like height, or intelligence as measured by an IQ test. Studies such as these indicate a substantial genetic contribution to the determination of IQ, for example. However, in man the similarity between close relatives such as brothers and sisters due to biological inheritance is always very difficult to separate from similarity due to sharing the same home environment. It does seem that genetic differences, like those that can be studied in the blood, also exist for other more complex attributes (Fig. 1.7). However, the relative contributions of genetics and environment to such traits may be hard to disentangle.

We have raised questions about the inheritance of genetic diseases, and about observed patterns of genetic variability, the factors that control them and, more generally, their evolution. These are questions we believe are very important for the study of all aspects of man, including cultural and biological factors. The transition to an agricultural society had important biological consequences and helped shape the future distribution of human races. What can be said about the much-discussed question of the contribution of genetic factors to racial differences? Approaches to curing genetic diseases by selective abortion of affected fetuses raise social and ethical questions that must be faced and answered. What will be the effect of curing genetic diseases on their incidence in the population? These and other questions form the subject of the last part of the book.

Medicine draws heavily on basic genetic knowledge for understanding human biology in general, as well as the more specific problems of genetic disease. Physical anthropologists use genetic differences as a major basis for defining population groups and their interrelationships. Social anthropologists, psychologists, and sociologists must be aware of the interrelations between biological and genetic factors on the one hand, and cultural factors on the other, in determining the societal structures and behavioral patterns they study. Our book is addressed to all of these needs for an understanding of genetics, population genetics, and the interactions between genetics and human welfare.

Figure 1.6
Identical twins. (*Top*) Bruno and Giorgio Schreiber at about two years of age. Even the adult twins cannot tell themselves apart in this childhood photo. (*Center*) As adults, the twins continued to look very much alike. (*Bottom*) A recent photograph. Bruno (left) and Giorgio (right) are now professors of zoology at the Universities of Parma (Italy) and Belo Horizonte (Brazil), respectively. (Courtesy of B. Schreiber.)

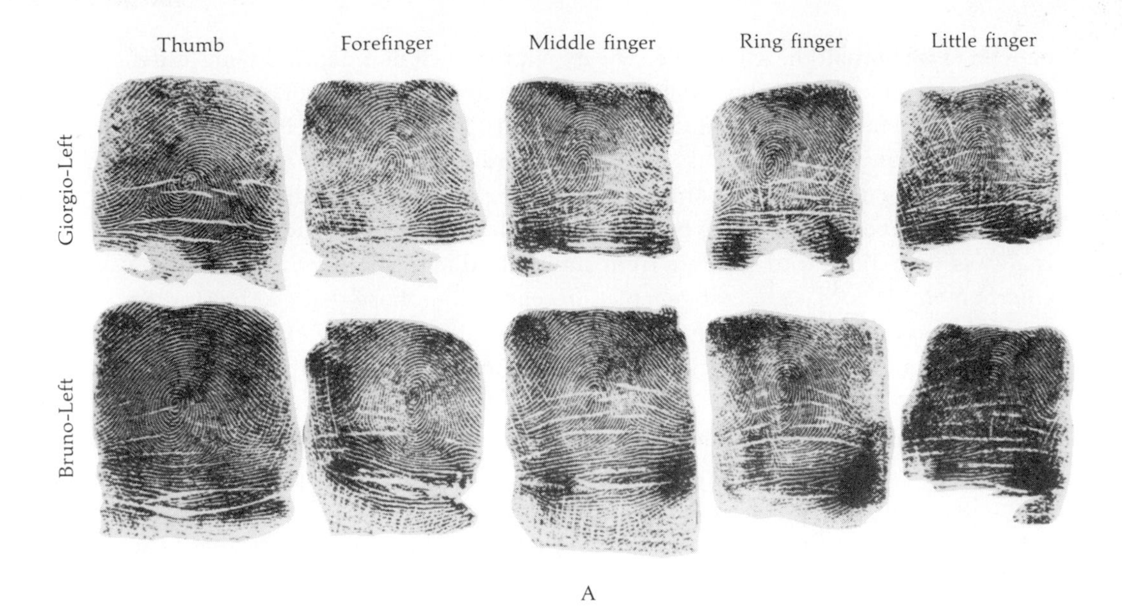

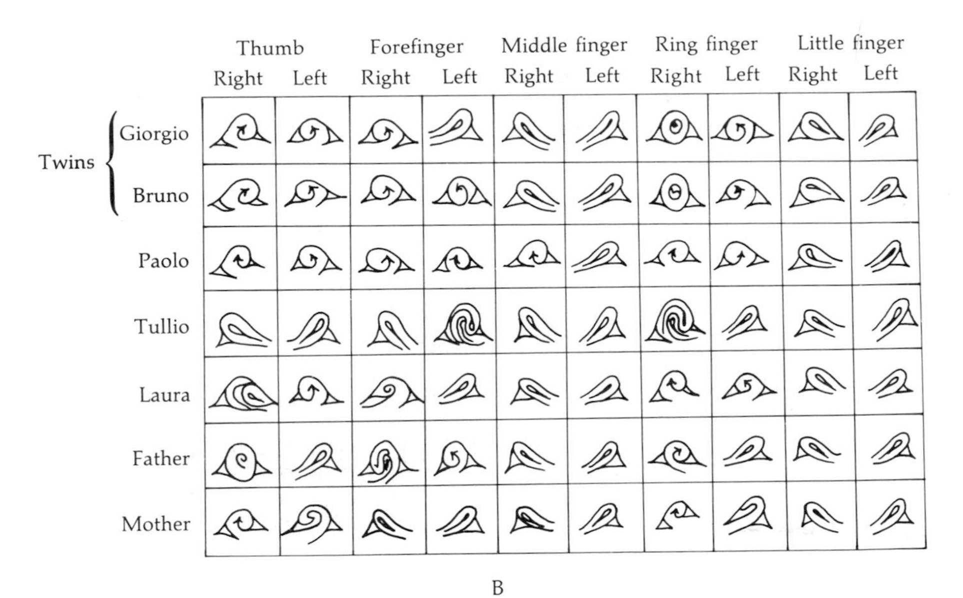

Figure 1.7
Fingerprints of identical twins. (A) Fingerprints of the left hands of the twins shown in Figure 1.6. (B) A schematic illustration of the major features of fingerprints from both hands of the twins, three of their sibs, and their parents. Note that, although the fingerprints of the twins are very similar, there is a clearcut discrepancy for the left forefinger—for which one twin resembles the father and the other resembles the mother. (From G. Schreiber, *Journal of Heredity*, vol. 9, pp. 403–406, 1930.)

References: Chapter 1

TEXTBOOKS ON GENERAL GENETICS

Goodenough, U., and R. Levine
1974. *Genetics.* New York: Holt, Rinehart and Winston.
Herskowitz, I. H.
1967. *Basic Principles of Molecular Genetics.* Boston: Little, Brown.
Srb, A. M., R. D. Owen, and R. S. Edgar
1965. *General Genetics,* 2nd ed. San Francisco: W. H. Freeman and Company.
Strickberger, M. W.
1968. *Genetics.* New York: Macmillan.

TEXTBOOKS ON HUMAN GENETICS

Cavalli-Sforza, L. L., and W. F. Bodmer
1971. *The Genetics of Human Populations.* San Francisco: W. H. Freeman and Company.
Harris, H.
1975. *Human Biochemical Genetics.* 2nd ed. New York: American Elsevier.
Levitan, M., and A. Montagu
1971. *Textbook of Human Genetics.* New York: Oxford University Press.
Stern, C.
1973. *Principles of Human Genetics,* 3rd ed. San Francisco: W. H. Freeman and Company.

COLLECTIONS OF IMPORTANT ORIGINAL ARTICLES

Boyer, S. H.
1963. (Ed.) *Papers on Human Genetics.* Englewood Cliffs, N.J.: Prentice-Hall.
Peters, J. A.
1959. (Ed.) *Classic Papers in Genetics.* Englewood Cliffs, N.J.: Prentice-Hall.

2

Cells and Chromosomes—The Laws of Inheritance

Living organisms are formed of cells. Cells have nuclei that contain chromosomes. Genes are chromosome segments to which specific functions can be assigned. Studying the behavior of chromosomes (in the reproduction of cells and of organisms) leads to an understanding of the laws of inheritance—that is, of the mode of transmission of inherited traits from parents to offspring.

2.1 Cell Division: Mitosis

The basic, self-sustaining unit of any living organism is the cell (Fig. 2.1A). This word was used by Robert Hooke in 1665 to describe structures he saw through one of his primitive microscopes. About 175 years later, Theodor Schwann and others formulated the cell theory of living matter, which simply states that all living organisms are complex collections of cells. The simplest organisms—such as bacteria, yeasts and amoebae—are single cells. However, they differ considerably in their shape, size, and internal organization. The most complex organisms are made up of very large numbers of cells. In the case of man, the number is estimated to be on the order of one hundred thousand billion—that is, one followed by fourteen zeros, or 10^{14}. All of these cells are derived from just one original cell, the fertilized egg.

There are, of course, many different types of cells in a human being: skin cells, nerve cells, several different types of blood cells, specialized cells in the kidney

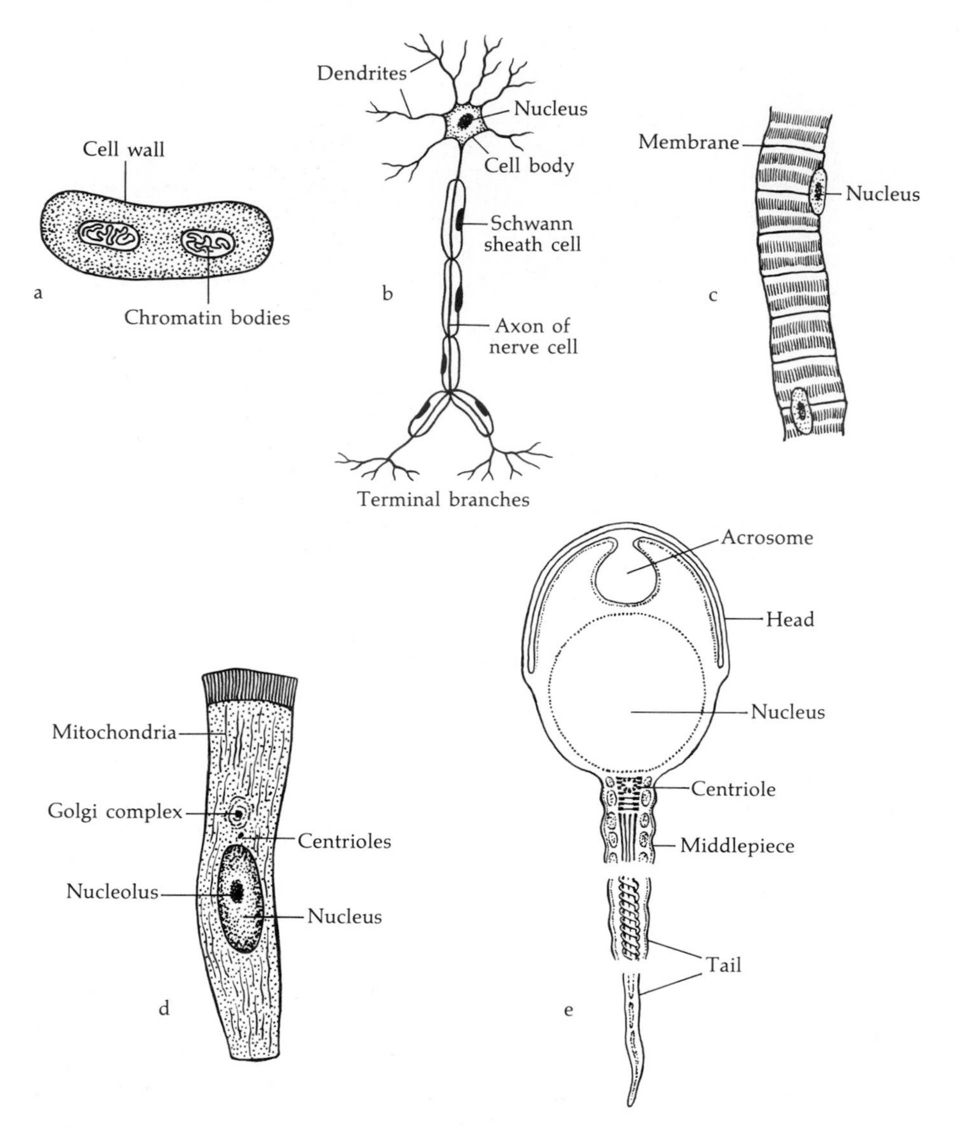

Figure 2.1A
Various types of cells: (a) bacterial cell, *Escherichia coli;* (b) nerve cell with surrounding Schwann sheath cells; (c) striated muscle fiber; (d) mammalian intestinal cell; (e) sperm cell. (Parts a through d after Loewy and Siekevitz, *Cell Structure and Function,* Holt, Rinehart and Winston, 1963; part e after Schultz-Larsen.)

and liver, to name just a few. Every different organ, in fact, seems to have its own characteristic types of cells, making their own special products, such as the red blood cell's hemoglobin. The fact that there are different cell types—whether among the different cells of higher organisms such as man or among the different single-celled organisms—means that cells when they divide must have the ability

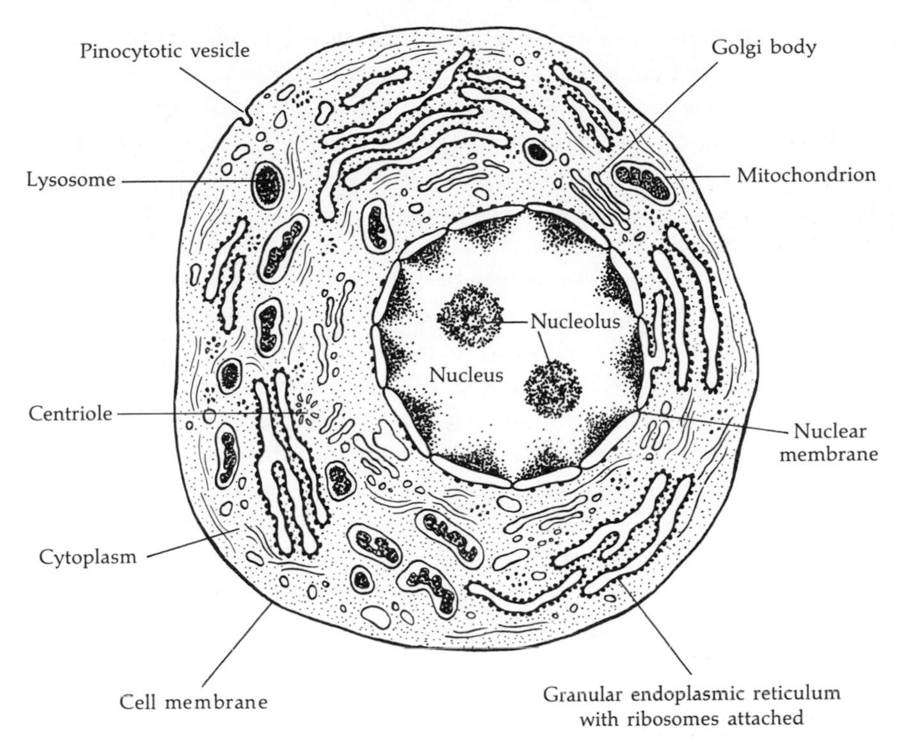

Figure 2.1B
Generalized diagram of a cell and its components. Some functions of the various components are as follows. Pinocytic vesicles: fluid ingestion. Lysosome: digestion of macromolecules. Mitochondria: energy production. Endoplasmic reticulum: protein synthesis. Centrioles: form poles for the division spindle. Golgi body: concentration and secretion of macromolecules. Nucleolus: synthesis of RNA. Nucleus: contains genetic material.

to reproduce their own special characteristics. This ability to reproduce like from like is, of course, the essence of living matter by which life itself must be defined. The study of inheritance is the subject of genetics. It concerns itself with the nature of the reproductive process and its consequences at all levels, from that of the single cell through that of the whole organism and of the groups of organisms which make up a population, to that of the populations which form a species.

The various types of cells of higher organisms, including those of man, differ in some respects but share many common features that reflect the same underlying biochemical and reproductive processes.

There is a well-defined region called the *nucleus* in the center of most cells (see Fig. 2.1B). The nucleus has a membrane around it during most of the lifetime of the cell. The region between this "nuclear" membrane and the cell's outer membrane is called the *cytoplasm.* Oskar Hertwig, in the second half of the nineteenth century, was the first to draw attention to the possible role of the nucleus in heredity. He pointed out that the sperm and egg are the sole contributors to the formation of a new individual, and therefore that they must contain all of the inherited information that is passed on from parent to offspring. Because the sperm is mostly nucleus, Hertwig argued that at least the genetic contribution of the male parent must reside in the nucleus.

For most of the time between one cell division and the next, there are few morphologically distinctive elements to be seen within the nucleus. Shortly before cell division, however, there appear in the nucleus a number of clearly visible, elongated structures called *chromosomes* (literally "colored bodies," because of their special staining properties). The number of chromosomes in the descendants of a given cell remains constant for nearly all types of cells. Thus nearly all the different types of cells within a given individual have the same number of chromosomes. These features of the chromosomes were noted by Weismann and others who began to study their behavior during cell division, late in the 19th century. Coupling this information with the fact that the chromosomes appear to be the major constituents of the nucleus, the investigators concluded that the chromosomes are the actual vehicles of inheritance. Each daughter cell must receive a complete and identical copy of the genetic information from the parent cell. Therefore, the chromosomes themselves must be duplicated at some time before cell division occurs, and during the division the two sets of chromosomes must be separated properly to end up with one complete set in each daughter cell.

Chromosomes become visible when the cell reproduces during the process called mitosis.

Mitosis (Greek, "formation of threads") is the name given to the period of a normal cell's lifetime when it divides. During mitosis, the chromosomes are clearly visible, having already duplicated, and their passage to the daughter cells can be traced (Fig. 2.2). Mitosis is conventionally divided into four major stages called prophase, metaphase, anaphase, and telophase. During the remainder of a cell's life cycle it is said to be in "interphase"; during this period the chromosomes are duplicated. The four stages of mitosis are defined by the state of the chromosomes.

Prophase is the time when the chromosomes first become visible as distinct entities. At first, they look like long, thin, somewhat knobbly, intertwined threads. Gradually they contract and become shorter and thicker, until they can be readily distinguished from one another. At this stage, they appear as relatively thick duplicate threads, called *chromatids*, held together at a region present in every chromosome and known as the *centromere*. This region is important at a later stage of mitosis for chromosome movement.

During *metaphase* the nuclear membrane, which has thus far remained intact, disappears. The chromosomes, each with its two chromatids, have condensed even more and are attached at their centromeres to fibers, which in turn are attached at either end to structures known as *centrioles,* located at diametrically opposite *poles* of the cell. The chromosomes align themselves with their centromeres lying in an equatorial plane between the centrioles; this plane is called the *metaphase plate.* The bundle of fibers that holds the chromosomes in their positions between the centrioles is called the *spindle*—hence the name spindle fibers. Metaphase is generally the best stage of mitosis at which to observe chromosomes, because they are then in their most contracted state and well distributed on the metaphase plate in the middle of the cell.

During *anaphase,* the duplicated chromosomes separate at their centromeres and begin to move toward the centrioles at opposite poles of the cell. They behave as if they were being pulled to the centrioles by the spindle fibers attached to their separate centromeres. This stage ensures that one copy of each chromosome goes to each of the two poles of the cell, later destined to become the nuclei of the two daughter cells formed by division of the single cell.

Telophase occurs after the chromosomes have reached their destinations at the poles, as nuclear envelopes form around each of the two resulting identical sets of chromosomes, thus creating the two new nuclei. Cell division then proceeds by an invagination (folding in) of the outer cell membrane at the site of the earlier metaphase plate. The chromosomes again become diffuse. Cell division is completed, yielding two daughter cells, each with a chromosome set that is identical to the parental one.

The behavior of a typical chromosome during these stages is illustrated schematically in Figure 2.2A; photographs showing a cell at various stages of mitosis are given in Figure 2.2B.

2.2 Gametes and Meiosis

We have just seen that mitosis is characterized by a regular, *equal* partitioning of the duplicated chromosomes between daughter cells in the course of normal cell division. Because the chromosomes are the carriers of the genetic information, this is the process that ensures that individual cells reproduce their like. What, however, is the fate of the chromosomes in the course of the sexual reproduction of the whole organism?

Sexual reproduction involves two important processes: fertilization and reduction, or meiosis.

The first clues to the transmission of the chromosomes between sexual generations of higher organisms came from the discovery that fertilization is the fusion of a male sperm cell with a female egg cell to form the primordial cell (*zygote*) of

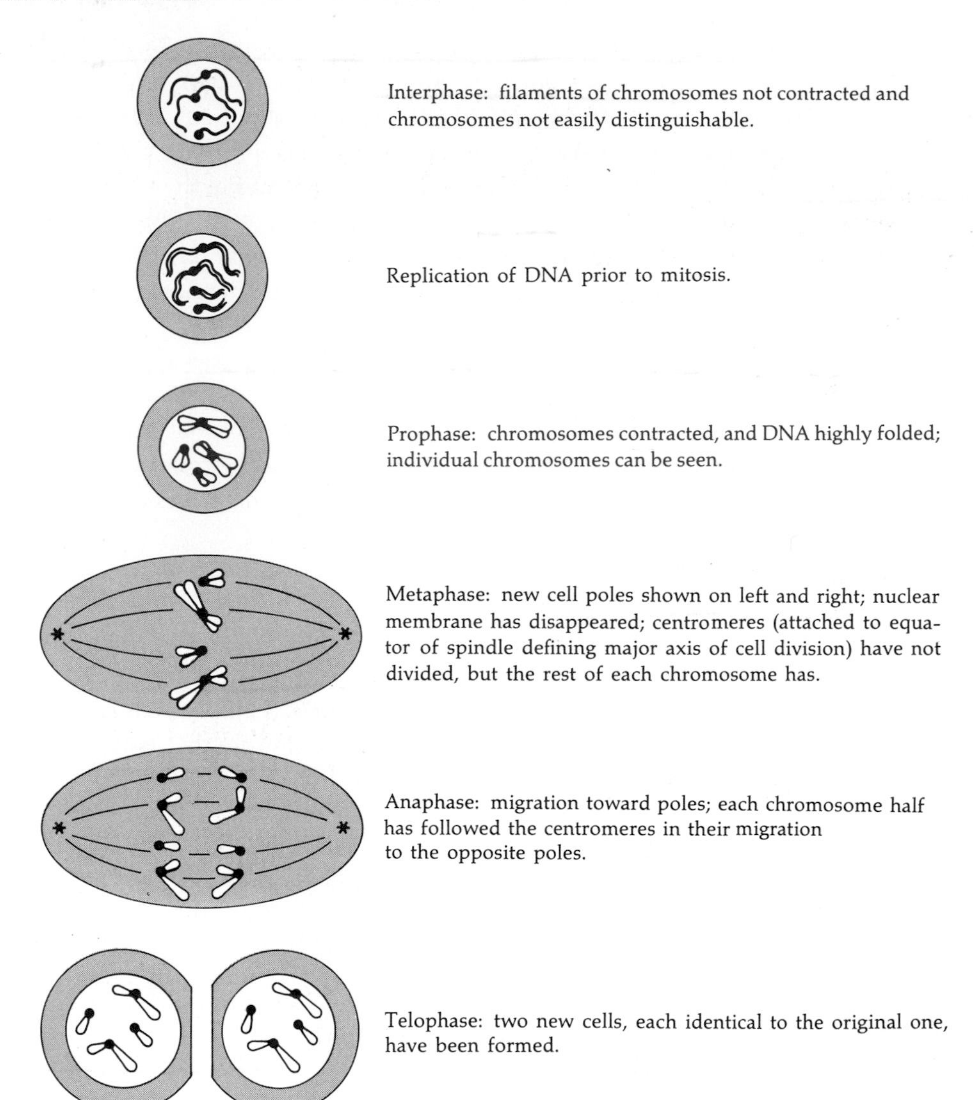

Figure 2.2A
Mitosis: diagrammatic representation of mitosis in a hypothetical organism that has one pair of short chromosomes and one pair of long chromosomes.

the new individual (Fig. 2.3). The sperm and egg cells are called *gametes.* It thus became clear that the chromosomes of the offspring must be derived from the parental nuclei of the two uniting gametes. Weismann, in 1887, correctly deduced that the production of gametes must involve a reduction in the chromosome number. Otherwise the number of chromosomes would increase from each generation to the next (Fig. 2.4). If each parent is to contribute equally to the zygote,

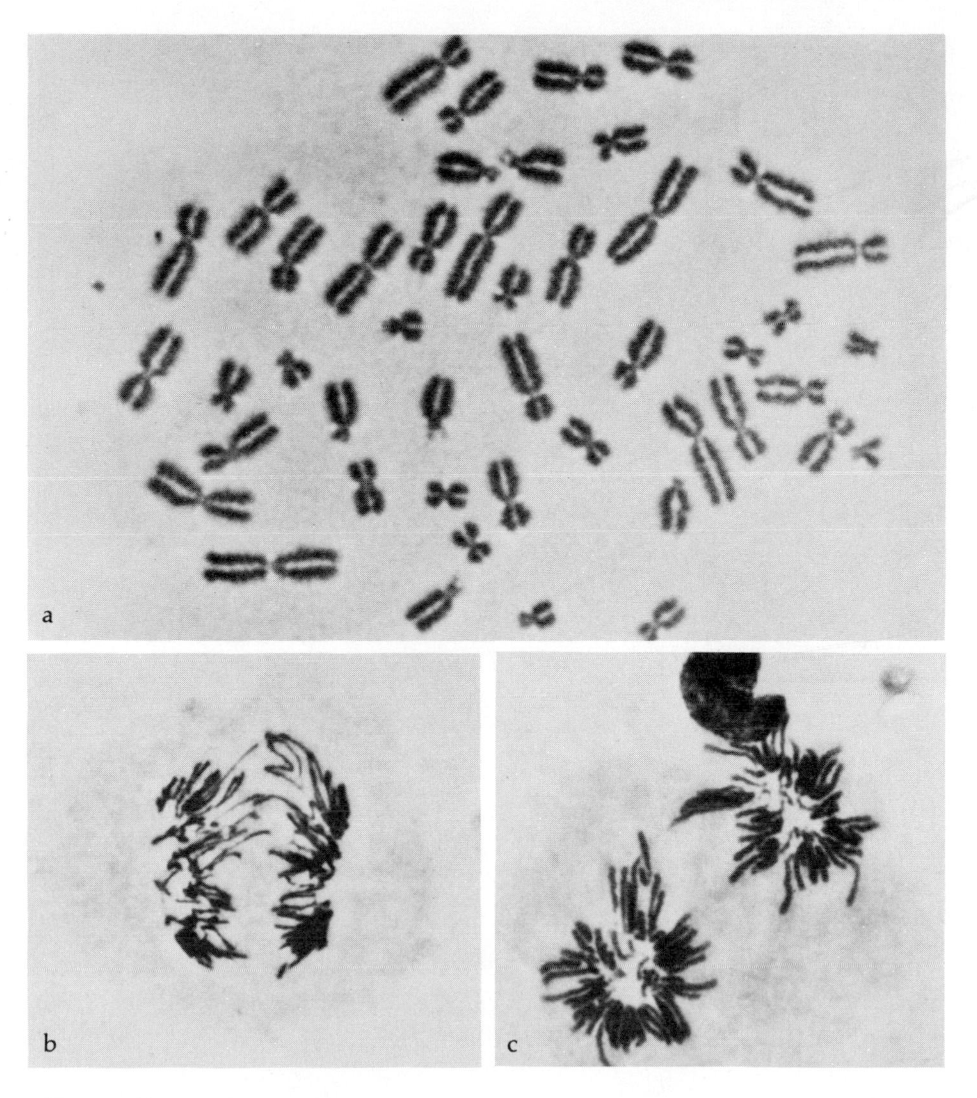

Figure 2.2B
Photographs of mitosis. (a) Metaphase spread from a normal human cell. (b) Early and (c) late anaphase in human cells. (Courtesy of M. Bobrow and D. Buck.)

and if fertilization is to restore the normal chromosome number in the zygote, the prefertilization reduction in chromosome number in gametes must be exactly by a a factor of two. That is, each gamete must have half as many chromosomes as does a normal cell.

The special reduction division which ensures that the gametes have half the number of chromosomes present in normal, nongamete cells is called *meiosis* (Greek, "reduction"). Nongamete cells are called *somatic* cells. There is a further

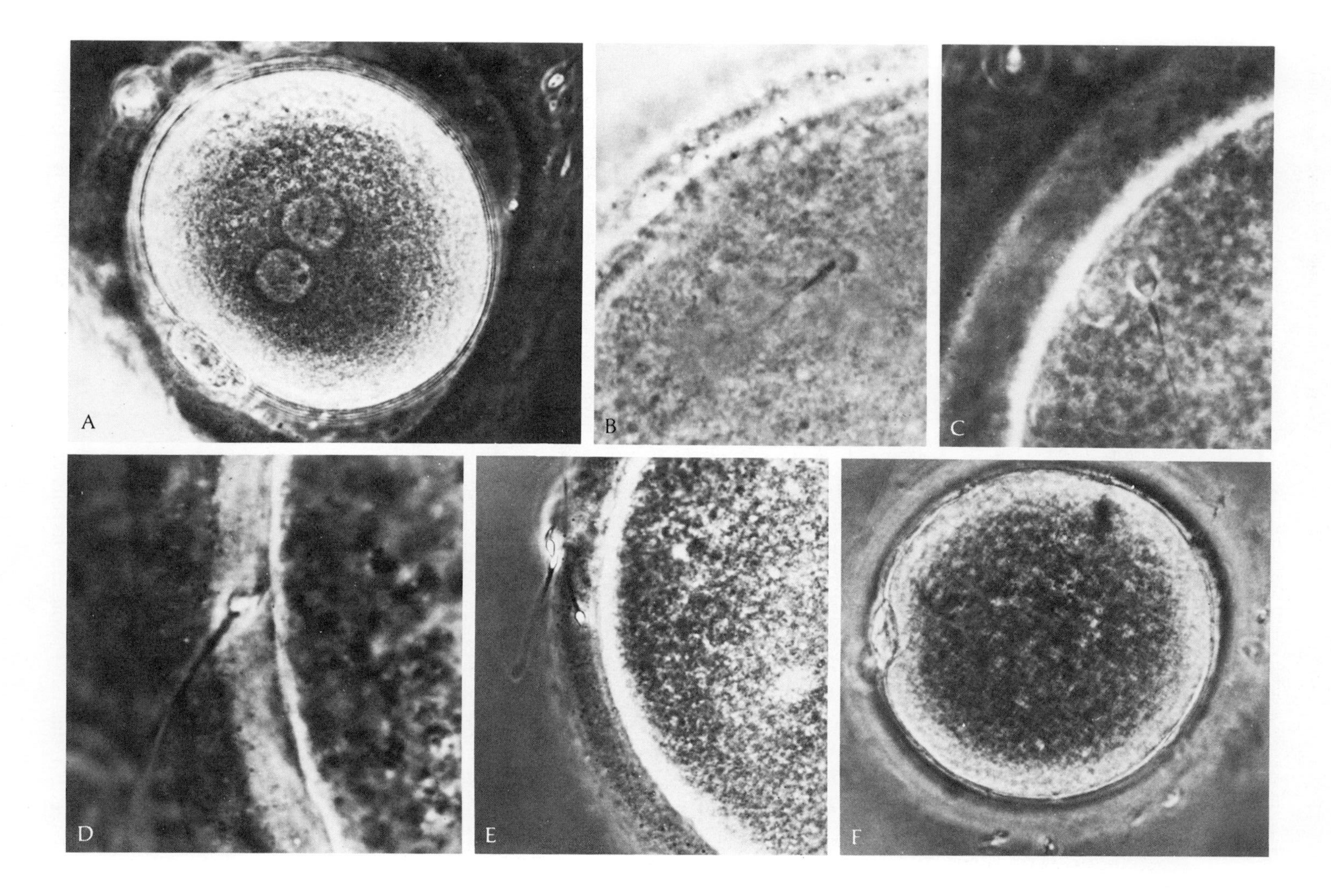
A
B
C
D
E
F

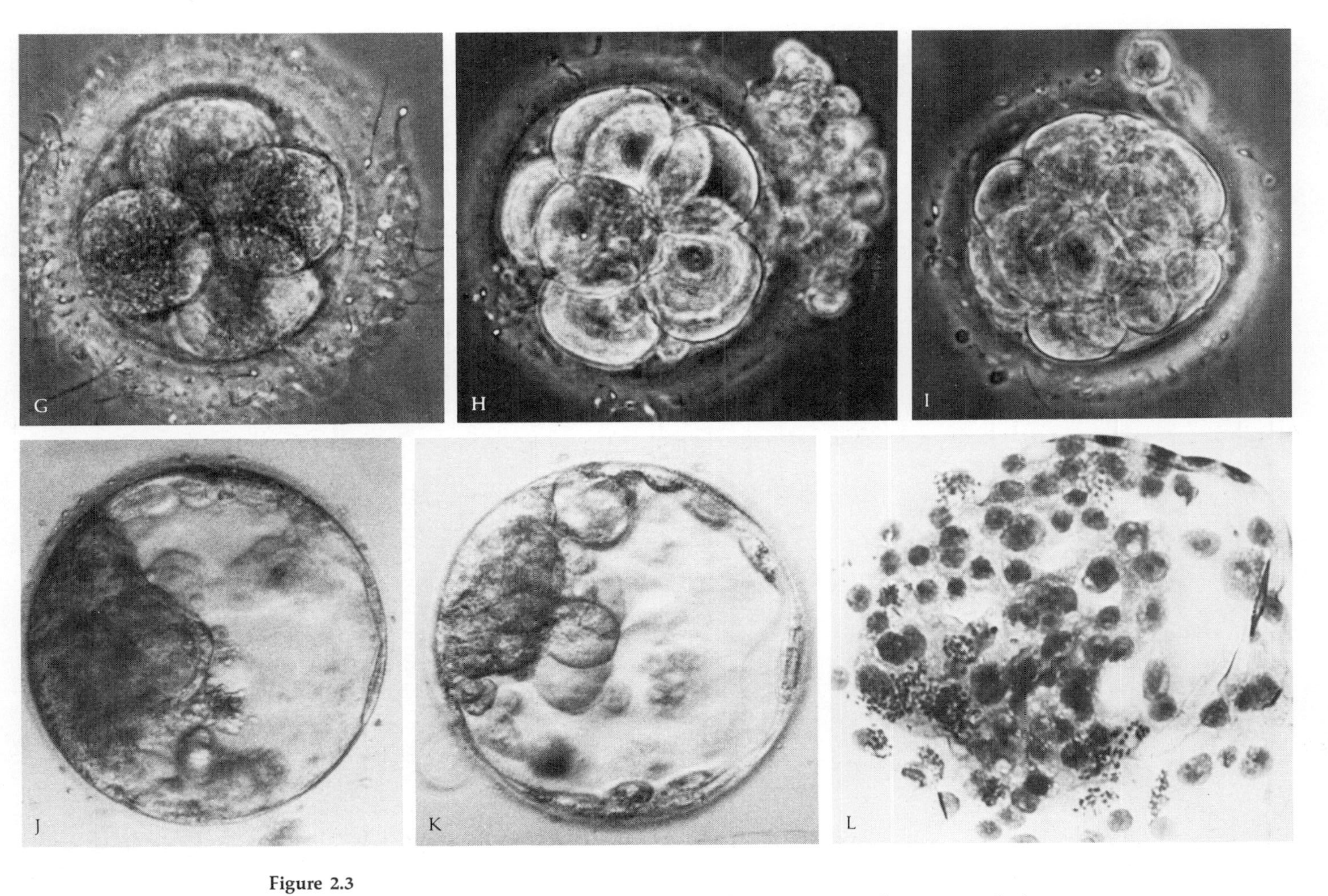

Figure 2.3
Fertilization and early development of human embryos. (A–F) Human eggs with sperm attached to membrane or inside egg. (G) Embryo after two divisions of egg (four-cell stage). (H–L) Successive stages of embryo development. (Courtesy of R. G. Edwards.)

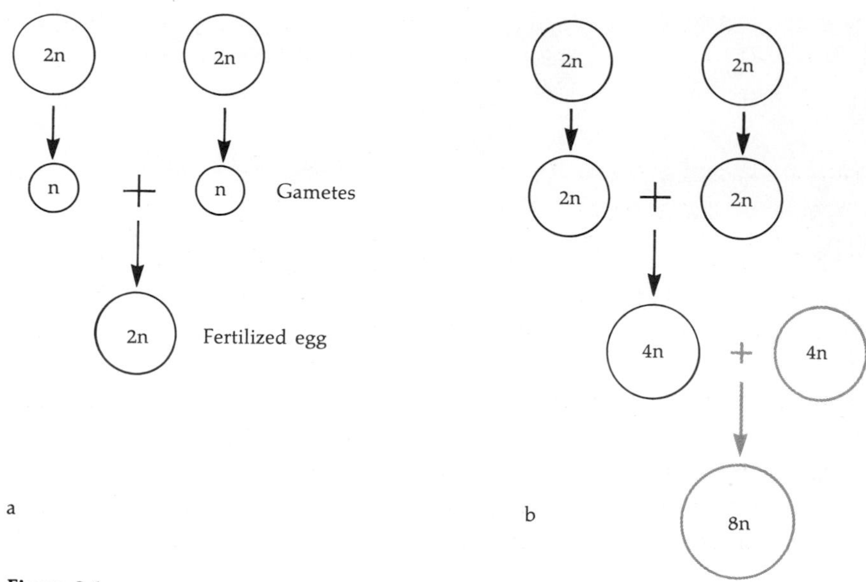

Figure 2.4
Significance of reduction division. Reduction division is necessary to maintain a diploid (double) chromosome set after fertilization. (a) The gametes, formed by meiosis, each receive a haploid (single) chromosome set. The fertilized egg, resulting from fusion of the gametes, has a normal diploid chromosome complement. (b) If reduction division did not occur, fusion of the gametes would lead to an increased number of chromosomes in the fertilized egg.

important requirement that meiosis must satisfy, in addition to reducing the number of chromosomes by one-half: the chromosome complement of parent and offspring must be equivalent. This means that the reduction in chromosome number cannot be arbitrary; it must be arranged in such a way that male and female parental contributions are not only equal in number, but are also equivalent in content. Sutton in the early years of this century realized that this requirement implies that the chromosomes in the normal somatic cells of an individual must occur in corresponding (homologous) pairs, one of which is derived from the father and the other from the mother. Meiosis must then insure that the gametes receive one member of each pair of homologous chromosomes. In this way, fertilization restores the normal double (*diploid*) chromosome complement. The single chromosome complement of the gametes is called *haploid.* If you look carefully at the picture of human chromosomes shown in Figure 2.2B, you will see that there are 46 chromosomes, 46 being the human diploid number, and that there are at least some morphologically distinguishable pairs, especially among the larger chromosomes. It was actually only in 1956 that the correct number of human chromosomes was firmly established, following the development of better techniques for looking at mammalian chromosomes. For many years before that, the human chromosome number had been thought to be 48.

There are two major features of meiosis that distinguish it from mitosis: (1) halving of the number of chromosomes and (2) pairing.

(1) Meiosis involves two cell divisions with only a single chromosome duplication. This leads to a reduction by one-half of the number of chromosomes in the gametes. (2) Duplicated homologous chromosomes, behaving as a unit, pair with each other during the first division of meiosis and then separate to opposite poles. This behavior of homologous pairs of chromosomes ensures that each gamete contains just one member of each homologous pair, as required if each gamete is to contain a balanced haploid set of chromosomes. Because there is no tendency for all the chromosomes derived from one parent to stay together, the haploid gametic chromosome set contains a mixture of chromosomes, some derived from the father and some from the mother. An important consequence of meiosis is, therefore, that it leads to a reassortment of the chromosome sets derived from the two parents.

The whole process of meiosis is illustrated diagrammatically in Figure 2.5; Figure 2.6 shows photographs of meiosis in the male. As in mitosis, the chromosomes are already duplicated by the time they become visible at the start of the first meiotic division. However, the metaphase of this first meiotic division is also preceded by pairing of the duplicated homologues. Each pair of duplicates is joined at the centromeres. Remember that a normal mitotic division involves the splitting of each duplicated chromosome, with an individual copy of each chromosome moving to each pole. In meiosis, however, the duplicated pairs remain joined throughout the first anaphase. The paternal homolog (a duplicated pair) moves to one pole; the maternal homolog (another duplicated pair) moves to the other. Careful study of Figure 2.5, and comparison with Figure 2.2A, should make the process clear.

The immediate products of the first meiotic division are two cells, each containing a diploid chromosome set. However, each homologous pair of chromosomes in one of these cells is a pair of maternally-originated chromosomes or a pair of paternally-originated chromosomes. (After mitosis, on the other hand, each daughter cell contains one maternally-originated and one paternally-originated homolog for each chromosome pair.) The assortment between the two cells is apparently random, with each resulting cell normally containing some chromosome pairs of maternal origin and others of paternal origin.

The second meiotic division is like an ordinary mitosis except that it is not preceded by chromosome duplication. Thus it is this second division which halves

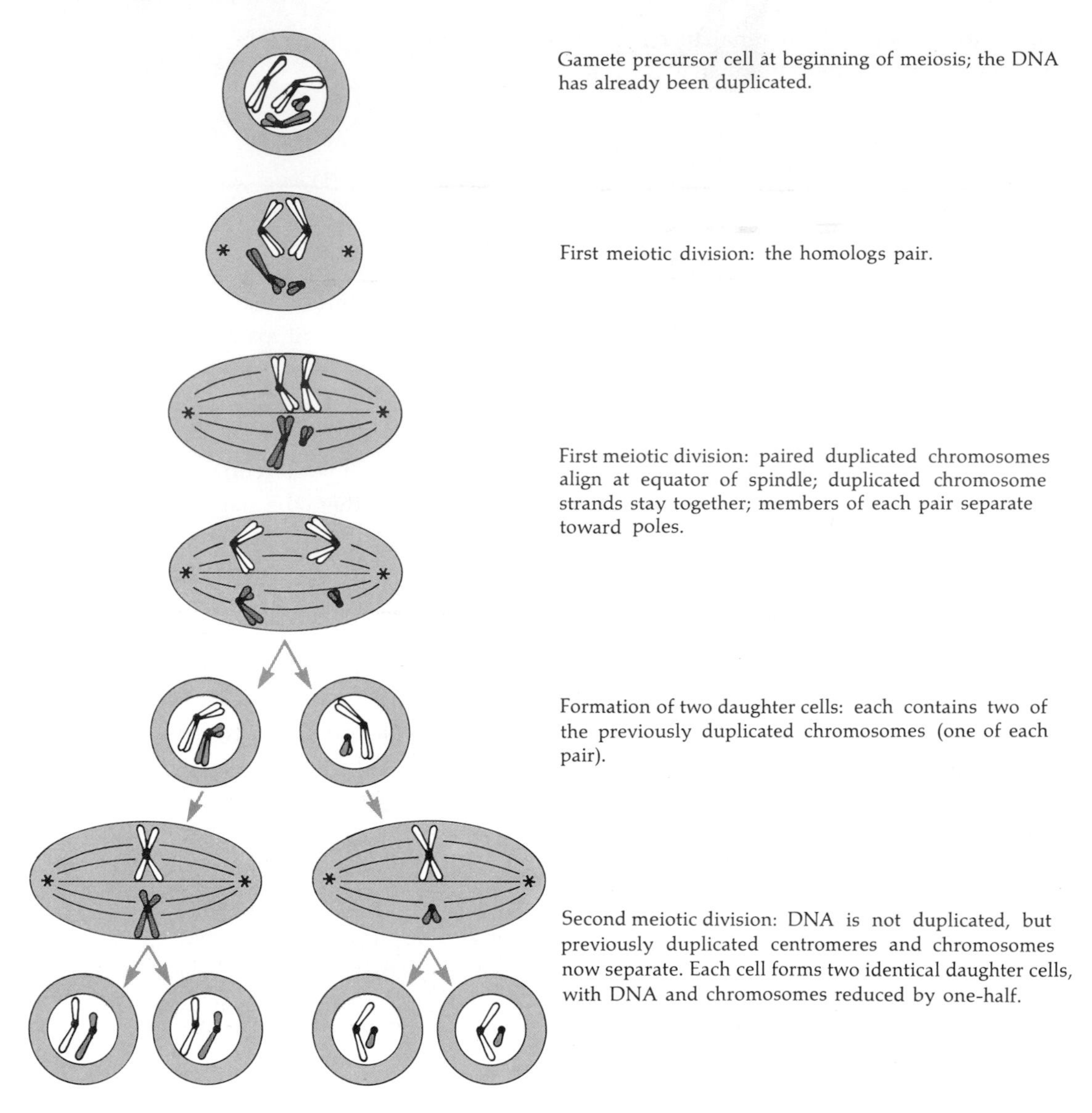

Figure 2.5
Schematic diagram of meiosis in a hypothetical male who has one pair of identical autosomes (white) and one dissimilar XY pair (shaded).

the number of chromosomes. In this division each of the two products of the first division produces identical daughter cells with half the usual number of chromosomes.

Each of the four products of a single meiosis in the male usually becomes a sperm. Each female meiosis, however, produces only one ovum or egg that de-

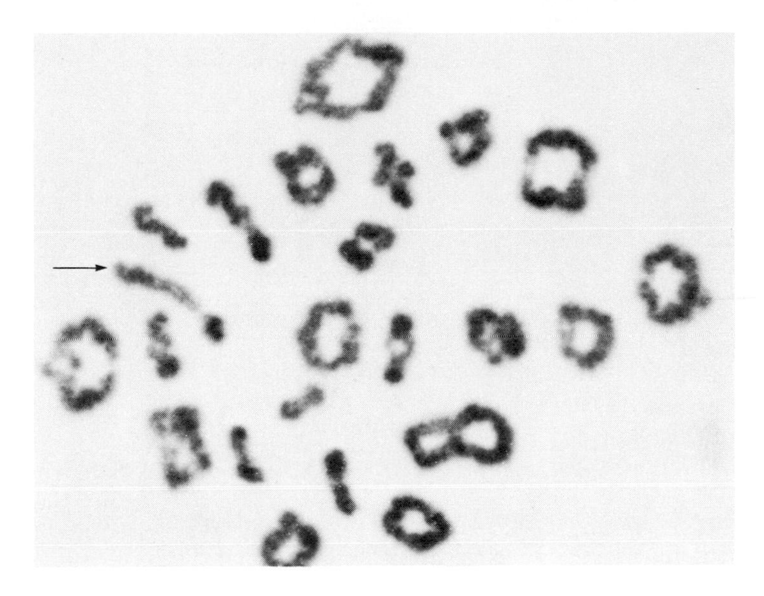

Figure 2.6
Meiosis in the male. Each separate structure is a pair of homologous chromosomes, photographed shortly before the pairs separate at the end of the first division. The X and Y chromosomes (arrow) are joined at their ends. (Courtesy of P. Pearson and M. Bobrow.)

velops at random from one of the four meiotic products (Fig. 2.7). The other three products remain in the periphery of the cell and form the *polar bodies.* Meiosis in the human female is not completed until after fertilization. All the cells in a female destined to become ova are already produced before birth. The oocytes, which are the precursors of the mature egg, remain in a suspended state in the middle of meiosis—either until they are fertilized, in which case meiosis is completed and followed by formation of the zygote, or until the oocyte is discarded at the end of a menstrual cycle. In the male, on the other hand, meiosis and production of sperm continue throughout adult life.

Until quite recently the only way to distinguish chromosomes was by their overall size and by the relative position of the centromere.

Using these two features of chromosome morphology, it was possible to identify unequivocally some of the homologous pairs, but many of them remained indistinguishable. The 46 chromosomes could, however, be arranged into seven groups, denoted by the letters A to G, and ordered according to their overall size. New methods of staining (called "banding") have now been developed that give rise to patterns of bands along the chromosomes.

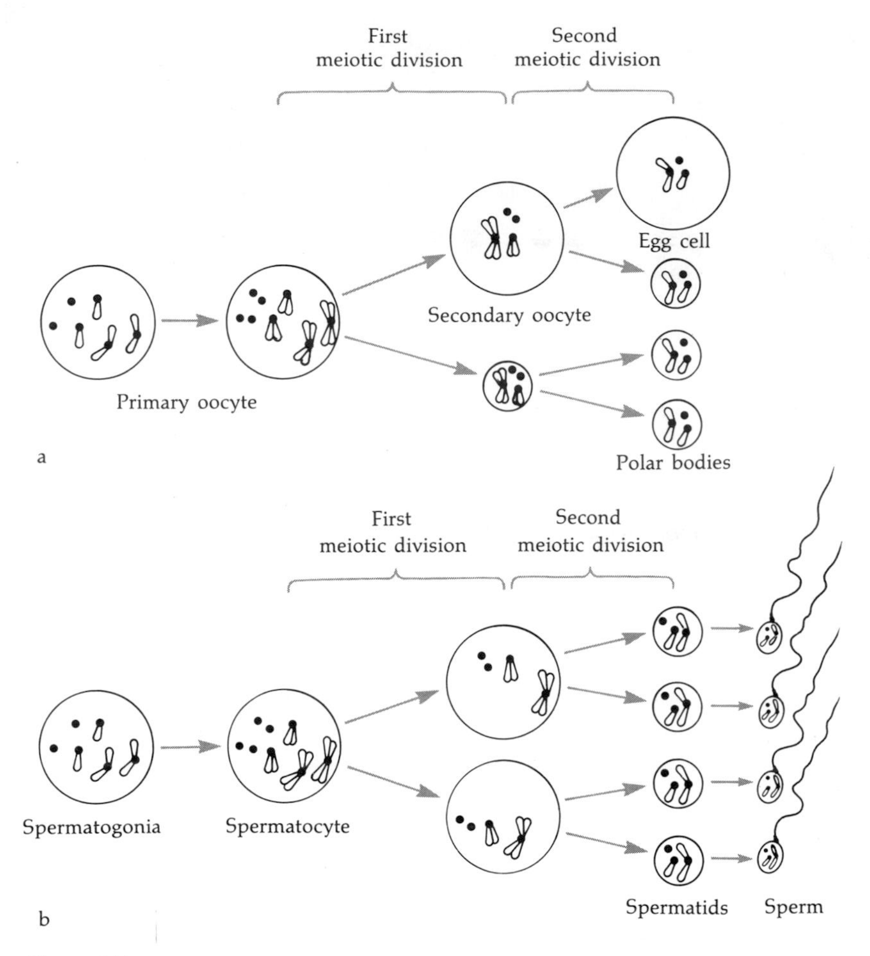

Figure 2.7
Meiosis in males and females. (a) In the female, one functional egg and three nonfunctional polar bodies are produced. (b) In the male, meiosis results in four functional sperm. (After H. Curtis, *Biology*, 2nd ed., Worth Publishers, 1970.)

Banding techniques reveal a unique and constant banding pattern for each chromosome.

All homologues can thus be distinguished without any difficulty. These relatively simple techniques have revolutionized human chromosome studies, because they enable the individual chromosomes to be identified with much greater precision than was previously possible. The chromosome set of a male cell, paired according to banding patterns and ordered by chromosome size, is shown in Figure 2.8. Chromosome arrangements such as these are called *karyotypes.*

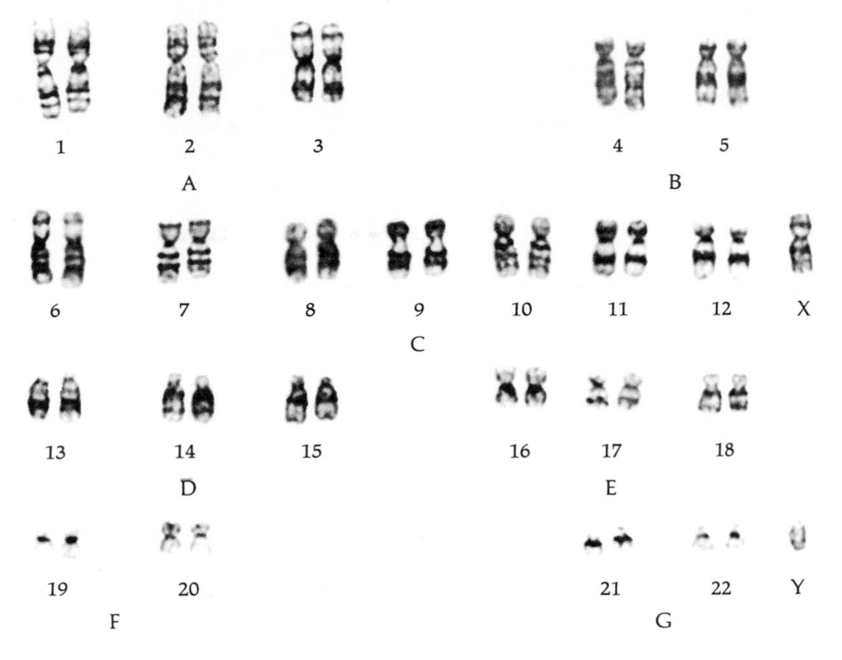

Figure 2.8
Normal male karyotype. The chromosomes are stained by a trypsin banding technique. Chromosomes are arranged according to their size and the position of the centromere. (Courtesy of M. Bobrow.)

There is one major difference between the male and female karyotypes.

Whereas the female has 23 clearly homologous chromosome pairs, the male has 22 such pairs and one dissimilar pair. One member of this nonidentical pair, called X, is just like one of the corresponding female pair. The male partner of the X chromosome, called Y, does not appear in the female karyotype. These two chromosomes are called the *sex chromosomes* because they are the genetic determinants of an individual's sex: XY individuals are males, XX are females.

Because X and Y are homologous chromosomes in the male, meiosis must lead to the production of equal numbers of X-bearing and Y-bearing sperm. Females, however, having the sex-chromosome constitution XX, can only produce X-bearing ova. On the assumption that each type of sperm has the same chance of fertilizing an ovum, male (XY) and female (XX) zygotes should be formed with equal frequencies, according to whether the fertilizing sperm carries a Y or an X chromosome. With respect to the X and Y chromosomes, the cross of a male by a female and its outcome can be represented as follows.

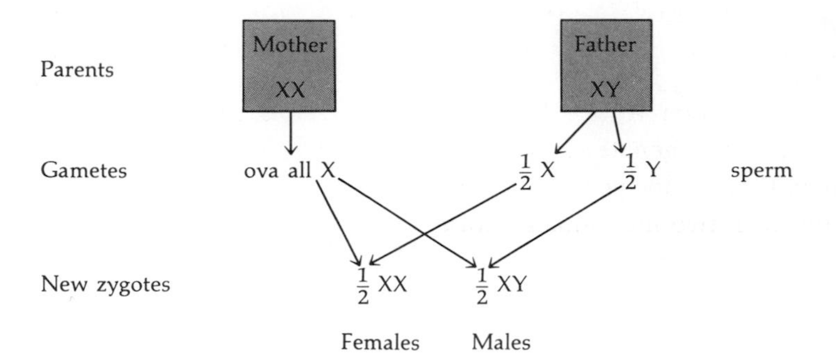

Note that a male always receives his Y chromosome from his father and his X from his mother. The consequences of meiosis, coupled with the X–Y sex-determination mechanism, readily explain the fact that the number of males and females born are approximately equal.

The sex ratio at birth (which is defined as the number of males divided by the number of females at birth) is generally slightly larger than one.

Though the departure from one is not large (for example approximately 105:100 in the white population of the United States), the large amount of available data on the sex ratio means that even such small differences are statistically very significant. It has been suggested that this departure from the expected 1:1 sex ratio may be due to different rates of production of functional X- and Y-bearing sperm, or to differences in the efficiency with which each type of sperm effects fertilization. An alternative suggestion is that there might be differential mortality of males and females after fertilization but before birth, resulting in a slightly greater chance of survival of males to the time of birth.

2.3 Mendel's Laws

The fact that one half of the sperm carry an X and the other half a Y chromosome is the direct result of meiosis. This distribution can readily be followed by the distribution of males and females among newborn offspring. Suppose now that a homologous pair of nonsex chromosomes, which are called *autosomes,* were somehow distinguishable by their effects (or morphologically, as implied in Figure 2.5). Then clearly these would also be equally distributed to sperm or egg, as are the X and Y chromosomes. This suggests that genetic differences, if they are carried by the chromosomes, should follow rules of inheritance that can be established from the behavior of chromosomes during meiosis, a fact first appreciated by Sutton and Boveri in 1903.

We can trace the expected pattern of inheritance of these traits from a knowledge of the mechanism of meiosis.

In the first chapter we used the sickle-cell anemia and its associated trait as model examples of genetically determined differences. The two characters (trait and anemia) are connected by the fact that on average $\frac{1}{4}$ of the offspring of a mating between two individuals with the sickle-cell trait have the severe sickle-cell anemia. How can we explain this fact in terms of our understanding of the mechanism of meiosis? Consider first the simpler situation of a mating between a normal person and one with the sickle-cell trait. Data from such matings show that, on average, half the offspring are normal and half have the sickle-cell trait—a 1:1 distribution, just the same as that expected from the mechanism of meiosis for the ratio of males to females.

Because the hemoglobin differences between the sicklers and nonsicklers are clearly inherited, we must assume that the genetic information for the determination of which type of hemoglobin is made by an individual is carried by some part of their chromosomes. We know, however, that chromosomes occur in homologous pairs which carry equivalent genetic information, one coming from the father and the other one from the mother.

The element of the chromosome that determines the hemoglobin type must, therefore, also exist in duplicate. We call such an element, which is the genetic determinant of an inherited character, a *gene.* In the next chapter, when we discuss the chemical basis of the genetic material of the chromosomes, we define a gene much more precisely in chemical terms. For the moment we must be satisfied with the definition of the gene just given, a definition used throughout the early period of the development of genetics and one close to that given by Mendel himself. However, Mendel used the word *element* instead of gene, the latter being a word coined in 1909 by W. L. Johannsen. The normal nonsickler has two equivalent normal hemoglobin genes that we shall call *A*. The sicklers, on the other hand, make both normal hemoglobin and the sickle-cell hemoglobin S. They must therefore have one normal hemoglobin *A* gene and, for their second hemoglobin gene, one that gives rise to the variant type of hemoglobin. We call this gene *S*. (Note that italics are used to distinguish the symbol for a gene, *S*, from that for its product, S.) The hemoglobin genetic formula for the normal individual can be written *AA*, while that for the individual with sickle-cell trait is *AS*. This means that the normal individual received *A*-carrying chromosomes from both his parents, but the carrier of sickle-cell trait must have received an *A* chromosome from one parent and an *S* chromosome from the other parent.

Alternative versions of genes, such as A and S, are called alleles.

Individuals, such as *AA*, who have the same allele of a given gene on both homologous chromosomes are called *homozygotes* ("homo" meaning same), whereas those who carry different alleles, such as *AS*, are called *heterozygotes* ("hetero" meaning other or different). An individual's genetic constitution, either in general or with respect to a specific set of genes, such as in the case of the hemoglobins, is called his *genotype.*

The consequences of meiosis, as we have already discussed, clearly lead to the expectation that, on average half of the gametes produced by the heterozygote AS will carry A, and the other half will carry S, depending on which chromosome is transmitted to the gamete.

The normal *AA* homozygote can only produce *A*-bearing gametes. Thus, exactly as in the case of the sex difference, we expect the cross *AS* × *AA* to produce half offspring *AS*, who will have the sickle-cell trait, and half who are *AA* (normal). Diagrammatically the cross can be represented as follows.

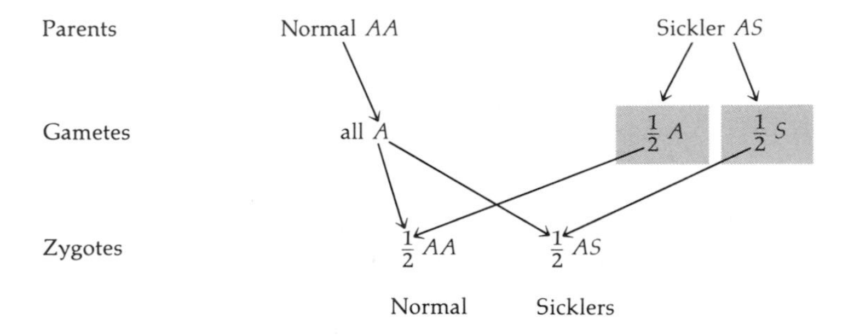

What about the cross between two sicklers, *AS* × *AS?* Here each parent produces gametes of which half carry *A* and the other half *S*. Fertilization can result from any pairwise combination of gametes, one taken from each parent (Fig. 2.9). Thus, if one parent contributes *A* and the other *S*, a heterozygote *AS* will be produced. If both parents contribute *A*, this will give rise to normal homozygote *AA*. If, on the other hand, both parents contribute *S*, a new type of homozygote, *SS*, will be produced. This homozygote carries only the gene *S* which produces the sickle-cell hemoglobin, and so the resulting cells can make no normal hemoglobin. It is the *SS* homozygous individual who therefore suffers from the severe sickle-cell anemia. (Note that homozygous and heterozygous are used as adjectives, homozygotes and heterozygotes as nouns.) If the various combinations of gametes are formed without regard to their genetic constitution, then the proportions of zygotes of a given type should just be the product of the frequencies with which the constituent gametes are produced. Thus an individual is *AA* only if each parent contributes an *A*-carrying gamete. The frequency with which each parent produces *A* gametes is $\frac{1}{2}$, so the chance of getting an *A* from each parent is $\frac{1}{2} \times \frac{1}{2} = \frac{1}{4}$. Thus, the chance of being *AA* is $\frac{1}{4}$ and similarly the chance of being *SS* is $\frac{1}{4}$. There are two ways in which the mating *AS* × *AS* can give rise to an *AS*

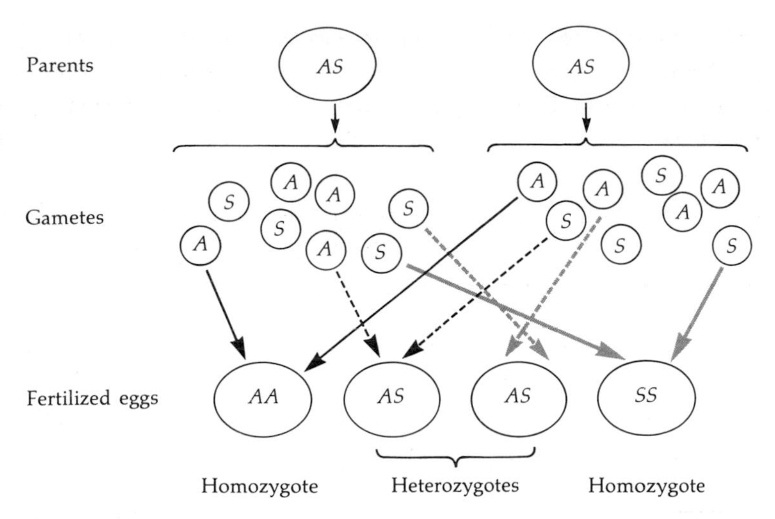

Figure 2.9
Offspring of mating of sickle-cell-trait heterozygotes. A heterozygote with sickle-cell trait produces normal and sickle-cell gametes in equal numbers. When two heterozygotes mate, any gamete of one parent may combine with any gamete of the other parent. One-fourth of the offspring thus can be expected to be normal homozygotes, one-half to be heterozygotes with sickle-cell trait, and one-fourth to be homozygotes with sickle-cell anemia. (A = normal allele; S = sickle-cell allele.)

offspring. Either the father's sperm carries A and the mother's egg S, or the father's sperm carries S and the mother's egg A. Each possibility has a chance of $\frac{1}{4}$, just as in the production of AA or SS. The two together therefore occur with a total frequency of $\frac{1}{4} + \frac{1}{4} = \frac{1}{2}$, which is the chance of an AS offspring from an $AS \times AS$ mating. The outcome of this mating can be represented diagrammatically as follows.

		Sperm (father AS)	
		$\frac{1}{2}A$	$\frac{1}{2}S$
Eggs (mother AS)	$\frac{1}{2}A$	$\frac{1}{4}AA$	$\frac{1}{4}AS$
	$\frac{1}{2}S$	$\frac{1}{4}AS$	$\frac{1}{4}SS$

This diagram gives the combined expected frequencies

$$\tfrac{1}{4}\,AA : \tfrac{1}{2}\,AS : \tfrac{1}{4}\,SS,$$

which is the 1:2:1 Mendelian ratio.

Our analysis of the cross between sicklers explains why we should expect an average of $\frac{1}{4}$ of their offspring to have the anemia, because only the homozygous genotype *SS* suffers from this disease. The other $\frac{1}{4} + \frac{1}{2} = \frac{3}{4}$ of the offspring, who are either *AA* or *AS*, are normal, at least as far as the anemia is concerned. The genotypes *AA* and *AS* can be distinguished only if one looks at the blood cells for evidence of sickling or at the hemoglobin to see if any of it is of type S.

When an allele A is such that genotypes AA and AS are indistinguishable, as they would be if we did not examine the blood, then allele A is said to be dominant *over allele S.*

Allele *S*, which only causes the anemia when in double dose in the homozygote *SS*, is said to be *recessive.* The reason that *AS* and *AA* are both clinically normal is because one *A* allele gives rise to enough normal hemoglobin A for the blood to function normally. Many inherited traits are known to be determined by a recessive allele, such as *S*, for which the two genotypes other than the recessive homozygote (in this case *AS* and *AA*) are not readily distinguishable.

The inborn error of metabolism, alkaptonuria, was one of the first to be described at the beginning of this century (see Chapter 1). Alkaptonuria is an example of a recessive trait. If we call the recessive allele *a* and the dominant allele *A*, the alkaptonurics are *aa* while genotypes *AA* and *Aa* are both normal and indistinguishable. (It is a common convention to use capital letters for dominant alleles and lower case letters for recessive alleles.) In this case the mating *Aa* × *AA* gives rise only to normal offspring. The mating *Aa* × *Aa* on the other hand gives, on average, $\frac{3}{4}$ normal offspring (*AA* or *Aa*) and $\frac{1}{4}$ alkaptonurics (*aa*), illustrating Mendel's 3:1 ratio (Fig. 2.10).

Mendel's breeding experiments, on the basis of which he formulated the laws of inheritance, made use of the garden pea. He worked with clearcut character differences that all turned out to be controlled by recessive genes. The cross *Aa* × *aa* is like that of *AS* × *AA*, and gives on average $\frac{1}{2}$ normal offspring (*Aa*) and $\frac{1}{2}$ recessive homozygotes (*aa*). The only way, therefore, that Mendel was able to distinguish genotypes such as *AA* and *Aa* was by testing them for example, in a cross with the recessive *aa*. If the offspring were all normal, the tested parent was *AA*, while if about $\frac{1}{2}$ were abnormal (recessive *aa*), then the tested parent must have been *Aa*. The expected results of all the possible crosses involving the three genotypes *AA*, *AS*, and *SS* are set out in Table 2.1. A cross between heterozygotes is often called an *intercross*, and that between a heterozygote and a homozygote a *backcross.*

The observed properties of an individual are called the phenotype *in contrast to the genotype, which is the genetic constitution.*

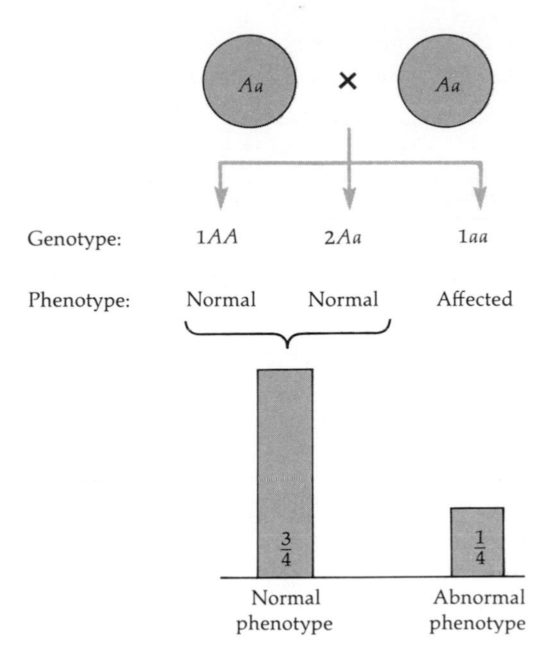

Figure 2.10
Distribution of genotypes and phenotypes in the mating of two heterozygotes.

Thus, when an allele such as that causing alkaptonuria is recessive, the phenotypes of the two genotypes *AA* and *Aa* are both the same, namely normal. In the case of the sickle-cell allele *S*, when we examine the blood, the phenotypes of all three of the genotypes *AA*, *AS*, and *SS* are distinguishable. In this case, alleles such as *A* and *S* both of which can be detected in the heterozygote are called *codominant*. When, however, we look just for the anemia, *AA* and *AS* have the same normal, nonanemic phenotype and then *A* is dominant over *S*. Thus whether one allele is dominant over another often depends on the level at which the phenotype is observed.

2.4. Pedigrees and Chance

Myotonic dystrophy is one of a class of rare, inherited diseases (the muscular dystrophies) in which the muscles gradually atrophy or degenerate so that they no longer function properly. The disease is also often associated with baldness and cataracts. There is, of course, no question of performing "human crosses" to establish a pattern of inheritance, as can be done with plants or animals. Appropriate data, equivalent to those that would be obtained from a planned cross, must therefore be sought in the population at large, usually by asking an individual who has a particular trait or disease about his or her relatives. Some typical

Table 2.1
The possible types of crosses involving the genotypes *AA*, *AS*, and *SS*, and their expected outcomes

Parents			Genotype proportions			Phenotypes: A normal, S anemic	
Father		Mother	*AA*	*AS*	*SS*	A	S
AA	×	*AA*	all	—	—	all	—
AA	×	*AS* }	1/2	1/2	—	all	—
AS	×	*AA* }					
AA	×	*SS* }	—	all	—	all	—
SS	×	*AA* }					
AS	×	*AS*	1/4	1/2	1/4	3/4	1/4
AS	×	*SS* }	—	1/2	1/2	1/2	1/2
SS	×	*AS* }					
SS	×	*SS*	—	—	all	—	all

Pairs of matings such as *AA* × *AS* and *AS* × *AA* are coupled because the result of the mating does not depend on whether it is the father or the mother who is *AA* or *AS*.

data obtained in this way, on the occurrence of myotonic dystrophy in two different families, is shown in Figure 2.11. The diagram is in the form of pedigree charts commonly used for the presentation of family data. All human genetic information is usefully presented in the form of pedigrees, and the analysis of pedigrees is often the best way to discover the mode of inheritance of a trait.

There are some simple rules for drawing and interpreting pedigrees.

Individuals on the same line are from the same generation, and individuals in each generation are numbered in sequence (such as II.1, II.2, II.3, and so on); circles indicate females, squares males. The symbols for wife and husband (such as I.1 and I.2 in Figure 2.11a) are joined by a line. The symbols for a sibship, the group of offspring derived from given parents, are attached to a common line (III.2, 4, 5, and 6 in Figure 2.11a). This line is itself attached to the line joining the parents (II.1 and II.2 in this case). Various other conventions are commonly used to symbolize the phenotypes and genotypes of the individuals in a pedigree. In this case, the open symbols represent normal members of the family and the shaded symbols those with myotonic dystrophy. The arrow indicates the patient from the family who first contacted a doctor and through whom the rest of the family was ascertained. Such an individual is called a *proband* or *propositus*. The small filled circle in the first position of generation IV indicates a stillbirth or

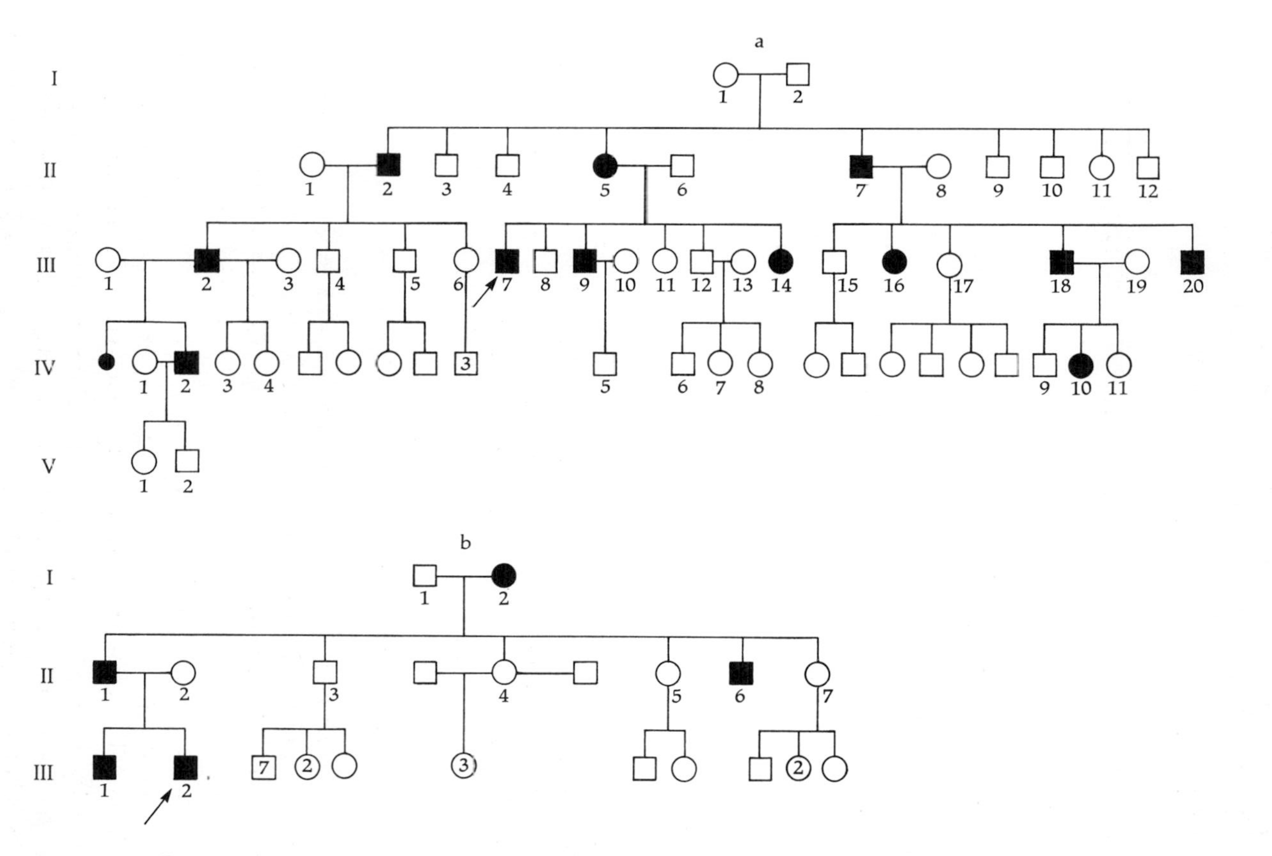

Figure 2.11
Pedigrees of muscular dystrophy: pedigrees for two different families, with shaded symbols representing individuals affected by myotonic dystrophy. This is a dominant disease with incomplete penetrance. See the text for discussion of pedigree diagrams, and see Figure 2.12 for a key to commonly used symbols. (Courtesy of P. Harper.)

abortion of unknown sex, and the number inside the square in generation IV (offspring of III.6) indicates the number of similar, normal males. A summary of some of the symbols commonly used in pedigrees is given in Figure 2.12, and a further discussion of pedigree analysis is given in the appendix.

One of the characteristics of a disease determined by a dominant allele is that it should appear in every generation of a pedigree.

Unless one or the other parent has the dominant gene, and so has the disease, none of the children will have the gene or the disease. Very often, however, the effect of such a gene in causing the disease may be quite variable. Thus, in the case of myotonic dystrophy, some people will show symptoms of muscle weakness at earlier ages than others and may eventually become much more severely affected. Occasionally, however, individuals carrying the relevant gene may be so mildly affected that their disease passes unnoticed. When this happens, we say that the gene shows *incomplete penetrance.* Variability in the expression and pene-

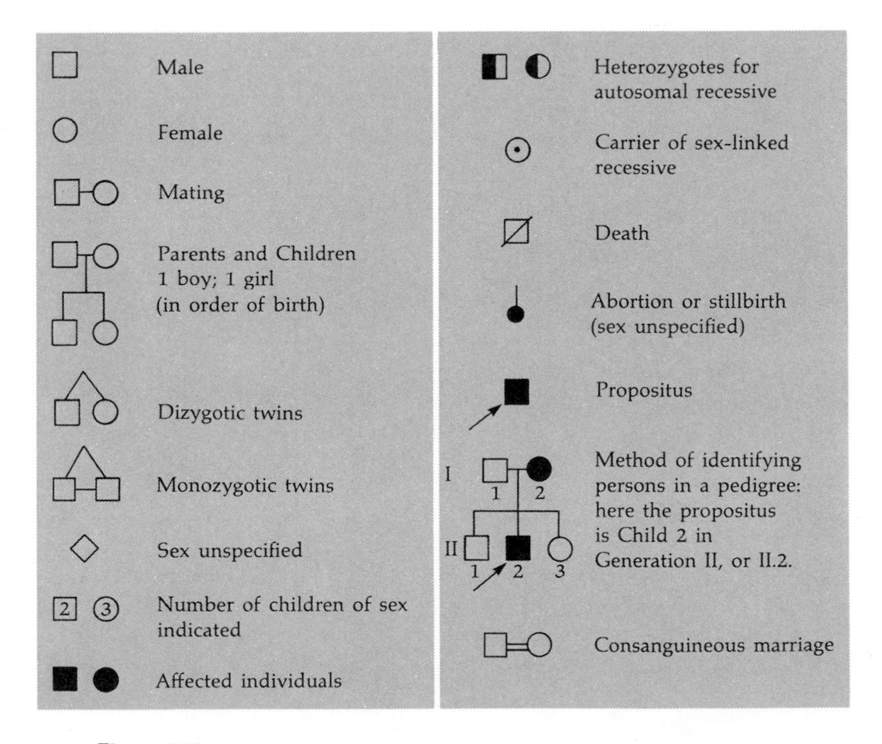

Figure 2.12
Symbols commonly used in pedigree charts. (From J. S. Thompson and M. W. Thompson, *Genetics in Medicine,* 2nd ed., W. B. Saunders Co., 1973.)

trance of genes is a common problem in the interpretation of pedigrees of dominant traits. In the pedigree shown in Figure 2.11a, neither parent I.1 nor I.2 is indicated as having the disease. One of them clearly must have carried the gene for myotonic dystrophy, as it has been passed down to three of their offspring and through two subsequent generations. Both of these individuals, however, were dead by the time the pedigree was ascertained, and so they could not be examined by a doctor. The presence of myotonic dystrophy, if mild, might easily have been overlooked by their surviving children, who would probably be the only available sources of information about their parents.

Cystic fibrosis is an example of a recessively determined inherited disease.

Among the classic features of this disease are chronic bronchial obstruction (due to accumulation of mucus, which often causes an infection of the lungs), malnutrition, growth failure, and increased salinity of the sweat. The disease is also often associated with a variety of other effects and has quite a variable expression. Affected individuals often survive into their teens but usually die before the age of 20. Until recent developments in the therapy for cystic fibrosis, such individuals would generally have died at a much younger age. Cystic fibrosis is the most frequent of severe recessive genetic diseases among Caucasians, occurring with an incidence of almost one in 2,000 live births. Though this might be considered rare from some points of view, it is about five times as frequent as any other severe inherited recessive or dominant disease found in Caucasians. Some of these diseases are very much less frequent and have been described in only one or two isolated families. Because almost 1,500 individuals with cystic fibrosis are born every year in the United States, the disease does pose a substantial health care problem (see Chapter 20). A typical pedigree of a family with cystic fibrosis is shown in Figure 2.13.

A characteristic of pedigrees of rare recessive diseases is that parents of the affected individuals usually are themselves normal.

This implies that the parents are heterozygotes for the recessive gene that causes the disease. This is common because heterozygotes for a rare gene occur with a much higher frequency in the population than do the affected homozygotes. The explanation for this lies in the laws that describe the behavior of genes in populations, a subject for the next part of the book.

From the rules of inheritance described in Section 2.3, we should expect that on average one-quarter of the offspring of a mating between two normal heterozygotes for the cystic fibrosis gene will be affected. In the pedigree shown in Figure

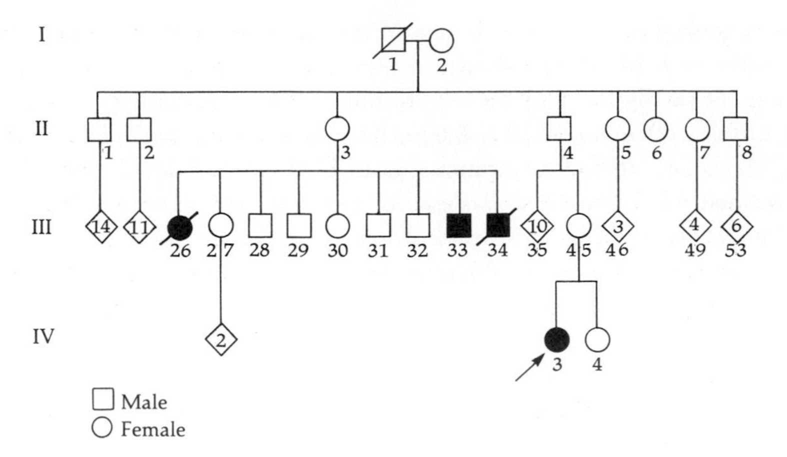

Figure 2.13
Pedigree of a kindred with cystic fibrosis. The propositus (patient) is IV.3. Individuals III.26 and III.34 died shortly after birth; III.33 has mild pulmonary and gastrointestinal disease at age 13. Parent I.1 died at age 65 of cancer; I.2 is living at age 80. Recessive inheritance is strongly suggested by the lack of affected individuals in generations I and II and in sibship III.35–45. (From J. B. Stanbury, J. B. Wyngaarden, and D. S. Fredrickson, *The Metabolic Basis of Inherited Disease,* 3rd ed., McGraw Hill Book Co., 1972.)

2.13, 3 out of 9 or $\frac{1}{3}$ of the children of individual II.3 are actually affected. Is this compatible with Mendelian expectations?

Let us consider again the myotonic dystrophy pedigrees of Figure 2.11, which show that the disease is caused by a dominant gene, say M. The offspring of affected individuals are not all affected, which is of course expected if the affected parents are heterozygotes (Mm) for the myotonic dystrophy gene. Their normal mates are presumably homozygous (mm) for the corresponding normal allele (m). Thus, matings between affected and normal individuals such as II.1 and II.2 are backcrosses ($Mm \times mm$) and should give rise, on average, to $\frac{1}{2}$ normal (mm) and $\frac{1}{2}$ affected (Mm) offspring. In this particular case, only one of five offspring is affected. On the other hand, three out of six of the offspring of II.2 and II.8 are affected. Once again, we can ask whether these observations are in reasonable agreement with Mendel's laws.

What do we mean when we say that "on average we expect" half of the gametes of an AS heterozygote to be carrying A and the other half to be carrying S; or that we expect one-half of the offspring of a mating between a normal individual and one with myotonic dystrophy to have the disease?

Clearly our expectations cannot always come true. There is no way of having 1.5 affected children out of a total of 3! To understand these questions, we must try to understand something of the rules of chance, frequencies, and probabilities—namely, the subject of statistics as applied to genetics.

Picking a gamete that may be either X or Y from a male (XY) is a little like tossing a coin. Heads it's an X, and tails it's a Y, with each possibility having the same chance of occurring. At least we assume the chance of heads and tails is the same, because it has to be one or the other and, if the coin is well balanced, there is no obvious factor that should make it more likely that a head rather than a tail will turn up. Another way of looking at the problem is to imagine that we have equal numbers of X and Y gametes in a sack, and when we pick out a gamete there is no way of telling which one has been picked until it has been looked at. If the gametes are all mixed up in the sack, then the chance of picking out an X is the same as that of picking out a Y, and since one or the other has to be picked, this chance must be $\frac{1}{2}$.

Another name for chance in this sense is probability.

Thus we can say that the probability of a sperm carrying a Y-chromosome is $\frac{1}{2}$. Because the female produces only X gametes, we can also say that the probability of a male offspring is $\frac{1}{2}$. What about the possible combinations that can arise if we toss a coin twice, or pick a gamete twice, or look at the sexes of families with just two children? There are two possibilities (male or female) for each child, and thus four combinations for the sexes of the two children, as can be seen from the following diagram (using M for male and F for female).

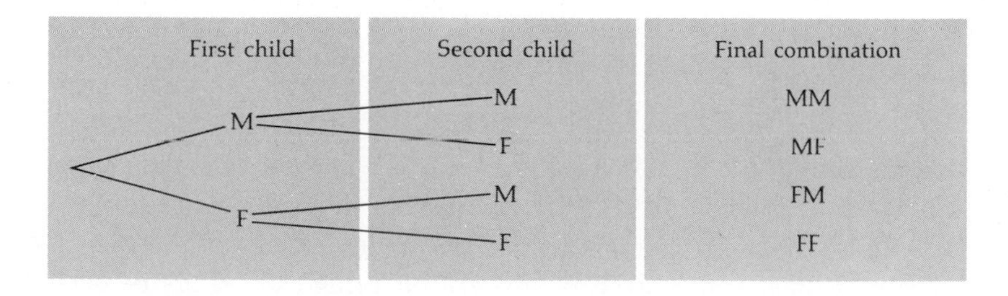

Thus, both children can be M, both F, the first M and the second F, or the first F and the second M. Clearly, each of the four combinations has the same chance, or probability, because at each stage there is nothing to make one combination more likely than another. Because just one of the four possible outcomes is two male offspring, we can say that the probability that the two offspring are both male is $\frac{1}{4}$. Another way of thinking about the probability of an event is, therefore, the number of ways that a particular chosen event can happen (in this case, one) out of the total number of ways from which the particular one in which we are interested is chosen (in this case, four).

Let us consider, then, the probability that of two offspring, one is male and the other female. This can happen in two ways out of four (MF or FM), so the probability is 2 out of 4, or $\frac{1}{2}$. Note that the probability of, say, two males is $\frac{1}{2} \times \frac{1}{2} = \frac{1}{4}$, which can be thought of as the probability that the first child is a male multiplied by the probability that the second one is a male. There is no simple way in which the sex of the first child should influence whether or not the second child is a male.

When two such events have no influence *on each other, then they are said to be* independent.

In the case of a family with two sons, the probability of both events occurring—namely, both children being male—is the product of the separate probabilities of the events. This rule for combining probabilities of independent events is a very useful one for working out the probabilities of compound events. Let us use it to find the probability that all of three offspring are male. There are now three successive events—namely, the births of three children. As far as sex is concerned, each is independent of the other two. Thus, the probability of three males is the product of the three separate probabilities that each birth is a male:

$$\tfrac{1}{2} \times \tfrac{1}{2} \times \tfrac{1}{2} = \tfrac{1}{8}.$$

We can summarize the results of our calculations on the sexes of two offspring in the following way:

Number of males	2	1	0
Number of females	0	1	2
Probability	$\frac{1}{4}$	$\frac{1}{2}$	$\frac{1}{4}$

Note that the sum of the probabilities is 1. This is a general rule: the sum of the probabilities of a set of mutually exclusive events (meaning that only one can occur at a time), which are also exhaustive (meaning that one of the events *must* occur), is always one. The analysis of sexes of offspring can readily be extended to families of three offspring. Following the same procedure we have outlined above for two-offspring families shows that there are $2 \times 2 \times 2 = 8$ different outcomes: one each for all males or all females, and three each for two females and one male or for one female and two males. These results can be summarized as follows.

Number of males	3	2	1	0
Number of females	0	1	2	3
Probability	$\frac{1}{8}$	$\frac{3}{8}$	$\frac{3}{8}$	$\frac{1}{8}$

Such analyses can be extended to any size of family and to any probability of outcome. Instead of $\frac{1}{2}$ for each type as for sex, the probability might be $\frac{1}{4}$, as it is for the expected frequency (or the probability) of a recessive cystic fibrosis offspring from a mating between two normal heterozygotes. The general formulas for this result, as well as the development of some other statistical ideas that are useful in genetics, are described more fully in the Appendix, which is intended for those who have some inclination toward understanding mathematical analyses. But what about the application of these ideas to our pedigrees?

We can now see that Mendel's laws are really framed in terms of probabilities.

The probability of a male offspring is $\frac{1}{2}$; the probability of a normal offspring from a mating between two cystic fibrosis heterozygotes is $\frac{3}{4}$ $(= 1 - \frac{1}{4})$. The laws also allow us to predict the probabilities of getting different combinations of types of offspring in families of given size. We can show, for example, that if myotonic dystrophy is determined by a dominant gene, then the probability of getting one affected offspring out of five from a mating between a normal and an affected person is $\frac{5}{32}$ (see the Appendix). Clearly, just looking at one or two pedigrees is not going to be enough to tell us whether the observations can be explained by Mendel's laws. For example, a family with one out of five affected might be explained better by a different probability for affected offspring, in this case a probability less than $\frac{1}{2}$. But the chance of observing such a result is certainly not unreasonably low if the segregation is 1:1. However, we should surely expect that, if we looked at further families, then the more families we had data on, the closer should be our results to those predicted by Mendel's law. This is just like saying that, if we keep tossing a coin and adding up the number of times heads comes up, then the longer we go on, the closer to $\frac{1}{2}$ should be the proportion of heads. In other words, as we go on tossing the coin, the observed proportion of heads gradually approaches the known probability of getting a head, namely, $\frac{1}{2}$. (Confirm this for yourself if you are not convinced that this must be the case.) This result is actually a most important and general law of probability and statistics.

The more data we get, the closer the results should be to what we expect.

At the same time, the smaller our body of data the greater will be the chance fluctuations from the expected result, and so the more difficult it will be to know whether data and model, or hypothesis, agree with each other. Some statistical procedures, the best known of which is the chi-square (χ^2) test for telling whether a given body of results are within the bounds of expectation, are also described in the Appendix.

One of the difficulties of doing human genetics is that family sizes are generally quite small.

Therefore data on a number of families must be accumulated even to establish a simple pattern of inheritance for a given trait. A simple way of looking at such a combination of data is to add up all the results for a given type of mating. For example, for the pedigrees of Figure 2.11, if we add up the total of affected and normal offspring from known matings between affected and normal parents, we get the result 13 affected and 20 normal, which is really not too far from a 1:1 ratio. Clearly, as in the case of penny tossing, the more families we could include in such an analysis, the closer the proportion of affected to normal should approach 1:1, which is the expected Mendelian ratio.

Mendel—in his description of his results of the crosses of pea plants that led to his formulation of the major laws of inheritance—was fully aware of the concept of chance variation around expected values. In his famous paper of 1865, he gave data from individual pea plants to illustrate the fact that this variation is greater, the smaller the number of progeny that are counted (Fig. 2.14). Mendel was also aware of another major factor that could cause departures from his expected ratios. This is the possibility that different genotypes may have different chances of surviving to the stage at which their phenotypes can first be scored. This means that, though zygotes may be produced in the expected proportions, mature individuals may deviate systematically from the expectations. This is a problem that is especially relevant to the interpretation of pedigree data on genetically determined diseases in man. These diseases are often associated with lowered probabilities of survival before and after birth, and so one should expect in such cases that the proportion of affected offspring observed in families might be significantly less than that expected on the basis of Mendel's laws.

2.5 Genes on the Sex Chromosomes

When we described the frequencies of different types of offspring expected from matings involving the three genotypes *AA*, *AS*, and *SS*, we could do so without taking account of which parent was male and which was female. For example, in the mating $AS \times AA$, the results are the same whether it is the mother or the father who is *AS*. Not all inherited traits show this lack of correlation with sex. Hemophilia is an inherited disease in which the blood does not clot properly; affected individuals continue to bleed when they are injured, hence the name "bleeders' disease." Though very rare, this disease did occur in one quite famous family—the British royal family. Ten of Queen Victoria's descendants were known to have the disease, and all were males. Hemophilia is actually a very old inherited disease, perhaps one of the first ever to be described. The earliest known

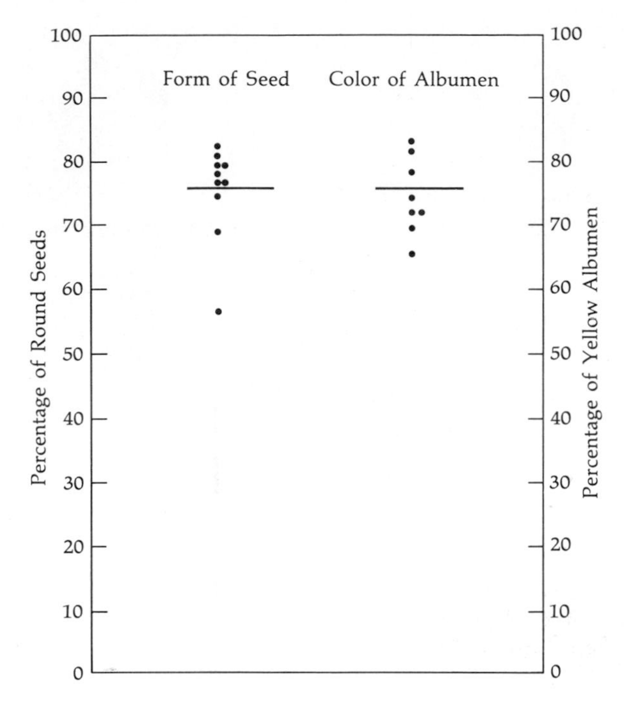

Figure 2.14
Variability in Mendel's data: each dot represents data from an individual plant. Among the traits that Mendel examined were shape of seed (round or wrinkled) and color of the albumen (yellow or green). Fortunately for future geneticists, each of these traits is under control of a single gene, and the phenotypes therefore segregate in easily recognizable patterns.

record of it was in a tract of the Talmud, written before the sixth century A.D. There it is recorded that a rabbi exempted from circumcision a boy whose brothers had bled profusely following this rite. This rule became part of the Jewish Talmudic law. The exemptions from circumcision extended to the sons of sisters of a woman who had given rise to male bleeder offspring, but not to the father's sons by other women! This clearly recognized the transmission of the inherited disease through the female but not through the male line. What can be the explanation for such a pattern of sex-linked inheritance?

Sex-linked inheritance follows the pattern expected for a gene located on the sex chromosomes.

This problem was first solved by T. H. Morgan, one of the most famous of the early geneticists and the leader of the school of genetics which worked with the small fruitfly called *Drosophila* (Fig. 2.15). He found that eye color in *Drosophila*

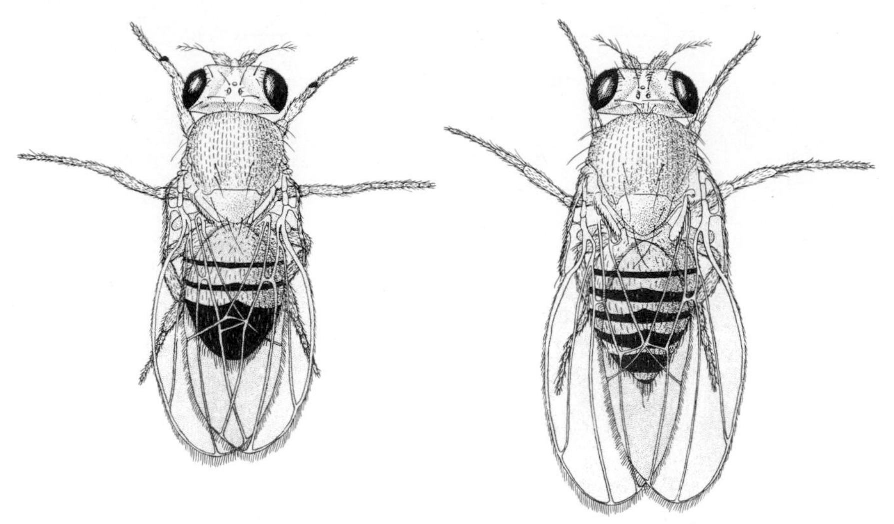

Figure 2.15
Drosophila melanogaster: male (left) and female (right). *Drosophila* is a favorite experimental organism of geneticists, rivaled only by bacteria and laboratory mice. (After A. H. Sturtevant and G. Beadle, *An Introduction to Genetics,* W. B. Saunders Co., 1940.)

followed a pattern of sex-associated inheritance similar to that we have described for hemophilia. Morgan interpreted the data as showing that such a pattern is caused by a recessive gene located on the X chromosome. This was, in fact, the first clearcut association of a gene, as defined by its pattern of inheritance, and a particular chromosome.

A part of Queen Victoria's hemophilia pedigree is shown in Figure 2.16. Fortunately for geneticists, she was quite prolific and had enough fertile offspring to show very clearly the pattern of inheritance expected for a gene on the X chromosome. The first feature of the pedigree is, of course, that all the affected individuals are males. Another point to notice from this pedigree is that none of Queen Victoria's daughters had hemophilia, though one of them, Princess Beatrice, had children and great grandchildren with the disease. Finally, Leopold's family shows how the disease was transmitted to a grandson through an unaffected daughter, Princess Alice. Unaffected women who transmit the disease to their sons are called *carriers,* as indicated in the figure. Unless a woman has children, there is no sure way of telling whether or not she is a carrier for hemophilia.

The existence of female carriers suggests a recessive mode of inheritance for hemophilia. However, none of Queen Victoria's normal male children had descendants with hemophilia, indicating that there is no such thing as a male carrier for hemophilia. Why is there this difference between the sexes? The answer lies in the fact that the Y chromosome is not really homologous to the X chromosome as

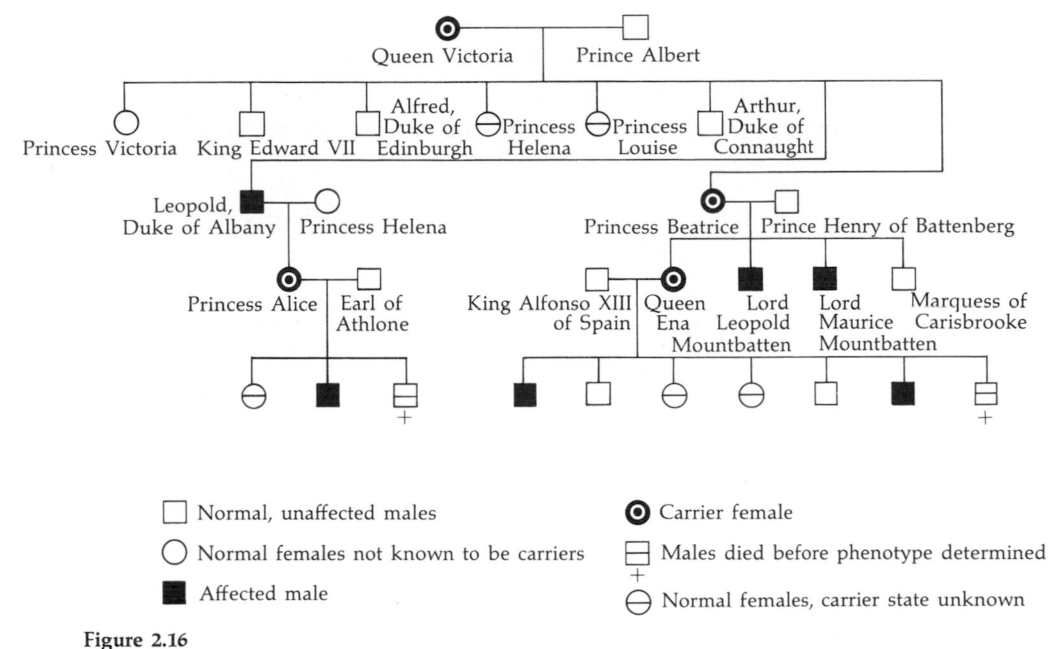

Figure 2.16
Queen Victoria's pedigree and the inheritance of hemophilia. X-linked inheritance is indicated by three facts: (1) all affected individuals are males; (2) none of Queen Victoria's daughters had hemophilia, but one of them (Princess Beatrice) had children and great-grandchildren with the disease; (3) in Leopold's family, the disease is seen to be transmitted to a grandson through an unaffected daughter (Princess Alice). Unaffected women with affected sons are called *carriers*.

far as its gene content is concerned. None of the genes on the X chromosome have yet been found on the Y. Thus, as far as the X chromosome is concerned, the male is genetically haploid and so can never be a heterozygote. Suppose we assume that there is a recessive gene on the X chromosome that is responsible for hemophilia. Let us call the X chromosome carrying this gene X^h and reserve X for the normal chromosome. Because we assume that X^h is recessive, only homozygous X^hX^h females would have hemophilia. Heterozygous X^hX females, however, are carriers of the hemophilia gene. There are only two sorts of males, the normal XY and those who are X^hY. The latter do not have the normal counterpart of the recessive hemophilia gene, and so they are effectively like homozygous X^hX^h females, and therefore are affected by the disease. Such males are said to be *hemizygous,* in contrast to homozygous.

Queen Victoria's pedigree shows two types of crosses that give rise to affected males or carrier females—first, those between carrier females and normal males and second, those between affected males and normal females.

Her own mating with Albert, who did not have hemophilia and so must have been a normal XY male, is an example of the cross of a carrier female by a normal male ($X^hX \times XY$). Following the rules of meiosis and Mendel, the expected outcome of this cross can be represented diagrammatically as follows.

Father (normal male, XY)

Mother (carrier female, X^h X)

Eggs \ Sperm	$\frac{1}{2}$ X	$\frac{1}{2}$ Y
$\frac{1}{2}$ X	$\frac{1}{4}$ XX (normal females)	$\frac{1}{4}$ XY (normal males)
$\frac{1}{2}$ X^h	$\frac{1}{4}$ X^h X (carrier females)	$\frac{1}{4}$ X^h Y (hemophilic males)

The male produces a normal population of sperm, half carrying X and the other half carrying Y. The female produces two types of eggs in equal proportions: those with an X and those with an X^h. Combining the gametes in fertilization according to these proportions, we expect $\frac{1}{2}$ of the male offspring ($\frac{1}{4}$ of the total offspring, of which $\frac{1}{2}$ are males) to be X^hY, and so to be hemophilic. All the females receive a normal X from their father and so are normal. One half of them, however, get an X^h chromosome from their mother and so should be carriers. These expectations are in accord with Queen Victoria's pedigree to the extent that she (and her carrier daughter Beatrice, and her carrier granddaughters Ena and Alice) produced both normal and affected males but only normal females—some of whom proved to be carriers.

What about the second type of mating involving hemophilia—namely that between Leopold, Duke of Albany, an affected male, and Princess Helena, a normal female? Because hemophilia is rare and Helena was not related to the royal family, it seems reasonable to assume that she was a normal XX female. Thus this cross is $X^hY \times XX$, and its outcome can be represented as follows.

Father (hemophilic male, X^h Y)

Mother (normal female, XX)

Eggs \ Sperm	$\frac{1}{2}$ X^h	$\frac{1}{2}$ Y
All X	$\frac{1}{2}$ XX^h (carrier females)	$\frac{1}{2}$ XY (normal males)

We see that in this case all the offspring are normal, but *all* the females are carriers, as was Princess Alice. An affected female must be X^hX^h and so must receive an X^h chromosome from both parents. Apart from the exceedingly rare possibility of a mating between hemophiliacs, only a mating between a carrier

Table 2.2
Patterns of X-linked inheritance. X^h represents an X chromosome carrying the recessive gene for hemophilia. Affected individuals are boxed.

Mating type	Parents		Offspring			
	Female	Male	Female		Male	
1	XX	XY	XX		XY	
2	XX^h	XY	$\frac{1}{2}XX$	$\frac{1}{2}XX^h$	$\frac{1}{2}XY$	$\boxed{\frac{1}{2}X^hY}$
3	$\boxed{X^hX^h}$	XY	XX^h		$\boxed{X^hY}$	
4	XX	$\boxed{X^hY}$	XX^h		XY	
5	XX^h	$\boxed{X^hY}$	$\frac{1}{2}XX^h$	$\boxed{\frac{1}{2}X^hX^h}$	$\frac{1}{2}XY$	$\boxed{\frac{1}{2}X^hY}$
6	$\boxed{X^hX^h}$	$\boxed{X^hY}$	$\boxed{X^hX^h}$		$\boxed{X^hY}$	

female and a hemophilic male ($X^hX \times X^hY$) could give rise to an affected female offspring. Although the British royal family did have a tendency to inbreed, they fortunately managed to avoid this particular combination!

The patterns of inheritance expected for all the possible matings involving recessive genes on the X chromosome are summarized in Table 2.2, using hemophilia as an example. Genes that are on the X chromosome are said to be *X-linked.* These patterns can readily be worked out from Mendel's laws, as we have already done for two of the types of matings. These two—carrier female by normal male, and hemophilic male by normal female (types 2 and 4 in Table 2.2)—are the ones that are most characteristic for an X-linked gene, and are generally the basis for diagnosing this mode of inheritance. For a rare disease, such as hemophilia, the frequency of the X^h chromosome in a population is so low that the other types of matings involving hemophiliacs or female carriers—types 3, 5, and 6 in Table 2.2—are rarely if ever observed.

Genes on the Y chromosome, if they are not also present on the X chromosome, should show a completely different pattern of inheritance from X-linked genes.

Such genes can be transmitted only from fathers to sons along with the Y chromosome. Daughters would never get a Y chromosome and so could never express a *Y-linked* trait (Fig. 2.17). Several pedigrees suggesting Y-linked inheritance have been reported but few, if any, are confirmed. Just one character—hairy ears, found with a relatively high frequency in parts of India—still retains a claim to being Y-linked. The striking pattern of inheritance expected for a Y-linked gene makes it very unlikely that any such differences would be overlooked.

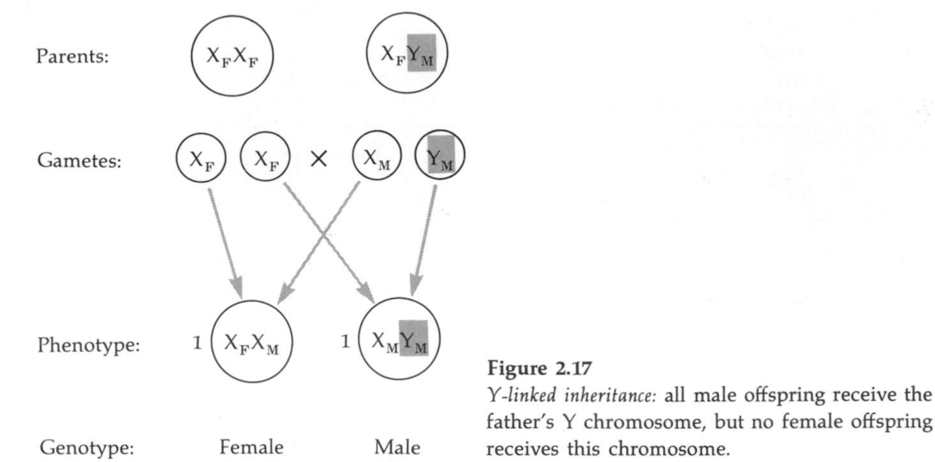

Figure 2.17
Y-linked inheritance: all male offspring receive the father's Y chromosome, but no female offspring receives this chromosome.

Sex-linked inheritance must be clearly distinguished from another mode of inheritance that also shows an association with sex, namely sex-limited inheritance.

A number of inherited traits are expressed quite differently in the two sexes, probably for physiological reasons that are not directly connected with genetic mechanisms. Pattern baldness—which in its milder forms is expressed as a thinning of the hair at the temples, and in its extreme form leaves the top of the head bald with hair remaining on the sides—is an extreme example of such a trait. It is expressed almost exclusively in males, but its pattern of inheritance shows that it clearly is not X-linked or Y-linked. An example of a pedigree of this trait is shown in Figure 2.18. Only males are affected, but the trait is transmitted from male to male. This clearly rules out X-linked inheritance. On the other hand, not all sons of affected fathers are bald, which rules out Y-linked inheritance. Because males in each generation are affected, it seems most likely that the trait is controlled by a dominant gene on an autosome (not on one of the sex chromosomes). Traits that show such a pattern of inheritance, with the trait limited almost exclusively to one sex but not controlled by X- or Y-linked genes, are called *sex-limited.*

Sex-linked and sex-limited inheritance may not always be so easy to distinguish.

This fact is well illustrated by the trait called testicular feminization, which provides an interesting paradox for sex-linked inheritance. Persons having testicular feminization are intersexes (that is, individuals having some of the characteristics of both sexes) with a normal XY karyotype and with well-developed female secondary sex characteristics. Most of these individuals regard themselves as females. They may have undescended testes, and they lack normal internal female

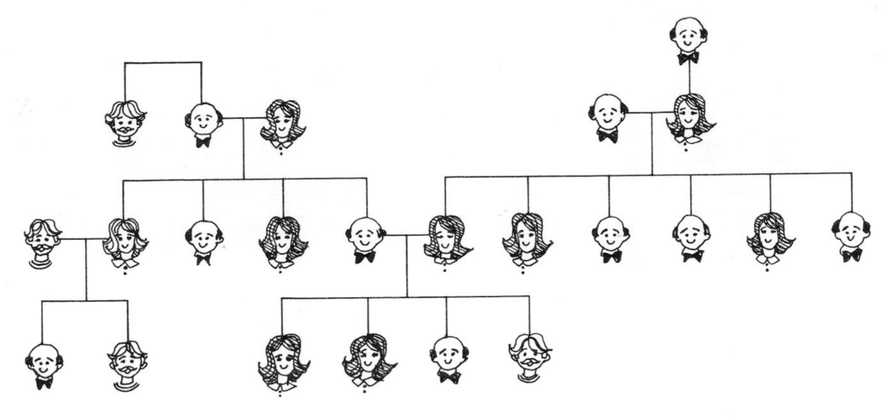

Figure 2.18
A pedigree of pattern baldness. Three types of individuals are represented here: females, bald males, and nonbald males.

genital organs, apart from a small "blind" vagina. Their "outward" phenotype is clearly female (a famous photographic model and twin airline stewardesses with the syndrome have been reported); many of them are happily married, though all of them are infertile. The frequency of the syndrome is estimated to be about one per 65,000 males born. An example of a pedigree showing the inheritance of testicular feminization is given in Figure 2.19. The pattern is that typically expected for an X-linked recessive. Carrier females can, as indicated in this pedigree, often be identified because they have almost no female pubic hair, though they are otherwise normal. The inheritance pattern can, however, be explained just as well by an autosomal dominant gene whose expression is essentially limited to the male. Recent data show that an exactly similar trait in the mouse is inherited through a gene on the X chromosome, suggesting that the same may be true in man.

2.6 Genes on Different Chromosomes

Thus far we have limited our discussion of inheritance to one character at a time—apart, that is, from sex. At the beginning of the last section, we pointed out that the inheritance of the hemoglobin types *AA* and *SS* is not influenced by the sex of the parent, nor is there any association between the types of offspring produced and their sexes.

The cross *AS* × *AA*, for example, produces (on average) equal numbers of males and females of both the offspring genotypes *AS* and *AA*. What this really means is that during meiosis the behavior of the X and Y chromosomes is in no way influenced by the behavior of the chromosomes that carry the hemoglobin

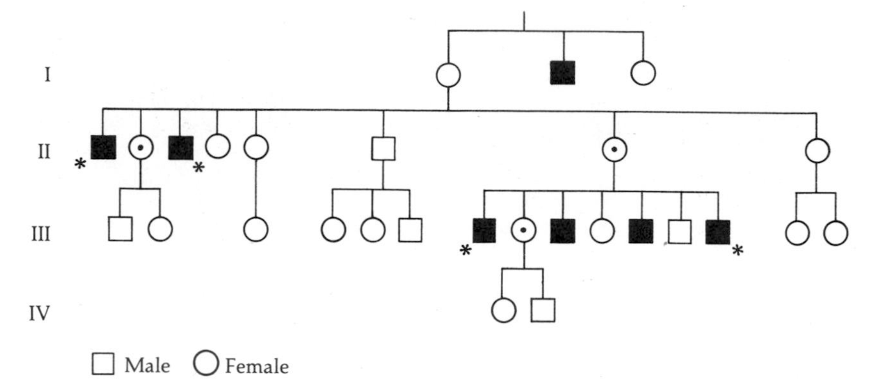

Figure 2.19
Pedigree of the testicular feminization syndrome. Note that the *prima facie* pedigree evidence can be explained either by an X-linked inheritance, or by autosomal dominant inheritance with male limitation. Black squares indicate men observed to have the testicular feminization syndrome; asterisks indicate cases of the syndrome that have been confirmed surgically. Circles with centered dots in them indicate women examined and found to lack pubic hair. (From V. A. McKusick, *On the X Chromosome of Man*, American Institute of Biological Sciences, 1962; adapted from W. E. Schreiner, *Geburten Frauenheilk.*, vol. 19, pp. 1110–1118, 1959.)

genes *A* and *S*. In statistical terms, the two pairs of chromosomes behave independently.

The behavior of chromosomes at meiosis allows one to predict the transmission of inherited traits determined by genes on different chromosomes.

Let us consider what happens during meiosis in a sickle-cell-trait male whose genotype is *AS*, XY. During the first division of meiosis, there are two equally probable ways in which the two pairs of homologous chromosomes can assort: either *A* with X and *S* with Y or *A* with Y and *S* with X. The first assortment gives rise to two *A*X and two *S*Y gametes, while the second produces two *A*Y and two *S*X. Thus the four combinations *A*X, *S*Y, *A*Y, and *S*X are produced with equal probabilities, as expected if the *AS* chromosome and the XY chromosomes behave independently (see Fig. 2.20). If the *AS*, XY male mates with an *AA*, XX female, they will produce four types of offspring as follows.

Father (*AS*, XY)

	Eggs \ Sperm	$\frac{1}{4}$ *A*X	$\frac{1}{4}$ *S*X	$\frac{1}{4}$ *A*Y	$\frac{1}{4}$ *S*Y
Mother (*AA*, XX)	All *A*X	$\frac{1}{4}$ *AA*, XX (normal females)	$\frac{1}{4}$ *AS*, XX (sickler females)	$\frac{1}{4}$ *AA*, XY (normal males)	$\frac{1}{4}$ *AS*, XY (sickler males)

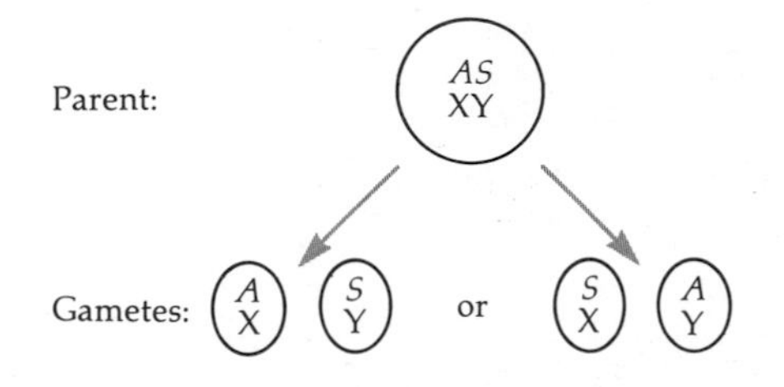

Figure 2.20
Consequences of independent assortment of chromosomes in meiosis. Each member of a gamete pair has complementary chromosomes to those of the other member of the pair. (*AS* and XY represent pairs of homologous chromosomes.)

We see that males and females are equally distributed between the genotypes *AA* and *AS*. Exactly the same arguments apply if we substitute for X and Y any pair of alleles on a different chromosome from that which carries the *A,S* pair.

Let us take as a second example of a genetic trait the MN blood-group system. This is one of a number of different sets of traits that can be detected on the surface of red blood cells, using appropriate techniques that we discuss in more detail later (Chapters 5 and 10). All we need to know at this stage is that three types of individuals can be distinguished: N, MN, and M. These three phenotypes correspond to three genotypes, *NN, MN,* and *MM,* determined by alleles *N* and *M* that are responsible for the production of the corresponding N and M substances on the surface of the red blood cells. Thus, if you have allele *N* your blood cells will be type N, and if you have *M* the cells will be type M. From this it follows that the heterozygote *MN* has both types N and M, and thus has the phenotype MN. Alleles such as *M* and *N,* which simply combine their phenotypes in heterozygotes, are (as mentioned earlier) called *codominant.* The MN blood system is convenient for genetic analysis because of the relatively high frequencies of the three phenotypes in most populations. In Caucasians, for example, the frequencies are typically about 30 percent M, 50 percent MN, and 20 percent N. The frequencies in Africans are not much different, so that it should not be too difficult to find a mating between an MN individual with the sickle-cell trait and a normal N individual: *AS MN* $\times$ *AA NN.* The expected outcome of this mating is just the same as *AS* XY $\times$ *AA* XX, with *N* substituted for X and *M* for Y. Thus, the four types *AA NN* (normal, N), *AS NN* (sickle trait, N), *AA MN* (normal, MN), and *AS MN* (sickle trait, MN) will each occur with a probability of one-fourth.

In the case of the MN blood groups, not all matings, of course, must be *MN* $\times$ *NN.* We can, for example, have a mating *AS MN* $\times$ *AA MN.* The first step in working out the expected consequences of such a mating is to work out the expected gametic output of each parent. Thus, *AS MN* produces $\frac{1}{4}$ *AM,* $\frac{1}{4}$ *AN,* $\frac{1}{4}$ *SM* and $\frac{1}{4}$ *SN,* while *AA MN* produces $\frac{1}{2}$ *AM* and $\frac{1}{2}$ *AN.* Now we work out the consequences of combining all possible combinations of gametes from the two parents in the proportions with which they are produced. This is best laid out as follows in the two-way table of the type we have already used a number of times.

		Parent (*AS, MN*)			
	Gametes	$\frac{1}{4}$ *AM*	$\frac{1}{4}$ *AN*	$\frac{1}{4}$ *SM*	$\frac{1}{4}$ *SN*
Parent (*AA, MN*)	$\frac{1}{2}$ *AM*	$\frac{1}{8}$ *AA, MM*	$\frac{1}{8}$ *AA, MN*	$\frac{1}{8}$ *AS, MM*	$\frac{1}{8}$ *AS, MN*
	$\frac{1}{2}$ *AN*	$\frac{1}{8}$ *AA, MN*	$\frac{1}{8}$ *AA, NN*	$\frac{1}{8}$ *AS, MN*	$\frac{1}{8}$ *AS, NN*

Finally, we add up the probabilities of similar offspring types to give $\frac{1}{8}$ *AA MM*, $\frac{1}{4}$ *AA MN*, $\frac{1}{8}$ *AA NN*, $\frac{1}{8}$ *AS MM*, $\frac{1}{4}$ *AS MN*, and $\frac{1}{8}$ *AS NN*. A similar, but somewhat more lengthy calculation can be done with the most complex mating type, *AS MN* $\times$ *AS MN*; more examples are given in the exercises at the end of the chapter.

2.7 Genes on the Same Chromosome—Linkage and Recombination

The gene for hemophilia is not the only one on the X chromosome; many traits in man are known to be X-linked. Clearly, one would not expect genes on the same chromosome to be transmitted to the gametes independently of each other as are genes that are on different chromosomes. Because there are only 23 chromosome pairs in man, there must be many genes on each chromosome. In fact, unless there were some special mechanism for separating genes that are on the same chromosome, such genes would always be inherited together.

Observations on the inheritance patterns of traits determined by genes on the same chromosome show that separation does occur by a process that is called crossing-over, *or* recombination.

Let us consider as an example the joint inheritance of two X-linked traits: color blindness and glucose-6-phosphate dehydrogenase deficiency (or Gd$^-$, for short). In Caucasian populations about 7 to 8 percent of males, but less than 1 percent of females, are color-blind. Family studies clearly show that this trait is controlled by an X-linked recessive gene. Its relatively high frequency makes it a good candidate for studies of inheritance involving the X chromosome. Glucose-6-phosphate dehydrogenase is an enzyme protein that is used by the body in the metabolism of the sugar glucose. (We shall be saying more about enzymes and proteins in the next chapter on the chemistry of heredity). A deficiency for this enzyme occurs with quite high frequencies in parts of Africa and in other areas where malaria is, or was formerly, prevalent. The enzyme-deficient, Gd$^-$ individuals, seem (just like the sickle-cell-trait *AS* individuals) to be relatively resistant to malaria. As expected for an X-linked recessive trait, Gd^- males are much commoner than Gd^- females. Color-blind (CB) males who are also Gd^- are found, for example, among Africans and American blacks. Suppose now that such a doubly affected (CB and Gd^-) male marries a normal female with no evidence of Gd^- or CB in her ancestry. A daughter from such a mating will receive from her father an X chromosome carrying both the genes causing Gd^- and CB, and from her mother the X chromosome carrying the corresponding normal alleles. A shorthand notation for the daughter's genotype is *cg/CG*—where *c* and *g* refer, respectively, to the recessive alleles for color blindness and Gd^-, and where *C* and *G* are the normal counterparts of the recessive alleles. Alleles that are inherited together on the same chromosome from one parent are separated by a line from those inherited together from the other parent. Thus, the original mating and its consequences for the daughter can be represented as follows.

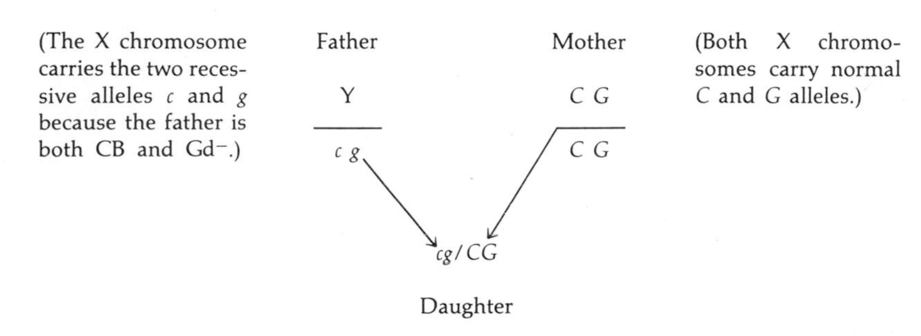

The daughter will be phenotypically normal because she is heterozygous for both the recessive alleles *c* and *g*.

But what will happen if the daughter has male offspring? A male must receive his single X chromosome from his mother, so we could expect half of her sons, on average, to be normal, and the other half to be both CB and Gd^-. In fact, observations of such cases show that two other types also are produced—namely, those that are only color-blind, and those that are only Gd^- (see Fig. 2.21). The normal and the doubly affected males presumably have received the intact *CG* and *cg* chromosomes from their mother. The only reasonable explanation for the other two types is that they have received a part of each of their mother's chromosomes. The color-blind non-Gd^- sons, must have received the recessive *c* allele with the normal *G* allele, while the Gd^- non-color-blind sons must have the reverse, *C* with *g*. In other words, the doubly heterozygous *cg/CG* female produces gametes carrying four types of X chromosomes: the *parental* types *cg* and *CG*, and the new or *recombined* types *cG* and *Cg* (which are called *recombinants*). Thus it appears that segments of homologous chromosomes carrying different genes can somehow be recombined during meiosis. This process is called *crossing-over*, or *recombination*. (Pairs of genes on different chromosomes, such as *A*, *S* and *M*, *N* discussed in the

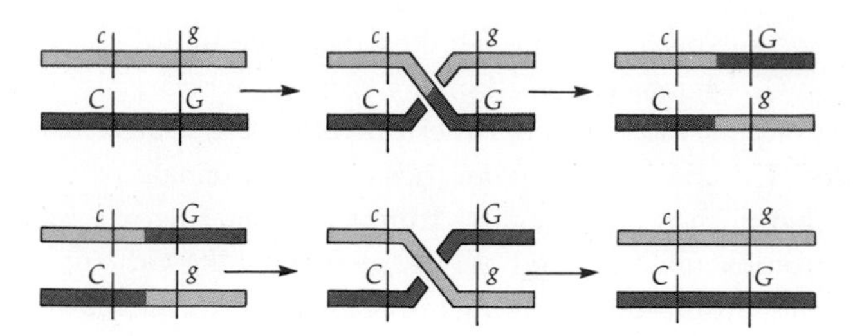

Figure 2.21
Recombination of genes on the X chromosome. The coupling heterozygote *cg/CG* produces *cG* and *Cg* as recombinant gametes. The repulsion heterozygote *cG/Cg* produced *CG* and *cg* as recombinant gametes. (*c* = color blindness; *C* = normal; *g* = G6PD deficiency; *G* = normal.)

previous section also, of course, produce recombinant gametes that carry the genes in a different combination than that transmitted by the parents. But this is expected from the independent assortment of different chromosomes at meiosis.)

There are two types of doubly heterozygous females for X-linked genes.

The first is the case we have just discussed, in which *C* and *G* are together on one chromosome and *c* and *g* are together on the other, namely *CG/cg*. The second, however, has *c* with *G* on one chromosome and *C* with *g* on the other, and so has a genotype that can be written as *cG/Cg*. In this case the parental or *nonrecombinant* chromosomes are *cG* and *Cg*, and these give rise in a mating with a normal male to singly affected CB and Gd$^-$ male offspring. The normal and the doubly affected CB,Gd$^-$ male offspring are now the products of the recombinant *CG* and *cg* chromosomes. The roles of recombinant and nonrecombinant chromosomes are thus interchanged when *cG/Cg* parents are compared to *CG/cg* parents (see Fig. 2.21). By convention, double heterozygotes such as CG/cg, in which dominant alleles are together on one chromosome and recessive alleles on the other, are said to be in *coupling*, while the other situation, *cG/Cg*, is called *repulsion*.

The frequency of recombination is independent of the combination of alleles.

Observations on the frequencies of the various genotypes produced by matings involving two genes on the same chromosome show two very important features of the recombination process.

(1) The two parental combinations (for example *cg* and *CG*) are, on average, produced in equal proportions (see Table 2.3). The two nonparental, or recombinant, combinations (in this case *cG* and *Cg*) are similarly produced in equal proportions. (Parental and recombinant combinations do not, however, occur in

Table 2.3
Expected outcome of coupling and repulsion double backcross matings for two autosomal genes on the same chromosome. *A,a; B,b* are two pairs of dominant and recessive alleles on the same autosome; *r* is the recombination fraction between them.

	Double heterozygote ◇		Double homozygote ◆	Offspring			
				Nonrecombinant		Recombinant	
Coupling	*AB/ab*	×	*ab/ab*	*AB/ab*	*ab/ab*	*Ab/ab*	*aB/ab*
				$\frac{1}{2}(1-r)$◇	$\frac{1}{2}(1-r)$◆	$\frac{1}{2}r$⬙	$\frac{1}{2}r$⬘
Repulsion	*Ab/aB*	×	*ab/ab*	*Ab/ab*	*aB/ab*	*AB/ab*	*ab/ab*
				$\frac{1}{2}(1-r)$⬙	$\frac{1}{2}(1-r)$⬘	$\frac{1}{2}r$◇	$\frac{1}{2}r$◆
			Totals	$1-r$		r	

Phenotype Symbols
◇ Normal
⬘ a
⬙ b
◆ ab

equal proportions.) The genotype *CG/cg,* for example, produces on average equal numbers of *CG* and *cg* chromosomes. The numbers of the two recombinant chromosomes *Cg* and *cG* are also, on average, the same. This suggests that recombination is a *reciprocal* process, the production of *Cg,* say, always being coupled with the production of *cG.* (2) The overall proportion of recombinant chromosomes out of the total produced is a feature of the particular genes being studied, and is the same whether the double heterozygote parent is in coupling or in repulsion. Thus, the coupling genotype, *CG/cg* produces *Cg* and *cG* recombinants with the same frequency, on average, that the repulsion genotype *Cg/cG* produces *CG* and *cg* recombinants.

The proportion of recombinant types produced by such double heterozygotes is called the recombination fraction.

Some examples of pedigrees showing the joint patterns of inheritance of CB and Gd$^-$ are shown in Figure 2.22. We shall concentrate our attention on the male offspring because they reveal directly by their phenotypes the types of chromosomes that they have received from their mothers. In family B, for example, the grandfather I.1 is CB and Gd$^-$. His daughter is normal, and therefore is presumably *cg CG,* while both her children III.1 and III.2 are CB Gd$^-$, and so must have received the parental, nonrecombinant, *cg* chromosome from their mother.

Family H illustrates a contrasting pattern of inheritance. Grandfather I.1 is CB but not Gd$^-$, so his X chromosome must be *cG.* His son II.2, on the other hand,

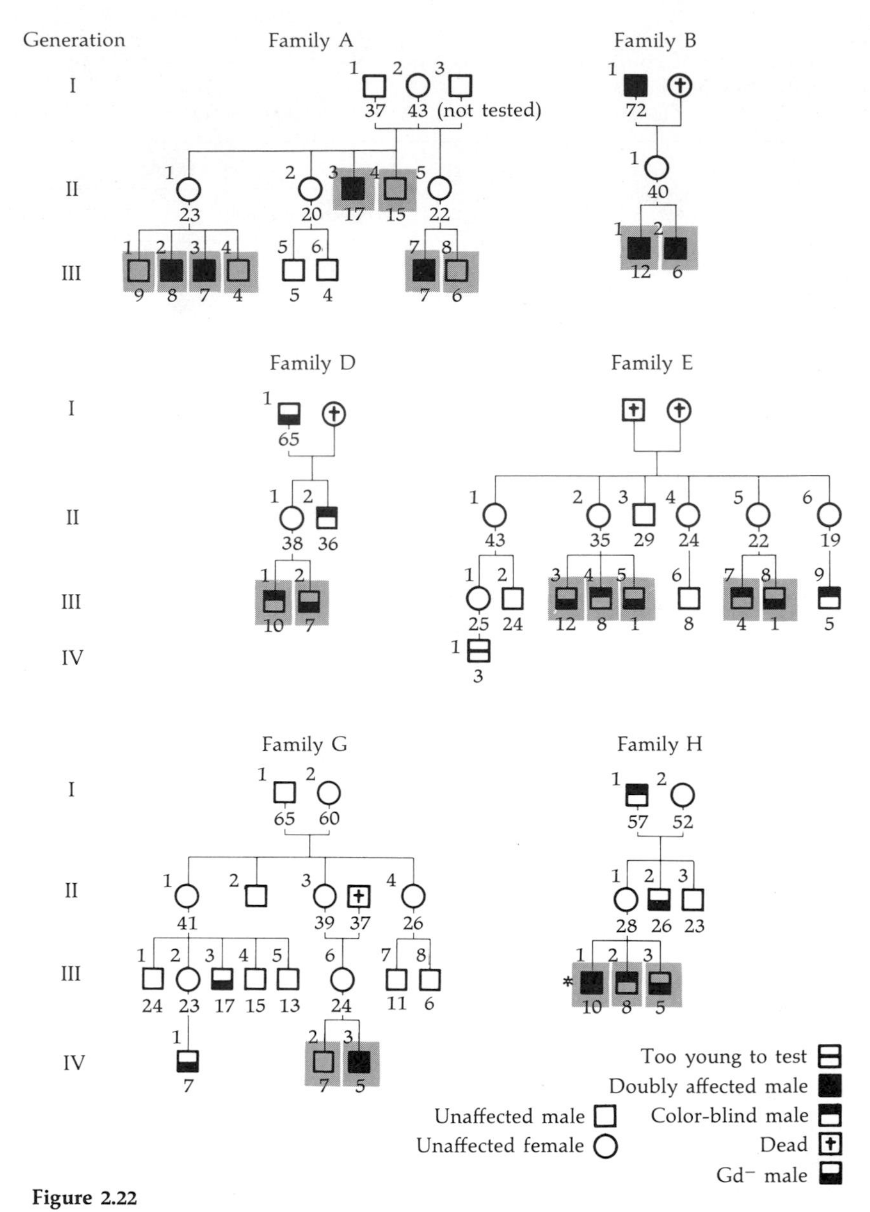

Figure 2.22

Linkage of color blindness and G6PD deficiency. Among 3,648 black schoolboys, color blindness was found in 134 (3.7 percent). All males (totaling 236) in the sibships of 106 of the color-blind probands were tested for Gd^-. In ten sibships, both traits were discovered; six of these families (having the deutan type of color blindness) are diagrammed here. Only one definite instance of recombination was discovered (in family H) out of 22 relevant offspring (indicated in the diagram by shaded squares). Numbers below the symbols indicate ages at time of testing; numbers above and to the left of symbols identify individuals within each generation. (After V. A. McKusick, *On the X Chromosome of Man*, American Institute of Biological Sciences, 1962.)

is Gd⁻ and not color-blind, and so must have received a *Cg* chromosome from his mother, showing that she was heterozygous at least for *G* and *g*. The daughter II.1 has three sons, two of whom are singly affected, one being color-blind and the other being Gd⁻; the third son is doubly affected, CB and Gd⁻. We know that she must have received a *cG* chromosome from her father, which accounts for the color-blind son. The Gd⁻ son indicates that she also got a *Cg* chromosome from her mother, as did her Gd⁻ brother. Thus, II.1 must be a repulsion double heterozygote *cG*/*Cg*. The doubly affected son, III.1 is therefore a recombinant who has received a *cg* chromosome from his mother. The phenotypes and genotypes of this family as we have just worked them out are illustrated schematically in Figure 2.23.

All the male offspring in families in which both CB and Gd⁻ occur are shaded in Figure 2.22. The presumed genotypes of the individuals in all of these families can be worked out just as we have done for families B and H. From this it turns out that individual III.1 of family H is, in fact, the only recombinant out of 22 male offspring from doubly heterozygous females. These data, therefore, give a recombination fraction of $\frac{1}{22}$, which is close to the value of 0.04 to 0.05 that is obtained from pooling all the available sources of data.

Thus far we have ignored the females in our analysis, because their phenotypes depend on the X chromosomes they get from their father as well as those from

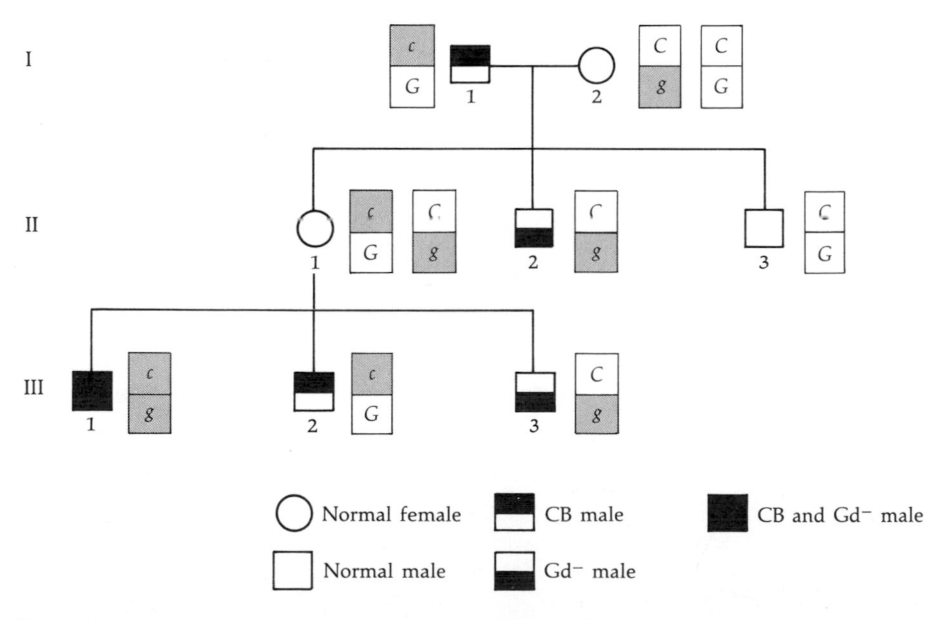

Figure 2.23
Schematic interpretation of linkage between two X-linked genes: a genetic analysis of family H from Figure 2.22. The individual alleles are shown, as well as the phenotype of each family member. III.1 is recombinant.

their mother. These phenotypes do not therefore in general indicate directly whether the female's X chromosomes are recombinant or not.

The female phenotypes indicate recombination only if their mothers are double heterozygotes (CG/cg or Cg/cG), and also their fathers are doubly affected CB and Gd⁻, namely of genotype cg/Y.

A schematic diagram illustrating the expected outcome of the cross ♀ *CG*/*cg* × ♂ *cg*/Y is shown in Figure 2.24. The frequencies shown are those expected for 100 offspring, assuming a recombination fraction of $\frac{1}{25} = 0.04$ (or 4 percent). The father either contributes a Y to his sons, or a *cg* X chromosome to his daughters. The mother contributes equal numbers of the two parental gametes *CG* and *cg*

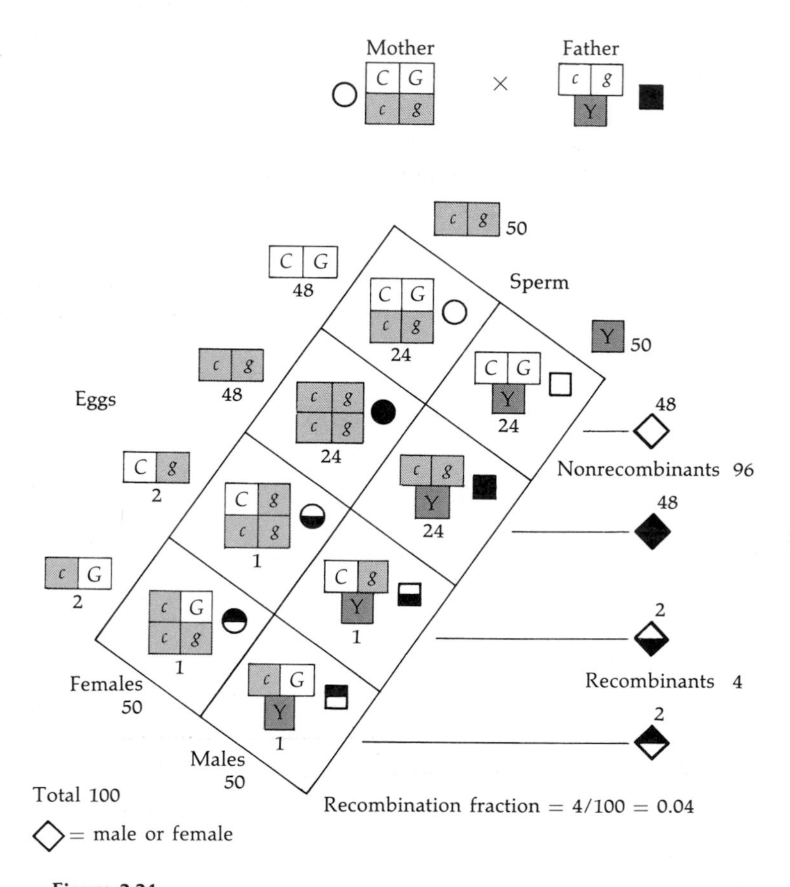

Figure 2.24
Expected outcome of the cross ♀ *CG*/*cg* × ♂ *cg*/Y. A recombination fraction of 0.04 is assumed. The gametes are represented along the edges of the rectangle.

and similarly equal numbers of the recombinant *Cg* and *gC* gametes. As before, the overall number of recombinant is much less than the number of nonrecombinant gametes. The phenotype of the males again is reflected directly by the type of X chromosome they have received from their mother. Now, however, the same is true for the females, because they all receive both the recessive alleles *c* and *g* from their father. Thus, in this case male and female phenotypes can be pooled as they give equivalent information, and the recombination fraction is, therefore, just the proportion of recombinant phenotypes (CB or Gd^-) out of the total number of offspring, namely $\frac{4}{100} = 0.04$.

Autosomal linkage can be analyzed in the same way.

We have, so far, used the X chromosome to illustrate the analysis of recombination because of the direct relationship in males between phenotypes and genotypes. As illustrated in Figure 2.24, females can be used for such analysis, if the father is hemizygous *cg*/Y for both the alleles causing color blindness and Gd^-. A simple extension of this principle is used in the analysis of recombination between autosomal genes (genes that are not on the X chromosome). Suppose we have two pairs of autosomal dominant or recessive alleles *A*,*a* and *B*,*b* such that homozygotes *aa* and *bb* determine distinguishable phenotypes a and b. The recombination fraction can then be directly determined in matings between coupling (*AB*/*ab*) or repulsion (*Ab*/*aB*) double heterozygotes and the doubly recessive "ab" phenotype, whose genotype must be *ab*/*ab*. This is possible because offspring from such matings will always receive both recessive alleles from one of their parents.

Table 2.3 illustrates the expected outcome of these *double backcross* matings, assuming an arbitrary recombination fraction, *r*. In each case, all the four distinguishable phenotypes correspond directly to the four genotypes that result from the four chromosome combinations produced by the doubly heterozygous parents. Thus recombinants and nonrecombinants can be counted directly. It is important to note that this analysis is possible only if we know whether the double heterozygote parent is in coupling or in repulsion. In the case of the X-linked genes, we were able to determine this from the ancestry of the double heterozygote females, in particular from their father's phenotype. However, in the case of autosomal loci it is usually more difficult to establish from earlier generations whether the double heterozygote is in coupling or repulsion.

What is the significance of the recombination fraction?

A comparison of Table 2.3 with the results expected for two genes on different chromosomes, as described in Section 2.6, shows that the two correspond when the recombination fraction is $\frac{1}{2}$. In this case the frequency of each of the four

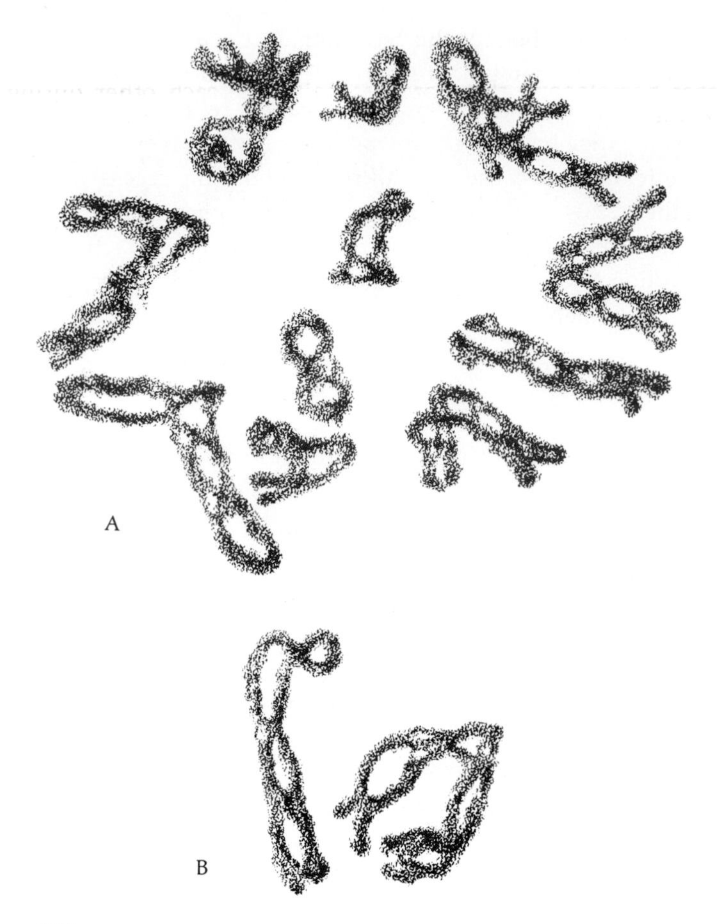

Figure 2.25
Meiosis in the male, with chiasmata and pairing: drawings from photographs of meiotic chromosomes in the spermatocytes of a species of salamander (*Batrachoseps wright*). Salamander chromosomes are easily observable and thus provide good photographs. In B, the four-strand structure is easily visible. Crossovers are clearly seen in A and B. (After photographs by J. Kezer.)

types of offspring is $\frac{1}{4}$, regardless of the origin of the doubly heterozygous parent. In other words, the recombination fraction for genes on different chromosomes is effectively $\frac{1}{2}$. Pairs of genes on the same chromosome that give rise to recombination fractions less than $\frac{1}{2}$ are said to be *linked*—hence the term *linkage,* and the use of *linkage phase* to describe the state of a double heterozygote, namely whether it is in coupling or in repulsion. The recombination fractions between linked genes range from zero (or almost zero) up to somewhat less than 0.5 (or 50 percent). To understand the significance of such differences, it is helpful first to consider what might be the physical basis for the production of recombinant chromosomes.

Pairing at meiosis is a prerequisite for crossing-over.

Recall that homologous chromosomes pair with each other during the first division of meiosis before they separate at anaphase. F. A. Janssens, in 1909, saw that such paired homologs were often connected with each other by cross-like figures which he called *chiasmata.* Figure 2.25 is a drawing illustrating chiasmata in meiosis. Janssens suggested that these chiasmata actually reflect the physical exchange of parts between homologous chromosomes that results in the production of a recombinant chromosome. The essentials of this suggestion remain valid to this day. The simplest model for recombination is that, during pairing in meiosis, chromosomes may occasionally be broken and rejoined with their homologs to form a physically recombinant structure as illustrated in Figure 2.26.

The "breakage and reunion" model of recombination assumes that duplicated pairs of homologous chromosomes are aligned during pairing at meiosis in such a way that breakage and rejoining of homologous strands can occur. The breaks in each strand must occur in exactly homologous positions; otherwise recombination would not be reciprocal, and one product of crossing-over would carry an excess while the other had a deficiency of genetic material. If the probability of breakage and rejoining were the same at all points on the chromosome then presumably, the farther apart two genes are, the more likely it is that a break and rejoin, or crossover, will have occurred between them. Because each crossover produces a recombinant, this means that, the farther apart two genes are on a chromosome, the larger will be the recombination fraction between them. Thus, as Thomas Hunt Morgan realized by 1911, the recombination fraction can be used as a measure of the distance between two genes on a chromosome. It was left to Morgan's student and close associate, A. H. Sturtevant, to construct the first *genetic map* based on recombination fractions, that of the X chromosome of *Drosophila* (Fig. 2.27).

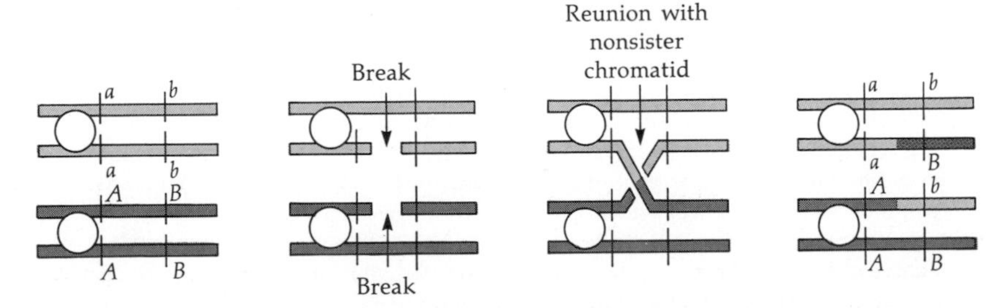

Figure 2.26
Schematic representation of the mechanism of single crossovers. Crossing-over occurs when chromatids break and the broken end of each chromatid joins with the chromatid of a homologous chromosome. The result is an exchange of alleles between chromosomes.

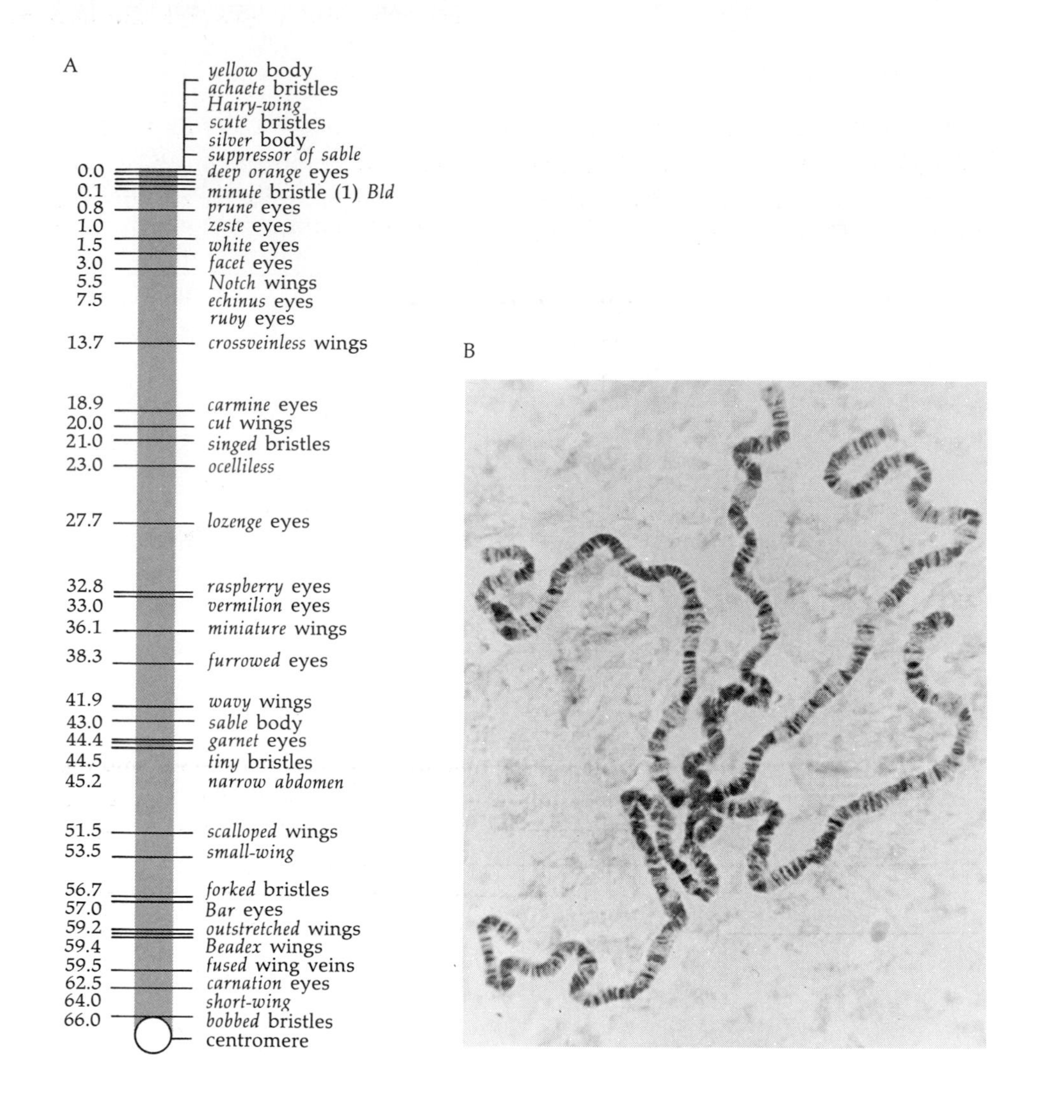

Figure 2.27

Genetic maps of Drosophila chromosomes. (A) Many of the phenotypic traits of *Drosophila* have been mapped to specific positions on the four chromosomes of this fruitfly. The map of the X chromosome is given here as an example of the extent to which particular genes have been mapped in this species. (B) Photograph of *Drosophila* salivary-gland chromosomes. These chromosomes are 100 times larger than those in ordinary body cells, a condition called polyteny. The light and dark bandings correspond almost exactly with the positions of specific sets of genes on genetic maps prepared through recombination analysis. (Part A after M. Strickberger, *Genetics*, Macmillan, 1968. Part B, photograph by B. P. Kaufmann.)

Genetic maps can be constructed on the basis of recombination analysis.

We shall illustrate the principle of genetic map construction, using the three X-linked genes for hemophilia (H), color blindness (CB), and glucose-6-phosphate dehydrogenase deficiency (Gd$^-$). Suppose the three pairwise recombination fractions between these genes are (in percent)

CB – Gd$^-$	:	4
H – Gd$^-$	:	2
H – CB	:	6

From this information it appears that H and CB are farther apart than either H and Gd$^-$ or CB and Gd$^-$. This suggests that Gd$^-$ lies between H and CB. Thus, if we use recombination fractions as measures of distance along a line, we get the following map.

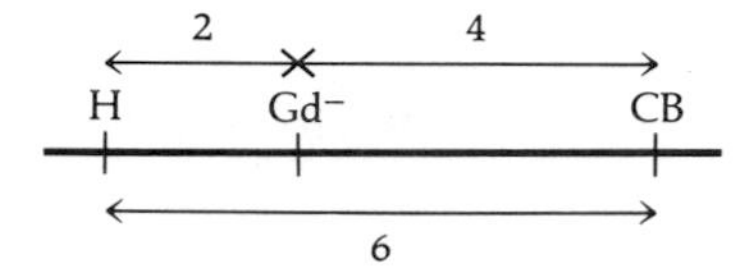

The recombination fraction for H-CB is equal to the sum of the H-Gd$^-$ and the Gd$^-$–CB fractions, just as expected from the map. Similar data should be obtained for any set of three genes, though the results may not always be as close to expectations as we have indicated. In particular, recombination fractions are not additive when the distance between the loci is much larger than that in this example. In fact, more than one crossover can occur within a given genetic region (Fig. 2.28). The triple heterozygote for the hemophilia, CB, and Gd$^-$ alleles *hcg*/*HCG* could,

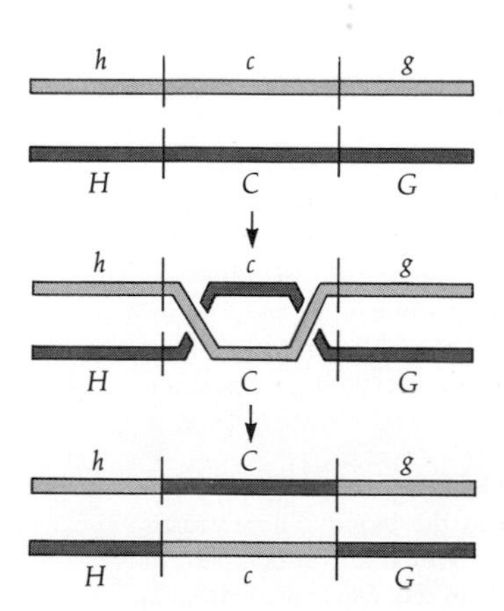

Figure 2.28
Schematic representation of the mechanism of double crossovers. Breaks occur between *H* and *C* (and between *h* and *c*), and also between *C* and *G* (and between *c* and *g*), producing the doubly recombinant gametes *hCg* and *HcG*.

for example, produce the doubly recombinant gametes *hCg* and *HcG* if recombination occurred in the *H*-to-*C* as well as the *C*-to-*G* region. By combining data from many sets of crosses involving two or more genes, we can create a linear map involving many genes. Such maps should be consistently linear in the sense that a cross involving any three genes always gives the sequence that would be predicted from the map. The position of a gene on its chromosome is often called its *locus.* The terms gene and locus are, in fact, often used interchangeably.

Extensive genetic maps have been constructed in a number of higher organisms, notably Drosophila and the house mouse.

Genetic mapping in man, however, has, until recently progressed much more slowly. Matings between suitable human genotypes cannot be arranged at will, as

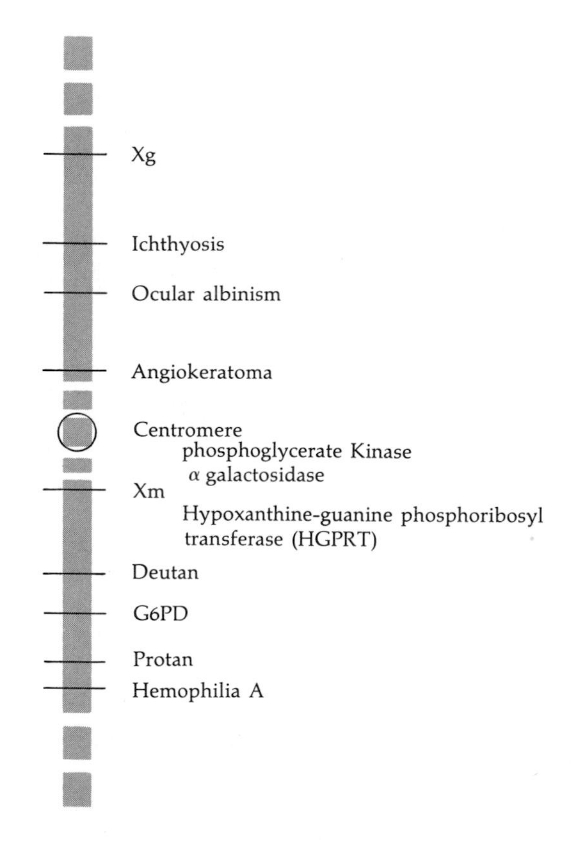

Figure 2.29
Map of the human X chromosome. Deutan and protan are two forms of color blindness. Xg is a blood-group protein, and Xm is another protein found in the blood. There are at least 93 loci known to belong to the X chromosome, most of which are unmapped, in addition to almost as many others that are believed to be X-linked. Several of the map locations given here are tentative.

can be done with experimental plants and animals. Appropriate matings must be sought in the existing population and may be difficult to find. Many of the possible genetic differences are, moreover, quite rare—so that families providing information on more than two markers at a time are not common. Clearly, in any case, only those families involving at least one parent who is a double heterozygote will be informative. Perhaps the most difficult problem, however, is the determination of the linkage phase of putative double heterozygotes. In the case of X-linked loci (Fig. 2.29), we have seen how information from three or more generations can be used to assign the linkage phase, but with autosomal loci this may not be possible. New techniques of genetic analysis using cells growing in the test tube have, however, led recently to major advances in our knowledge of the human gene map (see Chapter 5).

Summary

1. Nearly all somatic cells of an organism have the same number of chromosomes, which occur in pairs. In man, this diploid number is 46, or 23 pairs. Mitosis assures that daughter cells are identical, in their chromosomal constitution, to parental cells.

2. Gametes (sperm and eggs) have the haploid number of chromosomes (one-half the diploid number). Chromosome reduction occurs by a process called meiosis, which assures that every gamete receives one chromosome of each pair. The separation of homologues at meiosis is preceded by their pairing. Fertilization, or the fusion of a sperm and an egg, restores the diploid number.

3. The two different chromosomes, X and Y, form one of the 23 pairs. There are two types of sperm: those carrying an X and those carrying a Y. These determine, respectively, a female or a male zygote.

4. Individuals carrying the same gene (allele) in two homologous chromosomes are homozygous; the cross between two homozygotes for different alleles of the same gene gives rise to a heterozygote for the two alleles. Phenotypically the heterozygote may be similar to one or the other of the two parents, and in this case the parental type that is manifested in the heterozygote is called dominant. The heterozygote may alternatively be intermediate.

5. Homozygotes for a given autosomal gene form only one type of gamete for that gene; heterozygotes form two types, in equal proportions. Hence the mating of a homozygote to a heterozygote (a backcross) generates two types of progeny (identical to the two parents) in equal proportions.

6. The mating between two heterozygotes generates on average, out of four progeny, two heterozygotes and one each of the two types of homozygotes.

7. The result of any mating can be predicted if the genotypes of the two parents are known. The sequence of operations is the following: (a) write down the types of gametes that each parent can produce, and their relative frequencies for each parent; (b) combine each paternal with each maternal gamete, and obtain the expected frequency of each zygote generated as the product of the frequencies of the two corresponding gametes; (c) obtain the frequency of each genotype (or of each phenotype) by pooling zygotes with the same genotype (phenotype)—that is, by adding the relative frequencies.

8. Human inheritance is studied through pedigrees. Sometimes incomplete penetrance of a gene's effect may make interpretations uncertain. A fuller understanding of pedigree analysis requires a knowledge of population genetics (see Chapter 6 and following chapters).

9. Genes on sex chromosomes have special rules of inheritance. Those known are almost all on the X chromosome and have no counterpart on the Y. Fathers transmit their Ys only to sons and their Xs only to daughters. Sons obtain their Xs only from their mothers.

10. Sex may have an effect on manifestation of traits that depend on autosomal genes (sex-limited and sex-influenced traits).

11. Genes on different chromosomes behave independently in their inheritance.

12. Genes on the same chromosome, if they are not too far apart, show linked inheritance—that is, they tend to be transmitted together. They can be separated from each other by the process of crossing-over or recombination.

13. The frequency with which genes located on the same chromosome recombine in the progeny is a function of their separation on the chromosome. Recombination can thus be used to construct chromosome maps.

14. The linkage phase of a double heterozygote for two linked genes (coupling, where both dominant types are on one chromosome, and both recessives on its homolog; or repulsion, where there is one dominant and one recessive on each of the two homologs) needs to be known for the understanding of crossing-over. The phase can usually be assessed if the parents of the double heterozygote are known.

15. The best way to familiarize oneself with the laws of inheritance is to practice them on examples such as those in the following exercises.

Exercises

A number of exercises are provided at the end of each chapter. These fall basically into two categories. The first includes review questions that cover, more or less in sequence, the topics dealt with in each chapter. The second includes problems, often quantitative in nature. Some of the more difficult problems may appeal mostly to those who have some grounding in mathematics and statistics; further background information is given in the Appendix.

1. Review the functional aspects of mitosis and meiosis, in relation to hereditary transmission, and underline the most salient differences between the two processes.

2. Why are Mendel's laws expressed in terms of probabilities?

3. In which matings do we expect a segregation of 1:1? 3:1? (Include autosomal dominant, recessive, and X-linked.)

4. When do we expect two loci to segregate independently?

5. How is a percentage of recombination estimated, and what does it indicate?

6. Give at least three possible explanations why a cross between an individual believed to be homozygous for a trait A, crossed to one believed to be heterozygous for A and A′ (being of phenotype A′, and the progeny of a cross A × A′, where A′ is an inherited characteristic alternative to A), gives progeny A:A′ in proportions different from 1:1.

7. *Mitosis.* Indicate whether each of the following statements is true or false.
   a. Two daughter cells originating from a mitosis are genetically identical to each other.
   b. Two daughter cells originating from a mitosis are genetically identical to the parental cell.
   c. Barring exceptional events, all somatic cells of an individual are genetically identical.
   d. Barring exceptional events, a somatic cell of an individual is genetically identical to a somatic cell of another individual.

8. *Meiosis.* Indicate whether each of the following statements is true or false.
   a. If we take at random two of the four gametes originating by a meiotic process from one single diploid cell, we expect them to be identical for a given gene.
   b. Two gametes as above *can be* identical for a given gene.
   c. If two gametes as above are identical for a gene, the other two are identical to each other for the same gene.
   d. The formation of polar bodies is responsible for the fact that there are far fewer eggs (which are of the order of thousands) than spermatozoa (which are of the order of millions).
   e. If there were no meiosis, the number of chromosomes would increase in successive generations of a sexual organism.
   f. The formation of polar bodies might explain why there are a few more males than females at birth.
   g. If Y-carrying sperm is more motile than X-carrying sperm, the excess of males at birth might be explained.
   h. Occasional imperfections of the meiotic process may explain why there are individuals with an abnormal number of chromosomes.

9. *Autosomal genes.*
   a. What is the expected distribution of progeny from matings between an individual with the sickle trait and a sickle-cell anemic?
   b. Can a sickle-cell anemic who is married to a

"normal" individual have anemic progeny? have normal progeny?

c. In ABO blood groups, *A* and *B* are codominant alleles. What is the expected progeny of *AB* × *AB* matings?

d. In ABO blood groups there is also a third allele, *O*, which is recessive to both *A* and *B*. What can be the genotype(s) of an individual who is phenotypically A? of an individual who is phenotypically B? phenotypically O?

e. About 30 percent of Caucasians are unable to taste the substance PTC (phenylthiocarbamide, see Chapter 14), which the other 70 percent find bitter. The two types of individuals are called nontaster and taster, respectively. Assuming that tasting is dominant, so that tasters can be either homozygous or heterozygous for the gene in question, and nontasters are homozygous for the recessive gene, answer the following questions.
   1) Can there be tasters among the progeny of nontaster × nontaster?
   2) Can there be nontasters among the progeny of taster × taster?

f. If you have answered part e correctly, you can use the same reasoning inversely in the following problem. About 7 percent of Caucasians are anosmic (incapable of smelling) to the odor of musk. The children of an anosmic × anosmic mating are anosmic. Those of a normal × normal mating are usually normal, but a few of them are anosmic. What does this suggest about the inheritance of this type of anosmia?

g. Huntington's chorea is a severe and rare disease with a late age of onset (see Sect. 7.4 for a description). In many of the pedigrees in which a patient with this disease is found, one of the parents of the patient has the same disease. In others, a parent died well before the usual age of onset and so might have had the disease, had he or she lived long enough. What does this suggest with respect to the transmission of Huntington's chorea: is it dominant or recessive?

h. Albinism is a rare recessive condition. What is the expected proportion of albinos among the progeny of two heterozygous parents?

l. Maple-syrup-urine disease is a rare inborn error of metabolism. It derives its name from the odor of urine of affected individuals. If untreated, affected children die soon after birth. The disease tends to recur in the same family, but the parents of the affected are always normal. What does this suggest with respect to the transmission of the disease: is it dominant or recessive?

10. *Sex linkage.*
    a. Normal parents have a family with two hemophilic boys and one normal girl. What is the probability that the girl will transmit the disease?
    b. A color-blind man has a normal wife and one color-blind daughter. What can one say about the genotype of the wife? What is the expected incidence of color blindness among other daughters?
    c. A blood group called Xg is known to be X-linked. Xg^a is a dominant allele determining a positive reaction with a reagent called anti-Xg^a. If a woman is negative and her husband positive, how will their sons and daughters react to anti-Xg^a?
    d. A man who is color-blind has four normal sons from a first wife who died, and a color-blind daughter from a second wife, who is herself phenotypically normal. What can one say about the genotypes of the two wives?

11. a. Give the frequencies of all the genotypes that arise from crossing two individuals, each of which is heterozygous (*Aa Bb*) at a pair of unlinked loci.
    b. If the genes *A* and *B* are dominant, what are the phenotype frequencies expected from this mating?

12. *Linkage and recombination.*
    a. The mother of a family with ten children is Rh^+ (a dominant blood-group type, described in Section 10.2) and has a very rare condition with no adverse clinical effect in which the red cells instead of being round, are oval (elliptocytosis, E). The father is Rh^- (lacks the Rh^+ type) and has normal red cells (symbol: e), the children are 1 Rh^+e, 4 Rh^+E, 5 Rh^-e. Information is available on the mother's parents, who are Rh^+E and Rh^-e respectively. One of the ten children who is Rh^+E married

someone who is Rh^+e, and they had an Rh^+E child. First draw the pedigree of this whole family, and then answer the following questions.

1) Is the pedigree in agreement with the hypothesis that Rh^+ is dominant and that Rh^- is recessive?
2) What is the mechanism of transmission of elliptocytosis?
   Could the genes for E and Rh be on the same chromosome? If so, compute the map distance between them and comment on it.

b. More information on linkage is available from experimental animals than in man because one can breed special strains that can be used to make informative crosses, from which large numbers of individuals can be obtained. Remember that it is customary to indicate dominant alleles by capital letters, and recessive alleles by small letters. In a cross *AaBb* × *aabb* the following numbers of progeny were obtained: AB 110; Ab 5; aB 8; ab 91.
   1) The parent *AaBb* could be a double heterozygote in coupling (*AB*/*ab*) or in repulsion (*Ab*/*aB*). Which is the more likely possibility?
   2) What is the recombination fraction given by these data?

c. The cross *Ab*/*aB* × *aabb* gave offspring AB 2; Ab 37; aB 45; ab 0. Is this in agreement with the results of the cross given in part b? [For a quantitative answer, use the 2 × 2 table as described in the Appendix.]

d. A third locus *Cc* was segregating in the cross described in part b. The full genotypes of the parents were *ABC*/*abc* × *aabbcc* and the progeny were : ABC 95; ABc 15; AbC 5; aBC 1; Abc 0; aBc 7; abC 12; abc 79.
   1) Show that, ignoring *Cc*, these data are the same as those of part b, and hence give the same recombination fraction between A and B.
   2) Ignoring *Aa*, compute the recombination fraction between genes B and C.
   3) Ignore *Bb* and evaluate the last of the three recombination fractions.
   4) Construct the genetic map for genes A, B, and C, showing which of the three is in the middle and giving the map distances between them.
   5) Which, and how many, of the above progeny must have arisen from a minimum of *two* crossovers?

e. Repeat the same analysis as in part d for the following data from the cross *XYw*/*xyW* × *xxyyww*: XYW 4; xYW 114; XyW 69; XYw 325; xyW 301; xYw 81; Xyw 99; xyw 7.

f. In both parts d and e, the smallest progeny class arises from double crossovers. Justify this. Check the rule by which one can tell which of the three genes is in the middle, by simply looking for which gene must be exchanged in the triply heterozygous parent to obtain the double crossovers. (For example, if the rarest progeny from cross *DEF*/*def* × *ddeeff* are *dEF* and *Def*, then D is in the middle.)

g. An affective mental disorder (bipolar manic–depressive psychosis or bmdp, described in Section 17.4) is believed almost always to be due to a dominant sex-linked gene. The initial evidence for this was that affected males almost always have affected mothers and normal fathers. If the disorder is X-linked, it should show linkage with other X-linked genes that are not too far away. Which of the following matings will give the most clearcut results, and what is the expected distribution of progeny if there is a 10 percent recombination between bmdp and an X-linked marker gene *Mm* (which could, for instance be Xg, color blindness, or G6PD).
   1) father bmdp *M* × mother normal *m*
   2) father normal *M* × mother bmdp *m*
   3) father bmdp *m* × mother normal *M*
   4) father normal *m* × mother bmdp *M*

   Assume there is no information on the parents of the father and mother. Contrast the results expected from
   1) the case of a 10 percent recombination fraction between the loci;
   2) bmdp not being X-linked; and
   3) bmpd being X-linked, but with its location sufficiently far from that of the marker *Mm* that there is free recombination (50 percent) between them.

References

Stern, C., and E. R. Sherwood, eds.
1966. *The Origin of Genetics: A Mendel Source Book.* San Francisco: W. H. Freeman and Company. [This book includes the English translation of Mendel's papers and other papers and letters related to Mendel's original discovery. The reading of Mendel's 1865 paper, "Experiments on Plant Hybrids," is a unique and highly recommended experience.]

Swanson, C. P.
1967. *Cytogenetics.* Englewood Cliffs, N.J.: Prentice-Hall.

Whitehouse, H. L. K.
1973. *Towards an Understanding of the Mechanisms of Inheritance.* 3rd ed. London: E. Arnold.

3

The Chemical Basis of Heredity

We have seen how traits are inherited, and have clarified the modes of inheritance on the basis of the behavior of microscopically observable bodies, the chromosomes. Chromosomes must be made of genes, the elements that determine inheritance. We now proceed toward deeper understanding of the process by studying the chemical nature of the phenomena involved. Genes are made of DNA, which contains linear strings of "nucleotides," the genetic information. Duplication of DNA assures the transfer of genetic information to the progeny—that is, inheritance. The way in which this information is utilized to form an organism requires an understanding of how DNA can direct the synthesis of proteins, which in their turn carry out the life processes of the organism. The duplication of DNA, the synthesis of protein, and the regulation of the synthesis are the objects of "molecular" genetics.

3.1 Proteins and Gene Changes

Living organisms are made up of cells as their basic units. Cells, in their turn, are resolvable into their chemical constituents. Full understanding of many biological processes depends on unraveling their chemical basis. Some of the greatest advances in this direction have been in the study of the chemical basis of heredity. This third major avenue in the development of genetics (following chromosomes and Mendelism) is the subject of this chapter.

An extract of cells that have been broken open yields a bewildering variety of chemicals. Some are common, small molecules such as the various types of sugars

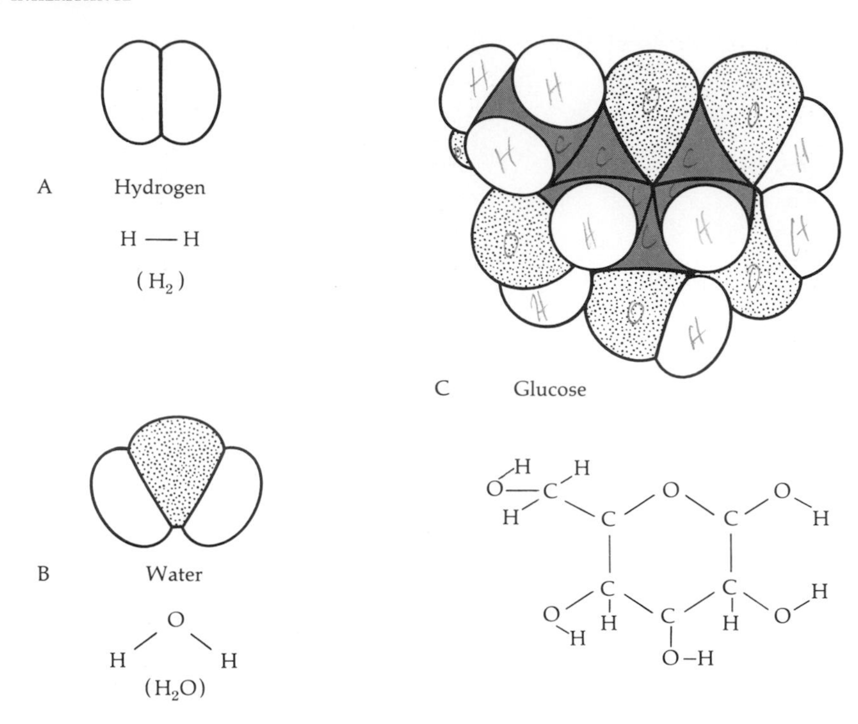

Figure 3.1
Simple molecules: representations of three of the very basic molecules of great importance to life. (A) Hydrogen. (B) Water. (C) Glucose (a simple sugar that is an important carbon source and building block for organic macromolecules); the structural formula is shown (somewhat distorted for correspondence) below the space-filling model.

and salts (Fig. 3.1). Many, however, are rather large from a chemical point of view, and are characteristic of living organisms. The chemicals typical of living matter are called organic (hence the term "organic chemistry"), and they have in common the fact that the carbon atom is one of their key components. There are four major categories of larger organic substances: the carbohydrates, the lipids or fats, the proteins, and the nucleic acids. Anyone who has ever worried about diet will be all too familiar with carbohydrates and fats and will be aware of the importance of proteins as a source of food. Carbohydrates (which are predominantly complex sugar molecules) and fats are major sources of metabolic energy and are important constituents of a cell's structure, especially of its membranes (Fig. 3.2).

The proteins, of which there are very many different types, are the most important functional molecules of the cell.

Proteins are responsible for carrying out most of the cell's metabolic activities, as well as contributing to cell structure. Hemoglobin, the molecule that carries

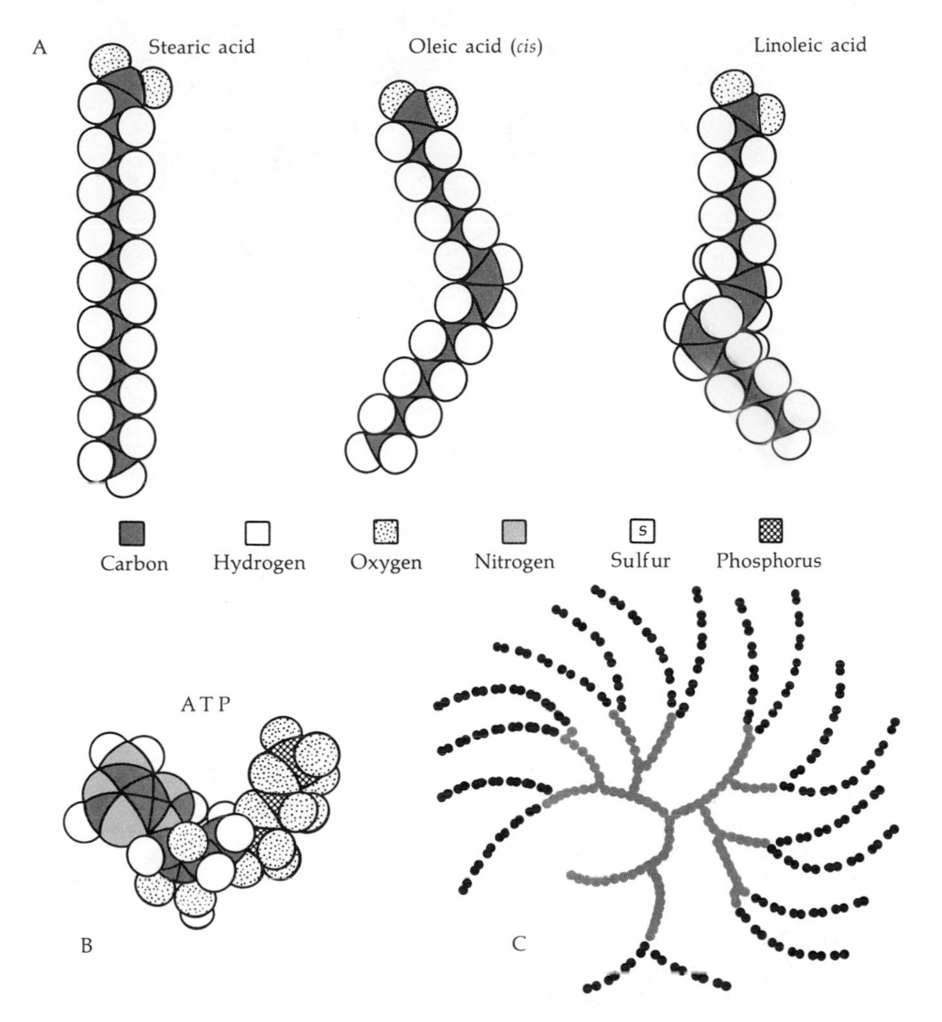

Figure 3.2
Representations of various organic molecules. (A) Space-filling models of three fatty acids. The long hydrocarbon tails are characteristic of fatty acids. Long-chain (C_{16} to C_{18}) fatty acids are essentially hydrophobic ("water-fearing," or water-insoluble); however, their Na^+ and K^+ salts (called soaps) will dissolve in water. (B) Space-filling model of ATP (adenosyl triphosphate), an important intermediate in energy metabolism. (C) Diagrammatic representation of the starch amylopectin. Each circle represents a glucose molecule. As shown here, amylopectin is a large, highly branched molecule. Its molecular weight may be as large as one million. (Parts A and C after A. Lehninger, *Biochemistry,* Worth Publishers, 1970.)

oxygen in the blood, is an example of a protein (Fig. 3.3). An average adult has a total of about one kilogram of hemoglobin. Another very abundant protein is actomyosin, which is a major component of muscle and which occurs in amounts of up to several kilograms per individual.

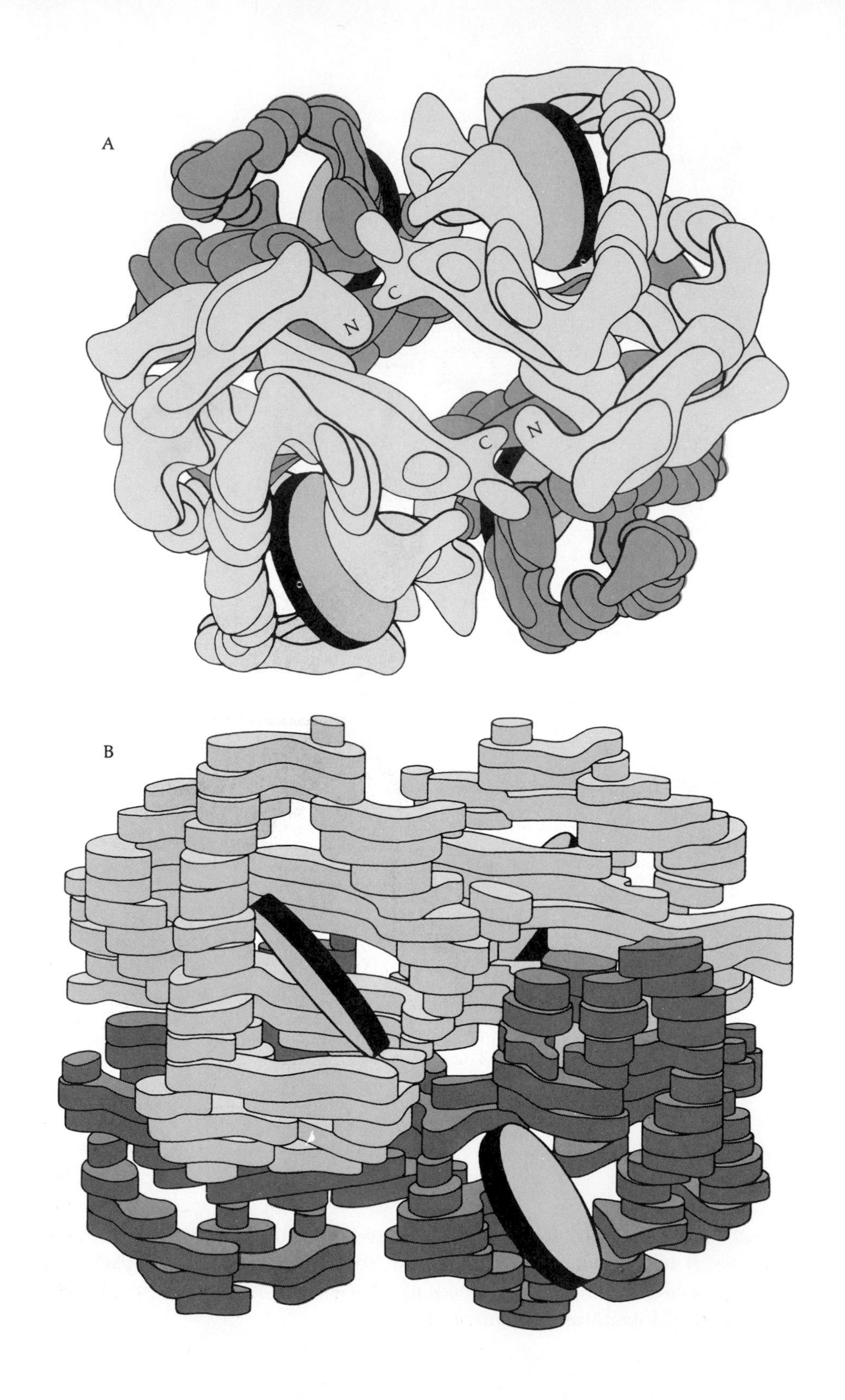
A
N
C
C
N
B

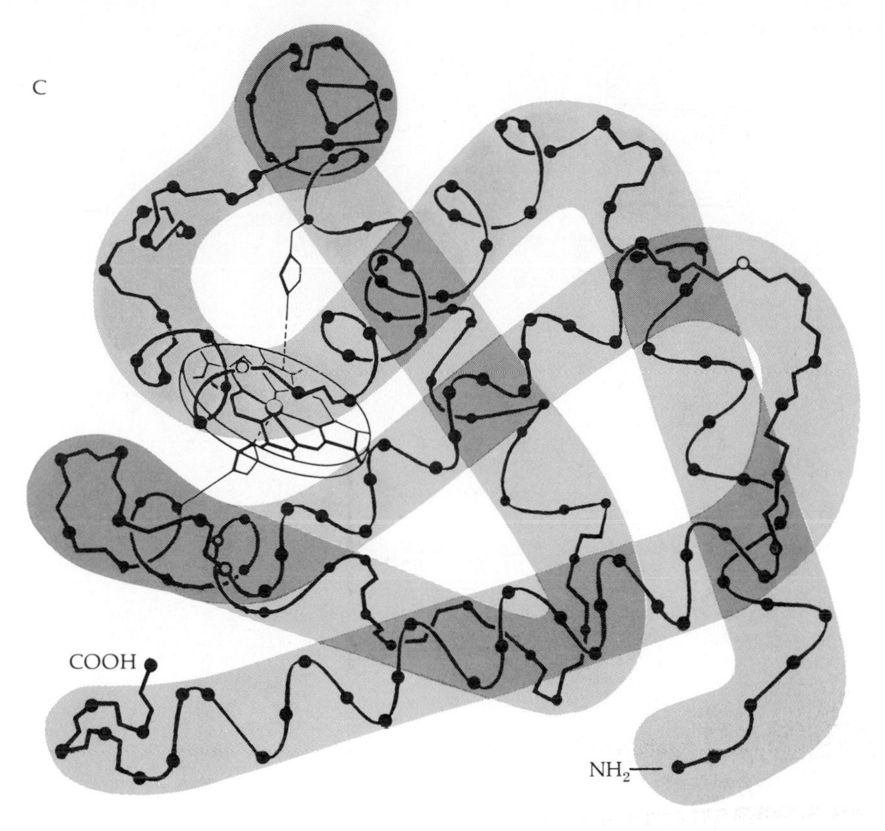

Figure 3.3
Protein molecules. (A) Hemoglobin (top view). (B) Hemoglobin (side view). The structure shown here was deduced from X-ray diffraction studies. The molecule is composed of two α chains and two β chains. Each chain enfolds a heme group (dark disk), which binds oxygen to the molecule. (C) Myoglobin (a molecule from muscles that is similar to one of the four chains of hemoglobin). (Parts A and B from "The hemoglobin molecule," by M. F. Perutz, *Scientific American*, November 1964. Copyright © 1964 by Scientific American, Inc. All rights reserved. Part C after A. Lehninger, *Biochemistry*, Worth Publishers, 1970.)

A major class of proteins is the *enzymes*—molecules that are responsible for carrying out most of the chemical reactions that take place in the cell. G6PD or glucose-6-phosphate dehydrogenase is an example of an enzyme. Its function is in one of the steps involved in using the sugar glucose as an energy source. There are many thousands of types of enzymes in a cell, each capable of carrying out a very specific metabolic function. The enzymes degrade and transform the substances that we ingest, as a prelude to their use as energy sources and building blocks for the cell. They thus mediate the transformation of food products into those substances required for the growth, function, and reproduction of cells. Enzymes are named after their specific function and characterized by the suffix -ase. In the course of this chapter, and throughout the rest of the book, we refer to many different types of enzymes and proteins.

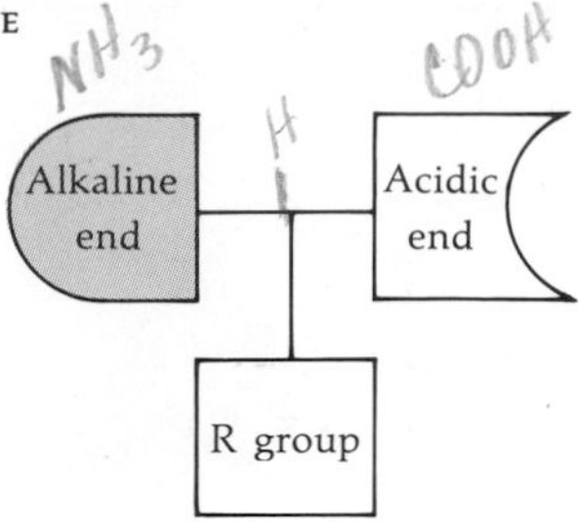

A A schematic amino acid

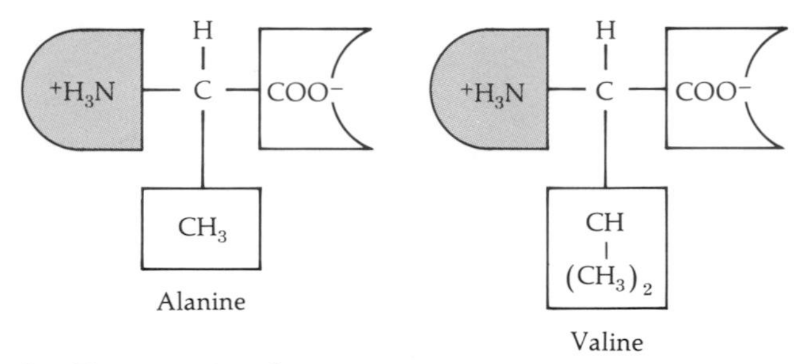

B Two examples of amino acids

Figure 3.4
Amino acids. (A) Schematic structure of a generalized amino acid. (B) Schematic representations of the structures of two amino acids.

Proteins, like many large organic molecules, are complex compounds made from relatively simple, small chemical building blocks.

The basic chemical subunits of the proteins are called *amino acids* (Fig. 3.4). There are twenty different types of amino acids (Fig. 3.5). Each has a short common element, one end of which is "acidic," while the other end is "alkaline" and contains nitrogen. This combination of properties accounts for the name amino acid (*amino* from ammonia). Amino acids can be joined, end-to-end, by their common elements, an acidic end fusing with an alkaline end to form what is known as a *peptide bond.* A linear string of amino acids connected together in this

---

Figure 3.5
Structural formulas and space-filling models of amino acids. Each amino acid except proline has a free ionized carboxyl group (COO^-) and a free ionized amino group (NH_3^+) on the α-carbon atom. Each amino acid also has a characteristic R group. The amino acids can be divided into four classes according to the polarity of the R groups. The amino acids with nonpolar or hydrophobic (water-repelling) R groups are alanine, isoleucine, leucine, methionine, phenylalanine, proline, tryptophan, and valine. The amino acids with polar but uncharged R groups are asparagine, cysteine, glutamine, glycine, serine, threonine, and tyrosine. Aspartic acid and glutamic acid have negatively charged R groups. Lysine and arginine have positively charged R groups, as does histidine at pH 6.0 to 7.0. (After A. Lehninger, *Biochemistry,* Worth Publishers, 1970.)

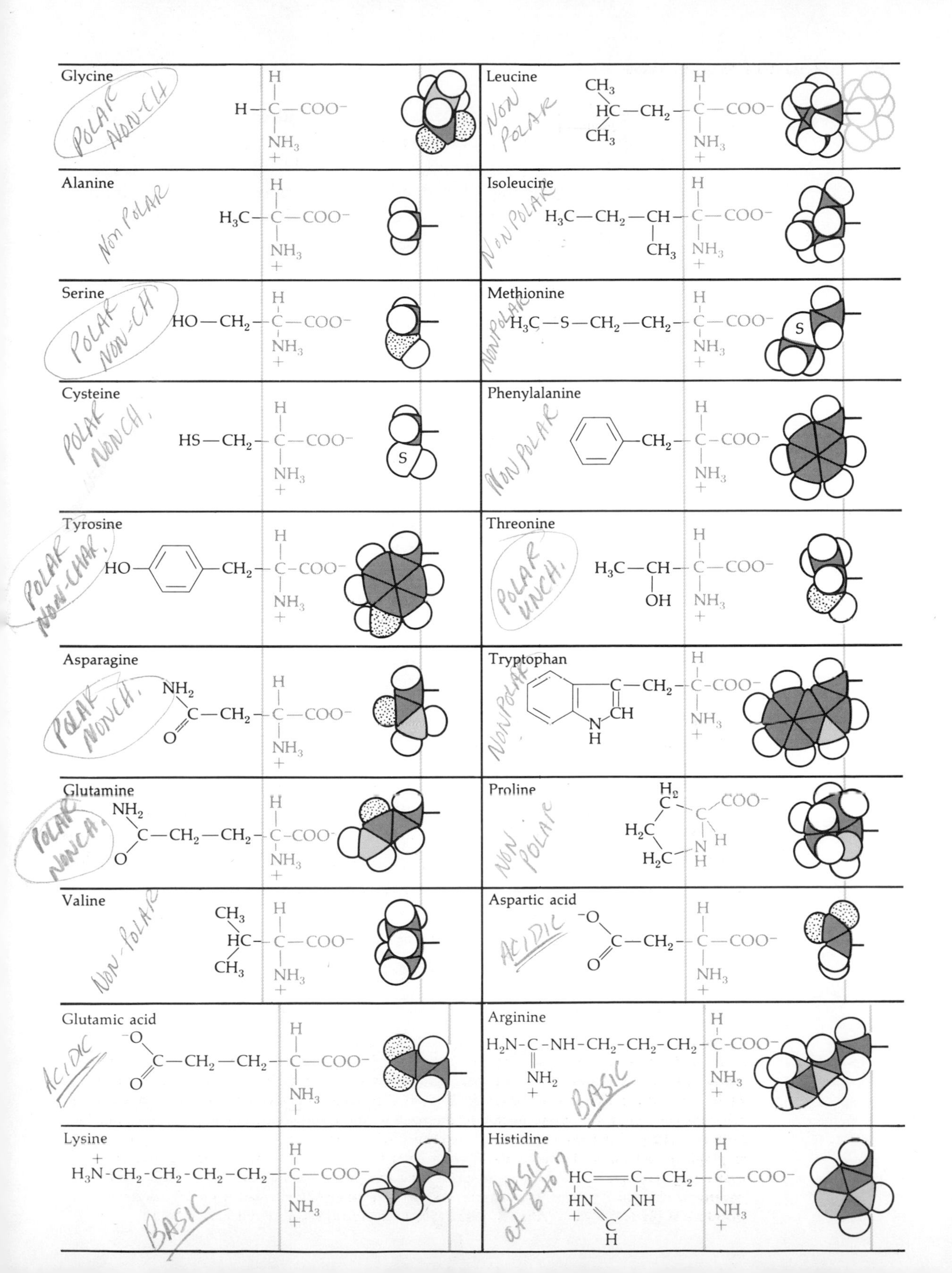

Glycine
POLAR NON-CH
Alanine
Non POLAR
Serine
POLAR NON-CH.
Cysteine
POLAR NONCH.
Tyrosine
POLAR NON-CHAR.
Asparagine
POLAR NONCH.
Glutamine
POLAR NONCH.
Valine
Non-POLAR
Glutamic acid
ACIDIC
Lysine
BASIC
Leucine
NON POLAR
Isoleucine
NON POLAR
Methionine
NON POLAR
Phenylalanine
NON POLAR
Threonine
POLAR UNCH.
Tryptophan
NON POLAR
Proline
NON POLAR
Aspartic acid
ACIDIC
Arginine
BASIC
Histidine
BASIC at 6 to 7

way is called a *polypeptide* and may contain several hundred amino acids (Fig. 3.6). Polypeptides form complex, three-dimensional, folded structures whose shape is determined by their particular sequence of amino acids. A protein is made up of one or more such polypeptide, folded subunits, generally held together by relatively weak chemical interactions. Most adult hemoglobins, for example, are made up of two different pairs of identical chains called alpha and beta (α and β), making four chains in all. A simplified chemical formula for such a hemoglobin can thus be written as $\alpha_2\beta_2$. As we shall see later, the change associated with the sickle-cell hemoglobin is in the β polypeptide chain.

The possible number of different sequences that can be formed from 20 different units arranged in sequences of length 100 is, for example, $20 \times 20 \times \cdots \times 20$ (one hundred times) $= 20^{100}$, or about 10^{130}, which is 1 followed by 130 zeros, a more than astronomical number. No wonder there can be so many different types of protein. The complexity of proteins actually led people to think at one time that genes must be made of proteins. We now know, of course, that this is not true.

The names of the 20 different types of amino acids are given in Figure 3.5. These can be thought of as letters of a twenty-letter alphabet, the polypeptide chains being long words of this special amino-acid language. Each word has its own particular meaning spelled out by its letter content. The shape of a protein

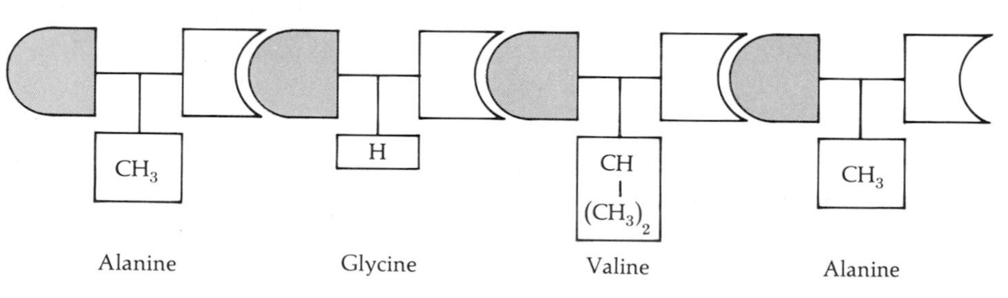

Schematic polypeptide

$$\mathrm{H_3N{-}\underset{CH_3}{\overset{H}{C}}{-}CO\cdots H_2N{-}\underset{H}{\overset{H}{C}}{-}CO\cdots H_2N{-}\underset{CH_2{-}CH_3}{\overset{H}{C}}{-}CO\cdots H_2N{-}\underset{CH_3}{\overset{H}{C}}{-}COOH}$$

Peptide bond

Chemical formula corresponding to the schematic diagram

Figure 3.6
Models for a polypeptide sequence. Amino acids are linked together to form polypeptides. A molecule of water is eliminated in the formation of the peptide bond (the hydrogen and oxygen coming from the carboxyl and amino groups).

determines what it can do, and this shape is determined by the amino-acid sequences of its constituent polypeptides. Enzymes, for example, have precisely shaped nooks and cranies into which fit the small molecules whose chemical reactions the enzymes are controlling. Hemoglobin, for example, has a special space for dealing with the oxygen molecules that it carries around the blood.

Electrophoresis, or the transport of molecules in an electric field, is of great utility in genetic studies.

How did Linus Pauling and his associates show that the abnormal hemoglobin S is associated with the sickle-cell trait and anemia, and in what way is this hemoglobin different from the normal hemoglobin A? There are many ways of distinguishing proteins from each other. All of them depend in some way on the differences in amino-acid contents and sequences of the proteins being studied. One technique, which has been especially useful particularly for genetic studies, is known as *electrophoresis.* This technique separates proteins mainly according to their average electrical charge properties. Some amino acids are negatively charged and some are positively charged; the average overall charge on a protein depends basically on its balance of positively and negatively charged amino acids. For electrophoresis, a solution of protein is placed in a marked position on a supporting medium, usually a gel (Fig. 3.7). An electric current is then passed along the gel. This current causes the positively charged proteins to move toward the negative pole and the negatively charged proteins to move toward the positive one. The proteins are thus separated on the gel by the electric current, the extent of the separation being a function of their charge differences. Clearly, only those proteins whose charges do differ will be separated in this way. The positions of the proteins on the gel after electrophoresis are revealed by appropriate staining techniques.

Electrophoresis was the procedure used by Pauling and his coworkers to distinguish hemoglobins A and S. Figure 3.8 is a picture of an electrophoresis illustrating the separation of these two hemoglobins. The hemoglobins of *AA* and *SS* homozygotes have different electrophoretic mobilities. The electrophoresis illustrates also very clearly that the heterozygous *AS*, sickle-cell-trait individuals have both hemoglobins A and S in substantial amounts. The electrophoretic difference between the two hemoglobins clearly suggests that they differ in some way with respect to their amino-acid content.

As already mentioned, there are two types of polypeptide chains in hemoglobin A, the α chain (known to be 141 amino acids long) and the β chain (146 amino acids). These chains can be separated from each other by suitable chemical procedures. In 1957 V. M. Ingram showed that hemoglobins A and S differ by just

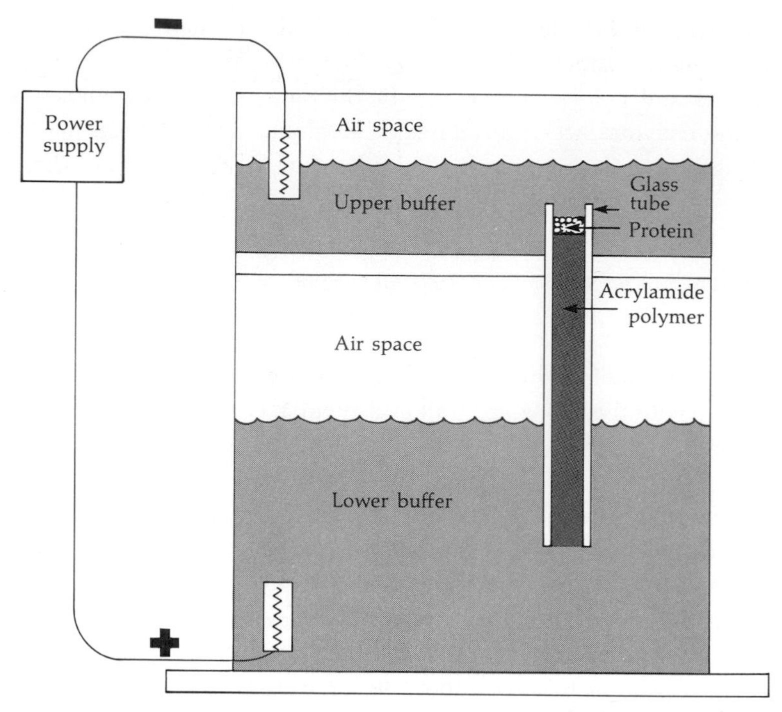

Figure 3.7
Electrophoresis: a diagrammatic representation of a polyacrylamide-gel electrophoresis apparatus. The apparatus is designed so that electric current is able to flow between the two buffers only through the gel tube. When a voltage gradient is established between the buffers, the proteins are pulled through the acrylamide. The more charged a protein, the faster it will migrate. The proteins are also segregated by size, because large proteins have their movement impeded to a greater extent by the acrylamide polymer. The buffers are needed to conduct the electricity and to ensure a constant pH; the pH of the buffer solution is chosen so that most of the proteins will be charged. The apparatus shown here is for disk-gel electrophoresis. In other kinds of apparatus (for slab-gel electrophoresis), the acrylamide gel is shaped as a slab.

one amino acid in their β chain. Where the normal hemoglobin A has glutamic acid in position 6, hemoglobin S has valine in its place (Fig. 3.9). Because valine is not negatively charged, while glutamic acid is, this change accounts for the electrophoretic difference between the two types of molecules. So we see that this difference in one amino acid out of 247 accounts for the hemoglobin change, for the sickle-cell trait and anemia, and for their associated phenotypes. This result provided the first direct link between a genetic change and the composition of a protein that could be pinned down to a specific amino-acid difference.

The connection between genetically inherited variations and biochemical deficiencies was clearly foreshadowed by Archibald Garrod in his description of inborn errors of metabolism. He realized that abnormalities, such as alkaptonuria, are due to blocks in the ability to carry out normal metabolic processes, and that

Phenotype	Genotype	Hemoglobin electrophoretic pattern Origin → +	Hemoglobin types present
Normal	*AA*		A
Sickle-cell trait	*AS*		S and A
Sickle-cell anemia	*SA*		S

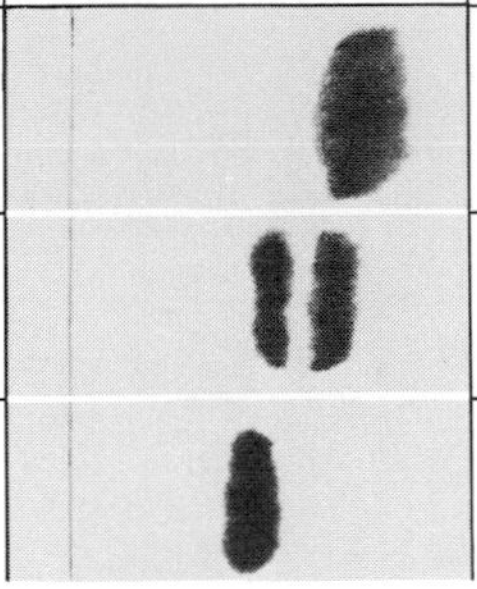

Figure 3.8
Electrophoretic patterns of hemoglobins A and S. (After I. M. Lerner, *Heredity, Evolution and Society,* W. H. Freeman and Company. Copyright © 1968. Photo by A. C. Allison.)

these blocks could be interpreted as enzyme deficiencies. Thus, he was actually the first to recognize the relation between gene and enzyme, though it was not until many years later—following the biochemical studies of George Beadle and Edward L. Tatum with the pink bread mold *Neurospora* (Fig. 3.10)—that the significance of this relationship was fully appreciated.

Many metabolic processes involve successive steps in the conversion of one chemical substance into another.

Each step carried out by a specific enzyme results in a small chemical change in a molecule that is being converted from one form (substrate or precursor) into another (the product). In this way, ingested foodstuffs are first broken down and then built up into the molecules that a cell needs to function, grow, and divide. An example of a metabolic sequence, which includes the step blocked in alkaptonurics, is shown in Figure 3.11.

Proteins taken in as food are broken down by the body into their constituent amino acids, of which phenylalanine is one. The sequence in the figure shows four consecutive steps in the metabolism of phenylalanine. Each of these steps is carried out by a corresponding enzyme. Thus, phenylalanine is converted to tyrosine (another of the 20 amino acids) by the enzyme phenylalanine hydroxylase. This is the enzyme whose function is missing in individuals with phenylketonuria (PKU). Tyrosine also plays a role in the synthesis of the dark pigment called melanin; albinism, another recessive disorder, is caused by a block in the conversion of tyrosine to melanin. Alkaptonuria is now known to be caused by

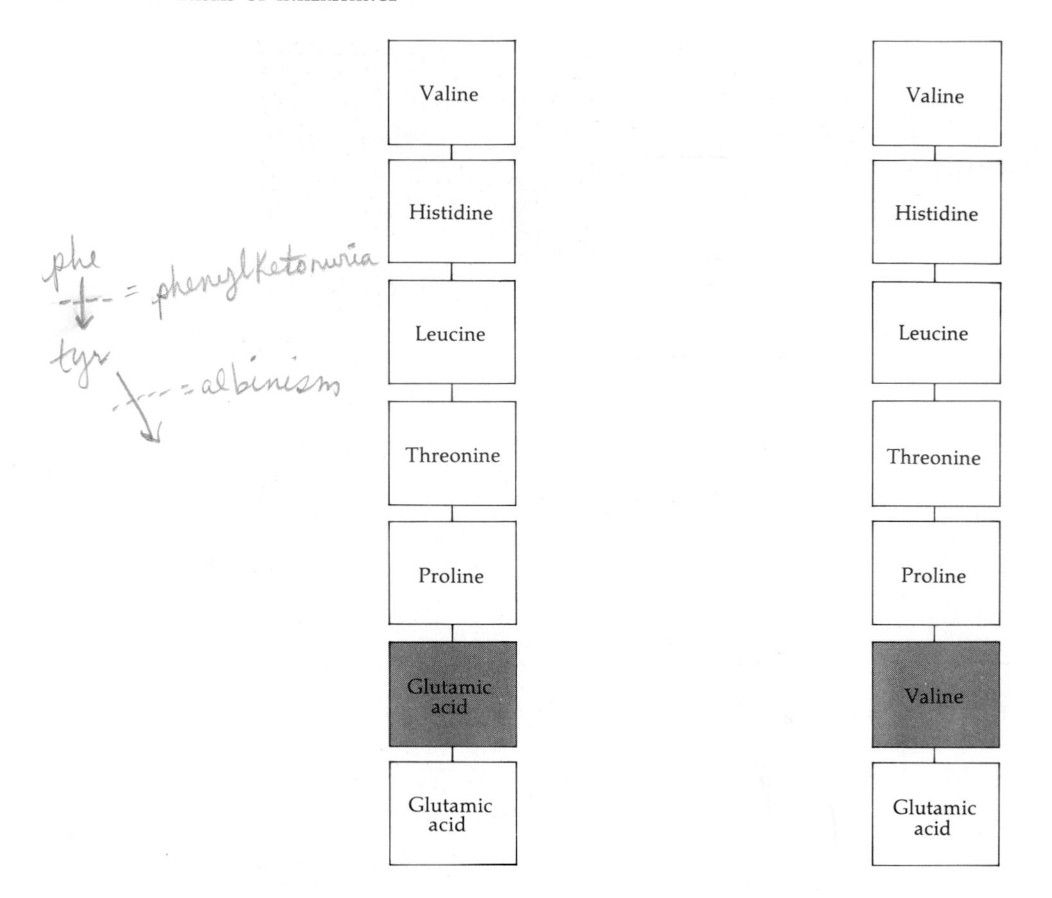

Figure 3.9
Amino-acid sequence of hemoglobin with sickle-cell mutation. A single-gene mutation is responsible for the difference between the β chain of a normal hemoglobin molecule (left) and that of hemoglobin S (right), the variant form responsible for sickle-cell anemia. Only the first seven of the 146 amino acids forming the β chain of hemoglobin are represented. The only difference between normal (A) and S hemoglobin is the substitution of valine for glutamic acid in the sixth position.

a deficiency for the enzyme homogentisic acid oxidase, which converts homogentisic acid to maleylacetoacetic acid. Tyrosinosis is another inborn error due to a block in the preceding step of this metabolic sequence.

The names of the enzymes are chosen to give a rough indication of their functions, but it is the principle of the metabolic sequence, rather than its details, that is important to grasp. Garrod actually managed to identify the position of the block in alkaptonuria by observing that patients with the disease excreted in the urine all the homogentisic acid fed to them, whereas normal individuals did not.

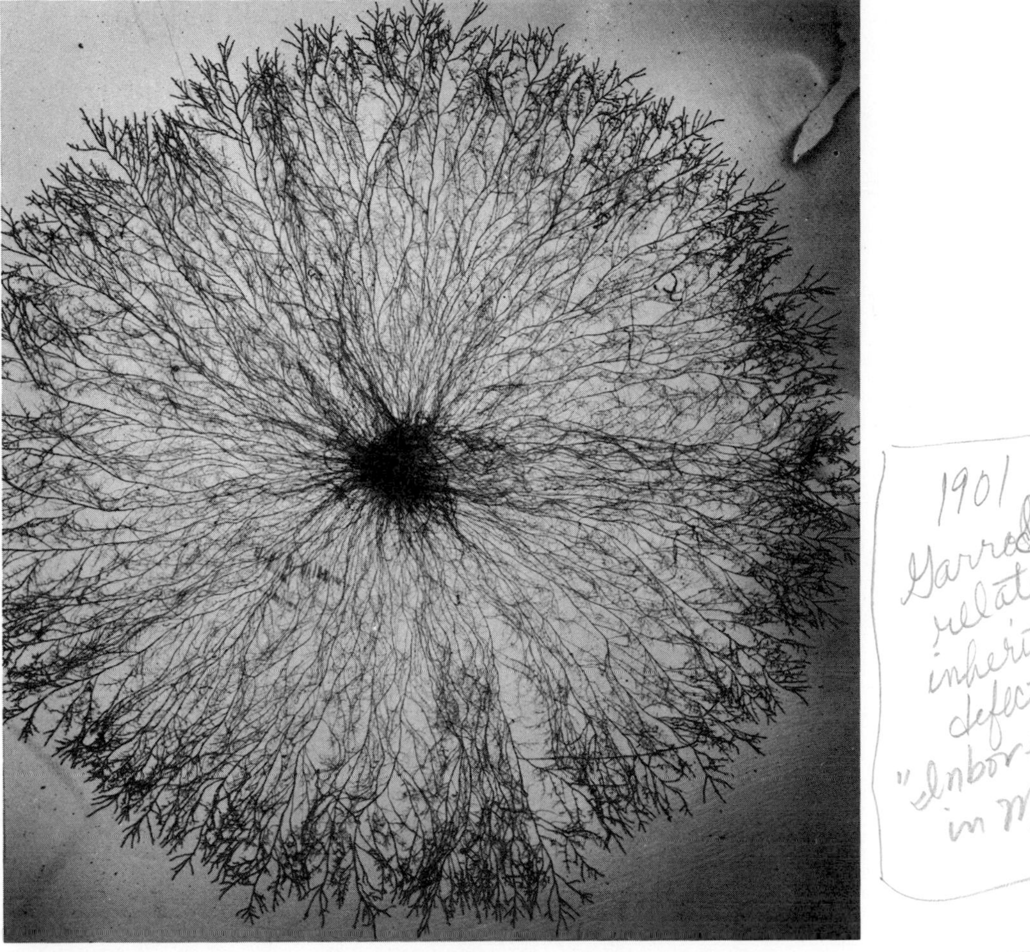

Figure 3.10
Neurospora: the vegetable form (mycelium) of the bread mold *Neurospora,* which can be grown in a simple synthetic medium. By altering the components of the medium, experimenters can select mutations concerned with the synthesis of particular substances of known biological importance. On the basis of their experiments with this mold, Beadle and Tatum proposed the theory that each gene has only one primary function. They suggested that this function most often is to direct the formation of a single enzyme (this proposal is often called the "one gene–one enzyme" theory). (Photo from Alfred Sussman.)

This is because normal individuals can metabolize homogentisic acid into its further breakdown products. Thus, Garrod realized that the position of a block could be determined by the fact that the substance immediately preceding it would be accumulated by affected individuals, who could metabolize it no further.

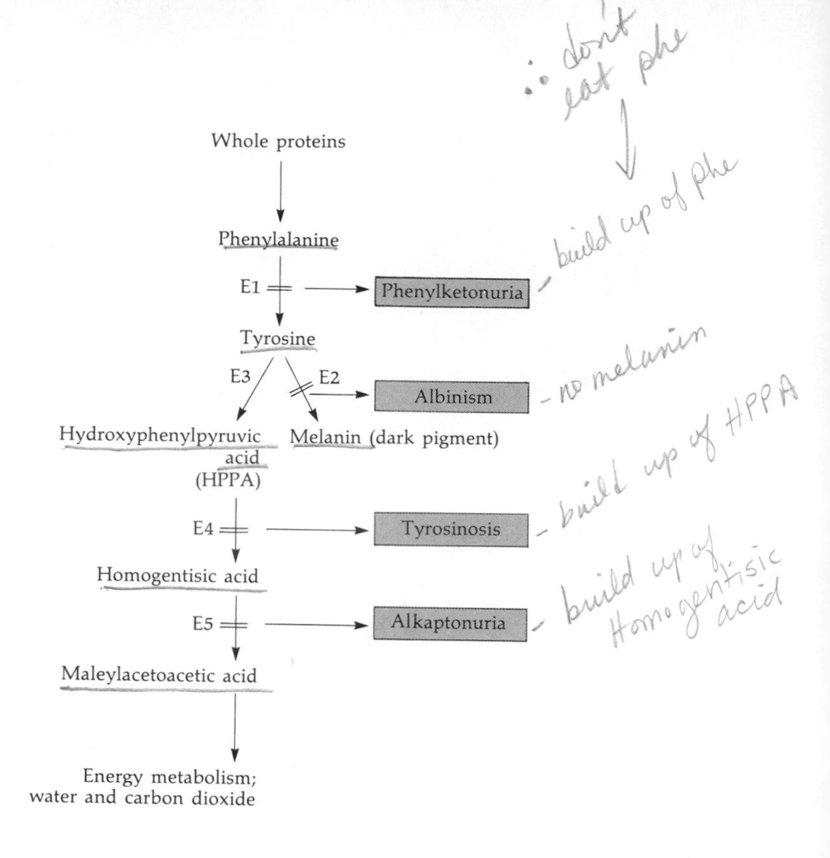

Figure 3.11
A metabolic sequence involving alkaptonuria and related defects. E1 through E5 represent enzymes, whose substrates and products are indicated above and below each enzyme, respectively. Beside each arrow is the name of the disease caused if that step in the process is prevented by a lack of the corresponding enzyme. For example, a lack of E1 (phenylalanine hydroxylase) prevents the conversion of phenylalanine to tyrosine, giving rise to the disease phenylketonuria. E2 is tyrosinase (plus other enzymes); E3 is tyrosine transaminase; E4 is HPPA oxidase; E5 is homogentisic acid oxidase.

This relatively simple principle has proved very important in studies on metabolic pathways. For example, PKU is most clearly identified by the accumulation of phenylalanine in the blood and urine, and it is the resulting excess phenylalanine (or its byproducts) that is toxic to the brain and causes mental retardation. Thus the dietary treatment for PKU, which is now proving quite successful, is to feed affected individuals on a diet that is deficient in the amino acid phenylalanine.

Beadle and Tatum's biochemical studies of genetically determined blocks in metabolic pathways were, of course, much more extensive than Garrod's. They involved the specific identification of many enzymes involved in a large number of biochemical pathways. Though the organism they studied was only a mold, the biochemical pathways they established were, in many cases, directly applicable to higher organisms and paved the way toward a much greater understanding of metabolic processes (Fig. 3.12).

3.2 DNA: The Genetic Material

Studies on gene–enzyme or gene–protein relationships brought the understanding of gene action down to a basic biochemical level. These studies clearly pointed to the idea that the most direct effect of a gene change was a change in the corresponding protein. These studies did not, however, identify the chemical nature of the gene itself. It took many years (as we have already mentioned) for the nature of DNA, the chemical substance of heredity, to be firmly established.

The nucleins that Friedrich Miescher first isolated in 1869 (from nuclei in pus cells) were later shown to be acids. They are unusual as biological substances in the fact that they contain substantial amounts of phosphorus. The nucleic acids are large molecules and, like the proteins, they are made up of a small number of relatively simple chemical subunits. In the case of DNA, there are four building blocks: the bases *adenine, guanine, cytosine,* and *thymine.*

Adenine (A) and guanine (G) belong to the class of substances known as *purines,* while cytosine (C) and thymine (T) are *pyrimidines.* The four bases are found in DNA as units called *nucleotides* (Fig. 3.13), which are combinations of a base, a phosphate residue, and a sugar called *deoxyribose*—hence the name *deoxyribonucleic* acid. The nucleotides are linked together in long chains, called *polynucleotides,* by their phosphate groups (Fig. 3.14). One of the reasons that DNA was for many years ignored as a candidate for the genetic material was that one of the pioneers of nucleic-acid chemistry had hypothesized that DNA is a regular alternating sequence of four nucleotides (such as AGCT AGCT . . .). This appeared to rule out the possibility that nucleotide sequences could carry any significant amount of information—as, for example, do the amino-acid sequences of a protein through their determination of a protein's three-dimensional folded structure. This "tetranucleotide" hypothesis was readily disproved when it was shown that DNAs vary appreciably in the relative amounts of the four bases A, G, C, and T that they contain.

While these developments in the chemistry of DNA were taking place, evidence for DNA's role as a carrier of genetic information was coming from an entirely different direction.

It had been known for some time that virtually all the DNA in the cells of higher organisms is contained in the chromosomes. However, the chromosomes also contain a lot of protein, so this fact did not firmly identify either DNA or protein as the genetic material. An English bacteriologist, Fred Griffith, obtained some

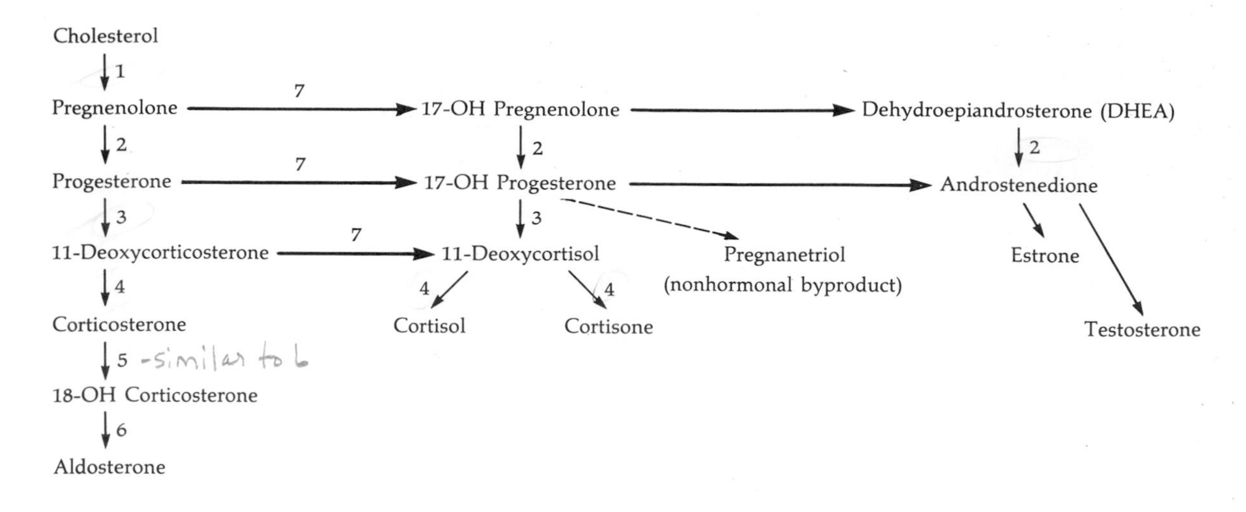

Effects of Enzyme Alteration on Hormone Production

	Altered Enzyme					
	1	2	3	4	6	7
Production of	Cholesterol desmolase	3-β Steroid dehydrogenase	21-β Hydroxylase	11-β Hydroxylase	18 Dehydrogenase	17 Hydroxylase
17-OH Pregnenolone	↓	↑	↑	↑	Normal	↓
DHEA	↓	↑	?	?	?	?
11-Deoxycortisol	↓	↓	↓	↑	Normal	↓
Aldosterone	↓	↓	↓	↓	↓	↑
Testosterone	↓	↓	↑	↑	Normal	↓
Pregnanetriol	↓	↓	↑	↑	Normal	Normal

Figure 3.12

Inborn errors of metabolism: congenital adrenal hyperplasia. There is a biosynthetic pathway that converts cholesterol into the "sex" hormones (estrone and testosterone). This pathway is also shared at several steps by the process leading to the production of other hormones called adrenal steroids. These steroids affect many parts of the body: muscles, liver, lymph nodes, fat cells.

In this particular pathway, the same enzyme may be involved in several different steps. Consequently, if an enzyme is altered, several hormones may be affected, and complications may arise in determining the sex of this individual. Also, because several different hormones share some steps of the pathway, blockage of production of one hormone (with consequent accumulation of the substance that would have been metabolized to form that hormone) may lead to production of an overabundance of another hormone.

Inborn errors of metabolism on this pathway lead to diseases called congenital adrenal hyperplasias. In the diagram, the enzyme controlling each step in the pathway is indicated by a number. The table gives the name of each enzyme and shows how alteration of each enzyme affects production of key hormones in the pathway (↑ = increased production; ↓ = decreased production). Thus, study of the hormone levels in an affected individual permits identification of the enzyme that has been altered and is now nonfunctional. Pregnanetriol is a nonhormone byproduct of the pathway that is readily identified and measured in the urine. Enzyme 5 is 18-hydrolase; alteration of this enzyme has effects similar to those of enzyme 6 on the concentrations of substances shown here. (Courtesy of R. Christiansen.)

The major purines

Adenine (6-aminopurine)

Guanine (2-amino-6-oxypurine)

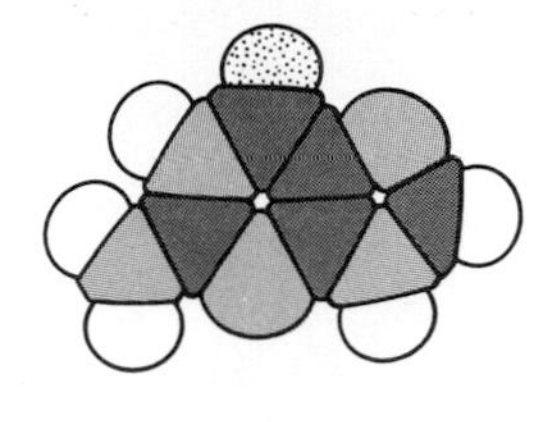

The major pyrimidines

Uracil (2,4-dioxypyrimidine)

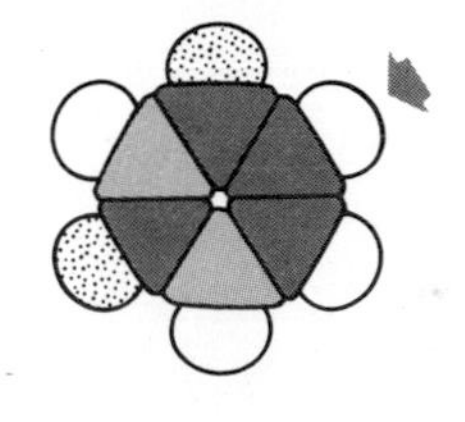

Thymine (5-methyl-2,4-dioxypyrimidine)

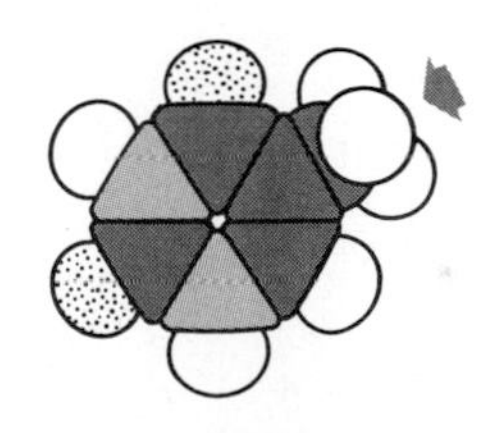

Cytosine (2-oxy-4-aminopyridine)

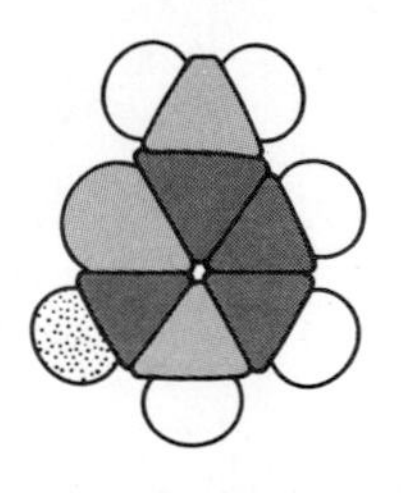

pyr 4C 2N
pur 5C 5N

Figure 3.13
Nucleic-acid bases: structural formulas and space-filling models. The structural backbone of a pyrimidine is a six-atom ring with four carbon and two nitrogen atoms. In a purine, the six-atom ring shares two carbon atoms with a five-atom ring that includes another two nitrogen atoms. Note that thymine and uracil are almost identical pyrimidines; a methyl group ($-CH_3$) in thymine replaces a hydrogen atom in uracil. Uracil is not found in DNA, but it is found instead of thymine in RNA (ribonucleic acid). A *nucleotide* is composed of a base (a purine or a pyrimidine) joined to a sugar (ribose in RNA or deoxyribose in DNA) and a phosphate group (see Fig. 3.14). (After A. Lehninger, *Biochemistry,* Worth Publishers, 1970.)

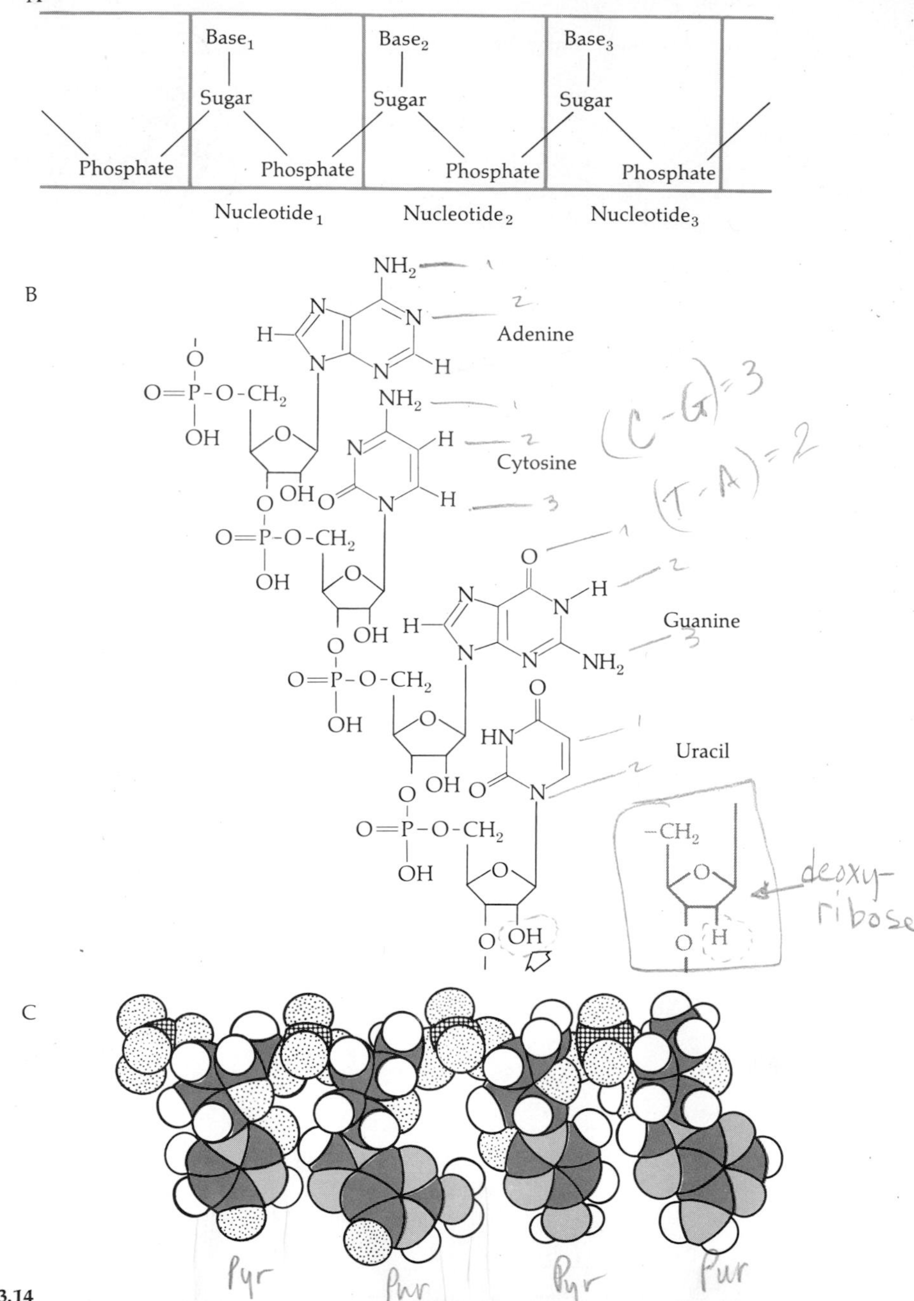

Figure 3.14
Polynucleotides. (A) Simplified diagram of a polynucleotide sequence, forming one strand of a nucleic acid. If the sugars are deoxyribose, the nucleic acid is DNA; if the sugars are ribose, the nucleic acid is RNA. (B) Structural formula for part of an RNA sequence. The arrow indicates where ribose (the sugar backbone of RNA) differs from deoxyribose (the sugar backbone of DNA). The structural formula of deoxyribose is shown in gray for comparison. Also see the structural formula for part of a DNA sequence in Figure 1.1. (C) Space-filling representation of part of an RNA sequence. (Part C after E. J. Du Praw, *Cell and Molecular Biology,* Academic Press, 1968.)

peculiar results working with the bacterium pneumococcus, which is one of the organisms that can cause pneumonia. He mixed heat-killed bacteria of one strain with live organisms of another strain and inoculated the mixture into a mouse. To his surprise, he obtained live bacteria that clearly combined some of the properties of both the live *and* the heat-killed bacteria in his mixture (Figs. 3.15 and 3.16).

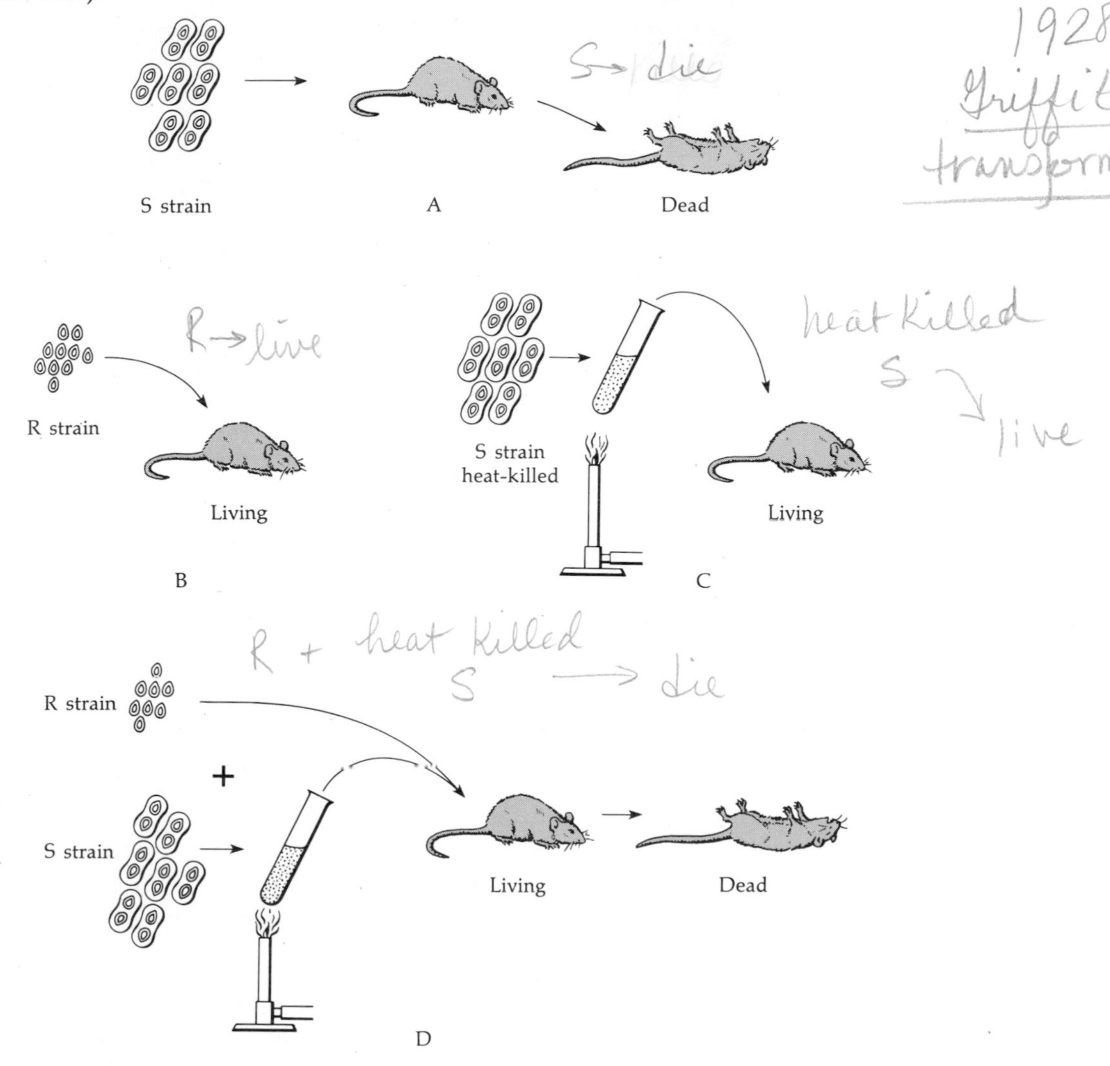

Figure 3.15
Schematic summary of Griffith's transformation experiment. (A) Mice injected with a culture of the S strain of pneumococcus die. (B) Mice injected with a culture of the R strain do not die. (C) Mice injected with a heat-killed culture of the S strain do not die. (D) Mice injected with a mixture of a living culture of the R strain and a heat-killed culture of the S strain die; living S bacteria are found in the blood from their hearts. (After G. Stent, *Molecular Genetics,* W. H. Freeman and Company. Copyright © 1971.)

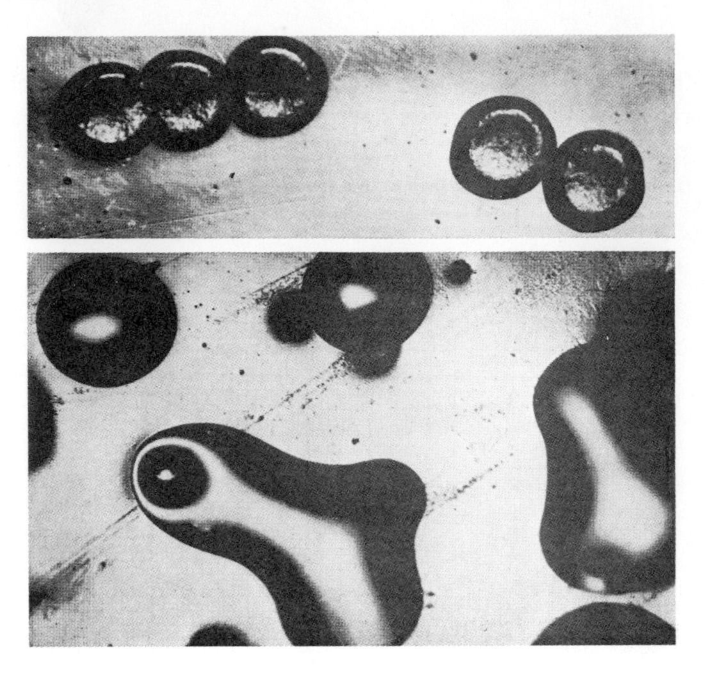

Figure 3.16
Strains of pneumococcus (species *Streptococcus pneumoniae*). The lower photo shows colonies of the normal, encapsulated bacterium; it is called the S strain because of the smooth appearance of the colonies. The upper photograph shows colonies of a nonencapsulated mutant strain of the bacterium. This strain is called the R strain because of the rough appearance of the colonies. The S strain is pathogenic—that is, it produces fatal pneumonia when injected into experimental animals. The R strain is nonpathogenic. (Photos by Harriett Ephrussi-Taylor.)

O. T. Avery and his coworkers at the Rockefeller Institute realized that Griffith's results imply that a chemical extract of one strain of bacteria can impart some of its properties to another strain with which the chemical extract has been mixed. They argued that this transformation of a recipient strain into a type corresponding to some extent to that of the donor of the extract is a form of genetic change. If they could identify the nature of the substance in the extract that is responsible for the transformation, then this material must be the genetic substance—at least for bacteria. In 1944, 16 years after Griffith's first description of his experiment, Avery and his coworkers announced that they had identified the "transforming" principle: it is clearly DNA. This result was reinforced some years later when Alfred D. Hershey and Martha Chase showed that the only part of a bacterial virus (or *phage*) that enters a bacterium upon infection is its nucleic acid (Fig. 3.17). Thus, they showed that the phage's genetic information also must be carried in its DNA.

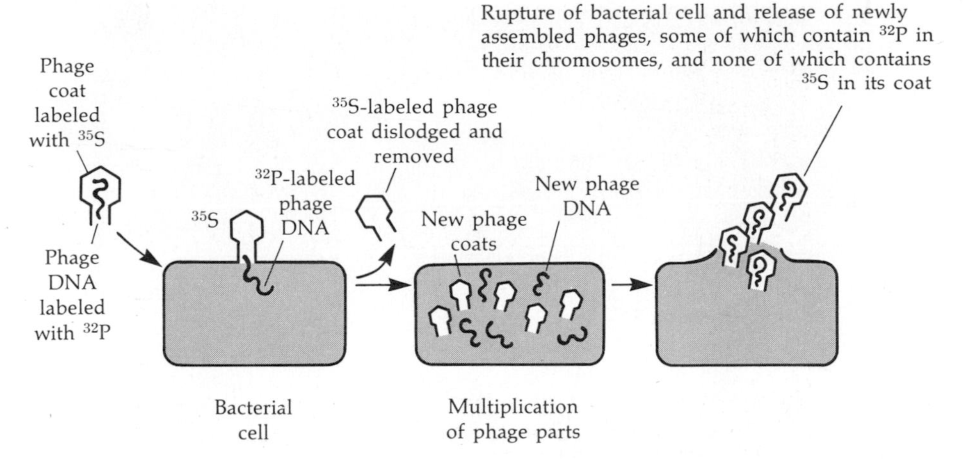

Figure 3.17
The Hershey-Chase experiment: confirmation of DNA's role as carrier of hereditary information. Bacterial viruses (called phage) are very simple organisms that consist of DNA wrapped in a protein coat. Hershey and Chase grew phage in media containing either radioactive sulfur (^{35}S) or radioactive phosphorus (^{32}P). DNA contains no sulfur, and proteins contain no phosphorus. Therefore, phage grown in the ^{35}S medium had radioactive protein coats but nonradioactive DNA, while phage grown in the ^{32}P medium had nonradioactive coats but radioactive DNA. Cells growing in a nonradioactive medium were infected with one or the other of these labeled phage. A series of experiments showed that ^{32}P found its way into the cell and became part of the next generation of phage, while ^{35}S was discarded with the coats of the original phage and never made its way into the infected cell. These results confirmed that it is the DNA—not the protein of the coat—that carries the information needed to produce new phage.

The elucidation of the structure of DNA by James Watson and Francis H. C. Crick in 1953 explained its function as genetic material with such conviction that there could be no lingering doubts about the fact that genes are made of DNA.

Watson and Crick's solution of the DNA structure depended on two major pieces of information. Erwin Chargaff—who was the first to show variation in amounts of A, G, C, and T between different DNAs—found that, in spite of this variation, the concentration of A is always the same as that of T, and the concentration of G is always the same as that of C (that is, $A = T$ and $G = C$ in terms of amount). Thus it appeared that a particular DNA can, in crude statistical terms, be characterized by its ratio of $C + G$ to $A + T$.

The second source of information came from X-ray pictures made from DNA crystals by Maurice Wilkins and Rosalind Franklin (Fig. 3.18). These pictures, whose detailed interpretation depends on complex mathematical theories, suggested to Watson and Crick that DNA has a helical structure, similar to that earlier proposed by Linus Pauling for some parts of protein molecules. Their

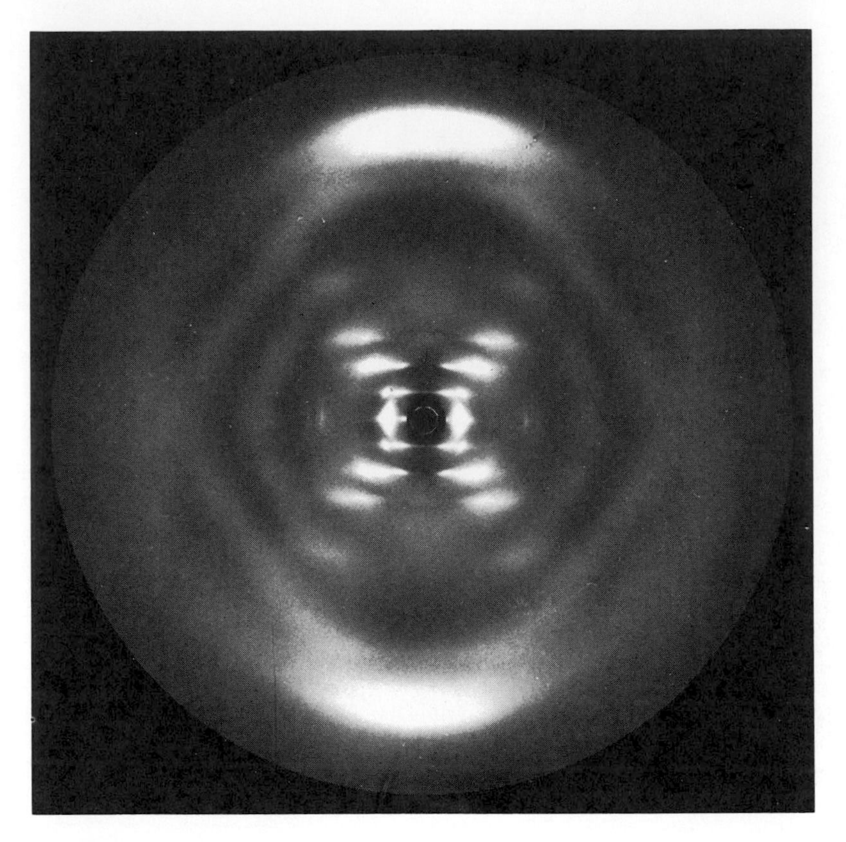

Figure 3.18
X-ray diffraction photograph of DNA. This famous picture, taken by Rosalind Franklin in the laboratories of Maurice Wilkins, indicated that the DNA molecule is a helix. (The pattern of reflections crossing in the middle is indicative of a helical structure.)

crucial suggestion, however, was that there are two paired helices in DNA, hence the famous *double helix* (Fig. 3.19). Each helical strand is a polynucleotide with bases joined by their sugars and phosphates as shown in Figure 3.14. The two strands are matched up in such a way that an A on one strand is opposite to a T on the other; similarly opposite a G always lies a C. Weak chemical interactions between these paired bases ("hydrogen bonds") are the forces that hold the two helices together.

As shown in Figure 3.19, the sugar–phosphate–sugar sequences of the two strands are like the two bannisters of a spiral staircase, while the paired bases that hold them together are the steps in between. The purines, A and G, are bigger molecules than the pyrimidines, T and C, but the pairing rule that always places a purine (A or G) opposite a pyrimidine (T and C) ensures that the widths of all

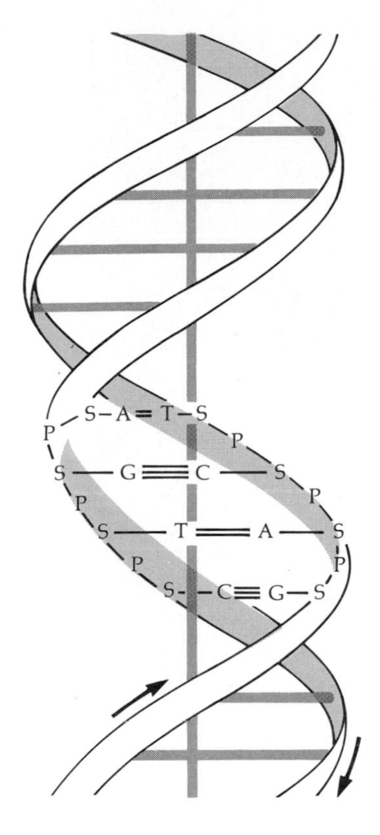

Figure 3.19
Schematic diagram of the DNA double helix. The backbone of each helical chain is made up of alternating deoxyribose sugar (S) and phosphate (P). The two chains are joined by weak interactions between the nucleotide bases adenine (A), thymine (T), cytosine (C), and guanine (G). Adenine–thymine pairs are held together by two hydrogen bonds, while cytosine–guanine pairs are joined by three such bonds. (Also see Fig. 3.20.)

the steps are the same (Fig. 3.20). Watson and Crick made detailed molecular models of their proposed structure; they predicted accurately the dimensions of the helix predictions that have since been more than amply confirmed.

The biological significance of the double helix was clearly pointed out by Watson and Crick in their original short note. The complementary structure of the double helix provides a simple answer to the problem of self-replication. Replication starts by separation of the two strands. Each resulting single strand then forms a template along which a new strand can be built by the complementary rules of an A opposite a T and a G opposite a C. Thus two identical daughter helices are formed, each of which has one old and one new strand (Fig. 3.21). The replication process is mediated by enzymes, called *polymerases,* and can actually be carried out in a test tube. The fine details of DNA replication are still being worked out and turn out to be more complicated than was thought to be the case a few years ago. The essential principle of complementarity remains, however, the simple basis for self-replication—which is, after all, one of the innermost secrets of life.

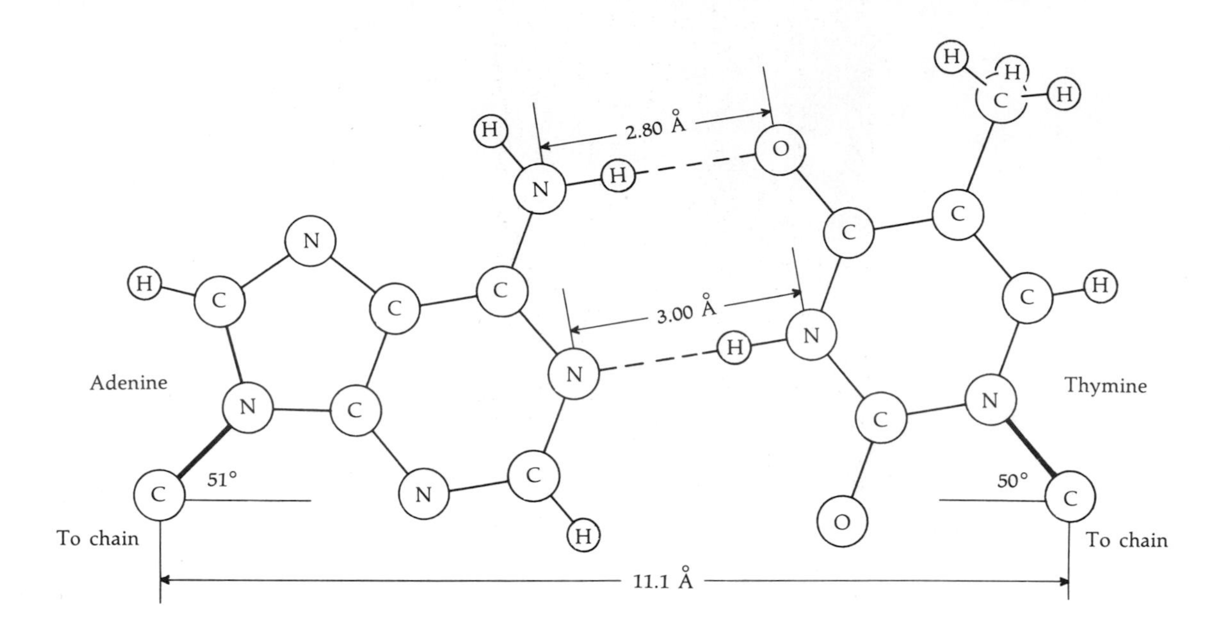
2.80 Å
3.00 Å
Adenine
Thymine
51°
50°
To chain
To chain
11.1 Å

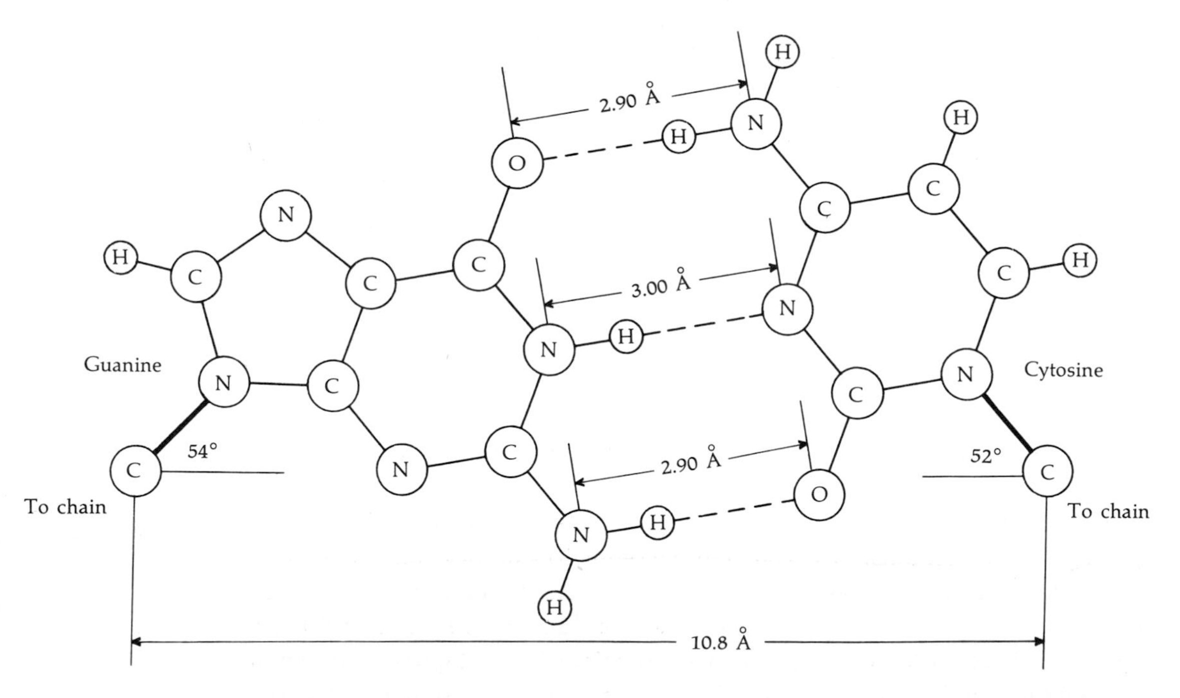
2.90 Å
3.00 Å
2.90 Å
Guanine
Cytosine
54°
52°
To chain
To chain
10.8 Å

A

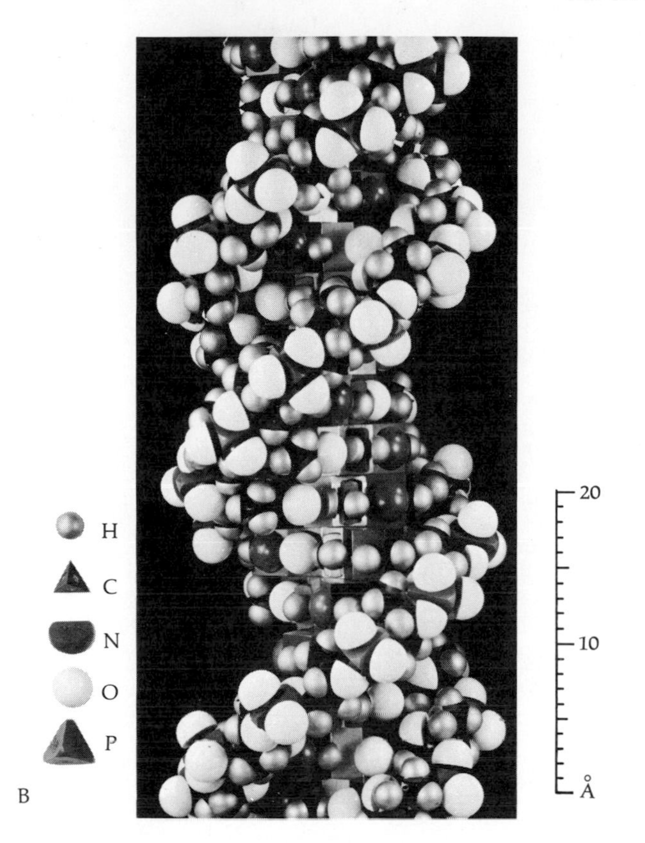

Figure 3.20
DNA base pairs. (A) Cytosine–guanine pairs are slightly closer together and more compact than adenine–thymine pairs. CG pairs are therefore slightly more dense than AT pairs. Note the two hydrogen bonds joining each AT pair, in contrast to the three such bonds joining each GC pair. (B) A portion of a space-filling model of a complete DNA double helix. The complexity of this molecule is readily apparent. (Part B from M. H. F. Wilkins.)

3.3 The Genetic Code: From DNA to Protein

The genetic language of DNA is written with four "letters": A, T, G, and C. The genetic information must be encoded in the sequences of these four "letters" that make up the DNA "sentences." The sequences of one strand, of course, determine the sequences on the other, so we need only consider one strand from the information point of view. Alternatively, if we consider *both* complementary strands forming the double helix, we think of the sequence in terms of nucleotide *pairs*—AT, TA, GC, and CG.

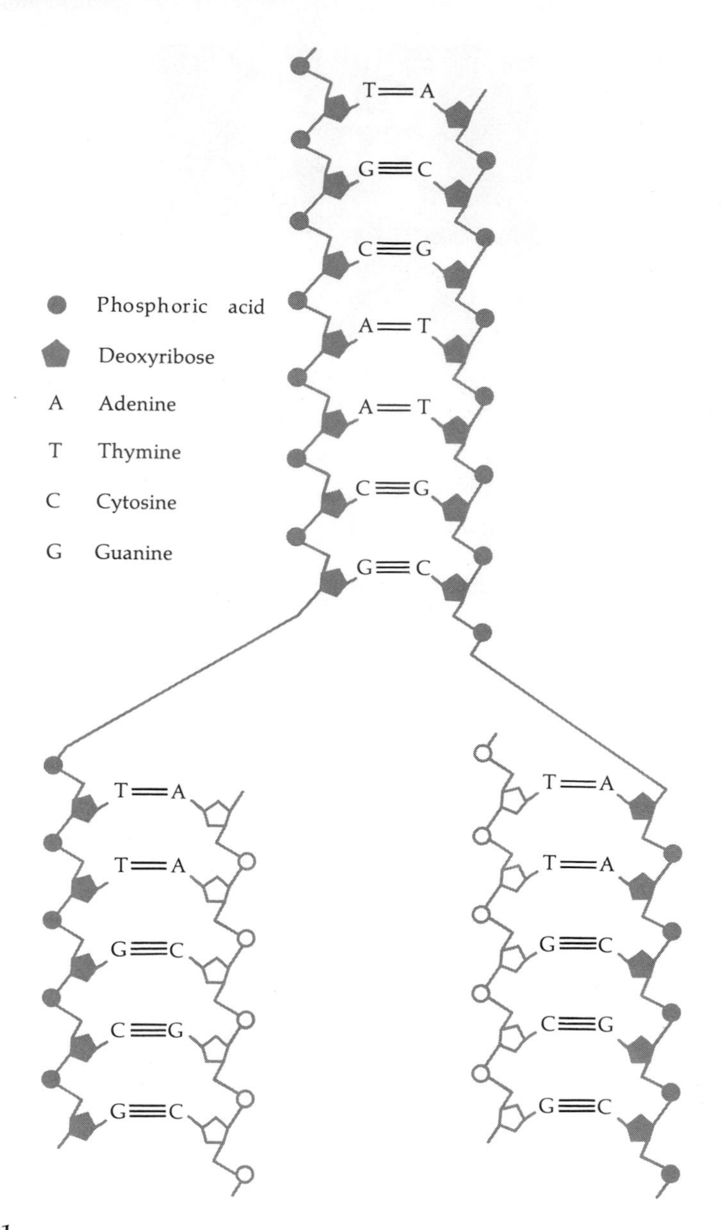

Figure 3.21
Schematic diagram of DNA replication. DNA undergoes "semiconservative" replication; the parental double helix separates into two strands, each of which serves as the template for assembly of a daughter strand. Two hybrid molecules are produced, each consisting of a parental strand and a daughter strand. The two hybrid molecules are identical to each other and to the original DNA molecule. Faithful reproduction is ensured by the pairing of A with T and of C with G. The two parental DNA strands are conserved, each one becoming part of a daughter molecule. Each daughter molecule is thus half composed of a strand conserved from the parent—hence the term "semiconservative replication."

DNA molecules can be very long; even the smallest of viruses has a DNA molecule containing several thousand nucleotide pairs. Because there are no obvious restrictions on the types of nucleotide sequences that can occur, the variety of types of DNA "words" can easily be made to match the variety of proteins which, as we saw earlier, is essentially unlimited. This ability to carry many different messages must clearly be a major feature of the genetic material.

What is the relationship between DNA and protein?

In the first part of this chapter we describe how protein differences can be closely tied to gene changes. Now we have seen the evidence that DNA is the genetic material. What is the relationship between the two? How is the four-letter language of DNA transformed into the 20-letter language of proteins? If a particular sequence of nucleotides in DNA corresponds to a given sequence of amino acids, then there must be a code for converting one into the other. Clearly there cannot be a one-to-one correspondence between nucleotide and amino acid, for then there could be only four amino acids. Even pairs of nucleotides would not be enough, because there are only $4 \times 4 = 16$ possible pairs—that is, only 16 two-letter words made up of A, T, G or C. The number of possible three-letter words is $4 \times 4 \times 4 = 64$, so three letters is the minimum length of nucleotide word needed to specify the 20 different types of amino acids. Thus, the simplest possible *genetic code* is one in which successive *triplets* of nucleotides along the DNA represent successive amino acids of a corresponding polypeptide. In this case, however, there might be some amino acids that are coded for by more than one three-letter word. All that is needed to complete the picture, at least formally, are some sequences to provide the punctuation in the DNA language—namely, to indicate where the sequence for a particular polypeptide starts and finishes.

The genetic code provides the formal answer for proceeding from DNA to protein, but what is the chemical mechanism through which the instructions embodied in the code are carried out?

There is no obvious chemical similarity between the structures of DNA and protein that could conceivably form a basis for translating one into the other. The answer to this puzzle involves a family of molecules that act as intermediates between DNA and protein: the ribonucleic acids (*RNA:* Figs. 3.13, 3.14).

RNA is a form of nucleic acid with many similarities to DNA. Although, like DNA, it is found in the chromosomes and the nucleus, much of a cell's RNA is found outside the nucleus in the cytoplasm. The basic chemical subunits of RNA are just like those of DNA with two important differences: (1) the sugar of RNA nucleotides is a ribose (not deoxyribose as in DNA), hence the name *ribo*nucleic acid; and (2) RNA has as its fourth base the pyrimidine uracil (U) instead of thymine (T). Thymine and uracil are very similar molecules and have essentially equivalent pairing properties with adenine (see Fig. 3.13). Thus, RNA polynucleotides are formed from the four bases adenine (A), guanine (G), uracil (U), and cytosine (C). The sugar–phosphate linkage between the nucleotides is just the same as that in DNA (except for the slight difference in the sugar structure). There is, however, one other major difference between RNA and DNA molecules. RNA is not a long double helix; most RNA molecules are relatively short and single-stranded. Even before the chemical structure of RNA was fully known, its involvement as an intermediary in protein synthesis had been suspected.

There are three main types of RNA that take part in protein synthesis. They are called messenger RNA (mRNA), transfer RNA (tRNA), and ribosomal RNA (rRNA).

The DNA nucleotide sequence is not "read" directly by the protein-making process. Instead, an RNA copy of a required DNA sequence is first made; this copy is called *messenger RNA* (mRNA) (Fig. 3.22). The copy is made by a polymerase, using just one strand of the DNA as its template and using the same pairing principle by which DNA itself is replicated, except that A now pairs with U. For example, the double-stranded DNA sequence

. . . -A-G-G-T-C- . . .
-T-C-C-A-G-

would be transcribed into a single-stranded RNA with sequence

. . . -U-C-C-A-G- . . .

if the upper strand of the DNA is used as the template. Note that the RNA sequence is just like that of the DNA strand complementary to the strand being used as a template, except that U replaces T.

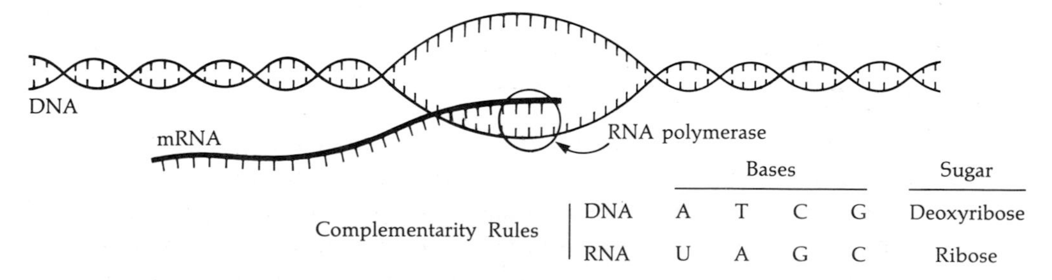

Complementarity Rules		Bases				Sugar
	DNA	A	T	C	G	Deoxyribose
	RNA	U	A	G	C	Ribose

Figure 3.22
Messenger RNA. The mRNA uses one of the DNA strands as a template. Therefore, the base sequence in the mRNA is exactly complementary to the original series of bases in the DNA.

The existence of mRNA was first recognized as a category of RNA with a base ratio ([C + G]/[A + U]) similar to that of the DNA in the same cell. However, the existence of a messenger RNA had already been postulated on the basis of some genetic experiments with bacteria. Special nucleotide sequences are needed to tell the RNA polymerase that makes mRNA where to start and where to stop copying.

Because there is no obvious chemical correspondence between nucleotide and polypeptide sequences, Crick had postulated that there must be some sort of adapter molecule that at one end "reads" a nucleotide triplet, and at the other end is attached to the corresponding amino acid (Fig. 3.23). This adapter turned out to be a relatively small RNA molecule—the *transfer RNA* (tRNA), which is generally about 70 to 80 nucleotides long. There is one transfer RNA for each nucleotide triplet that codes for an amino acid. A schematic diagram of tRNA is shown in Figure 3.24. The molecules typically have a partly double-stranded, "hair-pin" structure, which varies in its overall shape from one tRNA to another.

In the middle of the loop at the bottom end of the molecule, there is a triplet sequence that is complementary to the triplet code for the tRNA's amino acid. This is the region that fits, by its complementary nature, onto the triplet sequence of the mRNA. The free end of the tRNA, which in all cases has the sequence C–C–A, is attached to the corresponding amino acid. There is a specific enzyme for carrying out this attachment; the enzyme is different for each tRNA. This so-called *activating enzyme* must recognize the overall shape of its corresponding tRNA (determined by the tRNA's particular nucleotide sequence) and must then attach only the appropriate amino acid to the C–C–A end. Thus, the specificity of the correspondence between a tRNA and its amino acid depends on the activating enzyme.

The third type of RNA is a major constituent of the approximately spherical particles called *ribosomes,* from which *ribosomal RNA* (rRNA) gets its name. Ribosomes are present in most cells in large numbers, so that rRNA is usually the most abundant type of RNA in the cell. The ribosome is a complex particle containing two types of RNA and, in mammals, more than 50 different types of protein (Fig. 3.25). Ribosomes are now known to be the structures that mediate protein synthesis by holding together the mRNAs and tRNAs as one amino acid is joined to the next in the polypeptide chain. The detailed role of rRNA in the functioning of the ribosome is, however, not yet fully understood.

This description of the roles of the three types of RNA molecules in protein synthesis leads to a picture of the total process illustrated schematically in Figure 3.26. Messenger RNA copies are first made by the RNA polymerase enzyme from a defined DNA sequence. This copying process is called *transcription.* Ribosomes then attach themselves to the mRNA and move along it one triplet at a time. At each triplet position (called a *codon*), the corresponding tRNA, already charged with an amino acid, enters the ribosome and aligns its complementary sequence

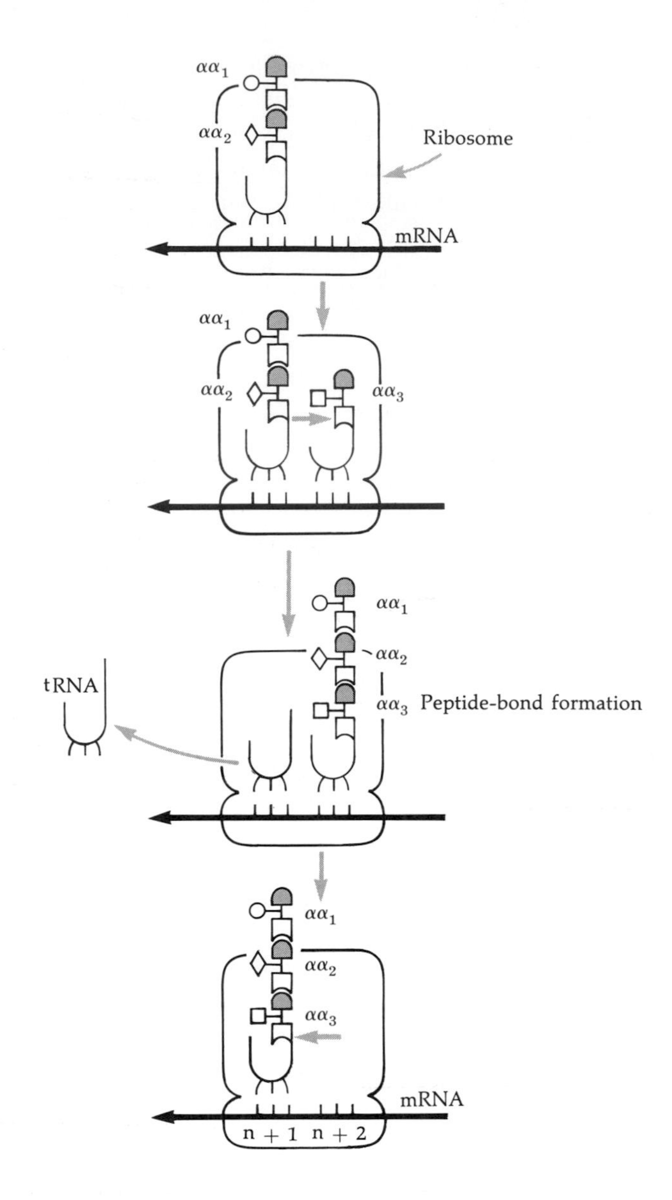

Figure 3.23
Transfer RNA. Each tRNA has a nucleotide triplet at one end and an amino acid at the other. The nucleotide triplet is complementary to some nucleotide triplet on the mRNA (which in turn was produced as the complement to a triplet in the DNA). As a ribosome moves along the mRNA, an appropriate tRNA moves into one of two sites in the ribosome, where it matches as complement to the next nucleotide triplet in the mRNA. The amino acid of the tRNA is joined onto the growing polypeptide chain, which is then detached from the previous tRNA in the other ribosome site. The previous tRNA is "evicted" from the ribosome, and the polypeptide–tRNA complex "bumps over" to the emptied second site, so that the process can begin again as a new tRNA is brought in to match up with the next mRNA triplet. The structure of tRNA is shown in more detail in Figure 3.24.

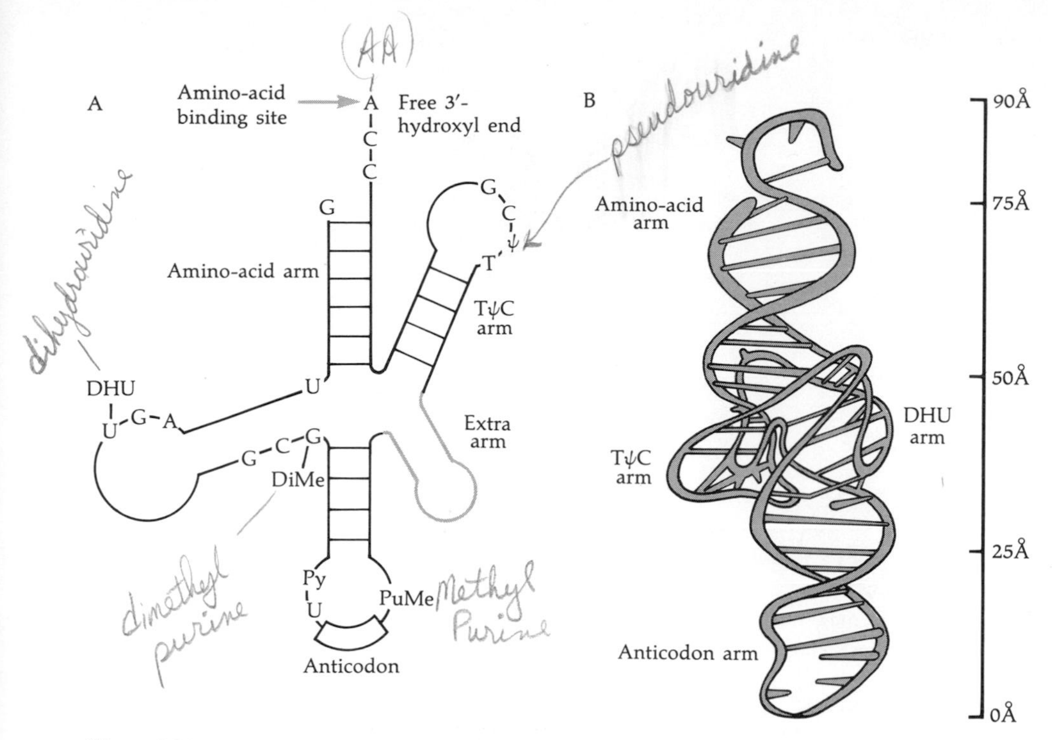

Figure 3.24
Structure of tRNA. (A) Schematic structural diagram. The extra arm (of varying length) is found only in some tRNAs, such as those for serine. The cloverleaf configuration gives maximal intrachain hydrogen bonding, but data from X-ray studies suggest that the lateral arms are folded closely alongside the vertical arms. The bases shown here are identical in all known tRNAs. In addition to the G, C, A, and T found in all RNAs, the tRNAs contain some unusual bases found only in tRNAs: ψ = pseudouridine, DHU = dihydrouridine, DiMe = a dimethyl purine, and PuMe = a methyl purine. Py represents a pyrimidine. (B) The three-dimensional conformation of tRNA, drawn from a model. (From W. Fuller, A. Hodgson, and M. Levitt, *Nature*, vol. 224, p. 759, 1969.) Recent more detailed data have suggested improved models of the structures given here.

(*anticodon*) with the current triplet being read on the messenger. The amino acid on the end of the tRNA is then transferred to the growing polypeptide, and the whole process is repeated for the next triplet along the mRNA. In this way amino acids are added one by one to a growing polypeptide chain, according to the nucleotide sequence dictated by the mRNA. The process by which the polypeptide corresponding to a given mRNA is synthesized is called *translation.*

The "cracking" of the genetic code that relates nucleotide sequences to amino-acid sequences was a clear and irresistible challenge.

Genetic studies were the first to confirm the proposed triplet nature of the code. However, the final solution to the problem (which has now been arrived at in

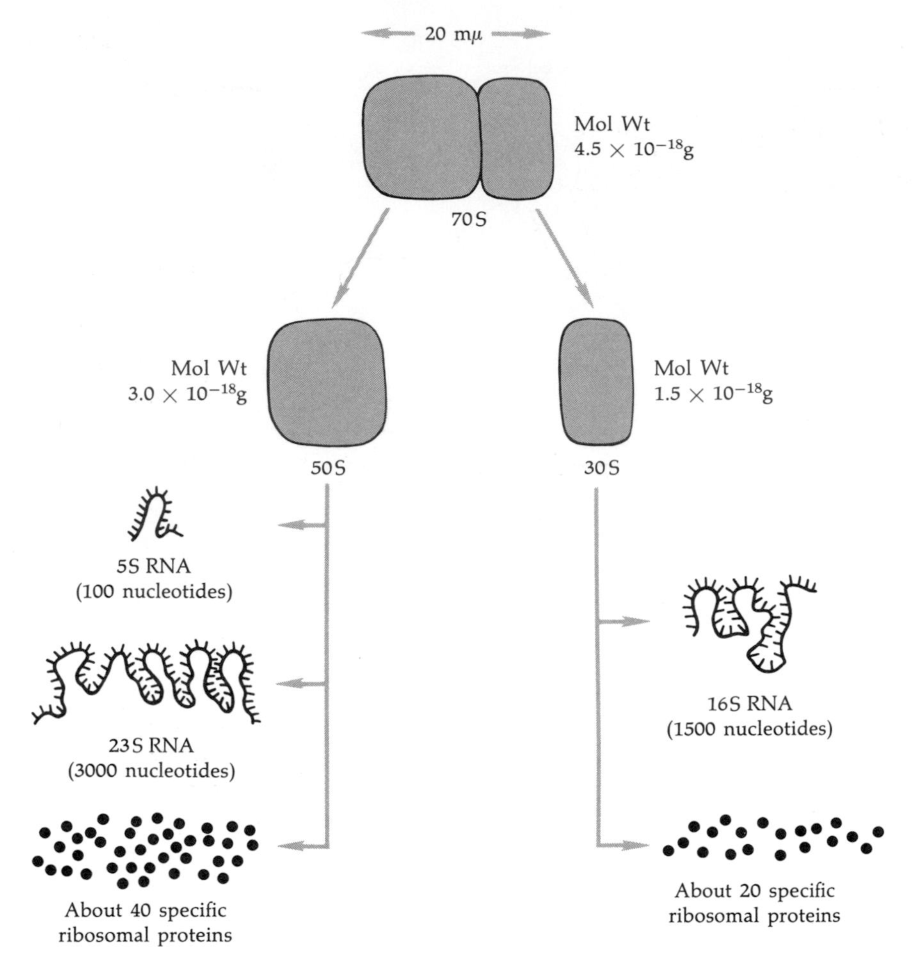

Figure 3.25
Ribosomes and rRNA. Ribosomes are made up of rRNA and ribosomal proteins. (From G. Stent, *Molecular Genetics,* W. H. Freeman and Company. Copyright © 1971.)

a number of ways) depended on a biochemical approach through the development of a test-tube *in vitro* system for studying protein synthesis. In spite of the complexity of the process, translation of mRNA can be made to take place in a test tube by using, under appropriate conditions, the various components of the system, including ribosomes, tRNAs, activating enzymes, and amino acids (Fig. 3.27). If a message of known sequence is added to the system, one can search for the corresponding polypeptide product and analyze its sequence.

The first application of this approach to cracking the code made use of one of the first simple RNA nucleotide sequences that could be made *in vitro:* a string of

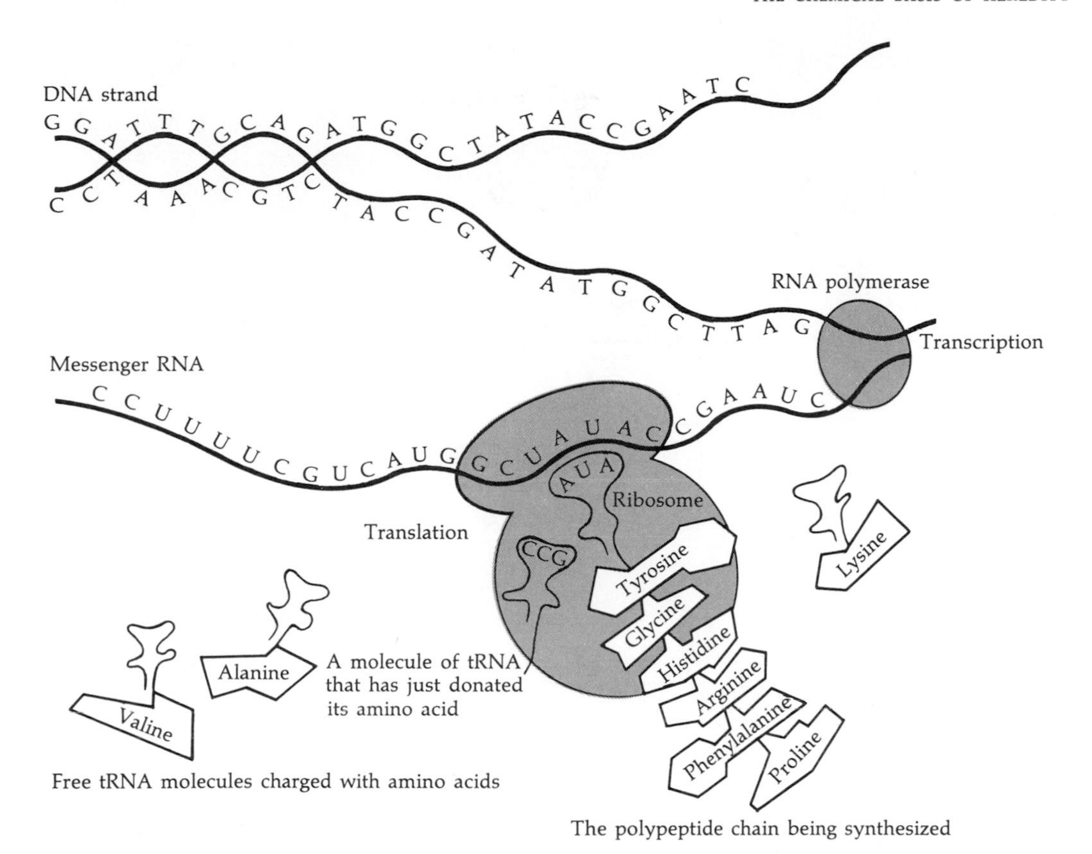

Figure 3.26
Scheme of protein synthesis. Strands of messenger RNA (mRNA) are formed on a DNA template in the nucleus. The mRNA travels to the cytoplasm, where it attaches to a ribosome. A tRNA-amino-acid complex—whose anticodon triplet matches the codon triplet being "read" (translated) on the mRNA—fits into place against the mRNA. As the mRNA strand is translated, the tRNA molecules are cleaved from the attached amino acids. The amino acids are joined together by peptide bonds to form the protein molecule encoded in the original DNA strand.

Us, U-U-U- . . . . When this message was added, the corresponding product was readily identified and turned out to be a sequence of phenylalanines. Clearly the triplet U-U-U is the code for the amino acid phenylalanine. Extensions of this approach have led to the complete identification of the code as given in Table 3.1. As can be seen from the table, all but three of the 64 triplets do code for amino acids and, of course, all the 20 amino acids have at least one corresponding triplet. The DNA triplets that do not code for amino acids—namely ATT, ATC, and ACT—have been shown to be codes for ending a polypeptide chain. When the ribosome comes to one of these "chain terminators," it leaves the messenger and

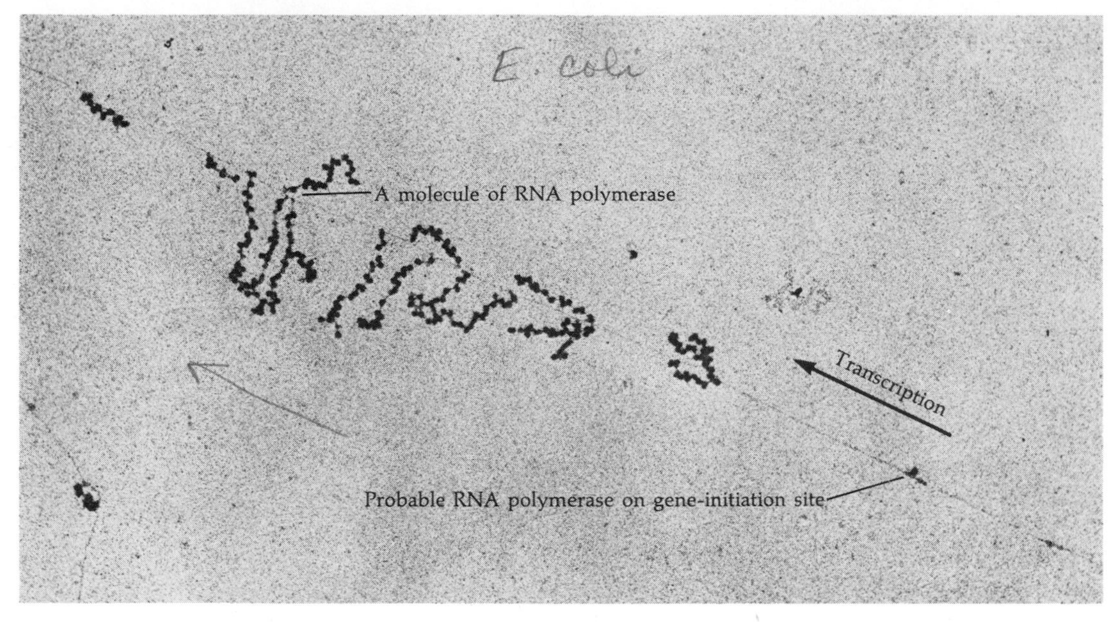

Figure 3.27
Electronmicrograph of transcription and translation in bacterial extract (*E. coli*). The straight line running diagonally across the picture is the bacterial DNA. The black dots are ribosomes; they are connected by the thin threads of mRNA to form polyribosomes. In this picture, transcription is thought to be proceeding toward the upper left, because the new mRNA molecules emerging from the upper portions of the DNA fiber carry more ribosomes than do those emerging from the lower portions. (Photo from O. L. Miller, Jr., Oak Ridge National Laboratory, and Barbara A. Hamkalo and Charles A. Thomas, Jr., Dept. of Biological Chemistry, Harvard Medical School.)

releases a completed polypeptide chain. Some of the amino acids (such as methionine and tryptophan) have only one corresponding triplet (TAC and TCC respectively), while others (such as leucine and arginine) have as many as six triplets each.

A striking feature of the code is that, in many cases, it is only the first two nucleotides of a triplet that are, in effect, needed to code for an amino acid. This feature of the code has led to the suggestion that the present version evolved from a more primitive version, which was a doublet rather than a triplet code. This evolution must have occurred very early—perhaps near the origin of all life—because subsequent studies have revealed that the code shown in Table 3.1 (which was worked out using the colon bacterium, *E. coli*) is the same for all living organisms so far studied, including man, and so is presumed to be universal. There could hardly be a more telling piece of evidence in support of the theory of evolution than this one striking fact of the universality of the genetic code.

Table 3.1
The genetic code: Nucleotide triplets and corresponding amino acids

DNA triplet	RNA triplet	Amino acid	DNA triplet	RNA triplet	Amino acid
AAA	UUU	Phenylalanine (phe)	ATA	UAU	Tyrosine (tyr)
AAG	UUC	Phenylalanine (phe)	ATG	UAC	Tyrosine (tyr)
AAT	UUA	Leucine (leu)	ATT	UAA	Chain end
AAC	UUG	Leucine (leu)	ATC	UAG	Chain end
GAA	CUU	Leucine (leu)	GTA	CAU	Histidine (his)
GAG	CUC	Leucine (leu)	GTG	CAC	Histidine (his)
GAT	CUA	Leucine (leu)	GTT	CAA	Glutamine (gln)
GAC	CUG	Leucine (leu)	GTC	CAG	Glutamine (gln)
TAA	AUU	Isoleucine (ileu)	TTA	AAU	Asparagine (asn)
TAG	AUC	Isoleucine (ileu)	TTG	AAC	Asparagine (asn)
TAT	AUA	Isoleucine (ileu)	TTT	AAA	Lysine (lys)
TAC	AUG	Methionine (met)	TTC	AAG	Lysine (lys)
CAA	GUU	Valine (val)	CTA	GAU	Aspartic acid (asp)
CAG	GUC	Valine (val)	CTG	GAC	Aspartic acid (asp)
CAT	GUA	Valine (val)	CTT	GAA	Glutamic acid (glu)
CAC	GUG	Valine (val)	CTC	GAG	Glutamic acid (glu)
AGA	UCU	Serine (ser)	ACA	UGU	Cysteine (cys)
AGG	UCC	Serine (ser)	ACG	UGC	Cysteine (cys)
AGT	UCA	Serine (ser)	ACT	UGA	Chain end
AGC	UCG	Serine (ser)	ACC	UGG	Tryptophan (try)
GGA	CCU	Proline (pro)	GCA	CGU	Arginine (arg)
GGG	CCC	Proline (pro)	GCG	CGC	Arginine (arg)
GGT	CCA	Proline (pro)	GCT	CGA	Arginine (arg)
GGC	CCG	Proline (pro)	GCC	CGG	Arginine (arg)
TGA	ACU	Threonine (thr)	TCA	AGU	Serine (ser)
TGG	ACC	Threonine (thr)	TCG	AGC	Serine (ser)
TGT	ACA	Threonine (thr)	TCT	AGA	Arginine (arg)
TGC	ACG	Threonine (thr)	TCC	AGG	Arginine (arg)
CGA	GCU	Alanine (ala)	CCA	GGU	Glycine (gly)
CGG	GCC	Alanine (ala)	CCG	GGC	Glycine (gly)
CGT	GCA	Alanine (ala)	CCT	GGA	Glycine (gly)
CGC	GCG	Alanine (ala)	CCC	GGG	Glycine (gly)

Abbreviations for amino acids are shown in the table. Abbreviations for nucleotide bases:
A = Adenine; C = Cytosine; G = Guanine; T = Thymine; U = Uracil.

Let us briefly consider the application of the code to the interpretation of the amino-acid difference between hemoglobins A and S.

Recall that, whereas the normal hemoglobin A has glutamic acid in the sixth position of its β chain, hemoglobin S has a valine in this place. From Table 3.1

we see that there are two possible triplets for glutamic acid, namely CTT and CTC. Valine, on the other hand, can correspond to any one of the triplets CAA, CAG, CAT, or CAC. Thus, one change, from T to A, in the middle nucleotide of the glutamic acid triplet could explain the change from glutamic acid to valine, whichever of the two triplets CTT or CTC is present in the DNA.

	Glutamic acid		valine
either	C[T]T	⟶	C[A]T
or	C[T]C	⟶	C[A]C

The simplest possible change in DNA is the substitution of one base for another, as in this case. Genetic changes generally are called *mutations,* which is just another word for alterations. Thus, we see that the simplest type of mutation that can be interpreted at the chemical level is a change from one base to another, and that this small difference in the large DNA molecule is ultimately responsible for all the consequences associated with the possession of hemoglobin S (see Fig. 3.9).

3.4 The Gene Concept at the DNA Level

Now that we have identified the genetic material and the way it works, we are in a position to redefine Mendel's "*Elemente,*" or genes, at a biochemical level. A gene is a segment of DNA recognizable by its specific function. In most cases the function is carried out by the polypeptide chain corresponding to the sequence of nucleotides in the DNA segments. Thus, the DNA sequence corresponding to the 146 amino acids of the hemoglobin β chain—which must contain at least $3 \times 146 = 438$ nucleotide pairs—is one example of a gene defined in this way. A gene, therefore, has its own complexity and is not (as was at one time thought) the minimal unit of genetic material; that unit is the nucleotide pair. Mutations can occur at any position along the sequence of nucleotides that comprise a gene. So it can be seen that there can exist many different versions, or alleles, of a given gene defined in this way.

The construction of genetic maps using recombination data is described in Chapter 2. There we point out that such maps, which are always consistently linear, seem to correspond—at least approximately—to the physical position of the genes on the chromosomes. The relation between nucleotide sequences and polypeptide sequences suggests that this concept of genetic mapping should apply even at the molecular level. In fact, genetic studies with bacteria have shown this to be the case. These studies have provided perhaps the most convincing evidence for the one-to-one correspondence between gene and protein.

Genetic analysis with microorganisms, including molds, bacteria, and bacterial viruses (phage), has contributed enormously to the development of biochemical genetics.

In fact, many of the most important advances in our molecular understanding of genetic processes could hardly have been achieved in any other way. The advantages of microorganisms for genetic studies are that they (1) divide quickly and so undergo many generations for study in short periods of time, (2) produce very large numbers of "individuals," and (3) can often be grown on simple, biochemically defined media so that the conditions of growth and reproduction can be closely controlled. Thus, for example, *E. coli*—the colon bacillus, which was the first bacterium used for genetic studies, and which is one of the most intensively studied of all organisms—can normally grow on a simple agar-based medium, containing sugar, salts, and ammonium as a source of nitrogen. Genetic variants, or mutants, can be obtained which require extra nutrients (for example single amino acids) in the medium in order to grow. Like the inborn errors of metabolism, these mutants are blocked in a biochemical pathway that leads to the production of an essential component such as an amino acid, and so cannot survive unless the missing essential component is provided for them in the medium.

Recombination in *E. coli* was first discovered through the observation that mixtures of two different mutant strains of bacteria that required different nutrients could produce at a low frequency normal bacteria which no longer needed any extras in the media in order to grow (Fig. 3.28). For example, suppose one strain requires the amino acid arginine, and the other requires a different amino acid, lysine. Neither strain will grow on a medium lacking both arginine and lysine. However, when the two are mixed under appropriate conditions, a small number of bacteria are produced that can grow on a medium lacking both arginine and lysine. These are recombinants that combine (1) the ability to grow without lysine derived from the arginine-requiring parent, and (2) the ability to grow without arginine from the lysine-requiring parent.

One of the most important features of this approach to picking out bacterial recombinants is that it is very sensitive; it is highly selective, because only the recombinants can grow.

Thus, one recombinant out of more than a million parents can readily be detected and it is not hard to obtain bacteria in the tens, hundreds, or thousands of millions. Genetic maps can be constructed in bacteria just as in higher organisms, though it is now known that mating (Fig. 3.29) and recombination do not take place in the orderly manner we have described for higher organisms in Chapter 2. Bacteria do not have chromosomes that are as highly organized as

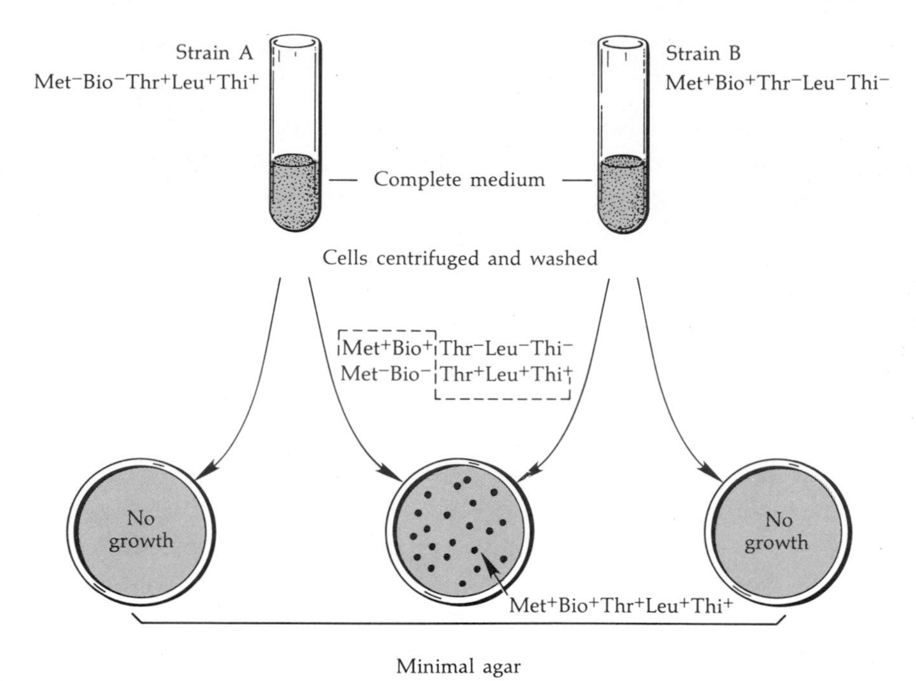

Figure 3.28
Recombination in E. coli: schematic diagram of an experiment performed by Joshua Lederberg and Edward L. Tatum. Strains A and B (obtained by mutation) grow only in the presence of nutrients not needed by normal strains: the amino acids Met, Thr, and Leu, or the vitamins biotin (Bio) and thiamin (Thi). The notation Met$^-$ indicates a requirement for that nutrient; Met$^+$ indicates a capacity to grow without it. Plating a mixture of the two strains on minimal agar (containing none of the nutrients required by strains A and B) gives rise to some colonies of recombinant bacteria. These bacteria possess all of the genes needed to produce the five nutrients absent in the minimal medium. (From G. Stent, *Molecular Genetics,* W. H. Freeman and Company. Copyright © 1971.)

those of other organisms, and so have no ordinary mitosis or meiosis. The bacterial "chromosome" seems to be just a very long DNA molecule with no other associated material. Recombination in *E. coli* occurs by the transfer of parts of the DNA molecule from one bacterial cell into another. Recall that transformation in pneumococcus was shown to be due to transfer of DNA, though in this case the donor DNA was in an extract and no longer in a cell.

Another way in which DNA can be transferred from one cell to another is by a virus or phage. Certain phages are known that can associate their DNA with that of the bacterium. They can pick up a part of the bacterial DNA, which is then incorporated with the phage DNA into a progeny phage particle. If such particles now infect another bacterium without killing it, they bring with them the DNA acquired from their previous bacterial host, which can be recombined into the new host's bacterial chromosome. Thus—if such a phage is grown on a normal strain

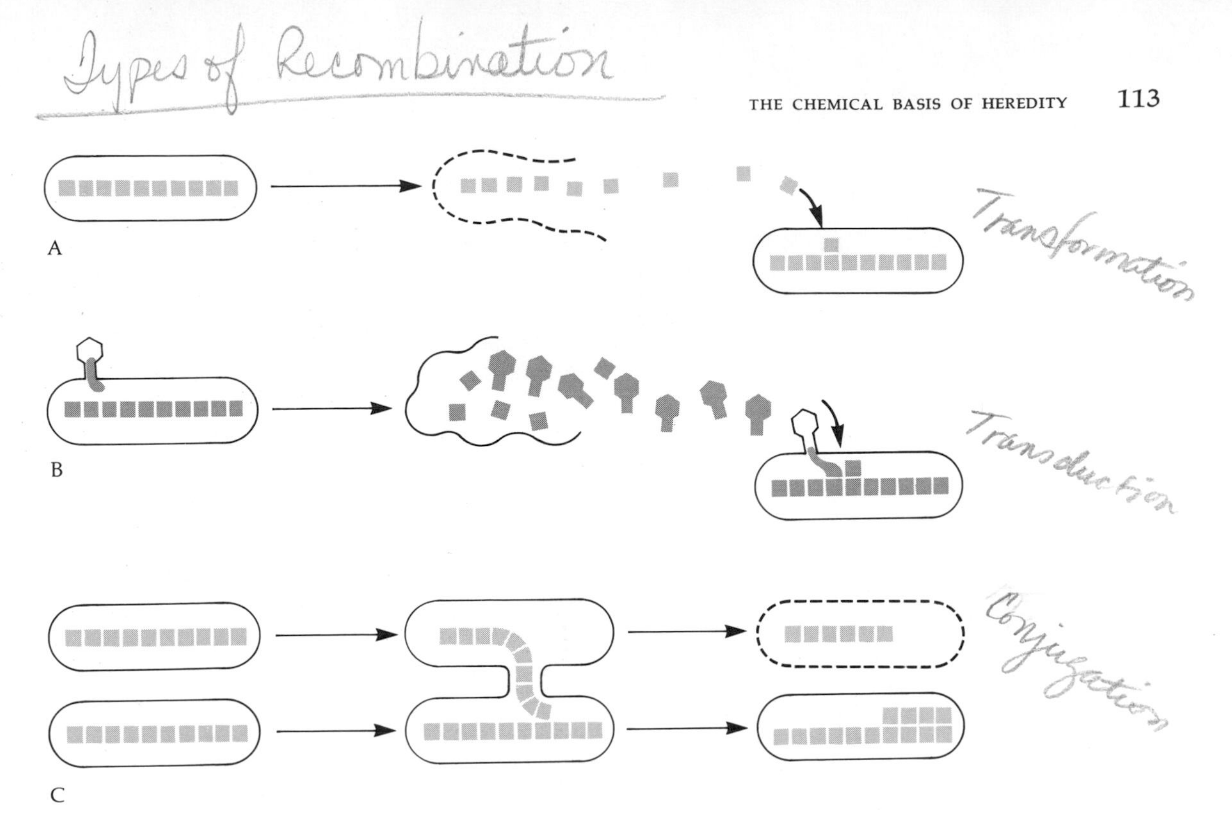

Figure 3.29
Gene transfer in bacterial cells. Genetic material can transfer from one bacterial cell to another by any of three methods. (A) Transformation: cells are disrupted, freeing the DNA; a small bit of DNA may then enter another cell. (B) Transduction: viruses infect the cells, carrying genetic material from one cell to another. (C) Mating: the DNA is passed directly from one cell to another. Greater amounts of DNA can be transferred by mating than by transformation or transduction.

of bacteria, and the resulting phage progeny grown on a strain requiring, say, the amino acid lysine—a small proportion of the new hosts are found that no longer require lysine for growth. They have received the normal gene for the blocked function via the phage. This process of phage-mediated transfer of genes (illustrated schematically in Figure 3.29B) is called *transduction.* These various approaches to bacterial genetics are important not only because of their contribution to biochemical genetics using bacteria, but also (as we shall see later) because they provide models for novel ways of doing genetics using cells of higher organisms.

We were led to this discussion of microbial genetics from a consideration of the construction of maps at the "intragenic" level. The best example of this mapping involves a series of mutants in the bacterium *E. coli,* mutants that are deficient with respect to the enzyme tryptophan synthetase (required in the final stage of the synthesis of the amino acid tryptophan). Recombinants between two different tryptophan mutants can be selected simply by growth on a medium lacking

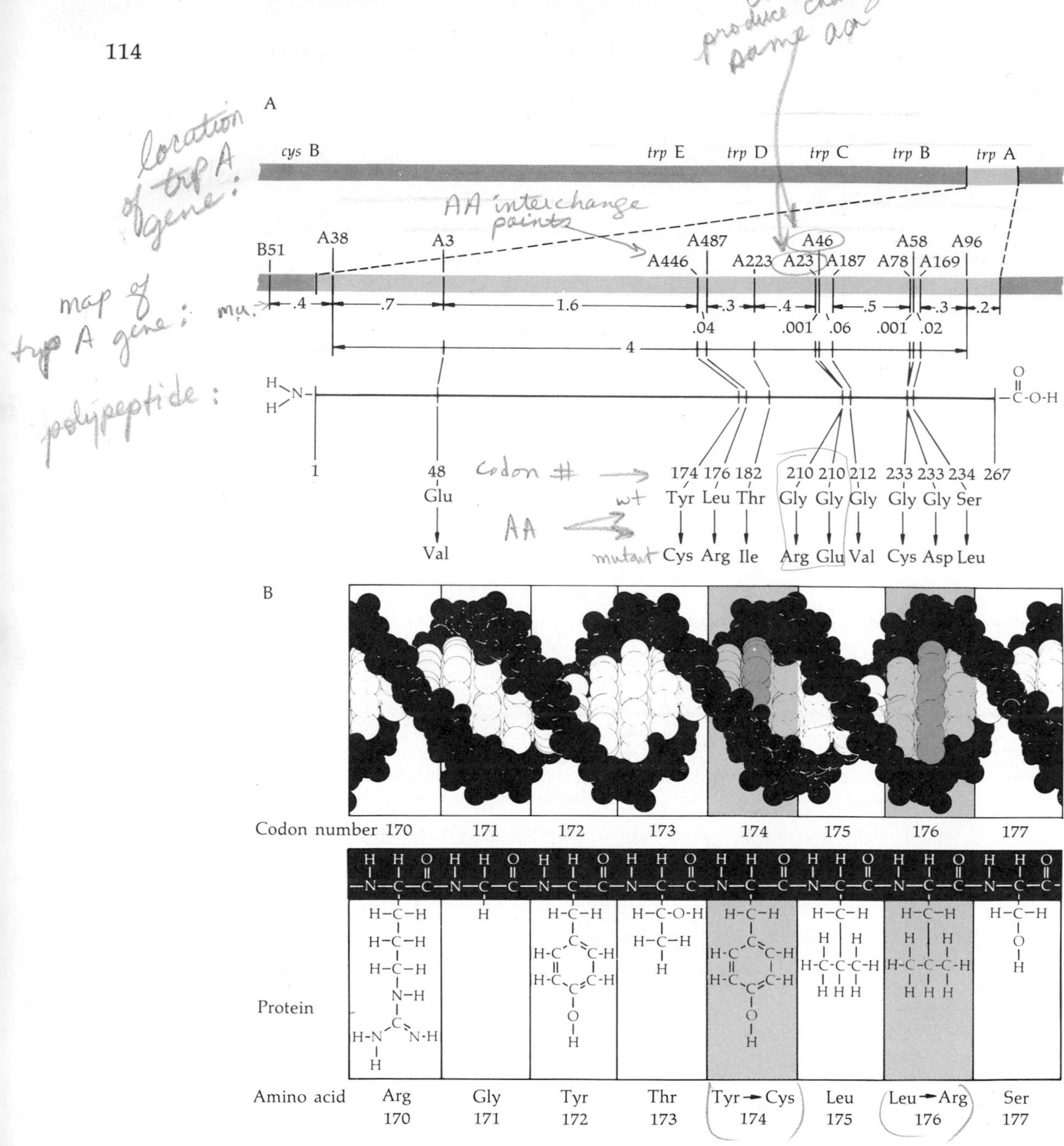

Figure 3.30

Colinearity of gene and protein. (A) Fine-structure map of a number of mutant sites in the *trp*A gene of *E. coli*. Numbers represent the probability (as a percentage) of a crossover between the mutant sites indicated by the arrows. The line drawn below the genetic map represents the tryptophan synthetase A protein polypeptide chain, whose production is controlled by this gene. The numbered amino-acid interchanges indicate the position and nature of the normal amino-acid

tryptophan. A genetic map can then be constructed from the pairwise recombination fractions between a series of such mutants, as described in Chapter 2. The difference here is that all the alleles we are studying are now at the same locus (mutations in the same gene). Moreover, the gene product is known, and its amino-acid sequence can be completely determined. Thus, each mutant that has been mapped can also be defined in terms of the amino-acid difference in the tryptophan synthetase protein with which it is associated.

The remarkable result of this research is that the genetic map obtained just from the recombination fractions corresponds precisely to the amino-acid map of the protein molecule. In other words, the sequence of the mutants from the map (and their distances apart) corresponds well to the sequence of amino acids changed in the protein (Fig. 3.30). Biochemistry and genetics have truly found their ultimate meeting point with this result. One interesting feature of this genetic map

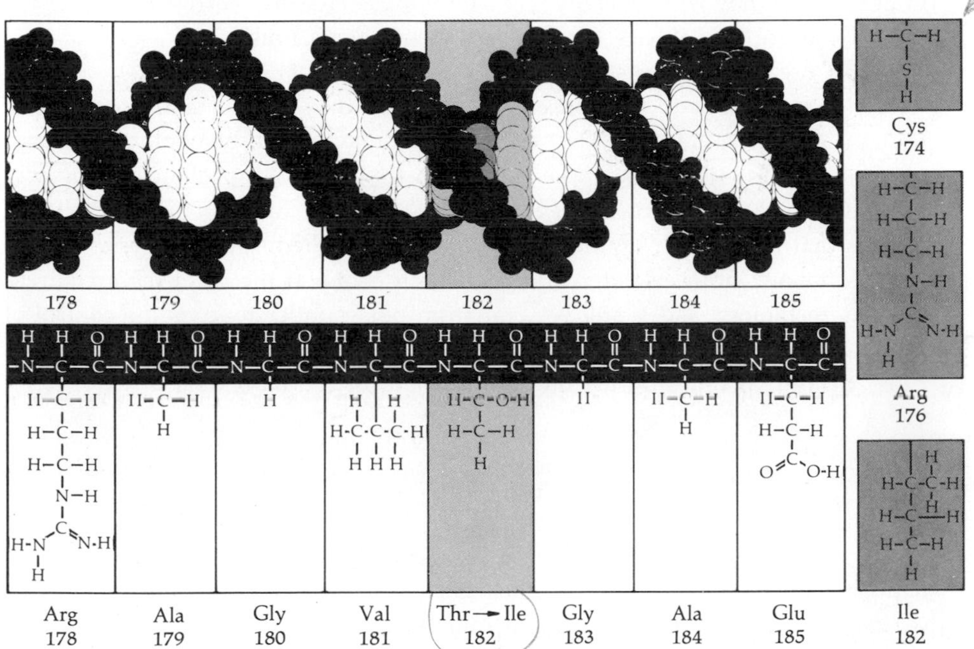

residues in that chain that have been replaced by an abnormal amino-acid residue as a result of each TrpA⁻ mutation. (B) Scale drawings of the double-helical DNA molecule of that part of the *trp*A gene in which amino-acid residues 170 to 185 of the A protein are encoded, and of the corresponding polypeptide chain. (After C. Yanofsky, "Gene structure and protein structure," *Scientific American,* May 1967. Copyright © 1967 by Scientific American, Inc. All rights reserved.)

is that two of the mutations, numbers 23 and 46, affect the same amino-acid position. Recombinants obtained from these two mutants have yet another amino acid in this same position in the protein, demonstrating that recombination can occur between neighboring nucleotide pairs. Thus the nucleotide pair is the minimal genetic unit, with respect to both mutation and recombination.

3.5 The Control of Gene Activity

An organism does not need all of its genes all of the time. Some enzymes—such as those involved in replicating DNA—may only be needed at certain times during the cell cycle. Other enzymes may only be required if certain nutrients are available in the environment in which the organism is growing. Bacteria, for example, can make use of a variety of different sorts of sugars as their basic energy source. They have genes for enzymes that can cope with each of the sugars they are able to utilize, but they will generally make only those enzymes corresponding to the sugars that are available in the medium at any given time.

Higher organisms show still more complex patterns of differential gene activity. Cells are specialized to such an extent that they may make mainly one sort of protein and few others. The red cell, for example, makes hemoglobin in large amounts, but no other cells of the body normally make this protein. The process by which the single fertilized egg diversifies into the variety of different types of cells that go to make up the whole organism is called *differentiation,* or *development.* A major feature of differentiation, which is one of the main research areas of modern biology and is still very far from being understood, is the control of gene activity such that different genes are working or "switched on" in different cells at different times during the development of the individual.

Important clues as to the possible mechanisms of control of gene activity have come from genetic studies with bacteria.

The problem is to determine how a cell decides when to make a given protein and when not to, and how much of the protein is to be produced. Switching on and off can, in principle, work at least at four different levels:

1. DNA: the production of RNA—namely, transcription can be controlled;
2. RNA: the production of protein—namely, translation can be controlled;
3. protein: the activity of the protein itself may be controlled by other molecules; and finally
4. the substrates on which the protein acts may be controlled separately.

One important system worked out in *E. coli* for the genes that make the enzymes that allow the utilization of the sugar lactose as an energy source is illustrated in Figure 3.31. Neighboring genes E_1, E_2, and E_3 make the enzymes needed for the utilization of lactose, while an unlinked gene R makes a protein whose only function is to regulate the activity of the enzyme gene. Its product, r, binds to a defined genetic region called the *operator* (O) just in front of the three enzyme genes in such a way as to block RNA polymerase attaching to the DNA, and so prevent the transcription of these three genes. The sugar lactose acts as an *inducer* (i); it binds to the *repressor* (r) and so stops the repressor from blocking transcription. Thus, in the presence of lactose, enzymes are made and the lactose is used up for energy metabolism. If there is no more lactose left, repressor r is free again to block transcription of mRNA and so switch off the production of the enzyme.

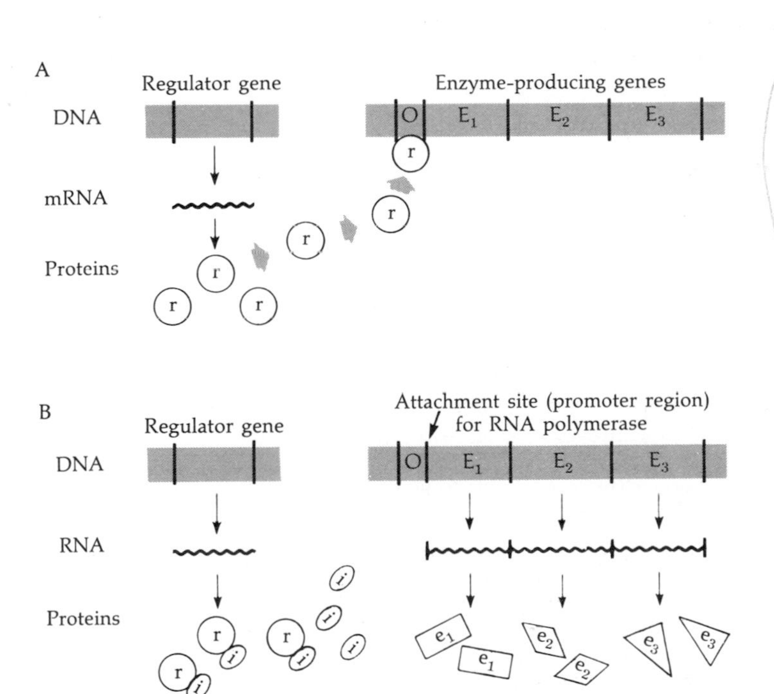

Figure 3.31
The operon model. E_1, E_2, and E_3 are neighboring genes that code for the enzymes needed to utilize lactose. R, an unlinked gene, makes a repressor protein (r) that (in the absence of inducer i) inhibits transcription of E_1, E_2, and E_3 by binding at O (the site of binding for RNA polymerase). The inducer (lactose in this particular case) binds to the repressor, preventing the repressor from binding to region O of the DNA, and thus permitting transcription of the genes to mRNA. (A) Substance i not present. (B) Substance i present.

The whole system is automatically self-regulating, rather like a toilet tank (Fig. 3.32). When you flush the toilet, water flows out of the tank, the float goes down, and water begins to flow back into the tank. When the tank is full, the float is up again and the flow of water into the tank is stopped. Such self-regulating systems are, in fact, the basis of automation, but nature discovered them first: they are also the clue to the regulation of living processes.

The control system just described works mainly at the level of the DNA. In Section 3.1, we describe a metabolic sequence (see Fig. 3.11) involving a number

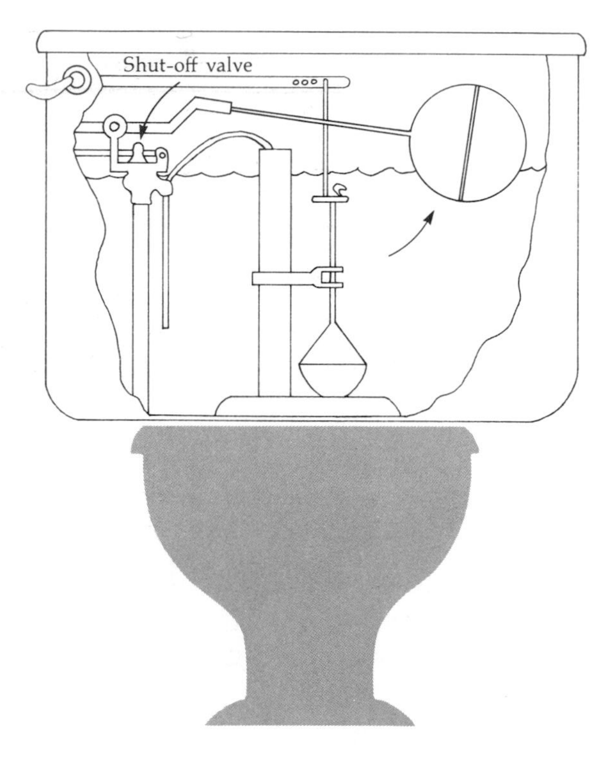

Figure 3.32
Feedback inhibition. A toilet tank is an example of a system working by feedback inhibition. When the toilet is flushed and water leaves the tank, the float moves down and opens the valve to cause water to flow into the tank. As the tank fills up, the float is raised, gradually retarding the flow of water until it is cut off completely at some threshold point. Thus, the presence of water in the tank inhibits the inflow of more water, while the absence of water acts as an inducer for the inflow of water. A thermostat also works by feedback inhibition: when the heat in the room exceeds a certain level, the thermostat turns off the heater to stop the production of further heat. The essence of a feedback-inhibition system is that the presence of a certain level of the product of the system leads to a shutdown of the system. In the example discussed in the text, the presence of a high level of lactose-utilizing enzymes leads to a depletion of lactose, which in turn leads to a shutdown of the enzyme-producing system. In this case, the absence of lactose also causes a shutdown of the system.

of successive steps, each carried out by a different enzyme. Many essential substances such as amino acids are synthesized step-by-step by a series of enzymes in such a metabolic sequence from simple precursors present in our food. Some control is needed to indicate to the organism when each of the essential substances has been synthesized. The level of, say, an amino acid may have to be quite closely regulated. Too little creates an obvious problem; too much—as for example in the case of PKU and phenylalanine—may be equally dangerous.

A common and important mechanism of control in such cases acts at the protein level and is also an example of feedback inhibition.

The end product of a metabolic sequence, producing say an amino acid, inhibits the activity of the first enzyme in the pathway leading to its formation. Thus, the more of the end product that is formed, the less active become the enzymes involved in making it from the available precursors, until eventually the whole pathway may be switched off. As the level of the end product goes down, the enzymes will once again be able to function and bring it up to the required level. Feedback inhibition in this case works at the protein level, rather than through DNA or RNA. Combinations of these various mechanisms and interactions between different metabolic pathways can give rise to very complex regulatory patterns.

The inborn errors of metabolism that we used as examples in our discussions of biochemical genetics are nearly all determined by recessive genes.

The simple explanation for this is that the homozygote for a deficient allele makes none of the required enzyme, and so is completely blocked in an essential pathway. The heterozygote has one normal functioning allele and this seems to allow for the synthesis of enough functional enzyme for normal development of the individual. Studies on enzyme levels, however, generally show that heterozygotes for a mutant allele, such as that causing PKU, have half as much of the corresponding enzyme—in this case phenylalanine hydroxylase (see Fig. 3.11)—as do individuals who are homozygous for the normal allele. This relation between the number of functional genes and the enzyme level is often called *gene dosage.* The lower amount of the enzyme in heterozygotes does not, in general, seem to cause any problems akin to the severe disease suffered by the homozygotes for the deficient allele. This situation is rather like the contrast between the sickle-cell anemic individual (*SS*) and the heterozygote (*AS*). The heterozygote (*AS*) makes enough hemoglobin (*A*) to function quite normally. Presumably the body's regulatory systems are able to compensate for the lower enzyme or hemoglobin levels in such a way that the individual is not affected, though heterozygous for a deficient allele. It is not the amount of enzyme present in an individual which is crucial but the amount of the product made by the enzymes.

The amount of product made by an enzyme is generally proportional to the amount of enzyme present. However, when there is an excess of the enzyme, the amount of product made follows a law of diminishing return. This is illustrated in Figure 3.33, which also shows how in some cases an inborn error of metabolism could be dominant. This happens when the amount of enzyme produced by the heterozygote is not enough to lead to an adequate level of enzyme product for normal functioning. Sometimes such a dominant effect may only show up under conditions of stress. This seems to be the case, for example, for some inborn errors involving enzymes of energy metabolism and the ability to utilize protein food-

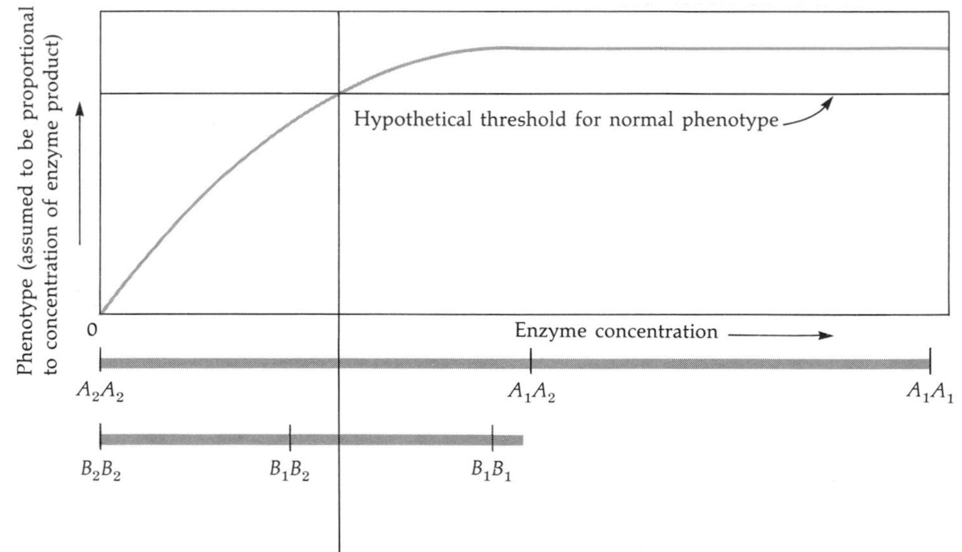

Figure 3.33
Effect of gene dosage on enzyme and enzyme-product concentrations. The concentration of enzyme is assumed to be the sum of the effects of two alleles; the heterozygote is therefore intermediate between the two homozygotes. The homozygote for the deficient allele is assumed to have none of the enzyme. The concentration of the enzyme product—on which the phenotype depends—follows a law of diminishing returns with respect to the enzyme concentration, as shown by the curve. Phenotypic effects are indicated for two types of genes (A and B). For gene A, both A_1A_2 (the heterozygote) and A_1A_1 (the enzyme-producing homozygote) genotypes produce enough enzyme to bring the product concentration above the threshold for a normal phenotype. Only the deficient homozygote (A_2A_2) is phenotypically abnormal; therefore, A_2 is a recessive allele with respect to A_1. For gene B, only B_1B_1 (enzyme-producing homozygote) has enough enzyme and enzyme product to achieve a normal phenotype. The heterozygote (B_1B_2) has an abnormal phenotype, so B_2 is a dominant allele with respect to B_1. Genes of type A are more common than those of type B. Also, in reality, enzyme production may not be zero even for a homozygous-deficient genotype, so that some enzyme and product may be present.

stuffs. A normal diet does not affect heterozygous individuals. However, when they take in an excess of protein, their enzyme levels cannot cope. They then suffer from severe headaches and vomiting due to the accumulation of toxic levels of byproducts of protein metabolism.

Different genes may complement each other to restore normal function.

The relationship between gene and enzyme for an inborn error may not always be obvious. Albinism is a recessive condition that occurs with a frequency of about one in 17,000 individuals, and is characterized by a complete (or almost complete) absence of the major pigment melanin (Fig. 3.34). Albino individuals have white (or almost white) hair, very light skin (making them sensitive to the sun), and pink eyes that are very sensitive to light levels (so that albinos often suffer from poor sight). Melanin is formed from the amino acid tyrosine (see Fig. 3.11), and so it is presumed that albinos have a metabolic block in the pathway leading from tyrosine to melanin, as first suggested many years ago by Archibald Garrod.

The notion that albinism is due to a single recessive gene was contradicted, however, by a report in 1952 of two albinos who met at a school for the partially sighted, married, and had three *normal* children. The simplest explanation for

Figure 3.34
Albinism: an African albino.

this unexpected result is that more than one gene locus is involved in albinism. Suppose for example that there are two enzyme steps in the pathway from tyrosine to melanin, each controlled by its corresponding gene. Recessive mutations that cause a loss of enzyme activity at either locus would lead to albinism. A mating between the two different types of albinos could then be written in the form

$$\frac{a1}{a1}\,\frac{A2}{A2} \times \frac{A1}{A1}\,\frac{a2}{a2}$$

where *A1* and *A2* are the dominant alleles of the two loci, and *a1* and *a2* are the corresponding recessive alleles. All the offspring would be double heterozygotes

$$\frac{A1}{a1}\,\frac{A2}{a2}$$

and so normal, having at least one normal functioning allele at each locus.

This phenomenon—the production of normal offspring from a mating between two apparently similar affected homozygotes—is called complementation.

The *"complementation test"* as we have just described it is in fact a functional test for whether two recessive mutations occur at the same gene locus. Some years after the description of the anomalous albino family this interpretation was confirmed when it was shown that some, but not all, albinos lacked the enzyme tyrosinase that is needed for the conversion of tyrosine to melanin. The pedigree of a mating between albinos in which such biochemical tests have been done is shown in Figure 3.35. The other biochemical lesion of albinos is not yet identified. In fact, there may well be more than two steps involved in going from tyrosine to melanin, leaving further forms of albinism still to be discovered.

Lack of complementation between alleles at the same locus is clearly illustrated by some of the hemoglobin variants. Thus, hemoglobin C is a variant that is found in relatively high frequencies in some of the same areas of the world where the sickle-cell hemoglobin S is prevalent. Homozygotes *CC* for the hemoglobin-C gene have a much milder anemia than do *SS* homozygotes. Heterozygotes *SC* are, however, not normal, but have an anemia whose severity is intermediate between that of *SS* and *CC*. The *C* allele is known to be associated with an amino-acid substitution in the same position (number 6) as that affected by the *S* allele, but the C substitution is from glutamic acid to lysine rather than to valine. Thus, biochemical studies in this case clearly demonstrate that *S* and *C* are alleles and, as expected, the heterozygote *SC* does not show complementation.

Many enzymes are made up of two protein subunits that, in contrast to the subunits of hemoglobin, are identical and are determined by the same gene locus.

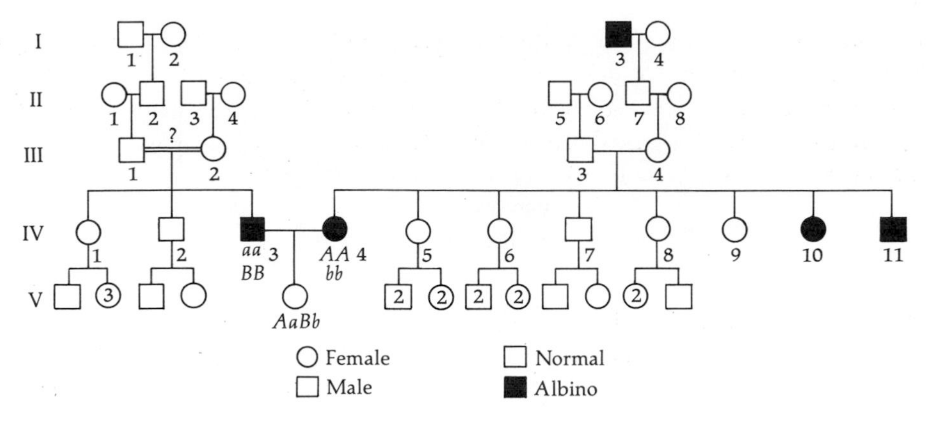

Figure 3.35
An albino pedigree illustrating complementation. This is the pedigree of a family where both parents (IV.3 and IV.4) are albino, but have a normally pigmented daughter. The father is tyrosinase-negative; the mother and her affected siblings are tyrosinase-positive. (From C. S. Witkop et al., *American Journal of Human Genetics,* vol. 22, p. 59, 1970.)

Heterozygotes for electrophoretic variants of such an enzyme show three bands of enzyme activity on electrophoresis. Two of these are the expected forms corresponding to the two types of bands observed in homozygotes for each of the two alleles, while the third band in between consists of one subunit from each allele. An example of this is shown in Figure 3.36 for the enzyme peptidase A, which is one of a class of enzymes that break the peptide bonds linking pairs of amino acids in a polypeptide. Peptidase A type 1 is obtained from a *1/1* homozygote and type 2 from a *2/2* homozygote. The heterozygote *2/1* has both these bands and has in addition a third, intermediate "heteropolymeric" band combining one subunit from the *1* allele with one subunit from the *2* allele. The occurrence of such intermediate bands in heterozygotes is in fact often the best evidence that an enzyme is made up of two identical subunits.

Sometimes it is found that alleles at the same locus controlling the same polypeptide chain do complement, in the sense that heterozygotes are normal or nearly so while homozygotes are deficient. This is thought to be due to compensating interaction between two subunits in a heteropolymeric molecule such as peptidase A 2-1. Complementation, therefore, does not always provide clearcut evidence that two genetic characters are concerned with separate functions, though it does always show that the complementing alleles are different.

Hemoglobin, which we have already used several times to illustrate the fundamental principles of genetics, provides some of the best examples of gene interaction and of the control of gene activity in mammals.

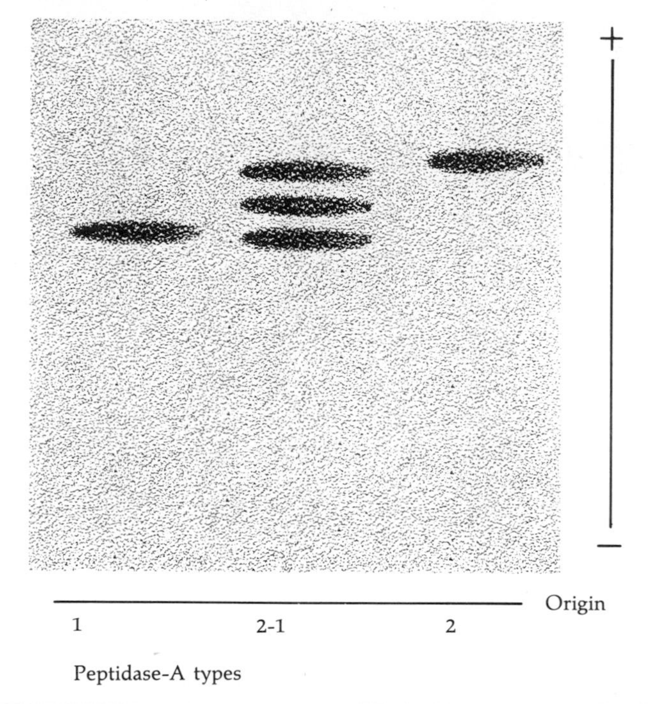

Figure 3.36
Electrophoresis of peptidase A, illustrating hybrid enzyme formation. The enzyme from the heterozygote (*2-1*) shows three bands; the middle band contains one subunit derived from allele *1* and one subunit coded by allele *2*. Such an electrophoretic pattern indicates that the enzyme normally is made of two identical subunits. (After H. Harris, *Principles of Human Biochemical Genetics,* North Holland Pub. Co., 1970.)

Recall that the main type of hemoglobin is made up of two different polypeptide chains (α and β) and that the differences between hemoglobins A, S, and C are due to changes in the amino-acid sequence of the β chain. The corresponding genes *A, S,* and *C* are thus alleles at the locus that codes for the β chain. What is the relation of the α-chain gene to the β-chain gene?

This question was answered when a family was found in which both hemoglobin S and hemoglobin HO2 (Hopkins 2, a variant of the α chain) were segregating. The pedigree of this family is shown in Figure 3.37. Its key feature is the occurrence of normal and doubly affected offspring from a parent carrying both the altered forms of hemoglobin. If the *S* and *HO2* genes were allelic then all offspring of the mating between double carriers and normals would either have hemoglobin S or hemoglobin HO2 and none would have both or neither. The

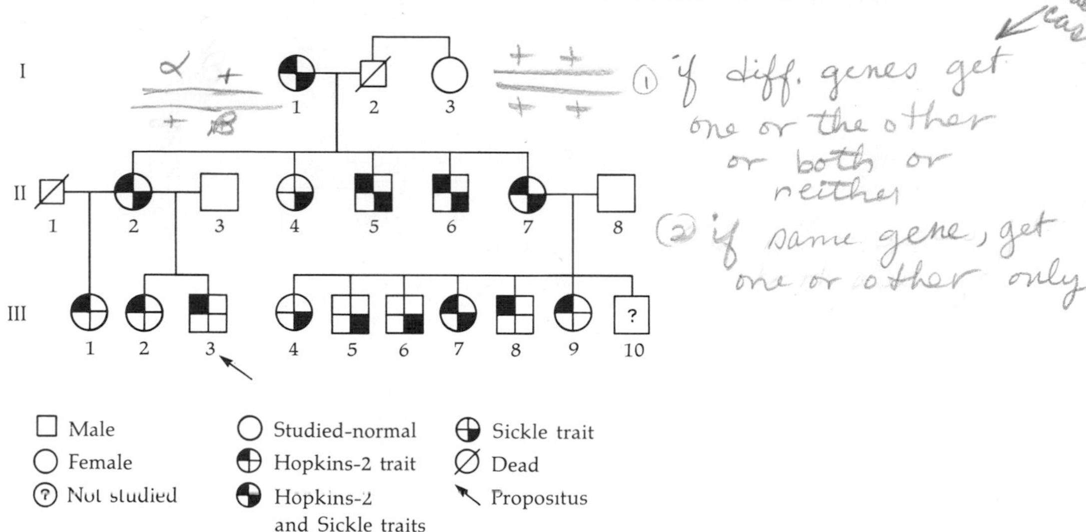

Figure 3.37
The Fuller kindred. In this family, both hemoglobin S (a β-chain mutation) and hemoglobin HO2 (Hopkins 2, an α-chain mutation) were segregating. A parent with both the altered forms of hemoglobin (II.7) had a doubly affected offspring (III.7). The genes for S and HO2—and therefore those for the β and α chains—cannot be allelic. If S and HO2 were alleles of the same gene, all offspring of a doubly affected parent would have either hemoglobin S or hemoglobin HO2, but not both. Other pedigrees confirm that the two genes for α and β chains segregate independently and most probably are on different chromosomes. (Pedigree by E. W. Smith and J. V. Torbert, *Bulletin of the Johns Hopkins Hospital,* vol. 102, p. 38, 1958.)

common occurrence of doubly affected and normal types in the family shows that the two genes are not allelic and may in fact be unlinked. The data obtained from the pedigree are quite consistent with the notion that the genes for the β and α chains are on different chromosomes.

A combination of genetic and biochemical data suggests that next to the β-chain gene there may be other genes for similar hemoglobin chains.

In addition to the main hemoglobin type A found in normal adults, there is a minor component called hemoglobin A_2. This hemoglobin is formed from the combination of two α chains with two chains called δ that are different from β. This hemoglobin thus has the formula $\alpha_2\,\delta_2$ in contrast to hemoglobin A, which is $\alpha_2\,\beta_2$. Rare families are found in which variants for both the β and the δ chain are segregating. In this case the two variant genes behave like alleles, as shown by the pedigree given in Figure 3.38. This shows that the β-chain and δ-chain genes are very closely linked.

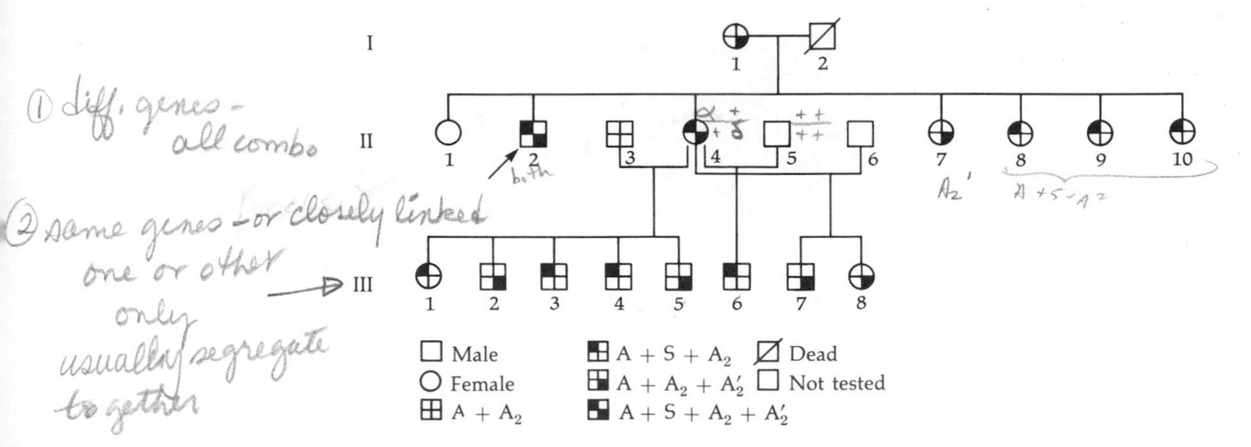

Figure 3.38
Segregation of hemoglobins beta and delta. In contrast with hemoglobins S and HO2 (Fig. 3.37), the δ-chain variant HbA_2 and the β-chain variant HbS segregate like alleles, indicating very close linkage between the genes for the two chains. (From B. F. Horton and T. J. Huisman, *American Journal of Human Genetics,* vol. 15, p. 395, 1963.)

The occurrence of certain very unusual hemoglobin chains that appear to be combinations of β and δ suggests that the two loci may actually be next to each other on the chromosome. These unusual "Lepore" hemoglobins (so called after the name of the family in which they were first found) have one end like a β chain and the other end like a δ chain, just as if they were the result of a crossover between the two chains. The β and δ chains each have 146 amino acids and are quite similar in their amino-acid sequence. This similarity presumably is reflected in the corresponding nucleotide sequence of the DNA. On the basis of this similarity, and on the assumption that the two genes are next to each other, it has been suggested that the Lepore hemoglobins are in fact derived from crossing-over of an unusual kind, called *"unequal,"* in which the β-chain gene on one chromosome pairs with the δ-chain gene on the other, instead of with its β homologue as is normally the case. Crossing-over under these circumstances, as illustrated in Figure 3.39, can give rise to Lepore-type recombinant molecules. Note that this type of unequal crossing-over can also increase the number of similar, adjacent genes—in this case, from 2 to 3.

There is a change in the types of hemoglobin produced during development.

There is a third type of hemoglobin (hemoglobin F) that in normal individuals only occurs at the fetal stage of life, and that is associated with another polypeptide chain called γ, which is quite similar to the β and δ chains. The molecular

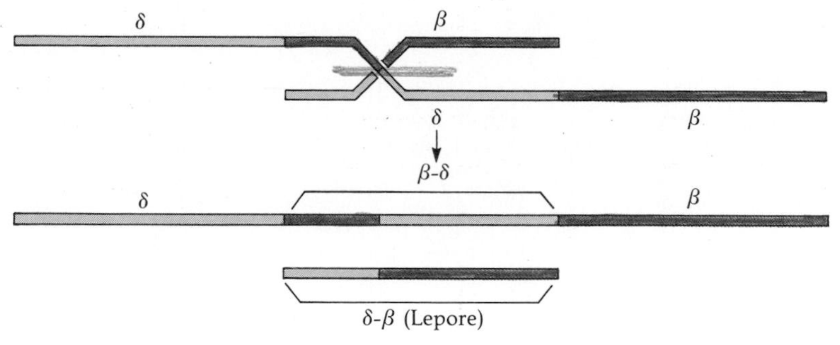

Figure 3.39
Lepore hemoglobins arise from unequal crossing-over. The β and δ hemoglobin loci may pair incorrectly and engage in unequal crossing-over. The Lepore hemoglobin produced in this manner is a combination of β and δ hemoglobins. (After H. Harris, *Principles of Human Biochemical Genetics,* North Holland Pub. Co., 1970.)

γ chain

formula for hemoglobin F is $\alpha_2 \gamma_2$. Thus each of these three types of hemoglobin—A, A_2, and F—have similar molecular structures. Each type has a pair of α chains; the types differ only in the other chains—respectively β, δ, and γ. It seems likely that the γ-chain gene is closely linked to the β-chain and δ-chain genes, possibly adjacent to them as indicated in Figure 3.40.

As already mentioned, hemoglobin F occurs predominantly in fetal life; as indicated in Figure 3.41, it is barely detectable by the age of 6 months. This figure also shows the existence of yet another hemoglobin chain ε that is probably similar to β, γ, and δ and that appears to be found only during early embryonic life. There are thus at least two major switchover points in hemoglobin synthesis during development. The first from ε to γ, and the second from γ to β and δ. Such differential gene activity at different stages of development (and of course in different tissues) is a major feature of the developmental process. The mechanisms that control such switches are not yet known and are the subject of intensive research, following the understanding of similar sorts of control mechanisms in bacteria as discussed earlier.

Some clues to these control mechanisms may come from the study of hemoglobin abnormalities resulting from a reduced rate of production of one or more of the hemoglobin chains.

There is a class of anemias called the *thalassemias* that are associated with a complete absence (or at least a very much reduced amount) either of β chains or of α chains. Individuals with "beta-thalassemia major," for example, are homozygous for a gene that appears to suppress the synthesis of β chains. This gene behaves as an allele of the β-chain locus and only affects the activity of the β-chain gene

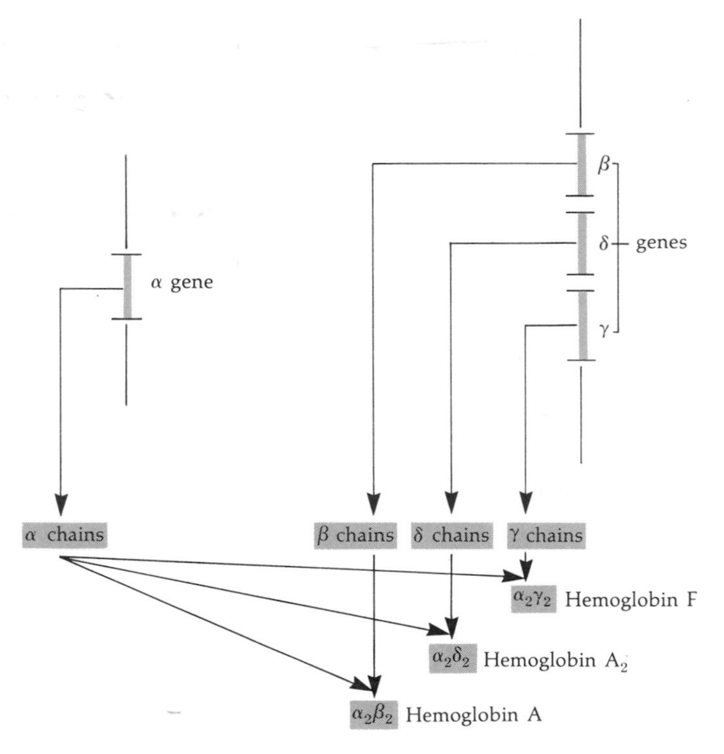

Figure 3.40
Genetic control of hemoglobins. The γ gene may be adjacent to the β and δ genes. The α gene is not linked to any of these genes.

on its own chromosome. Heterozygotes for the beta-thalassemia gene often show a mild anemia, which may not even be clinically apparent.

The mechanism by which the beta-thalassemia gene works is not yet fully understood, though it is known that there is no alteration in the β chain produced, and so the change is apparently not in the β-chain gene itself. Recent data suggest that the defect is at the level of production of messenger RNA from the β gene, so that the alteration may, for example, involve the DNA region where the RNA polymerase that makes the messenger is first attached. Individuals with beta-thalassemia major continue to produce fetal hemoglobin into adult life, presumably as a partial, though inadequate, compensation for their lack of the normal adult hemoglobin A.

The "α thalassemias," which are associated with a deficiency of α chains, are associated with apparent abnormal alleles at the α-chain locus, and they parallel in many respects the beta thalassemias. Homozygosity for an alpha-thalassemia gene is, however, associated with much more severe and earlier effects than is

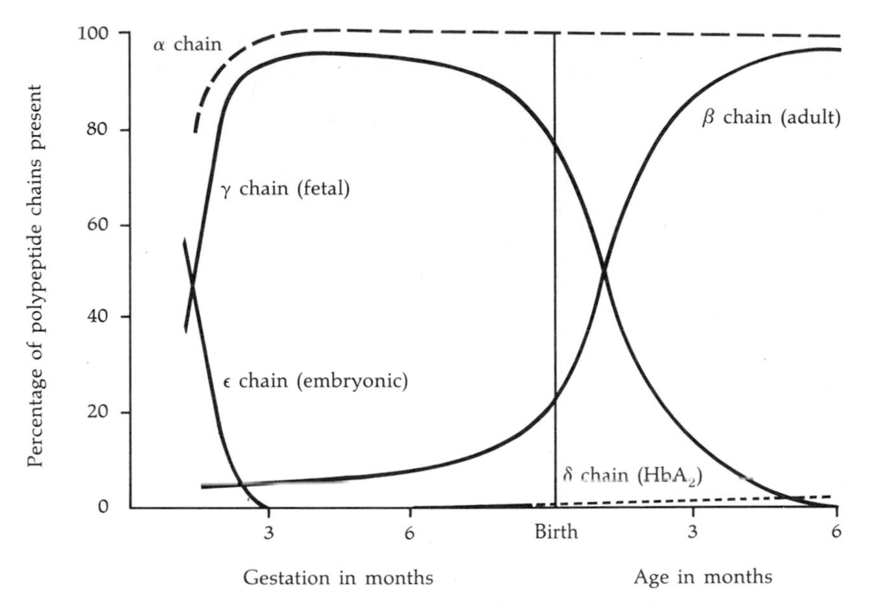

Figure 3.41
Synthesis of hemoglobin chains during development. Hemoglobin F ($\alpha_2\gamma_2$) is produced mainly during fetal life; it is almost undetectable by six months after birth. Hemoglobin chain ϵ is seen only during the first three months of gestation, after which time it is replaced by the γ chain. The γ chain, in turn, is replaced by the β and δ chains. The δ chain is present in very low concentration in the normal adult hemoglobin. (From E. R. Huehns et al., *Cold Spring Harbor Symposia on Quantitative Biology*, vol. 29, p. 329, 1964.)

homozygosity for a beta-thalassemia gene, as might be expected from the fact that α chain occurs in virtually all the normal forms of hemoglobin, in the fetus as well as in the adult. Homozygotes for alpha-thalassemia genes produce abnormal hemoglobins formed from four β chains (β_4, called hemoglobin H) or four γ chains (γ_4, hemoglobin Barts). Clinical and genetic data indicate that there are, most probably, several different types of alpha and beta thalassemias, and their study should certainly help to unravel some of the mechanisms of control of gene activity in mammals.

There are at least two classes of functional genes that do not code for proteins—namely the genes for the ribosomal RNAs and those for the transfer RNAs.

Recall that the ribosomal RNA is a major component of the ribosomes, which are the particles on which protein synthesis takes place, and that the tRNAs are the adaptor molecules that read the nucleotide triplets on messenger RNA and mediate the attachment of the correct amino acid to the growing polypeptide chain. These two sorts of RNA molecules are transcribed from DNA sequences just as if they were messenger RNA, but are not then translated into protein sequences. A particular feature of these genes, which seems to distinguish them from the protein-coding genes we have been discussing so far, is that they exist in multiple copies. Thus in mammals, including man, it is estimated that there may be several hundred copies of the genes coding for ribosomal RNA and at least ten or more copies of each of the tRNA-coding genes.

Biochemical techniques have led to detection of ribosomal-RNA genes near each of the centromere regions of the five human acrocentric chromosomes (numbers 13, 14, 15, 21, and 22). This is consistent with the fact that these same regions are associated with the "nucleolus," a nuclear structure in which the ribosomes are formed.

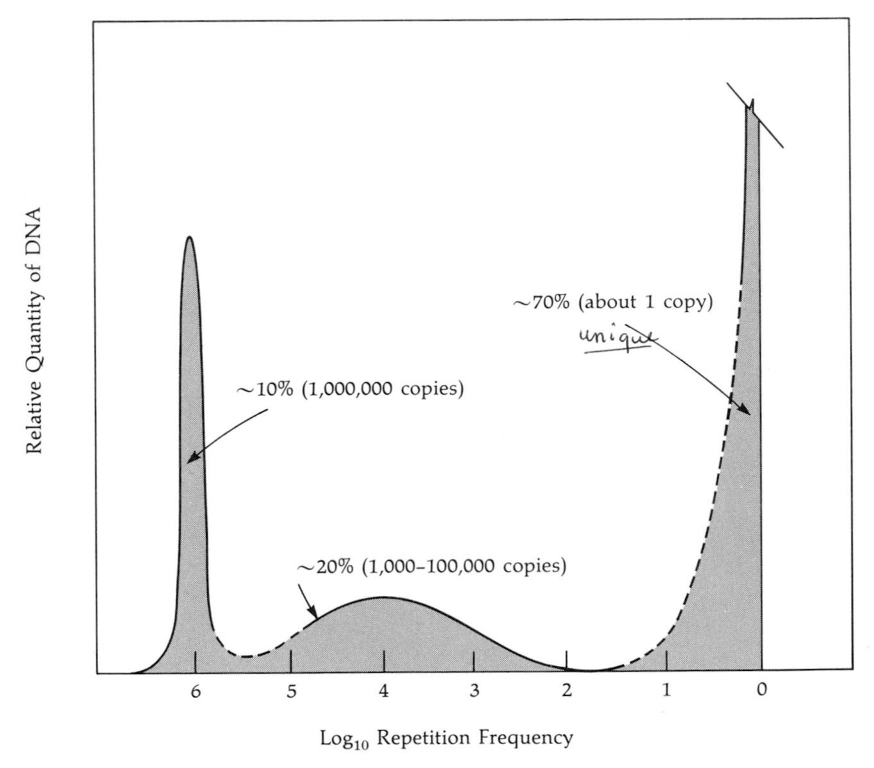

Figure 3.42
Distribution of repetitive DNA sequences in mouse DNA. The graph plots the relative quantity of DNA against the logarithm of the repetition frequency (6 = 1 million = 10^6 copies; 5 = 10^5 copies; and so on). The data are obtained by measuring the quantity and rate of reassociation of fractions of separated DNA chains. The dashed segments of the curve are regions of considerable uncertainty. (From R. J. Britten and D. E. Kohne, *Science,* vol. 161, p. 529, 1968. Copyright 1968 by the American Association for the Advancement of Science.)

A large fraction of the DNA of higher organisms is highly repetitious and has no obvious function.

This surprising fact has been established only quite recently, using sophisticated techniques of DNA biochemistry. Some DNA sequences, called "satellites," may be repeated up to 100,000 times or more (Fig. 3.42). This repetitious property enables them to be physically separated from the bulk of the DNA, making it possible to locate their position on the chromosome. Using staining techniques that are a variant of the banding procedures described in Chapter 2, the "satellite regions" show up as intensely staining bands that are usually located near the centromeres of chromosomes, as shown in Figure 3.43. No messenger RNA is apparently made from these satellite regions, adding to the puzzle of their possible function, about which there is much speculation but nothing established.

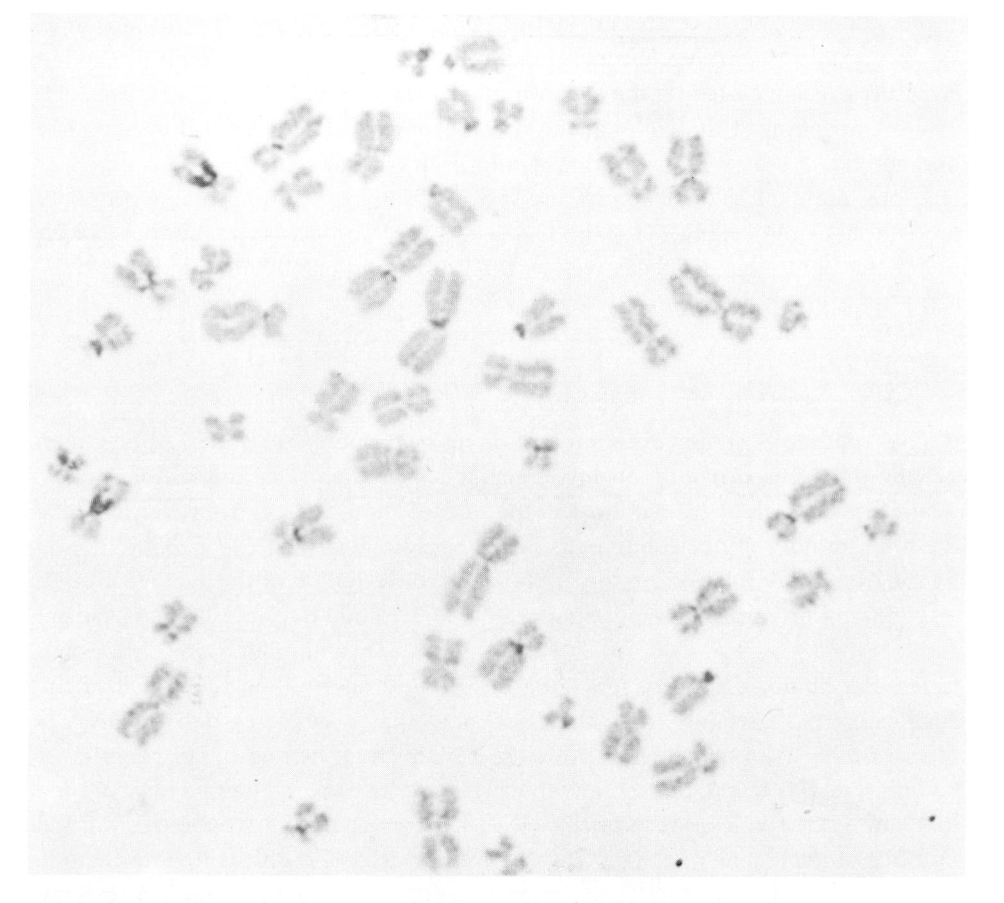

Figure 3.43
Centromeric heterochromatin and G-11 staining. The number-9 chromosomes stain most heavily; the number-1 chromosomes (the longest, above the center of the picture) also stain darkly. (Courtesy of M. Bobrow.)

Many lesser degrees of repetition than that represented by the satellite DNA also exist. Biochemical data suggest that at least 40 percent of the total DNA is repetitious to some extent, the remainder consisting of so-called "unique" sequences. Uniqueness is, however, very difficult to establish using these techniques, because differences between closely related sequences may not be readily detected. The repetitious sequences will therefore certainly include the ribosomal and transfer RNA genes, and they may also include some closely related genes such as those for the β, γ, δ, and ϵ chains of hemoglobin. Nevertheless, it seems that the bulk of the repetitious DNA is similar to the satellite DNA in that it is not transcribed into messenger RNA and has no obvious function.

The DNA content of the human genome, if fully stretched out in the form of a single continuous DNA double helix, would be about 2 meters long, or about two hundred thousand times the diameter of the average cell! This represents a total of about 6,000,000,000 (6×10^9) nucleotide pairs of DNA. Clearly the chromosomes as we visualize them in a karyotype must be highly condensed, multiply coiled structures, as suggested by the electron microscope picture shown in Figure 3.44. The organization of the DNA in the chromosome is at present still very poorly understood. It seems probable that within each chromosome the DNA exists as one long double helix surrounded by various proteins. These proteins probably mediate both the coiling up of the DNA into the compact organelle that is seen under the light microscope, as well as the unraveling of specific regions as their activities are expressed.

How many genes are there in the human genome?

This apparently simple question cannot yet be properly answered. At one time, genetic data, based mainly on linkage analysis in *Drosophila* and in the mouse, would have put the figure at most in the tens of thousands. If, however, we take the total number of nucleotide pairs in the haploid genome (3×10^9) and divide by the number of nucleotides that correspond to the length of the average protein (say 200 amino acids corresponding to $200 \times 3 = 600$ nucleotide pairs), we get a figure of $(3 \times 10^9)/600$, or 5,000,000 genes. This number is one thousand times that obtained for an average bacterium and one million times that of the smallest virus. Certainly, now that we know that a large fraction of DNA is repetitious and doesn't code for proteins, it seems that the number of genes must be less than 5,000,000, but by how much is uncertain. Some people have argued that only perhaps 2 percent of the DNA is functional, which would bring the number of genes down to 100,000—a figure that is not so different from the one assumed by earlier geneticists and is only 20 times that of an average bacterium. Biochemical data, however, suggest that as much as 50 percent of the DNA is functional, which would leave us with 2.5 million genes. The gap between these

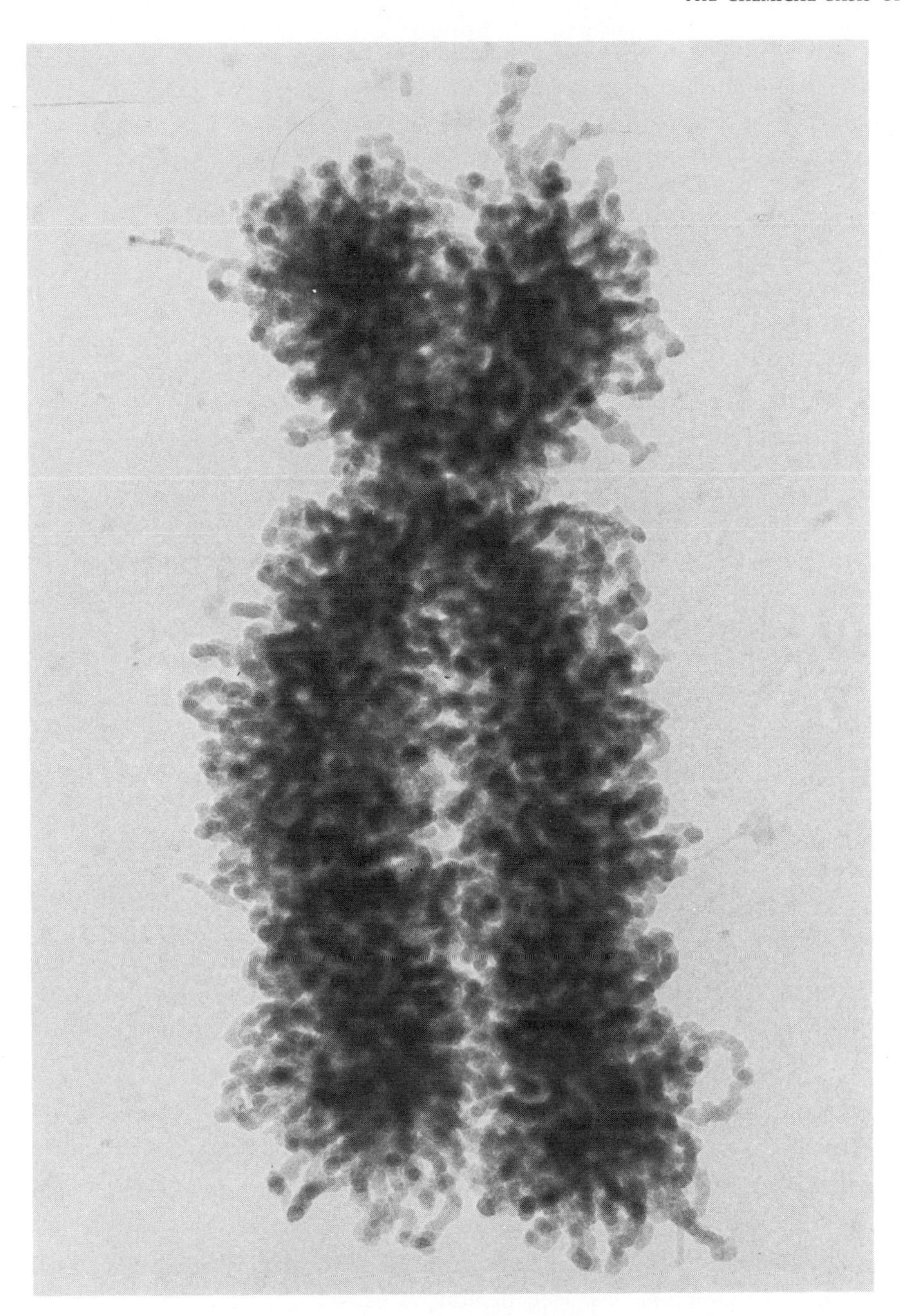

Figure 3.44
Electron micrograph of a human chromosome 12. (From E. J. Dupraw, *DNA and Chromosomes,* Holt, Rinehart and Winston, 1970.)

two extremes is still very considerable. Two points are clear, however. First, whatever the number of genes, it is very large. Second, it is extraordinary that so much of the DNA—whether 50 percent or 98 percent—has no obvious function in terms that can at present be fully understood.

Summary

1. Proteins basically are made of polypeptides (strings of amino acids), which owe their specific functions to their chemical compositions. Many of them are enzymes, which catalyze specific chemical reactions, thus utilizing nutrients to produce energy and materials necessary for growth and function. There are very many different types of proteins in an organism.
2. Proteins are synthesized by a mechanism—involving the ribosomes—that uses messenger RNA (mRNA) as the source of the information about the sequence of amino acids that forms a specific protein. The building blocks of the protein, the amino acids, are carried to the ribosomal site of synthesis by specific, relatively short RNA molecules called transfer RNA (tRNA). The whole process is called translation.
3. The production of the specific mRNAs, each forming a protein (or a series of them), is directed by a gene, a segment of DNA contained in chromosomes. The mRNA is complementary to one of the two DNA strands forming the double helix. This process of mRNA production is called transcription.
4. The genetic code is the dictionary stating which amino acid will eventually go into a protein molecule, in correspondence with a triplet of successive nucleotides in the DNA (and hence in mRNA). There are 64 possible triplets, 61 of which each specify one of the 20 amino acids. The other three triplets specify the termination of a polypeptide chain. The code is therefore redundant.
5. Genetic studies in microorganisms, made possible by special "crossing" techniques, have greatly helped the development of molecular genetics. They have proven that genetic information is contained in DNA; that DNA and proteins are colinear.
6. In bacteria, a mechanism of control of gene action through molecules that repress gene action has been demonstrated. This mechanism may also be important in higher organisms.
7. Other mechanisms of control of gene and enzyme action have been demonstrated which help us understand the development of multicellular organisms.

8. Inborn errors of metabolism are due to gene changes, whereby an enzyme is not produced or is inactive. A block thus occurs in a metabolic pathway at the step controlled by this enzyme. The product of this enzyme is not made available (or not available in sufficient amounts) and the precursor (substrate) and/or its derivatives accumulates. Usually an enzyme is produced in sufficient amounts by one single copy of the specific gene responsible for its synthesis, so that heterozygotes for a nonfunctioning allele have enough enzymes for normal function. Most, if not all, inborn errors of metabolism are therefore recessive.

9. When a protein is made by the products of more than one gene, changes due to defects of these genes may not appear in heterozygotes. This phenomenon, complementation, helps to show that more than one gene is responsible for a given effect.

10. Hemoglobin is made of two different types of subunits, each present in two copies. In addition, there are various types of hemoglobin in individuals, also depending upon the stage of development. This well-studied molecule gives examples of several important genetic phenomena: independent segregation, linkage, unequal crossing-over, complementation, duplication, and regulation.

11. Not all DNA codes for proteins. Some DNA codes for ribosomal RNA, which is not translated into protein. Other DNA may have other not-well-understood functions. In any case, the human DNA has room for hundreds of thousands or even millions of genes, conceived as functional units, many of which produce specific polypeptide chains.

Exercises

1. Review the general structure, mechanism of synthesis, and function of proteins.

2. Review the structure, synthesis, and functions of DNA.

3. In what ways is RNA an intermediate in protein synthesis?

4. Which were the major steps in cracking the genetic code?

5. What different mechanisms of gene regulation can be suggested to exist in higher organisms?

6. How many different genes might there be in man? What uncertainties are left due to our ignorance about regulatory genes?

7. Write the complementary strand, the mRNA, and the polypeptide sequences resulting from a DNA segment of the following constitution:

1	2	3	4	5	6	7	8	9	10	11	12	13	14	15
T	T	A	C	G	C	T	A	T	T	A	A	A	T	A

8. Rewrite the DNA code in the form of Figure 3.45, inserting in each case the abbreviation for the corresponding amino acid. Examine the table and say which of the three nucleotides of a triplet (first, second, or third) is *least* important in specifying an amino acid.

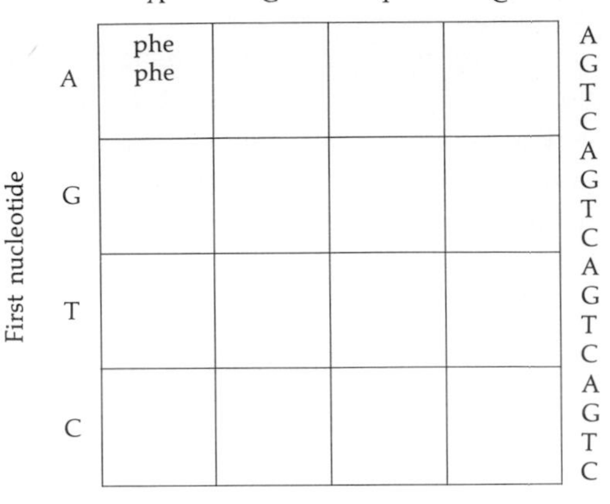
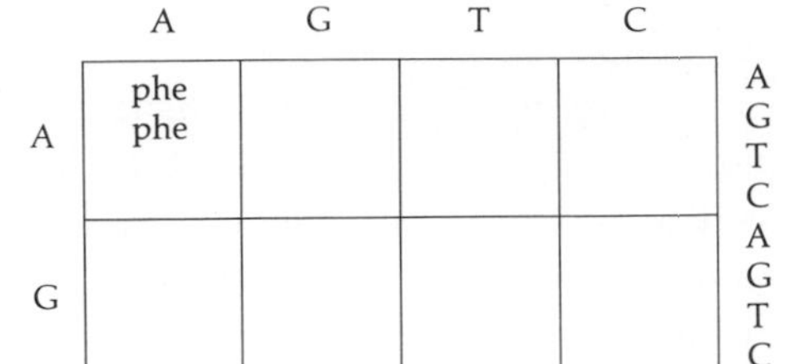

Figure 3.45
Genetic-code table for Exercise 8. Amino-acid abbreviations corresponding to triplets AAA and AAG have been entered as examples.

9. The enzyme lactate dehydrogenase is made of four subunits (it is a "tetramer"); there exist two types of subunits produced by two different genes, A and B. Assuming that the subunits A and B associate as in a tetramer, how many (and which) different enzyme forms can be distinguished? (These forms are called "isozymes.")

10. Once amino acids are joined in a polypeptide chain, most of them are "neutral"—that is, are neither acidic nor alkaline—because one acid (carboxyl) group and one alkaline (amino) group have been used in the formation of the peptide bond (Fig. 3.5). A few of them, however, have an acid or alkaline group ("radical") in the part labeled "specific portion" or R in the figure. Such a group remains free in the polypeptide chain. The following have an acid radical: glutamic acid and aspartic acid. The following have an alkaline radical: lysine, arginine, and histidine. The electric charge of a protein depends on the total number of such acidic and alkaline radicals in it because—unlike the neutral amino acids—they contribute, respectively, a negative or a positive charge. Substitution of a neutral—or a positive by a negative amino acid, or a positive by a neutral one, and so on—changes the charge of the protein molecule, and hence changes its electrophoretic mobility. Compute the probability that an amino-acid substitution in a protein changes its electrophoretic mobility, assuming that all codons are represented with equal frequency.

References: Chapter 3

Harris, H.
1975. *The Principles of Human Biochemical Genetics.* 2nd ed. New York: American Elsevier.

BOOKS ON MOLECULAR BIOLOGY

Crick, F.
1966. *Of Molecules and Men.* Seattle: University of Washington Press.

Hayes, W.
1968. *The Genetics of Bacteria and Their Viruses: Studies in Basic Genetics and Molecular Biology,* 2nd ed. New York: Wiley.

Haynes, R. H., and P. C. Hanawalt
1968. (Eds.) *The Molecular Basis of Life: An Introduction to Molecular Biology: Readings from Scientific American.* San Francisco: W. H. Freeman and Company.

McElroy, W. D.
1971. *Cell Physiology and Biochemistry,* 3rd ed. Englewood Cliffs, N.J.: Prentice-Hall.

Stent, G. S.
1970. *Molecular Genetics: An Introductory Narrative.* San Francisco: W. H. Freeman and Company.

Watson, J. D.
1970. *Molecular Biology of the Gene,* 2nd ed. New York: Benjamin.

4

Gene Mutations and Chromosomal Aberrations

Chapter 2 describes the inheritance of differences between people. All inherited differences arise from transmissible changes in the genetic material—changes that we call mutations. Some of them may be very small changes, perhaps in only a single nucleotide of the DNA. Others involve larger segments of DNA; in a few of these alterations, the changed DNA segments are so large that they can be detected microscopically as changes in the morphology or the number of chromosomes (chromosome aberrations).

This chapter is dedicated to a study of the various types of mutation and of the agents that induce these—in particular, radiation.

4.1 Mutation

Mutation in its broadest sense is, by definition, the origin of new hereditary types. It is the ultimate origin of all genetic variation; without it there would be no genetic differences, and so no evolution. The frequency of occurrence of new mutations is very low, as might be expected. If it were otherwise, there would be no continuity in biological inheritance, and Mendel's laws would simply be scrambled by a high rate of mutation. The rarity of mutation makes it a difficult phenomenon to study in mammals and especially in man.

The difference between the alleles determining the beta chains of hemoglobins A and S is an example of a change that must have arisen by mutation. As shown in Chapter 3, the change from glutamic acid in the beta chain of hemoglobin A to valine in that of hemoglobin S can be accounted for by one nucleotide or "base-pair" change—namely from T to A—in the middle of the glutamic-acid triplet. This substitution of one base for another is the simplest type of mutation that can occur. Any mutation must involve some change in the DNA nucleotide sequence. The identification of the precise nature of a mutation at the DNA level can, in general, only be achieved with knowledge of the gene product and its amino-acid sequence (or nucleotide sequence in the case of an RNA product). Few genetic differences in the human population can yet be characterized at this level.

Mutation is a random and rare event that may be difficult to document.

The hemoglobin-S allele is present in many individuals in various parts of the world, and it is not known exactly when or how it arose. To identify a mutational event, one needs evidence of a difference between parent and offspring that cannot be explained by the Mendelian laws of inheritance. In Chapter 2, hemophilia in Queen Victoria's family is used as an example of an X-linked recessive trait. None of Queen Victoria's ancestors were known to have had hemophilia, and so it might be thought that the hemophilia mutation in her family was first

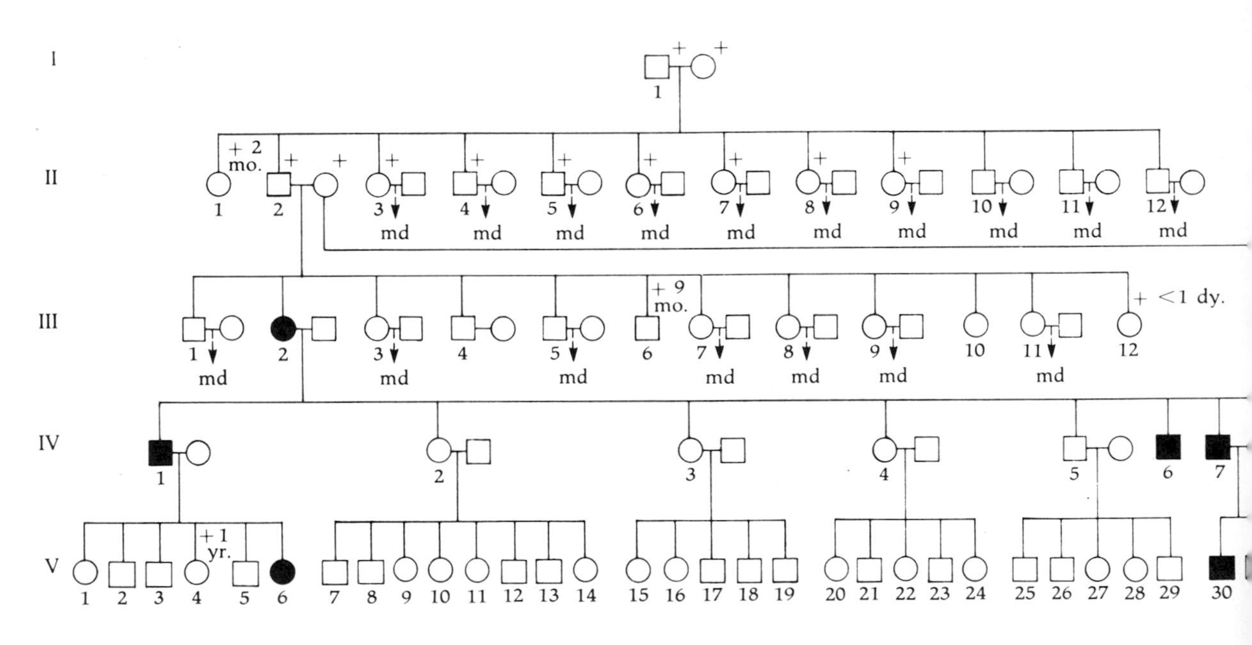

■ ● Spastic paraplegia
\+ Deceased
md Many unaffected descendants

carried by her. If so, the mutation presumably must have occurred in the formation of one of the two gametes contributed by her parents. The mutation could however have been carried for many generations in her female ancestors in the heterozygous state without ever having been exposed in a male, so its origin in Queen Victoria is not certain.

The problem of pinpointing the origin of an autosomal recessive mutation is still more difficult, as this mutation of course is only expressed in the homozygote of either sex. An autosomal recessive mutation may therefore be carried for many generations before it is exposed in the homozygous state, if it ever is. Only for clearcut dominant genes that are expressed unequivocally in the heterozygote can one be reasonably sure that a mutation has just occurred. In such a case, at least one parent of an "affected" individual must also be affected if no mutation has occurred. Mutational events are thus clearly identified by the occurrence of an affected offspring from normal unaffected parents.

An example of a pedigree documenting the occurrence of a mutation is shown in Figure 4.1. Affected individuals in this case have a spastic weakness of the legs called spastic paraplegia. The condition first occurred in this family in individual III.2, and was transmitted from her to both male and female offspring, some of

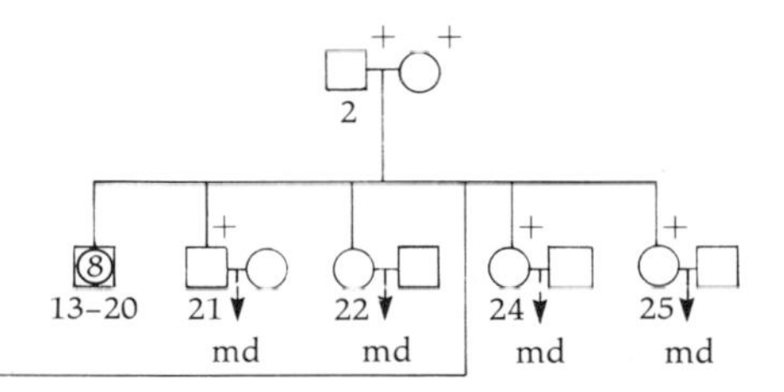

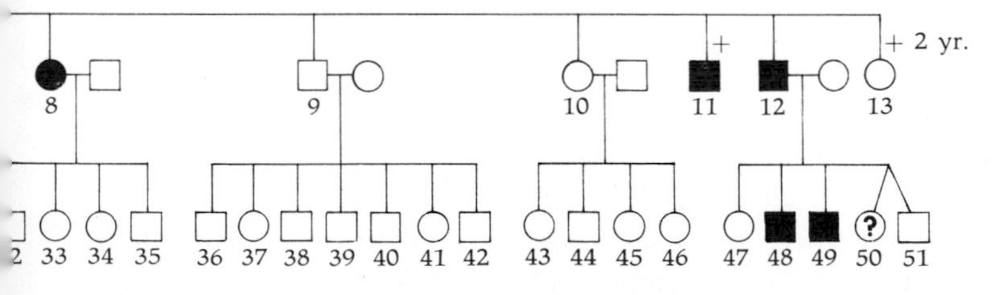

Figure 4.1
Dominant mutation. The mutation (for spastic paraplegia) first occurred in individual III.2. The transmission to male and female offspring, and through these offspring to a third generation, points up the dominant mode of inheritance. (After V. A. McKusick et al., *Cold Spring Harbor Symposia on Quantitative Biology,* vol. 29, p. 107, 1964.)

whom have passed it on to a third generation, showing clearly its dominant mode of inheritance. None of the many brothers and sisters of III.2 had the condition, nor did her grandparents. This pedigree comes from a closed religious community, the Amish, whose members are noted both for their fertility and their fidelity, so that illegitimacy can be confidently excluded. The study of transmission of other genetic markers (phenotypic differences whose mechanism of transmission is simple and well known) can also help in tracing possible illegitimacies. There seems thus little doubt that the mutation carried by individual III.2 was a new one produced in the germ cells of one or the other of her parents.

Because mutation is a rare event, many tens (or even hundreds) of thousands of pedigrees may have to be scanned to document a mutational event. Pedigrees such as that shown in Figure 4.1 are generally ascertained because the affected individuals come to the clinician's notice. Biochemical changes detected in heterozygotes, such as that from hemoglobin A to hemoglobin S in an AS heterozygote, would not be brought to the attention of a clinician because they have no obvious pathological connotations. In an analysis of a series of surveys for rare electrophoretic hemoglobin variants, only fourteen examples of unusual heterozygotes were discovered out of 10,971 individuals studied. Six of the variants occurred once only, four of them turned up twice, and none was observed more than twice. Thus, while the total variant frequency in the population was 1/784, the maximum frequency of any given variant allele was slightly less than 1/10,000 (2/21,942 because each individual carries two alleles). This frequency cannot, however, be directly related to the mutation rate because the parents were not studied, and so it is not known how many (if any) of these rare hemoglobin types were new mutations in the individuals being studied. The use of population frequencies such as this for the estimation of mutation rates depends on the principles of population genetics and is discussed in Chapter 9. In the present chapter we confine our attention to the types of mutation that can occur, the factors that influence the mutation rate, and some direct estimates of rates from animal studies.

The study of mutation rates in animals is done mostly in experimental organisms like Drosophila and mice.

The study of mutation as a process was initiated mainly by H. J. Muller in the 1920s working with the fruitfly, *Drosophila melanogaster.* This research depended on the development of *quantitative selective* techniques to identify mutations and so measure *mutation rates.* One of the first fruits of Muller's studies was the finding in 1927 that X rays greatly increase the rate of gene mutation. This finding underlies present-day concerns about the genetic effects of radiation, whether due to fallout from atomic bomb explosions, to byproducts of the generation of nuclear power, or to the medical uses of X rays. Some years later in 1941, as a byproduct

```
    H   H       H   H
    |   |       |   |
Cl—C—C—S—C—C—Cl
    |   |       |   |
    H   H       H   H
```

Figure 4.2
Structural formula of a chemical mutagen. Mustard gas (bis-2-chloroethyl sulfide) was the first known chemical mutagen.

of wartime research on the effects of mustard gas as a chemical warfare agent, came a demonstration that certain chemicals, such as mustard gas, could also greatly increase mutation rates (Fig. 4.2). Chemicals such as this are called *mutagens*. Nowadays there is concern that some of the chemicals used in food products might be significant mutagens. This concern—together, of course, with the need to monitor radiation effects—emphasizes the need to find ways of establishing mutation rates in man. To this end the mouse has been used extensively as a mammalian model because of its small size and comparative ease of breeding.

A few mutation-rate studies have been done in the mouse using a total of many hundreds of thousands of mice. Mutations to recessive alleles at specific loci where such alleles are already known (mainly for coat-color differences such as albino) are sought by crossing normal ("wild-type") mice with individuals homozygous for recessive alleles at a series of chosen loci. New recessive mutations of any of these loci will lead to homozygous mutant phenotypes in the offspring of such a cross, where only wild-type heterozygotes were expected (Fig. 4.3). The combined results of such an experiment, using wild-type males in the initial cross, carried out by the United States and United Kingdom Atomic Energy authorities gave, for the seven loci studied, 39 mutations out of 688,921 mice, or an average

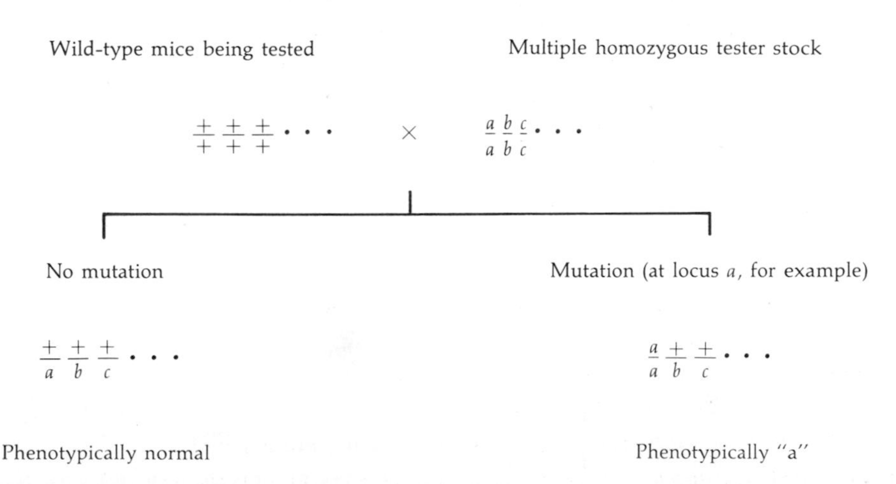

Figure 4.3
Detection of specific-locus, recessive mutations in the mouse. The recessive alleles at a series of loci are represented as *a*, *b*, and *c*; + represents the normal ("wild-type") allele for each locus.

rate per locus of 8.1 per million male gametes. The figure for females was somewhat lower at 1.4 per million gametes. Such sex differences are perhaps not unexpected in view of the radical differences between the sexes in their modes of gamete formation.

Some estimates of mutation rates to unspecified loci in the mouse have also been made. These come from routine examinations of inbred strains of mice for obvious phenotypic changes that are later shown to have a single-gene basis. Inbreeding allows one to pick up new recessive as well as dominant mutations. In one such study carried out at Jackson Laboratories at Bar Harbor (which is perhaps the foremost mouse-breeding establishment), 28 mutations occurring at 26 loci were found in approximately 3.5 million mice. Taking account of a necessary correction factor, this gave a frequency per locus of 0.67 per million. More than 14 million mice have been scored at the Jackson Laboratories for dominant mutations, and 54 of these have been found distributed among 12 loci. This gives a per-locus mutation rate for any dominant mutation of about 0.4 per million, which is much smaller than the estimates for recessive loci. These various figures are summarized in Table 4.1.

All these mutation-rate estimates suffer from at least two major deficiencies. In the first place the type of change detected is not well defined and is certainly not identified at the biochemical level. A subjective element must be involved in detecting the phenotypic changes that are counted as mutations. A second and perhaps even more serious problem is that these studies must inevitably pick up changes at those loci that have the highest mutation rates. Even with the data at hand, it is clear that mutation rates at different loci may vary quite considerably—

Table 4.1
Direct estimates of mutation rates in the mouse*

Number of mice studied	Mutation rate per locus
Recessives	
Specific loci	
Males: 688,921	8.1 per million
Females: 202,812	1.4 per million
Nonspecific loci	
Both sexes: 3,446,872†	0.67 per million
Dominants	
Any locus	
14,021,464†	0.4 per million (4.8 per million per gamete)

*Only mutations with visible phenotypic effects have been detected.
†In these studies, some mice were scored only for coat-color mutations.

perhaps by a factor of 10 or more. It can be shown that this factor of bias (necessarily identifying mutations at the most mutable loci) could lead to overestimates of mutation rates by a factor of at least 10, if not much more. Thus, while the method of detecting mutations may miss some, and so lead to underestimates, the bias due to picking up changes for the more mutable loci may lead to substantial overestimates. It at least appears from the mouse data, however, that the average rates of mutation to recessive alleles (which will include the majority of single-gene changes) are at most of the order of one per million per locus per generation, and could easily be tenfold lower than this.

At the DNA nucleotide-sequence level there are, in principle, three major basic types of changes that can occur: substitution of one base for another, deletion of one or more bases, and insertion of new bases.

The change from hemoglobin A to hemoglobin S, already discussed, is an example of a *substitution.* To illustrate the effect of a *deletion,* consider the following hypothetical DNA sequence of nucleotides:

AAA ACG AAA CCG AAG CAT CCT . . . .

and suppose that the fourth base starting from the lefthand end is deleted (Fig. 4.4). The new DNA message would then become

AAA CGA AAC CGA AGC ATC CT . . . .

Reading from the left end, the sequence of amino acids produced from the original DNA would be (following the genetic code given in Table 3.1)

phenylalanine–cysteine–phenylalanine–glycine–phenylalanine–valine–glycine

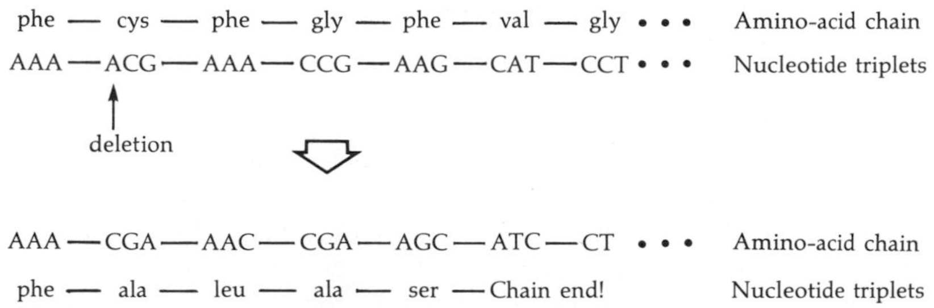

---

phe — cys — phe — gly — phe — val — gly • • •   Amino-acid chain
AAA — ACG — AAA — CCG — AAG — CAT — CCT • • •   Nucleotide triplets

deletion

AAA — CGA — AAC — CGA — AGC — ATC — CT • • •   Amino-acid chain
phe — ala — leu — ala — ser — Chain end!   Nucleotide triplets

Figure 4.4
Possible consequences of a deletion mutation. The "frame" is shifted by one base pair, resulting in a wholly different series of codons. In this example, amino acids would be altered, and transcription would be terminated early because one of the "new codons" is a chain-end signal. Such alterations are called frameshift mutations.

The corresponding sequence from the altered DNA is

phenylalanine–alanine–leucine–alanine–serine–chain end.

This new sequence shows two important features. First, the sequence is completely different for all the amino acids following the position of the mutation. Second, the mutated sequence ends prematurely because the sixth triplet (ATC) was turned into a "chain-end" signal.

Mutations leading to shortened polypeptide chains can, of course, also occur as a result of a base substitution. For example, changing the terminal G in the second triplet (ACG) of the original DNA sequence to a T would produce the chain-termination triplet ACT. Insertions, like deletions, also cause a complete change in amino-acid sequence after the position of the mutation. Such sequence changes are called "frameshift" mutations.

Because of their more marked effects on the amino-acid sequence of the corresponding proteins, deletions and insertions will generally have much more severe effects on protein activity than do substitutions. A recent review of all the known hemoglobin variants showed that 155 out of a total of 167 were due to substitution of one amino acid by another, and that all 155 could be explained by single DNA base substitution. Of the remaining 12 variants, 5 had missing amino acids, 2 had additional amino acids at the end of a chain, and 5 were "Lepore" type recombinant molecules (see Chapter 3). A much shortened molecule, due to a chain-end mutation occurring in the sequence, or a radically different molecule associated with a single nucleotide insertion or deletion, may of course never be detected. (However, hemoglobins that have undergone what appear to be terminal-frameshift mutations have very recently been described.) The frequency of observed variants, as discussed in Chapter 9, is a function of their survival value and so, as already emphasized, these observed frequencies do not properly reflect the patterns of occurrence of new mutations. Nevertheless, it does seem reasonable to suppose that substitution of one base for another is the most common type of primary mutational event at the gene level.

Because there are four different bases (A, T, G, and C), there are $4 \times 3 = 12$ different types of single-base substitutions, for each base can change to any one of the others. As discussed in Chapter 3, most of the amino acids have more than one corresponding triplet, and in a number of cases just the first two nucleotides of the triplet determine an amino acid's code. Thus a substantial fraction of base-pair substitutions do not result in a changed amino acid. There are some data, coming from an analysis of hemoglobin variants, that suggest that changes from C to T and from T to C may be substantially more frequent than other base-pair substitutions.

Most of the informative work on mechanisms of mutation at the chemical level has been done with microorganisms.

These studies used nutritionally deficient strains and the same types of selective techniques to identify mutations as those used for the discovery of recombination mechanisms in the bacterium *E. coli* (see Chapter 3). In this way, very rare mutational events can be readily and reproducibly identified, and so mutation can be studied as a process with comparatively little effort. These studies have identified various classes of chemical mutagens that can be characterized by their effects on the DNA molecule.

One class of mutagens, which includes nitrous acid, acts directly on the bases to cause "deamination," which involves the removal of a nitrogen atom from the outside of the molecule (Fig. 4.5). For example, deamination makes adenine behave like guanine, so that when the DNA is next replicated a C will be placed opposite the altered base, instead of the T that should have been there. In this way an AT base pair is replaced by a GC base pair.

Another class of chemicals, called "alkylating agents," includes the first discovered chemical mutagen mustard gas. These mutagens react mainly with guanine, causing it to be released from the DNA. The resulting gap is then filled by any one of the bases, usually causing a single-base substitution.

A third group of mutagens, related to the dye acridine orange, causes insertions and deletions of bases—probably because these chemicals can fit in between two

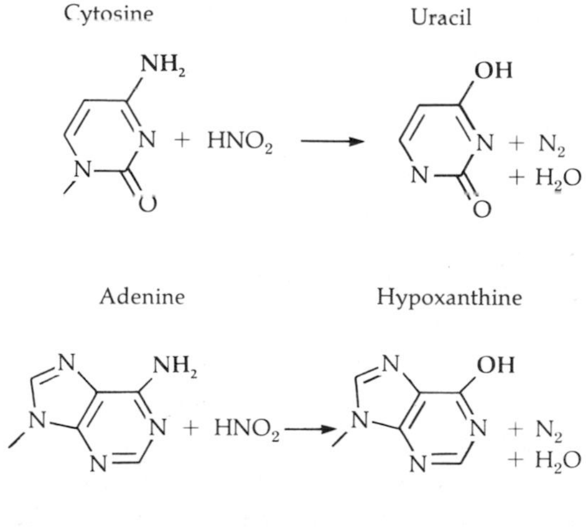

Cytosine + HNO_2 → Uracil + N_2 + H_2O

Adenine + HNO_2 → Hypoxanthine + N_2 + H_2O

Figure 4.5
Consequences of deamination. Nitrous acid (HNO_2) is a deaminating agent. Deamination of cytosine (normally forming three hydrogen bonds in base pairing) produces a base (uracil) with the hydrogen-bonding specificities of thymine (two hydrogen bonds). Deamination transforms adenine (two bonds) into a base (hypoxanthine) with the hydrogen-bonding specificities of guanine (three bonds). The result of such deamination is a nucleotide substitution during DNA replication. This substitution may cause an amino-acid substitution, or may even interfere with transcription.

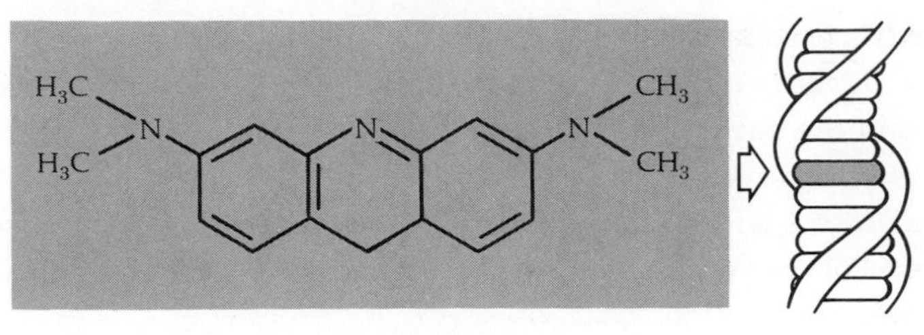

Acridine orange

Figure 4.6
Acridine insertion into DNA. Acridine dyes intercalate between the purine and pyrimidine bases in the double helix, causing frameshift mutations. Frameshift mutations can have drastic consequences during transcription (see Fig. 4.4). (After G. Stent, *Molecular Genetics,* W. H. Freeman and Company. Copyright © 1971.)

neighboring bases in the DNA double helix (Fig. 4.6). Yet another important group of mutagens is called "base analogs." These are substances that are sufficiently similar to the normal bases to be incorporated into DNA during replication, but different enough to cause mutations when the DNA is next replicated (Fig. 4.7).

Though mechanisms of mutation in higher organisms may differ in many details from those in microorganisms, they still must include chemical changes in DNA. Thus it is likely that chemicals which are mutagenic in microorganisms will also be mutagens for higher organisms; this is certainly true for some of the classes of chemicals just discussed. The studies with bacteria and molds thus provide important clues to the types of chemicals that are mutagens for man, as well as to their modes of action.

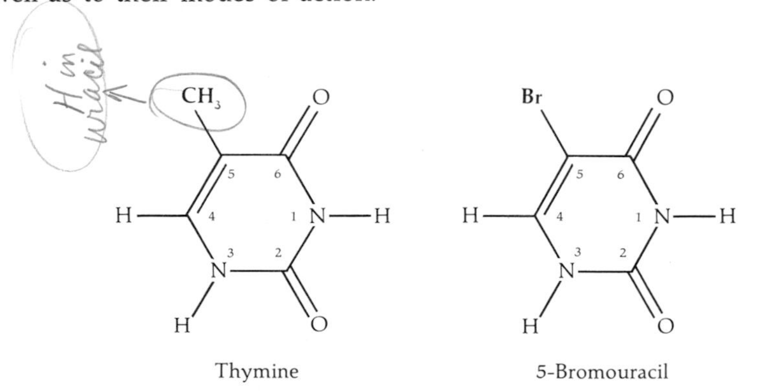

Figure 4.7
Comparison of thymine and 5-bromouracil (5-BU). In DNA biosynthesis, 5-BU can substitute for thymine. The substitution of Br for CH_3 changes the properties of the ring in such a way as to make base-pairing mistakes more likely.

4.2. Chromosomal Aberrations

The first new genetic variation to be analyzed in *Drosophila* proved to be due to the loss of a whole chromosome rather than to a change in a particular gene. Subsequent work, especially with *Drosophila*, established the existence of a wide range of genetic variations involving whole chromosomes or substantial segments of chromosomes. These are collectively called *chromosomal aberrations* to distinguish them from the single-gene mutations discussed in the preceding section.

Only after the development in 1956 of suitable techniques for looking at mammalian chromosomes did it become possible to search for human chromosomal aberrations.

The first abnormality in man that was shown to be due to a chromosomal aberration was Mongolism (now called Down's syndrome, after one of the men who first described it in the late nineteenth century). This is a congenital syndrome, usually recognizable at birth by an experienced physician, and one that is associated with severe mental retardation, characteristic facial features, and palmprint abnormalities (Fig. 4.8). It occurs with an overall incidence of about 1 in 700 births and so ranks as one of the commonest of the congenital abnormalities. The cause of Down's syndrome had been a puzzle for many years, though there had been some suggestions that it might be due to a chromosomal abnormality. It was clearly congenital and yet, as the affected individuals almost never reproduce, there was little chance of saying whether it is inherited in a simple Mendelian fashion. Down's syndrome also shows one other very striking feature. Its incidence increases some fortyfold as the age of the mother increases from the early 20s to the early 40s (Fig. 4.9). The age of the father and the number of previous offspring apparently have no effect on the incidence.

The riddle of Down's syndrome was essentially solved when J. Lejeune and R. Turpin showed in 1959 that it is associated with an extra chromosome 21. Thus affected individuals have 47 chromosomes instead of 46, with three copies of chromosome 21 instead of the usual two. Such individuals are said to be *trisomic* for chromosome 21. Chromosomal imbalance was known to cause abnormalities in other species, and so it was not surprising that this is also the case in man.

Trisomy (as in Down's syndrome) is the result of fertilization by a gamete carrying two copies of chromosome 21 instead of the normal one copy.

Such abnormal gametes are produced when homologous paired chromosomes fail to separate as they should at the anaphase stage of the first or the second division of meiosis, a phenomenon that is called *nondisjunction.* The expected consequences of nondisjunction for a particular chromosome are illustrated in

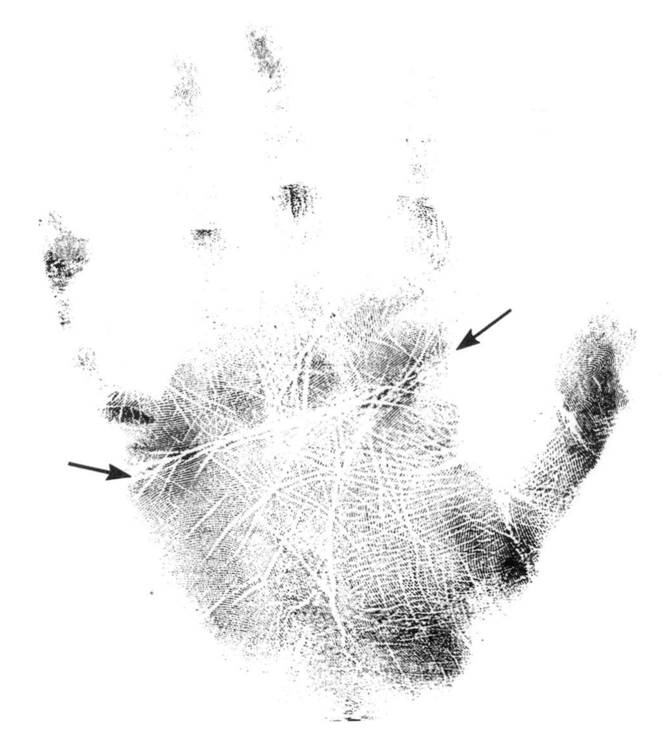

Figure 4.8
Palmprint in Down's syndrome. Two-thirds of the individuals affected with Down's syndrome (and occasionally those affected with other generalized congenital disorders) show the midpalmar crease (simian line) seen in this palm. (Courtesy of D. S. Borgaonkar.)

Figure 4.10. The first part of the figure shows what happens if the homologous chromosome pair does not separate at the first meiotic division. The products of this abnormal division are one daughter cell with both homologs and the other with neither. When followed by a normal second division of meiosis, the result (in the male) would be two gametes with both homologs and two with neither. (In the female there would, of course, be just one gamete that is chosen at random from these four possibilities, the other three products being extruded as polar bodies.) The two homologous chromosomes in the gametes in this case are different, one being maternal and the other paternal.

The consequences of nondisjunction at the second division of meiosis, shown in the second part of Figure 4.10, are somewhat different. Only one of the products of the first division will, in general, be affected so that two of the final nuclear products of meiosis will be normal. The other two are abnormal—one carrying two homologous chromosomes and the other none. In this case, however, the two

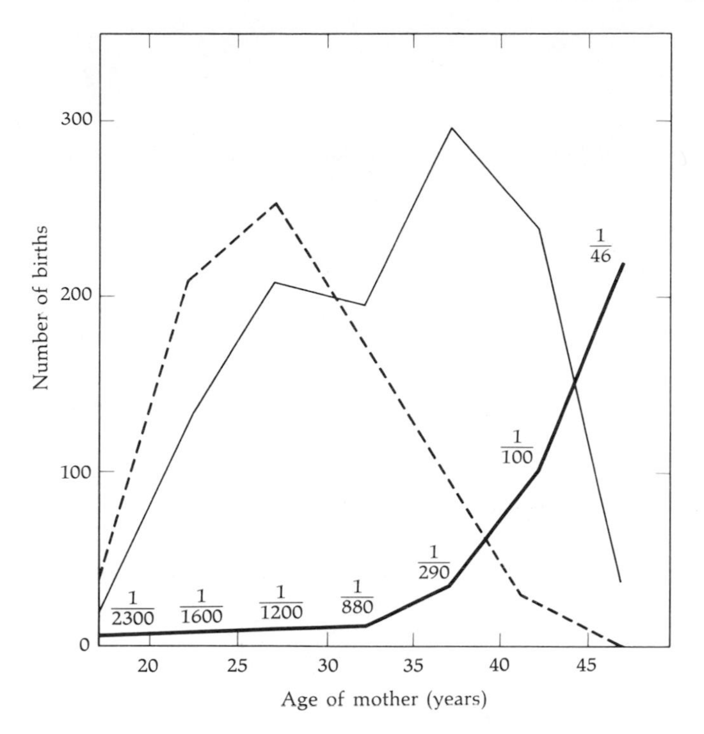

Figure 4.9
Incidence of Down's syndrome at birth, as a function of mother's age. The dashed line plots the total number of births (in thousands); the thin solid line plots the number of babies born with Down's syndrome. The thick solid line shows the incidence of Down's syndrome. All values are plotted as a function of the mother's age. (From L. S. Penrose and G. F. Smith, *Down's Anomaly,* Little-Brown, 1966.)

homologous chromosomes carried by the abnormal gamete (D) are duplicates of either the maternally or the paternally derived chromosome and so will not be different.

Union between a normal gamete and an abnormal one carrying two homologs instead of just one results in a trisomic zygote. Fertilization involving a normal gamete and one with no copy of the homolog gives rise to the complementary abnormality called *monosomy,* in which the zygote has only one instead of two copies of the chromosome.

The general phenomenon of variation in the number of one or more chromosomes is called *aneuploidy.* Monosomy or trisomy can in principle occur for any chromosome, though the phenotypic effects will be a function of which particular chromosome is involved. Nondisjunction is comparatively rare, though the rate of nondisjunction is much higher than that of single-gene mutations.

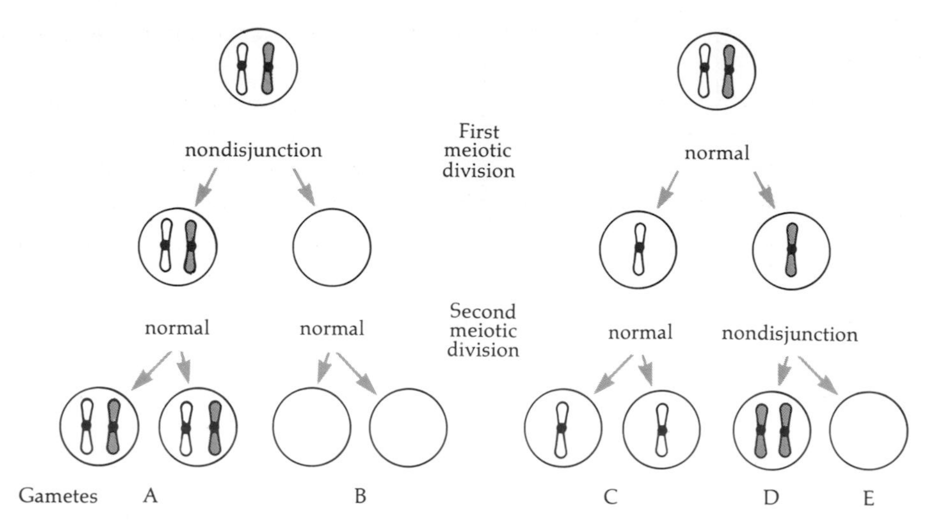

Figure 4.10
Consequences of nondisjunction at meiosis. Nondisjunction in the first meiotic division produces gametes having (A) both or (B) neither member of the homologous chromosome pair that is affected. Nondisjunction at meiosis II produces (D) gametes containing two identical chromosomes, both derived from the same member of the homologous pair, or (E) no chromosomes of that pair. Nondisjunction occurs more frequently at meiosis I than at meiosis II. Gametes lacking one autosome are apparently unable to form a viable zygote, for such genotypes have not been detected. A normal gamete (C) contains one chromosome from the homologous pair. The consequences of fusion with a normal gamete (N) at fertilization are readily apparent: A + N = trisomy, B + N = monosomy; D + N = trisomy, E + N = monosomy.

There are many known abnormalities of the sex chromosomes.

Very soon after the discovery that Down's syndrome is due to trisomy for chromosome 21, it was found that two well-known abnormalities of sexual development are also associated with chromosome abnormalities. In this case, however, it is the sex chromosomes that are affected.

Individuals with Turner's syndrome, the first of these two abnormalities, are basically female but have underdeveloped gonads, are generally infertile, and are mentally subnormal. They also have relatively short stature and frequently have web neck and wide-set nipples. Turner's syndrome is caused by monosomy for the X chromosome. Affected individuals have the sex-chromosome constitution XO, lacking the Y chromosome. They can presumably arise either through an "O" egg (lacking sex chromosomes) produced by nondisjunction for the X chromosome in the female and fertilized by a normal X-bearing sperm, or by an O sperm produced by nondisjunction in the male and fertilizing a normal X-carrying egg. Studies using X-linked genetic markers show that in nearly three-quarters of the cases of Turner's syndrome, the nondisjunction has occurred in the father, suggesting that nondisjunction for the X chromosome occurs less frequently in females.

The second of the two major abnormalities of sexual development, Klinefelter's syndrome, has the chromosome constitution XXY and so is a trisomy. Individuals with Klinefelter's syndrome are basically nonfertile males with small, underdeveloped testes and prostate, and with scanty secondary sexual hair. Though there is a significant increase in the incidence of mental retardation amongst XXY individuals, the majority of them fall within normal IQ ranges.

The fact that XO individuals who have Turner's syndrome are basically female, while XXY individuals with the Klinefelter's syndrome are basically male, demonstrates the male-determining capacity of the Y chromosome.

The same phenotypes for XO and XXY are observed in mice as in man, whereas the classical studies of sex-chromosome nondisjunction in *Drosophila* showed the reverse. The *Drosophila* XO is male and XXY is female. It is apparently the balance between the number of X chromosomes and autosomes that determines sex in *Drosophila,* the Y chromosome having no obvious effect. A wide variety of sex-determining mechanisms exists in plants and animals, ranging from single-gene differences to the X–Y chromosome mechanism, with a male-determining Y in all mammals.

An enormous variety of sex chromosome abnormalities, in addition to XO and XXY, has been described in man. The types of abnormalities expected on the basis of nondisjunction in the male or female are enumerated in Table 4.2. Notice that there are four types of abnormal sperm. Nondisjunction at the first division

Table 4.2
Abnormal sex-chromosome constitution due to nondisjunction

Sperm	Eggs	Normal	Abnormal	
		X	XX	O
Normal	X	XX^N	XXX	XO
	Y	XY^N	XXY	YO*
Abnormal	XY	XXY	XXXY	XY^N
	XX	XXX	XXXX	XX^N
	YY	XYY	XXYY	YYO*
	O	XO	XX^N	O*

*Lethal
^{N}Normal males or females

of meiosis will give rise to XY and O sperm (Fig. 4.11). Nondisjunction at the second division will, in addition to O, give XX or YY sperm depending on whether nondisjunction occurs in an X- or a Y-carrying first-division product.

There are three types of abnormalities in addition to XO and XXY that arise from nondisjunction in one or the other parent—namely, XXX, XYY, and YO.

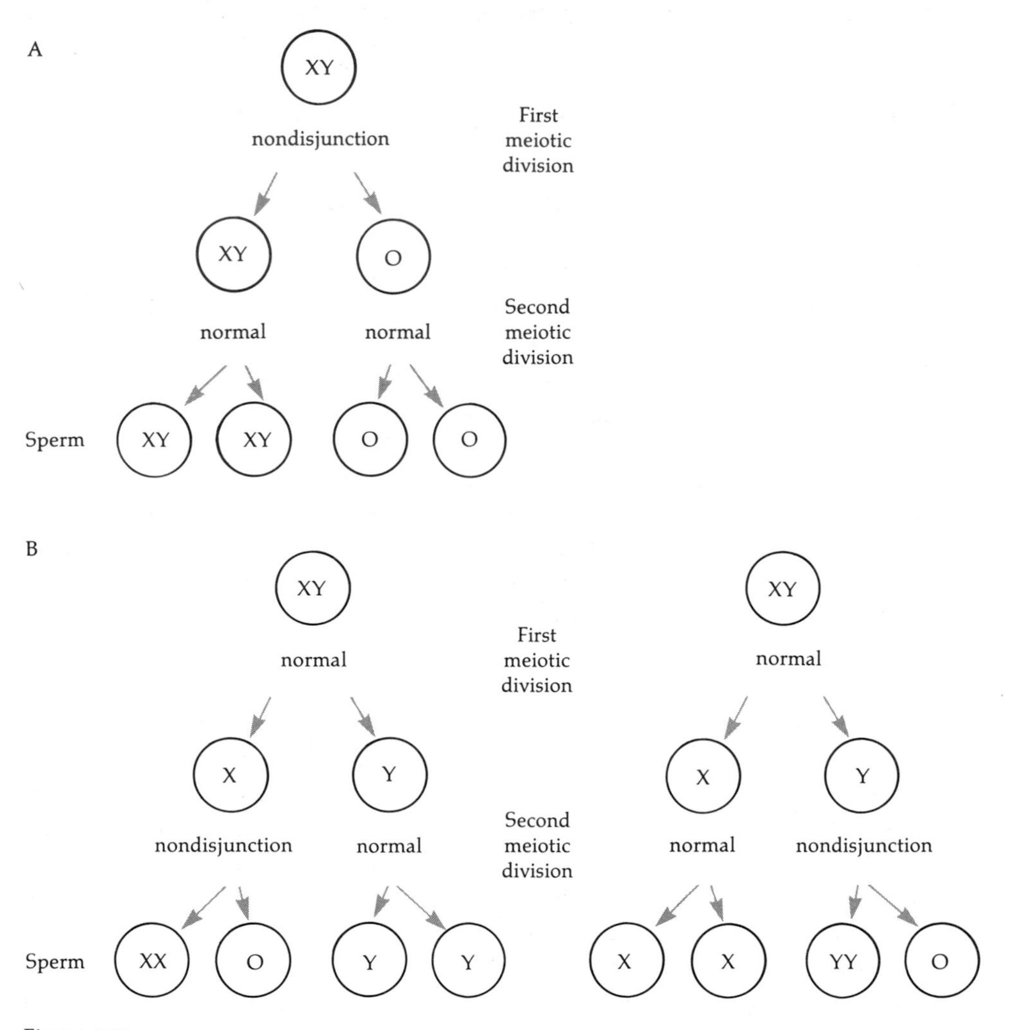

Figure 4.11
Consequences of sex-chromosome nondisjunction in the male meiosis. Nondisjunction in meiosis I produces abnormal gametes containing both X and Y chromosomes, and other abnormal gametes (O) containing neither sex chromosome. Fertilization of a normal X-bearing egg by these gametes will produce, respectively, XXY males (Klinefelter's syndrome) and XO females (Turner's syndrome). Nondisjunction in meiosis II produces abnormal gametes that are either XX, YY, or O. Fertilization of a normal egg will lead, respectively, to XXX females, XYY males, or XO females.

Of these, XXX is female and comparatively normal, but possibly with reduced fertility, while YO has never been observed and is presumed to be lethal at a very early stage of development. XO is in fact the only monosomy that has ever been observed in man beyond very early stages of development. Thus, while absence of the Y chromosome can be tolerated, and an X chromosome can function on its own (as it has to in the normal XY male, in any case), absence of a single copy of any other chromosome presumably causes very early death.

The XYY condition is exceptionally interesting, particularly from a behavioral point of view.

XYY individuals are, for the most part, normal males. However, in one of the first studies of the population incidence of XYY, an extraordinarily high frequency (7/197) of XYY individuals was found among a group of mentally retarded patients with criminal records who were in a maximum security prison hospital. Another feature of these XYY individuals was that they were up to six inches taller than average XY males in the same institution. Studies in the "normal" population suggest an overall incidence of XYY at birth of about 1 per 1,000.

Subsequent studies have tended to corroborate the initial finding of a concentration in mental and penal institutions, yielding on average an incidence of around 2 percent—some 20-fold higher than that in the control (normal) population. These studies have also confirmed that the XYY is taller than average. The few published studies suggest little if any departure from normality in other respects in the typical XYY male taken from the normal population, outside an institution.

It must be emphasized that—although the concentration of XYY is perhaps 20-fold higher in mental/penal institutions—only a small proportion of the population as a whole (perhaps 1 in 500) is in such an institution. This means that the overall proportion of XYY individuals who end up in mental/penal institutions is at most about 4 percent (Fig. 4.12). For example, consider a normal population made up of 100,000 individuals. We would expect about 100 of these to be XYY (0.1 percent of the population), and we would expect about 200 individuals from the total population to be in an institution (0.2 percent). If the incidence of XYY in the institutions is 2 percent, we would expect to find $0.02 \times 200 = 4$ XYY individuals in the institutions, so the other 96 XYY individuals would be among the "normal" population. Thus most XYY individuals (96 percent or more) are normal, and it is not known what other factors may lead to a criminal behavior associated with mental abnormality.

The Y chromosome is distinguishable on fluorescence banding by a particularly bright region near the end of its long arm.

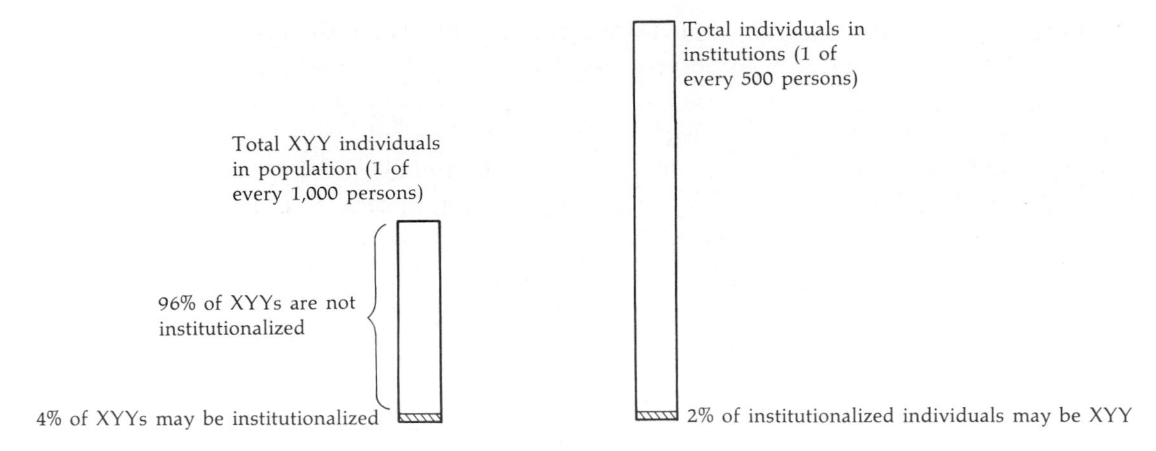

Figure 4.12
XYY incidence in general population and in mental/penal institutions. About one of every 500 individuals in the general population ends up in a mental or penal institution; 2 percent of these individuals may be XYY. The incidence of XYY in the general population is 1 in 1,000 (0.1 percent). These figures lead to the conclusion that about 4 percent of XYY individuals end up in institutions, which means that 96 percent of the XYY population does not behave so as to be institutionalized.

This banding of the Y chromosome is illustrated very clearly in the metaphase from an XYY individual shown in Figure 4.13. Such fluorescence can even be detected in interphase nuclei and in whole sperm (Fig. 4.14). A number of studies are now underway, aimed at using this technique to detect XYY individuals and follow up their development. The aim of these studies is to try and answer some of the questions about the overall phenotypic effects of an extra Y chromosome. In view of the possible association of XYY with abnormal behavior and the publicity given to this association, however, one may wonder what the effects on the parents and the affected child may be of knowing that the child is an XYY. It is difficult now to decide what advice to offer to the parents of an XYY child.

More complex sex-chromosome constitutions than those so far discussed can result from fertilization involving two abnormal gametes, as indicated in Table 4.2. The combinations YYO and O are clearly lethal, XXXY, XXXX, XXYY, and even XXXXY and XXXXX have been observed, though only very occasionally.

Nondisjunction also may occur in somatic development. A significant number of sexually abnormal individuals are found to have cells of two or more types, one or more of which are abnormal. For example, the commonest type is an individual with some XY and some XXY cells. Such individuals having cells with more than one genotype are called *mosaics.* Mosaics arise from nondisjunction that occurs during mitosis in one of the somatic cells very soon after zygote formation. Mitotic nondisjunction is essentially like the nondisjunction that occurs

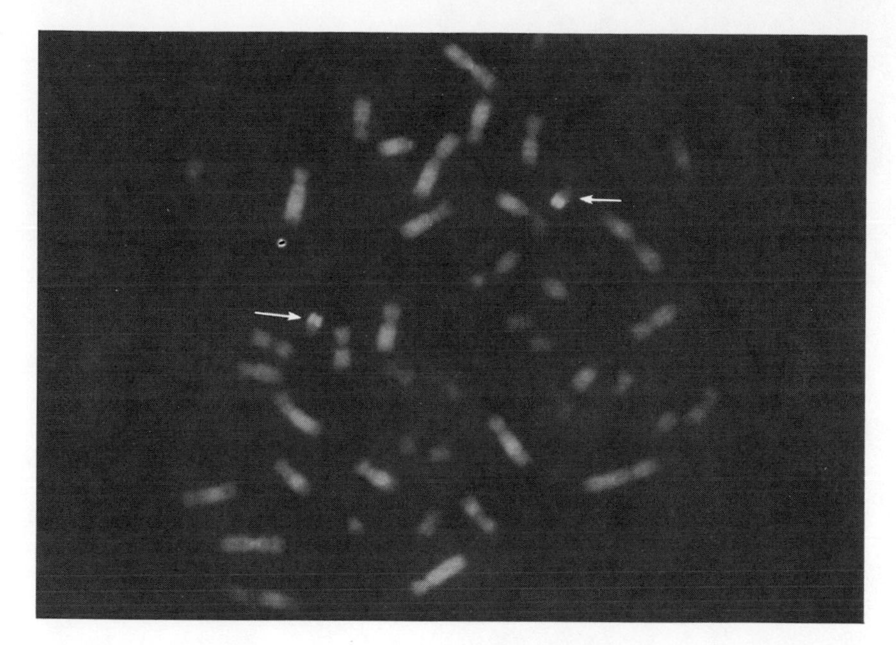

Figure 4.13
Fluorescence of the Y chromosome. This metaphase spread is from a 47 XYY individual. The two brightly fluorescent Y chromosomes are indicated by arrows. (Courtesy of M. Bobrow.)

during the first division of meiosis (Fig. 4.10). One of the resulting daughter cells has an extra chromosome and is trisomic (having three homologs instead of two for some chromosome pair), while the other daughter cell lacks one chromosome and is monosomic.

Suppose, for example, that one of the immediate products of the first division of mitosis in an XY male undergoes nondisjunction for the X chromosome, producing YO and XXY daughter cells. (The other cell resulting from the first mitotic division divides normally to produce XY daughter cells.) The result will be a mosaic individual having three types of cells: XY, YO, and XXY (Fig. 4.15). Because the YO cells will not survive, the result is an XY/XXY individual—which, as already mentioned, is the commonest type of mosaic observed. More complex combinations resulting from more than one mitotic nondisjunction (such as XY/XXY/XXYY) have also, though rarely, been found. In these cases the extent of the phenotypic abnormality observed depends on the relative proportions of the different types of cells. Clearly, it may be important in a case of suspected sexual abnormality to search carefully for abnormal cells, because the abnormality may be caused by a relatively unbalanced mosaicism with a small proportion of the unusual cell type.

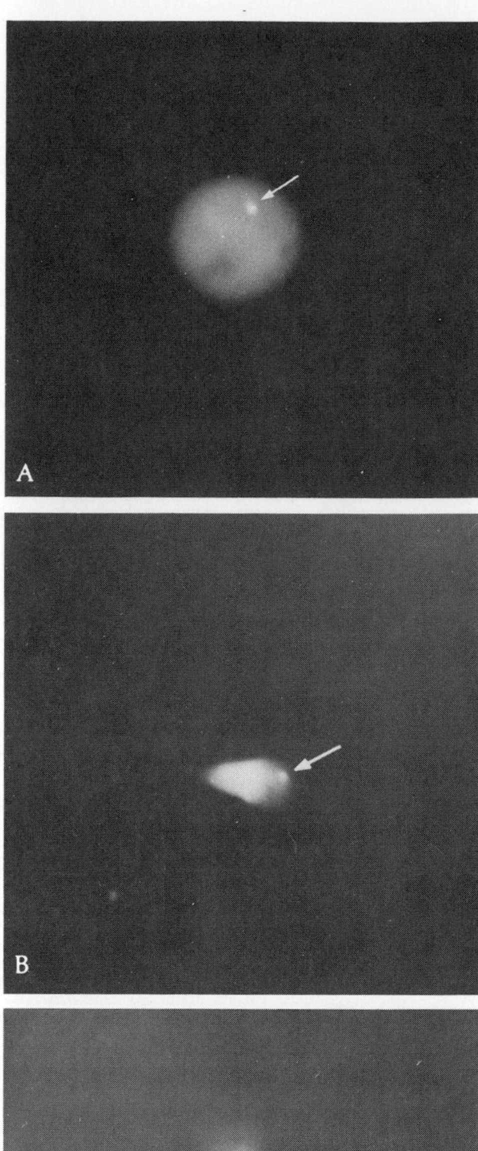

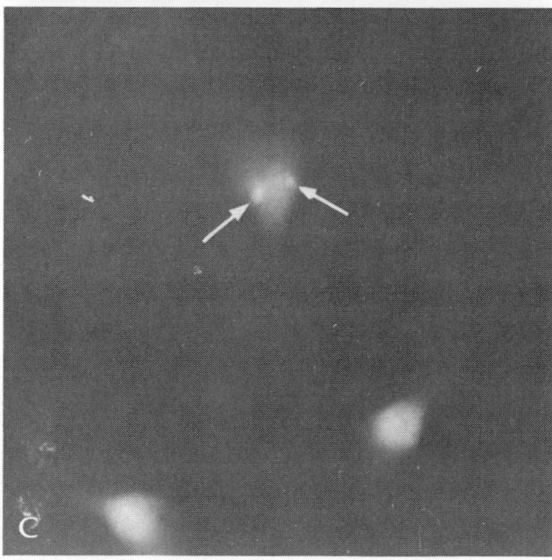

Figure 4.14
Y-chromosome fluorescence in nondividing cells. (A) One "Y body" in a lymphocyte. (B) Y body in sperm head. (C) Sperm head with two Y bodies. (Courtesy of P. Pearson and M. Bobrow.)

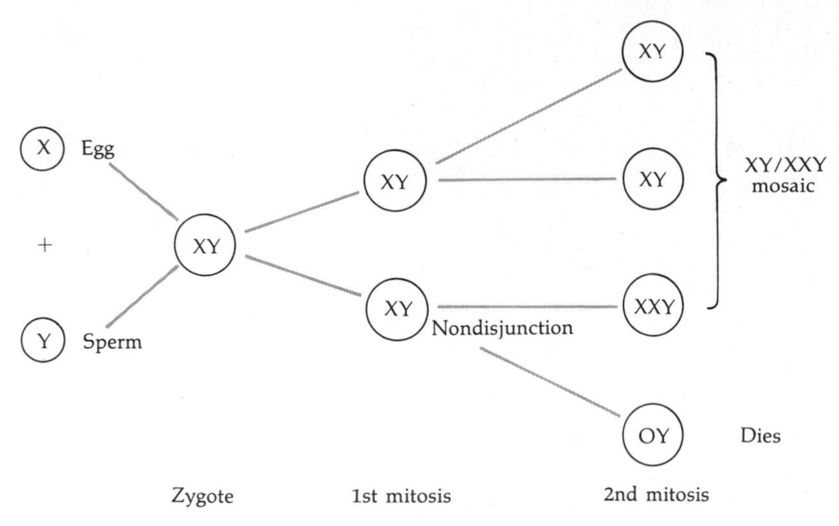

Figure 4.15
Production of XY/XXY mosaic. Nondisjunction may occur in mitosis, leading to two cell lines (XY and XYY); OY cells die. This example involves the X chromosome, but nondisjunction of the Y chromosome can also occur.

Apart from chromosomes X, Y, and 21, the only others for which trisomy has been observed so far are chromosomes of the D group (13, 14, 15) and chromosomes 16 and 18.

In the case of chromosome 16 hardly any liveborn infants have been observed, while for the others the effects are much more severe than trisomy 21 or the sex-chromosome abnormalities. Thus, individuals with trisomy 18 are severely mentally retarded, have characteristic heart malformations, and most die by the age of six months. Trisomy for D-group chromosomes, which is also associated with major central nervous system defects, is even more serious in its effects. This severity may, in part, account for the much greater rarity of observed cases of these forms of trisomy than that for the sex chromosomes or chromosome 21.

Sometimes instead of nondisjunction affecting just one chromosome of the set, separation of homologs fails completely for all chromosomes, resulting in a cell that has double the number of chromosomes that it should have. When this occurs during a mitosis, a daughter cell is produced with four sets of homologs instead of the normal two. Such a cell has 92 chromosomes (four times the haploid number 23) and is called *tetraploid.* If reduction fails during one of the divisions of meiosis, then a diploid instead of a haploid gamete will be produced. When this diploid gamete (with 46 chromosomes) is fertilized by a normal haploid gamete with 23 chromosomes, the result is a zygote with 69 chromosomes, or three times the haploid number. This zygote is called a *triploid* (Fig. 4.16).

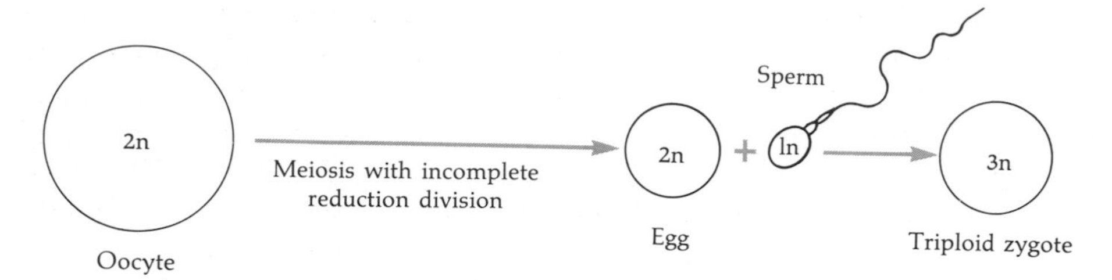

Figure 4.16
Origin of triploidy. If reduction division does not successfully occur in formation of the egg, then the egg may have two copies of a particular chromosome or of the whole chromosome set. In the latter case, fertilization by a normal sperm leads to triploidy. Errors in reduction for one chromosome may lead to trisomy or monosomy.

Polyploidy refers to the condition of having multiples greater than two of the haploid set of chromosomes.

Triploidy and tetraploidy are thus special cases of polyploidy. Though polyploidy is comparatively common in plants, insects, and some other groups of organisms (in fact, many of the cultivated crops are polyploids), it is very rare in mammals. In man triploids are known, but they hardly ever survive until birth. However, nearly 10 percent of spontaneous abortions are either triploid or tetraploid. About 15 percent of all pregnancies appear to end in spontaneous abortions, so as many as 1 percent of all fertilized zygotes in man may be triploid, indicating a remarkably high failure of reduction during meiosis. Note that, if we based an estimate of the rate of reduction failure on the incidence of triploids at birth, we would get a much lower and clearly incorrect figure. This is, of course, because so many of the abnormal zygotes die early and so are not counted among births, either stillborn or alive. We discuss in Chapter 8 how to take into account such viability effects in estimating mutation rates from observed incidences of genetic defects.

A small proportion of cases of Down's syndrome seem to occur on a familial basis, almost as if they were caused by a recessive gene.

This is certainly not expected if they are the result of trisomy due to nondisjunction in parental gamete formation. Trisomy is a comparatively rare event, and it would be most unusual for it to occur more than once in the same family. The occurrence of two or more cases of the syndrome in one sibship thus suggests that at least one of the parents has a hereditary tendency to produce affected offspring. This puzzle was soon solved when the chromosomes in such a family were studied carefully. The affected offspring had 46 chromosomes—not 47 as they should have if they were trisomic—but one of the chromosomes was quite different from all those seen in a normal karyotype. Both of the parents were phenotypically normal, but one parent had only 45 chromosomes, one of which was the same abnormal chromosome seen in the affected offspring. Clearly it is this abnormal chromosome that is the culprit.

Closer examination revealed that the abnormal chromosome was effectively a combination of two chromosomes—one being chromosome 21, which is usually trisomic in Down's syndrome, and the other being a member of the D group (13, 14, or 15). Concentrating for the moment only on the implicated chromosomes (21 and 14), the affected offspring had the constitution 21, 21, 14, and 14-21, the abnormal chromosome, which is called a *translocation*. This offspring therefore had three copies of chromosome 21, one of which was attached to the second copy of chromosome 14; this finding explains the Down's syndrome. The 45-chromosome parent was 21, 14, 14-21, having only one normal chromosome 21. This parent had the normal amount of chromosome-21 material, but in an unusual arrangement. He had a normal phenotype, because the rearrangement of the genetic material does not seem to affect its expression. As shown in Figure 4.17, the mating between a normal person with 46 chromosomes and a normal carrier of the 14-21 who has 45 chromosomes gives rise to three types of viable offspring. These are (1) normals, (2) carriers like the parents, and (3) individuals with the 14-21 translocation who have Down's syndrome.

Translocations between pairs of chromosomes arise by a combination of breakage and rejoining that is rather like recombination, but which takes place between nonhomologous chromosomes.

Thus, suppose that breaks have occurred simultaneously near the centromere on the very small short arm of chromosome 14 and on the long-arm side of chromosome 21 (Fig. 4.18). Now if the piece of chromosome 21 that has been detached from its centromere is joined onto the remaining part of the short arm of chromosome 14, a 14-21 translocation chromosome is formed. The reciprocal pieces when joined together (chromosome 14 short-arm fragment joined to centromere fragment of chromosome 21) form a small chromosome, which is generally lost. The fact that the 45-chromosome translocation carrier is normal indicates that the material in this small fragment can be lost without any untoward effect on the phenotype.

The presence of an odd number of chromosomes in the translocation carrier leads to complications of meiosis (Fig. 4.19). During pairing in meiosis, the translocation chromosome pairs in part with the normal chromosome 14 and in part with chromosome 21, forming a so-called "trivalent" combination. When the three centromeres in this "trivalent" separate at the first meiotic division anaphase,

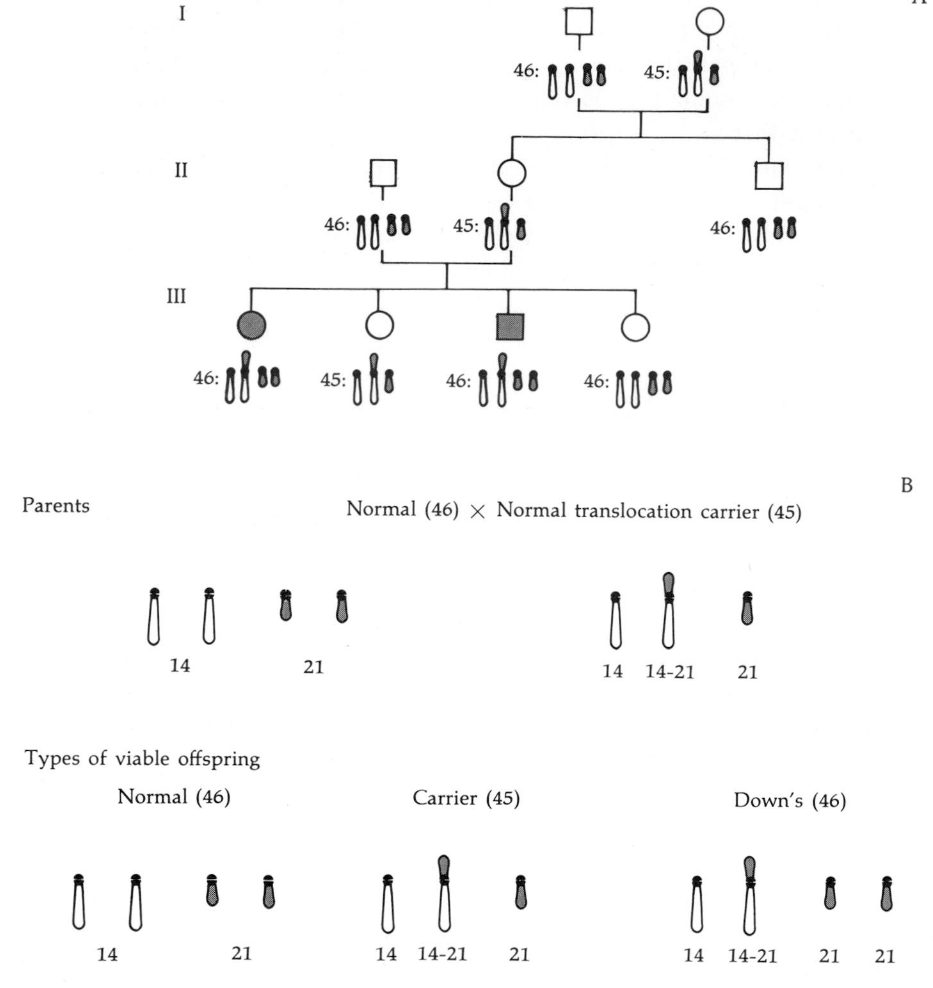

Figure 4.17
Pedigree and inheritance of Down's syndrome due to translocation trisomy 21. (A) Pedigree of a family where the phenotypically normal 45:14,14-21,21 mother (I.2) had two grandchildren with Down's syndrome due to a translocation trisomy 21. (B) Mechanism for inheritance of Down's syndrome due to a translocation between chromosomes 14 and 21. (Only chromosomes 14 and 21 are shown.) The total number of chromosomes is given in parentheses. (Part A after L. S. Penrose and J. Delhanty, *Annals of Human Genetics,* vol. 25, 1961.)

they can do so in three ways: 14,21 versus 14-21; 14,14-21 versus 21; and 14 versus 14-21,21. These give rise, when combined with the normal 14,21-bearing gamete, to six different types of fertilized zygotes. Three of these (numbers 1, 2, and 6 in Fig. 4.19) are the three observed types of viable offspring shown in Figure 4.17.

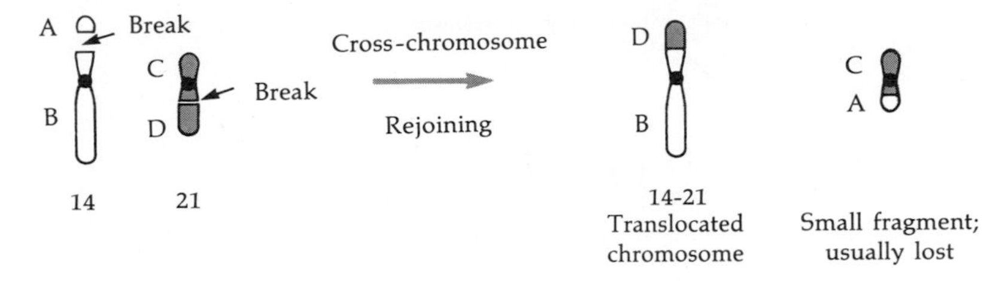

Figure 4.18
Origin of 14-21 translocation. A breakage of chromosomes 21 and 14 may lead to a shuffling of the genetic material. The small fragment usually is lost. The relative length of the short chromosome arms is exaggerated here and in Figure 4.19.

The other three types are either monosomic for chromosomes 14 or 21, or trisomic for chromosome 14, and they do not survive.

The expected proportions of the different types of offspring depend on the relative frequencies with which the three types of centromere separation occur. When the translocation carrier parent is a female, it is found that about 90 percent of the offspring (on average) are normal, and these are divided approximately equally between karyotypically normal and translocation carrier individuals. The remaining 10 percent of offspring have translocation Down's syndrome. However, when the translocation carrier parent is a male, only about 2 percent of the offspring have Down's syndrome. The reason for this difference between male and female carriers in the frequencies of the different types of meiotic segregation is not yet understood.

Translocations of the sort we have been discussing, which appear effectively to involve the joining together of the centromeres of two acrocentric chromosomes, are often called "Robertsonian" after the man who first described them. These are the commonest types of translocations observed at birth and account for more than half of the observed autosomal structural chromosome abnormalities. Typically, Robertsonian translocations in man involve combinations of D- and G-group chromosomes (as in the 14-21 case we have just discussed), as well as pairs of D-group (such as 14 and 15) or pairs of G-group chromosomes (21 and 22).

Many translocations have been observed in which the breakage points are not near the centromere, and also that involve biarmed as well as acrocentric chromosomes.

The new banding techniques described in Chapter 2 now make it possible to identify quite precisely the breakage points involved on two chromosomes that have taken part in producing a translocation. An example of banding patterns observed in a translocation between chromosomes 4 and 12 is shown in Figure

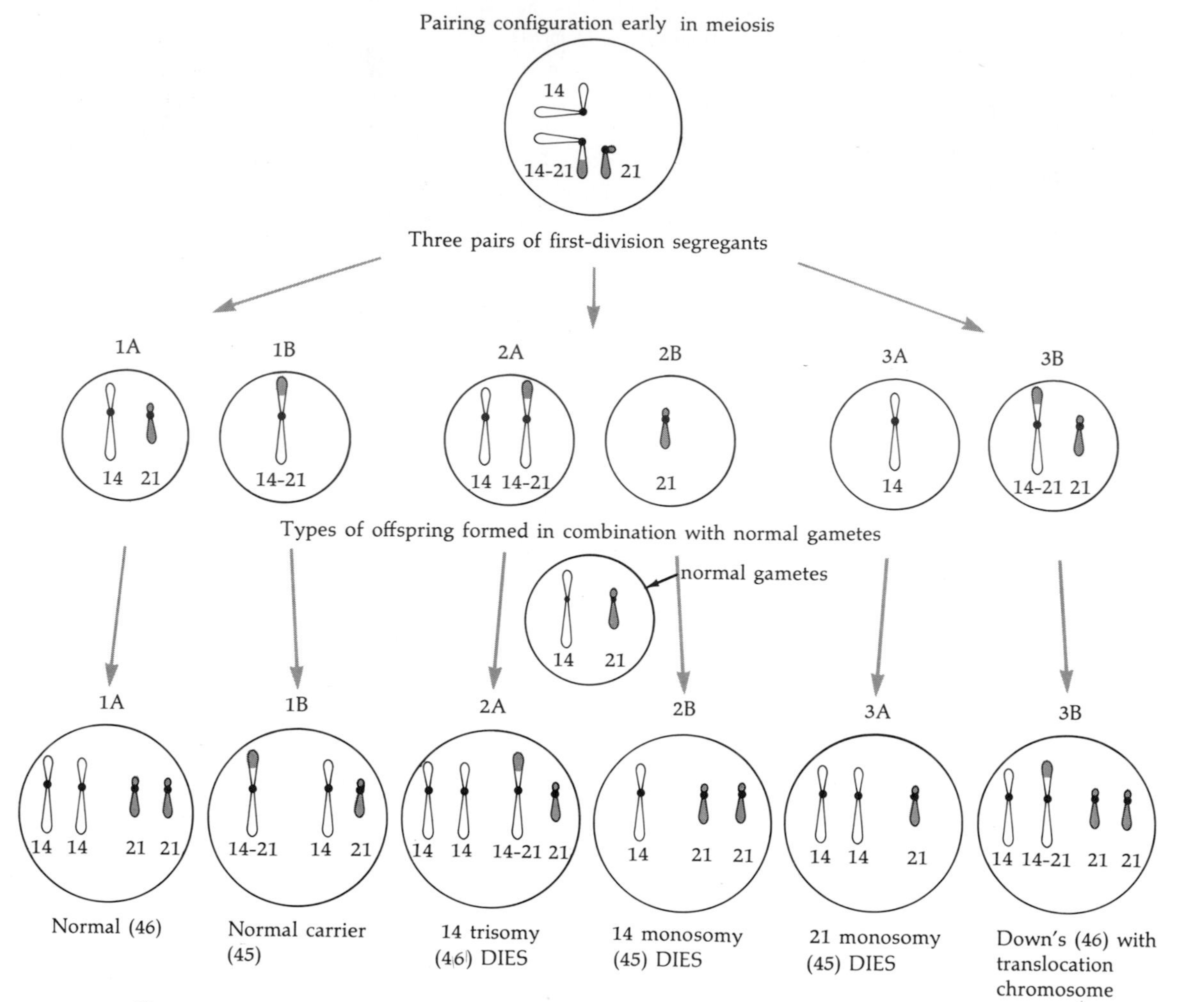

Figure 4.19
Three types of segregation in a phenotypically normal 14-21 translocation carrier. (A) Three different pairs of first-division segregants (1A + 1B, 2A + 2B, 3A + 3B) can arise during meiosis I. Each individual segregant (1A, 1B, 2A, 2B, 3A, 3B) forms a gamete that may combine with a normal gamete. (B) Combinations that may result from the fertilization of the normal gamete. Phenotypic consequences are also indicated.

4.20. Segments of the long arms of the two chromosomes have been interchanged exactly as indicated schematically in Figure 4.18 for translocation between 14 and 21, except that in this case the reciprocal piece (CA in Fig. 4.18) is quite sizeable and does not get lost. In fact, only the individual with both products of the translocation event is phenotypically normal. He carries a balanced translocation

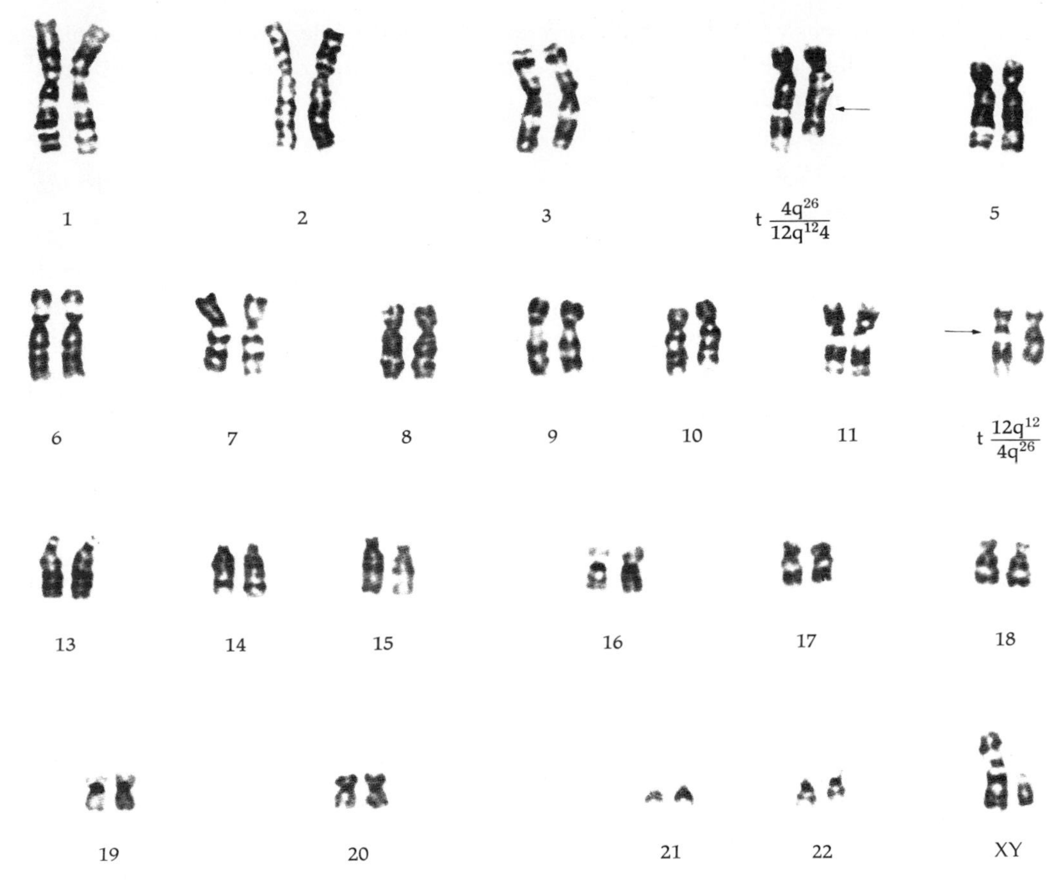

Figure 4.20
Banding patterns for a balanced translocation between chromosomes 4 and 12. The complete karyotype is shown. This karyotype can be represented by the notation 46,XY,t(4;12)(q26;q12); this notation means that there is a translocation (t) between chromosomes 4 and 12, involving the long arms (q). (After Hirschorn et al., *Annals of Human Genetics,* vol. 36, pp. 375–379, 1973. Cambridge Univ. Press.)

"complement" in which all the genetic material is there, though in a slightly different arrangement. His daughter (who has one normal chromosome 4, together with the 4-12 translocation chromosome and two normal chromosomes 12) has severe congenital anomalies, and she was the patient through whom the family was ascertained. She carries in triplicate a substantial region of chromosome 12 and misses a second copy of part of chromosome 4. This combination of partial trisomy for chromosome 12 together with partial monosomy for chromosome 4 undoubtedly accounts for her serious abnormalities. This pattern of observations for reciprocal translocations, as they are called, is quite general. The unbalanced combinations are always quite severely abnormal while the balanced combinations, who have both the reciprocal products of the translocation, are phenotypically normal.

Breaks in single chromosomes can occur, with the piece that is separated from its centromere simply being lost.

This results in a chromosome with one part of an arm simply deleted. An individual who has a normal and a deleted chromosome is effectively monosomic for the deleted region, and this again can have very severe consequences unless the deleted segment is quite small. In 1963 J. Lejeune and his colleagues described three cases of a very severe congenital syndrome associated with multiple effects on mental and physical development. This syndrome was distinguished by the fact that affected infants have a cry that sounds just like the plaintive meowing of a cat. They therefore called this condition *cri du chat*—"cat's cry." Examination of the chromosomes showed in all three cases that one of them, chromosome 5, had about half of its short arm missing. Since that time many cases of this same syndrome caused by deletion of part of the short arm of chromosome 5 have been described. Sometimes deleted chromosomes may originate from a translocation, in which case a balanced carrier of this translocation can give rise to offspring with the cri-du-chat syndrome (Fig. 4.21). Such individuals can also produce three other types of offspring: (1) normal, (2) a balanced carrier like the parent, and (3) an individual carrying the translocated 13-5 chromosome instead of the normal chromosome 13. Such an individual is trisomic for the part of the short arm of chromosome 5 that is deleted in complementary abnormality.

A third major type of chromosomal aberration, called "inversion," arises when two breaks occur in the same chromosome and the region between the breaks is reinserted after rotation through 180°.

Schematically, using letters to stand for segments of a chromosome, an inversion arises as follows.

Original chromosome		Chromosome with inverted segment
A B ↑ C D E F ↑ G	⟶	A B [F E D C] G
Breakpoints		Inverted segment

Because the only effect of an inversion is to change the sequence of the genes along the chromosome, inversions are not generally associated with any obvious growth abnormality. When the break points giving rise to an inversion occur in the same

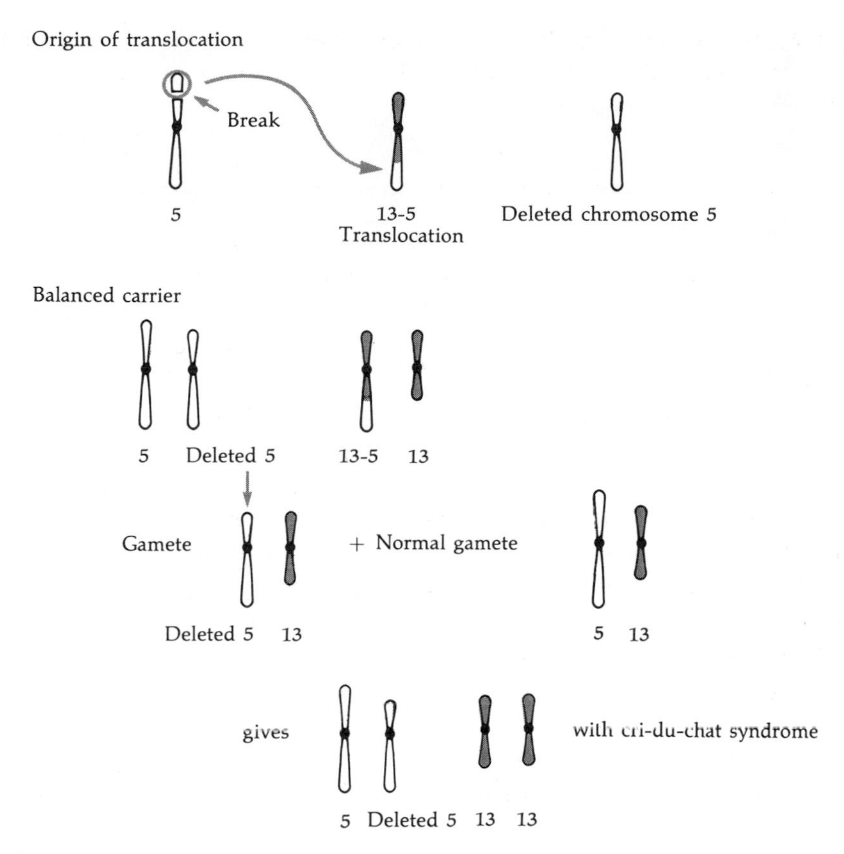

Figure 4.21
Inheritance of cri-du-chat syndrome from a balanced translocation carrier. Only chromosomes 5 and 13, which are involved in the translocation, are depicted. As illustrated here, deleted chromosomes may sometimes arise due to a translocation. Balanced carriers of this translocation can also produce offspring that are normal, have a balanced translocation, or carry the translocated 13-5 chromosome instead of the normal 13. In the latter case, abnormal development occurs due to trisomy for the extra part of chromosome 5.

arm of the chromosome, it is called a "paracentric" inversion and has no effect on the relative position of the centromere along the chromosome. The only way such an inversion can readily be detected is by a change in the pattern of banding along the chromosome arm in which the inversion has taken place. Because banding techniques were developed comparatively recently and inversions are rare, very few paracentric inversions have been described in man.

When two break points of an inversion occur on either side of the centromere, the inversion is called "pericentric" and may change the relative position of the centromere, giving rise to an easily recognizable new chromosome. One example

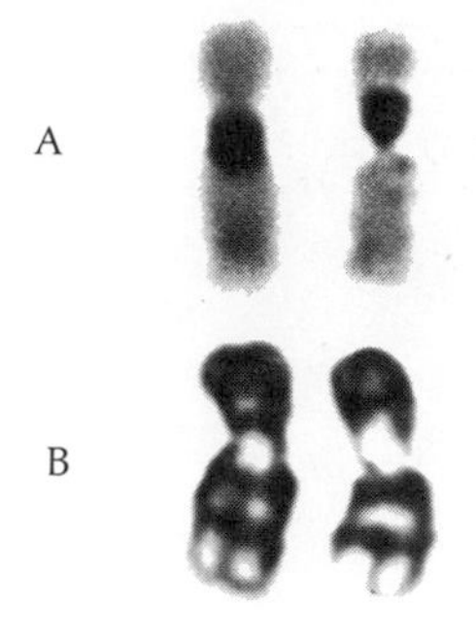

Figure 4.22
Apparent pericentric inversion of chromosome 9. "Pericentric" means "including the centromere" (see also Fig. 4.23). (A) G11 banding of number-9 chromosomes from a heterozygote for a pericentric inversion. The normal chromosome on the left has the heavily staining region near the centromere on the long arm, while the inverted chromosome on the right has it on the short arm. (B) Normal banding of these same chromosomes. (Part A from M. Bobrow et al., *Nature New Biology*, vol. 238, pp. 122–124, 1975; part B courtesy of K. Madan.)

of such an inversion for chromosome 9 is shown in Figure 4.22, in which the darkly staining region near the centromere has been moved from the long arm to the short arm. Figure 4.23 shows schematically the way that this probably occurred by a pericentric inversion including this region (also see Figs. 4.24 and 4.25). Variations in the amount of densely staining material near the centromere are quite common. They are found in about one to two percent of individuals, and so are more frequent than the other chromosome aberrations we have been discussing. Unlike those aberrations, the variations in centromere staining have not so far been found to be associated with detrimental effects.

4.3 Mutagenic Effects of Radiation and Other Agents

Following Muller's demonstration of the mutagenic effects of X rays, a great deal of work was done that was directed at elucidating the nature of the effect. What, for example, is the relation between X-ray dose and mutation rate? What information can this relation tell us about the nature of the gene and the mutation process? Much of this work was done before the chemical nature of the gene was known, when it seemed that quantification of the effects of radiation might give clues to the size and nature of the gene.

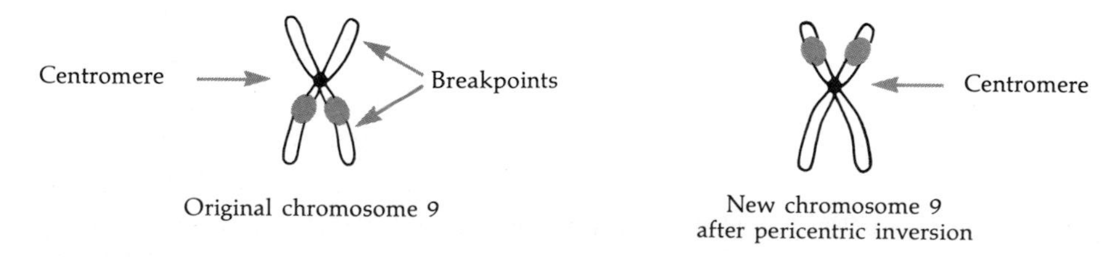

Figure 4.23
Schematic depiction of the origin of the chromosome-9 pericentric inversion. Figure 4.22A is a photograph of this inversion.

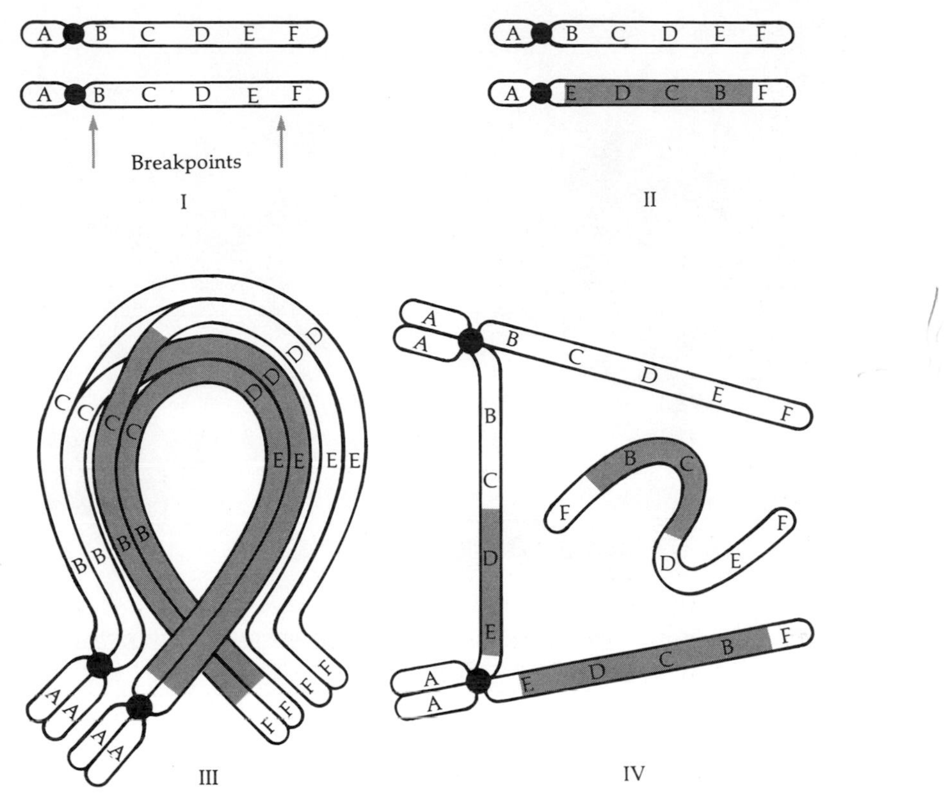

Figure 4.24
Genetic effects of a paracentric inversion. (I) Chromosome breaks occur between B and the centromere, and also between E and F. Both breaks are on the same side of the centromere, hence the name "paracentric." (II) A heterozygote for an inversion is thus generated. (III) At meiosis, pairing between homologous regions of a normal and an inverted chromosome generates a loop. A crossing-over is shown between two chromatids in the region between C and D. (IV) At the time the centromeres separate, it becomes clear that a chromatid bridge now joins them as a consequence of a crossing-over, and that there is in addition an acentric chromosome fragment (a fragment without a centromere). The bridge may break and the fragment is usually lost; thus gametes are generated that do not receive a regular haploid chromosome complement. These gametes are usually nonfunctional. Figure 4.25 shows a chromatid bridge. (After L. E. Mettler and T. G. Gregg, *Population Genetics and Evolution,* Prentice-Hall, 1969.)

More recently attention has been focused on radiation effects for practical reasons, because of concern for the effects of radioactive fallout from atom bomb tests, increasing use of X rays and other radiation in medicine, and the risks of exposure to byproducts of nuclear power stations. Work with chemical mutagens has focused more on the basic problem of the mechanism of mutagenesis in relation to DNA structure, as discussed earlier in this chapter. However, work is now also being done on the hazards arising from potential mutagenic chemicals that

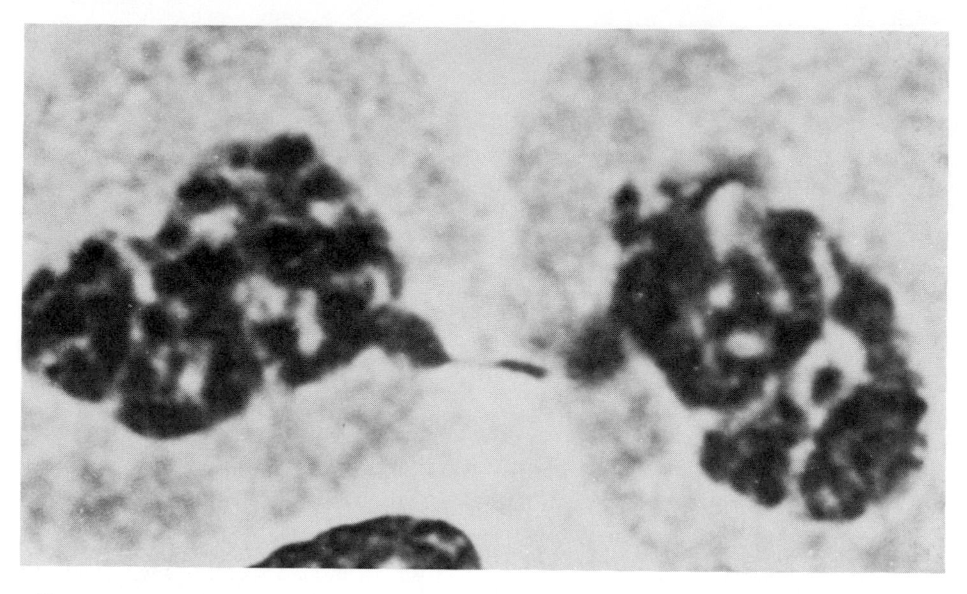

Figure 4.25
Chromatin bridge: the bridge formed during anaphase I in an inversion heterozygote (see Fig. 4.24IV). In this photograph, the bridge is still visible in interphase. (Courtesy of M. Bobrow.)

may be used, for example, in foodstuffs and as drugs. Concern about environmental mutagens is heightened by the fact that there seems on the whole to be a good correlation between agents that are mutagenic and those that cause cancers. In other words, most mutagens are also carcinogens. One of the best ways, in fact, of screening for carcinogens is to check their mutagenic activity on appropriately chosen and constructed strains of bacteria.

High-energy radiation comes in a variety of forms, the best known and so far most widely used being X rays. The primary effect of irradiation is due to collisions of the ray particles with atoms, which causes the release of electrons and leads to the formation of ions. An ion is an atom that has lost one or more of the electrons associated with it in its usual stable state. The instability of ions results in chemical reactions, which in turn cause mutations. This effect may be direct when an atom that is part of the DNA molecule is itself ionized, or indirect if the ions produced are not in the DNA itself, but cause mutations by interfering with DNA or with its replication.

Radiation dose can be measured by the amount of ionization produced in a given volume, the standard unit being the roentgen (r), named after Wilhelm Roentgen, who was the discoverer of X rays. A very similar alternative unit is the "rad" which measures the amount of radiation energy absorbed per gram of material.

The first and perhaps most basic fact of radiation genetics is that the mutation rate increases linearly with the radiation dose.

Much of the early work on induction of mutations by X rays was done with *Drosophila*, and this organism still provides the best data on the relation between mutation rates and radiation dose. A variety of sources of data illustrating the linear relationship is shown in Figure 4.26. The consistency of this relationship (extending over a very wide range of doses from 8 to 6,000 r and more) and the agreement between data from different sources are quite remarkable. There seems to be no lower threshold below which radiation has no effect on mutation rates. This point is very important in practice, as it indicates that any increase in radiation, however small, will give rise to a proportionate increase in the mutation rate. The classical interpretation of a linear relationship between dose and mutation rates is that one "hit" (ionization) in the neighborhood of the gene is enough to cause a mutation. Because radiation dose is measured by the number of ionizations, it is then to be expected that the mutation rate will increase directly in proportion to the dose. The *Drosophila* data, which measure lethal mutations on the X chromosome, suggest an increase of 3 percent in the mutation rate for every additional 1,000 r of radiation.

Analogous data on the induction of mutations by radiation have been obtained using the mouse. Being a mammal it should, hopefully, provide results more likely to be applicable to man. However, because experiments with the mouse obviously require much more effort than those with *Drosophila*, the available information is more limited. For 12 specific loci that have been studied, the average increase in mutation rate per locus per gamete per roentgen is 1.7×10^{-7}, which is less than one-tenth of the spontaneous mutation rate for these same loci (see Table 4.1). A commonly used yardstick for the effect of radiation on mutation is the dose that would lead to a doubling of the spontaneous mutation rate observed under normal conditions in the absence of specific exposure to radiation. A figure that is often quoted for the doubling dose on the basis of these mouse data is about 30 r. This figure is, however, subject to a great deal of uncertainty for a number of reasons.

Biological and physical factors affect the mutagenicity of radiation.

All stages of gamete formation are not equally sensitive to radiation effects, and there are significant sex differences. Thus, germ cells in the later stages of development, after meiosis, appear to be more sensitive than those in earlier stages, and spermatids (the developing sperm) are more sensitive than ova. It has also been shown that a given dose of radiation produces fewer mutations in younger than in older mice. Local environmental factors also seem to matter. The

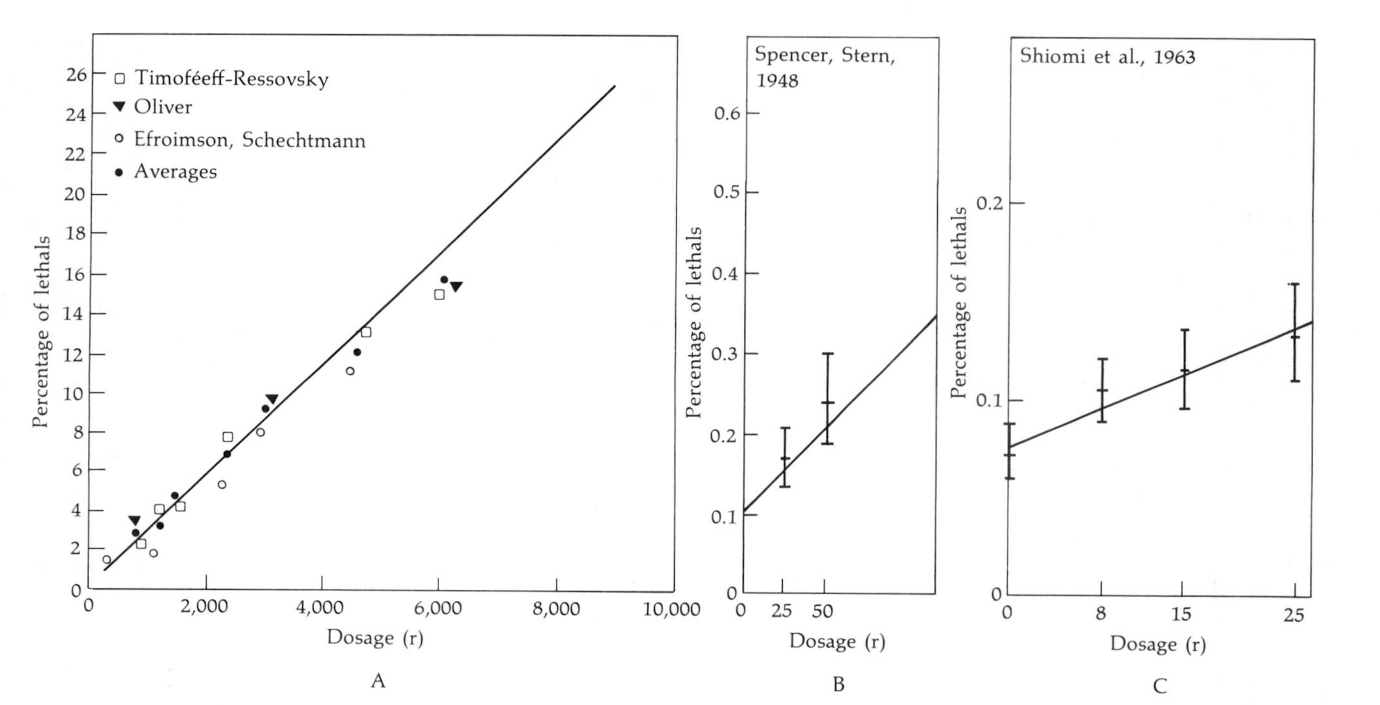

Figure 4.26
Linear relationship between mutation rate and radiation dosage. The graphs plot the frequency of lethal mutations induced in the X chromosome of *Drosophila melanogaster,* as a function of radiation dosage (in roentgens). (A) The relationship for high dosages. (B) The relationship for low dosages (25 and 50 r). The vertical lines represent the variation obtained in experiments at these dosages; the central "tick" on each vertical line represents the average mutation rate observed. The sloping line represents extension of the linear relationship observed for high dosages. (C) The relationship for dosages of 8, 15, and 25 r. (Part A after Timoféeff-Ressovsky et al., *Nachr. Ges. Wiss. Gottingen, N.F.,* vol. 1, 1935. Part B after W. P. Spencer and C. Stern, *Genetics,* vol. 33, 1948. Part C after Shiomi et al., *Journal of Radiation Research,* vol. 4, 1963.)

most notable such effect is the increase in mutation rate associated with an increase in oxygen concentration around the material being irradiated.

Perhaps the most important factor in assessing the overall mutagenic effects of irradiation arises from the observation that, in contrast to results with *Drosophila*, experiments with the mouse do suggest an effect of dose *rate* on mutation frequency. Thus for a given total radiation dose, a low rate of exposure (for example over days rather than in minutes) gives a much lower mutation frequency than that resulting when a whole dose is given in a short period of time, say a matter of minutes. The explanation for this effect lies in the discovery of enzymatic repair mechanisms that can correct radiation-induced lesions in the DNA. The enzymes involved can recognize breaks and other radiation-induced abnormalities in DNA, cut them out where necessary and then fill in the gaps by a form of "repair" DNA synthesis. When radiation is given at sufficiently low doses, the lesions it causes can be repaired before too much damage is done. When, however, radiation is given at too high a dose rate, the repair system cannot cope with the large number of simultaneously induced lesions. Thus, low dose rates allow much more opportunity for repair mechanisms to work than do high dose rates.

Many mutations that cause defects in repair mechanisms have been studied in bacteria. Such mutations can be identified by the fact that they increase the sensitivity of bacteria to ultraviolet light (UV). UV is very-short-wavelength light; it is the component of sunlight that causes sunburn and tanning of the skin. It is also a mutagen in bacteria. When UV is absorbed by DNA, lesions are formed that are in some respects similar to those caused by X rays and other sorts of radiation, and these lesions in turn give rise to mutations. The same types of mechanisms that repair radiation damage also lead to the repair of DNA damage caused by ultraviolet light (Fig. 4.27). A combination of genetic and biochemical studies has led to the identification of many of the specific enzymes involved in such repair processes in bacteria. Progress in higher organisms has so far been slower, but one striking example of a genetic disease caused by a DNA repair deficiency has been identified.

Xeroderma pigmentosum is a recessive disease associated with extreme sensitivity to sunlight, which causes the development of multiple cancers in affected individuals.

It has been shown that this disease is due to a deficiency in one step of the processes which repair DNA damage induced by the sun's ultraviolet light. The severity of the condition shows how important the DNA repair processes are in protecting us from the effects of ultraviolet light in the sun's rays, and no doubt also from the effects of other sources of radiation (Fig. 4.28). The existence of these repair mechanisms means that, in assessing the mutagenic effects of irradiation, both the *total* dose and the *rate* at which it is received must be taken into account.

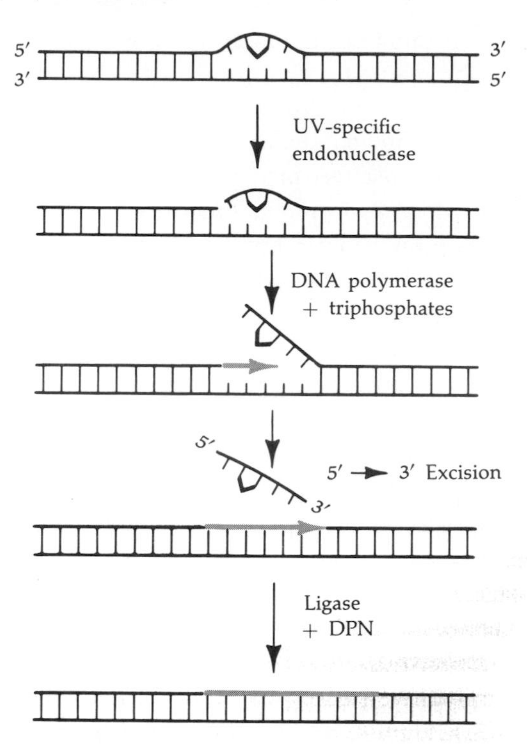

Figure 4.27
DNA repair synthesis. Much of the damage done to DNA by UV light, ionizing radiation, or chemical mutagens causes a structural distortion in the DNA molecule. (In the example shown here, UV light has led to formation of a dimer between two adjacent nucleotides.) Repair enzymes (endonucleases) recognize these distortions and excise them by nicking the DNA on the 5′ side of the lesion. New nucleotide bases are joined into the gap by the action of a DNA polymerase enzyme, and another repair enzyme (an exonuclease) cleaves the DNA on the 3′ side of the lesion. The newly made patch is then joined to the main body of the DNA, thus restoring a normal DNA strand. This scheme of excision repair is called "cut, patch, cut, and seal." (From P. C. Hanawalt, *Endeavor,* vol. 31, p. 83, 1972.)

Radiation not only induces mutations; it also causes chromosome breaks and rearrangements that can readily be seen in cytological preparations. Some examples of the types of abnormal chromosomes that are produced by radiation are shown in Figure 4.29. The effect of a break depends on whether it occurs before or after the chromosomes replicate, whether it is rejoined, and if so how. Studies on human material can be done by irradiating lymphocyte cultures *in vitro.* The frequency of abnormal chromosomes produced by single breaks appears to increase linearly with radiation dose, just as for mutations. A total dose of about 300 r is enough to produce many abnormal chromosomes per cell, which is a very high rate of abnormality, while doses as small as 10 to 20 r cause a detectable increase in abnormalities.

The induction of some chromosome abnormalities, such as translocations, requires two breaks to be produced simultaneously. In this case, as might be expected, the dependence of the frequency of production on dose is not linear but rather follows approximately a square law (Fig. 4.30). In this figure, data on the rate of induction of translocations by acute radiation doses (95 r per minute) are plotted against dose. The data were obtained by irradiating mouse spermatogonia, which are the premeiotic cells of the male germ cell line. Each point on the graph is based on the examination of a total of 800 spermatocytes from four

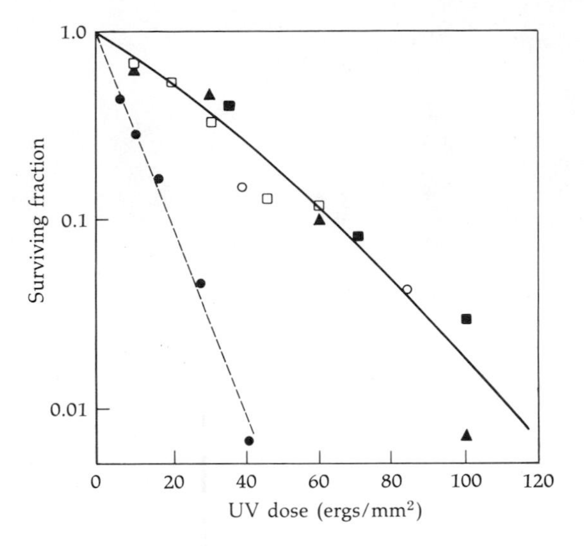

Figure 4.28
UV survival curves for xeroderma pigmentosum (XP) and normal fibroblasts. XP and normal fibroblasts (connective tissue cells from the skin cultured in test tubes) were irradiated with UV light. Black squares and triangles represent values obtained with normal fibroblasts; open squares and circles represent values obtained with XP variants having normal excision repair. The black circles represent values obtained with XP6, the common form of XP with reduced excision repair. As indicated by the dashed line, the survival curve for XP6 fibroblasts is significantly different from that for fibroblasts with normal excision repair. (After J. E. Cleaver, *Advances in Radiation Biology,* vol. 4, 1974.)

irradiated mice. The rate of production of translocations increases sharply as the radiation dose goes from 200 r to 400 r, in marked contrast to what is expected from a linear relationship as found for other types of induced damage. Radiation has also been shown to induce nondisjunction in the developing germ cell of the mouse. Fractionation of the radiation dose has the same ameliorating effect on the induction of chromosome abnormalities as it has on the induction of mutations, presumably for the same reason—namely the existence of repair mechanisms.

An increased frequency of abnormal chromosomes has often been found in lymphocytes taken from people who have been irradiated, either inadvertently or in the course of medical treatment.

Such chromosome abnormalities have sometimes been observed in lymphocytes many years after a known exposure to radiation. Because it must be presumed that the abnormalities were caused by the radiation, and so must have been produced at the time of exposure, this helps to show how long lymphocytes normally survive in the circulation. The rate of production of abnormalities by

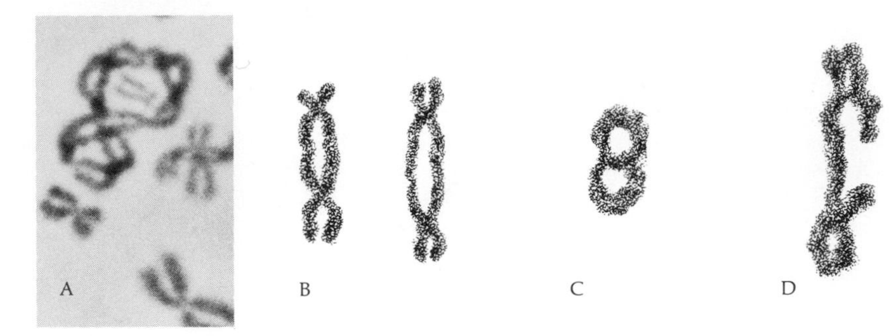

Figure 4.29
Types of chromosome breaks caused by radiation. (A) A ring chromosome formed as a result of multiple breaks. (B) Two dicentrics. (C) A ring chromosome. (D) A dicentric resulting from single-chromatid involvement. (Part A from M. Bobrow; parts B–D after A. Bloom in H. Harris and K. Hirschorn, *Advances in Human Genetics,* vol. 3, 1972.)

such *in vivo* exposures seems to be quite comparable to that found from *in vitro* studies. Diagnostic chest X rays are at too low a dose to produce detectable increases in abnormalities, but other more complex diagnostic procedures do involve doses that can be high enough to produce observable effects. The importance of these studies on chromosome breaks is that they provide a good way of monitoring for the effects of radiation. Inherited abnormalities will, of course, not be produced unless the radiation affects the gonads.

Induction of cancers is known to be associated with irradiation of somatic tissues.

An interesting and rather distressing example of this is provided by a study of radium-dial painters, who put luminescent paint on watch dials. They retain small amounts of radioactive radium, which accumulates in their bones. The resulting radiation more than doubles the frequency of chromosome breaks observed in their lymphocytes, and these individuals have a high risk of developing bone cancer.

The survivors of the atom bombs dropped on Hiroshima and Nagasaki at the end of World War II are another group of people who have been studied intensively in order to find out more about the genetic and other effects of radiation.

The Atomic Bomb Casualty Commission (ABCC), which is a joint United States and Japanese venture, has been studying these survivors and their children for more than 20 years. The dose received by an individual can be calculated from his position, and the distance from the place where the bomb fell, at the time of the explosion. In one study, about 1,400 children of individuals exposed to more than 100 r were examined for chromosome abnormalities. The results suggested a

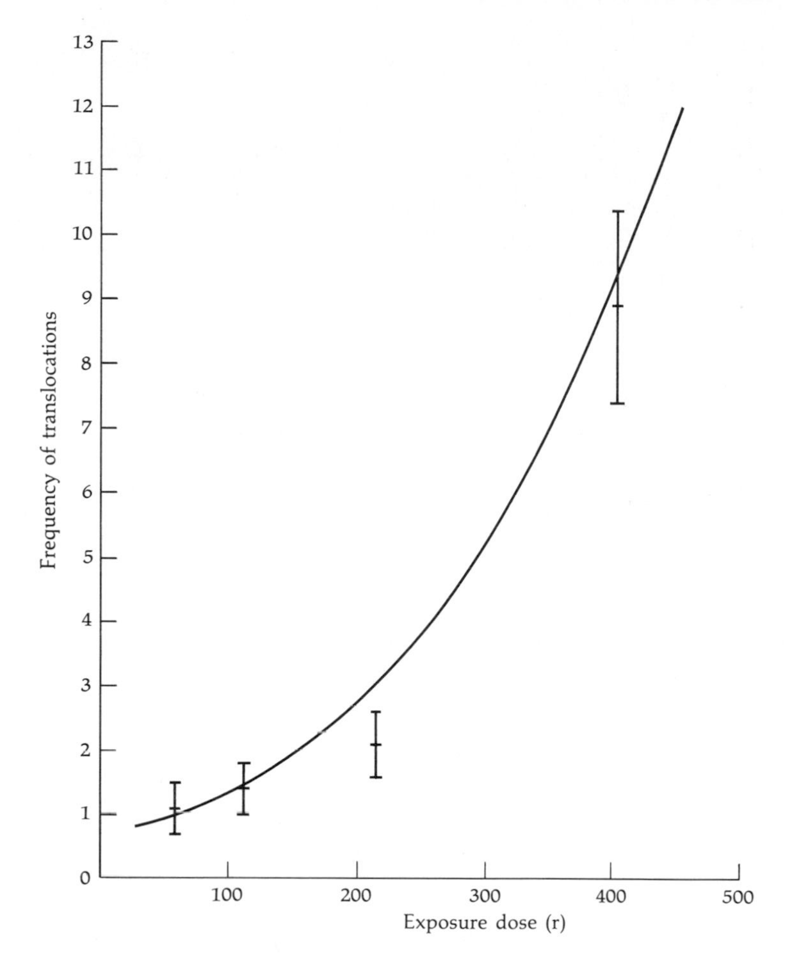

Figure 4.30
Square-law dose dependence for translocations. The frequency of translocations increases with the square of the radiation dosage.

possible two-fold increase in the incidence of chromosome abnormalities. This is not a very striking effect, and it emphasizes how much effort is needed to establish significant genetic effects even of known and major environmental agents. Other effects of irradiation, especially the induction of cancers, were much more striking. Thus, the incidence of leukemia in Hiroshima survivors who were within 1 kilometer of the center of the explosion was 1 in 60 after 12 years, an increase of almost 50-fold over control levels.

In order to assess the overall genetic effects of radiation and other mutagens on the population as a whole, we must know something about the behavior of mutant genes in populations. This is a subject treated in Part II of our book, and we shall therefore defer a discussion of these overall effects of environmental

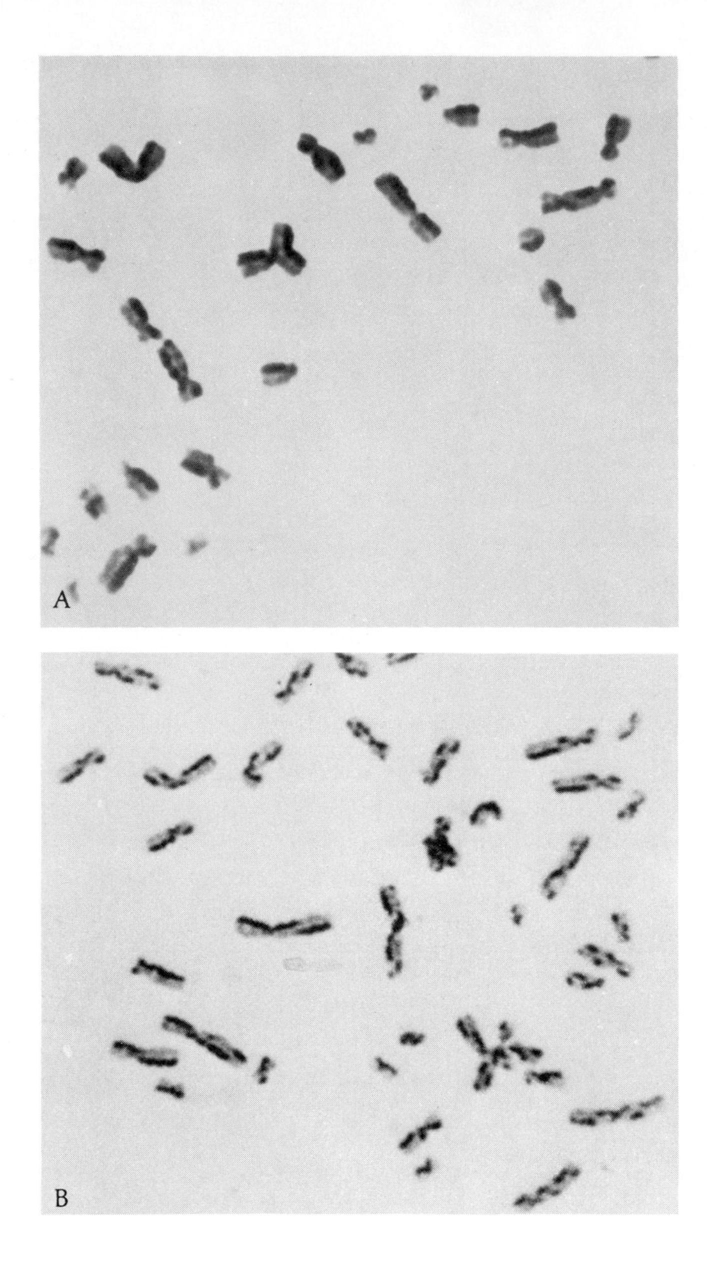

Figure 4.31
Sister-strand exchange. Recombination occurs by sister-strand exchange, which may be studied by fluorescence techniques. A fluorescent dye (which is quenched by heavy-metal atoms) is incorporated into the DNA during one replication. During the next round of semiconservative replication, 5BudR (which contains the heavy metal bromine) is added. Because replication is semiconservative, only one sister strand in each chromosome will contain 5BudR, and both will contain the fluorescent dye. (A) One strand of each chromosome is fluorescent, while the other is not. Sister-strand exchange is observable, meaning that crossing-over has occurred. (B) These cells have been treated with chlorambucil, which increases the frequency of sister-strand exchange. This technique may prove to be a simple and valuable procedure for detecting the effects of some types of mutagenic agents on chromosomes. (Courtesy of M. Bobrow.)

mutagens until Chapter 21, in which we discuss some of the social issues raised by genetic questions. There has rightly been much concern about the genetic hazards of radiation and other mutagenic agents, whether these are due to radioactive fallout from atom bomb tests or to medical uses, or are byproducts of the generation of nuclear power. It is undoubtedly most important that these hazards should be continually monitored.

Summary

1. Many mutations are, at the protein level, single amino-acid substitutions; they derive from single-nucleotide substitutions in DNA.
2. Deletions or additions of single amino acids are due to deletions or additions of nucleotide triplets in DNA.
3. Deletion or insertion of a number of nucleotides that is not a multiple of three results in more serious alterations of the proteins.
4. Various chemical mutagens have different modes of action.
5. Chromosome aberrations include aneuploidies, usually the consequence of nondisjunction at meiosis. Down's syndrome (or trisomy 21) and various sex-chromosome aneuploidies are the most commonly observed such abnormalities. Even triploidies and tetraploidies have been observed.
6. Other chromosome aberrations are due to breaks, followed by rejoining of the fragments in altered relations. Thus inversions, translocations, deficiencies, and duplications arise. Many of them determine serious pathological consequences.
7. Natural and man-made radiation accounts for a fraction of all mutations. Any increase in radiation causes an increase of mutation. Because most mutations are harmful, there is wide concern over possible increases of radiation and other mutagenic agents.

Exercises

1. Review basic types of mutation at the DNA level.
2. Which are the four basic types of chromosome mutations, and how are they detected genetically and cytologically?
3. Review the most frequent aneuploidies detected in man.
4. Which are the most important known mechanisms of action of chemical mutagens?

5. Why does mutation rate increase proportionately with dose of radiation?

6. How does one account for the effect of the intensity of administration of radiation?

7. Discuss relationships between
   a. mutagenicity and carcinogenicity;
   b. UV and X-ray mutagenesis;
   c. DNA repair mechanisms and mutagenesis.

8. Using the genetic code, write the polypeptide resulting from a segment of DNA of the following constitution:

1	2	3	4	5	6	7	8	9	10	11	12	13	14	15
T	T	A	C	G	C	T	A	T	T	A	A	A	T	A

   and those resulting from (a) a substitution of nucleotide 4 with G; (b) of 8 with C; (d) of 12 with G.

9. Considering the redundancy of the code, and assuming that all triplets are equally frequent (which is not exactly right, but is sufficiently correct for a first approximation), what is the probability that a single-nucleotide substitution will *not* cause an amino-acid substitution? Given the same assumption, what is the probability that a single-nucleotide substitution will cause the termination of the polypeptide chain?

10. For the DNA segment given in Exercise 8, introduce a thymine between the first and second nucleotides. Which polypeptide will result?

11. In hemoglobin (Hb) a great number of single-amino-acid substitutions have been found (Table 4.3). Can each of them be explained by a single-nucleotide substitution?

13. A patient with Turner's syndrome is found to be color-blind. Both her father and mother have normal color vision. How can this be explained? Does this tell us whether the nondisjunction took place in the father or in the mother?

14. A patient with Klinefelter's disease is found to be color-blind. Father and mother have normal color vision. How can this be explained? If the color-blindness gene were close to the centromere (it is not close), would the clinical data inform us whether the nondisjunction took place at first or second meiotic division?

15. A woman who worked as a nurse in a radiologist department has a child with a dominant autosomal achondroplasia. If she sues the hospital, believing that radiation induced the disease, what can a geneticist called by the judge say? (After C. Stern.)

16. A man working in a nuclear plant has a child affected by hemophilia. If he sues the nuclear plant for damages, what shall a geneticist say about the validity of his claim? (After C. Stern.)

Table 4.3
A sample of single-amino-acid substitutions known in hemoglobin (Hb)

Hb	Mutant	Chain	Position	Amino acid in normal chain	Amino acid in mutant chain
J	Torino	α	5	ala	asp
J	Oxford	α	15	gly	asp
G	Honolulu	α	30	glu	gln
L	Ferrara	α	47	asp	gly
	Sinai	α	47	asp	his
M	Kansas	α	87	his	tyr
S		β	6	glu	val
C		β	6	glu	lys
G	Galveston	β	43	glu	ala
D	Ibadan	β	87	thr	lys

References: Chapter 4

Bloom, A. D.
1972. "Induced chromosome aberrations in man." In H. Harris and K. Hirschhorn, eds., *Advances in Human Genetics,* vol. 3, pp. 99–153.

Hamerton, J. L.
1971. *Human Cytogenetics.* 2 vols. New York: Academic Press.

Neel, J. V., and W. J. Schull
1956. *The Effect of Exposure to the Atomic Bombs on Pregnancy Termination in Hiroshima and Nagasaki.* Publ. No. 461. National Acad. of Sciences–National Research Council. Washington, D.C.

United Nations Scientific Committee
1962. *Report on the Effects of Atomic Radiation.* General Assembly Official Records, 17th Session, Supplement 16 (A/5216). New York: United Nations.
1966. *Report on the Effects of Atomic Radiation.* General Assembly Official Records, 21st Session 14 (A/6314). New York: United Nations.
1972. Ionizing Radiations: Levels and Effects. New York: United Nations.

5

Cell Biology and Genetics

Major advances in our understanding of the way cells work and interact with each other have come from studies of cells cultured *in vitro* (''in the test tube''). The mammalian cell is itself a complex organism in comparison with viruses, bacteria, or fungi. The development of cell-culture techniques has now made it possible to work with mammalian cells using approaches that are modeled on the studies with microorganisms that have, over the last 20 to 30 years, laid the foundations of molecular biology, as described to some extent in Chapter 3. An aim of this chapter is to describe some of the more recent achievements of cell-culture studies using these approaches. We also describe an important phenomenon: immunity.

5.1 Cell Culture and Cancer

Mammalian cells are much more difficult to grow "in vitro" than are molds or bacteria. As a result it took some time before media were developed with the right combinations of vitamins, amino acids, and other components to support growth of cells outside the body. Even now such tissue-culture media still have blood serum, often from a fetal calf, as an essential component. The serum obviously contains one or more substances that are required for the growth of cells in culture but, in spite of much effort, comparatively little progress has been made in identifying what these essential growth factors might be. Despite the complexity of the media, tissue-culture techniques are now comparatively routine and are very widely used in biological and medical research.

If a small piece of skin is put in a dish with medium, cells will grow out from it after a few days and gradually spread over the whole surface of the dish (Fig. 5.1). The cells grow only if they can attach to the surface of the dish, and they will not grow over each other. When the surface of the dish is fully covered and each cell is in contact with other neighboring cells, there is no further growth. Cells can then, however, be detached from the surface on which they have been growing, usually using the enzyme, trypsin. If an aliquot of these detached cells is now transferred to another dish with medium, these cells will again grow to cover the surface of the dish and then stop, and the whole process can again be repeated. If only a small number of cells is transferred, the descendants of single cells form well-separated colonies that can be picked separately, so that mixtures of different types of cells can be resolved (Fig. 5.2). This process of *cloning* (a clone is formed by the descendants of a single cell, or a single individual) is an essential ingredient for any genetic study.

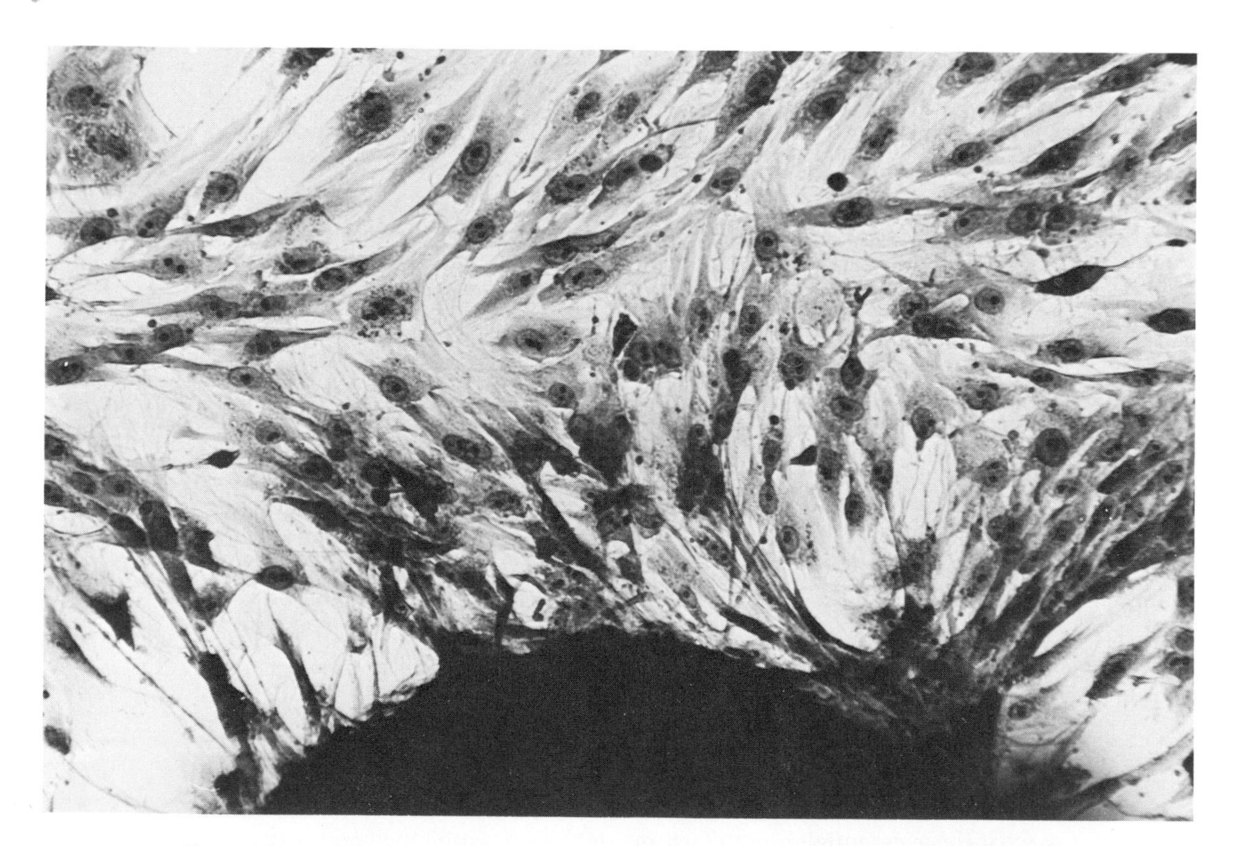

Figure 5.1
Cells growing out from a skin biopsy. Normal fibroblasts growing out from a small piece of skin in culture. (Courtesy of D. Buck.)

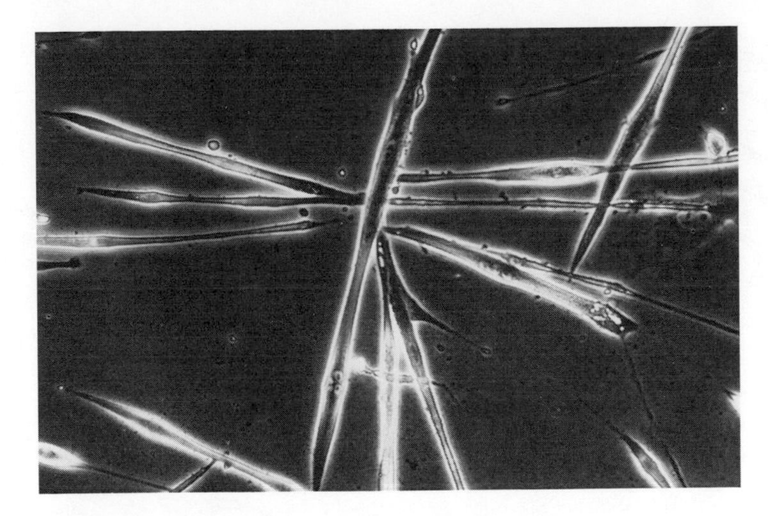

Figure 5.2
Muscle cells growing in culture. These cells were derived from the muscle tissues of a fetal mouse. (Courtesy of J. Peacock.)

The variety of cell types that can be grown in culture is still fairly limited.

The most common type that grows out from virtually any tissue is a long spindly-looking cell called a fibroblast (Fig. 5.3). This cell retains in culture many of the properties it is known to have in the organism. Unfortunately it has proved very difficult, for example, to grow cells from the liver that still manufacture in culture those products characteristic of liver cells. Many of the gene products, especially enzymes, that are produced by cells in culture can readily be studied using biochemical techniques. Where there are genetic differences between individuals these are, as expected, reflected in the cells cultured from their tissues.

One rather striking property of fibroblasts grown in culture is that after about 20 to 40 transfers they regularly die out, as if overcome by old age. This very intriguing phenomenon is now being studied as a model for the aging process. It is, however, a severe limitation to the use of fibroblasts for genetic and other studies. What sources of cells could be used that might not have this finite lifespan culture? The one clearcut answer that has been found to this question is cancer cells. The best known human-cell line is derived from a tumor; it has now been in culture for about 30 years and is grown in laboratories all over the world. It is called HeLa after Henrietta Lacks, the unfortunate donor of the cells, who died from a very malignant cancer soon after the cells from her tumor were put in culture.

Cell lines derived from tumors have another important property in addition to their unlimited life span. They are not usually inhibited in their growth by

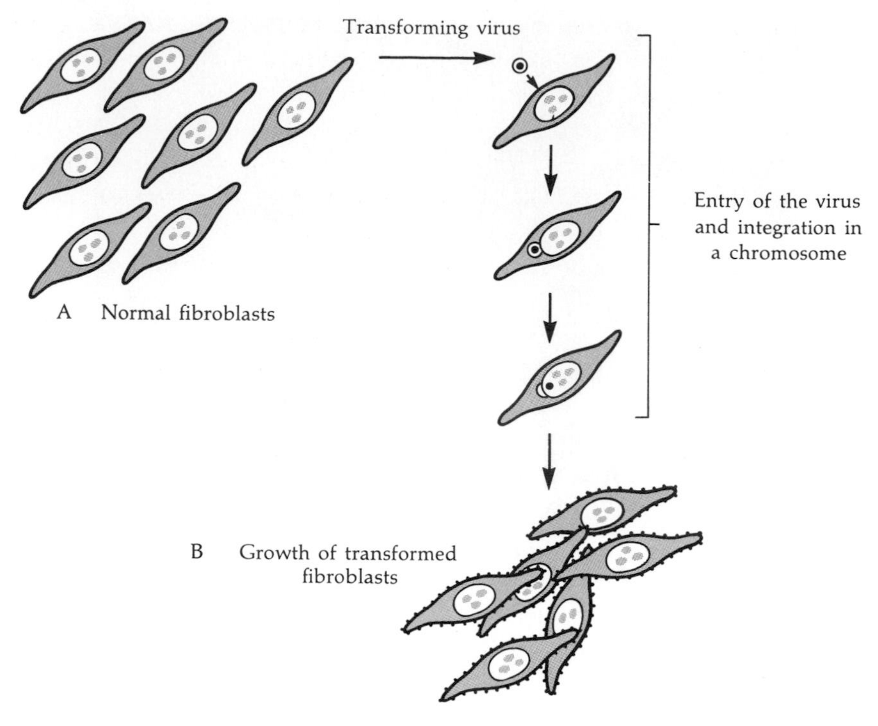

Figure 5.3
Normal and transformed fibroblasts. The normal cells form nonoverlapping networks, but the transformed cells (which are not responsive to contact inhibition as are the normal cells) pile up on top of one another. (After R. Dulbecco, "The induction of cancer by viruses," *Scientific American,* April 1967. Copyright © 1967 by Scientific American, Inc. All rights reserved.)

adjacent cells, and so can pile up on top of each other as they grow, reaching much higher cell concentrations than do fibroblasts (see Fig. 5.3). These properties of permanent cell lines derived from tumors seem to fit in very well with the uncontrolled growth of cancer cells when they are in their natural environment in the individual.

Cell cultures derived from normal tissue sometimes give rise spontaneously to permanent cell lines just like those that can be grown out from tumors.

This phenomenon seems to be an *in vitro* analog of the production of tumors in the animal. Cell lines that are produced in this way are called "transformed," because their properties are so different from those of the original cells from which they were derived.

The possibility that viruses may be one, if not the major, cause of cancer was first suggested many years ago, and there is no doubt that some cancers are caused by viruses. The discovery that certain viruses can cause transformation of normal cell cultures was therefore of great interest, because it provided an *in vitro* model for studying the induction of tumors by viruses. It is likely that all the lines that can grow permanently in cell culture are, in some sense, virus-transformed. The transforming virus seems to remain in the transformed cells in a latent state. In one or two cases it has been clearly shown that the DNA of the latent virus is integrated into its host's DNA, so that it actually becomes a part of the host's chromosomes. Some bacterial viruses behave in a very similar way, though in this case there is no apparent analog of the phenomenon of transformation.

Recently it has been found that permanent cell lines can be grown out from lymphocytes obtained from peripheral blood. These "lymphoid" cell lines appear to be lymphocytes that have been transformed by a virus called EB (after M. A. Epstein and Y. M. Barr, who first described it), which is the cause of mononucleosis or glandular fever. Lymphoid lines, instead of growing attached to a glass or plastic surface, grow in loose clumps and so are rather easier to handle than other types of cell culture. They are proving to be a most interesting and useful source of human-cell culture material, especially as they can be grown rather easily from any chosen individual.

It is very often found that permanent cell lines, just like tumors in the animal, have a quite abnormal chromosome constitution, which may itself be quite variable within the line or tumor.

HeLa, for example, has about 63 to 65 chromosomes instead of the normal 46, and many of them are quite unlike any of the normal human chromosomes (Fig. 5.4). These abnormal chromosomes appear to be derived from normal chromosomes by a series of inversions, translocations, and possibly deletions. At one time it was thought that an initial chromosome abnormality might itself be the cause of a tumor. Now, however, it seems more likely that the abnormal chromosome is produced during the outgrowth of a tumor, and may be selected for because it confers some form of advantage, in terms of growth rate, on the altered cell. It is also likely that some tumor viruses at least have a tendency actually to cause chromosome breaks themselves (Fig. 5.5).

Cell cultures, especially permanent cell lines, have been widely used for biochemical studies and for studies of cellular processes such as cell division, DNA replication, and the control of gene expression. Normal fibroblast cultures are also now used for growing viruses, such as polio, for the preparation of vaccines. As we shall now discuss, cell lines have in addition provided the basis for a completely new approach to human genetics.

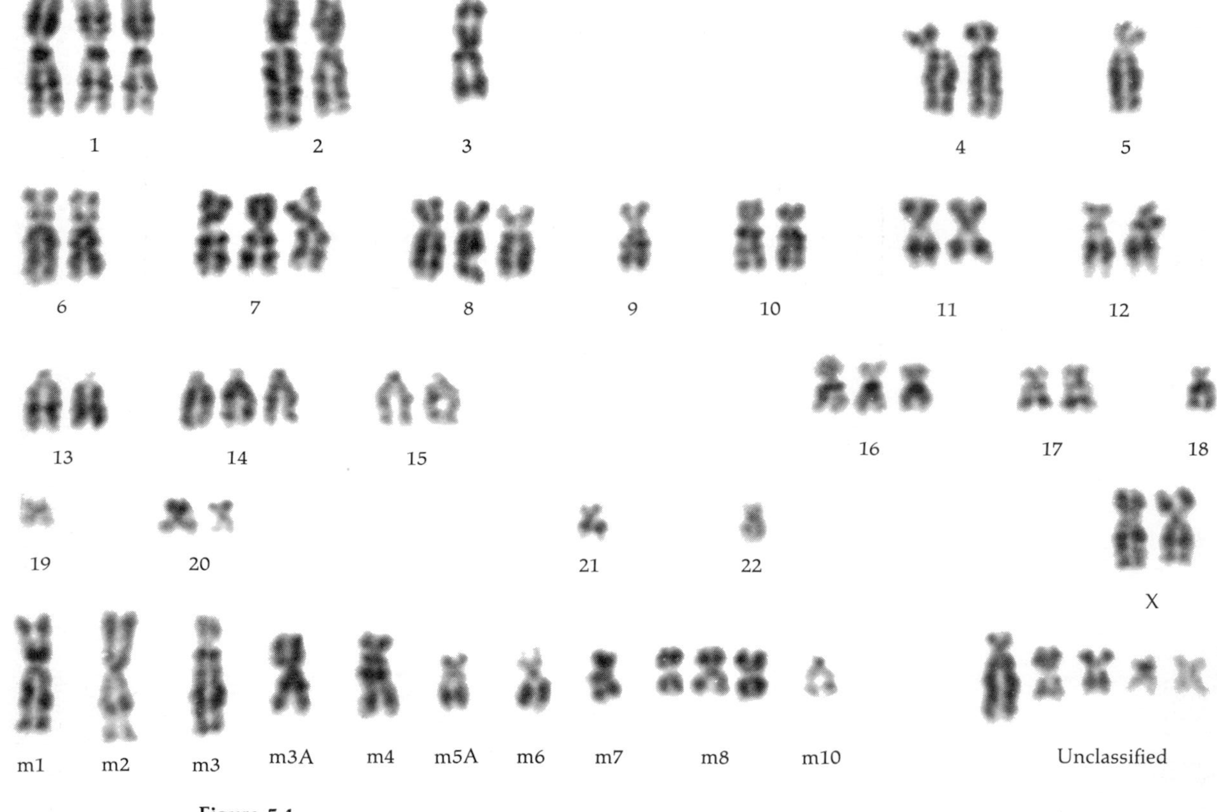

Figure 5.4
A HeLa karyotype. The normal chromosomes are listed first; these are present one, two, or three times each. Numbers m1 through m10 are abnormal chromosomes, but types that are consistently recognizable from one HeLa culture to another. The unclassified chromosomes are not readily characterized. HeLa karyotypes have an average number of 63 chromosomes; the range is from 62 to 65 chromosomes. (Courtesy of B. Bengtsson.)

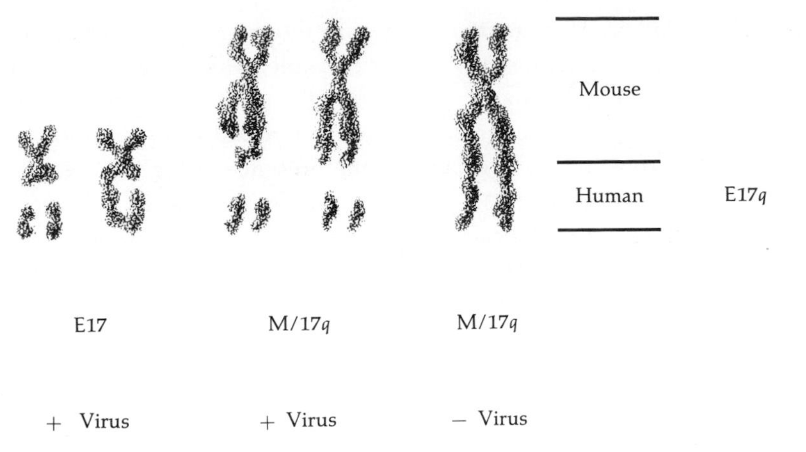

Figure 5.5
Chromosome breaks induced by adenovirus 12. Cells have been infected with adenovirus 12, which has induced a chromatid gap easily visible in the long arms of the translocated mouse chromosome 17 from a mouse–human hybrid cell line. E17 = human chromosome 17. M/17q = mouse–human translocated chromosome 17. (After J. K. McDougall et al., *Nature New Biology,* vol. 245, p. 173, 1973.)

5.2 Somatic Cell Genetics

There would clearly be great interest in using cell cultures for genetic studies along the lines of microbial genetics. Because cell cultures are somatic cells, not being derived from the germ cells, this approach to genetic analysis has been called *somatic cell genetics.* For such an approach to be possible, two requirements must be satisfied. First, it must be possible to identify and work with genetic differences in cell culture. Second, some means is needed of doing genetic analysis with different cell types, for example by making crosses.

Let us consider first the question of genetic markers.

As already pointed out, cells in culture seem in general to express the genes of the individuals from whom they are derived. Thus, for example, if there are genetically determined enzyme differences between individuals in a population that can be detected by the techniques of electrophoresis (described in Chapter 3), these same differences can also serve as genetic markers in cell culture. Mutations can also be selected for in cell culture, just as with microorganisms. Though some nutritionally deficient mutants have been produced, the opportunities for doing this are severely limited by the complex growth requirements of mammalian cells in culture.

The most widely used mutant cell lines are those that are resistant to certain drugs. The drug 8-azaguanine (8AZG), for example, which is an analog of one of the DNA bases, is highly toxic because, when incorporated into the DNA, it interferes with the normal processes of replication and transcription. Cell lines can be produced (by continued growth in the presence of high concentrations of this drug) that are resistant to its toxic effects. The resistance has been shown in most cases to be due to a deficiency in an enzyme called hypoxanthine-guanine-phosphoribosyl transferase (HGPRT), which the cell uses to incorporate 8AZG into its nucleic acids. When this enzyme is defective, the 8AZG can no longer be incorporated, and such cells are resistant to its toxic effects. This deficiency also prevents the cells from incorporating certain normal bases. They can, however, survive because they are able to manufacture these bases internally from more basic medium constituents.

The acquisition of resistance to such drugs has formed the basis for many studies of the effects of mutagens on cells in culture, paralleling similar studies with bacteria and fungi. The 8AZG-resistant cell lines turn out to be particularly useful for genetic studies because there is another drug, called aminopterin, that is toxic only to the resistant cell lines and not to the normal, 8AZG-sensitive cells. Thus, mutations can be selected in both directions. First, they can be selected to 8AZG resistance in the presence of 8AZG, and then selection back to sensitivity can be achieved in the presence of aminopterin.

Having answered the requirement for genetic markers, what about the means of genetic analysis?

For many years this problem was the major stumbling block to the development of somatic cell genetics. There seemed to be no counterpart in cell culture to the bacterial matings discovered by Lederberg and Tatum, though observations had often been made which suggested that neighboring cells might sometimes "fuse" to form double-sized cells. However, in 1960 G. Barski and his coworkers in Paris provided the first clearcut evidence for the occurrence of fusion between cells from different lines and the subsequent formation of a true hybrid cell. Here then was evidence for mating between cells in cultures. Two problems, however, remained. First, the frequency of fusion was quite low, and so the outgrowth of hybrids was difficult to recognize against a background of many parental cells. Second, these initial hybrids seemed to have simply the sum of the two chromosome complements obtained from the two parents. No segregation or reassortment of chromosomes had taken place, as happens of course in meiosis, and without this segregation there can still be no proper genetic analysis.

The first problem, that of the low frequency of production of hybrids, was solved in two main ways. First, techniques were devised for selectively picking out a very small proportion of hybrids from appropriate parental cell mixtures. Second, it was found that certain viruses could greatly increase the rate at which fusions took place between different types of cells. In fact, H. Harris and J. F. Watkins were able to show in 1965 that cells from species as far apart as man and chicken could be fused (Fig. 5.6).

The second problem—that of finding segregation in hybrids—was solved fortuitously when it was discovered that hybrids made between human and mouse cells lost most of the human chromosomes while retaining the mouse chromosomes. Different hybrid derivatives thus ended up with different combinations of human chromosomes, and this provided a basis for doing at least some genetic analysis with these rather novel hybrid cells.

The selective techniques used to pick out hybrids are based on the 8AZG-resistant and other similar drug-resistant cell lines that are sensitive to the aminopterin. Thus, in the experiments that led to the discovery of the loss of human chromosomes in man–mouse hybrids, a mouse line was used that is resistant

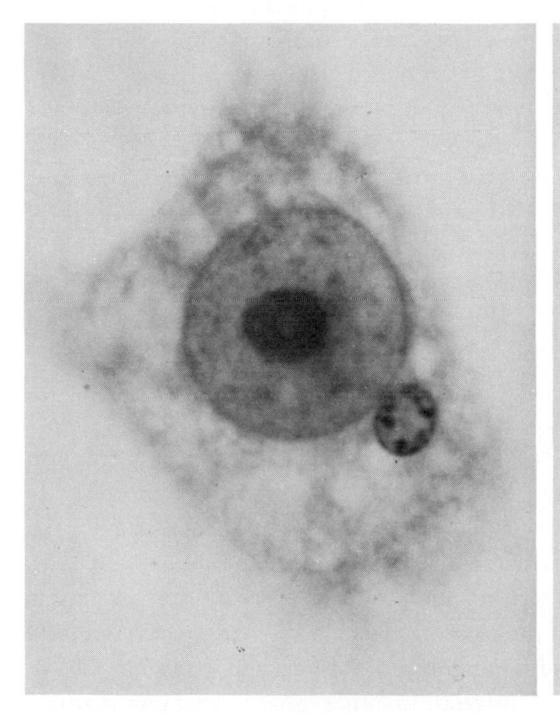

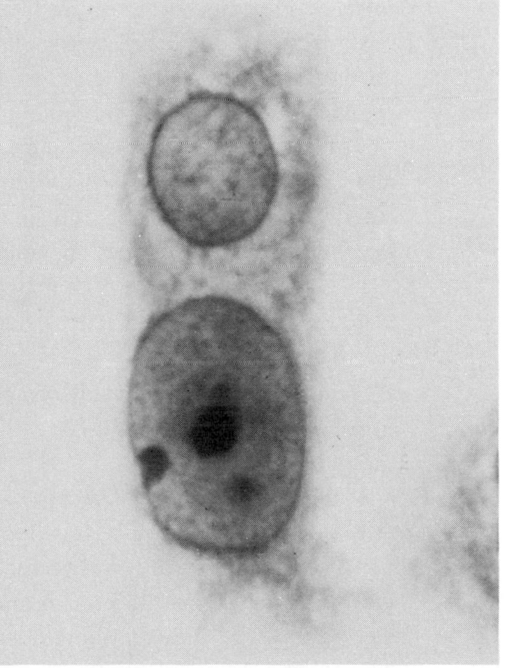

Figure 5.6
Products of fusion between a HeLa cell and a chicken red blood cell. (A) Photo taken immediately after fusion. The large nucleus is from HeLa, the small one from the chicken cell. (B) Photo taken 18 hours later. The dormant red-cell nucleus has enlarged as a prelude to being reactivated in the HeLa-cell cytoplasm. (Courtesy of Henry Harris.)

to the drug 5-bromo-deoxyuridine (5BudR). The resistance is due to a deficiency for the enzyme thymidine kinase (TK), which is used by normal cells for the incorporation of 5BudR and also of thymidine into DNA. Just as in the case of 8AZG resistance, the 5BudR-resistant cells are unable to grow in the presence of aminopterin. When drug-resistant mouse cells are fused with human cells that are not 5BudR-resistant, a hybrid is formed that contains an active human thymidine-kinase gene; this hybrid can therefore grow in the presence of aminopterin. Thus, if 5BudR-resistant mouse cells and normal human cells are fused and the mixture grown in aminopterin, only hybrid cells and the human parental cells can grow.

In these early experiments, a slowly growing human cell line was used; the hybrids could be picked out because they grow much faster than the human parent, while the mouse parent would not grow at all in the presence of aminopterin (Fig. 5.7). Subsequently, human peripheral-blood white cells, which do not grow in culture, have been used to make hybrids. In this case, selection just against the mouse parent yields only growing hybrid cells because the human parent cell, though it can fuse and form hybrids, is not able itself to grow in culture. When the chromosomes of hybrids between 5BudR-resistant mouse cells and normal human cells were observed, a number of cells were found that had only one remaining human chromosome (Fig. 5.8). Because the hybrids can only grow in aminopterin-containing medium if they have retained the normal thymidine kinase, the gene for this enzyme must be located on the one remaining chromosome, which has been identified as number 17. This, then, was the first achievement of somatic cell genetic analysis, namely to identify the human chromosome that carries the gene for thymidine kinase.

One can assign many genes to chromosomes by means of somatic cell hybridization.

The human and mouse forms of many enzymes can be distinguished by electrophoresis. All these differences can serve as genetic markers in studies with human–mouse hybrids. The electrophoretic patterns obtained for the enzyme G6PD from a series of hybrids between 8AZG-resistant mouse cells and normal human white blood cells, together with the parental mouse and human patterns, are shown in Figure 5.9. The first two positions indicate clearly the substantial difference between the human and mouse enzymes. The striking result with the hybrids is that they all show the presence both of human and of mouse G6PD, together with an intermediate band not found in either parent. This intermediate band is just like that found in heterozygotes for an enzyme made up of two subunits (see Chapter 3).

These results show that all the hybrids made with an 8AZG-resistant mouse cell retain the human gene for G6PD. Because G6PD is known from ordinary pedigree studies to be controlled by an X-linked gene, this suggests that all these hybrids

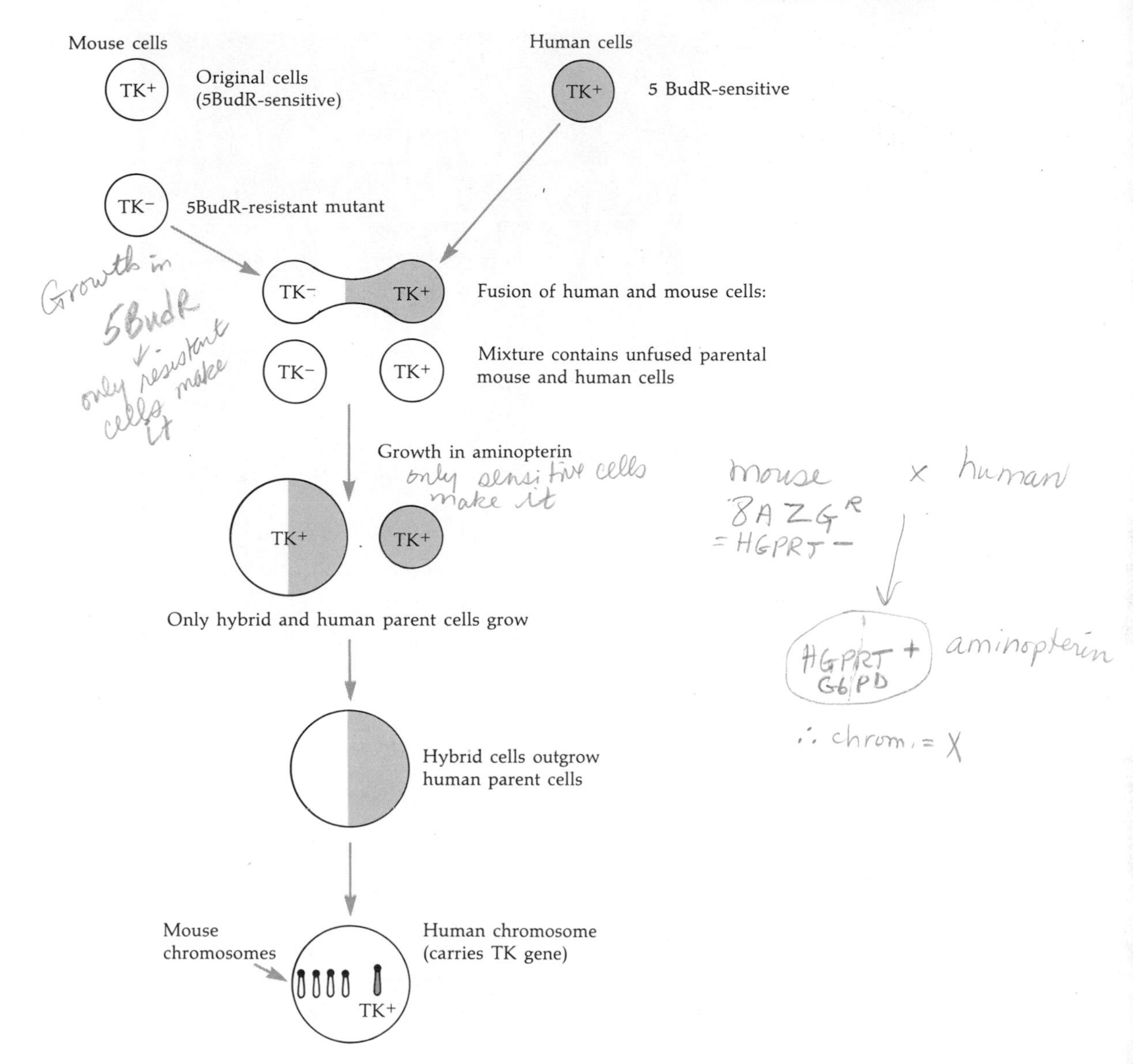

Figure 5.7
Scheme for selection of human–mouse hybrids. Growth in aminopterin eliminates the 5BudR-resistant mouse cells. Because a slowly growing line of human cells is used, the faster growing hybrid cells can be picked out; the hybrid cells are recognizable because they quickly form large colonies. TK = Thymidine Kinase; + = active; − = inactive.

have retained at least one human X chromosome. We know that each hybrid retains only a few human chromosomes, and so it would be surprising if each of the hybrids had kept an X chromosome just by chance. Much more likely is

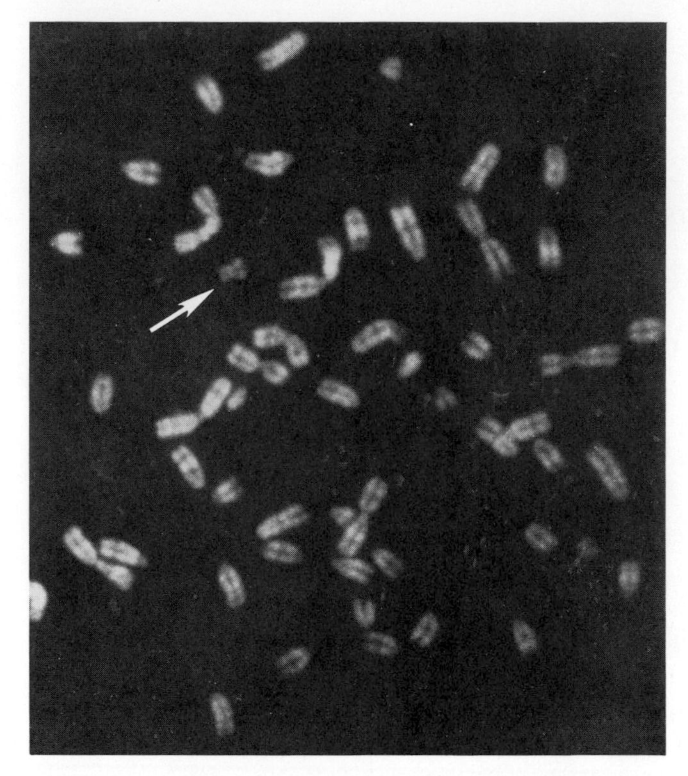

Figure 5.8
Human–mouse hybrid containing a single human chromosome. This photograph is of a metaphase cell obtained by hybridizing a 5BudR-resistant mouse cell with a human cell. In such hybrids, the human chromosomes are preferentially lost. By growing the hybrids under conditions requiring expression of a particular gene, it is possible to select for cells containing a particular chromosome. In this cell, the arrow indicates the single human chromosome 17, which codes for thymidine kinase (see Fig. 5.7). (From O. J. Miller et al., *Science,* vol. 173, p. 244, 1971. Copyright 1971 by the American Association for the Advancement of Science.)

the possibility that the human gene for the enzyme HGPRT, which is required by these hybrids for outgrowth in aminopterin, is on the X chromosome. In this case, the X must be retained by hybrids growing in aminopterin. This explanation for the data can be confirmed by isolating from the hybrids a series of clones that can again grow in 8AZG, as does the mouse parent. Such derivatives should have lost the human chromosome carrying the HGPRT because, so long as the human HGPRT gene is present and active, the cells will remain sensitive to 8AZG. Exactly as expected, it is found that all 8AZG-resistant derivatives from the 8AZG-sensitive hybrids have lost their human G6PD, and consequently their X chromosome. These experiments thus show that the human HGPRT gene is linked to G6PD and that both genes are on the X chromosome (Fig. 5.10).

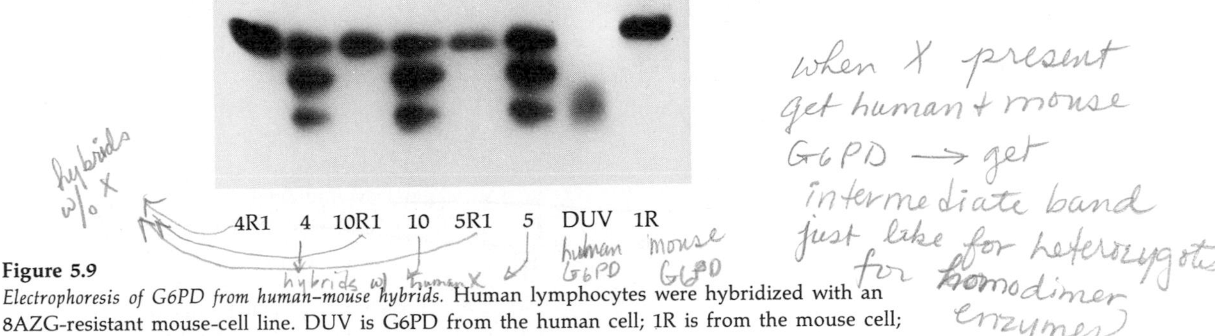

Figure 5.9
Electrophoresis of G6PD from human–mouse hybrids. Human lymphocytes were hybridized with an 8AZG-resistant mouse-cell line. DUV is G6PD from the human cell; 1R is from the mouse cell; 4, 10, and 5 are from hybrids containing the human X; 4R1, 10R1, and 5R1 are from hybrids lacking the X. The intermediate band is just like that found in heterozygotes for enzymes made up of two identical subunits (see Fig. 3.36). (Courtesy of S. Povey.)

Some years ago, a neurological abnormality was described whose most striking feature is that affected children have a compulsive urge to bite the ends of their fingers and lips and to hurt themselves in other ways, such as by banging their heads against hard objects. This urge is so extreme that patients with the disorder (called the Lesch–Nyhan syndrome after its discoverers) must be physically constrained. The syndrome is associated with a variety of other behavioral abnormalities, including mental retardation. Remarkably, it turns out that this complex neurological and behavioral abnormality is very simply explained at a biochemical level. It is due to a deficiency of the enzyme HGPRT, whose genetics we have been studying in cell culture. The cells of patients with Lesch–Nyhan syndrome are 8AZG-resistant just like the mouse parent cells used to make human–mouse hybrids. Pedigrees of the syndrome clearly show that it is controlled by an X-linked recessive gene (Fig. 5.11). Thus the experiments with hybrid cells show *in vitro* the same patterns of inheritance demonstrated in a more conventional, but less controllable, way over a period of many years by the individuals represented in the pedigree.

The mapping of the TK and HGPRT genes using hybrids was based on their association with the selective technique (growth in aminopterin) used to identify the hybrids in the first place. Many linkages have subsequently been established by cell hybridization that involve unselected markers. Thus, for example, if two human enzymes identified in man–mouse hybrids by electrophoresis are always either both present or both absent, then this suggests that they are controlled by genes on the same chromosome. This is because segregation in the hybrids occurs by the loss of whole chromosomes, as in the case of the X chromosome, there being no analog of the process of recombination that occurs in meiosis. Separation of linked markers would occur only if the chromosome was broken, which does

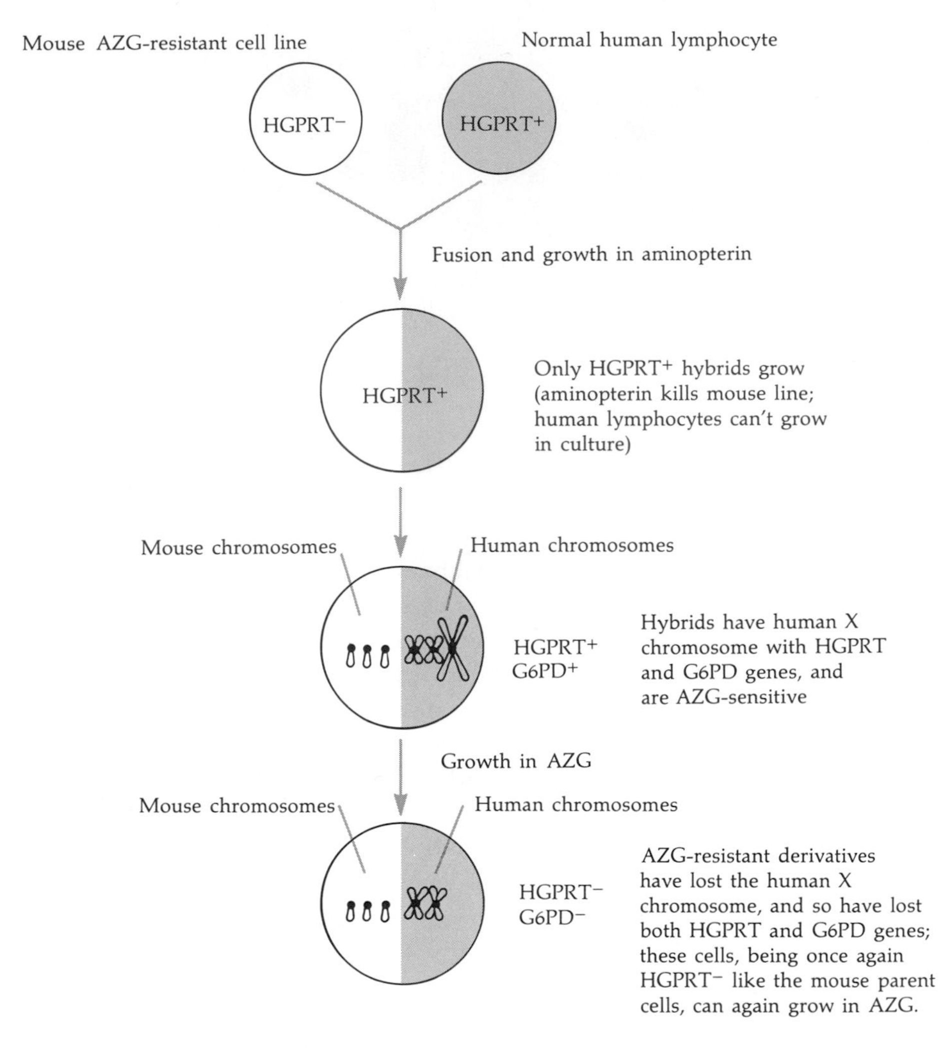

Figure 5.10
Mapping the HGPRT gene by cell hybridization. The human gene for HGPRT is needed for growth of the hybrids in aminopterin. The human lymphocytes do not grow in culture. Viable hybrids, which are HGPRT$^+$, are also G6PD$^+$ and have the human X chromosome; these cells are also AZG-sensitive. If hybrids are grown in AZG-containing medium, AZG-resistant segregants will arise. All of these segregants have lost the human X chromosome and are HGPRT$^-$ and G6PD$^-$.

occasionally happen. When banding techniques are used to identify the human chromosomes in man–mouse hybrids, the particular chromosomes associated with the human genetic markers present can be identified. In this way, human genes can be systematically associated in linkage groups corresponding to each of the human chromosomes (Fig. 5.12). Further refinements of these techniques making

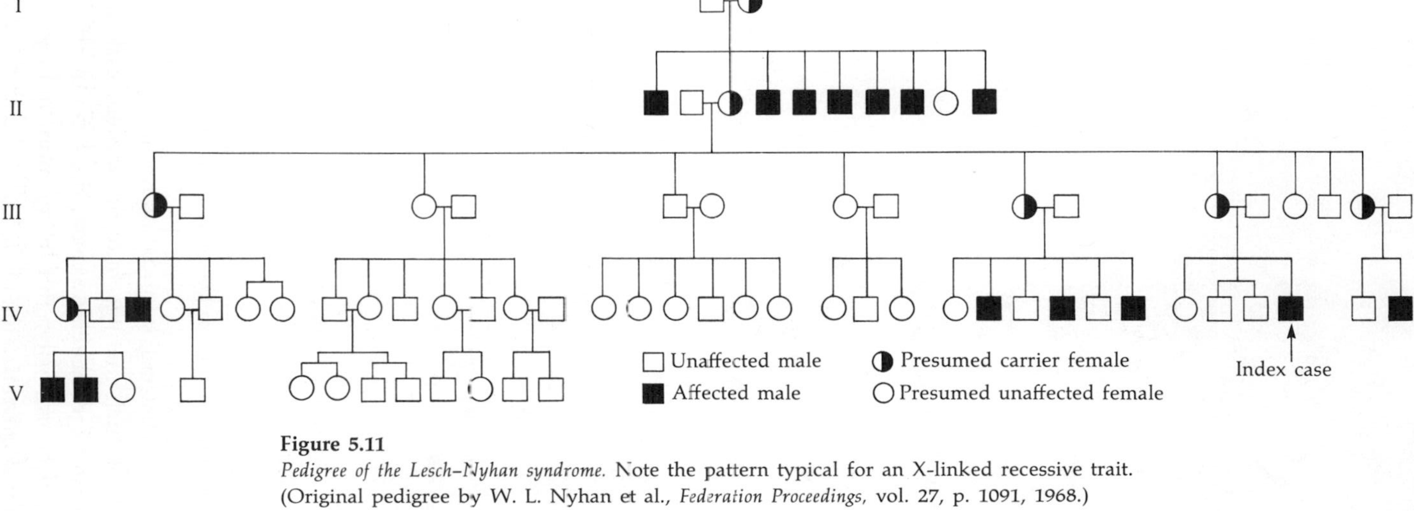

Figure 5.11
Pedigree of the Lesch–Nyhan syndrome. Note the pattern typical for an X-linked recessive trait. (Original pedigree by W. L. Nyhan et al., *Federation Proceedings,* vol. 27, p. 1091, 1968.)

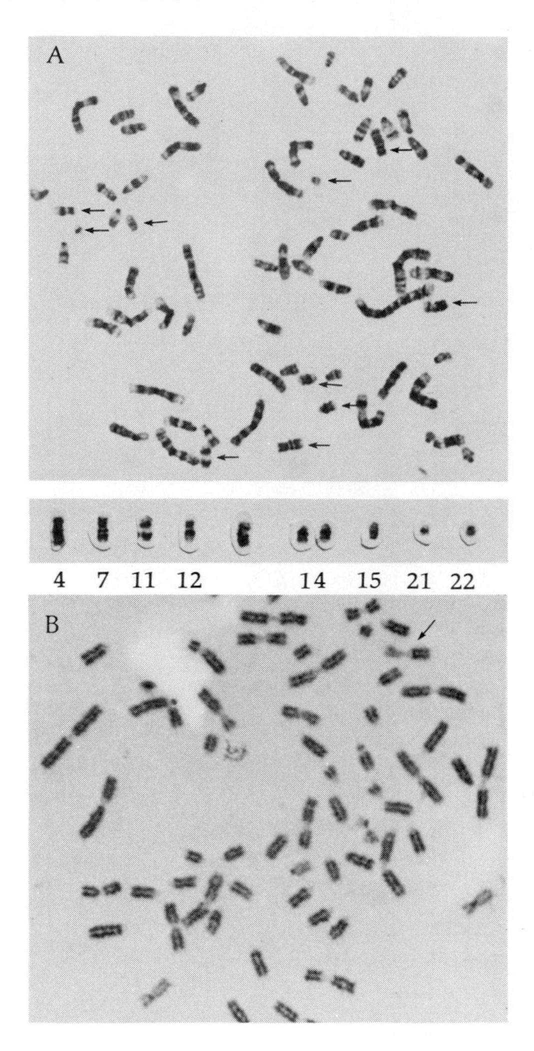

Figure 5.12
Karyotype of man–mouse hybrid cells. (A) Several of the human chromosomes are identified in this karyotype. (B) A metaphase spread showing a mouse–human translocation, as revealed by G11 staining (note the different staining of the mouse and human portions). (Part A from V. Van Heyningen et al., *Annals of Human Genetics,* vol. 38, p. 297, 1975. Part B from M. Bobrow and J. J. Cross, *Nature,* vol. 251, pp. 77–79, 1974.)

use of chromosomal aberrations can provide some information on the location of the genes along the chromosomes.

Until just a few years ago, only three or four autosomal linkages in man had been described.

Matings that can provide useful information for linkage studies are hard to find, and many offspring of such matings may be needed, especially to establish a comparatively loose linkage. In recent years, the level of effort devoted to such studies has increased enormously, and many more linkages have been found. With the advent of cell-hybridization techniques, the increase in information has been quite explosive, so that there are now one or more genes assigned to most of the human chromosomes, and some chromosomes (including the X) have quite extensive linkage maps. Table 5.1 and Figure 5.13 summarize the situation at the time of writing.

Complementation studies can be carried out in vitro.

Although intraspecific cell hybrids may be more difficult to use for genetic analysis because of their relative chromosomal stability and the more limited availability of genetic-marker differences within the species, they have proved very useful for complementation studies. As described in Chapter 3, the complementation test is a test for function in heterozygotes for apparently similar alleles. The somatic-cell hybrid formed by fusion between different cells that are from homozygotes for different (but similar) alleles is equivalent to the heterozygous individual, and such hybrids can therefore be used to test for complementation.

The first example of the application of this technique to a human disease involved the inborn error of metabolism, galactosemia. This is one of a class of diseases caused by an inability to utilize sugars found in milk; these diseases result from a deficiency in one of the enzymes used to convert these sugars into sources of energy. Galactosemic individuals lack a transferase enzyme needed for the utilization of the sugar galactose. They cannot tolerate normal milk, which contains a fair amount of galactose, but they manage quite well on a galactose-free diet. The disease is quite rare, occurring with a frequency that may be as low as 1 per 200,000, and it is not at all clear whether different cases are due to mutations to the same defective allele, or are even at the same locus.

When skin cells from seven different galactosemic individuals were fused in all possible pairwise combinations, three hybrids were found which complemented in the sense they had at least some of the enzyme activity needed for the utilization of the sugar galactose. The hybrid enzyme, however, was clearly distinguishable from that found in normal individuals. This result suggests that the complementation was between alleles at the locus for the transferase enzyme, and it shows that a substantial proportion of the mutant alleles causing galactosemia that are found in the population at large are in fact different in origin. A similar analysis of cells from different individuals with the repair-deficiency disease *xeroderma pigmentosum* (Chapter 4) has shown that there are at least two different loci that can give rise to mutants causing this disease. Complementation analysis thus proves to be useful for the detection of heterogeneity among genetic diseases.

The technique of somatic-cell genetic analysis based on hybridization is now being used to study a number of basic problems of cell biology.

D. Eisenhut
544-68-7279

Table 5.1
Key to human gene abbreviations used in Figure 5.13

Chromosome number	Abbreviation	Full name of gene locus
1	PPH	Phosphopyruvate hydratase
	PGD	6-phosphogluconate dehydrogenase
	Rh	Rhesus blood group
	E-1	Elliptocytosis-1
	UMPK	Uridine monophosphate kinase
	AK-2	Adenylate kinase-2 [located on *p* arm; exact position uncertain]
	PGM-1	Phosphoglucomutase-1
	Amy-S	Amylase, salivary
	Amy-P	Amylase, pancreatic
	Fy	Duffy blood group
	cat	Cataract, zonular pulverulent
	UDGP	Uridyl diphosphate glucose pyrophosphorylase
	Pep-C	Peptidase C
	5S	5S RNA gene(s)
	FH	Fumarate hydratase
	Ad	Adenovirus 12—chromosome modification site-1
	GUK	Guanylate kinase
2	MDH-1	Malate dehydrogenase-1
	AcP-1	Acid phosphatase-1
	Gal-1-PT	Galactose-1-phosphate uridyltransferase
	Gal + Act	Galactose + activator
	ICD-1	Isocitrate dehydrogenase-1
	MNSs	MNSs blood group
	Sct	Sclerotylosis
	Hbα,β	Hemoglobin alpha or beta
	If-1	Interferon-1
4	*AdeB*	Adenine B
	EstAct	Esterase activator
	Hbα,β	Hemoglobin alpha or beta
	PGM-2	Phosphoglucomutase-2
5	*HexB*	Hexosaminidase B
	If-2	Interferon-2
	DTS	Diptheria toxin sensitivity
6	SOD-2	Superoxide dismutase-2 (tetrameric or mitochrondrial; indophenoloxidase B)
	PGM-3	Phosphoglucomutase-3
	ME-1	Malic enzyme-1

Table 5.1 *(continued)*

Chromosome number	Abbreviation	Full name of gene locus
6	HLA	HLA histocompatibility region (includes *A*, first segregant series; *B*, second segregant series; *C*, third segregant series; *D*, mixed lymphocyte culture; Ir, immune response; Gb or Bf, glycine-rich β glycoprotein; Ch, Chido blood group)
	Pg	Pepsinogen
	Glo	Glyoxylase
7	SV-T	SV40 T antigen
	MDH-2	Malate dehydrogenase-2 (mitochondrial)
8	GR	Glutathione reductase
9	*AK-3*	Adenylate kinase-3
	AK-1	Adenylate kinase-1
	ABO	ABO blood groups
10	GOT-1	Glutamate oxaloacetate transaminase-1
	HK-1	Hexokinase-1
11	*AcP-2*	Acid phosphatase-2
	AL	Species antigen
	ES-A4	Esterase-A4
	LDH-A	Lactate dehydrogenase A
12	LDH-B	Lactate dehydrogenase B
	Pep-B	Peptidase B
	CS	Citrate synthase, mitochondrial
	Gly^{+} *A*	Serine hydroxymethylase (glycine A auxotroph complementing)
	TPI	Triosephosphate isomerase
13	RNr	Ribosomal RNA
14	RNr	Ribosomal RNA
	NP	Nucleoside phosphorylase
15	RNr	Ribosomal RNA
	MPI	Mannosephosphate isomerase
	BMG	β2-microglobulin
	PK-3	Pyruvate kinase-3
	HexA	Hexosaminidase A and/or C
16	LCAT	Lecithin-cholesterol acyltransferase
	Hpα	Haptoglobin, alpha
	APRT	Adenine phosphoribosyltransferase

Table 5.1 *(continued)*

Chromosome number	Abbreviation	Full name of gene locus
17	TK	Thymidine kinase
	GalK	Galactokinase
	AV-12	Adenovirus-12—chromosome modification site-17
18	Pep-A	Peptidase A
19	GPI	Phosphohexose isomerase (glucose-phosphate isomerase)
	PS	Polio sensitivity
	E11S	Echo 11 sensitivity
20	*ADA*	Adenosine deaminase
	DCE	Desmosterol-to-cholesterol enzyme
21	RNr	Ribosomal RNA
	AVP	Anti-viral protein
	SOD-1	Superoxide dismutase-1 (dimeric or cytoplasmic; indophenoloxidase A)
22	RNr	Ribosomal RNA
Y	TDF	Testes development factors
X	*Xg*	Xg blood group
	PGK	Phosphoglycerate kinase
	HGPRT	Hypoxanthine-guanine phosphoribosyltransferase
	G6PD	Glucose-6-phosphate dehydrogenase
	CB	Color blindness
	HemA	Hemophilia A
	(See Fig. 2.29 for other known loci on the X chromosome.)	

SOURCE: V. A. McKusick, *Mendelian Inheritance in Man*, 4th ed., Johns Hopkins Univ. Press, 1975.

What is the origin of the differences between the various types of differentiated cells in the body, all of which carry the same basic genetic information? What is the nature of the genetic control of fundamental cellular processes, such as protein synthesis and energy metabolism? What is the nature of the change in a cell that turns it into a potential cancer cell? The *in vitro* approach to cell biology and genetics, based on hybridization and other techniques, promises to be as important for the study of higher organisms, including man, as its precursor microbial genetics has been for the development of molecular biology.

5.3 Gene Dosage and the X Chromosome: X Inactivation

The key to the understanding of the inheritance of X-linked traits such as hemophilia lies in the fact that genes on the X chromosome have no counterpart on the Y. Males (who are XY) have only one copy of their X-chromosome genes, while females (who are XX) have two copies, just as for every other chromosome. This is in marked contrast to the gross phenotypic abnormalities that are caused by an imbalance in the number of autosomal genes such as, for example, trisomy for chromosome 21 and Down's syndrome. The contrast between the effects of different dosages of the X chromosome as compared to autosomes is further emphasized by the fact that trisomy for the X chromosome (XXX) has hardly any effect at all, whereas trisomy for most other chromosomes has not even been observed in liveborn offspring and is presumably incompatible with survival, let alone normal development.

Clearly there must exist a mechanism that can compensate for the differences between males and females in the dosage of genes on the X chromosome and so can prevent the metabolic imbalance that would otherwise be expected to result from the presence of different numbers of copies of X-linked genes in males and females. This "dosage compensation" effect can actually be observed at the level of the immediate gene product. Thus, persons who are heterozygous for autosomal recessive mutations causing inborn errors of metabolism due to enzyme deficiencies (such as PKU) generally have half the enzyme activity (in this case, phenylalanine hydroxylase; see Chapter 3) of normal homozygotes. Enzymes controlled by X-linked genes, however, such as G6PD (glucose-6-phosphate dehydrogenase) generally show the same level of activity in males and females.

The explanation for dosage compensation in mammals and in man that is now generally accepted is that only one of the two X chromosomes of the female is metabolically active in any given cell at a given time.

It appears that, relatively early in development, one of the X chromosomes in each cell of the female is inactivated. Either the paternally or the maternally derived X can be inactivated, probably at random, and this inactivated state persists stably in all of the cell's descendants. This explanation for dosage compensation, often called the Lyon hypothesis, is associated with the name of Mary Lyon, who was one of its main originators in 1961.

The history of the development of this hypothesis starts with the discovery in 1949 that most nondividing nerve cells of the female cat have a small dark-staining body in their nucleus that is never present in the cells of a normal male cat. This seemingly trivial observation was later extended to many other cell types in many

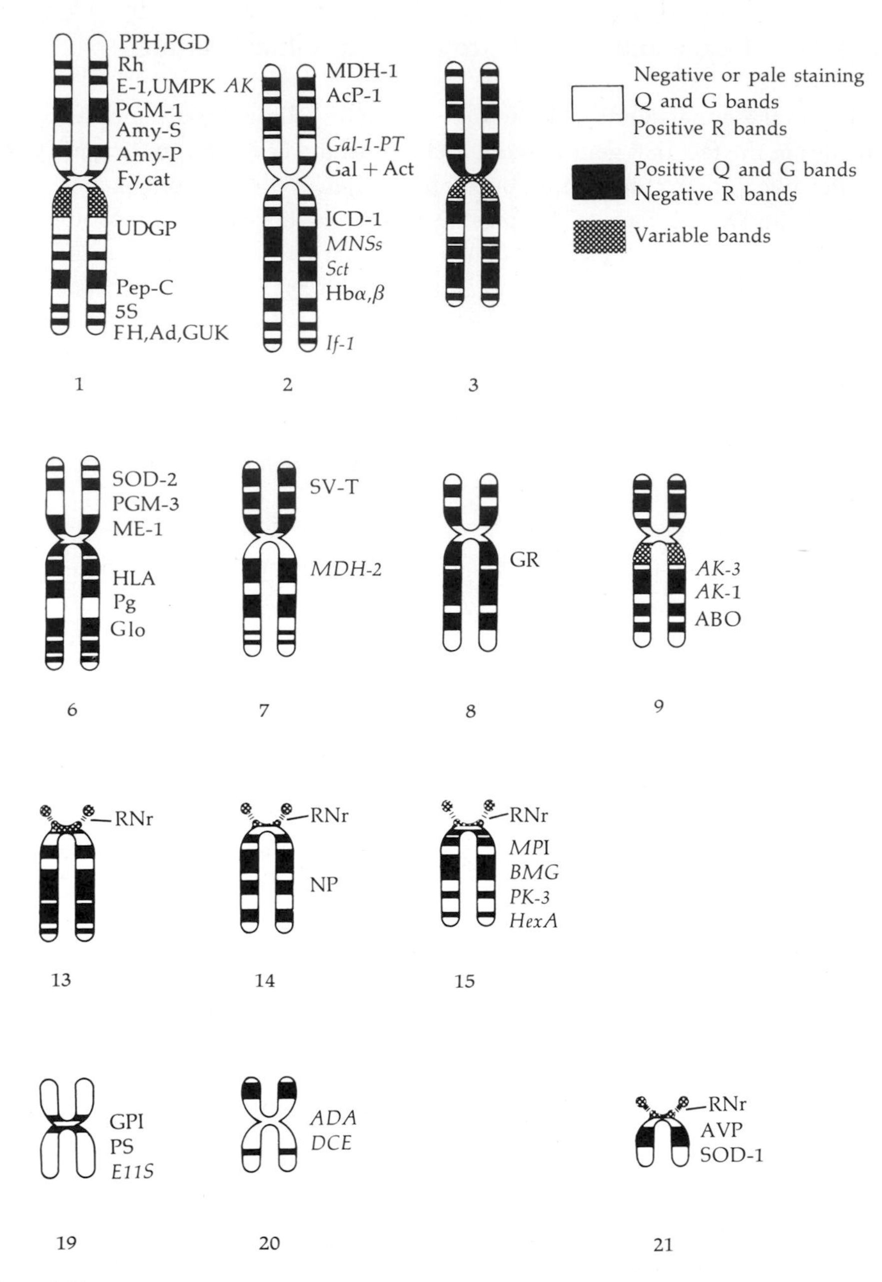

Figure 5.13
A human gene map. The diagram shows banding patterns produced by special staining techniques. Beside each chromosome are abbreviations of the names of genes that have been assigned to that chromosome; the information shown here is indicative of the state of knowledge about human gene locations as of September 1975. Italicized abbreviations indicate inconclusive location assignments. Table 5.1 is a key to the abbreviations used in this figure. Information about gene

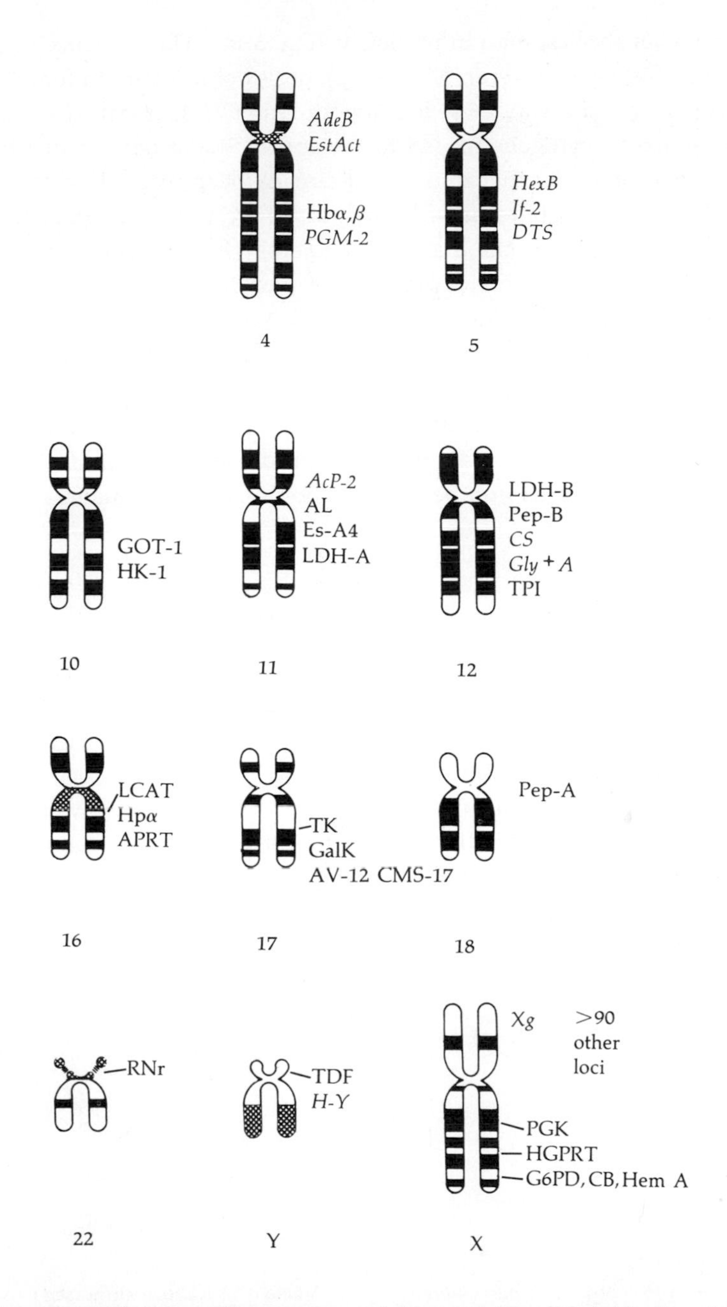

locations has come both from pedigree studies and from somatic-cell–hybridization studies. The two approaches often complement each other; linkages suggested by one method may be confirmed or extended through use of the other method. (After D. Bergsma, ed., *Human Gene Mapping,* National Foundation–March of Dimes, 1975; and V. A. McKusick, *Mendelian Inheritance in Man,* 4th ed., Johns Hopkins Univ. Press, 1975.)

other mammalian species, man in particular (Fig. 5.14). The presence of this Barr body (named after its discoverer, M. L. Barr), or sex chromatin, in female but not in male cells provided a way of sexing a population of cells before the development of mammalian cytogenetics led to the specific identification of the human X and Y chromosomes. At that time it had already been recognized that individuals with Turner's syndrome (now known to be XO), though basically female, lacked the Barr body, while the male-like Klinefelter's syndrome (now known to be XXY) was associated with the presence of one Barr body. Subsequent work showed more generally that individuals with *n* X chromosomes may have cells with up to a maximum of *n*–1 Barr bodies. Surveys for sex-chromosome abnormalities are still sometimes done by screening for Barr bodies in cells obtained from buccal smears from the inside of the mouth, as this is a much simpler procedure than karyotyping. The karyotype is then determined only for individuals who have cells with unusual numbers of Barr bodies, mainly females with none or males with one or more.

The sex chromatin or Barr body is an inert X chromosome.

Following the development of cytogenetic techniques for mammals, a careful survey of the origin of the Barr body throughout the cell cycle revealed that it is derived from one of the X chromosomes. Chromosome regions that stain heavily at some stage in the cell cycle but not at others, as does the Barr body, are called *heterochromatic* and are known in many organisms to be relatively inert from a genetic point of view. The identification of the heterochromatic Barr body with an X chromosome suggested that one of the X chromosomes in a cell with a Barr body is regularly heterochromatic and so probably inactive. This correlation was further strengthened by the observation that one X chromosome always tends to

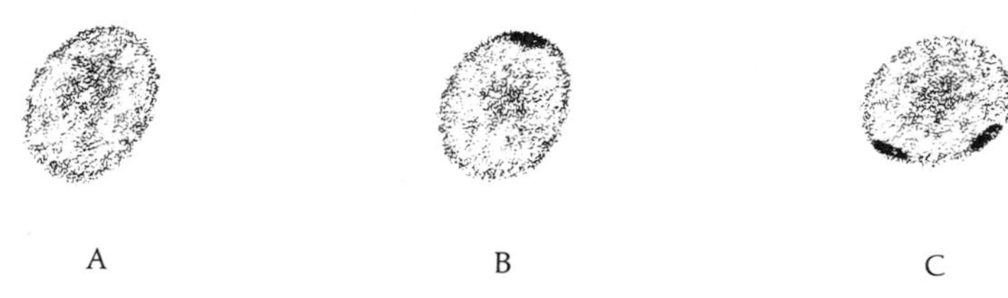

Figure 5.14
Human female cells with Barr bodies. (A) The nucleus of a cell with no sex chromatin. (B) Nucleus of a cell from a normal woman; one Barr body is seen at the periphery of the nucleus. (C) Nucleus of a cell from a triplo-X woman; two Barr bodies are present. (After photomicrographs by U. Mittwoch, *Scientific American,* July 1963.)

replicate later than the other, and this late X is the one that is destined to form the Barr body. Late replication during the cell cycle is another characteristic of heterochromatin.

These observations were united by Mary Lyon into her inactive-X hypothesis. This hypothesis states specifically that, at some stage in early development, one of the X chromosomes (chosen at random) of each cell of the female is inactivated, and that the same X remains inactive in all the descendents of any given cell. The hypothesis leads to the expectation that females heterozygous for X-linked mutants should be phenotypic mosaics—that is, mixtures of mutant and non-mutant cell lines according to whether it is the mutant or the normal X chromosome that is active in any given cell. Female mice heterozygous for X-linked mutations determining coat-color differences always look patchy. Their coat seems to be made up of normal and mutant patches, exactly as expected from the inactive-X hypothesis, and this was one of the clues that led Mary Lyon and others to their explanation for dosage compensation. The well known tortoise-shell cat (virtually always female) is an example of such patchy distribution in a female heterozygous for an X-linked coat-color mutation (Fig. 5.15).

Late replication of one X → Barr body

Figure 5.15
The tortoise-shell cat. The patches on the main part of the body and on the tail are black and brown, showing the patchy effect caused by the X-linked coat-color gene. The white areas are the result of another mutation on an autosome. (Courtesy of M. Lyon and D. Buck.)

The most convincing evidence for a single active X in any given female cell comes from studies of cultured cells obtained from females who are heterozygous for G6PD electrophoretic variants.

A relatively common electrophoretic variant of the X-linked enzyme G6PD is found in many populations of African origin. It is therefore quite easy to find females who are heterozygous *G6PD-A/B* where A and B denote the two electrophoretic forms of the enzyme (G6PD-A is the African-type variant, while G6PD-B is the type found in nearly all people of European origin). The electrophoretic patterns obtained from such heterozygous females show two bands, one for G6PD-A and the other for B (Fig. 5.16).

Using the cell-culture techniques described in Section 5.1, it is possible to grow out clones of skin cells where each clone originates from a single cell derived from the donor tissue. When such clones are obtained from a G6PD-A/B heterozygote, it is found that individually they are either G6PD-A or G6PD-B, but never both (Fig. 5.16). This is exactly as predicted from the X-inactivation hypothesis. Each clone is derived from a single cell from a heterozygous female, and in each cell only one X chromosome is active—either that carrying the *G6PD-A* allele or that carrying the *B* allele.

It is interesting to recall that, as described in Section 5.2 (Fig. 5.9), human-mouse cell hybrids that express both human and mouse G6PD give rise to a hybrid enzyme band. This is as expected from cells expressing two different electro-

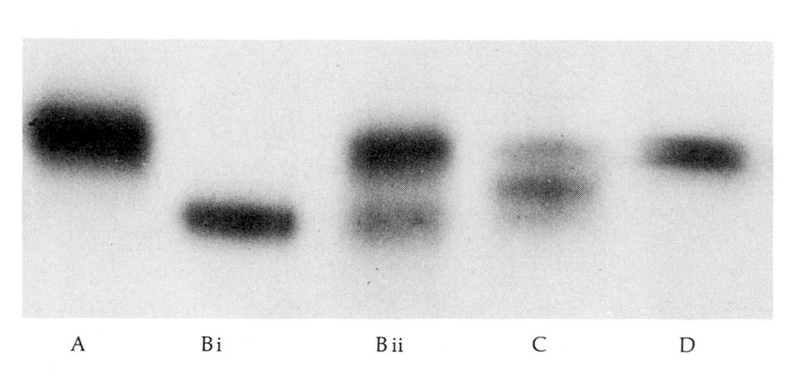

Figure 5.16
Electrophoresis of the X-linked enzyme G6PD. Extracts A and D are from a normal Caucasian, all of whose enzyme is G6PD-B. Extract Bi contains only the Negroid variant G6PD-A. Bii is an extract from a G6PD-A/B female heterozygote, showing a mixture of G6PD-A and -B enzymes, with no intermediate band. Extract C is from a hybrid cell, made by crossing G6PD-A and -B cells; extract C shows very clearly an intermediate heteropolymeric band of enzyme activity (as expected if both alleles are expressed in a single cell). The intermediate band does not appear in extract Bii because of X inactivation; each of the cells of the female heterozygote expresses either the A or the B allele, but never both. (Courtesy of S. Povey.)

phoretic forms of an enzyme that is made up of two subunits. Thus, individual cells that express both G6PD-A and G6PD-B should show an intermediate hybrid enzyme band. The fact that no such band is normally seen in extracts of cells from skin, blood, and other tissues taken from *G6PD-A/B* heterozygotes is thus by itself evidence that any given cell only makes one or the other form of the enzyme. Cell hybrids made by fusing G6PD-A and B fibroblasts do show such an intermediate enzyme band on electrophoresis, as expected if both the parentally derived X chromosomes remain active in the artificially produced cell hybrids (Fig. 5.16).

This result shows, incidentally, that the active and inactive states of an X chromosome are not affected by fusing with another cell. The single G6PD type for the clone from the G6PD-A/B heterozygote also, of course, shows the stability of the X-chromosome state through many divisions in cell culture. In spite of much effort, there is so far no evidence for the state of an X chromosome ever having been changed, at least in cell culture. X inactivation really seems to be an almost irreversible process. At least four other X-linked markers detectable in cell culture (including the HGPRT deficiency that causes the Lesch–Nyhan syndrome) have been shown to behave in the same way as G6PD in heterozygous females and so provide further direct evidence for X inactivation.

It seems likely that the majority of loci on the X chromosome are subject to inactivation.

There is one notable exception, however, and this is the X-linked blood group Xg, which does not appear to be inactivated, at least in cells from the blood and skin. Further evidence for incompleteness of X inactivation in man comes from the phenotypes of individuals with abnormal numbers of human chromosomes. Though XXX is a more-or-less normal female, individuals who are XO (Turner's syndrome) show a number of marked abnormalities and are invariably infertile. This suggests, for example, that there is some time during development when the overall activity of both X chromosomes is normally required, or that there is a region of the X that is not normally inactivated and that is therefore required in double dosage for normal development. Perhaps the Xg blood-group locus is located in such a region of the X chromosome.

Another exception to uniform X inactivation is the fact that female germ cells appear to have both Xs active. The most telling evidence for this is the fact that developing eggs from heterozygous G6PD-A/B females show the intermediate electrophoretic band (see Fig. 5.16) expected only when both the *G6PD-A* and *B* alleles are active in the same cell.

At what time during development does X inactivation occur?

Because in any given female cell it can be either the paternally or the maternally derived X chromosome that is inactivated, it would seem that there must be some stage at which both X chromosomes are active, even if only in the newly formed zygote. Suppose for the sake of argument that X inactivation occurred when there were only two cells, and let X_P and X_M stand for the paternally and maternally derived X chromosomes that formed the zygote. If inactivation really occurs at random, the chance that X_P is inactivated in one of the cells is one-half, and similarly for X_M. Thus the chance that, say, X_P is inactivated in both of the cells present at the time inactivation occurs is $\frac{1}{2} \times \frac{1}{2} = \frac{1}{4}$. This would mean that one-fourth of females heterozygous for an X-linked recessive mutation, such as that causing hemophilia, would behave as if they were homozygous because neither of the two cells in which inactivation occurred had ended up with normal X chromosomes being active. In this case, hemophilia would be nearly as common in women as it is in men, whereas in fact the frequency in females is at most 1 percent of what it is in males. These calculations suggest at least that X inactivation occurs when the developing embryo contains more than two cells.

It is perhaps surprising that a small piece of skin can give rise in culture to both types of cells—namely, those with a paternal and those with a maternal X active. On the assumption that inactivation takes place quite early, as seems to be the case, one might have expected there to be relatively large patches which would be all of one or all of the other type, as appeared to be the case in the tortoise-shell cat. Apparently, however, skin cells must migrate and so intermingle sufficiently after the time of inactivation to obscure any patches. Patch size is therefore not a good guide to the possible time when inactivation takes place.

Figure 5.17
Autoradiography of skin fibroblasts from an HGPRT⁺/HGPRT⁻ female. Black granules, indicating incorporation of tritiated hypoxanthine into nucleic acid, are seen over some but not all cells; this result is to be expected for X-linked genes in females. (The cells able to utilize the tritiated hypoxanthine, like those on the left of this figure, are $HGPRT^+$.) One of the female's X chromosomes is inactivated in each cell during embryonic growth. Inactivation is a random process and leads to mosaicism with respect to X-linked genes. (After W. Y. Fujimoto et al., *Lancet*, vol. 2, p. 512, 1968.)

There is another complication to these approaches to estimating the time of inactivation, and this is cell selection. Cells whose activated X chromosome carries a recessive mutation may not be able to survive as well as cells carrying a normal active X, and so may be selected against in the course of development. This has been observed very clearly in females who are heterozygous for the HGPRT deficiency that is the cause of the Lesch–Nyhan syndrome discussed in the previous section. Cells carrying normal and mutant X chromosomes can be distinguished by comparatively simple biochemical tests. Though skin cells from heterozygous females produce both $HGPRT^+$ and $HGPRT^-$ clones (Fig. 5.17), only $HGPRT^+$ cells can be found in the blood.

Further data on the timing of X inactivation have been obtained using a very elegant and delicate technique in the mouse. Single cells from one strain of mice can be injected into the developing embryo of another strain (Fig. 5.18), and the treated embryo can then be allowed to continue normal development to birth and adulthood. The resulting mouse is a chimera—that is, a mixture of cells of the two strains, donor and recipient. By injecting cells taken from appropriate heterozygous mice at different times, it can be shown that inactivation takes place some three to four days after fertilization, when the embryo consists of between 50 and 100 cells.

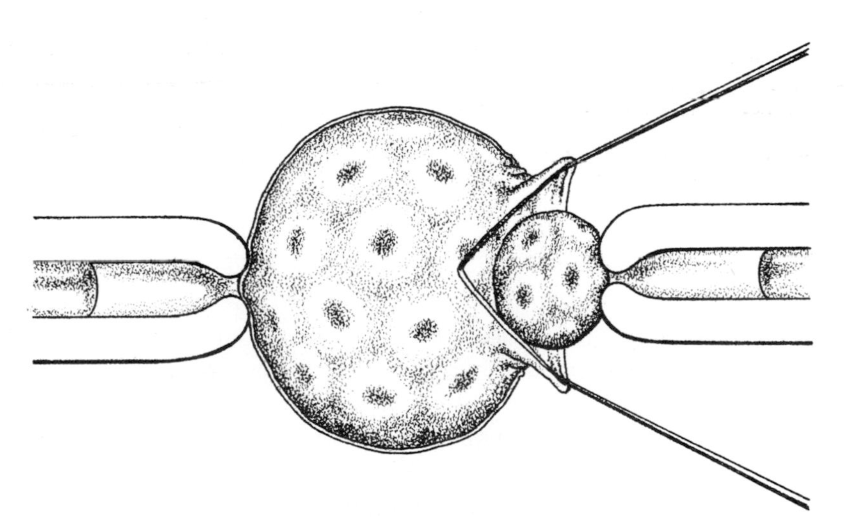

Figure 5.18
Injection of cells into blastocyst. Cells from one embryo may be transferred into another with this method. The blastocyst is held in place by suction on the tip of one smooth glass pipette (left) and by the donor inner-cell mass (right). A triangular hole is made in the wall of the host blastocyst, and the donor cell mass is introduced through this hole. Single cells may also be transferred with this method. (From R. L. Gardner, in *Embryonic and Fetal Development,* ed. C. R. Austin and R. L. Short, Cambridge Univ. Press, 1972.)

What might be the mechanism by which X chromosomes are activated or inactivated?

So far nothing is known about the mechanism, though there are many suggestions. One of the most extraordinary features of X inactivation is its stability. As already mentioned, no case of reactivation of the X chromosome has yet been documented. This stability sets X inactivation apart from the kinds of gene-control systems that have been studied so extensively in bacteria (Chapter 3). The stability suggests that inactivation perhaps occurs in some way at the level of the DNA, so that the inactivated state of the chromosome is passed on to daughter chromosomes along with the DNA, as an inherited property.

The phenomenon of X inactivation is a fascinating aspect of development about which much still remains to be discovered. It has also led to one important result in the field of cancer research. The question has often been asked: does a given cancer usually originate from just a single cell? In other words, are cancers clonally derived, perhaps from a cell that was, in some sense, a mutation? S. Gartler and his colleagues provided a simple answer to this question by studying the G6PD types of tumors obtained from heterozygous G6PD-A/B females. They found that nearly all the tumors they studied were either G6PD-A or G6PD-B, but not both, just like the clones obtained from skin cells in culture. Thus, they concluded that the origin of cancers was indeed clonal, at least for the cases they had studied, and perhaps quite generally.

5.4 Genes and Antibodies

When the body comes into contact with a foreign substance—which may be an infecting virus or a bacterium, a chemical taken in as food or drug, or even someone else's blood cells received in the course of a blood transfusion—it can often respond by manufacturing a special type of protein called an *antibody*. The main property of an antibody produced in response to such contact with a foreign substance is its capacity to bind specifically to that substance, usually in this context called the *antigen*. Thus an antibody is said to react or bind specifically to its corresponding antigen (Fig. 5.19). The range of antigens to which the body can respond in this way is in the thousands, if not the tens or even hundreds of thousands. This ability to respond quite specifically to such a large number of foreign substances, which is very widely spread throughout the animal kingdom, is one of the most remarkable and most widely studied biological phenomena.

The combination of an antibody with an invading organism such as a virus or bacterium enhances the body's ability to counteract an infection.

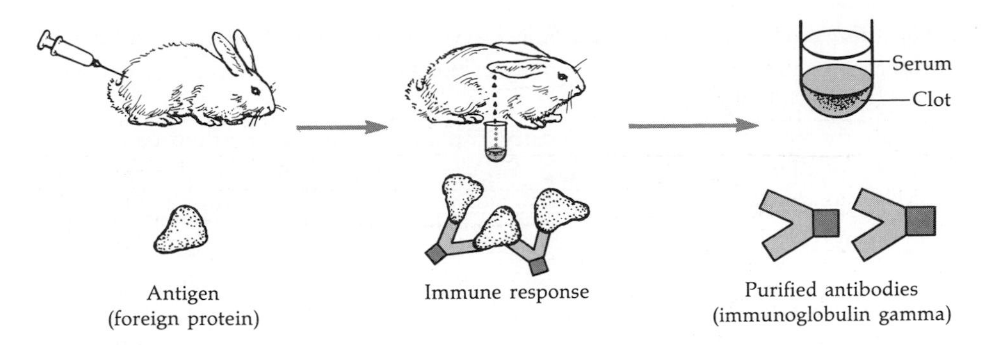

Figure 5.19
Antibody production. Antibodies are produced in response to stimulation by antigens (proteins foreign to the organism). The animal makes antibodies (primarily immunoglobulin gamma, or IgG) that bind specifically to the antigen. Antibodies are isolated from the serum fraction by first bleeding the animal and allowing the blood to coagulate. (From R. R. Porter, "The structure of antibodies," *Scientific American,* October 1967. Copyright © by Scientific American, Inc. All rights reserved.)

The antibody response is thus a most important mechanism for protection against infectious diseases, and this is undoubtedly one of its major functions and also perhaps the major reason for its evolution. Not all antibody responses, however, are beneficial. Excessive antibody responses to, for example, certain drugs or to pollen are causes of allergy and asthma. Certain severe diseases of the newborn are caused by a response of the mother to cells from her fetus, which escape across the placenta into the mother's blood. A number of chronic diseases, possibly even including some forms of rheumatism and also multiple sclerosis, are caused by abnormal reactions of the body to its own components, leading to the production of so-called auto-antibodies. Antibody response to mismatched blood given for transfusion can be fatal, and similar responses to transplanted kidneys and other organs are the reasons for their rejection.

Antibodies, because of their high degree of specificity, are often used as reagents for identifying particular substances, notably those determining the blood groups that have been so widely used in human genetics.

For all these reasons, it is important to understand the nature of the antibody response and to learn how to control it. The main aim of this section is to review briefly the nature of the antibody response and the antibody molecule itself, and also the uses of antibodies, especially for the identification of blood-group differences. The reasons for placing an emphasis on a discussion of genes and antibodies are, first, the importance of the phenomenon of antibody response and its basic interest as a fascinating biological system and, second, the importance of antibodies as a means of detecting many of the genetic differences that we discuss and use in following chapters.

Antibody molecules are found mostly in plasma, which is the fluid component of the blood. This can be obtained by preventing coagulation, usually by adding suitable chemicals and then removing the blood cells from a blood sample by centrifugation. More usually one allows the blood to clot and then removes the fluid, called *serum,* that can be separated from the blood clot. Serum (or plasma) contains many different sorts of proteins, including antibody molecules. All tests for antibodies depend in one way or another on their specific binding ability and usually make use simply of the whole serum containing the antibody, without any attempt to purify the antibody molecules from the serum. The specificity of an antibody–antigen reaction is such that other substances present in the serum mostly do not interfere with the reaction.

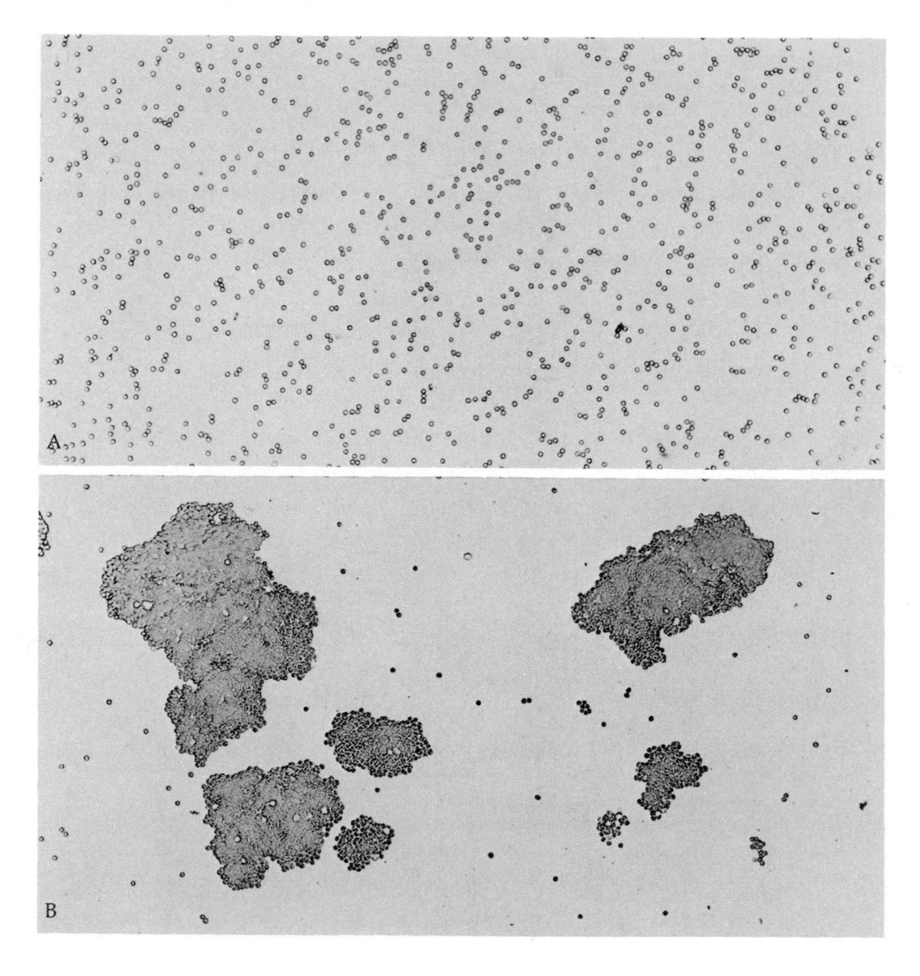

Perhaps the simplest reaction detected by an antibody is that called agglutination.

This is just the formation of clumps of cells held together by antibody molecules attached to antigens on the cell surface. This is the reaction used to detect the well-known ABO blood types and most of the other red-cell blood-group differences. The ABO blood types, which were the first discovered blood-group differences (described by Karl Landsteiner in 1900) form the basis for matching blood donors to their recipients for transfusion.

There are four common genetically determined ABO types—A, B, AB, and O—which are distinguished both by the antigens they have on their red blood cells and by the antibodies in their sera. Thus, for example, an antibody specific for type A, often called anti-A, will clump or agglutinate red cells of type A and AB but not those of type B or O, which lack antigen A. Similarly, an anti-B antiserum specific for type B will agglutinate B and AB red cells, but not O or A (Fig. 5.20). These patterns of reaction define the four ABO types.

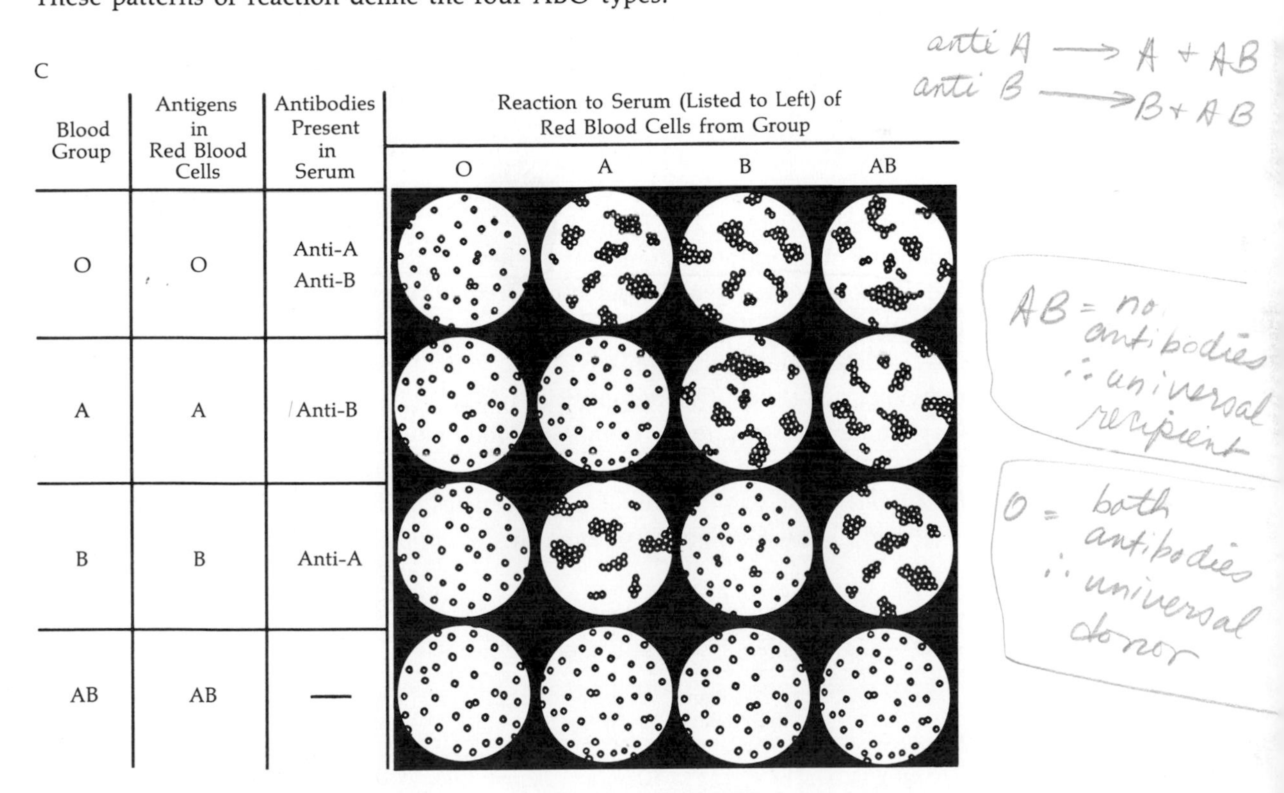

Figure 5.20
ABO agglutination patterns. (A) Nonagglutinated red blood cells. (B) Agglutinated cells. (C) Diagrammatic representation of reactions of red blood cells of O, A, B, and AB individuals to anti-A and anti-B antibodies. (Parts A and B courtesy of Dr. C. L. Conley. Part C after G. Hardin, *Biology. Its Human Implications,* W. H. Freeman and Company. Copyright © 1949.)

A peculiarity of the ABO system is that an individual has (naturally occurring in his serum) antibodies corresponding to the antigens other than those on his red cells. Thus, for example, a type-A individual has anti-B in his serum, while a type-O individual has both anti-A and anti-B. It is these naturally occurring antibodies that cause trouble if blood is not matched for transfusion. If, for instance, an O individual were to receive type-A blood, the O recipient's anti-A would combine with the incoming A donor red cells, eventually leading to the destruction of the donor cells. Byproducts of this destruction are responsible for the often fatal reaction that may follow transfusion of unmatched blood. The naturally occurring anti-A and anti-B antibodies are thought to be formed very early in life in response to A and B antigens, which are found on *E. coli,* a major constituent of the bacterial flora of the gut, on other bacteria, and in many foodstuffs.

The inheritance of the ABO blood types is controlled by three alleles—*A, B* and *O*—at a single locus, following the scheme given in Table 5.2. Thus allele *A* determines the presence of antigen A, and allele *B* determines the presence of antigen B, while the third allele, *O*, is "nul" and determines neither A nor B. (Note that italics, such as *A*, are used to designate the symbol for a gene corresponding to an antigen, such as A.) If an individual has a particular allele, say *A,* he always has the corresponding antigen A, whatever may be the second allele. Thus, type AB has both alleles *A* and *B* and so is heterozygous *AB* at the ABO locus. This independent behavior of dominant alleles, which is a characteristic of nearly all the blood-group antigens, is called codominance, as we have already mentioned. Individuals of type A can be of either genotype *AO* or *AA*, because the allele *O* has no detectable effect. Similarly, B individuals can be either *BO* or *BB*, but O individuals must be *OO*. Thus, only individuals of type O and AB have genotypes that can be inferred directly from their blood phenotypes. The frequencies of the four ABO types in a typical European population are also given in Table 5.2. All four types occur with appreciable frequencies, indicating that the three alleles, *A, B* and *O,* must be quite common in the population.

Table 5.2
ABO blood types, antigens, antibodies, and genotypes

Blood type	Typical frequency in European population	Genotype*	Antibodies in serum	Reaction† to Anti-A	Reaction† to Anti-B
O	45%	*OO*	anti-A & anti-B	−	−
A	41%	*AO* or *AA*	anti-B	+	−
B	10%	*BO* or *BB*	anti-A	−	+
AB	4%	*AB*	neither	+	+

**A, B,* and *O* are alleles at the ABO locus.
†No reaction = −; clumping or agglutination = +.

There are many other techniques (in addition to agglutination) for detecting antibody-antigen reactions.

For example, an antibody to a soluble antigen—say, another protein—can form a precipitate with the antigen bringing it out of solution (Fig. 5.21). Precipitation is a form of clumping between antigen and antibody at the molecular level that is analogous to agglutination at the cell level. Sometimes, the attachment of an antibody to an antigen on a cell surface can be detected by the fact that under appropriate conditions—namely in the presence of a series of proteins from fresh serum, collectively called complement—the antibody causes the cell carrying the corresponding antigen to burst open and be destroyed. Cells that do not have attached antibodies on their surface remain unaffected by the complement's activities. An antigen on the cell surface is thus detected by killing the particular type of cell that carries the antigen.

The amount of antibody activity present in a given serum is measured by the weakest dilution at which there is still a detectable reaction with the specific antigen. This dilution is called the serum's titer. Usually a serum is tested systematically in a series of one-in-two or one-in-ten dilutions and the weakest dilution still reacting is noted.

A key feature of the immune system, as the machinery responsible for an antibody response is called, is that it "remembers" previous contacts with an antigen.

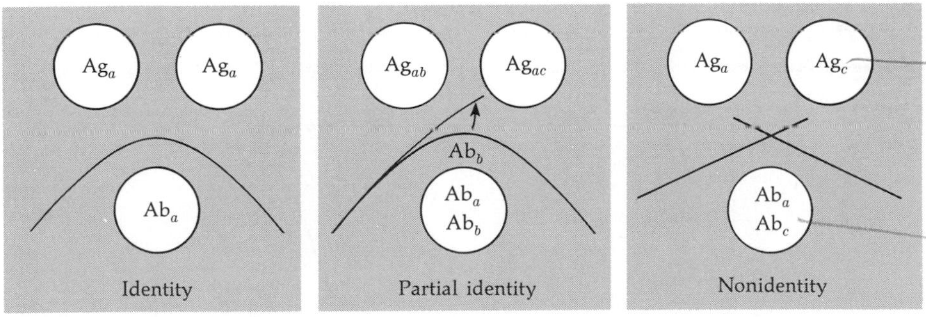

Figure 5.21
Immune precipitation in Ouchterlony plates. An Ouchterlony plate is a glass plate with a shallow central well and a number of similar wells arranged around it; in these diagrams, only the central well (at bottom) and two of the surrounding wells are shown. Antiserum is placed in the central well, and various antigenic proteins are placed in the surrounding wells. The proteins and antibodies diffuse toward each other; when they meet and react with each other, the antibody-antigen complex precipitates, forming a visible line across the plate. A preparation containing several antigens will yield multiple lines. The immunological relationship between the proteins can be assessed by observing the lines of precipitation. Completely confluent lines indicate immunological identity; a "spur" indicates partially related antigens; crossing lines indicate unrelated antigens. (After I. Roitt, *Essential Immunology*, Oxford: Blackwell, 1971.)

Thus, if an individual comes into contact with an antigen for a second time, even after a considerable interval, his response to the antigen is usually much more rapid and the resulting antibody titer appreciably higher. This is often true even when the second stimulus is given after all traces of antibody formed following the first stimulus have disappeared (Fig. 5.22). The heightened secondary response to an antigen shows that an individual's immune system "remembers" previous contact with the antigen. This is the basis for the efficacy of vaccination, which is the equivalent of the first contact with antigen, challenge with the disease itself generally forming the secondary stimulus. Immunization generally refers to the production of antibody in response to an infectious agent or its products, which results in an immunity to further infection. Much of this immunity depends on the fact that the immune system can "remember" previous contact with an antigen.

Another important feature of the immune system is that it incorporates the ability to recognize self.

This follows from the fact that, in general, antibodies are not made against substances naturally found in the body of the responding individual. Antigens will usually elicit an antibody response only if they are foreign—which means they are not normally found within the responding individual. Occasional breakdown of this recognition of self leads to the production of what are called auto-antibodies—namely antibodies against self components—and this, as mentioned earlier, can be a cause of serious diseases.

Any explanation for the mechanism of the immune system must at least account for its three major features, namely, (1) the extraordinary specificity of the re-

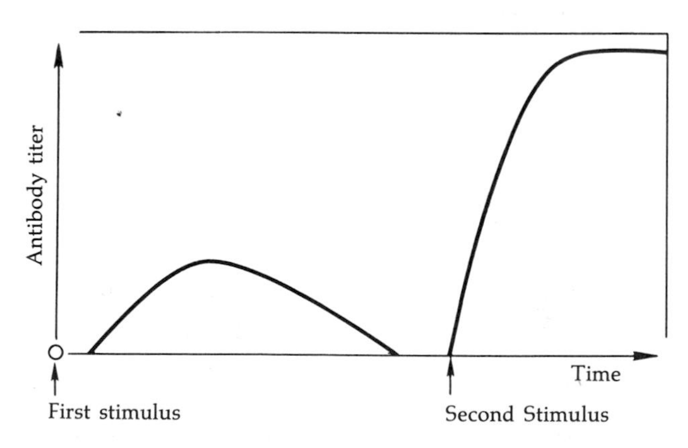

Figure 5.22
Schematic illustration of the time course of the primary and secondary responses to an antigenic stimulus.

sponse to an enormous, almost unlimited, variety of antigens; (2) the ability to remember very specifically previous contact with an antigen; and (3) the ability to distinguish between self and foreign antigens. Early discussion on the mechanism centered on the question of whether the antigen eliciting a response molds the antibody to fit it (instructional hypothesis), or whether the antigen picks out a closely fitting antibody molecule already being made by the responding individual (selective hypothesis). The knowledge that the antibody molecule is a protein and that therefore its shape is determined by its amino-acid sequence, which in turn is determined by the corresponding nucleotide sequence in the genetic material, now makes the instructional hypothesis untenable. Once a protein has been made, there is no known mechanism by which the protein can have its shape altered to fit any other antigen. This implies that each antibody corresponding to a different antigen must have, at least to some extent, a different amino-acid sequence to give it the capacity to bind specifically to its antigen. The selective hypothesis for antibody formation thus entails the production by each individual of a very large variety of different antibody molecules with correspondingly different amino-acid sequences.

Each of these sequences must, in turn, have its corresponding genetic information encoded in the DNA nucleotide sequence. There are two contrasting, still-unresolved hypotheses for the origin of the large amount of genetic information that must exist for making the antibody molecules (Fig. 5.23). One (the germ-line hypothesis) states that all the relevant genes are present in the fertilized zygote just like any other genes. In other words, there is a very large class of related genes in the genome, all there to make antibodies. The second (somatic hypothesis) supposes that the number of antibody genes is quite small but that these mutate in some way during development to give rise, in different somatic antibody-producing cells, to the large number of different sequences needed to account for the antibody response. Thus, the question that is not yet answered is whether the diversity of genetic information needed to code for this enormous variety of antibody molecules is included in the normal way in the genome, or is somehow generated by mutation during development.

Though this basic question of the source of antibody diversity is not yet finally resolved, many other aspects of the immune system have been elucidated in recent years. These developments have followed two main lines. One has been the discovery of which cells make antibodies and a clarification of the cellular aspects of the immune response. The other has been a detailed working out of the chemical structure of the antibody molecule and so of the basis for its specificity.

The clonal selection theory for antibody formation, first clearly outlined by F. M. Burnet in 1957, proposes that the cells that make antibodies are each committed to synthesizing just a single type of antibody molecule.

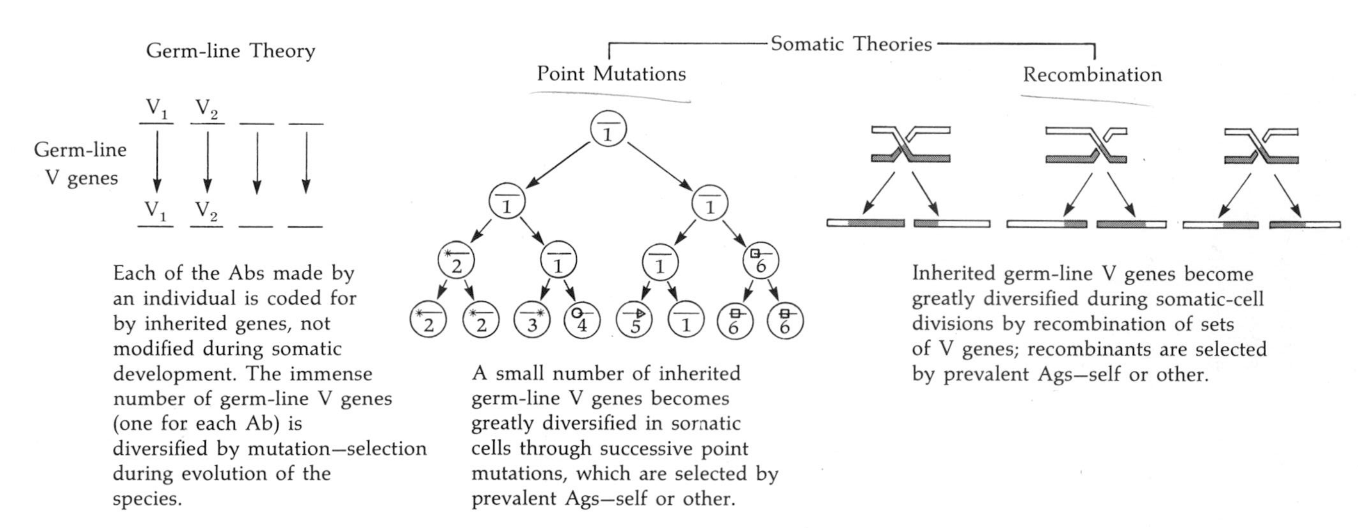

Figure 5.23
Origins of anitbody diversity. V genes code for the variable regions of the antibody molecule. (Abs = antibodies, Ags = antigens.) According to the germ-line theories, all of the possible V genes are present in the DNA and are transmitted through the germ line. The somatic theories are of two types: (1) mutation theories, which suggest that diversity of the V genes arises through point mutations occurring during duplication of a small number of inherited V genes; and (2) recombination theories, which suggest that V genes undergo recombination during mitosis to yield a large number of diverse antibodies. (After H. N. Eisen, *Immunology,* Harper & Row, 1974. Based on G. M. Edelman in *Neurosciences: Second Study Program,* ed. F. O. Schmitt, Rockefeller University Press, 1970.)

When such a cell comes into contact with antigen, it is stimulated to divide and synthesize large amounts of its single antibody, thus producing a clone of cells that are making antibodies specific to the antigen that elicited the response. The memory of the immune system resides in the cells that form such a clone. Next time an individual comes into contact with the same antigen, there is a larger number of cells making the specific antibody already present and able to respond once again, by division and antibody synthesis, to the antigen. The problem of recognition of self versus nonself is solved by having a system throughout development that kills those cells that produce an antibody to a self-antigen. Experimental evidence for this theory had to await the demonstration that the antibody-producing cell is actually the lymphocyte, one of the major types of white cell found in the blood.

The two major features of the clonal selection theory—namely production of only one type of antibody by a given lymphocyte and its clonal derivatives, and stimulation of lymphocyte division by antigens—are now well established. There is also evidence to support the idea that lymphocytes producing antibodies to self components are eliminated, though the mechanism by which this occurs is not yet known. However, as might have been expected, the details of the cellular steps involved in an antibody response turn out to be quite complex. One important discovery was the fact that there are two types of lymphocytes, called B and T, which cooperate in producing the immune response. B lymphocytes are the main antibody-producing cells. T lymphocytes direct the action of cells toward antigens, presumably via antibody-like molecules on their surface.

Antibody molecules were first identified by electrophoresis as a class of serum proteins called gamma globulins.

"Gamma" simply refers to their relative mobility on electrophoresis, and "globulin" is a general term for a protein with a basically globular shape. Subsequently, with the identification of several different types of antibody molecule, these have come to be known collectively as immunoglobulins. The first basic studies on the structure of gamma globulin molecules, by R. R. Porter and G. M. Edelman about 15 years ago, established that they are made up of two types of polypeptide chains, heavy (H) and light (L) chains. The most common type of immunoglobulin called IgG (for immunoglobulin G), which was the first to be studied, consists of two L chains and two H chains held together like the alpha and beta chains of hemoglobin molecules. It is now known that there are four other classes of immunoglobulin, in addition to IgG, that are called IgM, IgA, IgD, and IgE (Table 5.3).

These different classes of antibodies take part at different stages of the immune response or serve different immunological functions. Thus, IgM tends to be the

Table 5.3
Physical properties of major human immunoglobulin classes

Class designation	IgG (γG)	IgA (γA)	IgM (γM)	IgD (γD)	IgE (γE)
Molecular weight	150,000	160,000 and polymers	900,000	185,000	200,000
Number of basic peptide units	1	1 or 2	5	1	1
Valency for antigen binding	2	2, (? polymers)	5(10)	?	2
Concentration range in normal serum (mg/ml)	8–16	1.4–4	0.5–2	0–0.4	0.1–0.7
Percentage of total immunoglobulin	80	13	6	1	0.002
Major characteristics	NOTE 1	NOTE 2	NOTE 3	?	NOTE 4

NOTE 1: Most abundant Ig of internal body fluids, particularly extravascular, where combats microorganisms and their toxins.
NOTE 2: Major Ig in sero-mucous secretions, where it defends external body surfaces.
NOTE 3: Very effective agglutinator; produced early in immune response—effective first-line defence vs. bacteraemia.
NOTE 4: Raised in parasitic infections; responsible for symptoms of atopic allergy.
SOURCE: Adapted from I. Roitt, *Essential Immunology*, Oxford: Blackwell, 1971.

first type of antibody to be produced following an antigenic stimulus, while IgG is produced mainly as part of the secondary response to an antigen. IgA antibodies are found in secretions such as those in the gut, while IgE antibodies seem to remain attached to cells and are the class of antibodies mainly responsible for allergies. The particular function of IgD is not yet known. These various antibody classes are distinguished by their H chains, each having a unique type of H chain. The L chains, which can be of two types called kappa and lambda, are shared among all the classes of immunoglobulin.

The structural analysis of immunoglobulin was greatly helped by the existence of tumors of antibody-forming cells that are called myelomas.

People suffering from this form of cancer have in their serum a very high concentration of a *single* immunoglobulin, which is presumably the product of the myeloma, as might be expected if the tumor is derived from a single antibody-forming cell. Thus, on the clonal selection theory this original cell from which the tumor is derived is committed to making just one type of antibody, and this is the immunoglobulin found in high concentration in patients who have this type of cancer. In animals, especially mice, myelomas can be kept growing as tumors by passing them from one animal to another, or sometimes they can be maintained as cell lines in tissue culture. The importance of the myeloma proteins is that they are pure antibody molecules whose amino-acid sequence can be determined. They have, as a result, been intensively studied, and they provided the basis for much of the recently accumulated knowledge concerning the more detailed structure of antibody molecules.

Analysis of the amino-acid sequences of a number of myeloma-protein L chains has shown that they consist of a constant (C) and a variable (V) portion. The C regions all have exactly the same amino-acid sequence, with one exception. This exception is correlated directly with an inherited difference between individuals, associated with the kappa L chain. The V regions, on the other hand, show some differences for each myeloma protein that has been studied. These presumably are the differences that give the antibody molecule its specificity. Subsequent studies of H chains have shown that they too have constant and variable amino-acid-sequence regions. The variable regions of the H and L chains are aligned in

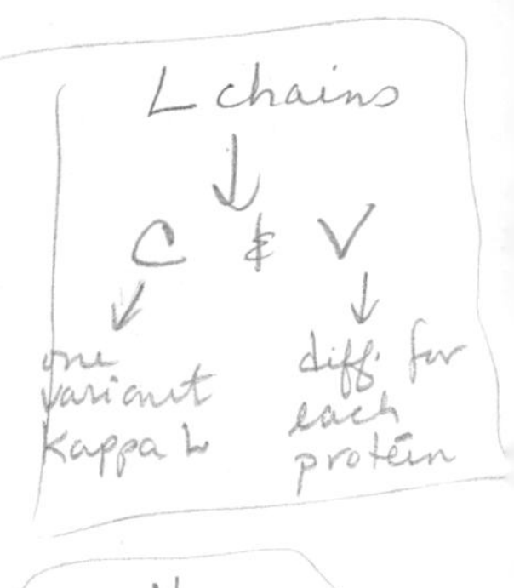

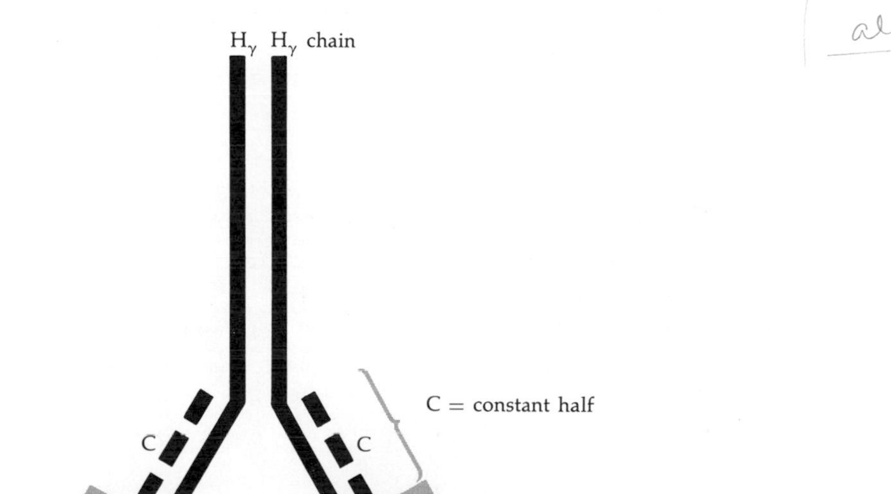

Figure 5.24
Schematic structure of γG antibody molecules. γG antibodies are identical dimers. Each monomeric component is composed of a heavy (H) chain and a light (L) chain. Each chain has a variable (V) region and a constant (C) region. Some variation may occur in C regions, but most variation occur in specific V regions called hypervariable regions. It is the changes in the V regions that alter the antigenic specificity of the antibody.

the complete IgG molecule, as shown diagramatically in Figure 5.24 to form a pair of variable ends of the molecule that are the sites where the antigen binds specifically to the antibody. Thus an antibody's specificity resides in the particular unique sequence of amino acid in its variable region, which presumably provides a shape for the molecule that can fit its corresponding antigen (Fig. 5.25). As in the case of the L chains, there are inherited differences on the constant parts of the H chains of the IgG molecule. These differences, which constitute the so-called Gm system, are themselves identified by immunological techniques.

The genes for the L and H chains seem not to be linked to each other. The genes for the different types of H chains, on the other hand, are closely linked if not adjacent, just like the β, γ, δ, and ϵ hemoglobin-chain genes (see Chapter 3). Though the constant and variable regions of the L and H chains are joined together to form a single polypeptide chain, there is good evidence to suggest that these two portions actually correspond to different genes. It is not yet clear whether the joining up of these two gene products occurs at the DNA level in the antibody-producing cell and presumably not in other cells, or at some later stage.

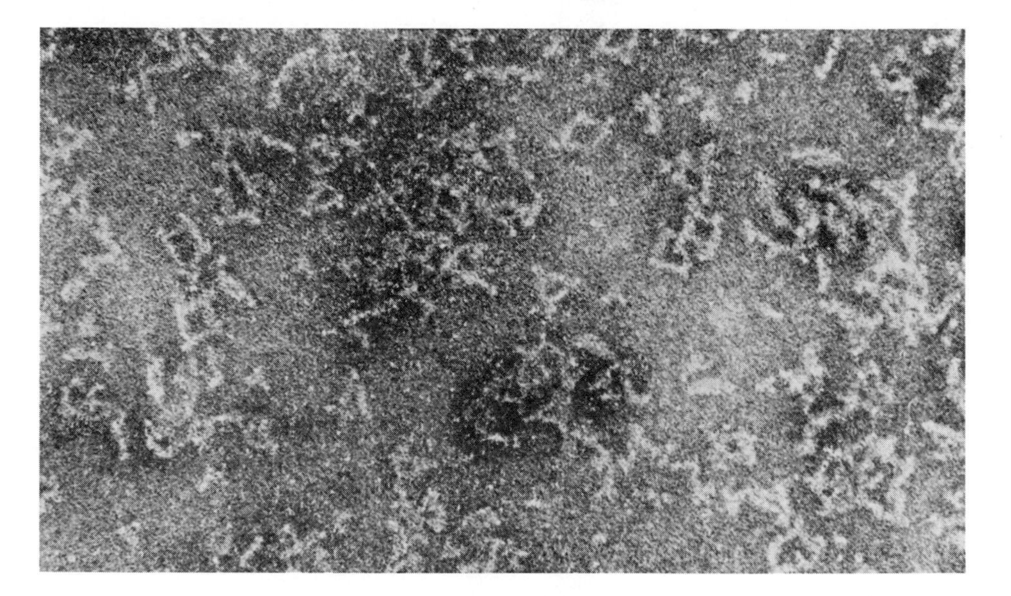

Figure 5.25
Electronmicrograph of antibody–antigen complexes. The specific shape of the antibody molecule, determined by the amino-acid sequence in its variable regions, fits with the shape of the antigen molecule to form a closely knit antibody–antigen complex. The formation of the complex inactivates the foreign protein (antigen), preventing damage to the organism. The antibody molecules are visible in this photograph; the antigen molecules are too small to be visible here. Each complex is made up of three or four Y-shaped antibody molecules, joined together at the arms of the Ys by antigen molecules. (Photo by M. Green and R. Valentine, National Institute for Medical Research, London.)

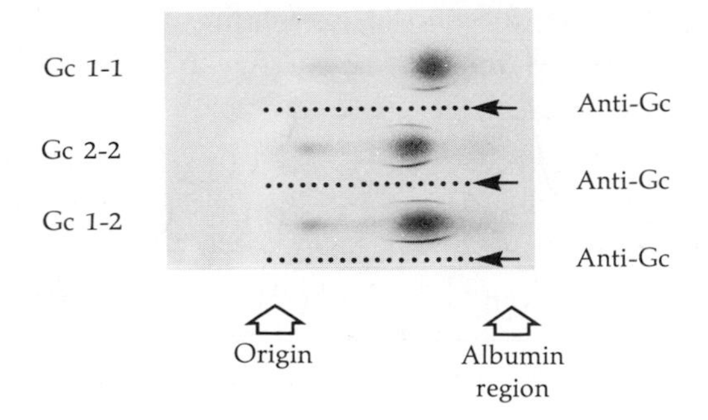

Figure 5.26
An experiment combining radioactive labeling, immunoelectrophoresis, and autoradiography. Gc (group-specific component) proteins in human plasma bind and transport vitamin D. There is an electrophoretic polymorphism for these proteins, coded for by two alleles, Gc^1 and Gc^2, which are found in all human populations. Possible phenotypes are Gc 1-1, Gc 1-2, and Gc 2-2.

If radioactive vitamin D (labeled with carbon-14) is added *in vitro* to plasma, it binds to and thus "labels" the Gc proteins. To distinguish between the two alleles, immunoelectrophoresis (IEP) is then conducted. (1) Small quantities of radioactive plasma from each individual being tested are inserted into cylindrical wells (labeled "origin" in the figure) in a horizontal agar gel. (2) An electric potential is placed across the agar gel and, over a period of several hours, proteins move to distinct positions determined by their electrical charges. In this specific experiment, Gc-1 proteins move slightly faster than Gc-2 proteins. (3) The voltage is turned off, and antiserum specific to human Gc is loaded in wells cut parallel to the direction of travel of proteins in the electric field (arrow in figure). (4) Antibodies and Gc protein diffuse toward each other and form precipitation arcs.

Following IEP, the agar gel (on a glass slide) is dried in warm air and is placed face-down on X-ray film; this process is called autoradiography. The developed X-ray film, as shown in the figure, then reveals all radioactive areas in the dried gel. This particular experiment demonstrates (1) that antibodies to human Gc precipitate a protein that has vitamin D associated with it (radioactive arcs), and (2) that labeled proteins from individuals of known Gc phenotypes 1-1 and 2-2 have different electrophoretic mobility. Plasma from Gc 1-2 individuals produces long arcs combining both proteins. (From S. Daiger et al., *Proceedings of the National Academy of Sciences*, vol. 72, pp. 2076–2080, 1975.)

The answers to these questions should resolve the fundamental issue as to whether the germ-line or the somatic theory is the correct explanation for antibody diversity.

In the first case, for the germ-line theory, there would have to exist a large number of genes corresponding to all the different L- and H-chain variable regions. For the somatic theory, however, there would be only one or perhaps a few V-region genes, and these would mutate in the course of lymphocyte development to produce a large variety of cells committed to synthesizing their own unique types of antibody molecules, identified by the particular V-region amino-acid sequences. The whole question of explaining the immune response at the molecular and cellular level remains one of the most tantalizing and actively studied problems in biology at the present time.

Summary

1. Cultures of somatic cells can be reproducibly established from certain tissues (also from the adult). In order to obtain indefinite multiplication, "transformation" of the cultivated cells by suitable viruses may be necessary.
2. There is a close resemblance between this type of cell transformation and cancer.
3. Drug resistances and enzyme deficiencies permit selection of hybrids that form spontaneously (or are induced by some viruses) between different types of cells, also from different species, such as man and mouse.
4. The study of hybrids (and of lines derived from them) that have lost some of their chromosomes permits assignment to specified chromosomes of genes whose products can be recognized in the cell cultures. This has greatly helped in building human chromosome maps.
5. In spite of the difference in number of X chromosomes between females and males, enzymes made by X-linked genes are in the same concentration in the two sexes.
6. This "dosage compensation" is explained by the fact that, in adult female somatic cells, one of the two X chromosomes (a randomly chosen one) is inactivated.
7. Vertebrates can produce molecules, called antibodies, that react specifically with other molecules, antigens (usually but not necessarily proteins). Antibodies are immunoglobulins (Ig) of which there exist many classes. They all contain in a part of their molecule a specific sequence of amino acids that is responsible for the specificity of their reaction with antigens.
8. Antibodies are produced by lymphocytes. A single lymphocyte can produce only one type of antibody. Presence of the antigen stimulates lymphocytes to multiply, thus producing greater concentrations of the specific antibody.
9. It is not yet clear if there is a great variety of genes—each of which produces a different antibody, and only one of which is activated in an antibody-producing lymphocyte—or if there is a single basic gene (or a few such genes) that can somehow mutate during development to produce a variety of cells, each making a different antibody.

Exercises

1. Describe types of problems of human genetics that somatic-cell culture can help to solve.

2. Which are the major developments that led to the formulation of the theory of X inactivation?

3. Outline the major properties of the immune system.

4. Discuss the importance of recognition of self, and the consequences of its breakdown.

5. Review the major aspects of the clonal theory of antibody production.

6. A rat liver tumor (hepatoma) that has been cultivated *in vitro* continues to produce albumin, a protein normally produced by the liver (and found in large amounts in normal blood). When crossed to a mouse-cell line that does not produce albumin, the hybrid produces both the rat and mouse albumin, as identified by electrophoresis. What simple explanation does this suggest about gene regulation?

7. What kind of experiments would permit mapping of the albumin-producing gene in rat or mouse chromosomes, using the system described in Exercise 6?

8. A blood-group system, Xg, is known from pedigree studies to be X-linked. Individual red cells of a heterozygote between two alleles show no difference in their antigens. What does this suggest about the Lyon effect?

9. Individuals have been found who are color-blind in one eye but not in the other. What would this suggest if these individuals were only or mostly females? What would it suggest if the phenomenon were found only or almost only in males?

10. In *myasthenia gravis,* a disease with a fairly high familial recurrence, there is a progressive weakening of muscles. Antibodies are found in these patients against components of the muscular cell. Drugs that inhibit antibody formation permit alleviation of symptoms. Which is the normal mechanism that is likely to have been altered in these patients?

11. If the hypothesis is correct, that each antibody is coded by a different gene, and if there are at least 100,000 different antibodies made, what fraction of the total genome should be devoted to antibody formation? Assume 5×10^9 nucleotide pairs in the whole genome; take the number of amino acids of H chains to be 500 and that of L chains as 200.

References

Bergsma, D.
1975. (Ed.) *Human Gene Mapping: Rotterdam Conference, 1974.* Birth Defects Original Article Series, vol. 11, no. 3. White Plains, N.Y.: National Foundation–March of Dimes.

Eisen, H. N.
1974. *Immunology.* Hagerstown, Md.: Harper and Row.

Ephrussi, B.
1972. *Hybridization of Somatic Cells.* Princeton, N.J.: Princeton University Press.
Ephrussi, B., and M. C. Weiss
1969. "Hybrid somatic cells," *Scientific American,* vol. 220, no. 4, pp. 26–35.
Harris, H.
1970. *Cell Fusion.* Boston: Harvard University Press.
Harris, M.
1964. *Cell Culture and Somatic Variation.* New York: Holt, Rinehart and Winston.
Hook, E. B.
1973. "Behavioral implications of human XYY genotype." *Science,* vol. 179, p. 139.
Jacobs, P. A., M. Bruton, M. M. Melville, R. P. Brittain, and W. F. McClemont
1965. "Aggressive behavior, mental subnormality, and the XYY male," *Nature,* vol. 208, p. 1351.
Lyon, M. F.
1961. "Gene action in the X-chromosomes of the mouse (*Mus musculus* L)," *Nature,* vol. 190, pp. 372–373.
Nesbitt, M. N., and S. M. Gartler
"The applications of genetic mosaicism to developmental problems," *Annual Review of Genetics,* vol. 5, pp. 143–162.
Roitt, I. M.
1971. *Essential Immunology.* Philadelphia: Davis.
Ruddle, F. H., and R. S. Kucherlapati
1974. "Hybrid cells and human genes," *Scientific American,* vol. 231, no. 1, pp. 36–44.

PART II

POPULATION GENETICS

In Part I we studied genes and chromosomes, their physical and chemical nature, and how they act in determining the development and the fate of an individual. We have also seen how genes and chromosomes change, by mutation. Mutations provide the source of differences between individuals, and we have discussed how these differences are inherited and how predictions can be made about the progeny expected from given parents. Mendelian genetics teaches us what to expect in crosses, families, and pedigrees. But individuals do not exist in nature in isolation. On the contrary, they usually occur together in large numbers, forming populations whose members are similar but show some differences and can, at least in principle, intercross. A population's genetic constitution changes with time, and this change is called evolution. We know what major forces determine these changes. The major factor is natural selection, which is the automatic preferential propagation of those types that are fitter in the environment in which they live. Chance phenomena also are of some importance—the more so, the smaller the population involved (random genetic drift).

These forces can be measured and their effects predicted. The analysis can usually be simplified by making resort to quantities called the gene frequencies, which are obtained from counts of the existing alleles at a locus. A genetic description of a population is therefore provided by its gene frequencies.

The events taking place in a population that are of interest from a genetic point of view form the subject of *population genetics,* to which Part II is dedicated. Population genetics is necessary to an understanding of evolution. Moreover—in the case of the special organism in which we are most interested—population genetics is essential to a full understanding of the laws of human inheritance. Many organisms can be bred in the laboratory, but it is not possible (or desirable) to set up special "crosses" of humans for purposes of genetic study. This limitation is a serious drawback for genetic analysis. Population genetics enables predictions to be made about observed matings, in spite of this limitation.

6

Genes in Populations

In contrast to genetic studies with many plants and animals, the study of human genetics cannot make use of experimental crosses between organisms of given types. Matings that occur spontaneously can, however, be observed and their progeny recorded. Are there simple rules for predicting the frequency of occurrence of spontaneous matings? The simplest assumption one can make is that they occur at random. If this assumption is valid, then it is easy to predict the behavior of genes in populations, and the relative proportions of various genotypes. Can one then test whether the randomness of mating is a reasonable assumption?

6.1 The Extent of Genetic Variation

Part I of our book deals mainly with the mechanisms of inheritance, the origins of new variations, and the chemical and cellular bases underlying these processes. Now we turn to the behavior of genes in populations and to the questions raised by the existence of so much genetic diversity among the individuals who make up a population. Chapter 1 begins with an emphasis on the fact that any two individuals (except perhaps for identical twins) look different from each other, and that this diversity is paralleled by genetic differences with respect to traits such as the blood groups that show simple patterns of inheritance.

The most direct estimate of the extent of genetic differences within populations comes from studies of enzyme variation using the technique of electrophoresis (see Chapter 3). Electrophoresis can be used to separate proteins that differ in overall electrical charge, as a result of differences in proportions of certain amino acids. Calculations show that, on the average, about one in three of all possible amino-acid substitutions in a protein should give rise to differences that can be detected by electrophoresis. The amino-acid sequence of a particular protein is determined by the nucleotide sequence in the corresponding gene. If any amino-acid differences appear among samples of the same protein taken from different individuals, the basic tenets of molecular biology imply that these differences must be genetically determined. Electrophoresis of enzymes and other proteins therefore provides a direct tool for detection of genetic differences between individuals in a population.

The electrophoretic analysis of a set of enzymes (chosen only because suitable techniques are available) has shown that in general at least 30 percent of such enzymes vary genetically within a population. This fact was first demonstrated in 1966 by two independent investigations: H. Harris working with human populations, and R. C. Lewontin and J. L. Hubby working with *Drosophila.* In Harris' study, only ten enzymes were examined; since that time, however, many more enzymes have been studied, and the overall proportion showing genetically determined electrophoretic variation remains at about 30 percent.

Similar data (though with some differences in the proportion of variants) have now been obtained in a considerable range of species, suggesting that extensive genetic variability in natural populations is a more-or-less general phenomenon, at least among sexually reproducing species. If we assume that the enzymes studied are controlled by genes that are a representative sample of all human genetic loci, then these data imply that at least 30 percent of all gene loci vary within the human population. Allowing for the fact that electrophoresis detects only a fraction of all possible protein variations, we conclude that the proportion of loci that vary must be even greater.

Although normal genetic variants like the blood groups have been known for many years, these variations were regarded as exceptions. Until comparatively recently, geneticists supposed that most genes would not vary significantly within a population. The potential for genetic variability revealed by the electrophoretic data is quite staggering. It now appears that a single individual has the potential to produce at least $2^{10,000} \cong 10^{3,000}$ different types of eggs or sperm. (That number, $10^{3,000}$, is 1 followed by 3,000 zeros!) For comparison, we estimate that the total number of sperm that have ever been produced by all human males who ever lived is around 10^{23} or 10^{24}. Clearly, only a very small proportion of the very large number of potential genetic combinations is ever realized.

The first task of the population geneticist is to devise ways to describe genetic variation in a population.

On the basis of these descriptions, models can be constructed to explain how genetic variability is maintained, and how genetic changes can be explained in natural populations. Of course, the models are inevitably over-simplified in comparison with the "real" situation, which is usually much too complicated to describe in all its detail. Nevertheless, these models have proved to be very useful for understanding the behavior of genes in populations. Much of Part II is concerned with the analysis of such models and with their application to particular genetic problems.

In this chapter, we begin with the most basic of all laws of population genetics—a law named after its two discoverers, the English mathematician G. H. Hardy and the German physician W. Weinberg, who worked out this relationship independently in 1908.

6.2 Random Mating, Genotype, and Gene Frequencies

In a survey carried out at an African university hospital, hemoglobin type was examined in 100 small children. Electrophoretic techniques were used, and the presence of hemoglobins S and/or A was recorded. In this sample of 100 individuals, only one (1 percent of the total) had only hemoglobin S, and therefore must have been of homozygous genotype *SS*. There were 18 heterozygous *AS* individuals (18 percent of the total) who had both types of hemoglobin, A and S. The remaining 81 percent were *AA* genotypes, having only hemoglobin A.

Why do we find these particular proportions of heterozygous and homozygous individuals in the sample?

We shall explain the proportions by a route that may at first seem somewhat devious, but that does allow considerable simplification in practice. We first assume that the individuals were a sample drawn from an identifiable population, in which there can be interbreeding without major internal geographic or other restrictions. The concept of an interbreeding population, though apparently simple, is fundamental to any population-genetic analysis that deals with frequencies of types in such a population, and then attempts to describe the way the frequencies change. More specifically, in this case we assume that parents of the individuals examined for hemoglobin *mated at random with respect to the genetic character* we are studying. This means that, for instance, *AA* individuals had no special preference for *AA*, *AS*, or *SS* individuals (especially since one cannot tell them apart without special tests) and thus that matings occurred in proportion to the frequencies with which the various types were present in the population.

This hypothesis of random mating may seem a little arbitrary. For some traits it may not hold. It is, however, easy to test, and is a very convenient assumption if it turns out to be correct.

G. H. Hardy, who was a professor of mathematics at Cambridge University, often used to meet his colleague, geneticist R. D. Punnett, over the dinner table at Trinity College. It is said that, on one such occasion, Punnett put to him the following problem: Why is it that a dominant genetic trait does not increase in frequency at the expense of the recessive, until the population is uniformly of the dominant type?

Hardy soon came up with a simple answer. As a mathematician, he considered the result so trivial that he was hardly even willing to publish it. This "simple" result, the cornerstone of population genetics, explains why (following Mendel's laws of inheritance) genetic differences tend to be retained in natural populations. Hardy may yet live to be remembered more because of his mathematically trivial genetic theorem, the now-famous Hardy–Weinberg law, than because of his fundamental contributions to mathematics.

It can be proved that the models of population genetics, which form the basis for evolutionary predictions, are greatly simplified if randomness of mating can be relied upon.

Fortunately, actual testing of the hypothesis shows that, in the great majority of instances, it is justified. The assumption of random mating thus plays an important role in all developments that follow, and we now discuss it in more detail.

The most direct way to test the randomness of mating would be to study the frequency with which various matings occurred. This is seldom done, however, because it is a laborious task and because it can in practice be replaced by a relatively simple shortcut. The shortcut makes direct use of the numbers of individuals of the three genotypes, (*AA, AS,* and *SS* in the example) existing in a population (or of their relative proportions). It proceeds through the following steps (see Table 6.1).

1. Count the number of genes of the two allele types, *A* and *S* in our example, in the population.
2. Determine their relative proportions, called *gene frequencies.*
3. Compute, on the basis of gene frequencies, the expected proportions of genotypes *AA, AS,* and *SS* that would be found in the population if mating were random. These proportions are obtained by the rule that goes under the name of Hardy–Weinberg.
4. Compare the proportions "expected" under random mating with those that were actually observed in the population. If there is a good correspondence

Table 6.1
Computation of gene frequencies by counting alleles and from genotype frequencies

	Genotypes				
	AA	*AS*	*SS*	Total	Gene Frequency
I. COUNTING ALLELES					
Number of individuals	81	18	1	100	
Number of *A* genes	162	18	0	180	180/200 = 0.9 = 90%
Number of *S* genes	0	18	2	20	20/200 = 0.1 = 10%
Total number of genes	162	36	2	200	
II. FROM FREQUENCIES					
Frequency of genotype	0.81	0.18	0.01	1.00	
Frequency of *A* genes	0.81	0.09	0.00	0.90	0.90 = 90%
Frequency of *S* genes	0.00	0.09	0.01	0.10	0.10 = 10%
Total frequency of genes	0.81	0.18	0.01	1.00	

This table is based on the numerical example given at the beginning of Section 6.2.

between observed and expected genotype frequencies, then we can rely on the hypothesis of random mating. It may seem strange that random mating is possible: does one really choose a husband or wife at random? That this is not absurd, is shown by the simple consideration that one often ignores, at marriage, the ABO blood group of one's spouse! Even if it were known, the choice of a husband or wife is not normally based on his or her ABO groups. Randomness means, in essence, absence of a specific choice.

Step 1. The numbers of *A* and *S* genes in our sample can be determined just by counting the alleles, and by remembering that every individual has two copies of each gene. If the hemoglobin genes of all the 100 individuals were pooled together there would be 200 genes in all. *AA* individuals, of whom there are 81, would each contribute two *A* alleles to the pool, or $81 \times 2 = 162$ *A* alleles in all. To these must be added 18 *A* alleles contributed by the 18 *AS* individuals. Thus, out of the pool of 200 genes, we should have a total of $162 + 18 = 180$ *A* genes. The number of *S* alleles can be counted similarly to give a total of 20: the sum of 18 from the 18 *AS* individuals, plus 2 from the single *SS* individual. Thus altogether out of the 200 genes in the pool from our 100 individuals, 180 are *A* and 20 are *S*.

Step 2. The relative frequencies of *A* ($180/200 = 0.9$ or 90 percent) and of *S* ($20/200 = 0.1$ or 10 percent) are the respective *gene frequencies* of *A* and *S* in the population of 100 individuals. This computation is summarized in Table 6.1 in the form of a balance sheet.

Step 3. The next step consists of computing from the gene frequencies the expected genotype frequencies, on the assumption that random mating has taken place. We shall first just give the procedure and leave its justification for Section 6.3. The expected proportion (or relative frequency) of a homozygote for an allele under random mating is the square of the gene frequency of that allele. Thus, for *AA* the proportion is the square of the frequency of *A* or $0.9^2 = 0.81$. Similarly, for *SS* the frequency is the square of 0.1, the gene frequency of *S*, or $0.1^2 = 0.01$. For heterozygotes the expected proportion is twice the product of the frequencies of the two alleles, and thus in this case it is $2 \times 0.9 \times 0.1 = 0.18$.

Step 4. These computed proportions are exactly the same as those of the "observed" sample—namely, in percentages, 81 *AA*, 1 *SS*, and 18 *AS*. The agreement between observed and expected is perfect; this is because we made up the "observed" numbers to simplify the problem. With actual observations, the agreement is usually not perfect because sampling (i.e., statistical) effects give rise to fluctuations of the observed frequencies. A statistical test (the χ^2 test, described in the Appendix) can then be used to evaluate the agreement between observed and expected numbers. An example of the computation of gene and genotype frequencies based on a real sample from Africa is given in Tables 6.2 and 6.3.

For two alleles, one gene frequency is enough to describe the state of the population.

In computing gene frequencies from genotype frequencies, we have already gained in simplicity. This is because only *one* gene frequency is really needed, for the other is determined by the fact that the two gene frequencies must add up to one. Thus, given the frequency of *A* as 0.9, we know that the frequency of *S* must be $1.0 - 0.9 = 0.1$. This one number, namely 0.9, can be used to predict the frequencies of all the three genotypes.

Table 6.2
Computation of hemoglobin *S* and *A* gene frequencies (from genotype frequencies for a sample of children from Accra, Ghana)

	Genotypes				
	AA	*AS*	*SS*	Total	Gene frequency
Number of individuals	701	135	4	840	
Number of *S* genes	0	135	8	143	$p = 143/1{,}680 = 0.085$
Number of *A* genes	1,402	135	0	1,537	$q = 1{,}537/1{,}680 = 0.915$
Total number of genes	1,402	270	8	1,680	

Data from G. R. Thompson (1962), *British Medical Journal* 1:682–685. (Hemoglobin C was also present in this sample, but for simplicity of this presentation it was pooled with A.)

Table 6.3
Prediction of expected genotype frequencies from observed gene frequencies

Genotype	Expected genotype frequency	Number of individuals in population of 840: Expected*	Observed
SS	$p^2 = (0.085)^2 = 0.0072$	6.05	4
AS	$2pq = 2(0.085)(0.915) = 0.1555$	130.62	135
AA	$q^2 = (0.915)^2 = 0.8372$	703.25	701
	0.9999†	839.92†	840

*The expected number of individuals is obtained by multiplying the population size (N = 840) times the expected genotype frequency.

†The sum of expected genotype frequencies should be equal to 1.000; the value differs slightly in this case because of rounding-off errors, which also account for the slight discrepancy from N in total numbers expected.

NOTE: The gene frequencies (p and q) and the observed numbers of individuals are taken from Table 6.2; see the footnote to that table.

6.3 The Hardy-Weinberg Equilibrium

The MN blood-group system was found about fifty years ago through preparation of appropriate antibodies against human red blood cells. (Human blood cells are injected into a rabbit; antibodies can then be purified from the rabbit serum after the rabbit's immune system has reacted to the invasion of foreign cells.) As discussed in Chapter 5, the antibodies can be used to detect genetic differences on red cells through use of the agglutination technique. For example, an anti-M serum agglutinates the red cells of about 84 percent of human individuals (this percentage varies somewhat in different populations). Under the same conditions, anti-N serum agglutinates the blood of about 64 percent of individuals. These two sera provide the basis for identifying the three MN blood types (M, MN, and N) described in Chapter 2.

The MN blood type of an individual is controlled by the alleles present at the MN-blood-group locus on the chromosome-2 homologous pair. If the *M* allele is present on one or both chromosomes, the red cells of the individual will be agglutinated by anti-M serum—presumably because of some substance (M) produced by allele *M*. Similarly, the presence of one or two *N* alleles causes red cells to be agglutinated by anti-N serum—presumably because of some different substance (N) produced by allele *N*.

All human blood samples (with very rare exceptions) are agglutinated either by anti-M or anti-N. Apparently an allele producing neither M nor N (equivalent to the *O* allele in the ABO system) does not exist or is very rare. Table 6.4 explains

Table 6.4
The MN blood group

Genotype	Typical genotype frequency	MN blood type (phenotype)	Reaction with serum*	
			Anti-M	Anti-N
MM	0.36	M	+	−
NN	0.16	N	−	+
MN	0.48	MN	+	+

*Agglutination = +; no reaction = −.

the basis for MN blood typing using these sera; the table also gives typical frequencies of the three blood types. These frequencies are expected to be the same in males and females, because the MN locus is not X-linked.

Assuming random fertilization of gametes, we can construct a simple mathematical model for the distribution of genes in a population.

Suppose we took a large sample of gametes (sperm and eggs) from 100 different individuals and mixed all the gametes together, allowing sperm to fertilize eggs at random (Fig. 6.1). The proportions of fertilized eggs that would be expected to be *MM*, *MN*, and *NN* could be computed from the frequencies of sperm or eggs that are *M* or *N*. (The gene frequencies will be the same for sperm or eggs, because the genotype frequencies are the same for males or females.) The first step thus must be to establish what proportion of sperm or eggs are *M* and what proportion are *N*.

This problem involves the same type of calculation that we did before for the gene frequencies of hemoglobin alleles *A* and *S* (Table 6.2). Of the 100 individuals in our sample, we expect 36 to be *MM*, 16 to be *NN*, and 48 to be *MN* (using the genotype frequencies from Table 6.4). All of the gametes from *MM* individuals (36 percent of the gametes) and half of the gametes from *MN* individuals ($48/2 = 24$ percent of the gametes) will be *M*, so the frequency of *M* gametes will be $0.36 + 0.24 = 0.60 = 60$ percent. Similarly, we can compute the frequency of *N* gametes to be $0.24 + 0.16 = 0.40 = 40$ percent.

The frequencies of M and N gametes are just the M and N gene frequencies.

The computation of gamete frequencies we have just carried out is identical to the computation of gene frequencies in the population, as illustrated in Table 6.2. Therefore, we see that the gene frequencies in a population are the same as the frequencies of various types of gene-bearing sperm or eggs to be expected in a pool of gametes obtained from all the members of the population.

The mixture of sperm and eggs shown in Figure 6.1 has the two types of gametes in the proportions of 0.60 *M* to 0.40 *N*. A homozygous *MM* individual will be formed only if an *M* sperm fertilizes an *M* egg. Let us assume that fertilization is at random—that is, any sperm in the population is equally likely to encounter and to fertilize any egg in the population, regardless of the MN type of sperm and egg. (This assumption would be invalid, for example, if *M* sperm tended to seek out or more successfully to fertilize *M* eggs. Such nonrandom factors seem to be extremely rare.) When a sperm fertilizes an egg, there is a probability of 60 percent that the sperm will be *M* and a probability of 60 percent that the egg will be *M*. Our assumption of random fertilization means that the two probabilities are independent, so the probability that both egg and sperm will be *M* is $0.60 \times 0.60 = 0.36$ (using the rule of multiplication of probabilities for the joint occurrence of independent events; see Section 2.4). Thus we expect 36 percent of the zygotes (fertilized eggs) to be *MM*.

To generalize this result, let p be the gene frequency of *M* (in this case, $p = 0.60$); then the genotype frequency of *MM* zygotes will be $p \times p = p^2$ (in this case, 0.36). Similarly, if q is the gene frequency of *N* ($q = 1 - p$, and in this case $q = 0.40$), then the frequency of *NN* zygotes must be q^2 (0.16 in this case). The remainder of the zygotes will be *MN* heterozygotes. Because the genotype frequencies must add up to one, we can compute the frequency of *MN* as $1.00 - p^2 - q^2$ (in this case, $1.00 - 0.36 - 0.16 = 0.48$, so 48 percent of the zygotes will be *MN*).

In fact, we could compute the heterozygote frequency independently. The probability of an *M* sperm fertilizing an *N* egg is $p \times q$, and the probability of an *N* sperm fertilizing an *M* egg is also $p \times q$. In either case, the resulting zygote will be *MN*, so the frequency of *MN* heterozygotes should be $pq + pq = 2pq$ (in this case, $2 \times 0.60 \times 0.40 = 0.48$).

This process of forming zygotes by random fertilization between sperm and eggs is illustrated graphically in Figure 6.2, where the sperm and eggs from Figure 6.1 have been arranged after fertilization so that eggs with *M* are at the left and sperm with *N* are at the bottom. Figure 6.3 shows that the frequencies we have just computed can be represented as rectangular areas within a square of side 1.0, after subdivision of the sides of the square in proportion to the gamete gene frequencies.

Random union of individuals gives rise to the same predictions as those obtained for random union of gametes.

Of course, fertilization in humans does not take place by random mixing of gametes taken from the whole population, but rather by formation of mating pairs (one male and one female), with each pair having a variable number of off-

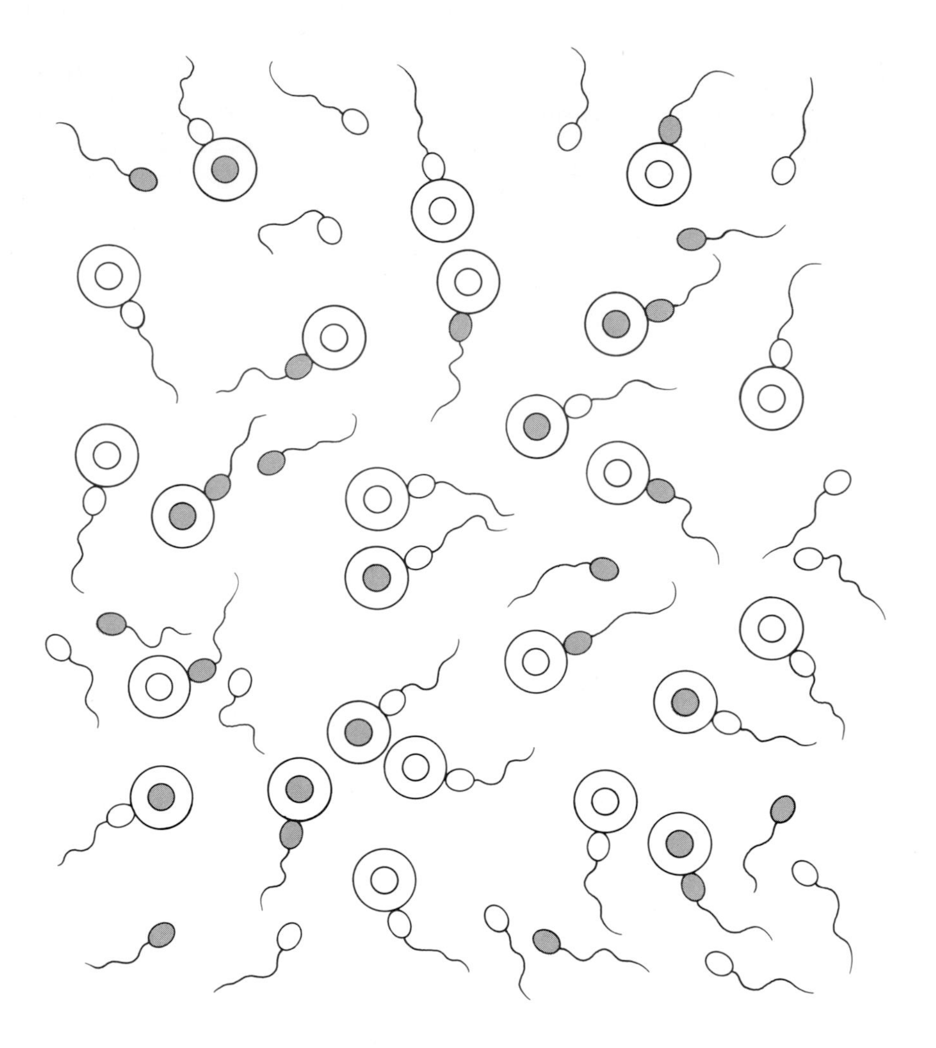

Figure 6.1
Random fertilization of eggs by sperm. Each sperm and each egg carries 23 chromosomes, each with a great many genes, but we consider here only one gene (as, e.g., the MN blood group). The head of the sperm or the nucleus of the egg is shaded if the gamete carries gene *N*; an unshaded (white) gamete carries gene *M*. Of 45 sperm shown in the figure, 27 carry *M* and 18 carry *N*. Of 25 eggs, 15 carry *M* and 10 carry *N*. Therefore, in both sperm and eggs, the frequency of *M*-carrying gametes is $\frac{27}{45} = \frac{15}{25} = 0.60$ and the frequency of *N*-carrying gametes is $\frac{18}{45} = \frac{10}{25} = 0.40$. After fertilization, the zygote's genotype for the MN locus is just the combination of the contributions from sperm and egg. This diagram is highly schematic. No account is taken of statistical fluctuations—the proportions shown here are exactly those expected in an ideal sample. Those of zygotes are: $MM\ \frac{9}{25} = 0.36$, $MN = \frac{12}{25} = 0.48$, $NN = \frac{4}{25} = 0.16$. Figures 6.2 and 6.3 show how to predict them.

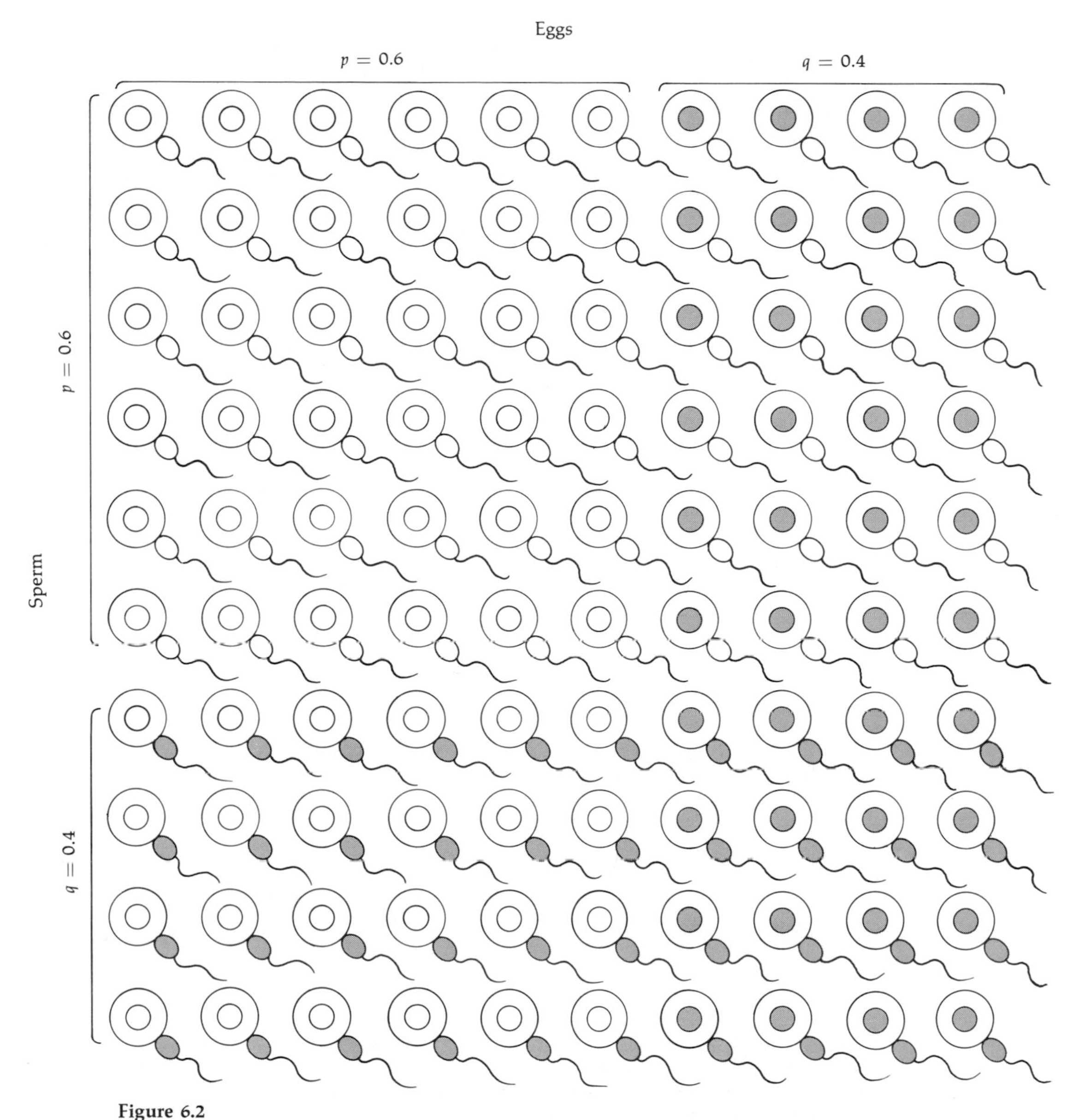

Figure 6.2
Proportions of zygote genotypes obtained from random fertilization: 100 zygotes obtained from a random fertilization (as in Fig. 6.1) with the frequency of *M*-carrying gametes (both sperm and eggs) equal to 0.60, and that of *N*-carrying gametes 0.40. The zygotes have been ordered after fertilization to put all *M*-carrying eggs at the left and all *M*-carrying sperm at the top of the diagram. The expected proportions of zygote genotypes (*MM, MN,* and *NN*) can be obtained by counting. Figure 6.3 shows a simplified geometric form of this diagram.

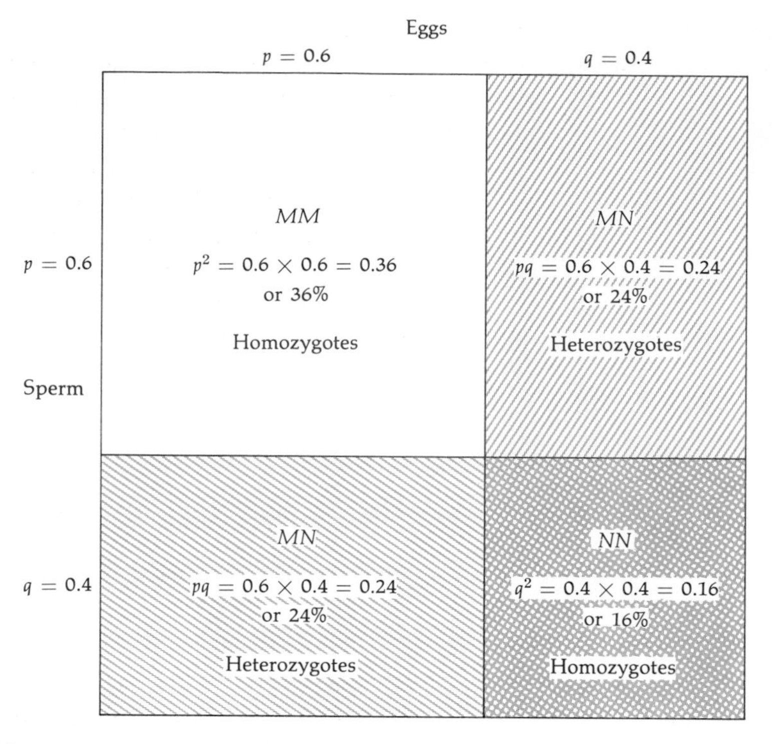

Figure 6.3
Geometric presentation of the Hardy–Weinberg law. The 100 zygotes of Figure 6.2 are represented as forming a square, divided into four parts. The frequencies of gametes of types M and N (equal to gene frequencies in the population) are shown on the sides. The four areas within the square represent the frequencies of the four types of zygotes—two of which are usually indistinguishable and can be considered together as the heterozygotes (resulting in this case from combination of an N egg and M sperm or from an M egg and N sperm). The area of the upper-left part of the square represents the frequency of MM homozygotes. This area is a square with sides of 0.6, so it gives a frequency of $0.6 \times 0.6 = 0.36$ for the MM homozygotes. Similarly, the area of the lower-right part of the square gives a frequency of $0.4 \times 0.4 = 0.16$ for the NN homozygotes. Each of the remaining parts is a rectangle with sides of 0.6 and 0.4, so the area of each of these parts is $0.6 \times 0.4 = 0.24$, and the frequency of heterozygotes (MN) is represented by the sum of these two areas, $0.24 + 0.24 = 0.48$. The sum of the expected genotype frequencies is 1.00, because the sides of the square are represented as having lengths of 1.0, giving a total area of $1.0 \times 1.0 = 1.00$. Note that this geometric representation accurately represents the Hardy–Weinberg law, predicting frequencies of p^2, q^2, and $2pq$ for the three zygote genotypes.

spring. However, if we assume that the pairs are formed at random (mates are selected without regard to their MN blood types) and that fertility averages out the same for any combination of blood types, then we can show that the predicted genotype frequencies for the offspring come out just the same (see Appendix).

Now let us summarize our results thus far. We assume a population in which two alleles are present for a single locus with gene frequencies p and q (where q must equal $1 - p$). We also assume (1) that mating occurs at random with respect to the genotypes being studied, (2) that fertility and chances of survival from fertilization to sexual maturity are not affected by the genotypes being studied. These are the assumptions that really matter in practice. The second, as we will more fully explain in the next chapter, is equivalent to absence of natural selection. Other assumptions—for instance that migration and mutation are rare enough to be neglected, and that heterozygotes produce gametes of the two types in equal numbers and with equal chance to fertilize (or be fertilized)—are less likely to be violated. Then, if the population is large enough and if statistical fluctuations can be neglected, we predict that the genotype frequency of heterozygotes among the offspring of this population will be $2pq$, the frequency of homozygotes for the allele with frequency p will be p^2, and the frequency of homozygotes for the allele with frequency q will be q^2.

The result summarized in the preceding paragraph is the relationship independently worked out by Hardy and by Weinberg in 1908, and it is now known as the *Hardy–Weinberg law.* Figure 6.4 shows genotype frequencies computed for various gene frequencies, using this law.

Gene frequencies are stable over time.

We have computed the genotype frequencies expected for the offspring of this population to be 0.36 *MM*, 0.16 *NN*, and 0.48 *MN*. Note that these frequencies are identical to the genotype frequencies among the original population (Table 6.4). This stability of genotype (and gene) frequencies from each generation to the next is quite general under the conditions that we have assumed. Under such conditions, the gene frequencies p and q will remain unaltered in the population from generation to generation indefinitely. Therefore, the proportions of the three genotypes (p^2, $2pq$, and q^2) will also remain the same indefinitely if the assumed conditions continue to hold true.

When gene and genotype frequencies do not change in successive generations, the population is said to be in *genetic equilibrium.* The Hardy–Weinberg law actually ensures that genetic variability will be maintained in a population, so long as there are no disturbing factors to upset the equilibrium. (Such factors could enter if one or more of the assumed conditions becomes invalid.) There is no tendency for less common (or for recessive) alleles to disappear from the population over time. The Hardy–Weinberg law is a striking example of how a little well-placed mathematics can lead to profound results in biology.

6.4 Composition of an Equilibrium Population

In Table 6.1, the genotype frequencies were 0.81 *AA*, 0.18 *AS*, and 0.01 *SS*. In the MN-blood-group example, the frequencies were 0.36 *MM*, 0.48 *MN*, and 0.16 *NN* (Table 6.4). The proportions of the three genotypes vary as a function of the gene

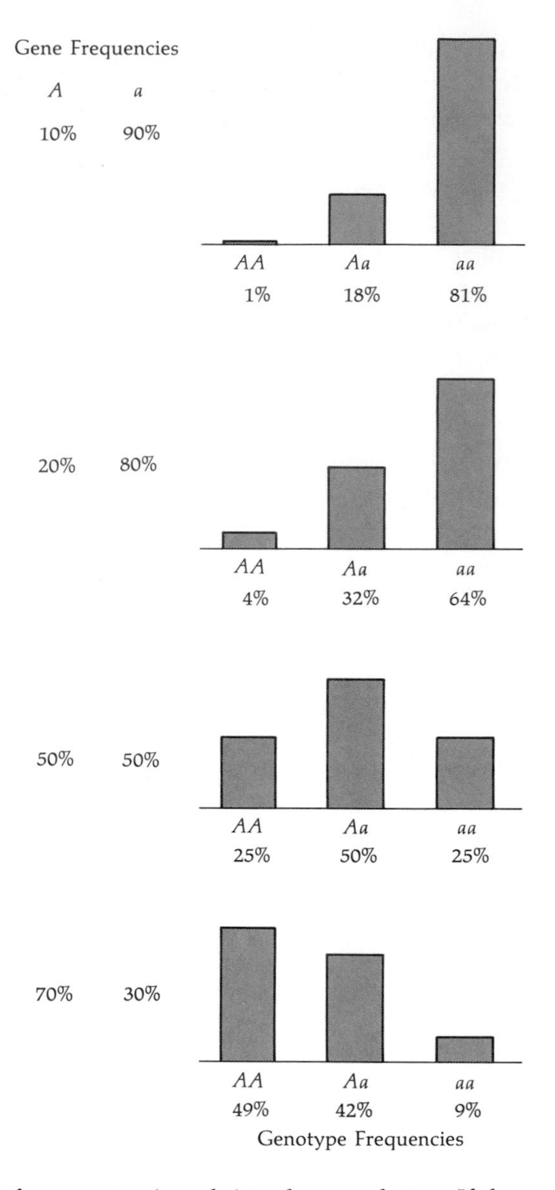

Figure 6.4
Hardy–Weinberg law: expected genotype frequencies for various gene frequencies, computed from the Hardy–Weinberg law.

frequencies (p and q) in the population. If there are just two alleles, it can be shown that the maximum possible frequency of heterozygotes is 0.50, and that this frequency will be obtained when $p = q = 0.50$. (Note that $2pq = 0.50$ in this case.) For any other values of p and q, the frequency of heterozygotes is less than 50 percent. The homozygote associated with the smaller of the two gene frequencies will be the rarest of the three genotypes.

Figure 6.5 gives a more complete picture of the relationship between gene and genotype frequencies. Note that the frequency of heterozygotes (*Aa*) is greatest and is equal to 0.50 when $p = 0.50$ (and therefore $q = 0.50$).

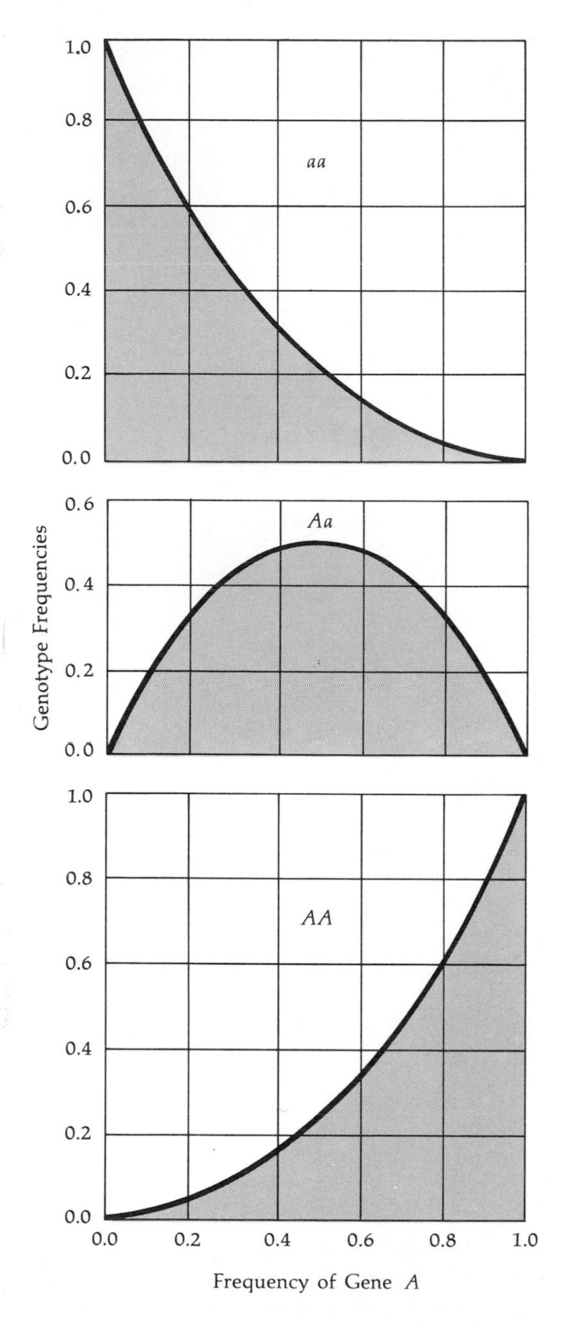

Figure 6.5
Frequencies of the three genotypes as a function of gene frequency. This graph shows the frequencies of the genotypes *aa*, *Aa*, and *AA*, plotted as a function of the gene frequency (p) of A. The values shown here are those computed from the Hardy-Weinberg law. The frequency (q) of gene a is, of course, equal to $1 - p$. NOTE: Use this graph, which is approximate, only to form an idea of the trend. For computation of expected frequencies use the formulas q^2, $2pq$, q^2 respectively.

When one allele is dominant, the Hardy–Weinberg law provides a simple way of computing the frequency of genes and of heterozygotes.

For example, consider the case of albinism—a rare, inherited, recessive defect discussed in Chapter 3. Albino individuals have essentially no pigment in their skin, hair, or eyes, and they are easily distinguished by their pale appearance. In many populations, albinism reaches a frequency of about 1 per 10,000 individuals. When dominance obscures the distinction between the heterozygote and one of the homozygotes, it is impossible to count cases of the three genotypes as we have done thus far in this chapter (Fig. 6.6). It is still, however, of interest in such cases to be able to predict the frequency of heterozygotes, and the Hardy–Weinberg law makes this prediction quite simple.

Albino individuals are homozygotes for a gene that blocks the ability to make pigment. The frequency of albinos in the population therefore must be equal to the square of the frequency of the albino gene. Therefore, the albino-gene frequency can be computed as the square root of the frequency of albino individuals: $q = \sqrt{1/10{,}000} = 1/\sqrt{10{,}000} = 1/100 = 0.01 = 1$ percent.

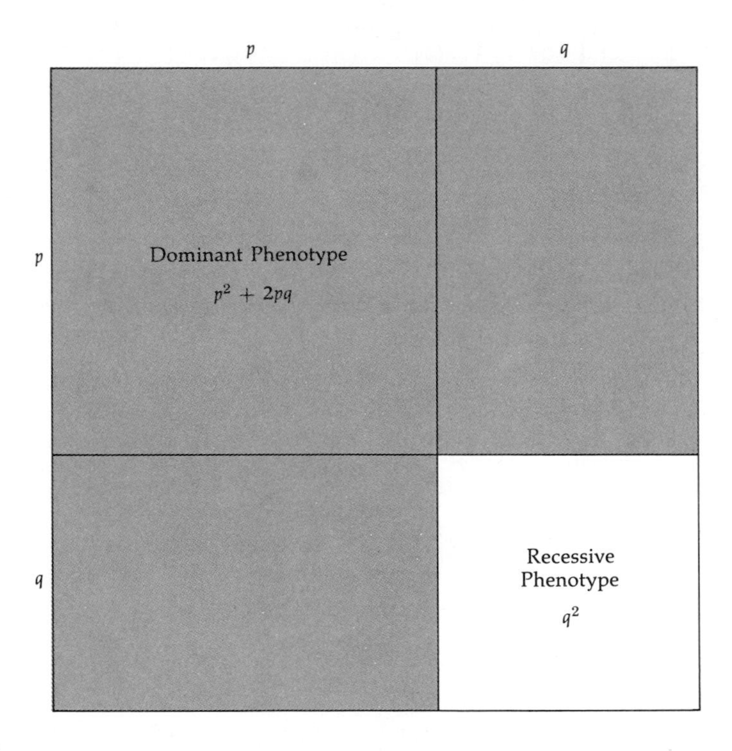

Figure 6.6
Frequency of the dominant phenotype. This geometric presentation is similar to that used in Figure 6.3. In this case, the entire shaded area represents the dominant phenotype, which includes both homozygotes and heterozygotes who cannot be distinguished phenotypically. (The frequency of the dominant gene A is p; the frequency of the recessive gene a is q.) However, because the frequency of the recessive phenotype *can* be observed and is expected to equal q^2, we can compute q and hence $p = 1 - q$. We can then estimate the frequencies of dominant homozygotes (p^2) and of heterozygotes ($2pq$).

We know that the frequency of the other allele must be $p = 1.00 - q = 1.00 - 0.01 = 0.99$. From the Hardy–Weinberg law, we know that the frequency of heterozygotes is $2pq = 2 \times 0.99 \times 0.01$. Because 0.99 is almost equal to one, we can simplify our computation by saying that the frequency of heterozygotes will be almost equal to $2 \times 1 \times 0.01 = 0.02$ or 2 percent. In other words, if q is very small, we can say that the frequency of heterozygotes will be very slightly less than twice q.

In the case of albinism, note that the frequency of albinos (recessive homozygotes) is only 0.0001 = 0.01 percent, while the frequency of heterozygotes is nearly 2 percent. The Hardy–Weinberg law thus explains why heterozygotes are relatively common, even when one of the homozygotes being considered is quite rare.

Suppose that we had used only the sickling test without electrophoresis in the sickle-cell anemia study of Table 6.2. In that case, it would have been impossible to distinguish between heterozygotes and homozygotes for the sickle-cell gene. We would have counted 139 carriers of hemoglobin S (either *AS* or *SS*) and 701 noncarriers (*AA*). We could still compute the gene frequency of *A* as the square root of the frequency of *AA* homozygotes: $q = \sqrt{701/840} \cong \sqrt{0.84} \cong 0.92$. We could then compute $p = 1.00 - 0.92 = 0.08$, the proportion of *SS* homozygotes as $p^2 = (0.08)^2 = 0.0064$, and the frequency of heterozygotes as $2pq = 2 \times 0.08 \times 0.92 \cong 0.15$. We would therefore have estimated the number of *SS* homozygotes in the sample to be $0.0064 \times 840 \cong 5.4$. This estimate is not very different from the more exact value of 6.05 computed from the three genotypes, and the agreement would be even closer if we had not rounded off our computations.

6.5 Extensions of the Hardy–Weinberg law

The Hardy–Weinberg law can be extended to multiple allelism.

For example, suppose that a given population has three alleles for a single locus—say, A_1, A_2, and A_3. These alleles give rise to three homozygotes (A_1A_1, A_2A_2, and A_3A_3) and three heterozygotes (A_1A_2, A_1A_3, and A_2A_3).

The ABO blood-group system (see Section 5.4) is controlled by three alleles (*A*, *B*, and *O*) in such a way that not all heterozygotes can be distinguished from homozygotes. *A* and *B* are both dominant over *O*, so that *AA* individuals are indistinguishable from *AO* (both being blood type A) and *BB* individuals are indistinguishable from *BO* (both type B). Individuals of type O must be homozygous *OO*, and individuals of type AB must be *AB* heterozygotes. Altogether, therefore, we can distinguish four phenotypes corresponding to the six genotypes (Table 5.2).

Of course, family analysis can sometimes distinguish homozygotes from heterozygotes. For example, an A individual who is the offspring of one A and one O parent must be an *AO* heterozygote. Also, if two individuals of type A have any O offspring, then both parents must be *AO* heterozygotes. Uncertainties do remain, however, even when family data are available. For example, suppose that an A individual marries an O individual, and that all of their offspring are type A. We cannot conclude with certainty that A is a homozygote. Given the relatively small size of human families, it is possible that all the offspring might be *AO* by chance, even though the A parent was *AO* and produced *O* gametes. It may have just happened that none of the *O* gametes ever chanced to combine with a gamete from the *OO* parent.

When the phenotypes in a population are enumerated using only standard blood-typing tests, homozygotes and heterozygotes among the A and B individuals are not distinguished. Nonetheless, we can extend the Hardy–Weinberg law to the case of three alleles and can use it to estimate all genotype and gene frequencies.

For example, suppose that in a Caucasian population the frequency of gene *A* is 0.3 (30 percent), that of gene *B*, 0.1 (10 percent), and that of gene *O*, 0.6 (60 percent). The generalization of the Hardy–Weinberg law to multiple alleles is then: (1) homozygotes have frequencies equal to the square of the respective gene frequencies: thus the frequency of *AA* is $(0.3)^2 = 0.09$, of *BB* $(0.1)^2 = 0.01$, and of *OO* $(0.6)^2 = 0.36$; (2) the frequency of heterozygotes is twice the product of the gene frequencies of the alleles forming the particular heterozygote being considered. For instance, the frequency of *AO* is $2 \times 0.3 \times 0.6 = 0.36$. The computation of the expected phenotype frequencies in terms of the gene frequencies is illustrated in Table 6.5 and Figure 6.7. It is also possible, assuming the Hardy–Weinberg law, to compute the gene frequencies given the phenotype frequencies, but the procedure is not simple because of the complications due

Table 6.5
Expected frequencies of phenotypes for the ABO system in Caucasians

Gene	Frequency	Genotype	Frequency	Phenotype	Frequency
A	$p = 0.3$	*OO*	$r^2 = (0.6)^2 = 0.36$	O	$0.36 = 0.36$
B	$q = 0.1$	*AA*	$p^2 = (0.3)^2 = 0.09$	A	$0.09 + 0.36 = 0.45$
O	$r = 0.6$	*AO*	$2pr = 2 \times 0.3 \times 0.6 = 0.36$		
		BB	$q^2 = (0.1)^2 = 0.01$	B	$0.01 + 0.12 = 0.13$
		BO	$2qr = 2 \times 0.1 \times 0.6 = 0.12$		
		AB	$2pq = 2 \times 0.3 \times 0.1 = 0.06$	AB	$0.06 = 0.06$

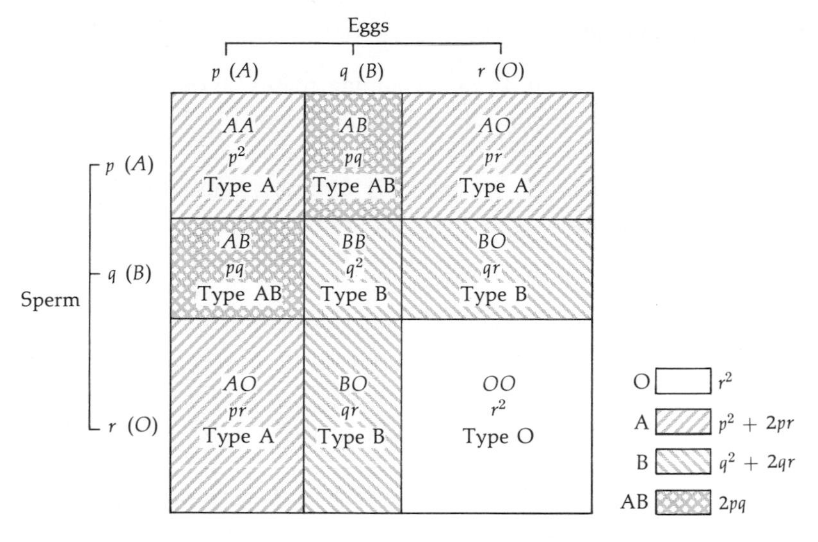

Figure 6.7
Geometric presentation of genotype frequencies for ABO alleles. This figure illustrates the computations given in Table 6.5.

to dominance of alleles A and B over O (more advanced books should be consulted for this purpose).

It is possible to predict the frequency of specific matings and their progeny.

The frequency of a given mating between two types, assuming randomness, is the product of the frequencies of the respective types. This is illustrated, with examples, in Exercise 10. The test using the Hardy–Weinberg rule, as in Sections 6.2 and 6.3, assumes not only random mating but, as we shall see, also absence of natural selection. If we want to test *only* the hypothesis of random mating, we must compare frequencies of observed matings with those expected. In a system like MN, there are three phenotypes (M, MN, N) and we can observe nine possible matings: M $\times$ M, M $\times$ MN, . . . , N $\times$ N. If we do not distinguish between husband and wife, we can pool matings such as M (husband) $\times$ MN (wife) and M (wife) $\times$ MN (husband), reducing the number of matings from nine to six. The frequency of M is p^2; that of the mating M $\times$ M is $p^2 \times p^2 = p^4$. The frequency of MN is $2pq$; that of mating M $\times$ MN is $p^2 \times 2pq = 2p^3q$; and that of MN $\times$ M $= 2p^3q$ also. The sum of the two mating frequencies then is $4p^3q$. Proceeding in this way and studying the progeny, one can demonstrate the Hardy–Weinberg rule in a more general form (see Appendix) than that given in Section 6.4.

X-linked genes require special treatment.

Color blindness is relatively frequent among males, about 1 out of 12 (some 8 percent) being affected among Caucasians. The defect is, however, much rarer among females.

Analysis of pedigrees has shown that color blindness is inherited as an X-linked recessive trait. It is in fact one of the commonest X-linked traits, especially in Caucasian populations. Recall that a male has only one copy of an X-linked gene, which of course always comes from his mother. A female, on the other hand, has two copies, one of paternal and the other of maternal origin (see Section 2.5).

Mating, of course, cannot be random with respect to sex! However, as far as the females are concerned, it seems reasonable to suppose that X-linked genes behave the same as autosomal genes and this can be proved rigorously. The frequencies of female genotypes expected are therefore given by the standard Hardy–Weinberg equilibrium. Males, however, each having only a single dose of X-linked genes, will have only one or the other of two phenotypes corresponding to the two genotypes associated with the two alleles. It can be shown that the frequencies of male phenotypes for X-linked genes are the same as the frequencies of the respective alleles in their mothers. This follows from the fact that males get their single X chromosome from their mothers. The Hardy–Weinberg equilibrium frequencies for an X-linked gene can thus be represented as shown in Table 6.6.

A recessive X-linked character is expected to be much more frequent in males than in females.

In Table 6.6, if a is an X-linked recessive, affected females are X^aX^a and occur with frequency q^2, while affected X^aY males occur with frequency q. Thus the frequency of homozygous recessives in females is the square of that in males. For example, color blindness in men would be expected to occur with a frequency of $q = 0.08$ (or 8 percent), while color blindness in women would be expected to occur with a frequency of $q^2 = (0.08)^2 = 0.0064$ (or 0.64 percent)—not far from the observed values for the female incidence. Hemophilia, a rare X-linked recessive disease that causes a delay of blood clotting (see Sections 2.5 and 5.3) has a frequency somewhat less than 1 in 10,000 in males and is almost unknown in females. This is not surprising, for we would expect the frequency in females to be less than $(1/10{,}000)^2$, or less than 1 in 100,000,000.

Although a higher frequency of a trait in males than in females is suggestive of X-linked inheritance, it is no proof that a trait is an X-linked recessive. Some conditions, such as pattern baldness, are *sex-limited* (or sex-controlled, or sex-influenced—see Section 2.5) rather than sex-linked. These traits are controlled by autosomal genes that are expressed differently in the two sexes. Sex-limited

Table 6.6
Hardy–Weinberg frequencies for X-linked genes

Sex	Genotype	Frequency
Males	X^AY	p
	X^aY	q
Females	X^AX^A	p^2
	X^AX^a	$2pq$
	X^aX^a	q^2

NOTE: A and a are alleles on the X chromosome with frequencies of p and q, respectively, in both sexes. Note that (1) the male genotype frequencies are just the gene frequencies of the two sorts of X chromosomes, X^A and X^a; and (2) female genotype frequencies are just the same as those for an autosomal locus with allele frequencies p and q.

characters occur with different phenotype frequencies in males and females, although they are not determined by X-linked genes.

If the genotype frequencies in a population are not initially at the values predicted by the Hardy–Weinberg law for the gene frequencies in the population, the genotype frequencies will approach the expected equilibrium frequencies under random mating (under the conditions assumed in Section 6.3). For an autosomal locus, it will take only one generation to reach the Hardy–Weinberg equilibrium. For X-linked genes, it will take more than one generation to reach the expected equilibrium values.

In Figure 6.3, the genotype frequencies for offspring were computed directly from the values of p and q from gametes—that is, from the parental generation. The expected genotype frequencies among the offspring would be the same, even if the genes had been distributed in some nonequilibrium pattern among the parents. For example, suppose that a group of people randomly selected from several different equilibrium populations were to establish a new settlement in some previously uninhabited area. The frequencies of genotypes for any particular locus would probably not show equilibrium values in this initial generation. However, there would be some particular values for the gene frequencies (p, q, . . .) in this generation. For an autosomal locus, the genotype frequencies among the *first* generation of offspring would reach the expected equilibrium values (p^2, $2pq$, q^2, . . .) if the Hardy–Weinberg conditions are valid.

The equilibrium frequencies of genotypes for X-linked genes, however, are not reached in one generation. If the gene frequencies differ initially in the two sexes, there will be changes in successive generations. The gene frequencies

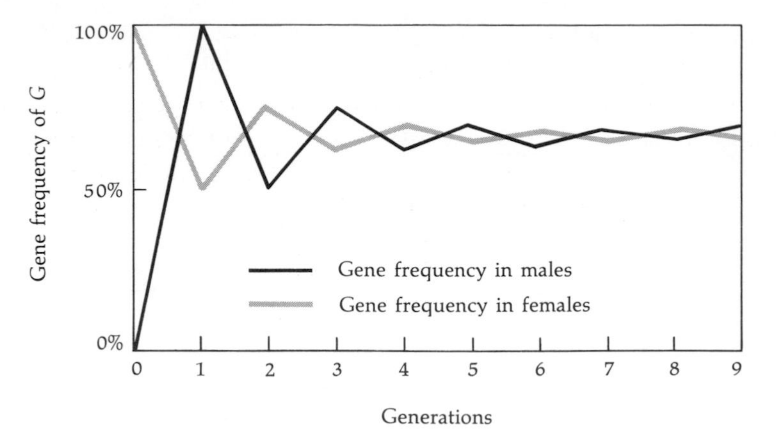

Figure 6.8
Approach to genetic equilibrium for an X-linked gene. The initial population is made up of females all carrying only *G* alleles for an X-linked trait G (p, the gene frequency of *G*, equals 1.00 for the females) and of males all carrying the allele *g* ($p = 0.00$ for males). There are equal numbers of males and females in the population, so the overall gene pool has twice as many *G* alleles as *g* alleles (because each female has *G* on both of her X chromosomes). Thus the overall gene frequency for the total population is $p = 0.67$. In the first generation of offspring, each male receives his single X chromosome from his mother, so males are all G and the value of p for males jumps abruptly from 0.00 to 1.00 in a single generation. Each female in generation 1 receives a *G* allele from her mother and a *g* allele from her father, so the value of p for females changes from 1.00 in generation 0 to 0.50 in generation 1. You can continue these computations for succeeding generations, noting that in each generation the males have a value of p equal to that of their mothers, while the females have a value of p equal to the average of the values for their mothers and fathers. The overall gene frequency remains $p = 0.67$ for each generation. After a few generations, the originally extreme difference between male and female gene frequencies becomes so small as to be negligible.

averaged over males and females remain the same at every generation, but the gene frequency in each sex considered alone changes from one generation to the next—with the frequencies in the two sexes gradually approaching the same equilibrium value by a series of damped oscillations (Fig. 6.8). This phenomenon may be of practical interest when there are massive hybridizations between populations that differ in the gene frequencies of X-linked genes.

6.6 Departures from Hardy-Weinberg Equilibrium

In the example introduced at the beginning of this chapter (the examination of Ghanaian children for sickle-cell anemia, Table 6.3), the agreement between the expected and observed values is so good that we would have very little doubt of the applicability of the Hardy-Weinberg law to those data. We can hardly, in general, however, expect such close agreement between observed and expected figures, even when the Hardy–Weinberg law holds. Observed numbers are always from limited—and usually small—samples, and in the sampling of a population, fluctuations due to chance inevitably arise. There are methods for deciding whether these fluctuations can be due to chance effects of sampling or not. The statistical procedure appropriate to this problem is the χ^2 test, which is discussed in the Appendix. Application of this test to the (real) sickle-cell anemia example of Table 6.3 would indicate a good fit. When this source of departure from expectation is taken into account, one finds that almost all real populations are apparently in Hardy–Weinberg equilibrium (Fig. 6.9). This may seem surprising in view of the long list of assumptions that must be made for the law to be valid in full rigor. Among the most important of them are the assumptions that mating takes place at random and that all pairs have the same probability of having children that reach sexual maturity.

Although significant observed departures from Hardy–Weinberg equilibrium are rare, there are many possible causes for discrepancies in either direction—toward a deficiency or an excess of heterozygotes.

The most obvious causes of discrepancy from Hardy–Weinberg equilibrium arise when the two assumptions just noted do not hold. *Assortative* mating (the choice of partners) as a cause of deviation from randomness in matings is very rarely observed for single genes. As we have already noted, the basis of attraction between a man and a woman can hardly be the genes that are most commonly studied, such as blood groups or enzymes. The individuals themselves are usually not aware of these properties, and their secondary effects (if any) are usually trivial. The other assumption—equal probability of all genotypes having viable progeny—can also be called the absence of natural selection, and it is discussed in the next chapter. However, it turns out that only rarely is natural selection of sufficient intensity to be detectable through deviations from Hardy–Weinberg equilibrium. Take the example of Ghanaian children (Table 6.3). As we shall see later, we know that sickle-cell anemia is under the effect of natural selection. Yet, in the sample of Ghanaian children we were unable to detect statistically significant deviations from Hardy–Weinberg expectation. This may be due to the fact that the sample is not sufficiently large; or that the children were too young. In fact, if we sample children at birth we do not find, in the case of sickle-cell anemia, deviations from Hardy–Weinberg; we may find them if we sample adults (and if a sufficiently large sample is investigated) because the three genotypes have a different probability of reaching adulthood.

In practice, deviations may be found also for other reasons—for instance, if the genetic determination of a trait is imperfectly understood, or if some individuals

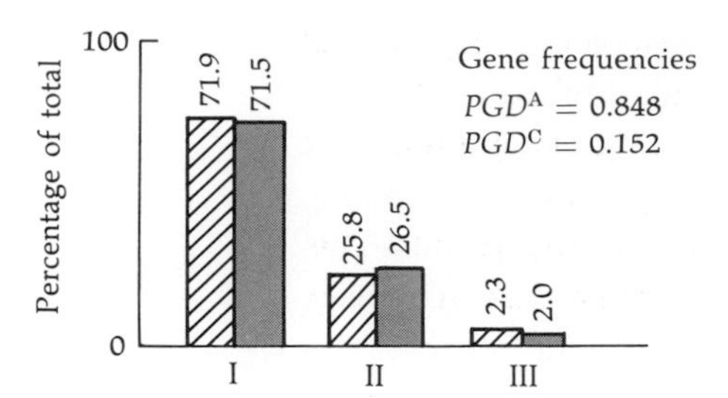

6-Phosphogluconate dehydrogenase phenotype in South African Bantus ($N = 200$)

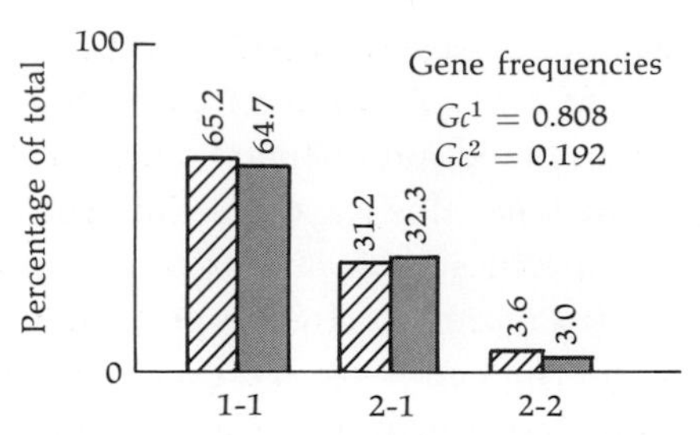

Group-specific component phenotype in Bangkok Thai ($N = 133$)

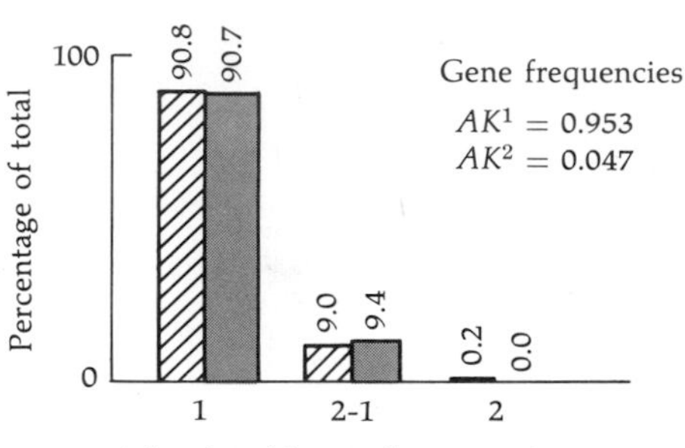

Adenylate kinase phenotype in Iraqi Jews ($N = 139$)

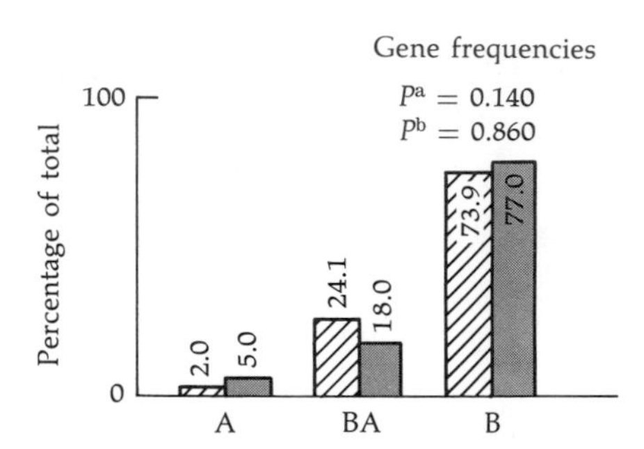

Red-cell acid phosphatase phenotype in Andean Indians of Peru ($N = 139$)

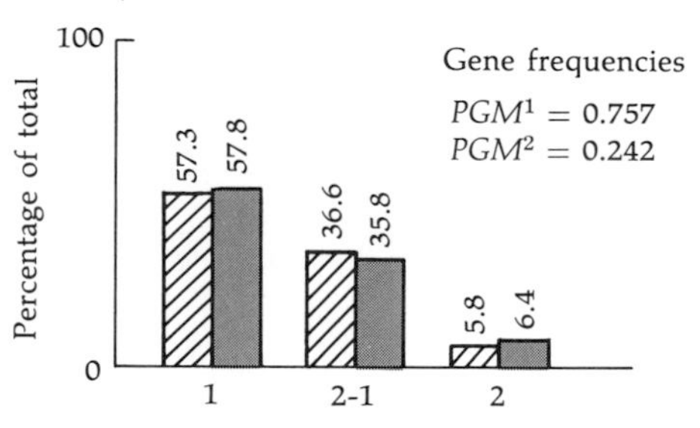

Phosphoglucomutase phenotype in Chinese ($N = 419$)

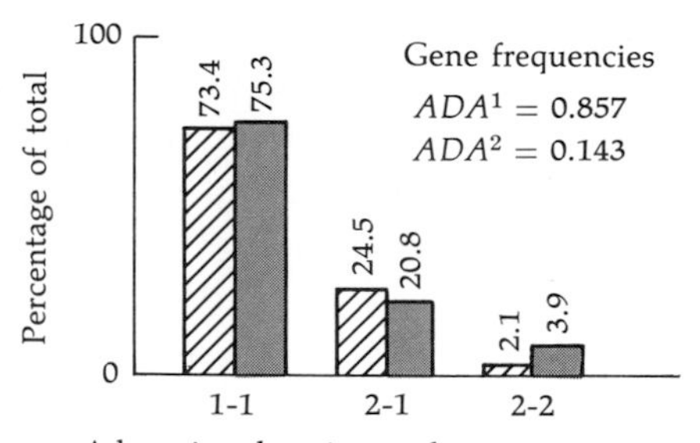

Adenosine deaminase phenotype in Ceylon Sinhalese ($N = 154$)

Figure 6.9
Examples of the application of the Hardy–Weinberg law. These bar graphs show expected and observed genotype frequencies for various human populations. Only codominant systems are shown, in which all genotypes can be identified and counted. Striped bars represent the observed frequencies; solid bars represent frequencies expected according to the Hardy–Weinberg law (using gene frequencies computed from the observed genotype frequencies). In five cases, the deviation between expectation and observation is small; a χ^2 analysis shows that each deviation can reasonably be explained by chance fluctuations for the size of the sample. Among the Andean Indians of Peru, there is a small shortage of heterozygotes.

are classified incorrectly into the various genotypes. Recognition of the cause of the deviation may then improve our knowledge of genetic determination of a trait. Thus, the analysis of the Hardy–Weinberg equilibrium can be a useful procedure for testing the validity of the interpretation of population-genetic data.

Another source of deviation arises when populations with different gene frequencies, each of which is in Hardy–Weinberg equilibrium, are pooled. This may well happen for instance in sampling individuals from a city in which recent immigrants from a distant, relatively different population are abundant. If immigrants are not distinguished, the population is heterogeneous. The outcome, as shown in Exercise 8, is a deficiency of heterozygotes due to heterogeneity. If the immigrants, however, have been present at least for one generation and have mixed with the earlier inhabitants, the deficiency of heterozygotes will disappear. Hardy-Weinberg equilibrium can in fact be established in one generation.

One of the results of the wide applicability of the Hardy–Weinberg law is that all the facts about the genetic behavior of a population can usually be summarized just in terms of the gene frequency or frequencies. The gene frequency is not expected to change in the absence of disturbing forces. These forces are described in subsequent chapters, but it may be useful to list here the major ones. They are (1) mutation (which, however, alters gene frequencies only very slowly), (2) natural selection (which alters the probability that individuals carrying one or the other allele leave progeny in the next generations) and, finally, (3) chance events in sampling of the gametes that form the successive generations (a phenomenon known as "random genetic drift"). Change in genotype frequencies is the basic evolutionary phenomenon. The fact that we can predict the frequencies of genotypes and phenotypes from gene frequencies, on the assumption of random mating following the Hardy–Weinberg law, is a major factor that simplifies the description of evolution and makes it possible to study many evolutionary phenomena simply as changes in gene frequencies.

Summary

1. It is advantageous for certain purposes to describe a population in terms of the frequencies of alleles at a locus.

2. Gene frequencies are the frequencies of alleles at one locus in a given population. They are also the frequencies of these alleles in the gametic pool.

3. When mating is random, the frequencies of genotypes are easily obtained as the product of the frequencies of the genes corresponding to the genotypes.

4. Therefore, a homozygote for an allele has a frequency equal to the square of the frequency of that allele, and a heterozygote frequency is twice the product of the two corresponding allele frequencies.

5. The above rule is part of the Hardy–Weinberg law, which also says that the expected proportions of genotypes reach equilibrium values (in one generation for autosomal genes), whatever the previous proportions, and do not change subsequently.

6. In the absence of mutation, natural selection, or chance variation, the frequencies of genes and genotypes remain constant, and genetic variability is maintained indefinitely (a state of equilibrium has been reached).

7. Deviations from the Hardy–Weinberg law are rare; they may be due to nonrandom mating (for example, assortative mating, or heterogeneity of the population) or error of the basic genetic hypothesis. Natural selection may also, though rarely, be detected as a deviation from Hardy–Weinberg equilibrium of the genotypes.

8. With X-linked genes, Hardy–Weinberg equilibrium may hold for females, but males' genotype frequencies are dictated by frequencies in the mothers' gametes, and at equilibrium they are equal to the gene frequencies.

Exercises

1. What are the main advantages of working with gene rather than genotype frequencies?

2. Outline the most important implications of the Hardy–Weinberg law.

3. What are the major causes of departure from Hardy–Weinberg equilibrium?

4. Indicate whether each of the following populations is in Hardy–Weinberg equilibrium.
   a. *AA* 0.30; *Aa* 0.42; *aa* 0.28.
   b. *BB* 0.09; *Bb* 0.42; *bb* 0.49.
   c. *CC* 0.9801; *Cc* 0.0198; *cc* 0.0001.
   d. *DD* 0.9216; *Dd* 0.0768; *dd* 0.0016.
   e. *EE* 0.0100; *Ee* 0.2678; *ee* 0.7222.

5. The following values are observed numbers of individuals with each genotype in various populations. For each population, use a χ^2 test (see Appendix) to determine whether the observed distribution represents Hardy–Weinberg equilibrium.
   a. *FF* 32, *Ff* 48; *ff* 20.
   b. *GG* 0; *Gg* 18; *gg* 35.
   c. *HH* 313; *Hh* 47; *hh* 21.
   d. *II* 27; *Ii* 105; *ii* 121.
   e. *JJ* 320; *Jj* 480; *jj* 200.
   f. *KK* 3,200; *Kk* 4,800; *kk* 2,000.

6. For each of the populations in Exercises 4 and 5 that is not in equilibrium, state whether there is an excess or a deficiency of heterozygotes.

7. Compare the results of parts a, e, and f in Exercise 5 and comment.

8. Two populations are examined for the same gene with the following results.
   Population 1: *AA* 4; *Aa* 32; *aa* 64.
   Population 2: *AA* 36; *Aa* 48; *aa* 16
   Now suppose that these two populations are combined to form a new population.
   Population 3: *AA* 40; *Aa* 80; *aa* 80.
   a. Determine if each of the three populations is in Hardy–Weinberg equilibrium. For any population that is not, indicate whether heterozygotes are in excess or in deficiency.
   b. What is the difference between populations 1 and 2?
   c. What is the effect of the admixture of the two populations on the genetic equilibrium?
   d. For how many generations will the effect of this single episode of admixture be detectable in population 3, if mating is at random?

9. Manic-depressive psychosis of the bipolar (bmdp) type is more frequent among females than among males. Indicate whether each of the following hypotheses is in agreement with the observed sex distribution:
   a. bmdp is X-linked and recessive;
   b. bmdp is X-linked and dominant;
   c. bmdp is sex-influenced but not sex-linked.

10. Suppose that a test of MN blood types is being carried out, but that no anti-N serum is available. The frequency of M^- individuals (giving no agglutination with anti-M) is found to be 0.20; the frequency of M^+ individuals (agglutination with anti-M) is 0.80.
   aa. What is the expected frequency of M^- children among the offspring of $M^+ \times M^+$ matings? Solution: The gene frequency of M is $1 - \sqrt{0.20} \cong 0.55$. The frequency of MM homozygotes is estimated to be $(0.55)^2 \cong 0.30$, and that of MN heterozygotes $2 \times 0.45 \times 0.55 \cong 0.50$. The frequency of $M^+ \times M^+$ matings should be $0.8 \times 0.8 = 0.64$, and that of matings between heterozygotes should be $0.5 \times 0.5 = 0.25$. Therefore, $0.25/0.64 = 0.39$ of all $M^+ \times M^+$ matings are between heterozygotes. These are the only matings between M^+ individuals that can produce M^- offspring, and we expect $\frac{1}{4}$ of the offspring of the heterozygous matings to be M^-. Therefore, the frequency of M^- offspring from $M^+ \times M^+$ matings should be $(\frac{1}{4}) \times 0.39 = 0.0975$.
   a. What is the expected frequency of M^- children among the offspring of $M^+ \times M^+$ matings if the observed frequency of M^- individuals among the parental generation is 0.04?
   b. What is the expected frequency of O children from A × O matings if the gene frequencies for the ABO system in the population are those given in Table 6.5?
   c. The Xg^a blood type is an X-linked, dominant phenotype; the Xg blood type is the homozygous recessive. What will be the frequency of Xg^a sons among the male offspring of $Xg^a \times Xg$ matings?

References

Hardy, G. H.
1908. "Mendelian proportions in a mixed population," *Science,* vol. 28, pp. 49–50. Reprinted in *Classic Papers in Genetics,* ed. J. H. Peters, Englewood Cliffs, N.J.: Prentice-Hall, 1959.

Weinberg, W.
1908. "On the demonstration of heredity in man." Reprinted in *Papers on Human Genetics,* ed. S. H. Boyer, Englewood Cliffs, N.J.: Prentice-Hall, 1963.

TEXTBOOKS IN POPULATION GENETICS

Cavalli-Sforza, L. L., and W. F. Bodmer
1971. *The Genetics of Human Populations.* San Francisco: W. H. Freeman and Company.
Crow, J. F., and M. Kimura
1970. *An Introduction to Population Genetic Theory.* New York: Harper and Row.
Jacquard, A.
1974. *The Genetic Structure of Populations.* New York: Springer-Verlag.
Lewontin, R. C.
1974. *The Genetic Basis of Evolutionary Change.* New York: Columbia University Press.
Li, C. C.
1955. *Population Genetics.* Chicago: University of Chicago Press.
Wright, S.
1968. *Evolution and the Genetics of Populations, Volume I: Genetic and Biometric Foundations.* Chicago: University of Chicago Press.
1969. *Evolution and the Genetics of Populations, Volume II: The Theory of Gene Frequencies.* Chicago: University of Chicago Press.

7

Natural Selection and Demography

Natural selection is the mechanism by which populations adapt to their environments. Those phenotypes that are transmitted to the progeny and confer an "advantage" will automatically expand within a population. The advantages that matter in terms of natural selection are the capacity to survive and to leave offspring. We are therefore led to consider questions of birth and death rates, fertility and mortality—in other words, the subject of demography.

7.1 Natural Selection Defined

There must have been a first person who carried a sickle-cell hemoglobin gene, which originated by a mutation in the sperm or egg that gave rise to this heterozygous carrier. We now know that, if he (or she) lived in an environment in which the disease malaria was common, he would have had a better chance than his "normal" peers of surviving through his early years (those years in which malaria takes its heaviest toll) and thus of reaching adulthood and having children. Half of these children would also carry the sickle-cell gene, and they would spread it further to their children. It is possible that there occurred only one mutation to sickle-cell hemoglobin in all of human history. Because of the advantage this mutation conferred on its carriers, the gene may eventually have spread through his descendants to those parts of the world—areas with heavy malarial infestations—where it would be advantageous to have such a gene. Other factors, however, that are inherent in the properties conferred by this gene on its carriers presumably limited its further spread and confined it to the malarial areas.

Sickle-cell anemia is one of the most thoroughly studied genetic conditions, and we consider it in detail in a later chapter when we discuss the factors that maintain genetic variability in a population. In this chapter we use sickle-cell hemoglobin as an example of *natural selection* to illustrate the modern interpretation of Darwin's theory of evolution through natural selection. Some genetic variants (that is, individuals carrying a new genetic mutation) may tend to have a different number of surviving offspring from that of pre-existing "normal" types. This number may be higher or lower: in the first case the genetic variant is selectively advantageous, while in the second it is disadvantageous with respect to the "normal" type. These differences in number of offspring usually depend on the environment in which individuals develop. For sickle-cell hemoglobin, the conditions that matter are the presence and concentration of the parasite that causes malaria. "Natural selection" means that some types are "chosen" or "selected" by the environment in which they are born, because they leave more descendants than do other types. The process will continue as long as the favorable environment is unchanged. Traits favored by the environment will automatically tend to increase in numbers as they are transmitted to their more numerous progeny; those that are disfavored will automatically tend to decrease. Eventually, the favored type may be the only one that remains.

There are two main mechanisms by which some genetic types may leave more (or less) descendants than other types.

First, an individual may leave more descendants because he is better able to resist an adverse environmental condition and so survive to adulthood or sexual maturity. In this case, the effect is manifested through differential *survival* (or its converse, differential *mortality*) of the various genetic types. Second, there may be differences in the number of children born—that is, differential *fertility*. Both of these mechanisms, survival and fertility, must be taken into account, for the number of descendants depends on both. Thus, the measurement of natural selection is based on measurements of differential survival and fertility of different genetic types.

It is important to remember that survival and fertility are measured in relation to a specific environment.

Natural selection is determined by the environmental conditions in which individuals live; if these conditions change, selection may be modified, or even reversed. A genetic type favored in one environment may be selected against in another. This is exactly what happens with sickle-cell hemoglobin, depending on whether malaria is present or absent in the environment.

Selection acts on the phenotypes of individuals.

We shall often refer to selection for or against certain genes and genotypes. Such reference, however, is an abstraction, for what survives and reproduces or dies are individuals (whole organisms), not genes. Selection acts on individuals (whole organisms), not genes. Moreover, selection acts on individuals on the basis of their phenotype. Genetic evolution is the change of genotypes. But phenotypes depend on genotypes; natural selection will thus inevitably lead to changes in the relative number of different genotypes and so bring about genetic evolution. Saying that "a genotype has a selective advantage (or disadvantage)" is really a shortcut for saying that, under given environmental conditions, the genotype determines a phenotype with an increased (or decreased) expected number of descendants relative to another genotype or genotypes. The relative increase (or decrease) in frequency of the genotype in future generations is determined by the joint effect of increased (or decreased) survival and fertility under the particular environmental conditions.

7.2 Demography

The study of populations—in particular how they survive, die, and reproduce or grow—is called *demography.* The measurement of natural selection, which as we have seen involves the chances of survival and reproduction of individuals that have different genotypes, follows concepts and methods closely akin to those of demography. The main difference is that demography involves the study of populations as a whole, whereas natural selection acts on groups of individuals, defined by their genotypes, who form only a part of a population. It is the comparison of the demographic properties of these groups within a population that forms the basis for the detailed investigation of natural selection.

The specific reasons why some genotypes have higher or lower mortalities or fertilities must be studied in a biochemical, physiological, and ecological context. But the measurement of the intensity of natural selection uses demographic techniques to estimate these mortalities and fertilities and does not usually depend on knowledge of the underlying mechanisms. In this and the next section therefore, we outline some of the basic principles of demography used in the measurement of natural selection.

The two basic factors that determine population numbers are birth and death.

The birth rate in a population is the number of individuals born in a given period, say a year, and is often expressed per thousand individuals living in the population, at that time. In present day human populations it varies (according to the country) from about 9 to 50 per thousand. The death rate is the number of individuals in a population who die in a given period, for example a year, and is also usually given per thousand living persons. It varies within limits similar to the birth rates. The growth of the population is measured by the growth rate, which (on a per-year basis) equals the number of births minus the number of deaths occurring in that year. Nowadays almost all populations are increasing because birth rates are higher than death rates. The rates of increase vary from country to country in the range 0 to 35 per thousand people per year.

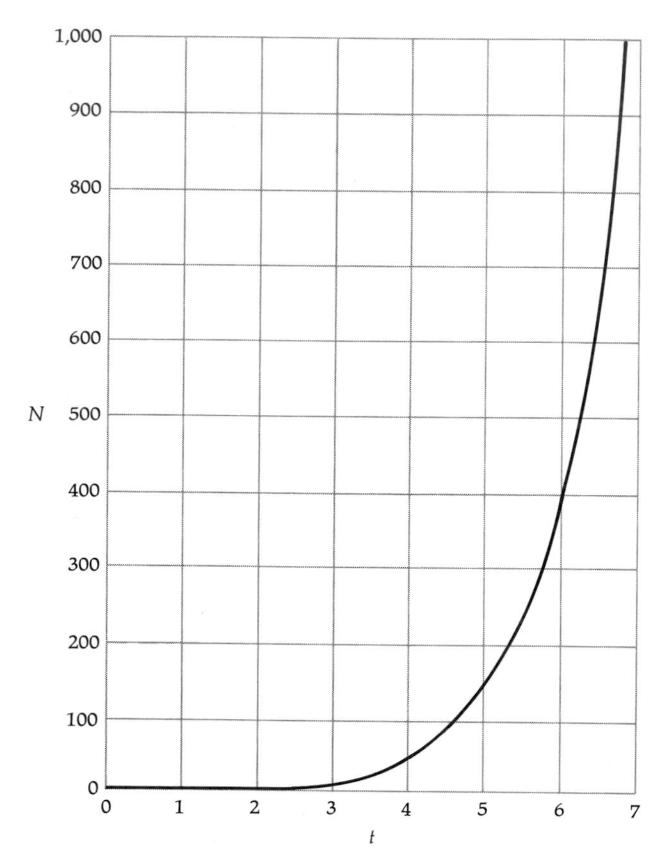

Figure 7.1
Exponential growth. This graph is a plot of the function $N = e^t$; the value of N is said to increase exponentially (or logarithmically) as a function of t. The logarithmic constant e has a fixed value of 2.71828 . . . (the exact value extends to an infinite number of places past the decimal). If t represents time in years, then this curve represents the growth of a population with an initial size of 1 individual and with a growth rate of 1.0 percent per year; the value of N gives the population size at each point in time. Note that the population grows very slowly while N remains small, but that the curve climbs ever more steeply as N increases.

If only birth and death rates are considered, the rate increase obtained is the "*natural*" or *intrinsic* rate. Actual changes in population numbers also include *migration* into or out of the population (immigration and emigration, respectively). The two rates of migration must be, respectively, added to and subtracted from the intrinsic rate to obtain the *actual* rate of population growth.

If the relative rate of increase of a population is constant (say, an increase of 3 percent per year), the size of the population grows geometrically, or *exponentially* (Fig. 7.1). Figure 7.2 shows the population growth that would occur over a century

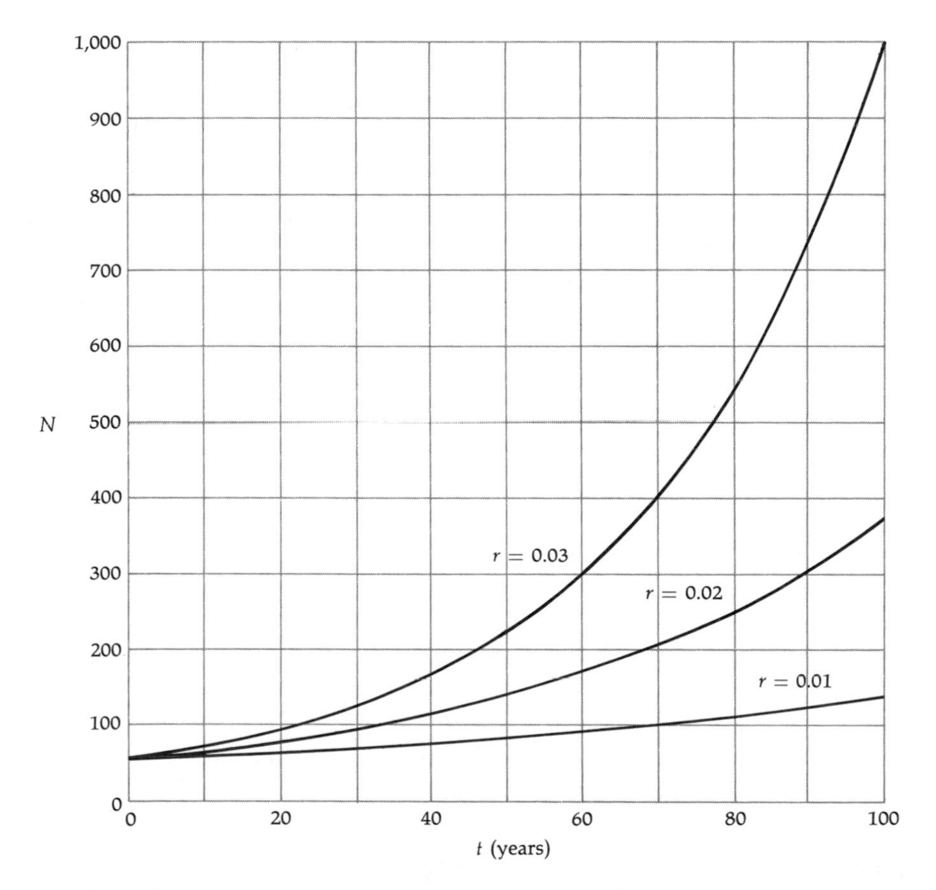

Figure 7.2
Exponential growth at different growth rates. The three curves show changes in population size at growth rates (r) of 1, 2, and 3 percent per year; each curve begins with an initial population size of $N_0 = 50$. The curves are plots of the general function $N = N_0 e^{rt}$. The same curves can be used to represent larger populations, because N could represent individuals, thousands of individuals, or even millions of individuals. In other words, the shape of the curve remains the same if we multiply N and N_0 by any number.

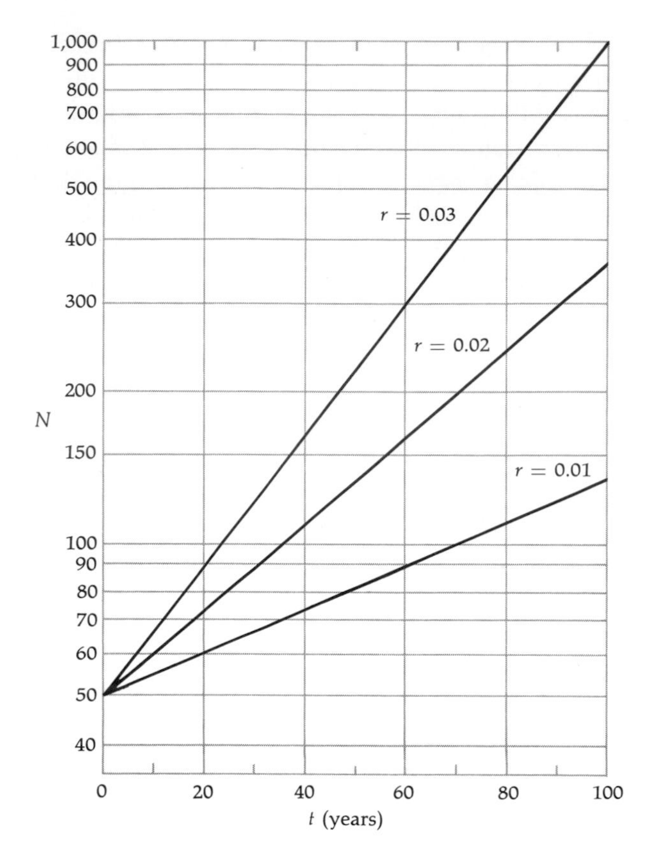

Figure 7.3
Exponential growth curves plotted on semilogarithmic graphs. This graph shows the same curves as those shown in Figure 7.2, but the vertical scale is proportional to the logarithm of N. Note that the curves become straight lines when plotted in this way. In other words, exponential growth causes N to increase in such a way that the logarithm of N increases by a fixed amount each year.

with growth rates of 1, 2, or 3 percent—in each case starting from an initial population of 50. Figure 7.3 shows the same growth rates plotted on a semilogarithmic graph. Here, the population size on the vertical axis of the chart is measured by the logarithm of the number of individuals in the population, and the growth curves appear as straight lines (see the Appendix for further discussion). If the growth rate is zero, the population remains constant in size; the population size is stationary, or constant.

Exponential growth, even with relatively small rates of increase, is fast. No population can grow exponentially for long because its size would soon become astronomical.

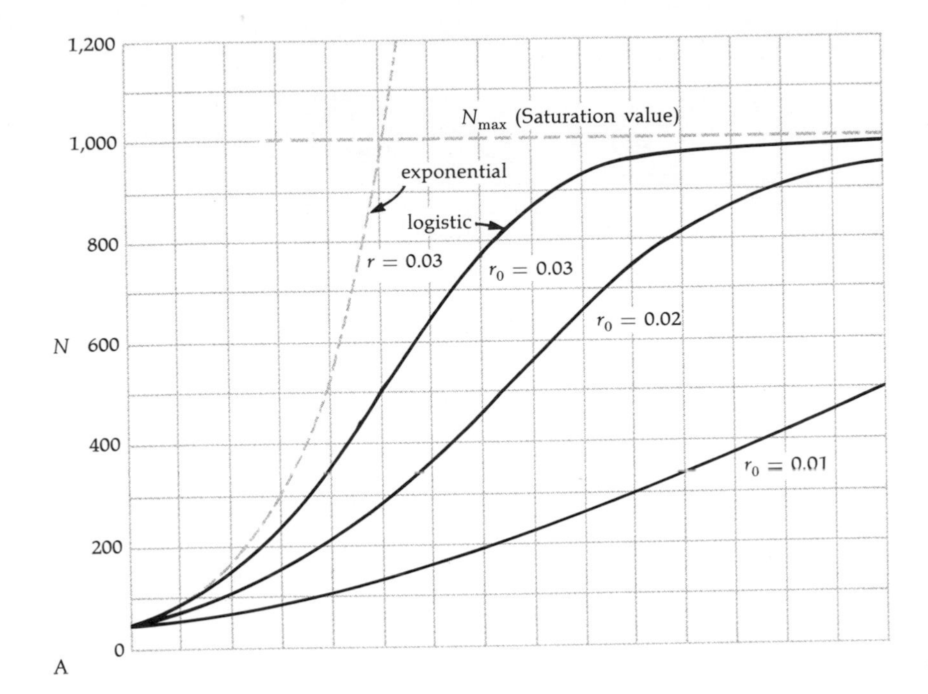

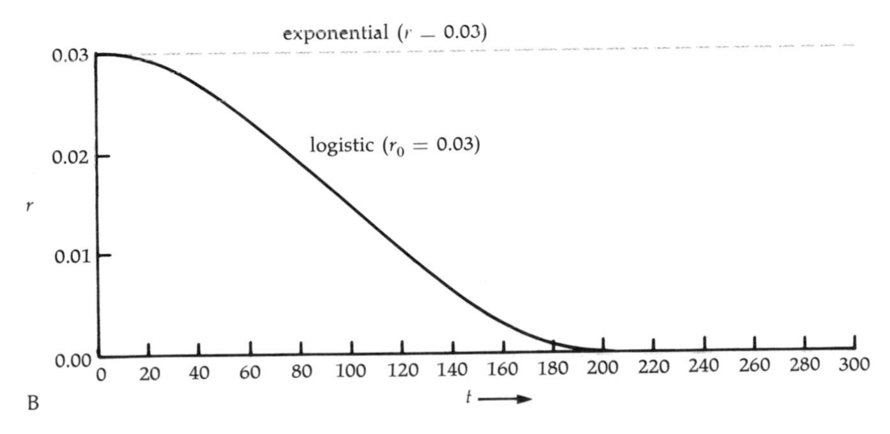

Figure 7.4
The logistic growth curve. (A) The solid lines show logistic growth curves with initial growth rates (r_0) of 1, 2, and 3 percent. In each case, $N_0 = 50$ and the maximum (saturation) population $N_{max} = 1{,}000$. The dashed curve shows exponential growth at a rate of 3 percent. Note that the exponential and logistic curves are very similar at first, but diverge more and more sharply as time passes. (B) This graph shows r as a function of time for the logistic and exponential curves with initial growth rates of 3 percent. For the exponential growth curve (dashed line), r remains constant at 0.03. For the logistic curve (solid line), r decreases as a function of time. The mathematical form of the logistic curve is quite complex; the expression for r can be written as $r = N_{max}r_0/(N_{max} - N)$, where N of course is a complex function of time.

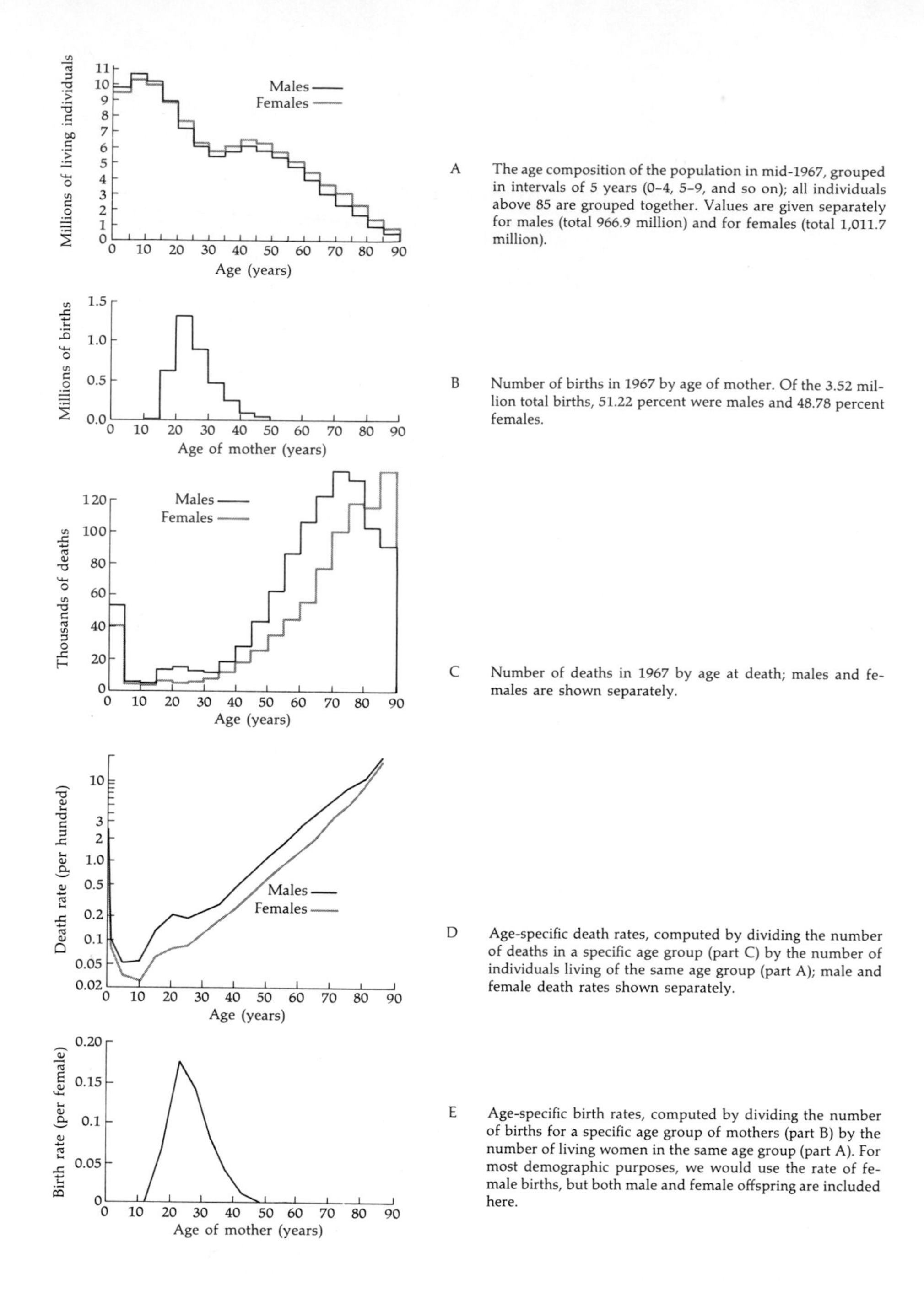

A The age composition of the population in mid-1967, grouped in intervals of 5 years (0–4, 5–9, and so on); all individuals above 85 are grouped together. Values are given separately for males (total 966.9 million) and for females (total 1,011.7 million).

B Number of births in 1967 by age of mother. Of the 3.52 million total births, 51.22 percent were males and 48.78 percent females.

C Number of deaths in 1967 by age at death; males and females are shown separately.

D Age-specific death rates, computed by dividing the number of deaths in a specific age group (part C) by the number of individuals living of the same age group (part A); male and female death rates shown separately.

E Age-specific birth rates, computed by dividing the number of births for a specific age group of mothers (part B) by the number of living women in the same age group (part A). For most demographic purposes, we would use the rate of female births, but both male and female offspring are included here.

Figure 7.5
Some standard demographic data for the United States. These graphs show data for the United States in 1967. Parts A, B, and C show information obtained by various sampling and estimation techniques. Parts D and E show age-specific death and birth rates computed from the other data. (Data from tables published by N. Keyfitz and W. Flieger, *Populations: Facts and Methods of Demography,* W. H. Freeman and Company, 1971.)

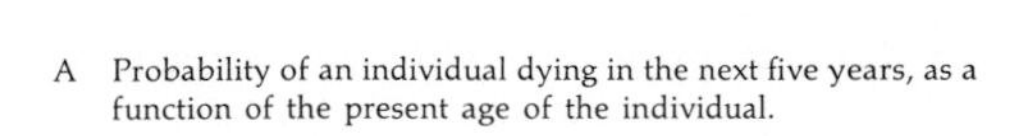

A Probability of an individual dying in the next five years, as a function of the present age of the individual.

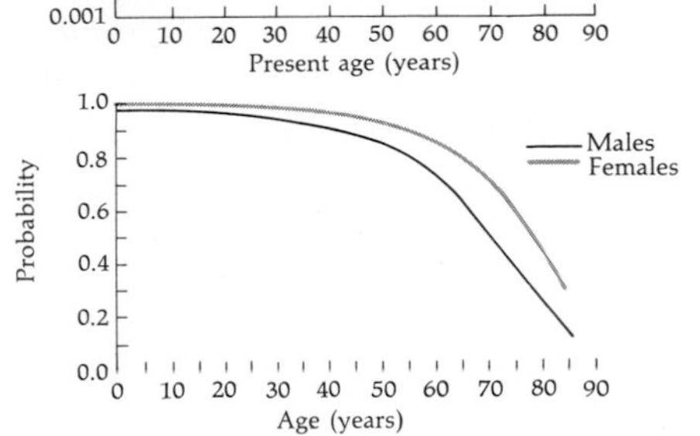

B Probability that a newborn individual will survive to a given age.

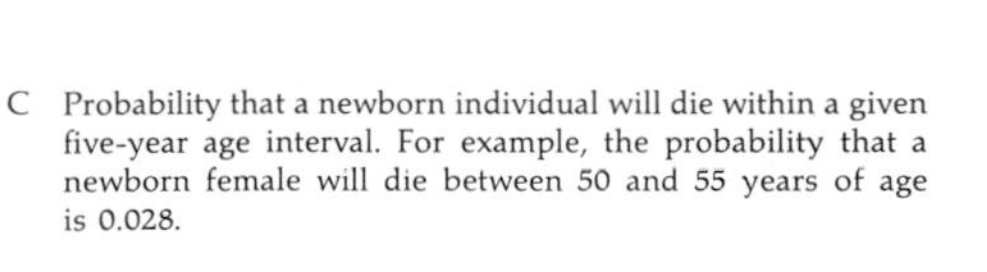

C Probability that a newborn individual will die within a given five-year age interval. For example, the probability that a newborn female will die between 50 and 55 years of age is 0.028.

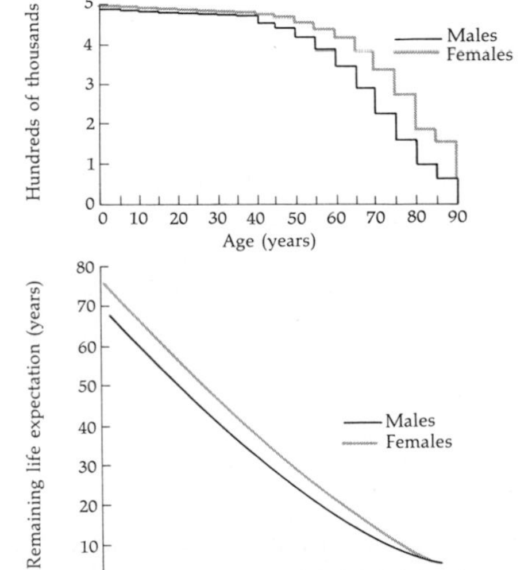

D Total years lived within a given five-year age interval, per 100,000 born. This curve also represents the age distribution the population would have if the overall birth rate equaled the overall death rate, if there were exactly 100,000 newborn each year, and if age-specific birth and death rates followed the same relative patterns as those now observed.

E Expectation of remaining life at a given age—that is, the average number of years lived beyond a given age by those who reach that age.

Figure 7.6

Some standard demographic probabilities and expectations. The graphs here exemplify the computation of projected future trends from the raw data of Figure 7.5 (for the United States in 1967). (Data from tables published by N. Keyfitz and W. Flieger, *Populations: Facts and Methods of Demography,* W. H. Freeman and Company, 1971. Procedures for obtaining these statistics from the raw data are explained by Keyfitz and Flieger in their book.)

Growth rates must, therefore, at some stage decrease; they must eventually go down to zero if population size is to remain constant. To obtain zero growth rate, either the birth rate must go down or the death rate up, or both, until the birth rate equals the death rate, and the population is neither increasing nor decreasing.

A growth rate that decreases with time produces a population growth curve that is sigmoid in shape (Fig. 7.4). A famous example of such a curve is called *logistic,* and its mathematical properties are explained in the Appendix. Initially, the logistic curve is like an exponential curve but, as numbers increase, the growth rate decreases until it becomes zero when the population size reaches its maximum or "saturation" level. The number of individuals increases more slowly than with exponential growth (the more so, the larger the number) until population size gradually levels out when saturation has been reached, and the growth rate becomes zero.

The logistic growth model is useful as an approximate description of the growth of some human and animal populations because it takes into account the fact that a given environment can at most support a certain maximum population size. The ecological term for this maximum is the "carrying capacity" of the land, which is of course limited.

We are in the middle of a "demographic transition."

Human population growth cannot be adequately represented by a simple logistic growth curve because it is complicated by a large number of different factors (Figs. 7.5 and 7.6). As we discuss more fully in Chapter 21, the world population is now in a transitory phase involving rapid and complex changes in birth and death rates. This may therefore be an especially difficult period for the study of population growth and, consequently, of selection. The major current trend is toward a radical decline in the relative number of deaths at early ages—namely, before and during the reproductive period of life. The decrease in death rate is accompanied, usually after some delay, by a decline in the birth rate; this whole phenomenon is called the "demographic transition." Its main features are similar in most countries, but detailed patterns (especially the approximate starting dates of the transitions) differ widely from one country to another. In some countries, the transition has barely begun, while in others it started over 100 years ago.

Figure 7.7 illustrates the demographic transition for Sweden. Birth rates tended to exceed death rates by a relatively small margin before 1800. However, there were substantial fluctuations from year to year because of pestilence, famine, and wars—the major "checks" on population size. Since 1800 the death rate has been decreasing quite steadily and it is still decreasing now, though somewhat more

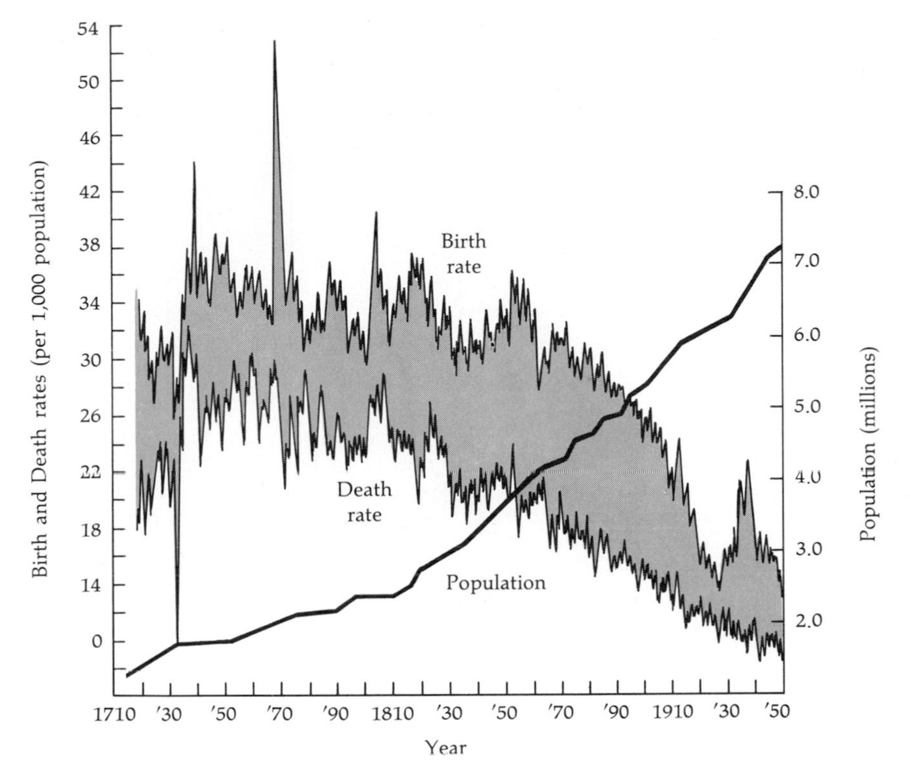

Figure 7.7
The demographic transition in Sweden. In the nineteenth century, death rates began to decrease, and exceptional peaks of mortality tended to disappear. Birth rates did not begin to decrease sharply until the second half of the nineteenth century. Population (dark line, scale at right) showed a higher rate of increase during the transition. (After E. Vielrose, *Elements of the Natural Movement of Population,* Oxford: Pergamon Press, 1965.)

slowly. The birth rate however started decreasing only around the 1850s and, as a result of this lag in the decline of the birth rate, the total population size has in the meantime been growing more-or-less exponentially.

The demographic transition seems to be associated particularly with economic development. As a result, there are at present considerable demographic differences between developing and developed countries. In the latter, hygiene and medicine have been very effective in reducing most postnatal causes of death, and thus in lengthening the expectation of life at birth. They have, however, been far less effective in reducing prenatal loss (especially spontaneous abortion) and postreproductive death rates. The contrasting patterns of death rates at the extremes of development are represented schematically in Figure 7.8. Various countries or regions show an almost continuous variation between the two extremes of difference. The trend for expectancy of life in Sweden is illustrated in

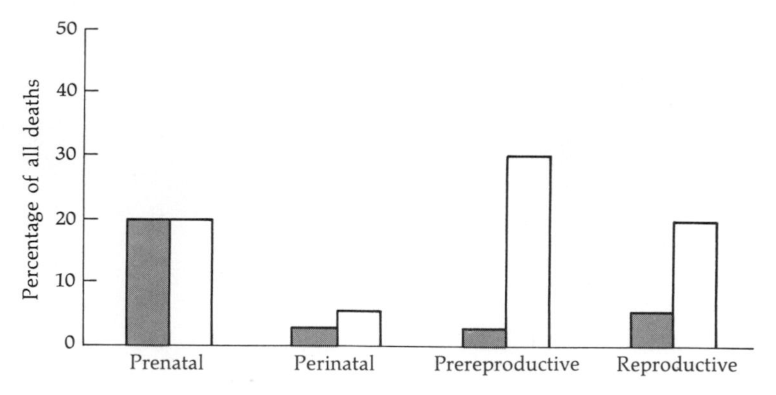

Figure 7.8
Death-rate analysis: typical values in highly developed countries (dark bars) and in under-developed areas (white bars). The bar graphs show the percentages of all deaths that occur within four age ranges: (1) *prenatal* (spontaneous abortions); (2) *perinatal* (stillbirths plus deaths in the first week of life); (3) *prereproductive* (ages zero to 15 years for women, zero to 20 years for men); and (4) *reproductive* (ages 15 to 45 for women, 20 to 50 for men). The remaining deaths (70.5 percent for developed countries, 25 percent for under-developed) occur in the postreproductive period. Because early spontaneous abortions are often undetected, the figures given for the prenatal period are highly approximate.

Figure 7.9. The fall in mortality at the younger ages has considerably increased the expectancy of life at birth, which has doubled in less than 200 years. But the expectancy of life at 50 years has only moderately increased (by less than 50 percent), and so has that at 70 years. Developing countries show patterns of mortality and, therefore, of life expectancy similar to those prevalent in Europe 100 or 200 years ago.

7.3 Sex and Age Distributions

There are approximately equal numbers of males and females in most human populations. The sex ratio—that is, the ratio of males to females—is however somewhat higher than 1:1 at birth and decreases gradually with age, because males have slightly higher age-specific death rates at all ages than do females. The deviation from a 1:1 sex ratio at birth, as discussed in Chapter 2, may be due to selection in favor of sperm carrying Y chromosomes, though other explanations are not entirely excluded. Various forms of *gametic selection* are known in other animals and in plants. Such selection in favor of Y-carrying sperm (if it exists) may have arisen by natural selection as a response to compensate for the higher mortality of males than females in the postnatal period. The numerical equality of sexes at the time of sexual reproduction is probably an advantage in a species

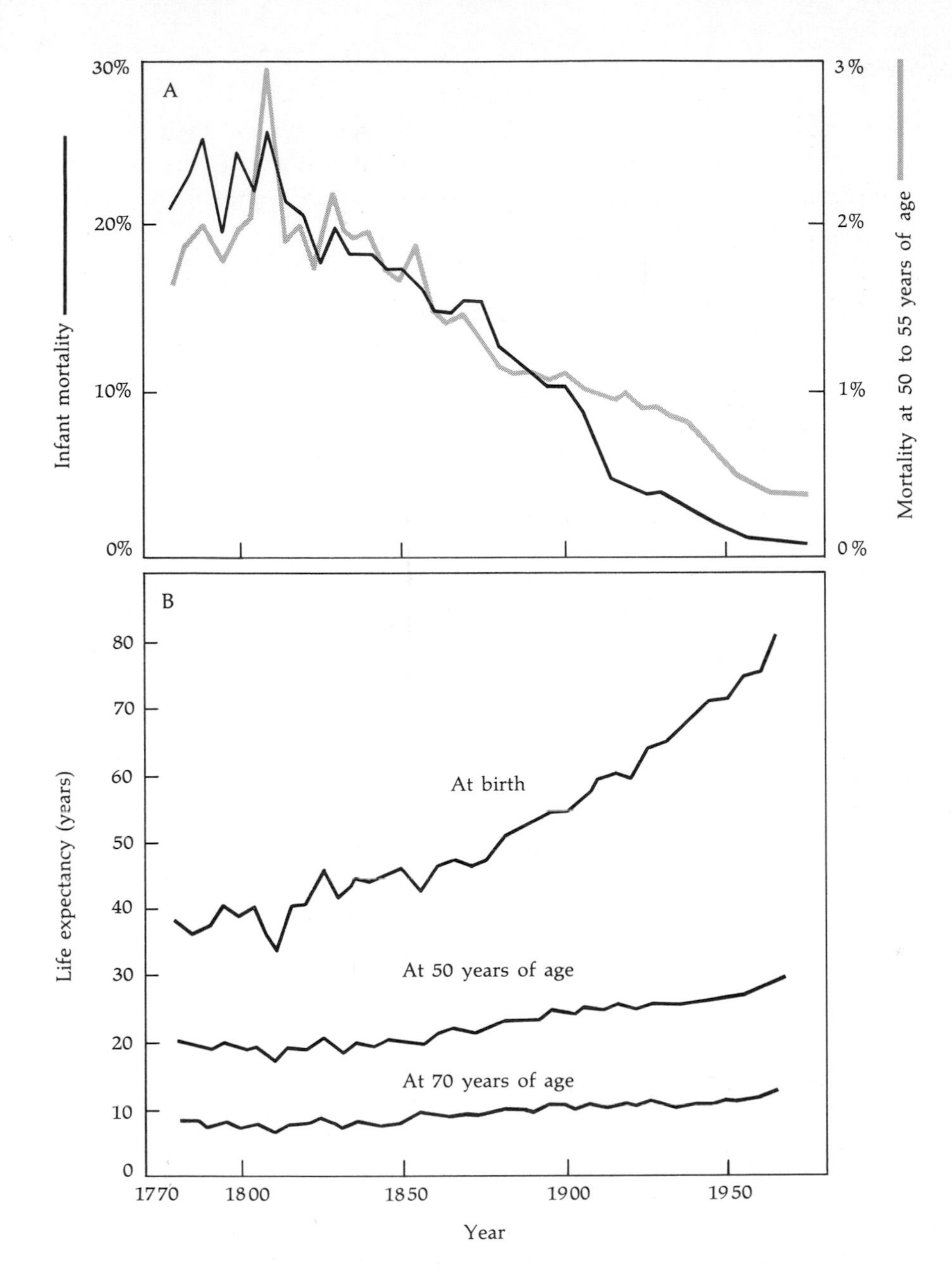

Figure 7.9
Mortality rates and life expectancies in Sweden over the past 200 years. (A) Mortality rates (percentage of newborns that die in a given age interval) for infants (first year after birth) on left ordinate and for females in the age range of 50 to 55 years on the right ordinate. Note that the infant mortality rate has decreased more rapidly than has the mortality rate for older individuals. (B) Life expectancy for females at various ages. Because of the greater decrease in mortality rates for younger ages (part A), life expectancy at birth has increased more strikingly than has life expectancy at later ages.

such as man, where both parents contribute to the upbringing of their progeny. The comparatively high sex ratio at birth balances almost exactly the postnatal selection against males, so that the two sexes occur in approximately equal proportions at the reproductive age.

The shape of the age distribution of a population—that is the proportion of people in the various age groups—is determined by the patterns of mortality and fertility as a function of age.

Age-specific fertility is the average number of children born per individual of a given age, and age-specific mortality is the number of people of a given age dying within a specified time interval (see Fig. 7.5). A fundamental mathematical theorem of demography (illustrated numerically in the Appendix) shows that, if age-specific fertilities and mortalities remain constant, the age distribution of the population soon stabilizes to a predictable shape. The theorem also predicts the rate of increase in population size in terms of the age-specific fertilities and mortalities. The age distribution in fact stabilizes sufficiently rapidly so that, even though age-specific fertilities and mortalities are changing quite quickly, they can nevertheless be used to obtain an approximate prediction of a population's age distribution, using current information on the natural or intrinsic rate of increase, r, at any given time.

Typical patterns of age-specific mortality and fertility for two extreme cases, representing the situation in developed as compared to developing countries, are

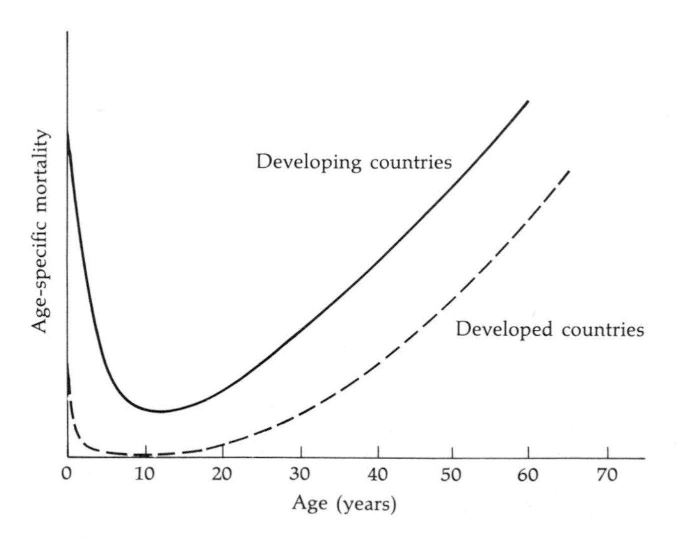

Figure 7.10
Age-specific mortality in two extreme situations: a schematic representation.

given in Figures 7.10 (mortality) and 7.11 (fertility). Higher mortality calls for higher fertility, and thus the number of children born is much higher in developing countries.

The actual age distribution is determined mostly by the intrinsic rate of increase of a population.

If a population has a growth rate of zero (is stationary), the age distribution expected is given by the life table in the sense that the fraction of the living population of a given age (Fig. 7.5A) is proportional to the probability that a newborn survives to that age. The latter value is l_x (Fig. 7.6B) in the life table. But in growing populations, as are most human ones, there will be a relative excess of younger people, and a relative deficiency of older ones—the higher the growth rate, the greater this discrepancy (Fig. 7.12).

7.4 The Measurement of Darwinian Fitness

A genotype associated with a disease that either reduces the chances of surviving to the end of the reproductive period or decreases the capacity to reproduce will be, at least to some extent, *deleterious.* Individuals with this genotype will be expected to have fewer children, on average, than those with other genotypes not

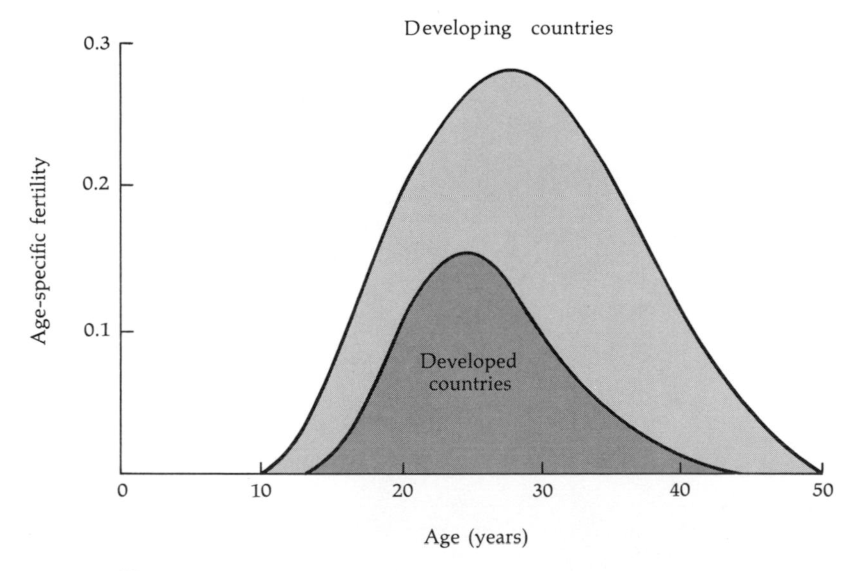

Figure 7.11
Age-specific fertility in two extreme situations: a schematic representation.

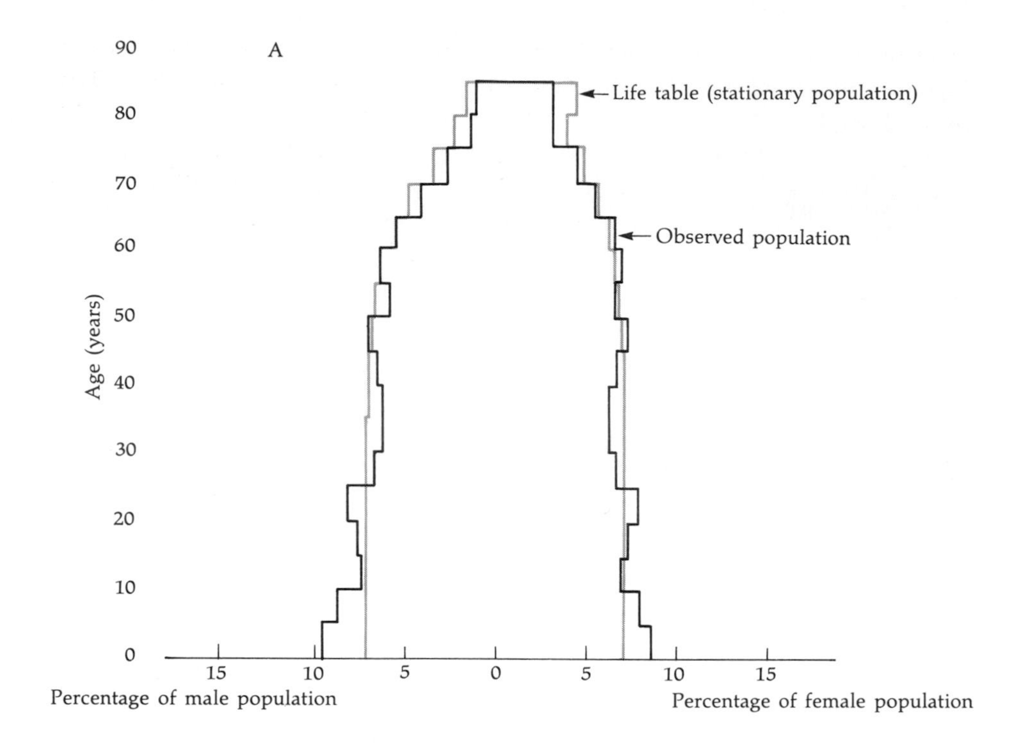
90
A
80
70
60
50
40
30
20
10
0
Age (years)
Life table (stationary population)
Observed population
15
10
5
0
5
10
15
Percentage of male population
Percentage of female population

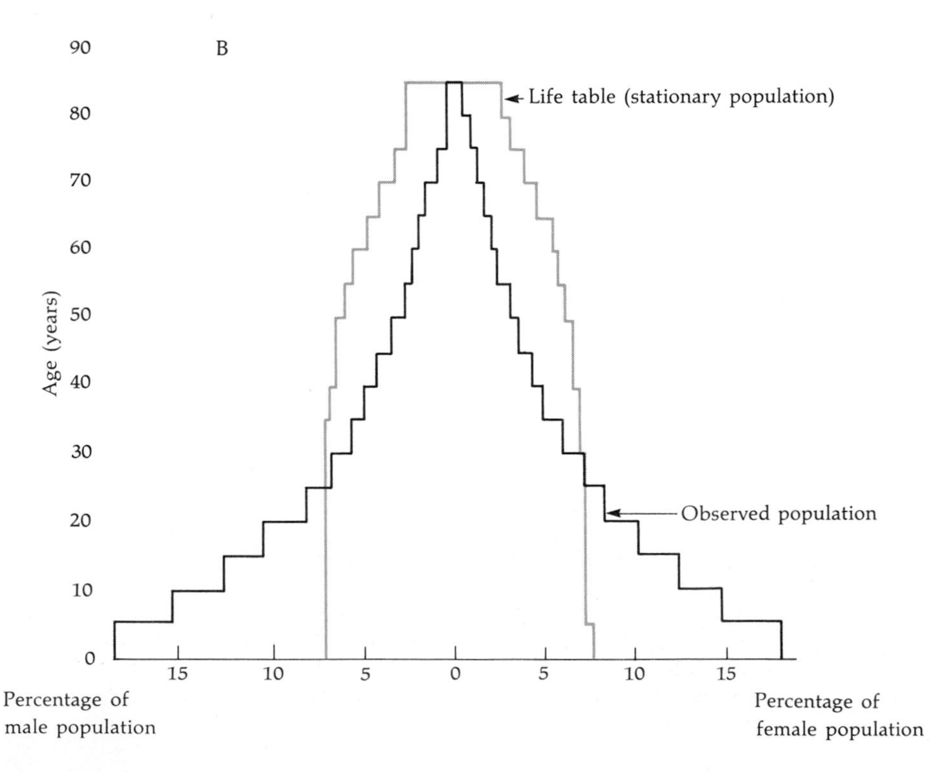
90
B
80
70
60
50
40
30
20
10
0
Age (years)
Life table (stationary population)
Observed population
15
10
5
0
5
10
15
Percentage of male population
Percentage of female population

associated with disease. Under these circumstances it is reasonable to expect that the deleterious genotype will decrease in frequency until eventually it disappears from the population. The deleterious genotype will clearly be expected to disappear more quickly, the smaller the average number of children born to affected parents. The frequency of genotypes in future generations can actually be predicted quite accurately given the average number of children produced by each genotype. The estimate of the expected number of children must take into account properly both survival and fertility.

The average number of children is a measure of a genotype's Darwinian fitness, which is the basic quantity for evaluating the intensity of natural selection.

The most refined way of measuring fitness is to consider each genotype as a separate part of the population, and then to evaluate the corresponding age-specific mortalities and fertilities as a demographer would for the whole population. This leads to an estimate of the intrinsic rate of increase, r, for each genotype. The relative frequencies of different genotypes in a population will clearly change if they have different values of r; those that have higher rates will become relatively more common. Rates of increase obtained in this way can be converted into conventional Darwinian fitness values (expected numbers of children). This method of measuring fitness is unfortunately not yet widely used and has, so far, been applied to only a few traits. The availability of much demographic data and of other statistics collected for public health purposes should, however, in the future allow considerable refinement in the measurement of the intensity of natural selection in man. In particular, it should even be possible to measure quite small selective differentials, which would need prohibitively large bodies of data if estimated directly.

Let us consider, as an example, the effect of ABO blood group on fitness. It has been shown that individuals of blood type O, for instance, are about 1.4 times more likely to get duodenal ulcers than are individuals of blood types A, B, or AB. As these ulcers may be a cause of death during the reproductive period, this should lead to some reduction in the Darwinian fitness of type-O individuals

---

Figure 7.12
Population pyramids. (A) England and Wales, 1968. (B) Mexico, 1966. Because the United Kingdom has a relatively low rate of growth, its age distribution (shown here in the form called a population pyramid) is not too far removed from that expected for a stationary population. (For such a population, the percentage of living individuals in a given age group would be proportional to the value given for that age group in a life table such as Fig. 7.6D.) In contrast, Mexico—which has a growth rate several times higher—has an age distribution deviating markedly from its life table. (After N. Keyfitz and W. Flieger, *Populations: Facts and Methods of Demography,* W. H. Freeman and Company. Copyright © 1971.)

relative to those of the other blood types. Using the known incidence of deaths due to duodenal ulcer at various ages and taking account of the age-specific fertility in the general population at those ages, it can be shown that the Darwinian fitness of type-O males is decreased, with respect to the other blood types, by about 0.01 percent. The decrease in females, whose death rate due to duodenal ulcers is less than that for males, is even smaller. The incidence of death due to the ulcers is so low that its selective effect on the ABO blood groups is indeed very small (see Chapter 10).

Genetic diseases that kill all carriers before reproduction or that prevent them from having any children have a Darwinian fitness of zero.

This was, for instance, the situation for hemophilic males until the availability of transfusion and other therapies for counteracting the delay in blood clotting. Many other genetic diseases still have zero fitness. Some genetic defects are not so severe, or else have on average a relatively late age of onset, so that their Darwinian fitness may be relatively high. Huntington's chorea, a dominant genetic disease in which there is progressive general mental deterioration, has an average age of forty years at onset (identified by lack of muscular coordination). No good estimates of the age-specific fertilities of patients with Huntington's chorea are available. Assuming, however, that their fertility is normal until the onset of the disease and is zero afterwards, it can be estimated that their Darwinian fitness is 26 percent lower than normal. It is possible that their fertility does not drop entirely to zero immediately after onset, though it will almost certainly do so when the patients are institutionalized, usually 6 to 7 years after onset. Assuming that fertility remains normal until institutionalization, the estimated reduction in the Darwinian fitness of individuals who get Huntington's chorea is only 3 percent. The correct figure for the fitness reduction must lie somewhere between these two extreme values of 26 and 3 percent.

Schizophrenia is a severe mental disease in which there is a significant genetic component (as described more fully in Chapter 17). It usually starts during early reproductive ages, and its onset considerably decreases the chances of marrying and having children. The decrease in fitness of schizophrenics with respect to "normals" was estimated, for 1940, to be 86 percent for males and 40 percent for females (in whom the disease has a later onset). The general decrease in the age at marriage during the last thirty years in the U.S. and elsewhere must, however, have reduced the selective disadvantage of schizophrenia. This change in social customs must mean that more schizophrenics marry and have children, as many will now do so before the onset of their disease. It can be estimated that the fall in the mean age at marriage had reduced the decrease in fitness due to schizophrenia by 1960 from 86 to 51 percent for males, and from 40 to 22 percent for females. There is, however, no defined genotype that causes schizophrenia, and the disease must also be considerably affected by environmental factors. It is the predisposition to schizophrenia that is genetically determined rather than the disease itself. Thus, these fitness estimates refer to a phenotype that actually may be quite heterogeneous, and not to a genotype. Selection acts on phenotypes, but it only has evolutionary consequences through its effect on genotypes. The lack of a clear-cut correspondence between genotypes and phenotypes thus makes it difficult to estimate precisely the selective effects of schizophrenia.

7.5 Selection Coefficients

The Darwinian fitness of a genotype is meaningful only when it is compared with that of other genotypes. Thus, only relative fitness values are usually of interest. It is common practice to assign an arbitrarily chosen genotype the fitness one, and to compute the fitnesses of other genotypes relative to that for the genotype taken as a standard. When, for example, a disease is involved, the "normal" genotype (homozygous for the allele not determining the disease) is the usual standard of reference. We have seen that Huntington's chorea, a dominant genetic disease, has a fitness that is, say, 26 percent lower than normal. This is the same as saying that 26 percent is the *selective disadvantage* associated with Huntington's chorea, relative to the normal nondiseased state. If we assign a fitness of one to normals, Huntington's chorea is associated with a fitness of 0.74 (a 26 percent reduction relative to 1.00). The quantity $1.00 - 0.74 = 0.26$ is then called the *selection coefficient* associated with the heterozygotes that get the disease.

The concept of "normal" in relation to different genotypes needs some further clarification. The number of possible genotypes in a population is enormous, as previously emphasized. If there were only two alleles, and hence three genotypes, at 10,000 different genes (and there are likely to be many more genes than that which are polymorphic in man), the number of possible genotypes would be $3^{10,000}$. This number is greater than 1 followed by three thousand zeros, which is very much larger than the number of electrons in the universe. The number of genotypes that actually exist in the human species is thus a very minute fraction of all those that are possible. The only pairs of individuals that have identical genotypes are monozygous or identical twins. Therefore, when we compare Darwinian fitnesses, we must restrict our analysis to a limited set of genotypes. This set is often those genotypes that can be formed by a series of alleles at a single gene locus: three for two alleles, and more when there are more than two alleles. We have to assume, as a first approximation, that other genes which affect fitness are equally distributed among the genotypes we are studying.

It is not always clear which genotype should be considered as the "normal" one.

In fact, only when referring to genes causing deleterious diseases does the concept of "normal" make sense. In most other cases the concept does not apply. Which, for example, is most "normal" among the *A*, *B*, and *O* alleles? The choice of a genotype to be used as a standard of reference for the calculation of fitness values is therefore usually arbitrary.

Selection coefficients are not necessarily constant.

Selection coefficients depend on age-specific fertilities and mortalities, and these may change not only in different environments, but also with time (in the same place), as they have in fact been doing dramatically in recent years. But there may, in addition, be more subtle changes. The fitness of genes may, for example, change with their gene frequency in the population, irrespective of the prevailing age-specific fertilities and mortalities. Examples of this are known in animals and in microorganisms, though none have so far been properly documented in man. There are, however, hints that such frequency-dependent selection may also operate in our species. One possible example is a special type of sexual selection, namely selection that favors certain types of mates. In countries where certain physical traits are particularly rare (for example, blue eyes or blond hair among southern Europeans, or dark eyes and dark hair among northern Europeans), these traits may have a special appeal because of their rarity.

7.6 The Rate of Change under Darwinian Selection

When man was still a hunter and gatherer, before the domestication of plants and animals, milk was drunk only during the first years of life because the only source was human. The adults of many ethnic groups are now known to lack an enzyme called lactase (produced in the stomach) that is necessary for the proper utilization of the sugar lactose found in milk. Absence of this enzyme leads to lactose intolerance, which is revealed by a variety of symptoms including nausea, vomiting, and abdominal pains. Other ethnic groups contain a high proportion of adults who are tolerant to milk. This proportion is a maximum (about 90 percent) among those populations, such as northern Europeans, that use the most fresh milk in their ordinary diets. The tolerant trait tends to be rare or absent in those populations, such as Oriental and most African ones, that do not usually use milk. The genetics of this trait is not yet fully understood because of the difficulties associated with assaying an enzyme that is present only in the wall of the stomach and the intestines. Nevertheless, there seems to be little doubt that the trait is inherited, and that it must have increased in frequency in those groups whose adults regularly use milk in their food. This habit could hardly have begun more than 10,000 years ago, because that is when the domestication of cattle and sheep is thought to have started.

Most probably, lactose tolerance represents a case of a mutant that became advantageous when dietary customs changed in certain human populations. We do not know the original frequency of the mutant, but we may assume that it was likely to have been very rare some thousands of years ago, before the change in diet took place. This provides an opportunity for estimating the intensity of selection needed to increase the mutant's frequency to the levels now observed in those populations in which it must have been at a selective advantage.

The speed of selection can be evaluated in terms of appropriate mathematical models.

Mathematical formulas given in the Appendix permit computation of *selection curves,* provided the fitnesses of all the relevant genotypes are known. For example, if we are dealing with the three genotypes formed by two alleles at a single locus and if we assume that the heterozygote is intermediate in fitness between the two homozygotes, we obtain curves such as those shown in Figure 7.13. The fastest selection process for advantageous gene A, shown in this figure, involves a selection coefficient of 10 percent which is high. Selection in favor of a new mutant of that intensity is perhaps rare.

Note that most of the curves, which plot the logarithm of the gene frequency against the time measured in generations, are approximately straight lines until the gene frequency is over 10 percent. They are, in fact, very close to logistic curves. Curves for selection coefficients other than those given in Figure 7.13 can be computed. If, for instance, we want to know the curve for a coefficient $s = 0.04$ (4 percent selective advantage of A over a), we take the curve for $s = 0.1$ and multiply the number of generations by the ratio of the two selection coefficients, in this case $0.1/0.04 = 2.5$. With $s = 0.1$, it takes about 115 generations to increase from a very low initial gene frequency of 1/100,000 to a frequency of 50 percent (which corresponds to 75 percent of individuals with the dominant, lactose-tolerant genotype). When $s = 0.04$, which is 2.5 times lower than 0.1, it takes 2.5 times more generations, namely about 290, to achieve the same change. Assuming an average generation time of about 30 years, 290 generations is about 9,000 years, which roughly corresponds to the time since the beginning of animal domestication. The calculation thus shows that in this interval, a selection coefficient of about 4 percent would have changed the gene frequency of lactose tolerance (assuming a single mutation is involved) from a very low value (comparable to that of just one new mutant in a large population) to the frequency now observed in northern Europeans (Fig. 7.14).

When selection coefficients are relatively large and so selection intensity is strong, genetic changes can occur comparatively rapidly and radical transforma-

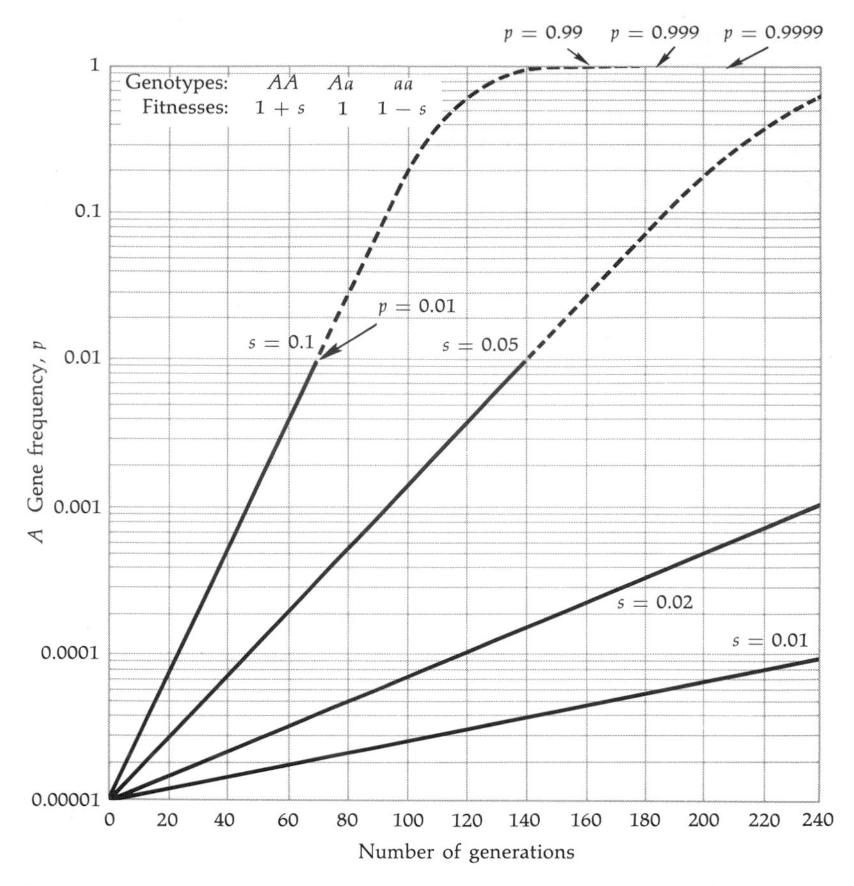

Figure 7.13
Selection in favor of a new advantageous allele. A single copy of mutant gene A appears in a population of 50,000 individuals at time zero. The initial gene frequency of A is therefore 1/100,000. The frequency of the gene increases according to the curves shown, depending upon the selection coefficient s. The fitnesses of the three genotypes are given in the upper left corner. Thus, for $s = 0.1$ (the highest curve), the fitness of AA is 10 percent higher and that of aa is 10 percent lower than the fitness of Aa. Time is given in number of generations (each generation being about 25 to 30 years). When the gene frequency is between 1 and 99 percent (indicated by the dashed portions of the curves), the gene is said to show a transient polymorphism. To estimate values for lower selection coefficients, multiply the number of generations by the same quantity used to divide the selection coefficient. For example, the curve for $s = 0.001$ can be obtained by using the curve for $s = 0.1$ and reading the time scale as hundreds of generations.

tions may take place in only hundreds of generations. Usually, however, selection coefficients are relatively small, and then natural selection results in slower changes. Information on selection intensities in man is, however, still very limited and mostly based on very indirect evidence.

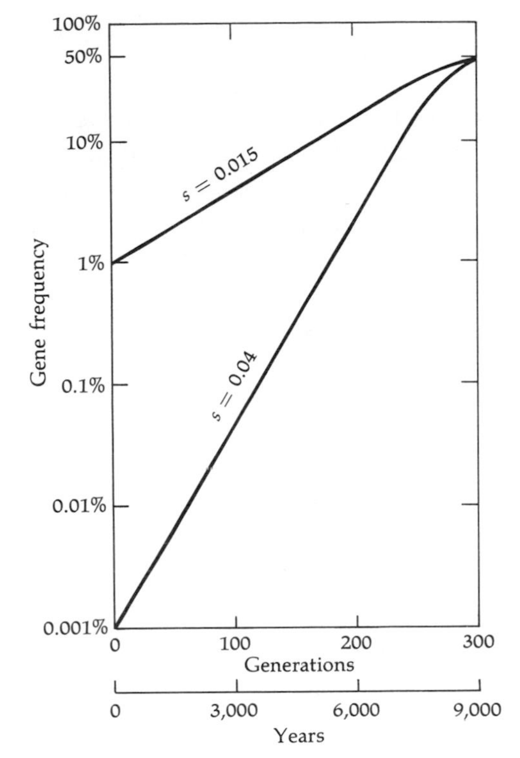

Figure 7.14
Selection for lactose tolerance of adults. Among modern northern Europeans, about 75 percent of individuals tolerate lactose as adults; this corresponds to a gene frequency of about 50 percent for the dominant gene. It is assumed that lactose tolerance first became advantageous about 10,000 years (300 generations) ago, when cattle were first domesticated. If the gene for lactose tolerance at that time had a frequency of 1/100,000 (comparable to that of a single new mutation), a value of $s = 0.04$ would lead to the present gene frequency. If the initial frequency was greater (say 1 percent), a smaller selection coefficient (0.015) would be sufficient.

7.7 Natural Selection and Genetic Adaptation

Natural selection ensures that genotypes which are more suited to a given environment will increase in relative frequency over the generations. The process is automatic, insofar as higher fitness means, in the Darwinian context, a greater capacity to leave descendants. In some cases the specific reason for a higher fitness may be quite clear, while in others the reason may be more subtle and is very often not known. One may for example speculate as to why man has lost most of his hair since his evolution from a presumably hairy common ancestor of man and the anthropoid apes, but these speculations are not readily supported by facts.

Natural selection brings about genetic adaptation, in the sense that the average fitness of the individuals forming a population increases and so the population as a whole becomes more suited, or adapted, to its environment.

Darwinian selection acts on individuals, whereas genetic adaptation takes place by changes in the composition of whole populations. Genetic adaptation by natural selection is thus adaptation of a population and not of individuals.

Individuals during their lifetime can, on the other hand, also adapt to environmental changes.

Such adaptation is not genetically transmissible; it may be called physiological, or individual, to distinguish it from the genetic or population adaptation that is the result of natural selection. The two phenomena may coexist as, for instance, with respect to skin color. Darker skin is due to increased formation of pigment in the skin, which offers protection against excessive solar irradiation. The skin darkens both in whites and in blacks a few weeks after exposure to the sun and this is a physiological, nontransmissible individual adaptation. But there has also been darkening or lightening of the skin by genetic adaptation, in those populations that have been exposed to environments with different intensities of solar irradiation long enough for selection to have had time to have a significant effect.

A tentative explanation of the mechanism by which skin color evolved has been advanced.

Man, like other mammals, requires vitamin D for calcium fixation and bone growth. Vitamin D is manufactured in the deep layers of the skin under the action of ultraviolet irradiation from the sun. An inadequate supply of vitamin D causes the bone disease called rickets. In high latitudes, the amount of ultraviolet light that penetrates to the appropriate skin layers is not sufficient to produce enough vitamin D if the skin is too highly pigmented. This is because the ultraviolet light is absorbed by the pigment in the skin before it can promote the synthesis of vitamin D. Hence, natural selection in favor of lighter pigmentation is believed to have taken place at higher latitudes because of its association with adequate vitamin D production. The theory goes so far as to assume that in tropical regions too much vitamin D may be produced in lighter skins, and hence that in these regions deeper pigmentation is favored by selection. These explanations are still quite speculative. It is usually not at all easy to understand why certain traits have a higher fitness than others.

Darwinian selection is sometimes called intragroup selection, because it acts on the individuals within a group.

There may, however, also be natural selection that favors one group relative to another, and this is called *intergroup selection*. Such selection is not necessarily due to direct competition or conflict between the groups. It may occur simply because certain groups are growing faster than others (*demic* selection). There are many possible explanations for the higher growth rate of some groups relative to others. If for any reason populations with different growth rates have different gene frequencies, the composition of the species as a whole will change with respect to those genes. The differences in gene frequency will usually have nothing to do with the differences in population growth rates. The ABO blood-group gene frequencies are, for example, different in the various ethnic groups. Figure 7.15C shows *A* and *B* gene frequencies for seven major geographic areas of the world. Taking into account the fact that there are big differences in the numbers of people in these areas (Fig. 7.15A), one can estimate that in the world as a whole the *A* gene has a frequency of 21.5 percent, the *B* gene 16.1 percent, and the *O* gene the remaining 62.4 percent. The population sizes of the various areas are, however, increasing at different rates (Fig. 7.15B). Thus, even if there were no changes in gene frequency within any area (that is, if there were no intragroup selection for or against any of the alleles), the world gene frequencies would change because of the differential population growth rates. If the present rates of growth persist, the *B* and *O* genes will increase in the whole human species at rates that have been calculated to be, respectively, 0.007 percent and 0.021 percent per year. The *A* gene will decrease at a rate equal to the sum of these two rates, namely 0.028 percent. If this process continues, the *A* gene may effectively decrease from 21.5 to 20.7 percent in one generation, solely as a result of intergroup or demic selection. This is a fairly important change, which would require a selection coefficient of about 5 percent in Darwinian (intragroup) selection. However, it is unlikely that the differences in growth rates between human groups observed today will persist for a long time.

Summary

1. The natural growth of populations depends on birth and death rates. When these are equal there is no growth. When the former is larger than the latter, the population grows; if the difference between birth and death rates remains the same, the population grows exponentially.

2. No population can sustain exponential growth indefinitely. A simple theoretical curve describing saturation of growth is the logistic curve.

3. The world is in the middle of a demographic transition.

4. Natural selection is the differential reproduction of genotypes. It can be due to differential fertility and/or mortality. It is measured using the techniques of demography.

5. Darwinian fitness of a genotype is its average number of progeny relative to that of another genotype taken as standard.

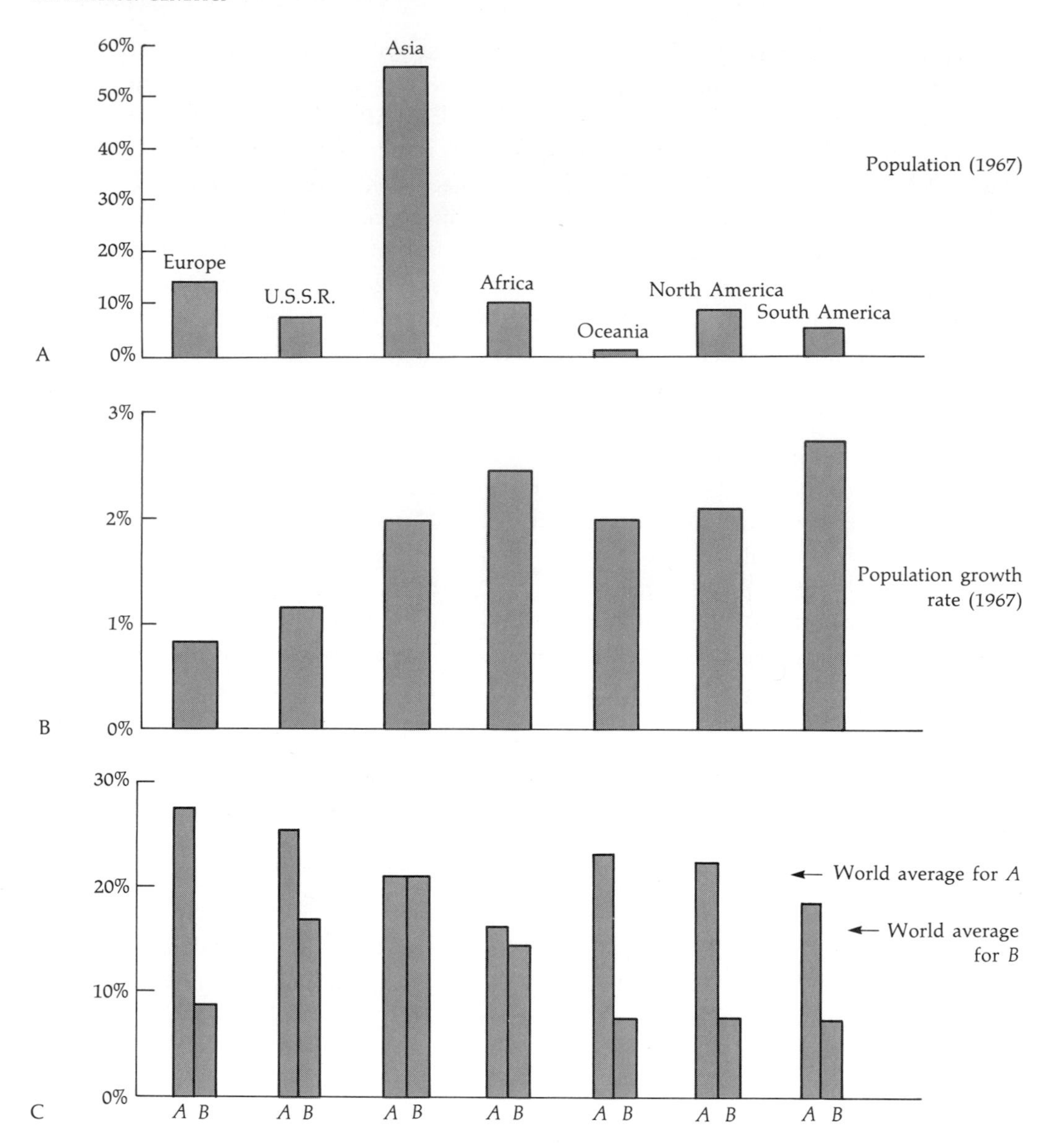

Figure 7.15
ABO gene frequencies and ethnic groups. (A) Percentages of the total world population found in various parts of the world. (B) Population growth rates for these areas. (C) Frequencies of *A* and *B* blood-group alleles in these areas. Because the populations have different gene frequencies and are growing at different rates, the gene frequencies of *A* and *B* alleles in the total world population are changing quite rapidly; the frequency of *B* is increasing (as is that of *O*), while the frequency of *A* is decreasing. (Data for part C from E. Thompson, *Annals of Human Genetics,* vol. 35, p. 357, 1972.)

6. A selection coefficient is the deviation of relative fitness from unity.
7. The rate of change of gene frequencies under natural selection depends on the magnitude of selection coefficients. Selection curves are easily computed if the fitnesses of all genotypes are known.

Exercises

1. What are the main components of biological fitness?
2. What types of information are needed to make population projections?
3. What is the basic course of the demographic transition?
4. Why has the fitness of schizophrenics increased in recent years?
5. Give some examples of factors that may change the magnitude and direction of natural selection.
6. What is group selection?
7. A population of bacteria grows exponentially, doubling every hour. Starting with one bacterium, how many are there expected to be after 10 hours? after 20 hours? Plot the curve of increase using time as the abscissa and the logarithm of the number of bacteria as the ordinate.
8. If a population grows at a constant rate r per year, it takes about $x = 0.69/r$ years for doubling (see Appendix for derivation). Compute the doubling time for populations growing at the rates $r = 0.005, 0.01, 0.02$, and 0.03 per year (also expressed as 0.5%, 1.0%, etc.). (The first and last values given are intrinsic rates of increase characteristic of human populations in developed and in developing countries, respectively.)
9. It is estimated by some that the carrying capacity of the earth is 50 billion people. Assume there are at present 4 billion people and that exponential growth continues at the present rate of about 2 percent per year. How many years will it take to reach the saturation limit? (Use the formula given in the Appendix.)
10. A type of bacterium, A, grows by doubling every hour. Another, B, grows half as fast—namely, doubles every two hours. At time 0 there are 100 times more bacteria B than A. What is the ratio between B and A bacteria after 2, 4, 6, 8, and 16 hours? Now, instead of the ratio, compute for each time the proportion of A among both, that is A/(A + B). Plot this proportion against the time. Do you recognize the curve that these data generate? If you are familiar with algebra you may obtain the theoretical curve, assuming two populations growing at exponential rates R_1 and R_2.
11. Assume that gene *a* has frequency 10 percent in a given generation, so that 1 percent *AA*, 18 percent *Aa*, and 81 percent *aa* are expected in the next generation under random mating and no selection. There is, however, selection favoring the *A* gene, so that one actually finds 5 percent more *AA* than there should be, and 5 percent less *aa* than there should be, the heterozygotes remaining unchanged. Compute the resulting gene frequency of *A* after one

generation. (Note that this corresponds to saying that the fitness of AA is $1 + s$, that of Aa 1, and that of aa $1 - s$, for $s = 0.05$. The curve for $s = 0.05$ in figure 7.13 was calculated in this way repeating the same procedure for some hundreds of generations.)

12. Construct a logistic curve using the equation $y = 1/[1 + e^{-kt}(q_0/p_0)]$, where p_0 is the initial frequency at $t = 0$, and $q_0 = 1 - p_0$. As t increases, y approaches a value of one. Plot y against t, first using an arithmetic scale for y and then again using a logarithmic scale. Note the similarity to the selection curves given in Figure 7.13. The curve is also identical to that obtained in Exercise 10, with k equal to the difference between the growth rates of the two bacterial strains. Comparing the results of Exercises 10 and 11, explain in general terms why the curves of Figure 7.13 are so similar to the logistic curve. What value of k in the formula of the logistic curve used here gives a curve similar to that for $s = 0.1$ in Figure 7.13?

References

Bajema, C. J.
1971. (Ed.) *Natural Selection in Human Populations.* New York: Wiley.

Cavalli-Sforza, L. L., and W. F. Bodmer
1971. *The Genetics of Human Populations.* San Francisco: W. H. Freeman and Company. [Chapter 6 of this text gives a more extensive treatment of the topics of this chapter.]

Dobzhansky, Th.
1971. *Genetics of the Evolutionary Process.* New York: Columbia University Press.

Fisher, R. A.
1958. *Genetical Theory of Natural Selection.* 2nd ed. London: Dover.

Haldane, J. B. S.
1932. *The Causes of Evolution.* New York: Harper. Reprinted, 1966, New York: Cornell University Press.

Keyfitz, N.
1968. *Introduction to Mathematics of Populations.* Menlo Park, Calif.: Addison-Wesley.

Keyfitz, N., and W. Flieger
1971. *Populations: Facts and Methods of Demography.* San Francisco: W. H. Freeman and Company.

Kretchmer, N.
1972. "Lactose and lactase," *Scientific American,* vol. 227, no. 4, p. 70.

Loomis, W. F.
1970. "Rickets," *Scientific American,* vol. 223, no. 6, p. 76.
1967, 1968. *Evolution and the Genetics of Populations.* 2 vols. Chicago: University of Chicago Press.

8

Deleterious Mutations

Many genetic diseases are now recognized, some dominant and some recessive. Almost all of them are rare, some extremely so. This suggests that we may have seen only the tip of the iceberg for genetic diseases. What explanations and predictions can be offered for these observations? Most such diseases seem to be maintained in the population by the balance between mutation and selection.

8.1 Dominant Mutations as a Cause of Disease

A particular type of dwarfism, called achondroplasia (one form of chondrodystrophy), is found at a very low frequency in nearly all human populations. The growth of the long bones in these dwarfs is abnormal, and as a result their limbs are short, squat, and often deformed, their skull bulges, and the bridge of their nose is "scooped out." The disease is not incompatible with life, but the dwarfs have, for obvious and unfortunate reasons, a rather low chance of marrying and having children. When one parent is chondrodystrophic and the other normal, one-half of the progeny on average have the disease; the trait is a clearcut dominant, and essentially all affected individuals are therefore heterozygotes for the dominant dwarfing gene. Only when two chondrodystrophics marry one another are homozygous abnormal progeny produced. The few such cases known have died *in utero* or shortly after birth with an extreme form of the disease.

The incidence at birth of achondroplasia is about 1 in 10,000. Natural selection is clearly very effective in keeping the frequency of the disease down by eliminating homozygotes and by decreasing the chances of heterozygotes marrying and having progeny. What force then leads to the recurrence of this disease? It is impossible to know whether the frequency of the disease has always been as high as it is at present, but the fact that the incidence at birth is very similar in widely different ethnic groups makes it probable that the frequency of the disease may be at an equilibrium that is independent of the environment. Although there are no good estimates of its incidence in the past, there are various documents indicating that the disease did occur a long time ago. Chondrodystrophic dwarfs were often buffoons at royal courts, as can be seen in some classical paintings (Fig. 8.1).

One can estimate directly the mutation rate for dominants.

Analysis of parentage, as discussed in Chapter 4, can provide evidence for the occurrence of new mutations. In the absence of mutations, all affected individuals should have either an affected father or mother, because one of the two parents must have carried the dominant gene responsible for the dwarfism. However, about four out of every five chondrodystrophic dwarfs have perfectly normal parents. Illegitimacy can, among other reasons, be excluded because other dominant traits do not show such irregularity, and because illegitimacy would have to be very frequent to account for such a discrepancy.

Figure 8.1
An achondroplastic dwarf, from a painting by Diego Velazquez, Spanish court artist, painted in the 1640s. (Prado, Madrid.)

Among other possibilities to be considered is a low *penetrance* of the gene—that is, the gene may not show its effect in every heterozygote. If this were the case, however, a generation might be skipped occasionally, but the disease would be likely to be found in a grandparent even if not in a parent, and chondrodystrophy is such a striking defect that it would hardly have escaped record in the family history.

The obvious remaining possibility is that a new mutation has occurred in a gamete of one of the parents, giving rise to the first chondrodystrophic dwarf recorded in the pedigree. Counting all such cases of births in which people born with the disease have normal parents gives (as discussed in Chapter 4) a *direct estimate of the mutation rate.* The estimate for chondrodystrophy has been found to be of the order of 1 in 20,000. Mutation must therefore be the factor that leads to the recurrence of the disease in the face of adverse selection.

A number of other dominant genetic diseases are known for which the mutation rate can be estimated in a similar way. The mutation rate for achondroplasia, however, turns out so far to be one of the highest (Fig. 8.2).

8.2 The Balance Between Mutation and Selection

The example of achondroplasia shows that there are two opposing forces which maintain the low frequency of this disease: mutation (which produces new alleles) and natural selection (which weeds them out). In this case, natural selection acts

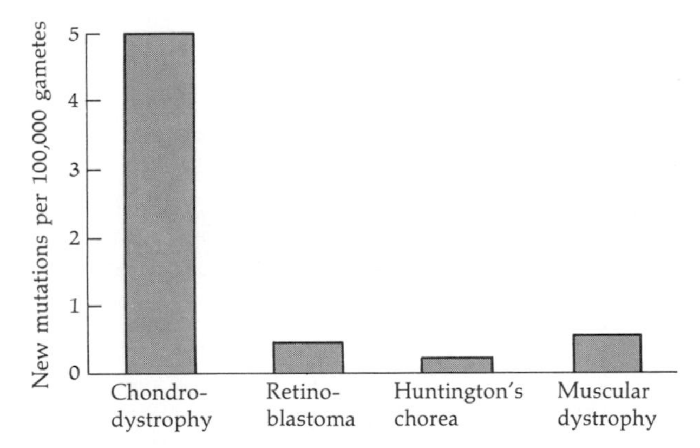

Figure 8.2
Direct estimates of mutation rates for some dominant deleterious mutations. Chondrodystrophy is achondroplastic dwarfism. Retinoblastoma is a tumor of the retina. Huntington's chorea is a neurological disease (see Section 7.4). Muscular dystrophy is a group of diseases that cause progressive deterioration of the muscles; the mutation rate shown here is that for facioscapular muscular dystrophy, which affects mostly the facial and upper trunk muscles.

not through premature death of the carrier of the mutation, but through a relative inability to leave progeny. In many other cases, natural selection acts by increasing mortality rather than decreasing fertility. When, in the extreme case, a genetic disease kills all carriers before the reproductive age, it is called a lethal. Mutations that increase mortality or reduce fertility are (as discussed in Chapter 7) called deleterious. Deleterious mutations can be either dominant or recessive. We shall consider first the fate of deleterious dominants—that is, genes that are deleterious when present in a single dose, and thus to heterozygotes.

Production of new copies of the deleterious gene by mutation in each generation should increase the frequency of the corresponding disease in the population (Fig. 8.3). If the new mutants produced each generation because of a constant mutation rate could accumulate in the population in the absence of counter selection, there would be a steady increase with time in the frequency of mutants (curve a in the figure). The rate of increase is actually only twice the mutation rate, which is in general small. The increase in frequency of the mutant will thus be very slow if it is selectively neutral—that is, neither at a selective advantage nor at a disadvantage.

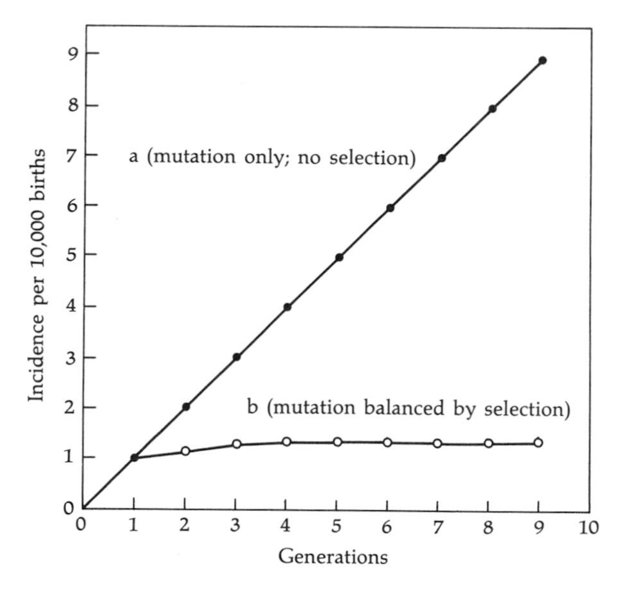

Figure 8.3
Mutation and selection. The graph represents a hypothetical population that includes no achondroplastic dwarfs in generation zero, and in which mutations for this gene occur at a rate of 1 per 20,000 births in each generation. If dwarfs reproduce at the same rate as normals (curve a), there will be a steady increase in the frequency of dwarfs, due to the accumulation of the new mutants produced by mutations occurring in each generation. If the Darwinian fitness of the dwarfs is 20 percent of that for normals (as apparently is the case), adverse natural selection counterbalances the production of mutants, and so their frequency in the population soon reaches a constant value that is somewhat higher than the mutation rate (curve b).

Natural selection eliminates carriers of deleterious mutations, thus preventing the accumulation of mutants beyond a certain incidence in the population.

If the mutants are selected against because they are deleterious to their carriers, their increase in frequency by mutation is checked by natural selection, which eliminates a substantial fraction of the mutant heterozygotes who have the trait. At some point in time, the two forces will balance each other (curve b in Fig. 8.3). The incidence of the disease at birth stabilizes at a level that is somewhat higher than the mutation rate, as we now prove (Table 8.1). In fact, the higher the fitness, the higher the ultimate frequency of the mutant.

We assume that the mutation rate and the intensity of natural selection remain the same each generation. The measure of natural selection is, as discussed in Chapter 7, the Darwinian fitness, represented by the expected number of progeny. We assume, for simplicity, that there are no mutants present in the initial population to begin with, and that an average of one out of every 20,000 gametes mutates in every generation to a genotype associated with chondrodystrophy, so that zygotes arising from them will become dwarfs.

If now we consider 20,000 sperm fertilizing 20,000 eggs at random (as in Fig. 6.1) to produce 20,000 zygotes with a mutation rate of 1/20,000, we expect one sperm and one egg to be mutant. The mutant sperm will almost certainly fertilize a normal egg, and similarly the mutant egg will be fertilized by a normal sperm, so that 2 out of the 20,000 zygotes will be heterozygotes. Thus the proportion of dwarfs due to mutation in the gametes forming this generation is 2 out of 20,000, or 1 in 10,000—that is, twice the mutation rate. If the mutation rate remains constant, two new mutants will be formed for every 20,000 births each generation.

Table 8.1
Expected incidence of dwarfs in a stationary population of 20,000

Generation	0	1	2	3	4	. . . Equilibrium
New mutants at birth (2μ)	—	2	2	2	2	2
Dwarfs in parental group (b)	—	0	2	2.4	2.48	2.5
Dwarfs reproducing from parental generation (wb)	—	0	0.4	0.48	0.496	0.5
Total dwarfs ($2\mu + wb$) (This value becomes b for the next generation)	0	2	2.4	2.48	2.496	2.5
Incidence of dwarfs in population (dwarfs per 10,000)	0	1	1.2	1.24	1.248	1.25

NOTE: The mutation rate $\mu = 1$ mutation per 20,000 gametes. It is assumed that no dwarfs are present in generation 0. The fitness (w) of the dwarf type is 0.2 (20 percent). The dwarf type is produced by a dominant allele. The last column shows conditions after several generations, when equilibrium has been reached. The incidence of dwarfs at equilibrium can be shown (see text) to be $b = 2\mu/(1 - w) = (2/20{,}000)/(1 - 0.2) = 1/8{,}000 = 1.25/10{,}000$.

In subsequent generations, some of the mutants born earlier will reproduce and have children. Now we know that, if the total population size remains constant, every individual has, on average, one child who becomes adult; if children are counted per married couple, every couple has two children. A population of 20,000 adult individuals forms 10,000 couples. Two of these matings in the first generation will be between a dwarf and a normal individual. From these, a total of four children are expected, of which only two are expected to be dwarfs because of the Mendelian segregation of a dominant gene. However, the fitness of these two dwarfs is only 20 percent of normal, and so only 20 percent of 2, or 0.4 dwarfs will be added to the second generation.

Fractions of individuals nearly always arise when we compute expected numbers, as we do here. Numbers of individuals can of course only be integers, but we could always multiply the population size by a factor of, say, ten to remove fractions, and leave relative numbers the same. Thus, in this case, if the population size were 200,000, then the 0.4 dwarfs would become four dwarfs, resolving the paradox of the fractions. If the population size is kept small, then a fractional average of expected dwarfs actually means that in some cases no dwarfs are produced, sometimes one, sometimes two, and so on, with different probabilities. The average over a number of generations can then, for example, remain at 0.4.

The second generation again has two new mutants, and the 0.4 dwarf individuals produced by those already present in the population, to give a total of 2.4 dwarfs in generation 2. As before, because of selection, these dwarfs contribute only 20 percent of 2.4 to the third generation, or 0.48 dwarfs, which must be added to the two new mutants to give 2.48 dwarfs in generation 3. Continuing the process to later generations, as shown in Table 8.1, we find that at each generation, there are two dwarfs (out of 20,000) from new mutations, and a fraction more who have been reproduced from the preceding generations. This addition gradually approaches 0.5, to give a limiting total of 2.5 dwarfs as can be seen in the table. We shall now give a more general approach to the problem.

The total incidence of mutants at birth, including new mutations and those transmitted from earlier generations, approaches the limit $2\mu/(1 - w)$, where μ is the mutation rate and w the fitness of the dominant mutants.

An alternative way of predicting the outcome of the balance between mutation and selection is to make use of the fact that the frequency of mutants at birth will neither increase nor decrease when the number of new mutants arising per generation equals the number lost because of adverse selection. This state is called a "genetic equilibrium," because at this point gene frequencies do not change any more. Recall from Chapter 6 that the Hardy–Weinberg law predicts an equilibrium state after one generation of random mating in the absence of disturbing forces. The equilibrium we are now discussing is reached gradually and is the result of the opposing and counterbalancing action of two of these forces, mutation and selection.

We can predict the equilibrium frequency as follows. The proportion of dwarfs arising per generation because of mutation is 2μ, expressed as a proportion of the whole population. Suppose that the total proportion of dwarfs in the previous generation, is b. Then, given that the fitness of the dwarfs is w, a proportion (wb) of these reproduce, while the rest, $(1 - w)b$, do not and represent the loss of dwarfs from the previous generation due to selection. Note that in Table 8.1 the number (wb) that reproduce is given in the third line of the table, while the number lost is the total minus the number reproducing ($2 - 0.4 = 1.6$; $2.4 - 0.48 = 1.92$; and so on). When the loss, $(1 - w)b$, is equal to the new additions by mutation each generation (2μ)—that is, when

$$2\mu = (1 - w)b$$

—the process is at equilibrium, and the frequency of dwarfs at birth, b, does not change any further. At this point we find, by rearranging the terms,

$$b = 2\mu/(1 - w),$$

as already given. In terms of the example in Table 8.1 the loss at equilibrium is $(1 - 0.2) \times 2.5 = 2$, which is equal to the number of new mutants produced each generation. The equilibrium total frequency 2.5 is simply the number that, when multiplied by the selection coefficient $s = 1 - w = 0.8$, gives 2, the number of new mutants.

The equilibrium formula confirms that the incidence of a mutant at birth will be higher, the higher the mutation rate, and the higher the fitness of the mutant. Thus, for example, if $w = 0.9$ instead of 0.2 for the dwarfs, their equilibrium frequency would be eight times higher—namely, $2\mu/0.1$ rather than $2\mu/0.8$.

The formula for the mutant equilibrium frequency under mutation selection balance can also be used to estimate the mutation rate μ, if the frequency (b) and fitness (w) are known.

The dominant disease, Huntington's chorea, for example, whose fitness we discuss in Chapter 7, has an incidence of about 1/10,000. If we take the minimum fitness estimate for Huntington's chorea of 0.74, then the equilibrium relation between μ (the mutation rate), b (the incidence of 1/10,000), and w (the fitness of 0.74) must be

$$b = \frac{1}{10{,}000} = \frac{2\mu}{1 - w} = \frac{2\mu}{1 - 0.74} = \frac{2\mu}{0.26} = \frac{\mu}{0.13}.$$

Rearranging it, this relation gives

$$\mu = \frac{0.13}{10{,}000} = 1.3 \text{ per } 100{,}000$$

as the estimate of the mutation rate for the Huntington's chorea gene. If we had used instead the maximum fitness estimate for Huntington's chorea (0.97), the estimated mutation rate would be 1.5 per million, which is nearly ten times lower. The trouble with these estimates is the uncertainty surrounding the values used in the equation, especially for the fitness w.

In the case of achondroplasia it has been noted that most fresh mutants are the progeny of older fathers. There is a fairly regular relationship between age of the father and incidence of mutations in the children (Fig. 8.4). This relationship has also been found for some other mutations, but not for all, and is not completely understood.

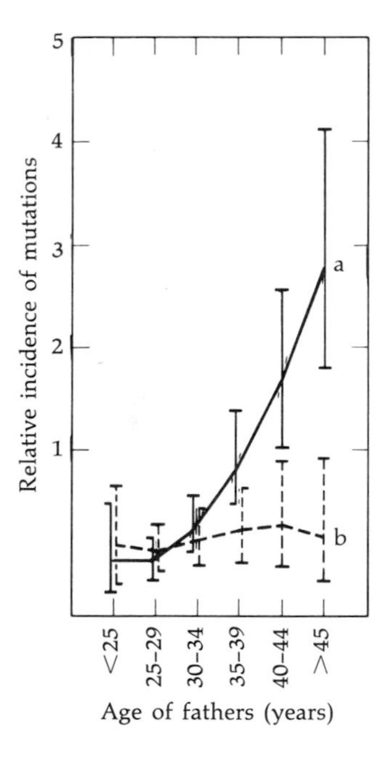

Figure 8.4
Parental age and mutation rate. This graph shows the relative incidence of some dominant mutations among individuals, grouped according to the father's age at the birth of the mutant offspring; the vertical bars represent intervals of error. Curve a shows the incidence of achondroplasia; curve b shows the incidence of neurofibromatosis and some other dominant genetic diseases. Curve a shows a striking effect of the father's age on the rate of occurrence of the dominant mutation for achondroplasia; curve b is given for comparison. (From F. Vogel, *Genetics Today,* vol. 3, p. 838, Oxford: Pergamon, 1965.)

8.3 Chromosomal Aberrations as Deleterious Dominants

Chromosomal aberrations such as Down's, Turner's, and Klinefelter's syndromes can, as far as their behavior in populations is concerned, formally be considered as deleterious dominants. Their fitness is generally close to zero, and so estimates of mutation rates can be derived fairly accurately from their incidence at birth. In the formulas discussed in the Section 8.2, if the fitness w is zero, it can be seen that the incidence of a given defect is just equal to twice the mutation rate. Conversely, of course, the mutation rate can be estimated as one-half the incidence. If the mutation rates in males and females were different, then the incidence would actually be equal to the sum of the two rates. Many chromosome mutations do seem to be produced at different rates in male and female gametes. From a social point of view, it is also important to remember that the rate of production of Down's syndrome in older women is much higher than in younger women (see Chapter 4). The mechanism responsible for this striking difference is not known, but one (so far unsubstantiated) suggestion, for example, is that it may be related to hormonal changes taking place with age. Sex and age differences in the rates of production of chromosomal aberrations other than Down's have also been described. Changes in the pattern of age at reproduction can therefore alter substantially the overall incidence of chromosome abnormalities.

Down's syndrome and sex-chromosome aberrations constitute a large part of all chromosome mutations.

Incidences of chromosome abnormalities at birth from a survey of 24,468 consecutive hospital births in the United Kingdom, United States, and Canada are presented in Table 8.2. There was a total of 126 cases, constituting 0.51 percent of all births. Nearly half of them were sex chromosome abnormalities, mostly XYY (22 cases), XXY (17 cases), and XXX (9 cases). Of those three abnormalities, as discussed in Chapter 4, only Klinefelter's syndrome (XXY sterile males) has a marked effect on reproductive capacity, though not necessarily on other aspects of the overall phenotype. Individuals who are XYY or XXX have more-or-less normal fertility, and so in those cases we cannot use the simple estimate of mutation rate being one-half the incidence.

Turner's syndrome (XO) was much rarer in this survey (only one case), though other studies suggest a higher incidence of about 0.02 percent. However, the incidence of XO among spontaneous abortions is at least 7 to 8 percent and thus is very much higher than that among live births or stillbirths. Clearly the majority of XO zygotes do not survive beyond the early stages of gestation, and so their rate of production as estimated from birth incidence may be only $\frac{1}{10}$ to $\frac{1}{50}$ of the true rate of production. The overall incidence of chromosomal abnormalities

Table 8.2
Chromosomal aberrations in 24,468 consecutive hospital births in the U.K., U.S., and Canada.

	Percentage of liveborn population	Mutation rate per gamete per generation*
Sex-chromosome abnormalities	0.20%	–
Autosomal trisomies	0.13%	0.065%
Autosomal structural abnormalities	0.18% (0.04%†)	0.02%
TOTAL	0.51%	

*Average for the two sexes, assuming fitness zero.
†New mutations
SOURCE: P. A. Jacobs, *Humangenetik*, vol. 16, pp. 137–140, 1972.

among spontaneous abortions is at least 30 percent which is really remarkably high. Putting it the other way round, chromosomal aberrations are actually a major cause of spontaneous abortion. Most of them do not survive to birth because they cause such gross derangements of development. The observed incidence of spontaneous abortion is about 20 percent of all births. This means that at least 7 percent of all gametes that effect fertilization have a detectable chromosomal aberration, corresponding to an overall mutation rate of 3.5 percent, which is of course much higher than the figures given in Table 8.2.

The major contribution to the autosomal trisomy incidence at birth is, of course, Down's syndrome (27 out of 32 cases), and nearly half of the autosomal structural abnormalities are D/D, most of the remainder (15 of 36) being a miscellaneous collection of reciprocal translocations involving other chromosomes. Down's syndrome remains, perhaps, the greatest problem of any of the chromosomal abnormalities, mainly because of its comparatively mild effect and hence relatively long average survival time.

8.4 Recessive Deleterious Genes

Dominant genes cannot escape selection upon first appearance, because they manifest their effects in the heterozygote. Their frequency thus usually remains very low, as we have seen. Genes whose deleterious effects are recessive, and therefore are revealed only in homozygotes, can accumulate in a population to higher frequencies before they are checked by selection (Fig. 8.5). This follows from the Hardy–Weinberg law, which shows (among other things) that a rare gene is present much more often in heterozygotes than in homozygotes (see

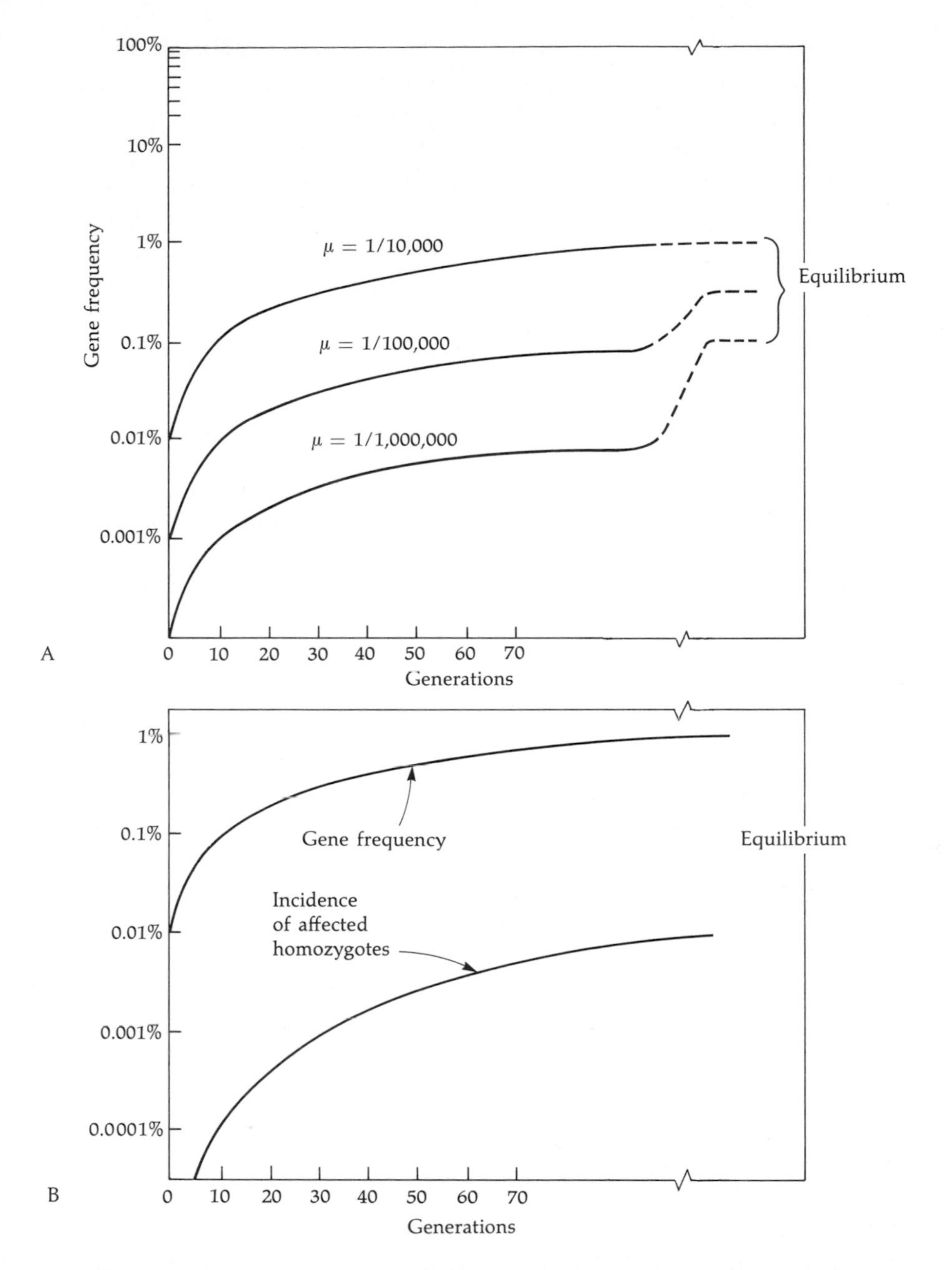

Figure 8.5

The increase of a deleterious recessive gene with time. At time zero, no copy of the deleterious gene is present. (A) Three values of the mutation rate (μ) are shown. The recessive gene is taken to be fully lethal, and the fitness of the heterozygote is assumed to equal that of the normal homozygote. (B) Gene frequency and frequency of affected individuals for $\mu = 1/10,000$. The incidence of affected individuals increases more rapidly, but starts at a much lower level, than the gene frequency.

Fig. 6.5). For example, in a population of 10,000 individuals mating at random, a gene frequency of 1 percent should produce just one homozygous individual, because the square of 1 percent is $(0.01)^2 = \frac{1}{10,000}$. The heterozygote frequency, on the other hand, is about twice the gene frequency, or $\frac{2}{100}$.

The approximate frequency at birth of the recessive defect phenylketonuria is of the order of 1 in 10,000. This disease, as discussed in Chapter 3, is caused by an inability to convert the amino acid phenylalanine into tyrosine, resulting in the accumulation of phenylalanine, which is associated with the development of severe mental defects. If the defect is discovered early enough, the child can be given a phenylalanine-poor diet and can develop more or less normally.

Mutations can lead to accumulation of relatively high frequencies of recessive genes in populations before natural selection can act to check their increase.

The increase in the frequency of a deleterious allele g (the symbol used for mutant genes) to the relatively high value of 1 percent must be due to its continued production by mutation. In Section 8.2 we derived the equilibrium frequency of a deleterious dominant gene subject to the opposing forces of mutation and selection. This was done by equating the loss (per generation) of deleterious alleles by selection to the number of new mutants arising (per generation) by mutation. The loss per generation for a deleterious recessive can be computed from the Hardy–Weinberg law on the assumption of random mating. If the frequency of g, the gene for phenylketonuria, is $q = 0.01$ (or 1 in 100) among the gametes that form a given generation, the expected proportion of homozygous recessives is $q^2 = 0.0001$ (or 1 in 10,000). A fraction, s, of these is lost, where $s = 1 - w$ is the selection coefficient, and w is the fitness of the homozygous affected genotype gg relative to the other two "normal" genotypes, GG and Gg. The loss of recessives per generation is thus $sq^2 = (1 - w)q^2$.

The rate of production of new g genes by mutation is the mutation rate (μ) as before. Thus, for the equilibrium condition, we can equate the loss by selection to the gain by mutation as before, to obtain

$$(1 - w)q^2 = \mu.$$

The "loss of recessives" that we have used here is the proportion of homozygous recessive *individuals* lost per generation because of adverse selection. But it is also equal to the relative loss of recessive *genes*, because recessive homozygotes carry only recessive genes. Therefore it is legitimate to equate the loss of recessives (as a loss of recessive genes) to the mutation rate (which approximately equals the relative increase of recessive genes due to mutation).

We can now use this equation to predict the expected proportion of individuals affected at birth. These individuals are the recessive homozygotes, whose frequency is q^2. Rearranging the terms in the equilibrium equation, we find that the frequency of recessive homozygotes at birth is given by

$$q^2 = \mu/(1 - w)$$

for the equilibrium condition.

With equal w, incidence of recessive defects is half that of dominants.

This equilibrium frequency is just one-half that computed for dominant mutations, which is $2\mu/(1 - w)$ (see Table 8.1); the fact that a deleterious mutation is recessive rather than dominant thus halves the proportion of affected individuals. If w is zero (the gene is a recessive lethal), then the proportion of affected individuals is just equal to the mutation rate μ. If we assume, for example, that PKU has a zero fitness (and if equilibrium between mutation and selection has been reached in the population), then the mutation rate for the PKU gene is 1 in 10,000—the frequency of affected individuals at birth. The gene frequency at equilibrium is the square root of the homozygote frequency of 1 in 10,000, and so is 1 percent. For a recessive, the gene frequency thus is much higher than the frequency of affected individuals. The frequency of heterozygote carriers of the PKU gene (which is approximately twice the gene frequency) is thus 2 percent, and so deleterious recessive genes can achieve much higher gene frequencies than can deleterious dominants, even though the proportion of affected is somewhat smaller. This is because most of the deleterious recessive genes occur in the normal heterozygous carriers.

The estimation of mutation rates for recessives is unsatisfactory.

It has been argued that one should be able to estimate the mutation rate for deleterious recessive genes given the incidence of the defect at birth (or in general, before selection) and its fitness as we did, in fact, for phenylketonuria. Several difficulties, however, make such estimates rather unreliable. First of all, the approach to equilibrium takes a longer time for recessives than it does for dominants, and so it is difficult to say whether a population is effectively at equilibrium (Fig. 8.5). More importantly, the hypothesis is made in deriving the equilibrium formulas that the heterozygote *Gg* has *exactly* the same fitness as the normal homozygote, *GG*. It is true that frequently we cannot easily distinguish heterozygotes from normal homozygotes, or we must employ delicate laboratory tests to make that discrimination. Fitness of heterozygotes, however, may depend on factors other than those that we can measure in the laboratory. We cannot assume, just because *we* have difficulties in distinguishing between the behavior of heterozygotes for a gene and the normal homozygotes, that *nature* cannot make the distinc-

D. Eisenhut
544-68-9279

tion. Moreover, quite small fitness differences in the heterozygote, which may not even be detectable, can have very marked effects on the equilibrium gene frequencies.

For instance, assume as before that PKU is at equilibrium with a gene frequency of 1 percent, giving the observed frequency of 1 in 10,000 homozygotes, and that these have fitness zero. Assuming complete recessivity—namely, *no* fitness difference between heterozygous carriers and normal homozygotes—we calculated that a mutation rate of 1 in 10,000 would maintain this equilibrium. However, if now we assume that the heterozygote has a fitness slightly higher or lower than that of the normal (which would be difficult to detect by ordinary measurements of fitness), it can be shown that the mutation rate now needed to maintain the observed equilibrium would be lower or higher (easily by an order of magnitude)

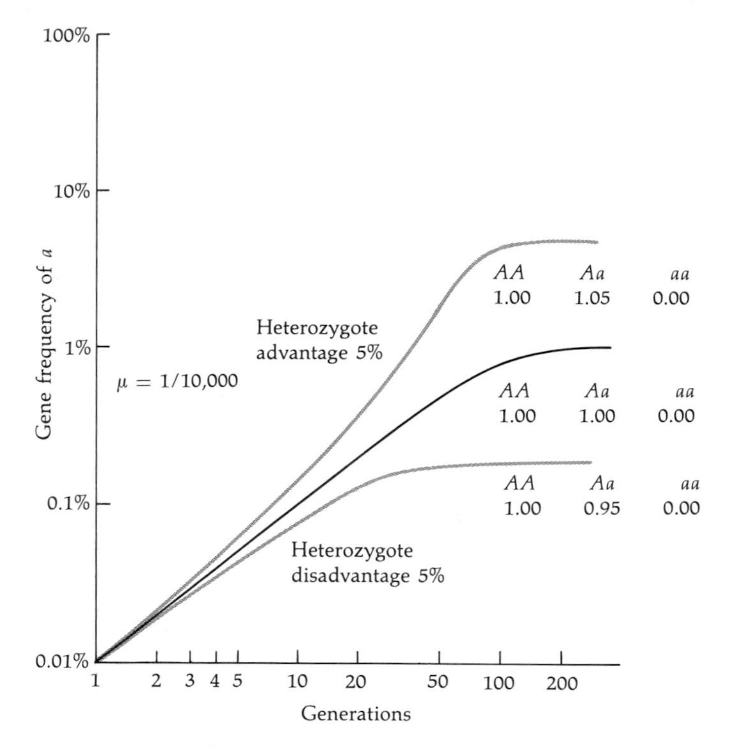

Figure 8.6
The overriding effect of the fitness of the heterozygote for a recessive lethal. The mutation rate is 1 in 10,000. The dark line represents the case in which the fitness of the heterozygote is equal to that of the normal homozygote, as in Figure 8.5. The curve above the dark line shows the change in gene frequency when the fitness of the heterozygote is greater than that of the normal homozygote by 5 percent. The lower curve represents the case where the heterozygote fitness is 5 percent lower than the fitness of the normal homozygote. (Note that, unlike Fig. 8.5, this graph shows both time and gene frequency on logarithmic scales.)

than that estimated assuming complete recessivity (Fig. 8.6). Clearly in the face of difficulties in estimating accurately the fitness of carrier heterozygotes, little credence can be given to mutation-rate estimates obtained in this way of deleterious recessive genes.

An X-linked deleterious gene subject to mutation–selection balance has a behavior intermediate between that of autosomal recessives and that of dominants.

Even though recessive (in females), such a gene is exposed to selection in males as if it were dominant. An X-linked deleterious recessive therefore behaves in a way that is closer to that of an autosomal dominant than to an autosomal recessive gene. A similar formula to that we obtained for deleterious dominants can be derived for X-linked recessives and used to estimate their mutation rates (see also Table 8.3). The X-linked genes have been extensively studied because they are so readily identified by their characteristic inheritance (see Sections 2.5 and 6.5).

A survey has been done of incidences of serious defects known to be X-linked (mostly recessive). From this survey it was possible to evaluate approximate mutation rates, and the distribution of values obtained is given in Figure 8.7. The average mutation rate can be computed from these data to be of the order of one in a million per gene per generation, or even lower. This shows, among other things, that the smaller sample of Figure 8.2 represents a collection of especially high mutation rates.

As mentioned when we first discussed mutation in Chapter 4, there are a number of important sources of bias involved in estimating mutation rates. One of these is the tendency of research workers to select unconsciously for study those genes that have especially high mutation rates, because it is easier to obtain data for such genes. Another source of bias is due to the fact that only defects that have been identified as inherited by earlier studies can be included. This inevitably selects for traits that have already been classified, and therefore selects for

Table 8.3
Comparisons of gene frequencies and incidence of affected individuals at birth for dominant, recessive, and X-linked deleterious mutations (at equilibrium)

	Dominant		Recessive		X-linked recessive	
Mutation rate	Gene frequency	Incidence at birth	Gene frequency	Incidence at birth	Gene frequency and Incidence in males	Incidence in females
1/10,000	1/10,000	1/5,000	1/100	1/10,000	3/10,000	$\sim 10^{-7}$
1/100,000	1/100,000	1/50,000	1/316	1/100,000	3/100,000	$\sim 10^{-9}$
1/1,000,000	1/1,000,000	1/500,000	1/1,000	1/1,000,000	3/1,000,000	$\sim 10^{-11}$

NOTE: Fitness w of affected is assumed to be zero; for $w > 0$, multiply all incidence values given in the table by $1/(1 - w)$.

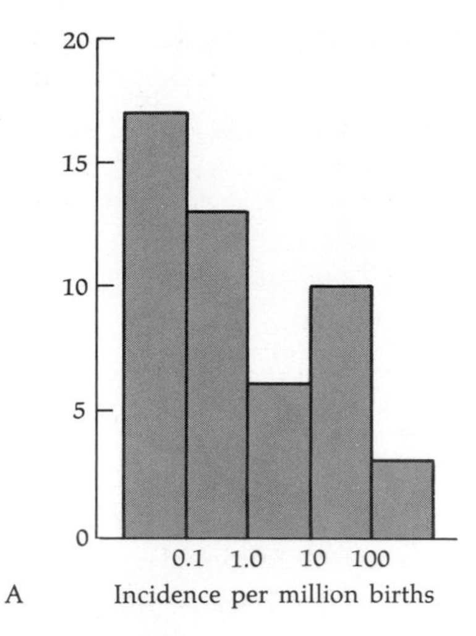

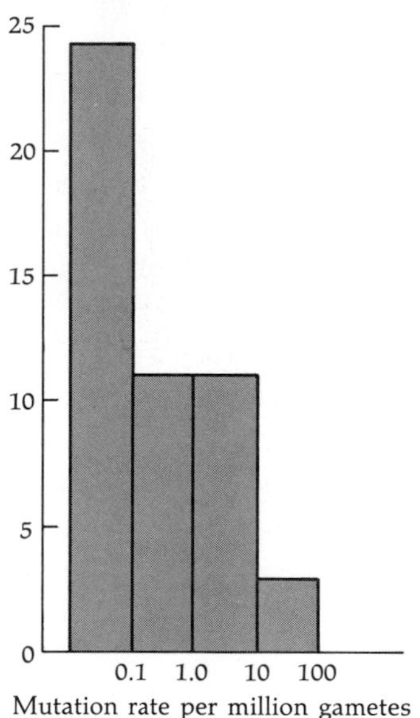

Figure 8.7
Distribution of incidence at birth and mutation rate for 49 X-linked traits. The height of each bar indicates the number of disorders (out of the 49 investigated) that have the incidence (part A) or the mutation rate (part B) shown on the horizontal axis. (Data from A. C. Stevenson and C. B. Kerr, *Mutation Research,* vol. 4, pp. 339–352, 1967.)

frequently occurring, more mutable traits. It can be shown that taking this fact into consideration may reduce the estimate of the average mutation rate of deleterious mutations by as much as a factor of ten, to give a value even lower than one in ten million per gene per generation.

The total incidence of X-linked defects at birth other than color blindness is of the order of 0.1 percent.

This estimate is, however, considerably affected by uncertainties as to the frequency of the few more frequent defects, which may themselves make up a large fraction of that total incidence. The X chromosome is only one-twentieth of the total genome in size; and so, if the other chromosomes contribute genetic defects at birth at the same rate as the X chromosome, we should expect about twenty times more, or some 2 percent total defects at birth due to deleterious gene mutations. This value is, however, very difficult to establish directly with any precision. One element of uncertainty, which is not always easy to quantify, is the degree of severity of the diseases included in such a survey. Estimates of incidence may vary widely depending on the level of severity chosen for the cut-off point between normal and defective. Everyone probably suffers from some genetic defect if one takes into account relatively minor problems such as short or long sight and other minor eye abnormalities, predisposition to allergy, and idiosyncratic reactions to drugs—to name just a few of the more common minor ailments that are believed to have, for the most part, a genetic basis.

A further complicating factor is that for many defects it is not easy to establish the extent to which, if at all, they are genetically determined. Some, such as cataracts (an opacity of the eye lens) may be either genetic, or the consequence of a viral infection during pregnancy. Other abnormalities, especially of the fetus, may be due to the adverse effects on the developing fetus of drugs taken by the mother during pregnancy. A tragic example of this was the effects of the drug thalidomide, which when taken by pregnant women led to their babies being born without limbs, though mentally normal.

8.5 Mutations Are Rare

New alleles are produced by mutation. Because they represent changes in a complex, highly organized structure, which has evolved through a long process of adaptation, mutant alleles are usually deleterious. It is hardly likely that changing a connection in a computer at random would improve its performance. The same is true for the living organism, if we consider changing a gene as the analog of changing a computer connection. The organism needs almost all of its parts to be functioning for perfect performance. Failure of even a single part may be disastrous. Death due to a genetic defect may occur early in development. If it occurs sufficiently early *in utero,* the pregnancy may remain undetected.

Most genetic lethals have their own characteristic distributions of age at death. Tay–Sachs disease, a recessive causing mental deficiency and blindness, is a lethal for which death occurs by four years of age, while cystic fibrosis, a relatively common recessive disease whose main effects seem to be malabsorption of food and obstruction of the bronchial tubes and other tissues, nowadays causes death by the late teens. Huntington's chorea, on the other hand, as discussed in Chapter 7, does not lead to death until the late forties or early fifties. Some genetic defects may of course decrease the chance of producing progeny without actually affecting the duration of life. Their effects may be due either to infertility or to a reduction in the probability of mating.

Genes are relatively stable, and mutations of a given gene, as we have seen, are usually quite rare.

Not all gene mutations will be deleterious, but the fraction that are disadvantageous is not known. The proportion of affected is, as we have shown, as high as the mutation rate for recessives and twice as high for dominants, when the fitness is sufficiently low for a mutation to be considered lethal. Thus, the frequency of a genetic disease that is highly deleterious is close to the mutation rate, and therefore is very low. It may of course be somewhat higher if the fitness is not too low, but its frequency still will in general be quite small.

The number of genes that can mutate to give rise to deleterious effects is certainly very large, and perhaps is as large as the total number of genes in the genome.

To date more than 1,000 different genetic diseases have been described, and this number is increasing rapidly. Many of these diseases are heterogeneous, being due to mutations of a number of different gene loci that have phenotypically similar effects. Thus, even if the frequency of affected for each disease is small, the overall frequency of all serious diseases may be quite considerable. The total estimate, including mental deficiency, may be as much as a few percent of all births. *Thus, though each genetic disease will tend to be rare, the aggregate of all genetic diseases is significantly common.*

It is therefore likely that there are many genetic diseases that have not yet been discovered, simply because they are rare. There are often difficulties in deciding whether a disease is genetic before enough cases have been studied. It is thus not unlikely that we have seen only the tip of the iceberg, as far as genetic disease is concerned. Clearly, deleterious mutations, just because they are deleterious, must form an important part of doctors' problems. There may also be special difficulties in devising therapies for diseases hitherto unseen, or for those of which there is very little experience because only a few cases are on record. Another serious complication arises from the fact that successful therapy increases the fitness of the genetically diseased. Thus, it may with time increase the population burden of genetic disease or, as is sometimes said, have substantial "dysgenic" effects. We return to some of these problems in Part IV.

Summary

1. Most genetic diseases are rare. They are maintained by the balance between mutation and selection. Because mutation is a rare event, and because selection against genetic diseases is usually strong, the low frequency of most genetic diseases is readily accounted for.

2. The balance between mutation and selection is established rapidly for dominant (and more slowly for recessive) diseases.

3. For dominant mutations, one can estimate mutation rates from data on incidence—either directly, when the incidence of cases with normal parents is known, or indirectly, when the general incidence and the fitness of the affected are known.

4. The estimation of mutation rates for X-linked recessive genes is also fairly satisfactory, but that for autosomal recessives is not.

5. There is considerable variation in the mutation rates of individual conditions.

6. Most chromosome aberrations of clinical importance behave as dominant lethals. Many are lost in spontaneous abortions.

7. Some mutation rates are strongly affected by the age of the parents.

8. Even though each genetic disease tends to be rare, their aggregate frequency is appreciable, but it is hard to obtain reliable estimates of their overall incidence.

Exercises

1. What is the basic explanation for the fact that most deleterious genetic traits have a very low population incidence? Why are they not completely eliminated from the population?

2. What problems can arise in the direct estimation of dominant mutation rates from pedigree data?

3. Why is it so difficult to obtain reliable estimates of mutation rates for recessive genes?

4. Why are the population frequencies of recessive deleterious genes generally much higher than those for dominant deleterious genes?

5. What reasons can you give to explain the fact that the birth incidence of Turner's syndrome (XO) is much less than that of Down's syndrome?

6. Why should X-linked deleterious recessive mutations generally have a lower frequency in the population than autosomal deleterious recessives?

7. What factors may bias the estimation of mutation rates?

8. Why is it to be expected that many mutations are deleterious?

9. If a disease, never described before, appears in a family, is it possible to recognize it as genetic (a) if it is dominant? (b) if it is recessive? In the latter case, what information might give a hint that the disease might be genetic? (For an extension of this reasoning, also see Exercise 1 in Chapter 11.)

10. Suppose a successful therapy is found for achondroplasia, whose present incidence is, say, one per ten thousand. With a mutation rate of one in twenty thousand, what will be the proportion of new cases that will present themselves for therapy a generation from now? two generations from now?

11. A successful prophylaxis has been found for PKU. Its incidence is now, say, one in ten thousand. Assume the mutation rate is one in ten thousand. What will be the incidence one generation from now? two generations from now?

12. Down's syndrome can be recognized *in utero,* and the affected fetus can be aborted without serious danger or burden for the mother. In Part IV we discuss the practical and ethical aspects of this. Under what circumstances would you suggest carrying out the necessary diagnostic procedure?

References

Carr, D. H.
1965. "Chromosome studies in spontaneous abortions," *Obstetrics and Gynecology,* vol. 26, pp. 308–326.
1967. "Chromosome anomalies as a cause of spontaneous abortion," *American Journal of Obstetrics and Gynecology,* vol. 97, pp. 283–293.

Cavalli-Sforza, L. L., and W. F. Bodmer
1971. *The genetics of Human Populations.* San Francisco: W. H. Freeman and Company. [Chapter 3 gives a more extensive treatment of the topics of this chapter.]

Haldane, J. B. S.
1935. "The rate of spontaneous mutation of a human gene," *Journal of Genetics,* vol. 31, pp. 317–326.

Penrose, L. S.
1955. "Parental age and mutation," *Lancet,* vol. 11, p. 312.

Penrose, L. S., and G. F. Smith
1966. *Down's Anomaly.* Boston: Little, Brown.

Russell, W. L.
1965. "Evidence from mice concerning the nature of the mutation process," in *Genetics Today,* vol. 2, ed. S. J. Geerts. [Proceedings of the 11th International Congress of Genetics, The Hague, Sept. 1963.] New York: Pergamon.

Stevenson, A. C., and C. B. Kerr
1967. "On the distribution of frequencies of mutations in genes determining harmful traits in man," *Mutation Research,* vol. 4, pp. 339–352.

Vogel, F.
1965. "Mutations in man," in *Genetics Today,* vol. 3, ed. S. J. Geerts. [Proceedings of the 11th International Congress of Genetics, The Hague, Sept. 1963.] New York: Pergamon.

9

Balanced Polymorphism

When the heterozygote is at an advantage over both homozygotes, a situation results in which both alleles tend to remain at substantial frequencies in the population. This situation is called balanced polymorphism, of which sickle-cell anemia was the first well-investigated case. The driving force in this case is malaria, a disease that also is responsible for other genetic variation.

9.1 Heterozygote Advantage and the Sickle-cell Trait

Sickle-cell anemia is responsible for almost 100,000 deaths per year throughout the world. In some populations—ranging from Western, Central, and Eastern Africans to Greeks and Asiatic Indians—the frequency of the gene for hemoglobin S reaches values as high as 10 percent. This high frequency occurs despite the fact that the homozygote for the *S* gene (who has the severe anemia) has only a small chance of surviving to the reproductive age.

Could the hemoglobin *S* allele be maintained at its high frequency by the balance between mutation and selection, as in the case of the much rarer genetic diseases discussed in the last chapter? If so, we should have to postulate that, in those populations where the gene is frequent, the mutation rate to *S* is inordinately high, while in other populations it is very small. Moreover, the mutation rate needed to keep the gene frequency as high as that observed in some populations would be so large that new mutations should surely have been discovered in family studies. Thus, a high mutation rate cannot account for the high frequency of the hemoglobin *S* allele in certain populations, and some other explanation is needed.

When two or more alleles at a given genetic locus occur with appreciable frequencies in a population, the locus is said to be polymorphic.

Thus, the hemoglobin locus is polymorphic in those populations in which sickle-cell anemia is a relatively common disease and in which the hemoglobin *S* allele has a frequency of at least a few percent. In other populations, such as those of northern Europe, where almost all individuals have the normal type-A hemoglobin, the locus is not polymorphic. We have already referred to a number of other examples of polymorphisms, such as the MN and ABO blood groups. These polymorphisms are not, however, associated with obvious differences in fitness such as those found in the case of sickle-cell anemia.

The definition of how large a gene frequency must be in order to be "appreciable" and so define a polymorphism is fairly arbitrary.

In practice, taking into account the size of readily available samples of human populations, the level used to define a polymorphism is of the order of one percent. That is, a locus is considered polymorphic if the second-most-frequent allele has a relative frequency greater than one percent. Polymorphism was, at one time, defined in terms of the selective mechanisms responsible for maintaining relatively high gene frequencies of two or more alleles at a locus. We now know, however, that these mechanisms may be very difficult to determine, so that a definition of polymorphism based on them would be of little value.

What mechanism maintains the sickle-cell polymorphism?

The clue to this puzzle was provided by J. B. S. Haldane's suggestion, made in 1949, that heterozygotes for thalassemia (another form of hemoglobin disease; see Section 3.5) might be more resistant to malaria than are individuals with normal amounts of hemoglobin. Many years earlier, in 1922, R. A. Fisher had shown with a simple mathematical model how a selective advantage of the heterozygote for two alleles at a locus over both homozygotes could give rise to a balanced polymorphism. The term "balanced" was used to describe the fact that such a polymorphism is maintained by a balance of selective forces acting on the three genotypes formed by two alleles at a single locus. We shall consider further in the next section this fundamental mechanism for explaining the maintenance of a balanced polymorphism. Let us return first to the evidence for selection in the case of the sickle-cell hemoglobin.

Haldane's suggestion was based on a comparison of the geographic distribution of malaria with that of the hemoglobin disease. Malaria is due to an infection by the protozoan parasite *Plasmodium,* the severest forms of the disease being associated with the species *Plasmodium falciparum* (see Fig. 9.1). The geographic distributions of the hemoglobin *S* gene and of *falciparum* malaria are illustrated in Figures 9.2 and 9.3. A comparison of the two maps shows that there is a fairly clearcut correlation between the two distributions. This correlation could hardly be expected to be perfect, because changes in the local frequencies of both

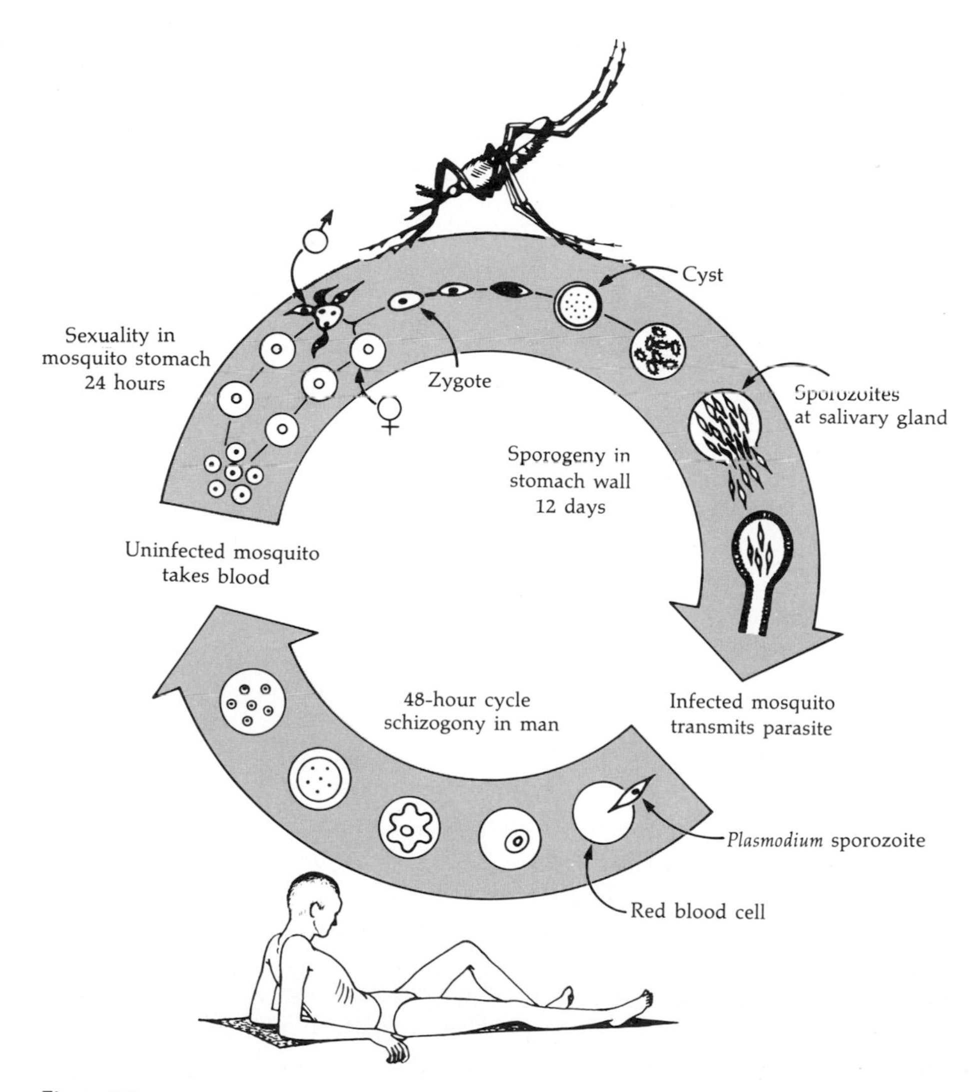

Figure 9.1
The life cycle of a malarial parasite. (After P. B. Weisz, *The Science of Biology,* McGraw-Hill, 1966.)

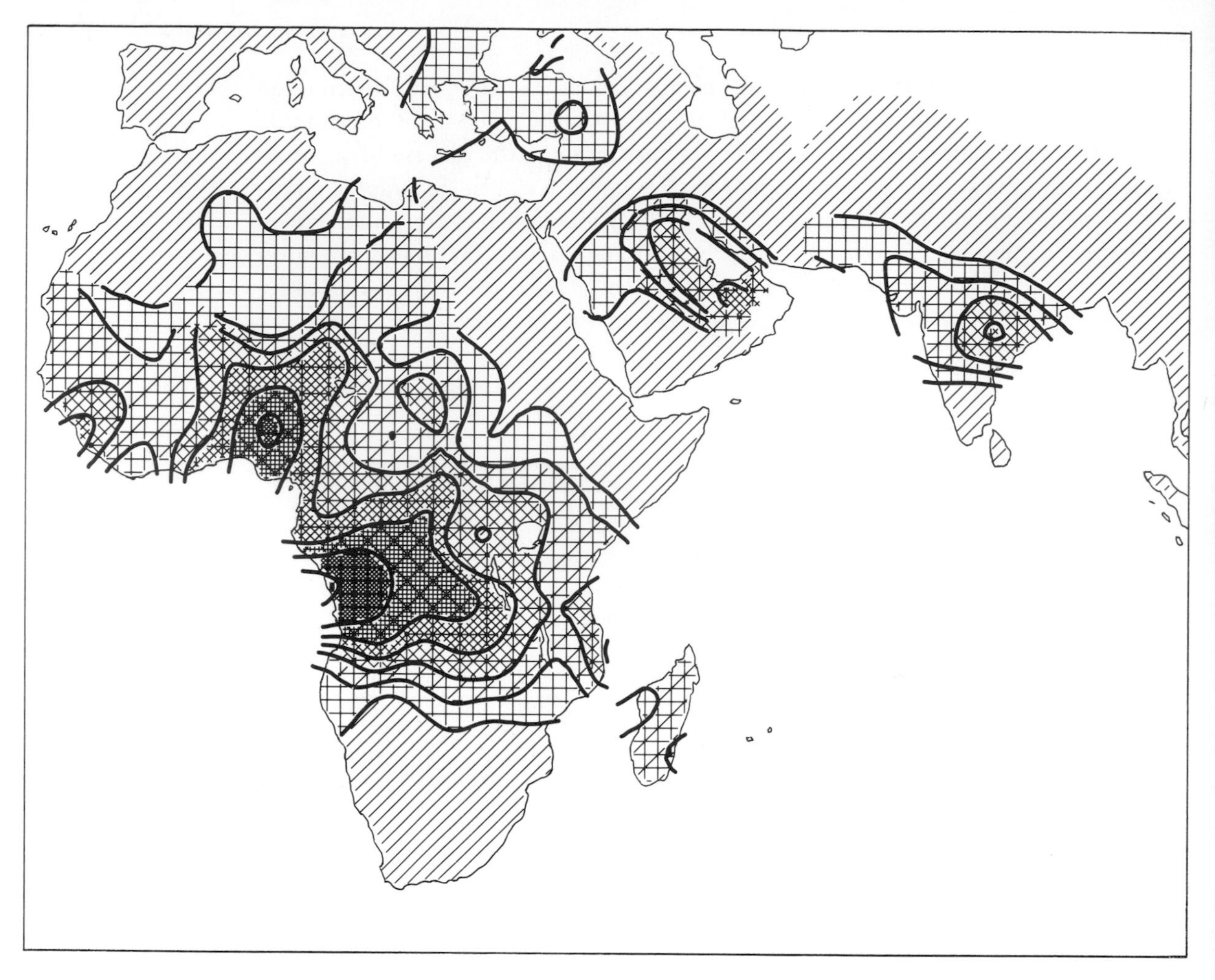

Figure 9.2
Distribution of hemoglobin-S gene in the Old World. This computer-generated map shows the frequency of the sickle-cell gene *S* as indicated by the key. Data from compilation of F. B. Livingstone. (Courtesy of D. E. Schreiber, IBM Research Laboratory, San Jose.)

phenomena may have occurred over the centuries. Still, the geographic distribution correlation is not sufficient proof of the malaria hypothesis, and further evidence is needed.

Direct experiments were carried out on volunteers to compare the resistance to malaria of sicklers (heterozygotes for allele *S* who have the trait but not the

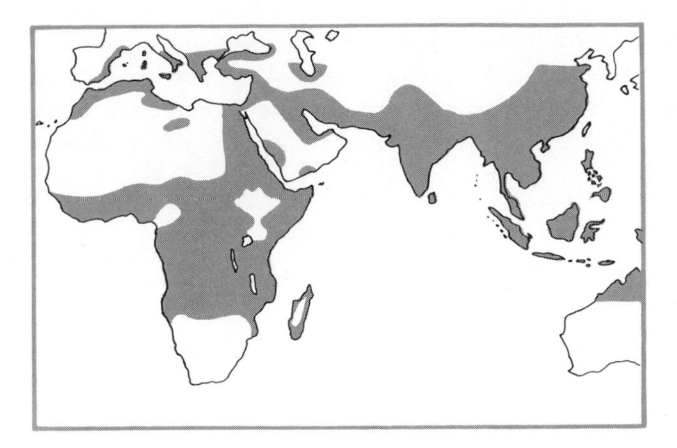

Figure 9.3
Distribution of malaria in the Old World. The map indicates the areas where there is a significant incidence of *Pl. falciparum* malaria. (From M. F. Boyd, ed., *Malariology*, Saunders, 1949.)

severe anemia) and homozygous normals. Fifteen sicklers and fifteen normals from the Luo population of East Africa were injected with *Pl. falciparum.* Malarial parasites were subsequently found in the blood of fourteen of the normals but only in two of the sicklers—a highly significant difference. (These experiments were not dangerous because malaria can readily be cured under these conditions.)

Even more striking evidence for the malaria hypothesis comes from the fact that, in hospital patients, the relative incidence of severe or fatal malarial infection is almost nil among sicklers.

Because the homozygote for allele *S* has the severe anemia, these data show that, in a malarial environment, the heterozygote *AS* is fitter than either the normal homozygote *AA* or the sickle-anemia homozygote *SS.* There are good reasons why the malarial parasite cannot thrive well in sicklers. Hemoglobin S tends to crystallize at the low oxygen pressures that occur in the narrow blood capillaries. This leads to a tendency of the red cells to break open (or lyse), so causing the hemolytic anemia (hemolysis effectively means lysis of red cells). The malarial parasite, which multiplies in the red cells and probably derives most of its nutrition from hemoglobin, therefore finds a less satisfactory environment when hemoglobin S is present. Red cells containing the parasite, and lysing because of the sickling, are also more likely to be phagocytosed (engulfed by other scavenger cells), leading to their preferential removal and so also to the removal of the parasite from the blood circulation.

Thus, there are very good reasons for accepting the idea that heterozygotes for the sickle-cell hemoglobin gene are more resistant to malaria. This resistance may apply not only to the chances of surviving a malarial attack, but also to the chances for survival of a fetus in an infected mother. Is the magnitude of the resistance sufficient to explain the fact that the *S* gene often reaches a high frequency when malaria is prevalent? We will explore this problem further in the next section.

9.2 Balanced Polymorphism Due to Heterozygote Advantage

A simple extreme example, which *Drosophila* geneticists call a *balanced lethal,* illustrates how heterozygote advantage can lead to a balanced polymorphism. A recessive lethal gene on the *Drosophila* second chromosome, called Curly (symbol *Cy*), can be recognized in the heterozygote by the fact that flies have curly wings. Another recessive lethal, Plum (*Pm*), which is also on the second chromosome, shows up in the heterozygote by the plum color of the eyes. By crossing the two strains, we can obtain *Cy/Pm* individuals with both traits. In the absence of crossing-over between *Cy* and *Pm,* which can be suppressed by the use of appropriate additional genetic markers, the only possible segregants from crosses among *Cy/Pm* flies are *Cy/Cy, Pm/Pm,* and *Cy/Pm.* The first two die because both *Cy* and *Pm* are homozygous lethals, and thus the only survivors are *Cy/Pm.* Such a stock can be maintained indefinitely with all individuals heterozygous (barring a rare crossing-over) because the homozygotes always all die. The gene frequencies of *Cy* and of *Pm* thus remain at exactly 50 percent in each generation (Fig. 9.4).

The case of a population with the sickle-cell gene in the presence of malaria is not so extreme as the Drosophila balanced lethal, but the principle is similar.

Not quite all of the *SS* homozygotes die because of sickle-cell anemia. Not all of the *AA* or normal homozygotes—in fact only a slightly higher fraction than that of heterozygotes—die from malaria. Heterozygotes do not have as complete an advantage as in the case of balanced lethals. How does this give rise to an equilibrium, and what is the expected gene frequency at equilibrium?

The answer to these questions comes from a mathematical analysis that is given in the appendix. Here we illustrate the model using numerical examples. Let us consider again the case of the sickle-cell gene *S* and its normal counterpart *A.* We assume random mating, so the Hardy–Weinberg law applies, and we assume that the fitnesses of the three genotypes *AA, AS,* and *SS* are $\frac{8}{9}$, 1, and 0, respectively (these values are chosen to simplify the numerical illustration). Suppose, as indicated by Table 9.1, that we start with an adult population with 80 people of type *AA* and 20 of type *AS.* The *S* gene frequency, which is just one-half the frequency of *AS* heterozygotes when there are no *SS* homozygotes, is 10 percent.

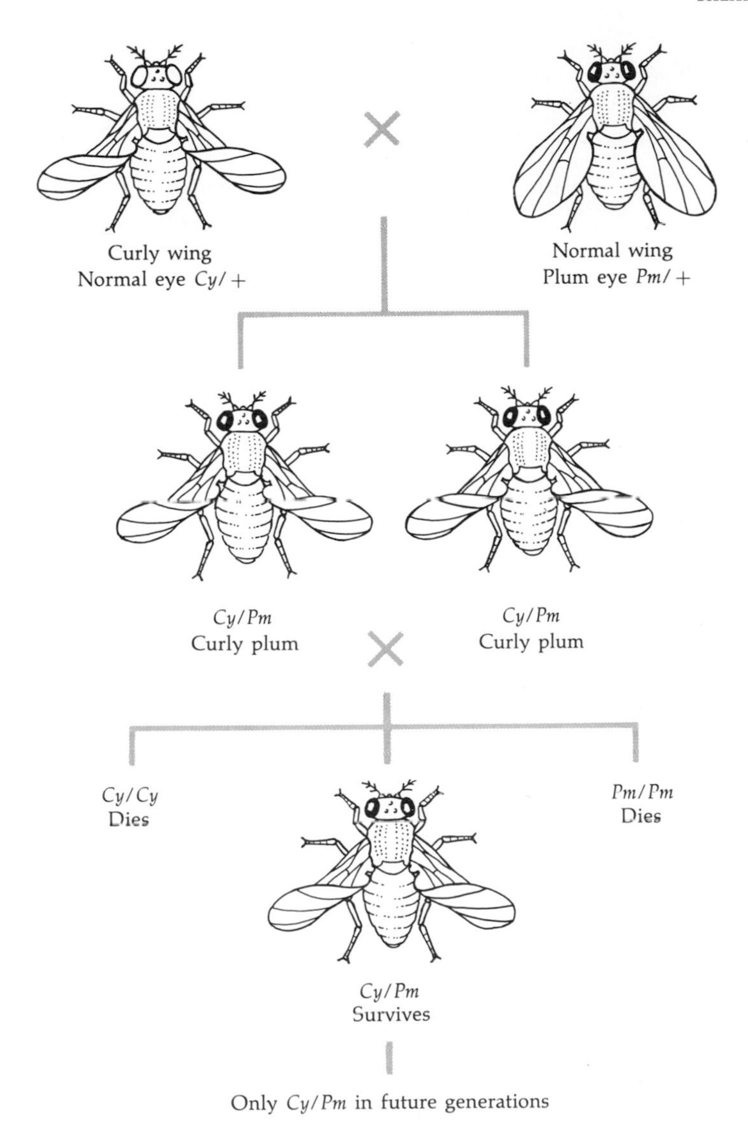

Figure 9.4
A balanced lethal situation in Drosophila. Because all homozygotes die, the gene frequencies of *Cy* and of *Pm* will remain balanced indefinitely at 50 percent each.

If these adults now mate at random, they will produce offspring in Hardy-Weinberg proportions, as indicated in Table 9.1: $(9/10)^2$ *AA*, $2 \times (9/10) \times (1/10)$ *AS*, and $(1/10)^2$ *SS*. These, however, are the expected frequencies with which zygotes should form, before natural selection has had an opportunity to act. To obtain the frequencies in the second parental generation (after selection), we must

Table 9.1
Balanced polymorphism due to heterozygote advantage.

	Genotypes			Alleles	
	AA	*AS*	*SS*	*S*	*A*
Fitness	8/9	1	0	–	–
Initial adult population (100 individuals)	80	20	0	–	–
Gene and genotype frequencies	$80/100 = 8/10$	$20/100 = 2/10$	$0/100 = 0$	$(1/2) \times (2/10) = 1/10$	$8/10 + (1/2) \times (2/10) = 9/10$
Zygote frequencies after random mating (by Hardy–Weinberg)	$(9/10)^2 = 81/100$	$2 \times (9/10) \times (1/10) = 18/100$	$(1/10)^2 = 1/100$	–	–
Zygote population (100 zygotes)	81	18	1	–	–
Population after selection* (total 90 survivors)	$(8/9) \times 81 = 72$	$1 \times 18 = 18$	$0 \times 1 = 0$	–	–
Gene and genotype frequencies among survivors‡	$72/90 = 8/10$	$18/90 = 2/10$	$0/90 = 0$	$(1/2) \times (2/10) = 1/10$	$8/10 + (1/2) \times (2/10) = 9/10$

*Computed as product of fitness times number of zygotes.
‡Computed dividing population after selection by total number of survivors: $72 + 18 = 90$.
NOTE: The population is in equilibrium; calculations for succeeding generations are the same as those just given for the second generation.

multiply each Hardy–Weinberg expectation by the corresponding fitness. Thus, the 81 *AA* zygotes leave only $(8/9) \times 81 = 72$ survivors, while all of the *AS* and none of the *SS* zygotes survive. Of 90 survivors, 72 are *AA* and 18 are *AS*. These values represent genotype frequencies of $72/90 = 80$ percent *AA* and $18/90 = 20$ percent *AS*—just the same frequencies as those in the original population.

This example shows that an equilibrium can be maintained with appropriate fitnesses corresponding to an advantage of the heterozygote over each of the homozygotes. In this case, the fitnesses were chosen to correspond to an equilibrium gene frequency for *S* of 10 percent. Table 9.2 shows the formula for computing equilibrium gene frequencies from known genotype fitnesses. This formula takes its simplest form (shown here) if we assume that the fitness of the heterozygote is one, while the two homozygotes have fitnesses less than one by the amounts s for the *AA* homozygote and t for the *SS* homozygote. Then, for example, the expected equilibrium gene frequency of allele *S* is $s/(s + t)$. For the example of Table 9.1, $s = \frac{1}{9}$ (because $1 - \frac{1}{9} = \frac{8}{9}$, the assumed fitness of *AA*) and $t = 1$ (because the fitness of *SS* is assumed to be $1 - 1 = 0$). The formula

Table 9.2
Equilibrium gene frequencies under heterozygote advantage

If genotypes	*AA*	*AS*	*SS*
have fitnesses	$1 - s$	1	$1 - t$
then, at equilibrium,			
alleles	*A*	*S*	
will have frequencies	$t/(s + t)$	$s/(s + t)$	

NOTE: For heterozygote advantage, both s and t must be positive.

from Table 9.2 then yields an expected equilibrium frequency for allele *S* of $(1/9)/(1 + 1/9) = 1/10$.

What happens to the gene frequencies if the frequencies in the initial population differ from the equilibrium values?

Table 9.3 shows an example in which the initial frequency of allele *S* is $\frac{1}{100}$. The table shows the same computations as those in Table 9.1, but the initial set of zygotes has the Hardy–Weinberg proportions $(99/100)^2$ *AA*, $2 \times (99/100) \times (1/100)$ *AS* and $(1/100)^2$ *SS*. Now the composition of the next generation changes to give an increased gene frequency for *S* of about 1.1 percent instead of the initial 1 percent. If we continued this process step by step for each generation, keeping the genotype fitnesses the same, the gene frequency would increase steadily until it approached its equilibrium value. Had we started with a gene frequency for *S* that was *above* rather than below the equilibrium value of 10 percent, then we should have noted a *decrease* after one generation. Repeating the process would then lead to a gradual decrease to the equilibrium frequency. This illustrates the essential property of a true stable equilibrium—whatever may be the gene frequency in the population to start with, it always moves toward the equilibrium value. Some examples of the approach to equilibrium calculated by computer, using different fitnesses, are given in Figure 9.5.

The stability of the equilibrium illustrated in Figure 9.5 depends on the fact that the heterozygote is fitter than both homozygotes.

Other fitness relationships between the genotypes will lead to quite different results. For example, if *AA* were fitter than *AS* and *SS*, then the *A* gene would be at an overall advantage over the *S* gene. Essentially, *S* would then be a deleterious partially dominant gene, maintained in the population at a low frequency by mutation–selection balance. This is the situation that might effectively arise when malaria is eradicated from the environment of a population carrying the *S* gene.

Table 9.3
Gene-frequency change under heterozygote advantage

	Genotypes			Alleles	
	AA	*AS*	*SS*	*S*	*A*
Fitness	8/9	1	0	–	–
Initial adult population (100 individuals)	98	2	0	–	–
Gene and genotype frequencies	98/100	2/100	0/100	(1/2) × (2/100) = 1/100	98/100 + (1/2) × (2/100) = 99/100
Zygote frequencies after random mating (by Hardy–Weinberg)	$(99/100)^2$ = 9,801/10,000	2 × (99/100) × (1/100) = 198/10,000	$(1/100)^2$ = 1/10,000	–	–
Zygote population (10,000 zygotes)	9,801	198	1	–	–
Population after selection (total 8,910 survivors)	(8/9) × 9,801 = 8,712	1 × 198 = 198	0 × 1 = 0	– –	– –
Gene and genotype frequencies among survivors	8,712/8,910 ≅ 97.8/100	198/8,910 ≅ 2.2/100	0/100	(1/2) × (2.2/100) = 1.1/100	98.9/100

NOTE: The gene frequency of *S* has increased 10 percent (from 1/100 to 1.1/100) in the first generation. Such increases will continue in future generations, until the gene frequency of *S* closely approaches the equilibrium value of 1/10 = 10/100.

It can be shown that if, for some reason, the heterozygote were less fit than either homozygote, an unstable situation is produced. The population will move toward being either all *AA* or all *SS,* depending on the starting gene frequency. The starting gene frequency that divides values into those moving to *AA* from those moving to *SS* is an "unstable" equilibrium point. This result is important because it shows that selection *against* heterozygotes can never (in the absence of other factors) lead to a balanced polymorphism. These various possibilities are summarized in Table 9.4. (Note that the counterpart to *A* having an unconditional advantage occurs when *SS* is fitter than *AS* and *AA;* then allele *S* has an unconditional advantage and the population always moves toward being all *SS.*)

When, in Chapter 6, we discussed the Hardy–Weinberg law, we used as an example (Table 6.2) data on the distribution of the genotypes *AA, AS,* and *SS* among a group of very young children. The fit with the Hardy–Weinberg law, including that for the genotype *SS,* is very close. This is expected because, by examining children, we have obtained the frequencies before selection has had much time to act. If we look at adults instead of children, the fit with the Hardy–

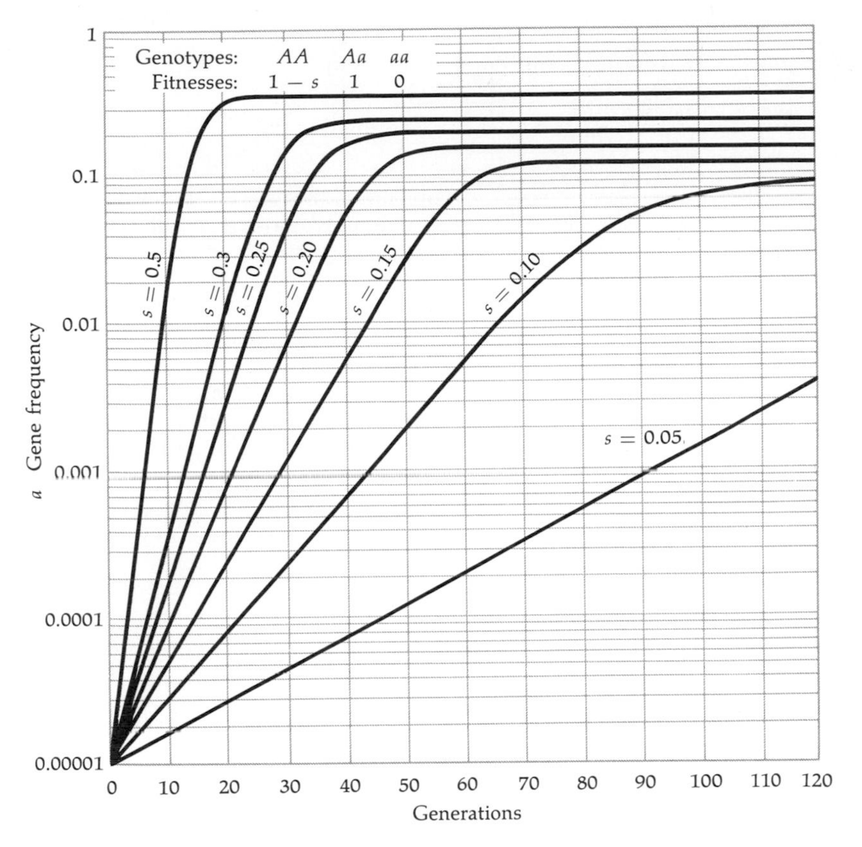

Figure 9.5
The approach to equilibrium. The graph shows the approach to equilibrium for the frequency of a gene *a*, a recessive lethal, for different fitness values of the dominant homozygote. In each case, the heterozygote is assumed to have a selective advantage ($s > 0$), and the initial frequency of *a* in the population is assumed to be 1 in 1,000,000. Note that smaller values of *s* (smaller advantages for the heterozygote) lead to slower approaches toward equilibrium and to smaller equilibrium values for the gene frequency of *a*.

Weinberg law is not good, and now we can use the difference to estimate the fitness of the various genotypes, as illustrated in Table 9.5. First, from the adult frequencies, by counting *A* and *S* genes, we obtain the gene frequencies for *S* (0.123) and for *A* (0.877). Then we can compute the expected frequencies of genotypes *AA*, *AS*, and *SS* from the Hardy–Weinberg law as $(0.877)^2$, $2 \times (0.877) \times (0.123)$, and $(0.123)^2$ respectively, and so obtain expected numbers by multiplying these by the total sample size (12,387). The ratio between the observed numbers and these expected numbers is then an estimate of the fitness, at least on the assumption that selection is acting mainly through differential mortality.

Table 9.4
Equilibrium situation for different fitness relationships

Fitness relationship	Consequence
Heterozygote advantage	
Aa fitter than AA and aa	Population always moves toward stable polymorphic equilibrium in which both A and a genes are represented.
Heterozygote disadvantage	
Aa less fit than AA and aa	Population always moves away from unstable intermediate gene frequency toward either all AA or all aa, depending on initial frequency.
*Gene A at unconditional advantage**	
$AA > Aa \geq aa$ or $AA \geq Aa > aa$	Population always moves towards all AA.
*Gene A at unconditional disadvantage**	
$aa > Aa \geq AA$ or $aa \geq Aa > AA$	Population always moves towards all aa.

*Here, the symbol $>$ means "fitter than" and the symbol $\geq$ means "fitter than or as fit as."

Note that, as expected, the most striking deficit is with respect to the homozygotes *SS*, where the ratio of observed to expected is only $29/187.4 = 0.155$, indicating that only perhaps about 15 to 16 percent of newly formed *SS* zygotes survive to be adults. To obtain the fitnesses in standard form (with the fitness of *AS* equal to one), we divide each of the fitness values (for *AA*, *AS*, and *SS*) by the value for *AS* (1.12). This procedure yields the final fitness estimates of 0.14 for *SS*, 1.00 for *AS*, and 0.88 for *AA*. To make use of the expression for equilibrium gene frequency shown in Table 9.2, we compute s and t as $s = 1 - 0.88 = 0.12$, and $t = 1 - 0.14 = 0.86$. We then compute the expected equilibrium frequency of *S* as

$$s/(s + t) = 0.12/(0.12 + 0.862) = 0.12/0.982 = 0.122,$$

a value that is remarkably close to the observed frequency of 0.123. This result shows that the data really agree very closely with the expected equilibrium gene frequency based on heterozygote advantage, using the calculated fitness values.

Given the fitness values that we have estimated for the three genotypes, would the equilibrium gene frequency for S be reached within a reasonable period of time?

Table 9.5
Genotype fitnesses for sickle-cell anemia and sickle-cell trait, computed on the basis of departures from expected genotype frequencies among adults

Genotype	Observed frequency among adults (*O*)	Expected frequency from Hardy-Weinberg (*E*)	Ratio *O/E*	Standardized fitness
SS	29	187.4	0.155	$0.155/1.12 = 0.14 = 1 - t$
SA	2,993	2,672.4	1.12	$1.12/1.12 = 1.00$
AA	9,365	9,527.2	0.983	$0.983/1.12 = 0.88 = 1 - s$
Total	12,387	12,387		

NOTE: From the observed genotype frequencies, the gene frequency of *S* is computed to be $\{29 + [(1/2) \times 2{,}993]\}/12{,}387 = 0.123$, and so the *A* gene frequency must be $1.00 - 0.123 = 0.877$. These gene frequencies are then used to compute the genotype frequencies expected under the Hardy-Weinberg law as $(0.123)^2 \times 12{,}387$, $2 \times 0.877 \times 0.123 \times 12{,}387$, and $(0.877)^2 \times 12{,}387$, respectively. The computation of fitness values given in the table assumes that all selection acts by mortality between zygote formation and adulthood, and thus ignores the effects of fertility. The selection coefficients obtained from these calculations are $t = 1.00 - 0.14 = 0.86$ (or 86%) and $s = 1.00 - 0.88 = 0.12$ (or 12%).

The data given in this table apply to the Yorubas of Ibadan, Nigeria (from G. M. Edington, 1957; as cited by F. B. Livingstone, *Abnormal Hemoglobins in Human Populations*, Aldine, 1967). For simplicity, hemoglobins C and G have here been pooled with A.

This question can be answered by using the type of graph shown in Figure 9.5. Suppose that the *S* gene were present at the beginning of the selection process at a very low frequency—as might happen if only one mutant had originally been produced in a large population, or alternatively if just one immigrant carrying the gene had entered the population from outside. In this case, a frequency very near the equilibrium frequency would be reached in about 100 generations. Assuming a time per generation of about 30 years, this means that equilibrium could be closely approximated within about 3,000 years.

We know that malaria was probably introduced (or at least became much more prevalent) as a result of the introduction of agriculture. When the forest is cleared and irrigation systems are introduced, small pools of stagnating water are easily created. The larvae of the mosquito that is the vector of the malarial parasite thrive particularly well in such pools. In addition, agricultural development led to the formation of larger conglomerates of people, thus increasing the chances of vector-man-vector transmission. Thus, agriculture is though to have helped greatly in promoting the diffusion of malaria. In many African (and other) populations, agricultural development originated more than 3,000 years ago, thus providing enough time for equilibrium to be reached for the *S* gene under the impact of natural selection in favor of the *AS* heterozygotes. However, malaria may well have antedated the origin of agriculture.

The case of the hemoglobin *S* gene has been discussed in detail because it is such a good example of the chain of inference leading from a changed nucleotide in DNA to an understanding of a balanced polymorphic situation. No other polymorphism in any species has so far been understood as completely as the sickle-cell polymorphism.

9.3 Other Polymorphisms That May Reflect Adaptations to Malaria

Thalassemia *is a genetic disease which is found especially in the Mediterranean region—Italy, Greece, and elsewhere—and also in Africa, India, and many other areas.*

There are, as explained in Section 3.5, at least two thalassemia genes on different chromosomes, one linked to the gene determining the α chain of hemoglobin and another linked to the gene determining the β chain. These genes appear to cause a decrease, or even a complete absence, of the synthesis of the α or β hemoglobin chains. The result is a very serious anemia in the homozygote, whether for the α or for the β thalassemia gene, leading to early death in practically 100 percent of the cases. The heterozygotes in both cases have at most very minor clinical signs, but can be detected on the basis of a hematological examination.

The α and β thalassemias are thus recessive lethals, whose homozygous effects are even more serious than that of the sickle-cell gene. Nevertheless, they are widespread in malarial areas and most probably represent another example of a genetic adaptation to malaria, associated with an increased resistance of the heterozygote. In some formerly malarial areas, such as Sardinia, the genes reach frequencies as high as 10 percent, with about 20 percent heterozygotes and with some 1 percent of all births affected by the severe homozygous disease (Fig. 9.6). A careful examination of the data shows that the gene frequency varies considerably from village to village, and that the ecological conditions also differ enough to have led to radical differences in malarial prevalence in different villages. As migration between the villages is limited, and genetic adaptation is comparatively rapid, one might expect (and indeed does find) a fairly good correlation between the prevalence of malaria and the frequency of thalassemia observed in the villages sampled (Fig. 9.7).

Figure 9.7 also illustrates a correlation between malaria prevalence and the frequency of yet another gene, the X-linked glucose-6-phosphate-dehydrogenase (G6PD) deficiency, which we have mentioned several times in connection with studies on the X chromosome. This sex-linked deficiency was first discovered because its symptoms include a pronounced sensitivity to certain antimalarial drugs and also to eating the fava bean, *Vicia faba.* The G6PD-deficiency gene frequency varies considerably. Values observed in various Jewish groups who lived in widely different environments after the diasporas (dispersions of the Jews that occurred 2,000 to 2,500 years ago) are especially informative (Fig. 9.8).

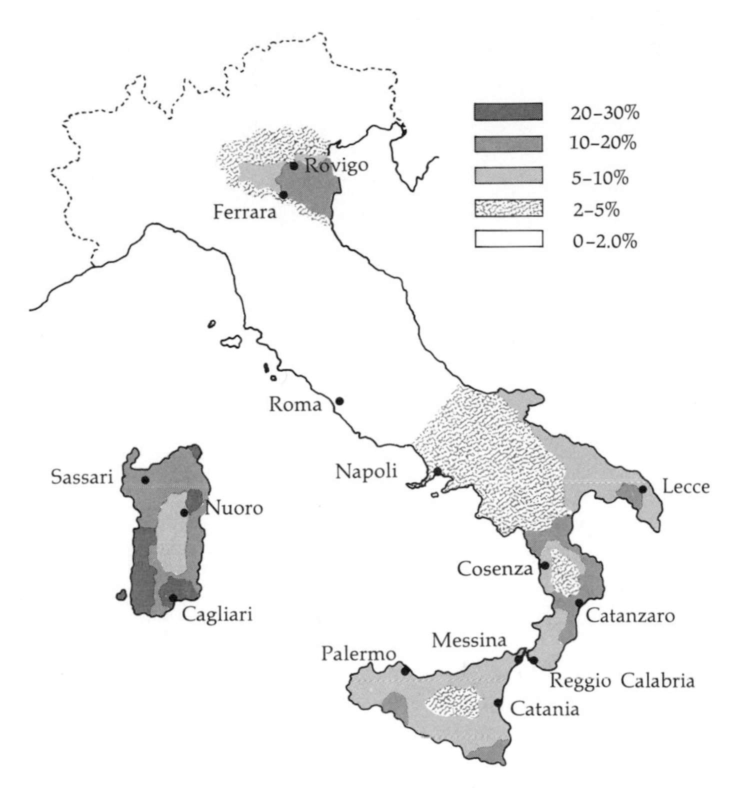

Figure 9.6
Incidence of thalassemia in various regions of Italy. Frequencies shown are those of heterozygotes for the gene; gene frequencies are approximately half of the values shown here. (After E. Silvestroni and I. Bianco, *American Journal of Human Genetics*, vol. 27, p. 198, 1975.)

Gene frequencies of up to 60 percent are found in some populations, indicating how rapidly a very strong selection pressure may have changed gene frequencies over a period of about 100 generations, starting presumably from very low frequencies. There are, in fact, a number of different G6PD-deficiency alleles, involving various degrees and modes of enzyme changes.

For G6PD deficiency there is a direct demonstration of the resistance of carriers to the malarial parasite.

This is based on the relative preference of the malarial parasite for a red cell containing G6PD enzyme activity (Fig. 9.9). It is not, however, entirely clear whether the female heterozygote is at an advantage, or whether it is only the homozygous females and affected males that have an advantage over normals,

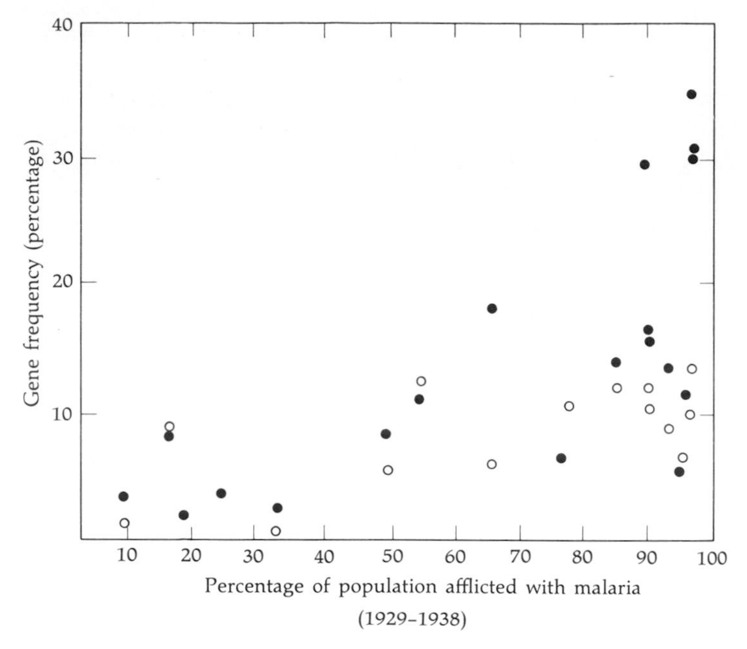

Figure 9.7
Correlation between malarial incidence, thalassemia, and G6PD deficiency. Solid circles show the relationship between malarial incidence and the frequency of the gene for G6PD deficiency in Sardinian villages; open circles indicate the relationship between malarial incidence and the frequency of the genes for thalassemia. (From M. Siniscalco, L. Bernini, B. Latte, and A. G. Motulsky, *Nature,* vol. 190, p. 1179, 1961.)

at least for some alleles. As G6PD is sex-linked, there are no heterozygous males. (The male condition for an X-linked gene is sometimes referred to as hemizygous.)

In addition to the abnormal hemoglobin S, there is another abnormal hemoglobin, called C, which is due to an allele of S and is also common in at least some parts of West Africa.

As we have already seen in Section 3.5, *CC* homozygotes have a much milder anemia than do *SS* homozygotes, but *SC* heterozygotes have an anemia whose severity is intermediate between those two homozygous genotypes. It is presumed that the hemoglobin *C* allele is also maintained polymorphic because of an association with resistance to malaria, but the evidence for this is much less clearcut than that for hemoglobin *S*. The analysis of a polymorphism involving three alleles (*A*, *S*, and *C*) is, of course, more complicated than that for two alleles described in the previous section. There is, however, one very interesting feature of the distribution of the *C* allele, illustrated in Figure 9.10. The gene frequencies suggest

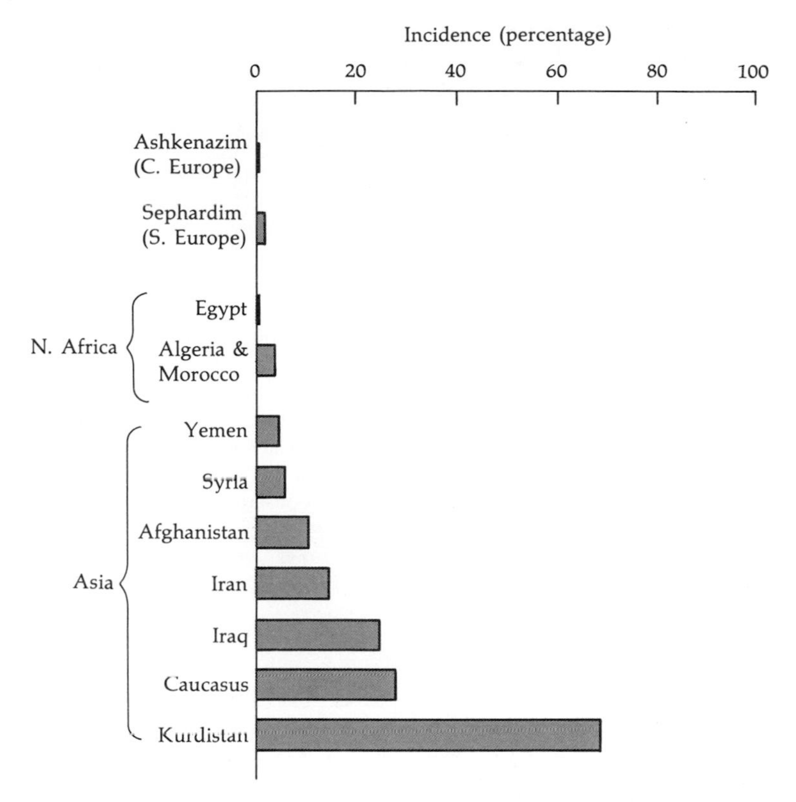

Figure 9.8
Incidence of G6PD deficiency in various Jewish populations. The incidence is shown as a percentage in the male population. Because the gene is X-linked, the incidence is equal to the gene frequency. Note the large frequencies in some Jewish populations from the Middle East. (Data from A. Szeinberg, in *The Genetics of Migrant and Isolate Populations,* ed. E. Goldschmidt, Williams and Wilkins, 1963.)

that the *C* allele is or was spreading from a center of origin somewhere near the Upper Volta region. The figure shows the observed distribution of *C* allele frequencies in West Africa, and some curves drawn to depict the expected frequencies for a gene spreading from such a center of origin.

A further analysis of this complex situation has suggested that it may be possible that the *S* allele diffused more widely and more rapidly because it was subject to stronger selective pressures in the presence of malaria. The *C* allele, however, may have been increasing in frequency slowly and steadily for similar or other reasons. This discussion emphasizes the difficulty in deciding whether an observed polymorphism is truly "balanced." The data for the *S* allele suggest that its frequency may be close to equilibrium, while the data for hemoglobin *C* show that

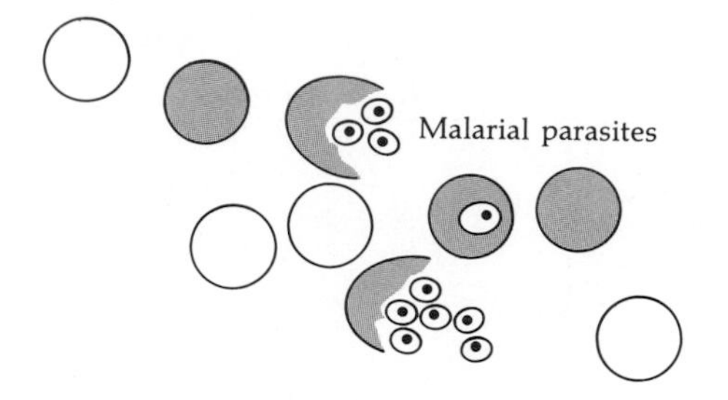

Figure 9.9
Direct demonstration of the malaria resistance conferred by G6PD deficiency. The diagram represents a slide of blood taken from a woman who is heterozygous for the G6PD deficiency (Gd^-) and who is suffering from a spontaneous malarial attack. Because of the inactivation of one X chromosome in each cell (see Section 5.3), half of the red cells (in which the Gd^+-carrying chromosome is active) produce the enzyme G6PD, while the other half (in which the Gd^- chromosome is active) do not. Those cells that contain G6PD are stained dark, and those that do not contain G6PD are not stained. The small organisms depicted in some cells are malarial parasites (illustrated schematically), and they are found to multiply well only in the cells with G6PD enzyme activity. Parasitization of Gd^- cells does take place, but at much lower rates than those for Gd^+ cells. (From an experiment by L. Luzzato, E. A. Usanga, and S. Reddy, *Science*, vol. 164, pp. 839–842, 1969.)

this is clearly not the case for allele *C*. The latter represents a possible example of a *transient* polymorphism, namely a gene caught in the midst of the process of increasing in frequency to almost 100 percent. In practice it may often be very difficult to decide whether an observed polymorphism is balanced or transient. This problem is further discussed in Chapter 13, where we consider the question of molecular evolution.

It is perhaps not too surprising that so many polymorphisms connected with resistance to malaria have been demonstrated, and still others are suspected. Malaria is a very widespread and highly debilitating disease. In areas where malaria is endemic, practically everybody is affected from an early age and has recurrent episodes of the disease throughout life. Death need not always be due to malaria itself, but may be caused by other diseases to which an individual has been made less resistant by the malarial condition.

9.4 Other Mechanisms That Can Lead to Balanced Polymorphism

Heterozygote advantage is the simplest example of the way in which a balance of selective forces can give rise to a balanced polymorphism. There are, however, many different combinations of balancing forces that can give rise to stable polymorphisms. The following are a few examples of models that have been suggested.

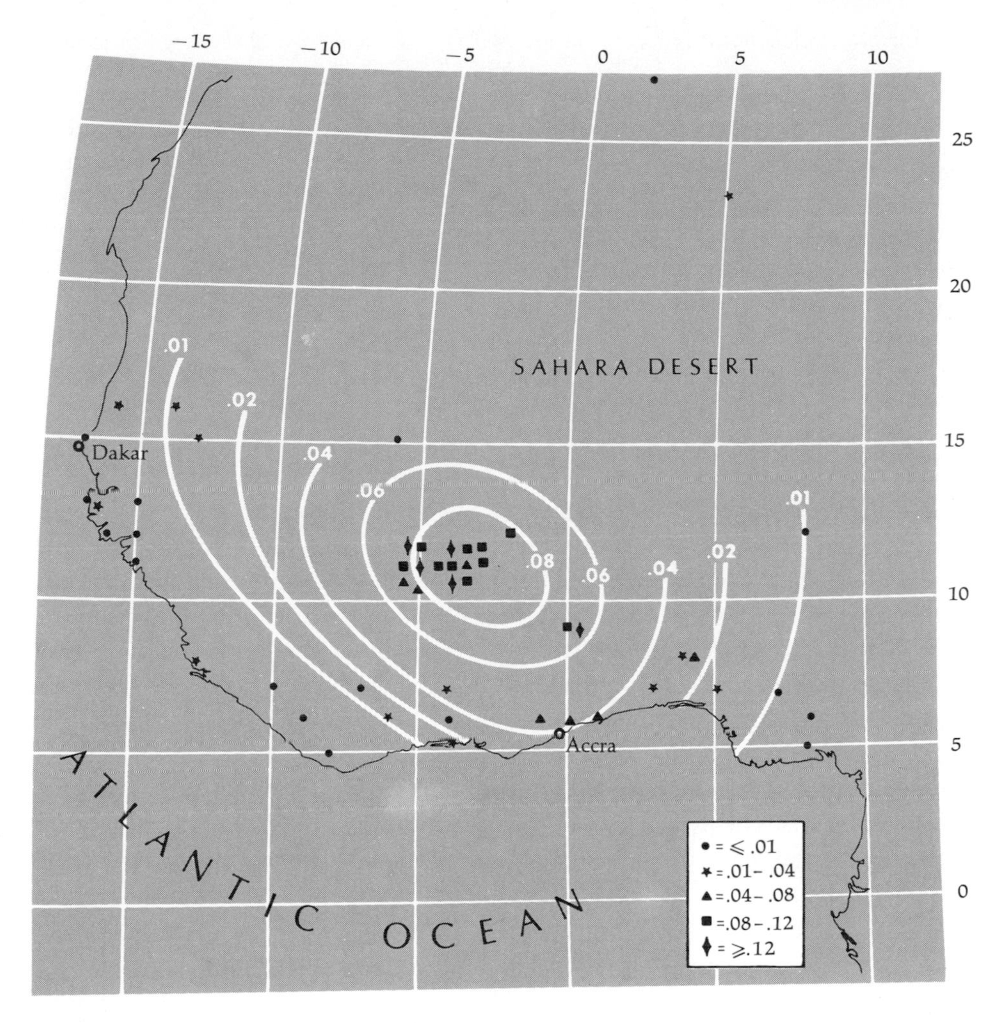

Figure 9.10
Gene frequencies of hemoglobin C in West Africa. Hemoglobin S is also found in the same area. The relationship between hemoglobin C and malarial resistance has not been fully confirmed.

Selection may vary with the frequency of the genotypes in a population.

We discuss in Chapter 7 the possibility that a genotype—such as that conferring blue eyes, or fair hair—may sometimes be at an advantage only when it is comparatively rare. Then it can be shown that a polymorphism can be maintained in the absence of any heterozygote advantage, provided only that the rare genotypes are always at an advantage over the commoner ones.

Another possible balancing situation arises if viability and fertility effects act in opposite directions.

For example, a given genotype may increase the probability of survival but, at the same time, decrease the chances of having children. It can be shown that such a balance can, under appropriate conditions, give rise to a stable polymorphism. A similar situation can arise if selection acts in opposite directions in the two sexes—one genotype, for example, being favored in males while another is at an advantage in females.

Selection intensities may vary in different generations or in different ecological niches.

This situation also can give rise to balanced polymorphisms. Epidemic infections, droughts, and floods, for example, could give (and certainly must in the past have given) rise to substantial temporal and geographical variation in selection intensities. From some points of view this may perhaps be the most important mechanism for the maintenance of polymorphisms because on an evolutionary time scale of hundreds of thousands or millions (rather than just thousands) of years, it is hard to consider any polymorphism as absolutely stable.

Many other examples of complex selection models can be constructed that may lead to balanced polymorphism, and many of these will be combinations of the types of situations we discuss in this chapter. In the next chapter we consider in more detail some other polymorphisms connected with the blood groups and related systems. The explanation of the maintenance of balanced polymorphisms is a major aspect of the explanation for the overall maintenance of genetic variability in populations.

Summary

1. An advantage of heterozygotes over both homozygotes leads to a balanced polymorphism. The gene frequencies tend to be maintained around a value that depends on the selective disadvantages of the two homozygotes.
2. The approach to equilibrium is relatively fast.
3. The best example of a balanced polymorphism is that of sickle-cell anemia in malarial regions. Here one can compute fitnesses directly from comparisons of observed genotype frequencies with their expected values according to the Hardy–Weinberg law. These direct estimates of fitness lead to expected equilibrium gene frequencies that are very close to those observed.

4. Thalassemia and possibly G6PD deficiency are similar examples of balanced polymorphisms maintained by malarial selection.

5. Mechanisms other than heterozygote advantage can also lead to balanced polymorphism.

Exercises

1. Why is it very unlikely that the sickle gene *S* is maintained in West Africa by mutation-selection balance?

2. What are the main arguments for the relation between the sickle-cell trait and resistance to malaria?

3. What is a balanced lethal?

4. Why can the Hardy-Weinberg law be used for the analysis of balanced polymorphisms, even though it is derived on the assumption of no selection?

5. What is it that distinguishes a stable from an unstable equilibrium?

6. Does the existence of natural selection necessarily invalidate the application of the Hardy-Weinberg law to adult frequency data?

7. What types of selective mechanisms, other than simple heterozygote advantage, may give rise to balanced polymorphisms?

8. What is the incidence of thalassemia at birth in a population in which 20 percent of the adults are heterozygotes? Calculate the frequency (a) assuming that all thalassemia is of the β type, and (b) assuming half of the thalassemia genes are of type α, and half β, and that the α and β genes show perfect complementation.

9. If the incidence at birth of sickle-cell anemia is 0.25 percent and if people decide to avoid having children born with the disease, (a) which marriage partners should heterozygotes *AS* choose? (b) Which should anemics choose? (c) What limitation would this restriction generate on their chances of marriage, in terms of the probability of finding the right partner?

10. Predict the ultimate fate of alleles *A* and *a* given each of the selective situations (genotype fitnesses) indicated.

	AA	*Aa*	*aa*
a.	1	0.98	0.10
b.	0.98	1	0.10
c.	0.85	1	0
d.	0.8	1	1

11. In many Sardinian villages the frequency of heterozygotes for thalassemia was of the order of 20 percent and is believed to have been at equilibrium. Compute the fitness of the normal homozygote.

12. Malaria was eradicated in Sardinia shortly after World War II. What should have happened to the gene frequencies—given a frequency of heterozygotes of 20 percent in 1945—in one generation since that time (which is about now)? Compute the gene frequency assuming that the fitnesses of heterozygotes and normal homozygotes have, since eradication, been equal, but that the fitness of the homozygote for thalassemia has remained at zero. Use the same scheme as in Table 9.3 with the appropriate gene frequencies and fitnesses.

13. Repeat the computation of Exercise 12 assuming that the heterozygote has, after eradication of malaria, a fitness which is 90 percent of that of the normal homozygote.

References

Allison, A. C.
1964. "Polymorphism and natural selection in human populations," *Cold Spring Harbor Symposia on Quantitative Biology,* vol. 29, pp. 137–149.
Cavalli-Sforza, L. L., and W. F. Bodmer
1971. *The Genetics of Human Populations.* San Francisco: W. H. Freeman and Company. [Chapter 4 contains further discussion of this topic.]
Livingstone, F. B.
1967. *Abnormal Hemoglobins in Human Populations.* Chicago: Aldine.
White, J. M.
1972. "Hemoglobin variation," in *The Biochemical Genetics of Man,* ed. D. J. H. Brock and O. Mayo, New York: Academic Press.

10

Blood Groups and Other Immunologically Detected Polymorphisms

Immunological techniques have played a major role in the detection of human polymorphisms, including especially the various blood-group systems. Several of these polymorphisms—in particular the ABO and Rh blood-group systems and the HLA histocompatibility system—are of clinical significance. Important questions, both from the medical and the biological points of view, have been raised by the discovery of the HLA, Gm, and other similar systems detected by immunological techniques.

10.1 Blood Groups as Polymorphic Systems

The ABO blood-group system, discovered by Karl Landsteiner in 1900 (the year Mendel's work was rediscovered) must rank as the first genetic polymorphism defined in man, or for that matter in any organism. Since that time blood groups (and other similar differences detected by the use of immunological techniques discussed in Section 5.4) have played a prominent role in the study of human polymorphisms.

In addition to their use as genetic markers, however, blood groups are of great importance medically for blood transfusion, and they have been shown to be involved in hemolytic disease of the newborn, a disease of the blood caused by an immunological reaction of the mother against her fetus. This is the disease associated with the well-known rhesus (Rh) blood factor. More recently, a blood-group-like system called HLA, which is detected on white rather than red blood cells, has proved of value in matching donors and recipients for kidney and other organ transplants. Some of the factors of this HLA system have also recently been shown to be associated with a number of important genetic diseases including, for example, the skin disease psoriasis and the severe degenerative nervous disease, multiple sclerosis.

Blood is the most readily available of the human tissues, and this no doubt explains why it has been studied so intensively from a genetic point of view.

Many of the best known polymorphisms pertain to components of the blood, though there is no reason to expect that these should show more genetic polymorphism than components of other tissue. In this chapter we first discuss the three blood-cell polymorphisms that seem to be of most importance medically (the ABO, Rhesus, and HLA systems) and then review briefly the wide range of other systems known so far. We choose these systems as important examples of polymorphisms that illustrate many principles of human and population genetics. Their discussion is meant to illustrate the types of data that can be collected and the way they are collected, the use that can be made of the data and the problems that they pose.

10.2 Rhesus Blood Groups and Hemolytic Disease of the Newborn

In 1939 P. Levine and R. E. Stetson made a simple clinical case report of an unusual complication during childbirth. Their report paved the way toward an understanding of the basis for hemolytic disease of the newborn (in which the infant's red blood cells are destroyed).

This discovery of the Rh system still remains one of the most important contributions of genetics to medical practice.

The case involved a woman whose second pregnancy terminated with a stillbirth, and who reacted severely to transfusion with her husband's blood. Experiments showed the presence in her serum of an antibody that agglutinated her husband's blood cells—in fact, this antibody agglutinated red blood cells from about 85 percent of a random sample of American Caucasians. The reactions of the antibody showed no correlation with the then-known blood groups ABO, MN, and P (a blood group defined at about the same time as the MN system). Levine and Stetson correctly hypothesized that the antibody had been formed by the mother during pregnancy in reaction to the paternal antigen carried by the fetus (an antigen, of course, absent from the mother). The antibody from the mother's blood then caused destruction of fetal tissue, leading to stillbirth.

In the following year (1940), Karl Landsteiner and A. S. Wiener reported the discovery of an antibody (produced by immunizing rabbits with blood from the rhesus monkey) that also reacts with the blood of about 85 percent of the human population and is not related to the ABO, MN, or P systems. They called this antibody anti-Rh (Rh standing for rhesus). It was soon discovered that the antibody earlier described by Levine and Stetson has essentially the same specificity as anti-Rh, and that many cases of hemolytic disease of the newborn can be attributed to its activity (Fig. 10.1). People who react to this antibody are called Rh-positive (Rh^+) and those who do not are Rh-negative (Rh^-). The Rh^+/Rh^- difference is inherited as a simple dominant determined by alleles *D* and *d;* genotypes *DD* and *Dd* are Rh^+, and *dd* is Rh^-. Many other specificities that belong to the rhesus system have been found subsequently, but the difference first described remains the one of greatest significance for hemolytic disease of the newborn—accounting for about 90 percent of the observed cases of this disease.

Only about 5 percent of offspring with hemolytic disease are firstborns, and 40 to 50 percent of first appearances of hemolytic disease in a family occur at the third or later births.

These facts suggest that, on the average, more than one stimulus from an Rh^+ fetus is needed to produce enough antibody to give rise to clinically observable

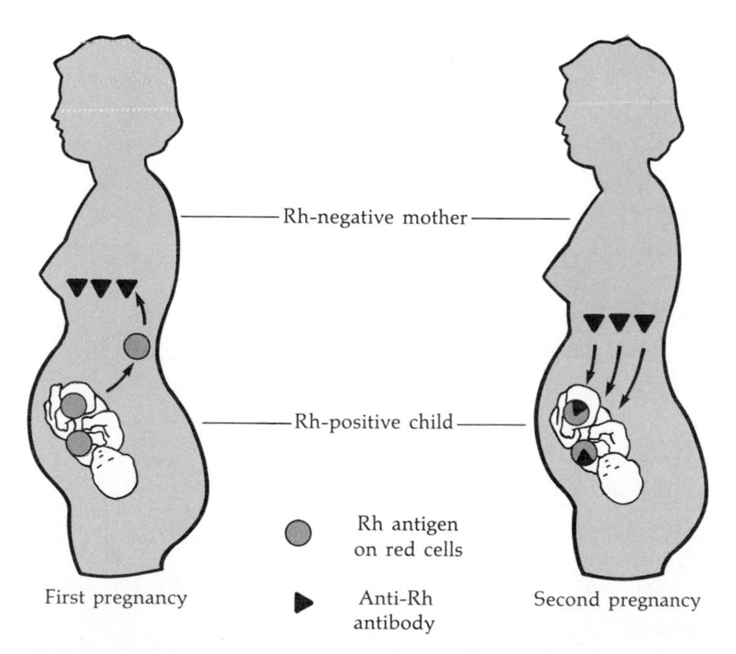

Figure 10.1
Hemolytic disease of the newborn and the rhesus group. An Rh^+ child born to an Rh^- mother immunizes her against the Rh antigen. If the mother later carries another Rh^+ child, the fetus will receive anti-Rh antibodies, causing hemolytic anemia of the newborn.

disease. Once an Rh$^-$ mother has had an affected offspring, the chances of subsequent Rh$^+$ offspring being affected are very high. Moreover, these later offspring are usually affected more severely. This is expected from the fact that the secondary response to an antigenic stimulus is usually much more rapid and much stronger than the initial response.

Only matings in which the mother is Rh$^-$ and the father Rh$^+$ are at risk with respect to hemolytic disease of the newborn. In terms of genotypes, there are two such matings: ♀ *dd* × ♂ *Dd*, and ♀ *dd* × ♂ *DD*. The first produces $\frac{1}{2}$ *Dd* and $\frac{1}{2}$ *dd* offspring, of which the *dd* (being Rh$^-$) are not affected nor do they immunize the mother. The second mating produces only *Dd* offspring, which are all Rh$^+$ and so are all potentially at risk (Table 10.1). Of the 84 percent of the population that are Rh$^+$, the Hardy–Weinberg computations indicate that 48 percent are heterozygous *Dd* and 36 percent are homozygous *DD*.

Selection due to incompatibility tends to eliminate the rarer allele.

The hemolytic disease of the newborn that occurs in a proportion of the Rh$^+$ offspring of Rh$^-$ mothers clearly represents a form of natural selection acting on the rhesus system. This selection is, however, quite unusual for at least two reasons. First of all, it is a form of selection *against* heterozygotes, rather than for heterozygotes as in the case of the sickle-cell trait discussed in the previous chapter. The selection is against heterozygotes because all individuals who are affected have Rh$^-$ *dd* mothers, but the affected individuals must be Rh$^+$ and so must have the genotype *Dd*. Secondly, natural selection acts on individuals not only as a function of their own genotype (*Dd*), but also as a function of their mother's genotype, which must be *dd*.

As we have seen in the last chapter, constant selection against heterozygotes can never give rise to a stable, balanced polymorphic situation. It can be shown that the same is true for the Rh$^+$/Rh$^-$ difference, even though the situation is complicated by the mother–child interaction. Thus, in the absence of any other factors influencing the situation, it would be expected that either the *D* or the *d*

Table 10.1
Matings at risk for rhesus hemolytic disease

Mother Rh$^-$	Father Rh$^+$	Offspring
Genotype *dd*	Genotype *DD*	All Rh$^+$ *Dd* and at risk
Genotype *dd*	Genotype *Dd*	{1/2 Rh$^-$ *dd*, not at risk {1/2 Rh$^+$ *Dd*, at risk

NOTE: None of the other possible matings (Rh$^-$ mother and Rh$^-$ father, or Rh$^+$ mother with either Rh$^+$ or Rh$^-$ father) give rise to offspring at risk with respect to hemolytic disease.

gene (whichever gene is less common) would disappear from the population. A simple explanation of this is the following. Suppose in a population there are, say, 60 Rh^+ genes and 40 Rh^- genes. Their proportions remain constant except for elimination of heterozygotes due to incompatibility. The elimination of one heterozygote brings the Rh^+ and Rh^- genes down to 59 and 39, respectively, shifting the ratio slightly in favor of Rh positives. The elimination of another heterozygote would bring the ratio to 58:38. Continuing the process after elimination of the 40th heterozygote, there would remain only Rh^+ genes. If instead we began with 40 Rh^+ genes and 60 Rh^- genes, the elimination of the heterozygotes would change the ratios to 39:58, 38:58, . . . , 0:20, thus eventually leaving only Rh^- genes. Of course, this analysis oversimplifies the computations, but a more complete analysis using the Hardy–Weinberg law and fitness coefficients leads to the same conclusion: eventually the allele that is initially less common is removed from the population.

By far the highest frequency of Rh^-, about 16 percent, occurs in populations of European type.

Africans have Rh^- frequencies of about 1 percent, and this genotype is virtually unknown in most Oriental populations. Rhesus hemolytic disease of the newborn thus is essentially a Caucasoid disease. An Rh^- frequency of 16 percent, or 0.16, corresponds (following the Hardy–Weinberg law) to a *d* gene frequency of $\sqrt{0.16} = 0.4$, or 40 percent, so that even in European populations *d* genes are less common than *D* genes. Therefore, in the absence of other selective factors, the *d* gene would be expected to disappear eventually. Is there some other factor that maintains the high level of Rh^- genes in European populations?

One suggestion that has been made is the possibility that Rh^- mothers who have lost babies because of hemolytic disease may tend to overcompensate for their losses by having even more surviving children than do other mothers. This explanation does not, however, seem to be the whole story, and there may well be other factors (as yet unknown) that favor those *Dd* heterozygotes who do not get hemolytic disease. Such selection in favor of the heterozygotes would at least partially counteract the selection against heterozygotes caused by hemolytic disease. In any case, the frequency of *serious* hemolytic disease in the absence of treatment—even for Rh^+ offspring of Rh^- mothers—is only about 6 percent, which does not represent very strong selection against the heterozygote.

Several factors are now leading to a marked reduction in the incidence of the disease, and so presumably to a reduction in the magnitude of selection against the *Dd* heterozygote.

The first of these factors is the reduction in the average family size that results from present reduced fertility rates, especially in the United States and in Europe.

This reduction in family size reduces the chances that an Rh$^-$ mother will be immunized against the Rh$^+$ antigen before carrying an Rh$^+$ child, because she is likely to have fewer offspring. The second factor reducing the incidence of the disease is the existence of medical treatment (mainly exchange blood transfusion) that can cure all but the most severe forms of the disease. The final factor is the existence of a novel form of prophylactic treatment that can prevent immunization of the Rh$^-$ mothers who are at risk. In this treatment, anti-Rh$^+$ antibodies are injected into the Rh$^-$ mothers within 72 hours after the delivery of an Rh$^+$ baby (usually their first). Apparently, the fetal Rh$^+$ red blood cells that have passed into the mother's blood circulation are coated by the injected Rh$^+$ antibodies, so the mother does not develop many of her own anti-Rh$^+$ antibodies and does not become immunized. Incidentally, these studies also verify that the leakage of fetal blood into maternal circulation *at the time of delivery* is the major source of immunization of Rh$^-$ mothers.

The Rh system as a whole involves a complex pattern of specificities. The interpretation of this pattern is based on a hypothesis suggested by R. A. Fisher.

Soon after the original description of the rhesus antigen, new antigen specificities were discovered, either from other cases of hemolytic disease of the newborn or from cases of reactions to blood transfusion. These specificities were identified with the rhesus system because they proved to be closely associated with the original Rh$^+$ type, both in family and in population studies. The main specificities are usually labeled by the symbols D (for Rh$^+$), C, c, E, and e. The specifications C/c and E/e are paired because they are inherited as "alleles," in the sense that it is very unusual to inherit both C and c, or both E and e, from the same parent. This pairing of specificities was first pointed out by the famous English population geneticist, R. A. Fisher, who did a great deal to promote early work on the rhesus and other blood groups in Britain. Fisher pointed out that the inheritance of the various Rh specificities could be simply described in terms of eight complex allelic combinations (*CDE, CDe, CdE, Cde, cDE, cDe, cdE,* and *cde*), formed by taking one each of the letters *C* or *c, D* or *d,* and *E* or *e.*

Such allelic combinations are usually called *haplotypes,* a shorthand word for "haploid genotypes." The presence of a letter in a haplotype indicates the genetic determination of the corresponding antigenic specificity—except that no antiserum is yet known to detect *d.* Genotypes are formed, of course, by taking pairs of haplotypes. For example, the genotype *CDe/cde* has the phenotype C, c, D, and e; the genotype *cde/cde* (the main Rh$^-$ European type) has just the antigens c and e. Figure 10.2 shows the typical frequencies of these various haplotypes in European populations.

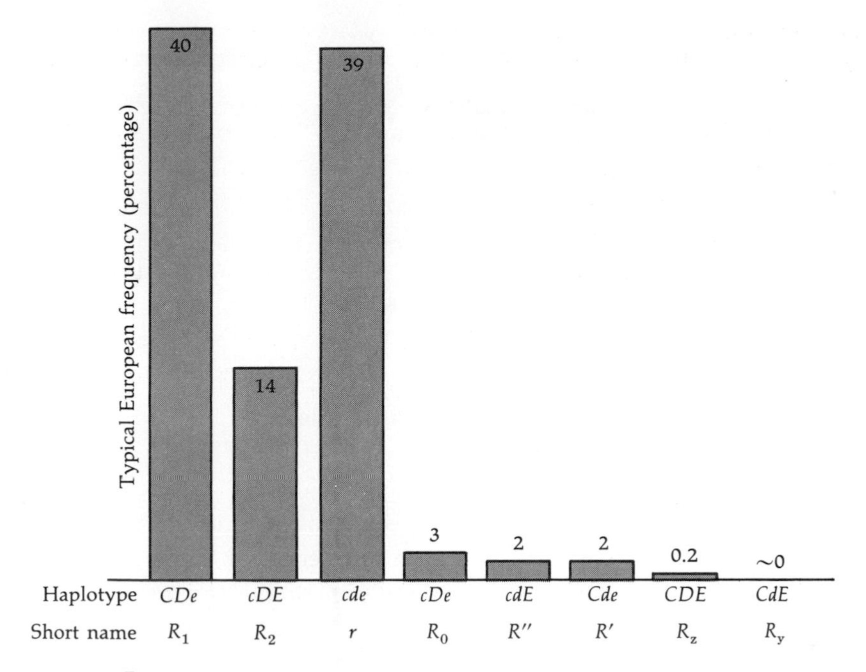

Figure 10.2
The complex Rh haplotypes and their proportions in European populations.

Note that three haplotypes (*CDe, cDE,* and *cde*) predominate, so that, for example, most Rh^- individuals are homozygous *cde/cde.* Following the notation used by Alex Wiener (codiscoverer of the Rh system), the three common haplotypes are often denoted as R_1 (for *CDe*), R_2 (for *cDE*), and *r* (for *cde*). Using this notation, we can say that most Rh^- individuals are homozygous *rr.* Of the 36 possible genotypes that can be formed with the eight possible haplotypes, only five genotypes (*rr*, R_1R_1, R_2r, R_1r, and R_1R_2) occur with frequencies greater than 3 percent, and these five between them account for nearly 90 percent of the European population.

Fisher suggested that the pairwise combinations *D/d, C/c,* and *E/e* might actually correspond to alleles at three closely linked loci, and that the sequence of loci is *D/d–C/c–E/e.* In this case, rare haplotypes might be derived from more common ones by recombination. For example, the relatively common heterozygote R_1R_2 (or *DCe/DcE*) could produce the rare haplotype *DCE* by recombination between the *C/c* and *E/e* loci (Fig. 10.3). The global distribution of the various haplotypes, together with other considerations, suggests that the combination R_0 (or *Dce*) may have been the original haplotype. Mutation, followed by recombination, could have given rise to all the other haplotypes roughly in proportion to their frequencies in Caucasians (Fig. 10.4).

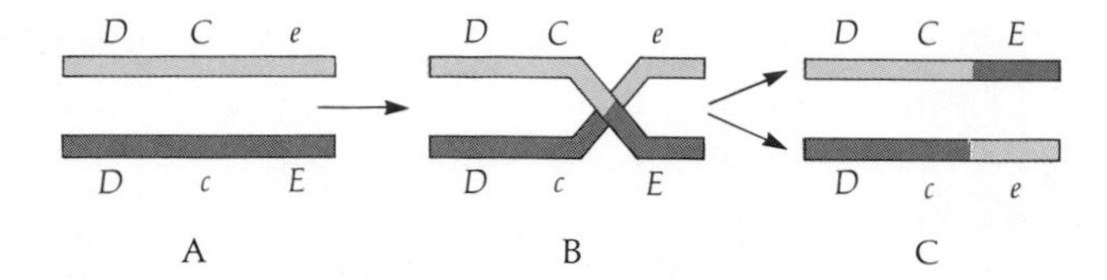

Figure 10.3
Formation of rare haplotypes by recombination. (A) The relatively common *DCe*/*DcE* (R_1R_2) heterozygote. (B) Occurrence of crossing-over. (C) After division, the new haplotypes *DCE* (R_z) and *Dce* (R_0) have been formed.

10.3 ABO Blood Groups, Selection, and Disease

The early discovery of the ABO blood groups and their significance for blood transfusion have placed this system at the center of attention for many decades. It was soon recognized that there are ethnic differences in the frequencies of the *A*, *B*, and *O* alleles. (We now know that there are also allele frequency differences for many other genes that have been investigated subsequently.) For example, the *B* gene is twice as frequent among Africans and three times as frequent among Orientals as it is among Caucasians, where it has its lowest frequency. But in

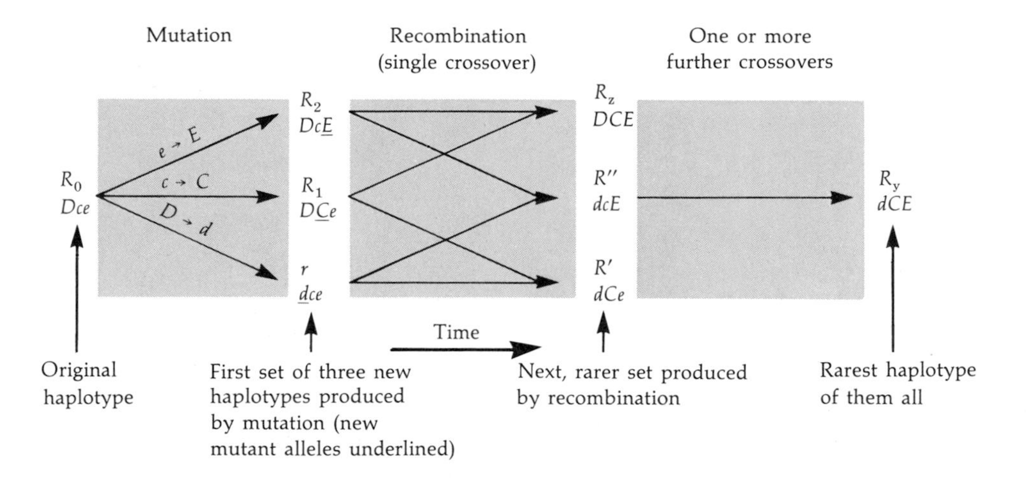

Figure 10.4
Suggested evolutionary origin of the major rhesus haplotypes. If the original haplotype is *Dce* (the most common haplotype in Africa), mutation at each of the three sites may have produced the three haplotypes *DcE, DCe,* and *dce.* The first two of these are common in most of the world, and the third is very common among Caucasians. The other haplotypes (relatively rare in the world population) could then originate by recombination. (After M. W. Feldman et al., *American Journal of Human Genetics,* vol. 21, pp. 171–193, 1969.)

all American Indians (despite their recognized Oriental origin), the *B* allele is almost entirely absent; in the majority of tribes, the *A* allele is also absent, frequently leaving the *O* allele as the only one represented (unless there has been admixture with Caucasians).

The A type can actually be subdivided into subtypes A_1 and A_2, controlled by corresponding alleles A_1 and A_2. When these subtypes are considered, there are six A_1A_2BO types: O, A_1, A_2, B, A_1B, and A_2B. There is no type A_1A_2 because the A_1A_2 genotype has the A_1 phenotype, allele A_1 being dominant over A_2. Allele A_2 is present in approximately equal frequencies in Caucasoids and Africans, but is absent from Oriental populations (Fig. 10.5).

These ethnic variations in the ABO gene frequencies lead naturally to the following question: what selective forces, if any, are acting on this genetic system? The possible association between ABO blood groups and disease has been a subject for much discussion and investigation ever since their discovery. A number of significant associations have been found (as we shall now discuss), but they do not really explain the maintenance of the ABO polymorphism.

ABO differences can sometimes be a cause of hemolytic disease of the newborn.

The ABO hemolytic disease is rare (occurring at most in 0.1 percent of the newborn) and seems to be limited mainly to the combination of an O mother and

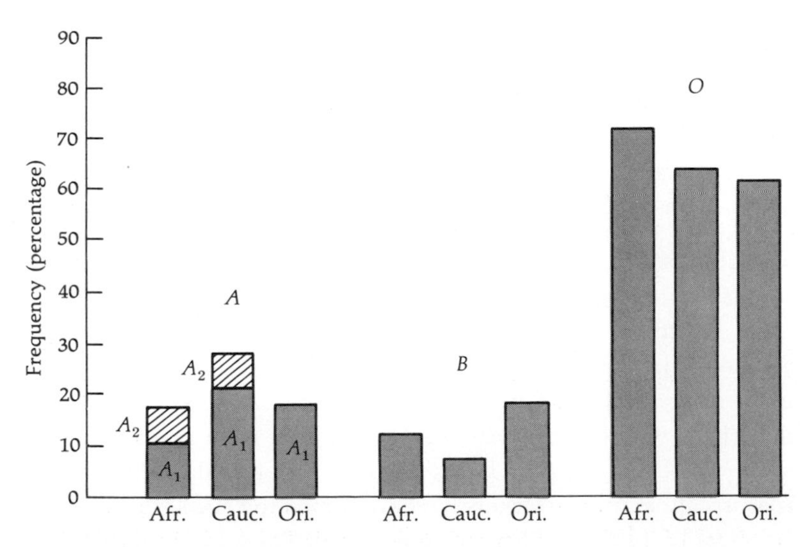

Figure 10.5
Frequencies of ABO alleles. The chart shows frequencies of the A_1, A_2, *B*, and *O* alleles of the ABO blood-group system among Africans, Caucasians, and Orientals.

offspring who are of type B or of the stronger A subtype, A_1. In marked contrast to the rhesus disease, the ABO hemolytic disease occurs among firstborn offspring. This seems to be because the A or B stimulus that is responsible for the abnormally high level of maternal antibodies needed to produce the disease in the fetus can come from A or B antigens present in bacteria or in normal food components. The stimulus need not come from previous offspring, as it does in the case of the Rh system.

The ABO hemolytic disease is much too rare to be significant as an overall selective effect on the ABO system. From time to time, however, there have been suggestions that ABO differences between mother and offspring could be responsible for a significant proportion of early fetal deaths that do not show up as hemolytic disease of the newborn. Quite early in the history of ABO studies, it was claimed that fewer A offspring are produced by ♀ O × ♂ A matings than by ♀ A × ♂ O matings. The mating ♀ O × ♂ A is said to be "incompatible," in the sense that the father carries the A-antigen gene, which the mother lacks; therefore the mother can respond immunologically to the presence of A antigen in the fetal blood. The suggestion was made that this response could cause early deaths of A-type fetuses—rather than only hemolytic disease, which is detected much later on in pregnancy. The compatible mating ♀ A × ♂ O does not provide this opportunity for an immune attack by the mother on her offspring.

Some subsequent studies have reported significant incompatibility effects, but the overall picture is not clearcut. The present information suggests either heterogeneity in the effects or biases in the sources and methods of analysis of the data, which in any case must be collected on a large scale if they are to show significant effects for relatively small differences.

The one really significant effect of ABO incompatibility is its relation to Rh hemolytic disease of the newborn.

Following his discovery of hemolytic disease, Levine showed that parents of diseased offspring are more often ABO compatible than incompatible. A mating is generally said to be compatible when the father has no antigens (in this case A or B) that the mother lacks, and to be incompatible if the father does have antigens not present in the mother. Thus, in the case of the ABO system, the following matings (written with the mother first) are the only ones that are incompatible: O × A, O × B, O × AB, A × B, B × A, A × AB, and B × AB. Since Levine's original observation, many workers have confirmed the strong protective effect of ABO incompatibility against Rh incompatibility. Thus, the reported frequencies of ABO-incompatible combinations among Rh^- mothers of infants with hemolytic disease are between two- and four-fold less than the frequencies expected in the population as a whole. The same sort of interaction has been observed when Rh^+

blood is injected into volunteer Rh^- recipients. If the blood is also ABO incompatible, anti-Rh^+ antibodies are much less likely to be formed than if it is ABO compatible. Presumably (in both the pregnancies and the transfusions), the naturally occurring anti-A or anti-B antibodies in the recipient coat the incoming A, B, or AB cells and in some way prevent them from providing an immune stimulus with respect to the Rh^- difference. Figure 10.6 summarizes the factors controlling the incidence of Rh hemolytic disease, taking into account the effects of ABO incompatibility.

Associations do exist between ABO and diseases.

Whatever their magnitudes, the incompatibility effects always involve selection against heterozygotes (in one form or another), and so these effects cannot account for the maintenance of ABO polymorphism in the population. Prompted by the

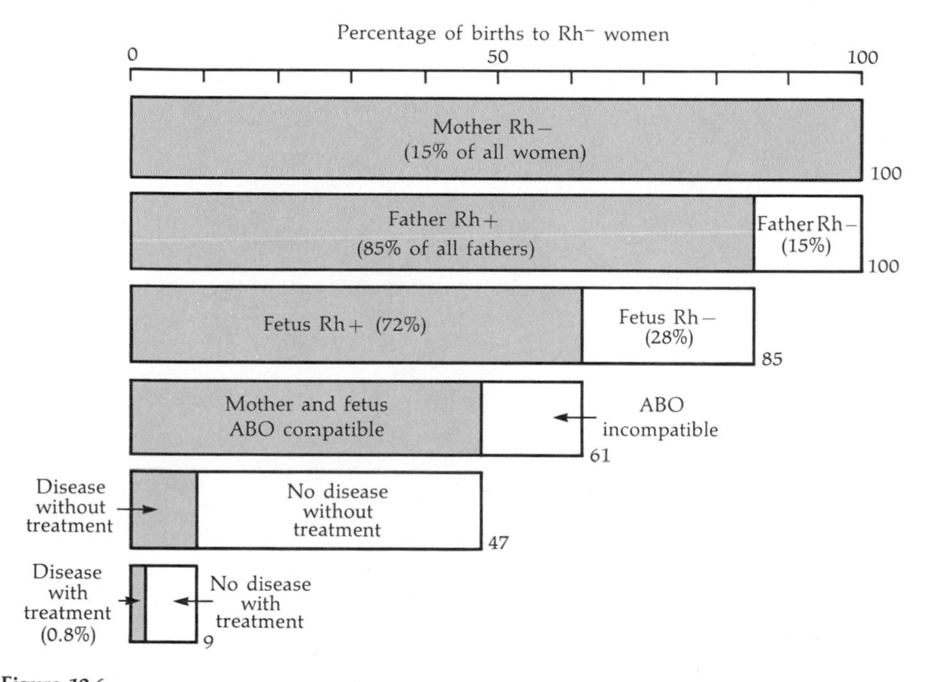

Figure 10.6
Some of the factors explaining the lower-than-expected incidence of Rh hemolytic disease. Of all Rh^- mothers (making up some 15 percent of all mothers), only about half produce an Rh^+ fetus with which they are ABO compatible and therefore are at risk of becoming isoimmunized. Of these mothers, only one-sixth will actually show detectable antibodies by the end of a second Rh^+ pregnancy, even without treatment. With treatment, the incidence of immunization drops by about 90 percent. (Adapted from C. A. Clarke in *Medical Genetics,* ed. V. A. McKusick and R. Claiborne, H.P. Publishing Co.)

apparent existence of some significant selection associated with the ABO system, the English population geneticist E. B. Ford in 1945 urged a search for associations between blood groups and specific diseases. Since then, a number of chronic diseases have been shown to have significant associations with one or another of the ABO blood types. These associations, however, are also unlikely to be of much selective importance because they involve diseases either that are very rare or that occur fairly late in life, after the end of the reproductive period.

One of the most significant associations (now confirmed by at least 44 different studies on a total of over 26,000 patients) is that between type O and duodenal ulcer. Type O occurs about 1.3 to 1.4 times more frequently among duodenal ulcer patients than among the normal population. However, as discussed in Chapter 7, the selective effect of this association is essentially negligible. Table 10.2 gives data on some of the more significant associations between ABO blood groups and chronic diseases.

A wide variety of cancers and ulcers, and even such diseases as diabetes, rheumatism, and heart disease have been implicated. In all cases, however, the effects (though possibly significant) are very slight, and so they do not offer much help in explaining the maintenance of the ABO polymorphism. Eventually, however, they may be of help in understanding some of the factors influencing the diseases involved.

Some associations have been claimed between ABO blood types and infectious diseases—notably leprosy, tuberculosis, syphilis, and smallpox. Because of the severity and worldwide prevalence of these infectious diseases (at least until quite recently), these associations could have much more selective significance than do those with chronic diseases. Unfortunately, however, these claims have on the whole not yet been substantiated.

Table 10.2
The association between ABO blood groups and some diseases.

Disease	Mean relative incidence
Duodenal ulcer	1.35 O:A
Pernicious anemia	1.25 A:O
Cancer of the stomach	1.22 A:O
Stomach ulcer	1.17 O:A

SOURCE: F. Vogel, *American Journal of Human Genetics*, vol. 22, pp. 464–475, 1970.

NOTE: The mean relative incidence is a measure of the relative increase in frequency of a blood type in diseased as compared to control (normal) individuals. It is computed by taking the ratio of blood types (for example, O to A) in diseased patients and dividing that value by the ratio obtained in a control series.

The ABO blood groups and two related systems called secretor and Lewis are the only blood-group systems whose chemistry is understood in some detail.

The secretor polymorphism separates individuals into two categories: (1) those who secrete in their saliva and other body fluids A and B substances that are similar to those on the red blood cells, and (2) those who do not secrete such substances. The substances are detected by their ability to inhibit agglutination either of type-A cells with anti-A, or of type-B cells with anti-B antibodies. The difference between secretors and nonsecretors is controlled by a dominant gene *Se*, so that only the recessive homozygotes *se se* are nonsecretors.

The second polymorphism associated with ABO is the Lewis blood-group system, which classifies individuals into three categories—Le(a^+b^-), Le(a^-b^+), and Le(a^-b^-)—according to the presence or absence of the antigens Lewis a and Lewis b. Note that there are no Le(a^+b^+) individuals. It turns out that all Le(a^-b^+) individuals are also secretors, whereas all Le(a^+b^-) individuals are nonsecretors. Le(a^-b^-) individuals, who have neither of the Lewis antigens, are homozygotes *ll* for a recessive gene *l* that seems to prevent the formation of either Lewis antigen. Table 10.3 illustrates these various interrelations between the ABO, secretor, and Lewis systems.

Using data such as those given in Table 10.3, R. Ceppellini in 1959 realized that the patterns of inheritance could be explained by a series of interacting metabolic steps. At the same time, W. T. J. Morgan and W. M. Watkins showed how the various substances are related chemically to each other. The first step in the chemical analysis was to show that the specificities of the A and B antigens correspond to certain complex sugars. The antigens themselves, as detected in secretions (on which most of the chemistry has been done), are combinations of sugars attached to a protein backbone (Fig. 10.7). The sugars at the end of the

Table 10.3
Interrelationships among the ABO, secretor, and Lewis systems

Genotype at		Approximate frequency in caucasoids	Red cell antigens		Secretion of A and B antigens
Secretor locus	Lewis locus		Lewis a	Lewis b	
Se Se or *Se se*	*LL* or *Ll*	72%	−	+	+
se se	*LL* or *Ll*	24%	+	−	−
Se Se or *Se se*	*ll*	3%	−	−	+
se se	*ll*	1%	−	−	−

NOTE: The symbol + indicates presence of antigens in red cells or secretions, and − indicates absence of antigens. The overall frequency of secretors is about 75 to 80 percent in all populations. The frequency of Le(a^-b^-) (who must be *ll* homozygotes) ranges from about 4 percent in Caucasoids to more than 40 percent in Africans.

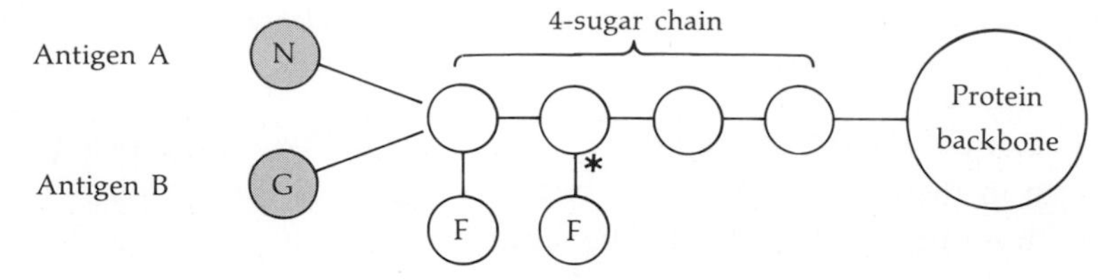

Figure 10.7
Chemical structure of ABO and Lewis antigens. When sugar N (N-acetyl galactosamine) is attached to the end of the four-sugar chain, the antigen has specificity A. If sugar G (galactose) is attached, the antigen has specificity B. Attachment of sugar F (fucose) to the second sugar from the end of the chain (marked *) creates the Lewis-b antigenic specificity.

chain determine which of the specificities (A or B) is expressed: either galactose for B, or a galactose derivative called N-acetyl galactosamine for A. The actual combined protein–sugar structure is quite complex, involving many such sugar chains attached to a given protein backbone. Thus, people of type AB may have mixtures of A- and B-determining chains attached to the same protein backbone. The *A* and *B* genes themselves determine enzymes, called transferases, that add the corresponding sugar to the end of the chain. The difference between the *A* and *B* alleles is presumed to reside in a difference in the specificity of the corresponding enzymes, directed in the case of *A* at N-acetyl galactosamine and in the case of *B* at galactose.

The Lewis-b specificity, as shown in Figure 10.7, depends on the addition of the sugar fucose to the penultimate position of the four-sugar side chain. The action of the dominant *Se* apparently enables the product of the dominant Lewis gene *L* (presumably also a transferase enzyme) to act by adding the fucose sugar. The complete system shows, in an interesting way, how a series of genes that are not linked to one another still work together in making a complex product. The true functional significance of all these activities, however, still remains to be discovered.

10.4 Histocompatibility and the HLA System

Records exist of tissue grafts carried out as long ago as 3500 B.C. by the Egyptians, but it was not until sixteenth century that the first proper description of some of the techniques of transplantation was made by Gasparo Tagliacozzi, a professor of anatomy at the University of Bologna. He was, perhaps not surprisingly, bitterly attacked by the theologians of the time for interfering with the handiwork of God, and some concern about grafting has persisted up to the present day.

Grafts are ordinarily rejected.

In the absence of appropriate medical treatment, any organ such as a piece of skin or a kidney grafted arbitrarily from one person to another will survive for only a short time. Skin grafts, for example, will last for only 10 to 15 days unless they are made between identical twins, when the graft will survive indefinitely. Genetic differences between people result in any transplanted tissue or organ being recognized as a foreign body by the recipient's immune system (see Section 5.4). It was a simple detailed case report by T. Gibson and P. B. Medawar in 1943, of skin grafting carried out on a patient suffering from severe burns, that first established clearly the immunological basis of normal tissue transplantation. Grafts of the patient's own skin, taken from another part of her body, survived indefinitely and became incorporated as a part of her own skin. The first graft taken from her brother, however, was rejected after about 20 days, while a second graft from the same individual never even took properly and was rejected in a few days. This rapid rejection of the second graft reflects the immune system's memory of the first graft and its recognition of the genetic differences between donor and recipient, which form the basis for rejection.

The nature of the genetic control of graft rejection has been elaborated after many years of study with experimental animals, mainly the mouse. The animals can be bred easily under controlled conditions and made more-or-less genetically homogeneous by extensive inbreeding, usually by many successive brother-sister matings. These studies have established that graft rejection is determined by genetically controlled antigenic differences, similar to the red-blood-cell antigens that are responsible for transfusion reactions and hemolytic disease of the newborn. The relevant antigens are called *histocompatibility antigens.* A histocompatibility antigen will elicit graft rejection if the donor has the antigen while the recipient does not, and such combinations are said to be *incompatible.* As for other antigens, such as the blood groups, histocompatibility antigens are determined by codominant alleles. Only those antigens controlled by loci that are *polymorphic* will be detectable histocompatibility antigens.

Intuitively we expect that the chance that two individuals are compatible will be higher for closely related individuals (Fig. 10.8) and will decrease as the number of polymorphic histocompatibility antigens increases. At least 30 specific histocompatibility loci have been identified in the mouse and more certainly still remain to be worked out. One system, however—the H-2 locus—seems to be much more important than all the others. In man there are, most probably, about the same number of histocompatibility loci, though only the ABO locus and a complex antigenic polymorphism on white blood cells called HLA, which is the human equivalent of H-2 in the mouse, have so far been unequivocally identified as histocompatibility systems.

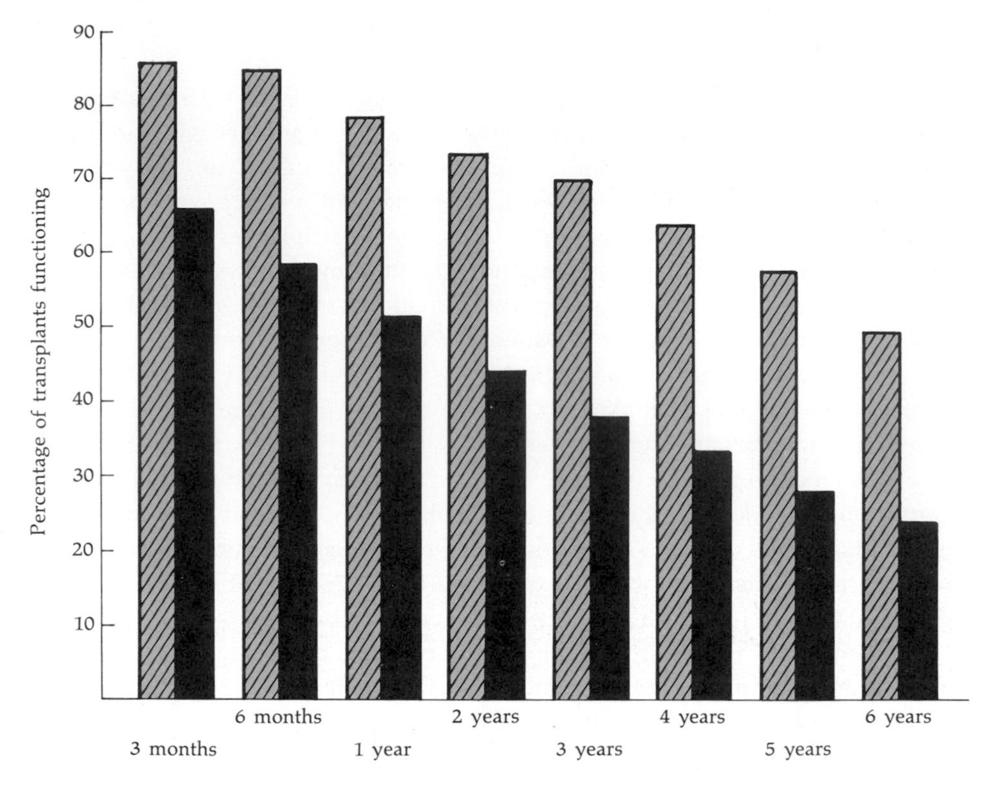

Figure 10.8
Survival time for kidney transplants carried out between 1967 and 1971. Striped columns represent transplants from sibling donors (1,170 cases); black columns represent transplants from cadavers (4,438 cases). (From A. Jones and W. F. Bodmer, *Our Future Inheritance: Choice or Chance,* Oxford Univ. Press, 1974.)

Two approaches have been followed in attempts to maximize the chances of a graft taking successfully.

The first approach is to try to inhibit the immune system by treating patients with drugs, appropriately called "immunosuppressors." The second is to try to match donors and recipients, as far as possible, for their histocompatibility antigens. Though neither of these techniques is yet completely successful, a combined approach has served to make kidney transplantation a reasonable and valuable therapy for most forms of severe kidney malfunction.

It turns out that matching for the ABO red-cell blood groups is more or less essential for successful kidney transplantation. A great deal of attention, however, has been focused on the major complex system, called HLA, which has now been shown also to be important in connection with a number of chronic diseases, as well as in tissue matching for transplantation. In the remainder of this section we describe the HLA system and its association with transplantation and disease.

Early work with transplantation in animals suggested that histocompatibility antigens might be more readily detected on the white cells of the blood rather than the red cells. The search for specific histocompatibility-antigen systems in man was thus concentrated on the white blood cells—more specifically, the lymphocytes. The problem of finding a suitable source of antibodies for studying the white-blood-cell groups was solved when it was found that at least 20 to 30 percent of women who have had two or more pregnancies have antibodies that react with their husband's (and children's) white cells and with those of a significant fraction of the population. These antibodies are formed by fetal-maternal stimulation, just as in the case of the rhesus system, but do not seem, in general, to have any detectable clinical effects on the fetus. Agglutination assays were at first used in studying the antigens of the HLA system but these have now, for the most part, been replaced by cytotoxicity assays on lymphocytes, routinely carried out on a micro scale (Fig. 10.9).

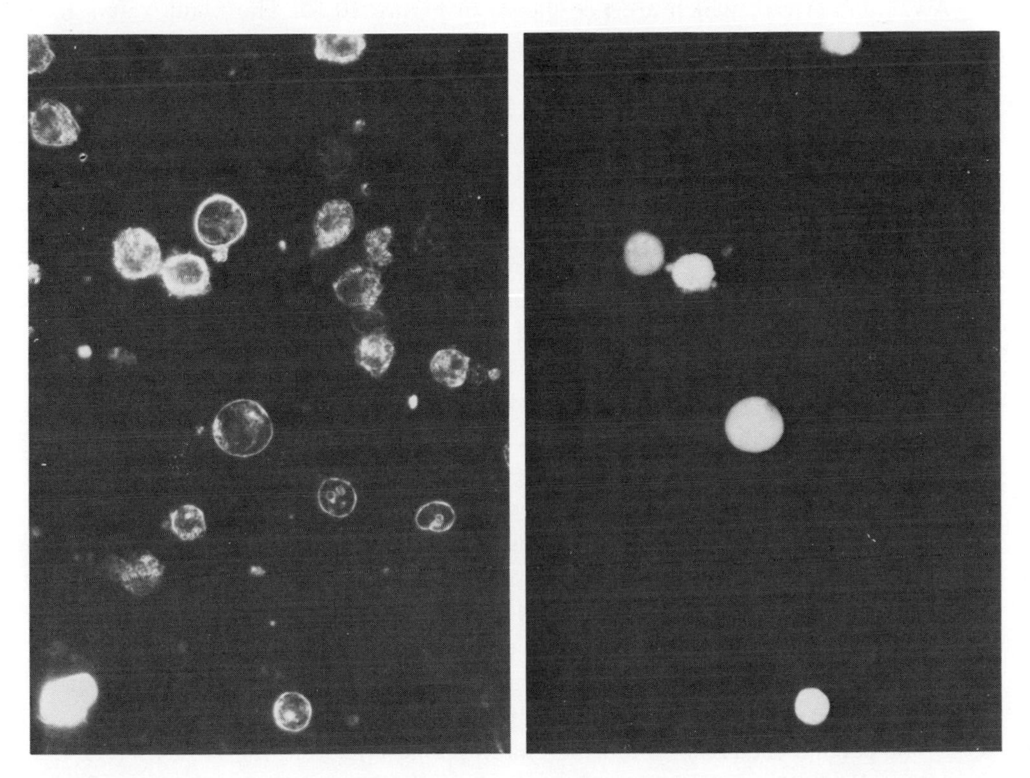

Figure 10.9
Living and dead lymphocytes: the reaction used for HLA typing. Living cells fluoresce under blue light after uptake of the dye FDA; dead cells do not. The photo on the left shows a group of cells under white (nonfluorescent) light. On the right are the same cells under fluorescent light; only the live cells are seen here. The proportion of cells killed is a measure of the reaction of lymphocytes with antibodies used as reagents for HLA typing. No kill = no reaction; 100 percent kill = complete reaction.

More than thirty different antigens which can be detected on lymphocytes using these techniques have now been defined.

These can be separated into two series, called A and B, which between them constitute the major antigens of the HLA system. The antigens of each series behave as if they are controlled by a set of multiple alleles at a single locus. This means that an individual can have at most two antigens from each locus (one inherited from his father and one from his mother), or four in all. Pedigree data show that the A and B loci are quite closely linked. Figure 10.10 shows typical frequencies for most of the various alleles in the three major world population groups. The numbers 1, 2, and so on are the names of the allele, those with w in front being provisional designations. The complexity of the system is such that alleles are sometimes split (rather like A_1 and A_2 of the ABO blood groups) to form two or more new alleles. For example, HLA-A10 has now been split into Aw25 and Aw26, which are not shown in Figure 10.10. The blank refers to a residual, as yet undefined, which is in some ways similar to the *O* allele of the ABO blood groups, though much less frequent.

The most striking feature of the HLA gene frequency distributions is the extent of polymorphic variation.

Very few alleles have frequencies much higher than average. Nearly 40,000 genotypes and more than 16,000 corresponding phenotypes can be formed with the alleles referred to in Figure 10.10. It can be shown that more than 75 percent of the individuals in a typical Caucasoid population have four different HLA antigens, being heterozygous at both the A and B loci, while less than 2 percent have only two antigens and are homozygotes at both loci. By these criteria the HLA system is undoubtedly by far the most polymorphic system known in man.

Because of the close linkage between the A and B loci, the parental antigen combinations are generally inherited as pairs kept together on the same chromosome. Such pairwise combinations, consisting of an antigen-determining allele from the A locus and one from the B locus, are haplotypes that are analogous to the rhesus combinations (such as *CDe*) discussed earlier in this chapter. Table 10.4 shows an example of the expected pattern of inheritance for a typical mating between two individuals who are double heterozygotes at the A and B loci.

The parents each have four different antigens—two from the A series (A1 and A3 for the first parent) and two from the B series (B7 and B8 for the first parent). In the absence of recombination, these antigens are passed on to the offspring in pairwise combinations that correspond to the parental haplotypes. Thus, the first parent passes on either A1 with B8 (corresponding to the haplotype *A1,B8*) or A3 with B7 (corresponding to the haplotype *A3,B7*). In the absence of recombination, therefore, only four types of offspring are possible—corresponding to the four possible haplotype combinations that can be formed by taking one of the two haplotypes from each parent (*A1,B8/A2,B12,* and so forth, as in Table 10.4). Each of these combinations is produced with an expected frequency of $\frac{1}{4}$, following a standard Mendelian pattern of inheritance.

Of course, the pattern of antigens inherited by the offspring can be used to determine the haplotype combinations inherited from the parents. Suppose that the following four phenotype combinations were observed in the progeny of matings similar to that illustrated in Table 10.4; (A1,A2;B7,B12), (A1,A9;B5,B7), (A2,A3;B8,B12), and (A3,A9;B5,B8). In this case, the genotype of the first parent must have been *A1,B7/A3,B8* rather than *A1,B8/A3,B7*.

However, suppose that the four offspring combinations shown in the table are observed, together with a single individual of type A1,A2;B7,B12. This observation could be explained by the occurrence of a crossover between the A and B loci during production of a gamete by the *A1,B8/B3,B7* parent, leading to formation of a *A1,B7* haplotype. From the observed incidence of such exceptions to the pattern of inheritance expected in the absence of recombination, it appears that the recombination fraction between the A and B loci is about 0.8 percent.

The HLA gene frequencies given in Figure 10.10 take no account of the linkage between the A and B loci. In order to correct for this linkage, we must determine haplotype frequencies rather than just the separate frequencies of the alleles at each of the loci. When this is done, we find that some pairs occur with a frequency substantially higher than that to be expected if the loci behave independently of one another.

As an example, consider the haplotype *A1,B8,* which has an average frequency of about 9 percent (or 0.09) in populations of Northern European origin. The separate frequencies of the alleles are 0.17 for the A-locus allele *A1* and 0.11 for the B-locus allele *B8*. If the two loci behaved independently, the expected frequency of *A1,B8* would be the product of the frequencies of the constituent alleles—namely, $0.17 \times 0.11 = 0.0187$, or a little under 2 percent. The observed

Table 10.4
Expected pattern of inheritance of HLA haplotypes

Parents	Phenotypes	A1,A3;B7,B8				A2,A9;B5,B12		
	Genotypes	*A1,B8/A3,B7*				*A2,B12/A9,B5*		
Expected	Phenotypes	A1,A2;B8,B12	:	A1,A9;B5,B8	:	A2,A3;B7,B12	:	A3,A9;B5,B7
Offspring*	Genotypes	*A1,B8/A2,B12*	:	*A1,B8/A9,B5*	:	*A3,B7/A2,B12*	:	*A3,B7/A9,B5*
	Frequencies	$\frac{1}{4}$	:	$\frac{1}{4}$	:	$\frac{1}{4}$	:	$\frac{1}{4}$

*The offspring shown are those expected in the absence of recombination. For example, recombination in the *A1,B8/A3,B7* parent could produce an *A1,B7* haplotype, giving rise to either of the phenotypes A1,A2;B7,B12 or A1,A9;B5,B7—neither of which are expected in the absence of recombination.

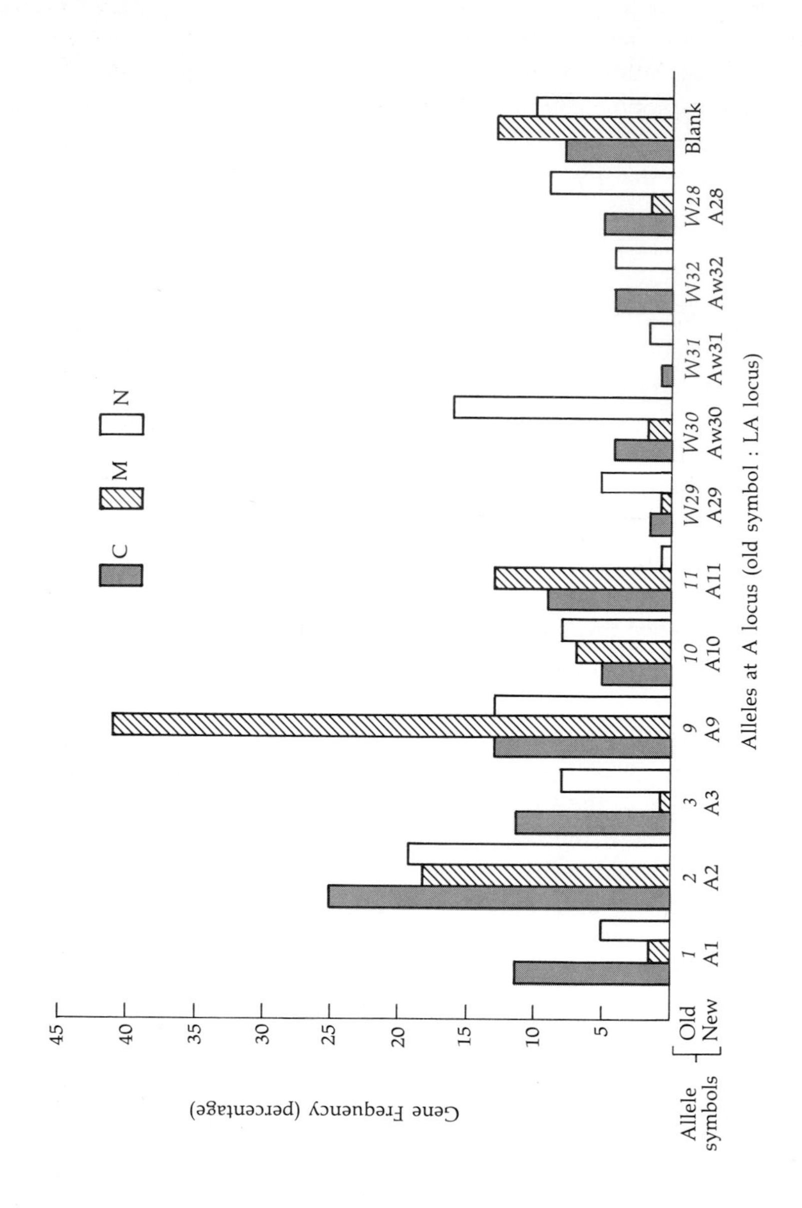

C
M
N
Gene Frequency (percentage)
45
40
35
30
25
20
15
10
5
Allele symbols
Old
New
1
A1
2
A2
3
A3
9
A9
10
A10
11
A11
W29
A29
W30
Aw30
W31
Aw31
W32
Aw32
W28
A28
Blank
Alleles at A locus (old symbol : LA locus)

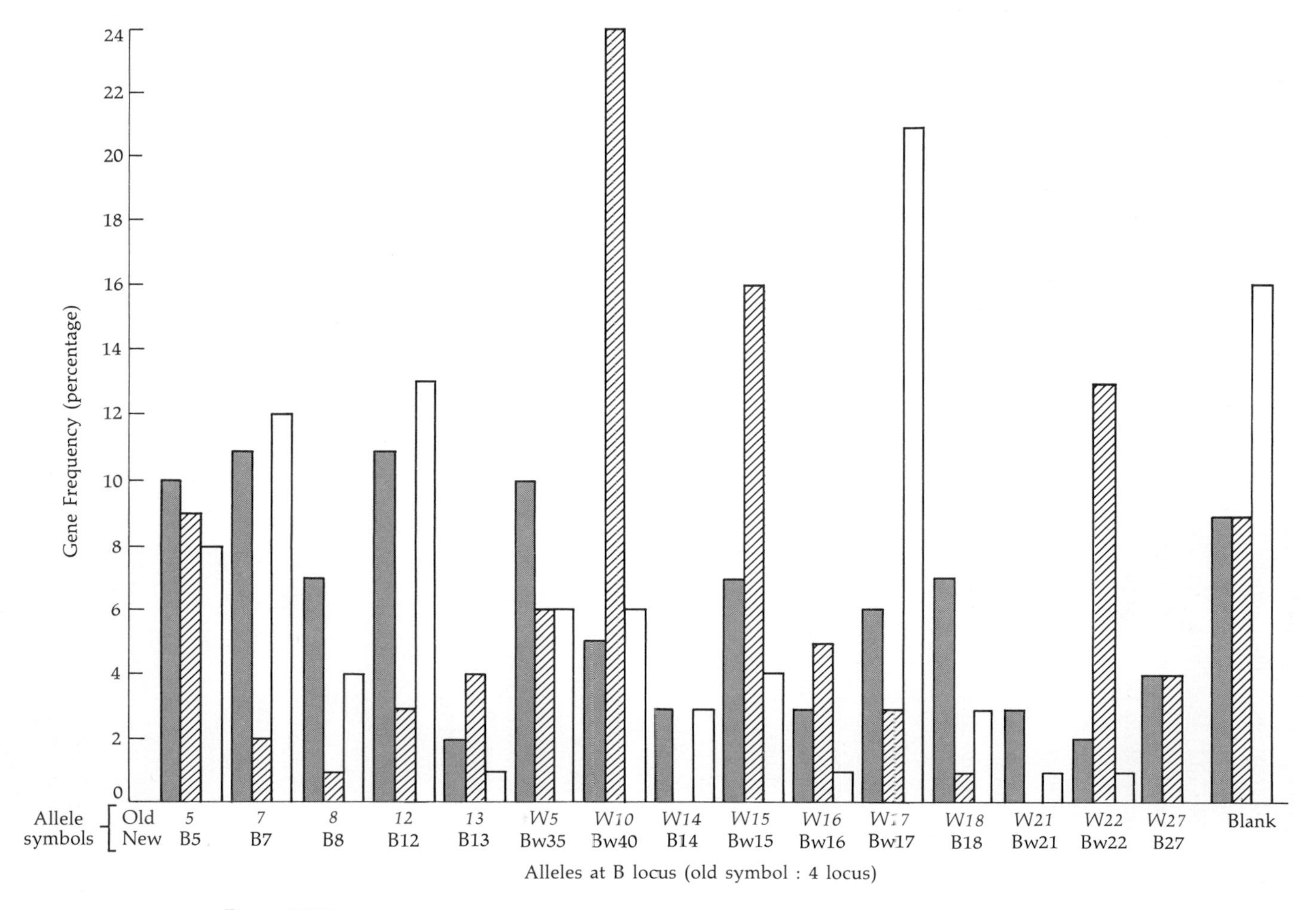

Figure 10.10
HLA gene frequencies in the three major human racial groups. C = Caucasian, M = Oriental (Mongoloid), N = African (Negroid). (Data from J. Bodmer and W. F. Bodmer, *Israel Journal of Medical Sciences,* vol. 9, pp. 1257–1268, 1973.)

frequency of the *A1,B8* haplotype thus is more than four times as high as the expected frequency. This genetic tendency for the *A1* and *B8* alleles to occur together on the same haplotype (a tendency that has been called *linkage disequilibrium;* see Table 10.5) leads to a phenotypic association in the population between the presence of the A1 and B8 antigens. These antigens occur together much more often than would be expected if they were associated at random.

Linkage disequilibrium may indicate selection.

Theoretical studies of the population genetics of two linked loci have shown that, in the absence of selection and given enough time for an equilibrium situation to have been reached, no associations between alleles due to linkage disequilibrium are expected. When, therefore, they do occur, as in the case of A1 and B8, they are an indication that natural selection may be favoring the combination of A1 with B8, and so in turn favoring the *A1,B8* haplotype, at least in Northern European populations. The majority of HLA haplotypes do not actually show evidence of linkage disequilibrium, though each population group tends to have a few associated pairs, like A1 and B8, and these pairs differ as one goes from one part of the world to another.

The best evidence for HLA as a histocompatibility system comes from a comparison of the survival of kidney grafts exchanged between HLA identical compared to nonidentical sibs. As shown in Figure 10.11, the HLA matched siblings do much better than their unmatched counterparts. In fact, their graft survival is almost as good as that for identical twins. Note that the expected proportions of sib pairs who are HLA identical is, from the pattern of inheritance illustrated in Table 10.4, just $\frac{1}{4}$. The results of HLA matching for unrelated donor–recipient pairs are, unfortunately, much less clearcut. The first problem is that, because of the extremely high degree of polymorphisms for the HLA system, it is very hard to find a well-matched cadaver donor for any given recipient. However, even

Table 10.5
Linkage disequilibrium between HL-A1 and HL-A8 in Northern European countries

	Frequency
A-locus allele *A1*	0.17
B-locus allele *B8*	0.11
Expected frequency *A1,B8* haplotype if no linkage (0.17 × 0.11)	0.0187
Observed frequency *A1,B8* haplotype	0.09
Expected frequency less observed frequency (0.09 − 0.0187)	0.0713

NOTE: The difference between expected and observed frequencies (0.0713) is a measure of the extent of linkage disequilibrium.

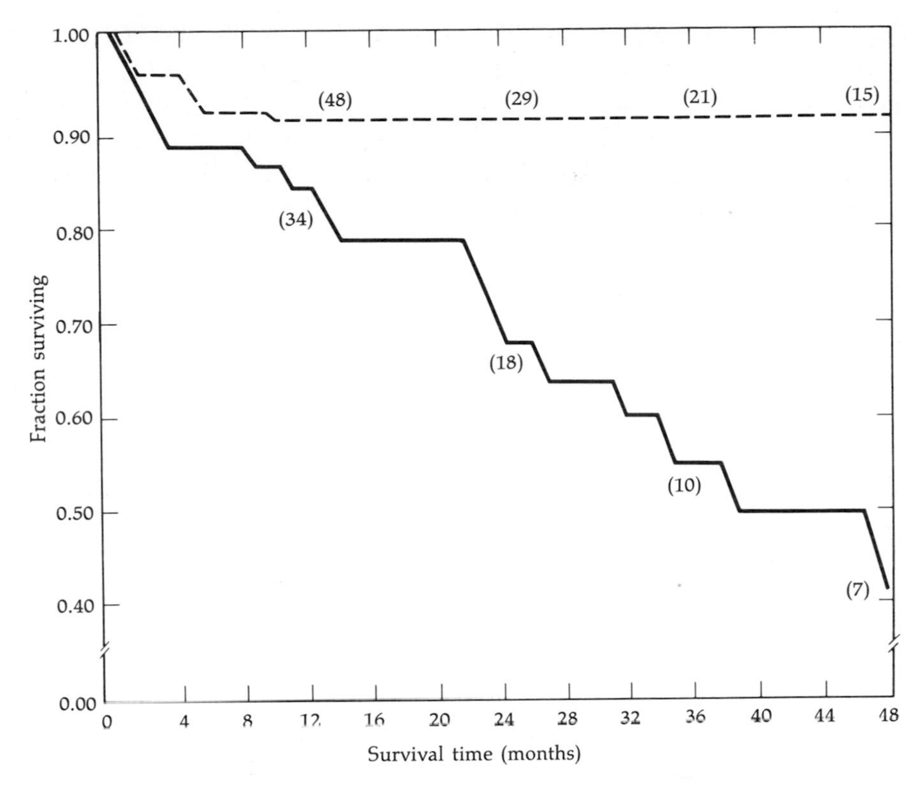

Figure 10.11
Survival curves for kidney transplants between HLA-matched and HLA-mismatched sibling pairs. The numbers along each curve indicate the number of patients at risk. The dashed line shows survival of matched transplants, the solid line that of mismatched transplants. (From D. P. Singal et al., *Transplantation,* vol. 7, pp. 246–258, 1969.)

when such a donor has the same HLA type as the recipient, though kidney survival is somewhat better, the results are not nearly as striking as those for HLA matched sibs as illustrated in Figure 10.11. It appears that there are other genetic factors, both within and outside the HLA system, that lead to this difference between matched sibs and matched unrelated donor–recipient pairs.

Although the separation of the alleles of the A and B loci occurs less than 1 percent of the time by genetic recombinations in families, it can be shown that this separation may allow room for hundreds, if not a few thousand, of other genes in the genetic region between the two loci. A number of the functions of these many genes are now being identified. So far they all seem to be connected with the cell surface and involved in the immune system including the mechanisms by which tissue grafts are rejected. Perhaps the most interesting of these functions are those discovered in studies with the mouse and other animals. These functions are connected with the ability to respond to a given immune stimulus and with resistance to viruses that cause leukemia. Thus, it has been shown in the mouse that there are single dominant genes, located in the H-2 region (which is the mouse equivalent of HLA), that control susceptibility to virus-induced leukemias. There are also other similar genes that control the ability to make antibodies to given chemically defined antigens. These findings in the mouse have stimulated the

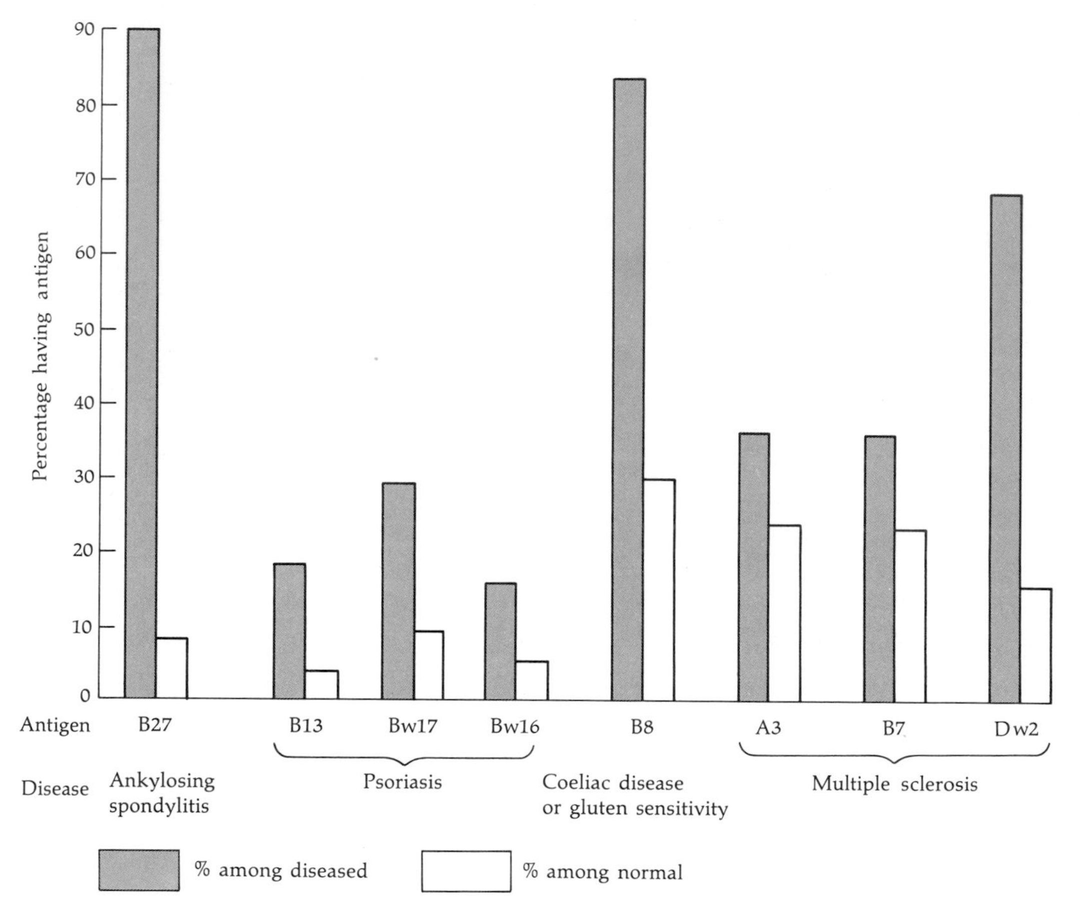

Figure 10.12

Some associations between HLA and disease. Ankylosing spondylitis is a form of rheumatism giving rise to lower back pains caused by fusion of the lower vertebrae of the spine; incidence is about one per thousand. Psoriasis is a chronic recurrent disorder of the skin associated with reddish patches of skin covered with silvery white scales; incidence is about 1 to 2 percent. Coeliac disease, or gluten sensitivity, is a form of sensitivity to gluten (a mixture of proteins occurring in the grain of wheat and other cereals); incidence is about one per 3,000. Multiple sclerosis is a progressive degenerative nervous disease caused by destruction of the "myelin sheath" that normally protects the nervous fibers. This disease is particularly distressing because it can, over a period of 25 years or even more, lead to almost complete physical incapacitation without severe impairment of mental functioning. The incidence of multiple sclerosis is about one per thousand. (Data based on H. O. McDevitt and W. F. Bodmer, *Lancet,* vol. 1, pp. 1267–1275, 1974.)

search in man for associations between the HLA system and a wide variety of diseases. Figure 10.12 summarizes data on some of the diseases associated with HLA antigens.

There are striking associations between HLA antigens and certain groups of diseases.

The most striking association is that between ankylosing spondylitis, a form of rheumatism causing lower back pains, and the antigen B27. This antigen occurs in 90 percent of the people with the disease, as compared to 7 percent in control (European) populations. Typing for B27 is actually now used as an aid to diagnosing the disease. The associations involving multiple sclerosis, which is a very distressing progressive degenerative nervous disease, are particularly interesting. There is a relatively slight, though undoubtedly significant increase in the antigens A3 and B7 in people with the disease. There is however a very marked increase in the frequency of Dw2, which is an allelic product of another gene in the HLA region. An interesting feature of most of the diseases associated with HLA, which include hayfever caused by allergy to ragweed pollen, is that they appear to be connected with abnormal immune responses to "self" antigens, and so come into the category of autoimmune-related disorders.

It thus seems likely that the associations with the HLA antigens are really due to associations with alleles of immune response genes in the HLA region that are controlling susceptibility to the diseases through differences in the ability to respond to various immune stimuli. These immune-response-gene alleles must be in linkage disequilibrium with the corresponding alleles of the B or A loci, for otherwise there would be no association between the antigens and the disease. The Dw2 and multiple sclerosis association is especially interesting because it shows how the products of other genes in the HLA region may show more significant associations with a disease than do the products of the B and A loci themselves. It seems probable that the HLA linked immune response genes, and genes with other related functions, will be shown to be important genetic factors predisposing to resistance or susceptibility to a variety of cancers, autoimmune disorders, and infectious diseases in man. These associations should eventually contribute to an understanding of the evolution of the HLA system and its extraordinarily high level of polymorphism.

10.5 Other Blood Groups and Immunologically Detected Systems

At least fifteen red cell blood-group systems have now been described, many of which (such as ABO, MN, Rh, secretor, Lewis, and the X-linked blood group Xg) we have already referred to. A list of these blood groups is given in Table 10.6,

Table 10.6
Time of recognition of human blood groups

Time of discovery	Major polymorphic systems
1900	ABO (51%)
1901-25	none
1926–30	MN (70%), P (50%)
1931–35	Se (50%)
1936–40	Rh (66%)
1941–45	Lu (8%)
1946–50	Kell (11%), Fy (51%)
1951–55	Jk (50%), Le (30%), Di (0%)
1956–60	Yt (8%)
1961–65	Au (49%), Xg (45%), Do (49%), Sc (2%), S (27%)
1966–70	Co (8%), Sd (32%), Bg (27%), Ch (23%)
1970–	none

NOTE: The percentages in parentheses refer to the frequencies of heterozygotes in N. Europeans. Di (Diego) is polymorphic only among Orientals and American Natives. In addition to the above 21 blood-group polymorphic systems, 41 nonpolymorphic ones are known for which only rare variants have been detected.
SOURCE: Courtesy of R. R. Race and R. Sanger.

together with their times of discovery and the overall proportion of people (in the English population) who are heterozygotes for each system.

Much less is known about most of these blood groups than about the ABO and rhesus systems, which we have described in some detail. Medically they are, in general, of little importance—though the Kell system can occasionally give rise to severe hemolytic disease of the newborn. The Xg blood group has been especially valuable for studies of the X chromosome. Very recently, evidence has been presented that antigens of the Duffy blood-group system (symbol Fy) may act as "receptors" for malarial parasites at the surface of the red cells. Thus the Fy antigens may be important for the entry of the parasites into the red cells, where they will multiply. A Duffy allele, Fy^0, determines the recessive phenotype Fy(a^-b^-), which gives negative reactions with the antisera against other alleles of the Duffy system. One might then expect that individuals with this phenotype are endowed with higher resistance to some forms of malaria. This hypothesis offers an explanation for the fact that most people of African origin have very high frequencies of the Fy^0 allele; the frequency reaches 100 percent in populations from Central Africa, such as the African pygmies.

In addition to the red cell blood groups and to the HLA system, there are a number of other immunologically detected polymorphisms, mostly concerning the immunoglobulins.

Undoubtedly, the most important of these is the Gm system, which consists of a series of alleles determining differences in the constant regions of various immunoglobulin heavy chains. In Section 5.4 we describe how the immunoglobulins, which are the antibody molecules themselves, are made up of two types of chains, light (L) and heavy (H) chains. Each chain has a "variable" region that is part of the antibody recognition site, and a constant region that is, in general, the same for all antibody molecules of a given type found in a given individual. There are different types of heavy chains, each with their own characteristic constant regions, which determine the different types of antibody molecules, such as Ig G, and Ig M. Many different Gm types, corresponding to different variations in the heavy chains of immunoglobulin G molecules have been described. These occur together in haplotypes, analogous to those we have discussed for the Rh and HLA systems. The major Gm types found in Caucasoid populations are usually called a, x, f, b, and n, and these occur in the combinations a, ax, fb, and fbn with frequencies as indicated in Figure 10.13.

There are a number of rarer haplotypes not shown in the figure, which occur with an overall frequency of at most 2 percent. The Gm system is very polymorphic, though still not nearly as much so as the HLA system. However, there are extraordinarily large differences in the frequencies of Gm haplotypes of human populations, and combinations are quite commonly found in one major group of populations that are not found anywhere else. These differences suggest that the

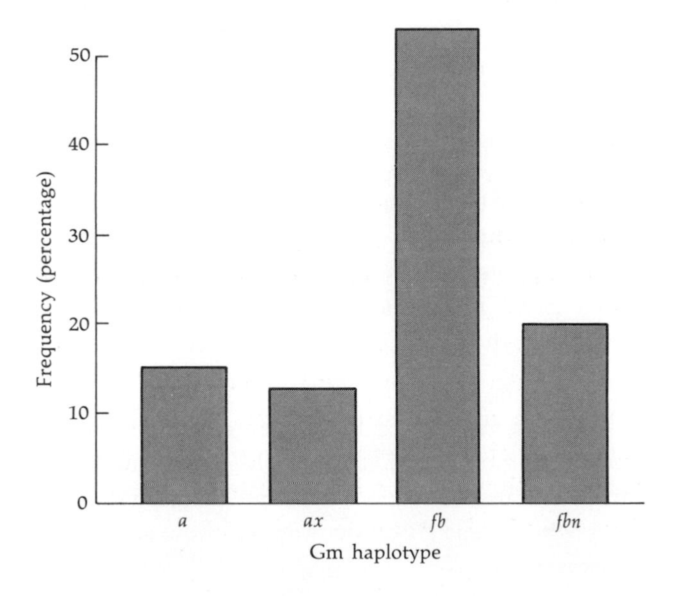

Figure 10.13
Typical Gm haplotype frequencies in a Caucasian population.

Gm system is associated with considerable selective effects. So far, however, no clearcut functional significance has been ascribed to the Gm polymorphism, though there is reason to expect that, as in the case of the HLA system, it may be associated with immune response differences. The system may thus be subject to selection, perhaps especially through differential resistance or susceptibility to infectious diseases.

There is at least one other polymorphic system involving the immunoglobulins, which is called Inv. The two alleles of this system, Inv_1 and Inv_2, control differences on one of the immunoglobulin light chains. The frequency of allele Inv_1 is about 10 percent in Caucasoid and 30 percent in Negroid and Mongoloid populations.

Antigen polymorphisms have been described that are specific to the granulocytes, one of the major types of white cells other than lymphocytes, and to platelets, the small elements of the blood that are involved in the formation of blood clots.

Immunological techniques have also been used to detect polymorphic differences in some proteins found in the serum, in addition to the immunoglobulins. There can be no doubt that immunological techniques used on components of the blood have uncovered an important and fascinating range of polymorphic systems in the human population.

Summary

1. Immunological techniques (including especially red cell agglutination) have permitted the discovery of many polymorphisms for antigens found on the red cell surface, of which the ABO and Rh systems are most important.

2. The Rh blood-group system is responsible for hemolytic disease of the newborn due to incompatibility between mother and fetus. The clinical consequences, however, are nowadays almost completely under control, and are relevant almost only to Caucasians. In addition to the main Rh^+ antigen, which is the cause of hemolytic disease, there are many other antigens of the Rh system.

3. The ABO was the first discovered blood-group system and, in fact, the first described genetic polymorphism in animals. It is important clinically for blood transfusion and for transplantation, and is an occasional cause of hemolytic disease of the newborn. The ABO blood types show some weak associations with certain diseases.

4. The biochemical genetics of ABO and related polymorphisms is fairly well understood. Their evolutionary significance still, however, defies our understanding.

5. The HLA antigens of the major human histocompatibility (or tissue matching system) are found on lymphocytes and most other cells. There are at least two major loci, A and B, which control the serologically detected antigens of the HLA system, and these loci are closely linked. There are many alleles at each locus. HLA is of importance in transplants and shows very close associations with several diseases, some of which are autoimmune or may be related to chronic virus infections.

6. The Gm and Inv types are polymorphic antigen differences detected on the antibody molecule itself.

Exercises

1. Why have blood groups played such a prominent role in the study of human polymorphism?

2. What were the arguments that led to the identification of the Rh^+ factor as a cause of hemolytic disease of the newborn? What are the major factors that limit the incidence of this disease?

3. What are the arguments for suggesting $R_0 = Dce$ as the possible "original" Rh haplotype?

4. Why is it difficult to accept the idea that the ABO-and-disease associations observed so far are responsible for maintaining the ABO blood-group polymorphism?

5. Outline the relationship between the ABO, Lewis, and secretor systems. In what way does a knowledge of the chemistry of these systems help understanding their interrelationships?

6. What are the main arguments for associating transplantation rejection with genetic differences?

7. Why is it likely that there are hundreds, if not a few thousand, of genes in the HLA region?

8. What is the significance of linkage disequilibrium for the interpretation of associations between HLA and disease?

9. Why is HLA matching for tissue transplantation so much more effective when donor and recipient are sibs, than when they are unrelated?

10. What is the relation of the Gm polymorphism to the basic structure of the immunoglobulin molecule?

11. What is the probability that the first child of an Rh^- woman married to an Rh^+ man is Rh^+? Assume the gene frequency of the Rh-negative gene is 0.4. If necessary, follow the lead given in Exercise 10 of Chapter 6.

12. Use the list and the frequencies of haplotypes of Figure 10.2 to answer the following questions.
    a. What are the possible genotypes corresponding to an Rh phenotype reacting positively with antisera to C, c, D, and e, and negatively with E? Recall that there is no known anti-d serum, and that alleles *C*, *c*, and *E*, *e* are known to be codominant.
    b. Which of the possible genotypes is the most likely?
    c. What is the probability that an individual of this phenotype is heterozygous for *D* and *d*?

13. Incompatible marriages in the ABO systems are defined as those in which the wife does not have antigen A or B, while the husband has it. Using

the data of Table 6.5, give the probability of each type of incompatible mating and the total frequency of such matings.

14. An extract from a plant agglutinates the red cells of all O individuals, and also of about 20 percent of all A individuals among Caucasians. Could it be that it is able to detect the *O* gene in *AO* heterozygotes? [HINT: Compute the expected frequency of such heterozygotes; assume the gene frequency of *A* is 30 percent and that of *O* 55 percent.]

15. The same extract as in the previous exercise also agglutinates the red cells of about 20 percent of AB individuals, but never those of B individuals. Does this reinforce the conclusion reached in Exercise 14?

16. The Gm alleles affect the constant portion of the heavy chain of the Ig G molecules. Does this suggest that individuals of given Gm types can only or more readily make antibodies against some types of antigens? Does it make it impossible that this explanation is correct?

17. There is an X-linked recessive gene that determines agammaglobulinemia—the incapacity to produce most types of immunoglobulins. Given what we have seen of the genetics of immunoglobulins in Chapter 5 and 10, is this sex-linked gene more likely to be coding directly for immunoglobulins or to be a regulatory gene? Why?

References

Bodmer, J., and W. F. Bodmer
1973. "Population genetics of the HL-A system: A summary of data from the Fifth International Histocompatibility Testing Workshop," *Israel Journal of Medical Science*, vol. 9, pp. 1257–1268.

Bodmer, W. F.
1972. "Evolutionary significance of the HL-A system," *Nature*, vol. 237, pp. 139–145.

Cavalli-Sforza, L. L., and W. F. Bodmer
1971. *The Genetics of Human Populations*. San Francisco: W. H. Freeman and Company. [Chapter 5 treats the topic of this chapter more thoroughly.]

Clarke, C. A.
1973. "The prevention of Rh isoimmunization," in *Medical Genetics*, ed. V. A. McKusick and R. Claiborne, New York: H. P. Publishing Co.

Dausset, J., and J. Colombani
1973. (Eds.) *Histocompatibility Testing 1972*. Copenhagen: Munksgaard.

Fudenberg, H. H., J. R. L. Pink, D. P. Sites, and A. Wang.
1972. *Basic Immunogenetics*. New York: Oxford Univ. Press.

Grubb, R.
1965. "Agglutination of erythrocytes coated with 'incomplete' anti-Rh by certain rheumatoid arthritic sera and some other sera: The existence of human serum groups," *Acta Pathologica et Microbiologica Scandinavica*, vol. 39, pp. 195–197.

McDevitt, H. O., and W. F. Bodmer
1974. "HL-A, immune-response genes and disease," *Lancet*, vol. 1, pp. 1269–1275.

Porter, R. R.
1973. (Ed.) *Defense and Recognition.* Baltimore: Univ. Park Press.
Race, R. R., and R. Sanger
1975. *Blood Groups in Man.* Philadelphia: Davis.
Terasaki, P. I.
1970. (Ed.) *Histocompatibility Testing 1970.* Copenhagen: Munksgaard.
Vogel, F.
1970. "ABO blood groups and disease," *American Journal of Human Genetics,* vol. 22, pp. 464–475.
Watkins, W. M.
1966. "Blood-group substances," *Science,* vol. 152, pp. 172–181.

11

Inbreeding and Its Consequences

Inbreeding is defined as mating between close relatives. In plants and animals inbreeding has been used experimentally to produce "pure lines"; in man, social customs mostly prevent inbreeding, though it does occur to some extent. The main genetic consequence of inbreeding is an increase in the proportion of homozygotes. Through inbreeding, recessive genes are more easily brought to the fore. One can thus obtain an estimate of the amount of hidden genetic variation. As some of this variation is detrimental, it is sometimes called the "genetic load."

11.1 Inbreeding, Consanguinity, and Incest

The aim of animal and plant breeders is always to improve their stock. In addition, they constantly try to make their stock homogeneous—that is, homogeneously good. Improvement is achieved by choosing the best individuals for the purpose of reproduction: this is "artificial" (as contrasted with "natural") selection. Homogeneity is increased by mating together close relatives, the process called inbreeding. In plants, one can carry inbreeding to an extreme degree: in many species one can mate an individual to itself. Only a few animal species however (mostly invertebrates) are hermaphroditic—that is, consist of individuals that are bisexual and can reproduce by mating to themselves. But mating between sibs or between parents and offspring is possible, and is often practiced. To increase genetic homogeneity, inbreeding is repeated generation after generation. One can thus reach a point at which most or all the individuals in a stock are homozygous for the same allele at a locus, and this is true of most loci.

Usually, inbreeding causes deterioration and outbreeding causes improvement of most characters.

Breeders soon noticed, however, that inbreeding practically always leads to a deterioration in many important qualities: fertility for instance, tends to decrease, and many an inbred stock has been lost because the fertility level became too low for maintenance of the line to be possible. In addition, some traits, such as overall general size also decrease. In almost all instances of practical breeding these are serious drawbacks, so that a compromise has to be reached between the desire to make a stock homogeneous and that of improving its quality. This phenomenon of deterioration on inbreeding is known as *inbreeding depression* (Fig. 11.1).

In contrast to inbreeding depression, if two independent inbred lines are crossed, the hybrids between them (at least in the first generation) mostly show a considerable increase in size, fertility, and many other desirable traits. This has been called *hybrid vigor,* or *heterosis,* and clearly has a great potential for applications in agriculture and animal husbandry. The first practical application of hybrid vigor as a technique for crop improvement was to corn, and it led to a very significant increase in productivity. The practice is now being extended to other plants and animals. It requires, however, a complex organization; the traditional system for plants of setting aside seed to be sowed next year must be abandoned. This is because hybrids can be used with maximum utility only in the first generation. If they are allowed to reproduce, their progeny shows great variability due to Mendelian segregation, and production drops drastically. Hybrid seed must be made by specialists using the appropriate inbred stocks. Knowledge and application of inbreeding depression and hybrid vigor rest on firm experimental data in animals and plants. We shall now see to what extent these conclusions extend to man. It is interesting to note, at this stage, that social customs in every human society serve to avoid inbreeding, at least to some extent.

All human societies in existence prohibit mating between first degree relatives.

"Incest" is mating between relatives as close as sibs or parents and offspring. The incest taboo is one of the strongest taboos in man. Even though incest sometimes takes place, especially in very isolated communities, it is quite infrequent. Avoidance of incest was at one time believed to be one of the unique characteristics of man; but recent observations on other mammals (especially the higher primates) indicate that some tendency toward avoiding incest may also be present in animals.

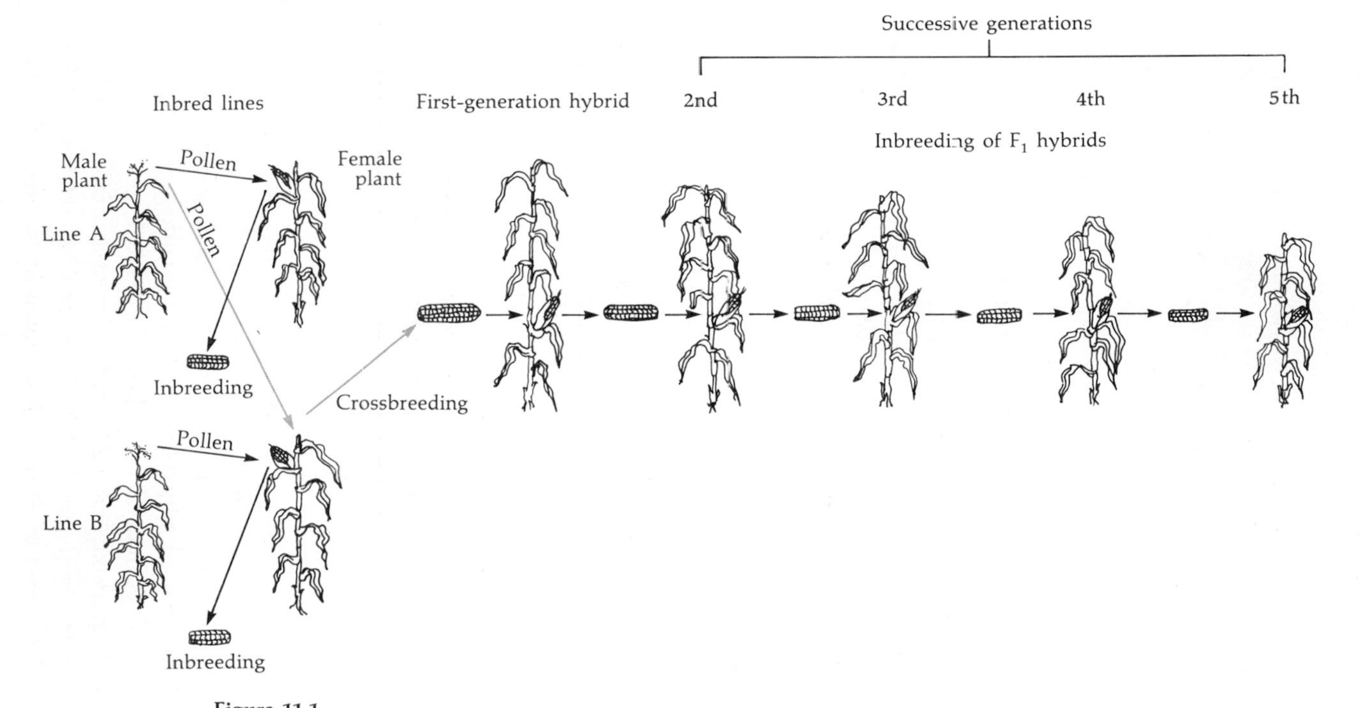

Figure 11.1
Hybrid vigor and inbreeding depression in corn. The plants of inbred lines A and B have been inbred for many generations. Each line is highly homogeneous genetically and each is highly homozygous—though not necessarily for the same genes in the two different lines. Overall size and yield are low in the inbred lines. When plants from the two lines are crossbred, the resulting hybrid (F_1, or first-generation hybrid) is much larger. If F_1 individuals are crossed among themselves, the next generation (F_2, or second-generation hybrid) is somewhat smaller and more variable, as homozygotes appear. Further generations of inbreeding of F_2 plants lead to increasing homozygosity and decreasing size and yield. Commercially available hybrids are usually the result of double crosses, in which four lines (A, B, C, and D) are first mated two by two (A × B and C × D), and then the offspring of these matings are crossed: (A × B) × (C × D).

Marriage between relatives less close than sibs or parents and offspring is not necessarily outlawed; but the dividing line between legal and illegal varies somewhat between countries. Thus in about half of the United States, uncle–niece, aunt–nephew, and first-cousin matings are forbidden by law. In many other countries they are not illegal; while in some, they can tàke place only if special permission is acquired from state authorities. Several religions also have precise rules about "consanguineous" matings.

Consanguinity is the name given to close blood relationships (as distinct from relationships by marriage).

Consanguineous individuals have at least one not-too-remote ancestor in common; how remote the ancestor should be for the definition to apply is somewhat vague. In practice, the limit often is set by the number of generations for which pedigrees can be reconstructed. The number is not often greater than two or three, corresponding to going no farther back than to grandparents or great grandparents. In some societies, however—ranging from Icelandic to Chinese and the Touaregs in North Africa—there has been a considerable interest in genealogies, so that pedigrees can be reconstructed, in part at least, for ten or even twenty generations. Documents such as "parish" books of births, deaths, and marriages also exist that allow pedigree reconstruction to extend back for ten or more generations.

Many religions require that a "dispensation" be requested before a marriage between consanguineous individuals of a certain degree can take place. Figure 11.2 shows types of relationship for which a dispensation is (or was until a few years ago) to be requested, according to the Roman Catholic church. Dispensations were very rarely granted until the beginning of the eighteenth century. In Roman Catholic Europe, first-cousin marriages—once very rare and strongly discouraged—then began increasing in frequency and continued to increase until the middle or the end of the last century. This was probably a response to changes in the laws of inheritance of property. Napoleon's abolition of the right of primogeniture caused excessive splitting of land property, which could to some extent be counteracted by marriage between close relatives. Later, in most places, the frequency of consanguineous marriages started declining. This decline was presumably a response to the industrial revolution, which caused an increase of geographic mobility. Extensive internal migrations, caused by work opportunities offered in distant places, separated individuals who were related and thus decreased the chances of a consanguineous mating. This happened in almost every country and led to a decrease in the frequencies of marriages between relatives. This phenomenon is sometimes referred to as the "breakdown of isolates" (Fig. 11.3).

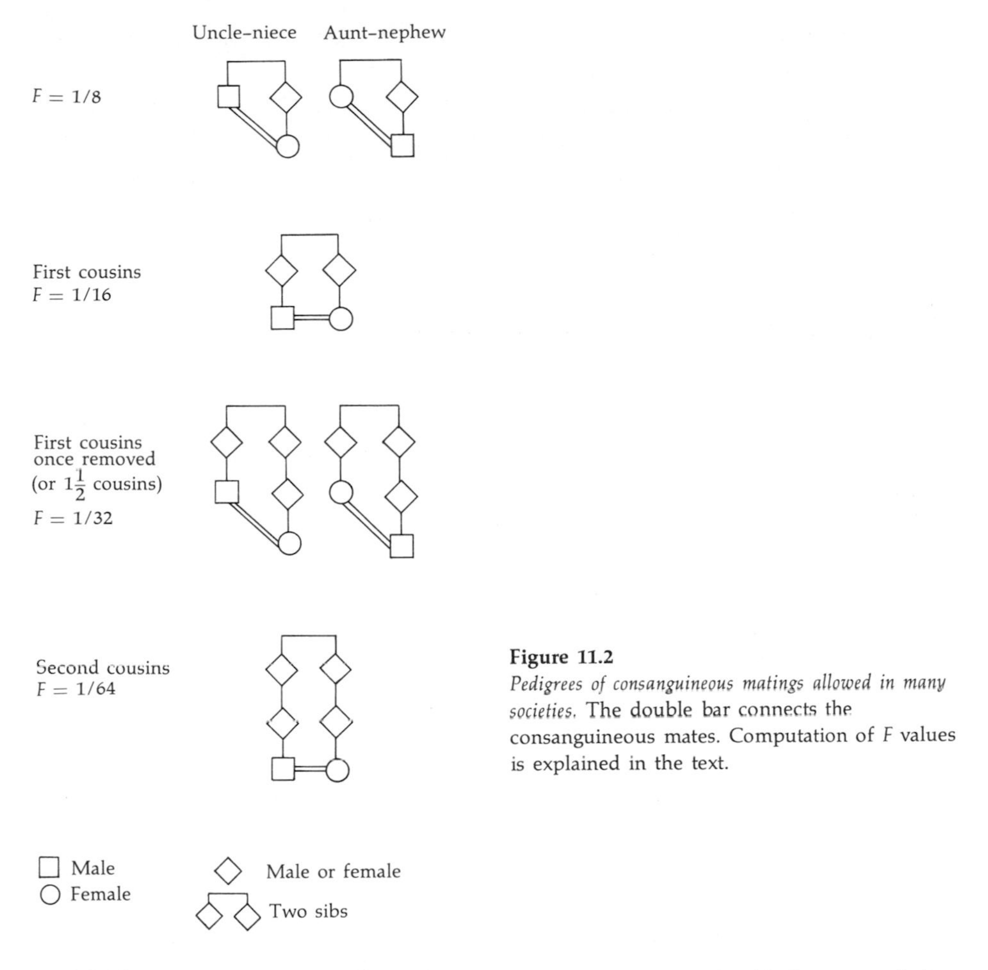

Figure 11.2
Pedigrees of consanguineous matings allowed in many societies. The double bar connects the consanguineous mates. Computation of F values is explained in the text.

The attitudes toward less close consanguineous matings vary considerably between societies.

In most traditional African societies, consanguineous marriage is not allowed, or is allowed only to a limited extent. Mating usually occurs only outside a very wide circle of relatives. Such social practices, apart from their biological advantages, may have considerable advantages in widening the network of relatives acquired by marriage, and thus that of individuals pledged to mutual support. In other societies, however, the opposite tendency may prevail—involving, for example, in some cases a direct preference in favor of certain kinds of consanguineous marriages. Thus in Japan, first-cousin marriage is encouraged and, in certain areas or social strata, up to 10 percent of the marriages are between first cousins. In Andhra Pradesh (India) certain castes favor uncle–niece marriages, which form more than 10 percent of all marriages. Clearly, such preferences or aversions have

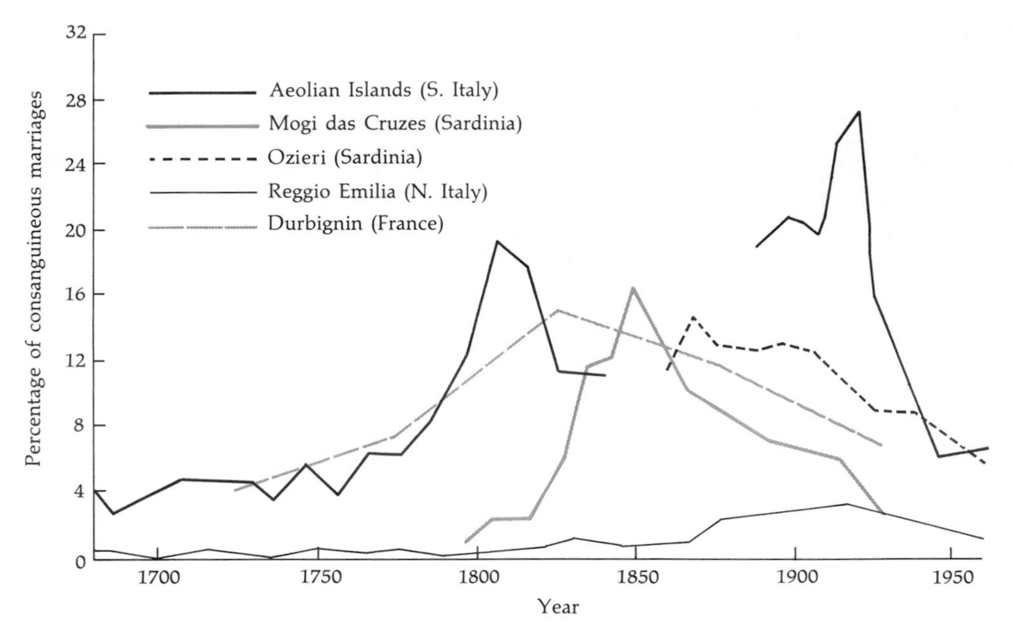

Figure 11.3
Frequency of consanguineous matings in some European populations over time. The frequency in some populations increased at the beginning of the nineteenth century, and then decreased more recently. The decreasing trend (the "breakdown of isolates") is due to the increased migration that accompanied industrial development. (After A. Moroni, *Historical Demography, Human Ecology, and Consanguinity,* International Union for the Scientific Study of Population, 1969.)

a predominantly social origin. In fact, many social scientists now believe that all restrictions on incest and consanguineous marriage exist because of social advantages to such restrictions, and that avoidance of biological effects of inbreeding is a secondary benefit that plays no major role in maintaining the taboos.

The biological effects of consanguinity can be traced to the fact that inbreeding can unmask a hidden recessive gene.

Consanguineous individuals have a recent common ancestor: first cousins, for instance, have two grandparents in common (Fig. 11.4). If one of these common ancestors had a single dose of (was heterozygous for) a recessive gene, the consanguineous parents may *both* be heterozygous for that same gene. As a consequence, their progeny can be homozygous for that gene.

A recessive gene carried in a single dose in a common ancestor may well remain hidden until it comes to light for the first time in an inbred descendent. Intuitively

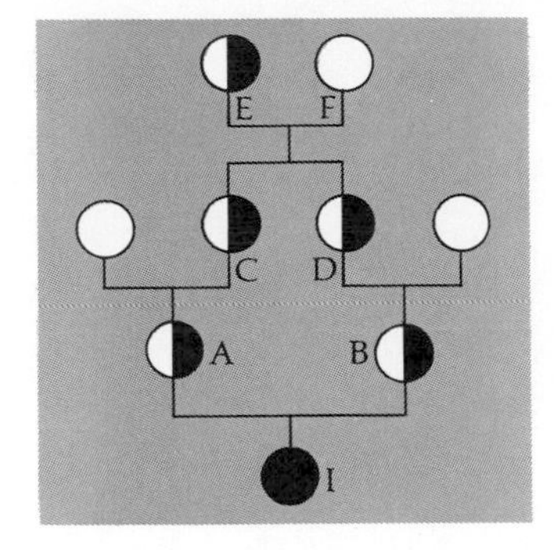

Figure 11.4
Pedigree of the offspring of two first cousins. The individual I is the offspring of parents (A and B) who are first cousins. The shaded half of each circle indicates the recessive gene transmitted to A and B by common ancestor E.

we expect, therefore, that recessive traits will occur with increased frequency in the progeny of consanguineous mates. This was, in fact, the first clue used by Archibald Garrod in establishing the inheritance of his "inborn errors" of metabolism (see Chapter 3). Thus, consanguinity is important in the study of recessive inheritance and of recessive traits. Clearly, the closer the consanguinity, the larger its effect is likely to be. Therefore, it is useful to evaluate the closeness of a relationship, with a view to predicting the genetic consequences of inbreeding.

11.2 The Inbreeding Coefficient

A coefficient, F—first proposed by the American geneticist Sewall Wright, one of the founders of population genetics—was devised to measure the genetic consequences of inbreeding.

The coefficient of inbreeding, F, of an individual is the probability that this individual receives at a given locus two genes that are identical by descent—that is, are copied from a single gene carried by a common ancestor.

Let us compute as an example the inbreeding coefficient of the progeny of a mating between two sibs. Let us call the two alleles at one locus in the sibs' father a and b, while c and d are the corresponding alleles in the sibs' mother. The genotypes of the two parents will therefore be ab and cd; the genotypes of their progeny are thus: $\frac{1}{4}$ ac, $\frac{1}{4}$ ad, $\frac{1}{4}$ bc, and $\frac{1}{4}$ cd. We are interested in the probability that children of the sibs are homozygous for any one allele of the sibs' father or mother—that is, are either aa, bb, cc, or dd. These homozygous children then have two copies of one particular allele carried by the original parents—that is, each of their alleles is identical by descent. If we then add up the frequencies of all of these homozygous children, we have the total F, or the total probability of identity by descent. This

Table 11.1
Computation of inbreeding coefficient (F) for progeny of a mating between sibs

Father (*ab*) ═ Mother (*cd*)

Sibs 1/4 *ac* 1/4 *ad* 1/4 *bc* 1/4 *bd*

Random mating between sibs

Second sib

First sib		1/4 ac	1/4 ad	1/4 bc	1/4 bd	*Homozygotes
	1/4 ac	1/16 × { 1/4 aa* 1/2 ac 1/4 cc* }	1/16 × { 1/4 aa* 1/4 ac 1/4 ad 1/4 cd }	1/16 × { 1/4 ab 1/4 ac 1/4 bc 1/4 cc* }	1/16 × { 1/4 ab 1/4 ad 1/4 bc 1/4 cd }	2/64 aa 2/64 cc
	1/4 ad	1/16 × { 1/4 aa* 1/4 ac 1/4 ad 1/4 cd }	1/16 × { 1/4 aa* 1/2 ad 1/4 dd* }	1/16 × { 1/4 ab 1/4 ac 1/4 bd 1/4 cd }	1/16 × { 1/4 ab 1/4 ad 1/4 bd 1/4 dd* }	2/64 aa 2/64 dd
	1/4 bc	1/16 × { 1/4 ab 1/4 ac 1/4 bc 1/4 cc* }	1/16 × { 1/4 ab 1/4 ac 1/4 bd 1/4 cd }	1/16 × { 1/4 bb* 1/2 bc 1/4 cc* }	1/16 × { 1/4 bb* 1/4 bc 1/4 bd 1/4 cd }	2/64 bb 2/64 cc
	1/4 bd	1/16 × { 1/4 ab 1/4 ad 1/4 bc 1/4 cd }	1/16 × { 1/4 ab 1/4 ad 1/4 bd 1/4 dd* }	1/16 × { 1/4 bb* 1/4 bc 1/4 bd 1/4 cd }	1/16 × { 1/4 bb* 1/2 bd 1/4 dd* }	2/64 bb 2/64 dd

Total homozygotes = 4/64 *aa* + 4/64 *bb* + 4/64 *cc* + 4/64 *dd* = 16/64 = 1/4 = F.

NOTE: The letters *a*, *b*, *c*, and *d* indicate different alleles at one locus. If the father and mother are both heterozygotes with no duplication of alleles between them, there are four possible offspring genotypes, each with probability of 1/4. If there is random mating among these sibs, 16 different matings are possible, each with a probability of 1/16. The table shows the expected frequencies of various offspring genotypes from each possible sib–sib mating. The total probability of a homozygous offspring resulting from a sib–sib mating is 16/64 or 1/4. This value, by definition, is the inbreeding coefficient (F) for the sib–sib mating.

probability is the F value of progeny from a sib–sib mating, and Table 11.1 shows that its value is $\frac{1}{4}$.

Adding a generation to a consanguineous pedigree reduces by half the probability that an ancestral gene is found in the consanguineous mates.

This conclusion can easily be tested by extending the computation of Table 11.1 to the case of an uncle–niece mating. The value of F in this case proves to be $\frac{1}{8}$, or one-half the F value obtained for the mating between sibs. The value for an aunt–nephew mating is also $\frac{1}{8}$. Thus, addition of one generation on one side of the sib–sib pedigree reduces the value of F by a factor of one-half. Similarly, offspring of first cousins (adding one generation on the other side of the pedigree as well) have an F value that is one-half of the F value obtained for offspring of uncle–niece or aunt–nephew matings—namely, $\frac{1}{16}$.

If the sib–sib mating is between half sibs rather than full sibs, the F value again is halved (see Exercise 11). Figure 11.5 summarizes F values for some of the more common types of consanguineous matings. In more complex pedigrees, other methods can be used to estimate F.

11.3 Average Inbreeding Levels in Human Populations

The amount of inbreeding that takes place in most human populations is extremely small. The closest consanguineous marriages allowed in some populations have coefficients of inbreeding of the order of $\frac{1}{8}$ (uncle–niece, aunt–nephew, double first cousins) but, except in very special situations, such marriages are extremely rare. The two most common types of consanguineous marriages in most populations are those between first cousins ($F = \frac{1}{16}$) and between second cousins ($F = \frac{1}{64}$). The frequencies of these cousin marriages are usually low in most populations, so that the average degree of inbreeding in a human population is always very small and no genetic "purity" could ever be achieved.

It is customary to measure the average consanguinity in a population as the average inbreeding coefficient of its individuals.

This is defined by the quantity often called α, which is obtained by averaging the F values pertaining to the progeny of consanguineous matings. As an example, assume that of 100 couples forming a population, five are first cousins ($F = \frac{1}{16}$), seven are second cousins ($F = \frac{1}{64}$), and the others are "unrelated" (F effectively zero). Then the average value of F is the combination of the contributions of $(\frac{5}{100}) \times (\frac{1}{16}) = 0.0031$ from the five first cousins and $(\frac{7}{100}) \times (\frac{1}{64}) = 0.0011$ from the seven second cousins, giving a total value of $\alpha = 0.0042$. The average inbreeding coefficient (α) in human populations is generally around one per thousand (0.001), as shown in Figure 11.6.

There are a few areas or populations in which α is higher than 0.01, but these are exceptional cases. Most of them are "isolates"—relatively small populations that have little or no gene exchange with other populations. However, some large populations (such as in South India) do have high values of α because of a preference for close consanguineous matings. Moreover, even small isolates do not

Type	Symbol	Degrees of Relationship: Roman Catholic usage	Degrees of Relationship: Napoleonic Code	Inbreeding Coefficients (F)*: Full	Inbreeding Coefficients (F)*: Half
Uncle–niece; aunt–nephew	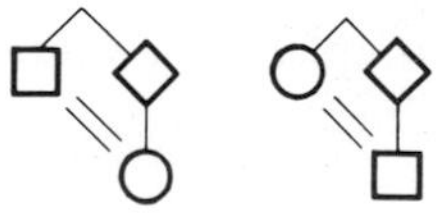	I in II	III	1/8	1/16
First cousins	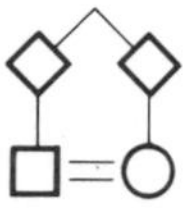	II	IV	1/16	1/32
First cousins once removed ($1\frac{1}{2}$)	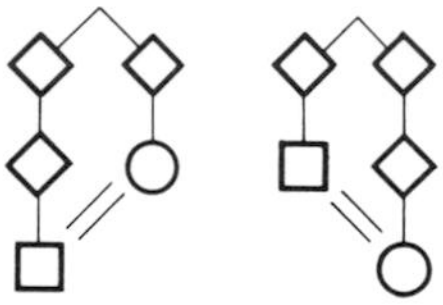	II in III	V	1/32	1/64
Second cousins	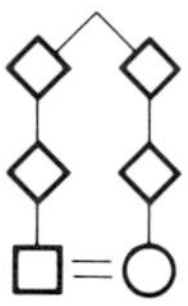	III	VI	1/64	1/128
Second cousins once removed ($2\frac{1}{2}$)	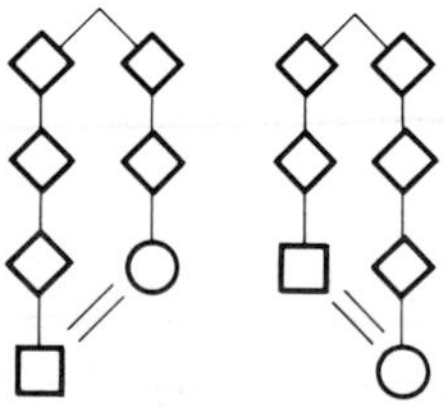	III in IV	VII	1/128	1/256
Third cousins	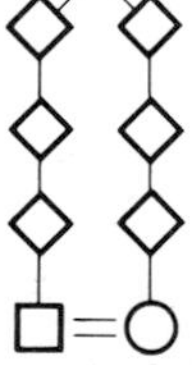	IV	VIII	1/256	1/512

Figure 11.5

The most common pedigrees of consanguineous matings in man and their inbreeding coefficients. "Full" and "half" refer to the two sibs starting the chains of descent, who are the top two individuals in each pedigree; full sibs have both parents in common, and half sibs only one parent. Diamond-shaped symbols in the pedigrees represent an individual of either sex.

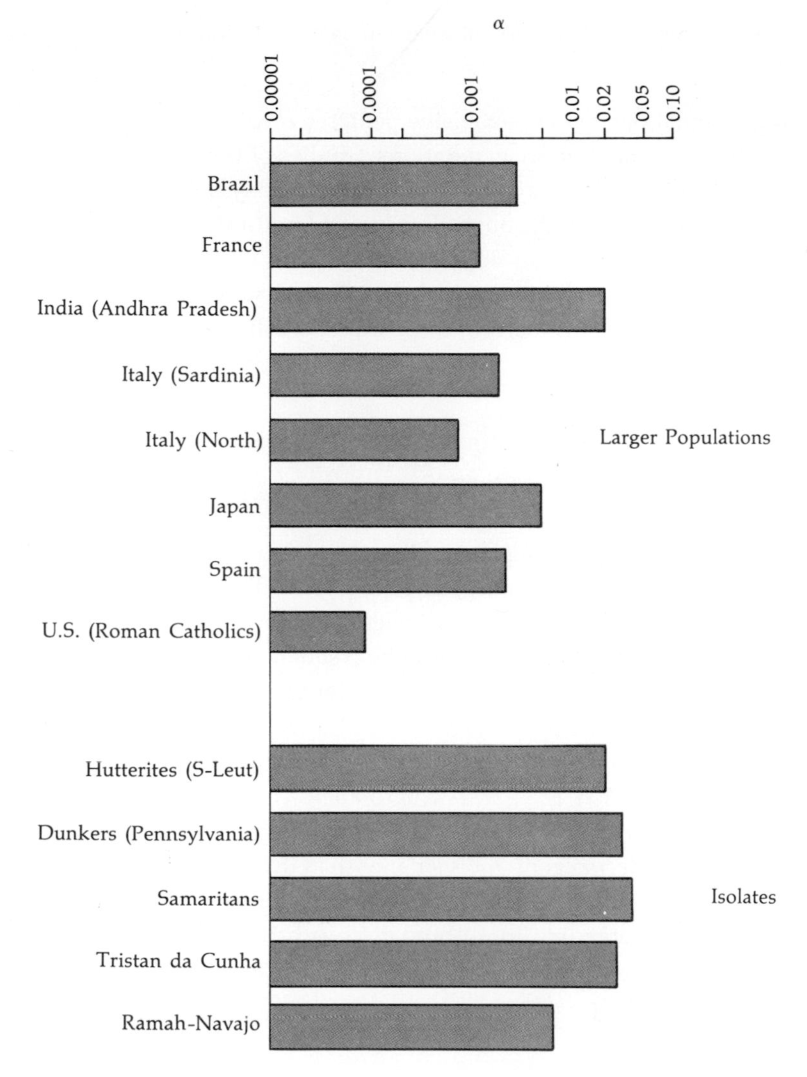

Figure 11.6
Average inbreeding coefficient (α): in some larger populations, and in some isolates. Note that α is given on a logarithmic scale.

always attain large average inbreeding coefficients. Polar Eskimos, because of a careful avoidance of consanguineous matings, have maintained a low value of α (<0.003) in a very small isolate. These inbreeding estimates, however, take into account only easily detectable consanguinity, which rarely includes relationships more remote than third cousins.

The α value, therefore, is usually an underestimate of the true level of inbreeding in a population.

In a few populations, where extensive pedigree records are available, it is possible to go back for many more generations. Studies of these populations indicate that the true value of α (at least in some instances) may be two or more times as large as the value obtained through the usual procedures. However, average inbreeding coefficients in human populations—even in the most extreme isolates—are still small when compared with those of artificially inbred populations of domesticated plants and animals. In these cases, α values are often above 0.1 and are sometimes close to one.

11.4 The Effects of Inbreeding

We have seen that inbreeding may make inbred individuals homozygous for genes carried by their common ancestor or ancestors. The coefficient of inbreeding is a measure of the probability that this event will happen. The Hardy-Weinberg law indicates that under random mating the frequency of homozygotes *aa* will be q^2 for an allele *a* that has frequency q in the general population. However, in an inbred individual with coefficient F, it can be shown that the probability of homozygosity has the higher value $q^2 + Fpq$. Similarly, the probability of the individual being homozygous *AA* is $p^2 + Fpq$, where allele *A* is present with frequency p in the general population. On the other hand, the probability that the inbred individual will be heterozygous *Aa* is smaller than the $2pq$ expected under random mating—namely, $(1 - F) \times 2pq$. The same expressions can be used to estimate the average effects of inbreeding in a population, substituting α for F to obtain the expected genotype frequencies in nonrandom mating. (See the Appendix for a derivation of these expressions.)

One consequence of these effects is that consanguineous parents are likely to be overrepresented among parents of recessive homozygotes.

This observation is the basis of the so-called "retrospective" analysis of the effects of consanguinity: parents of recessive homozygotes are more likely to be consanguineous than are couples chosen at random from the population. If a recessive is very rare, many or even most of the homozygotes will be the progeny of consanguineous parents.

Figure 11.7 gives some data comparing the frequencies of one type of consanguineous marriage (between first cousins) among parents of individuals with certain recessive traits and among the corresponding general population. For some

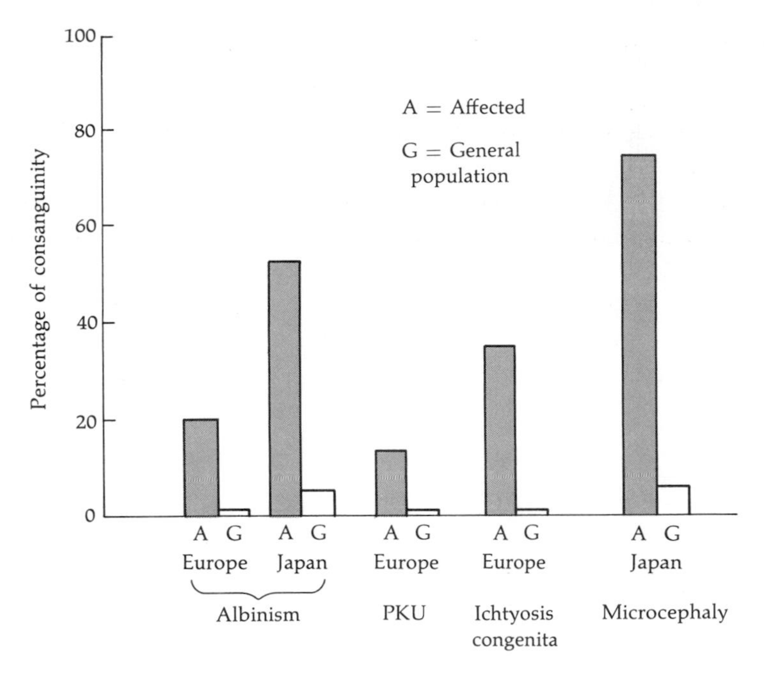

Figure 11.7
Consanguinity among parents of homozygotes for recessive conditions. The bar graph shows the frequency of first-cousin marriages (A) among parents of individuals affected by the recessive conditions listed below each pair of bars, and (G) in the general population. PKU is phenylketonuria; ichtyosis congenita is a skin disease; microcephaly is a condition involving an abnormally small head.

diseases, there is a very high proportion of first-cousin marriages among the parents of the afflicted. In general, the rarer the disease, the higher this proportion. Parental consanguinity can be a useful criterion in clinical diagnosis. When a doctor is confronted with a rare or previously unknown disease and he finds that the parents of the patient are consanguineous, the diagnosis of a recessive genetic disease is worthy of serious consideration.

The increased incidence of recessive homozygotes in inbred progeny causes an increase in mortality.

This effect is illustrated in Figure 11.8 with data from the cities of Hiroshima and Nagasaki, where careful genetic studies were conducted to ascertain the possible genetic effects of the atom bombs that destroyed these cities in World War II. No genetic effects were actually found because the population that could be investigated was too small. But in this investigation the effects of parental consanguinity

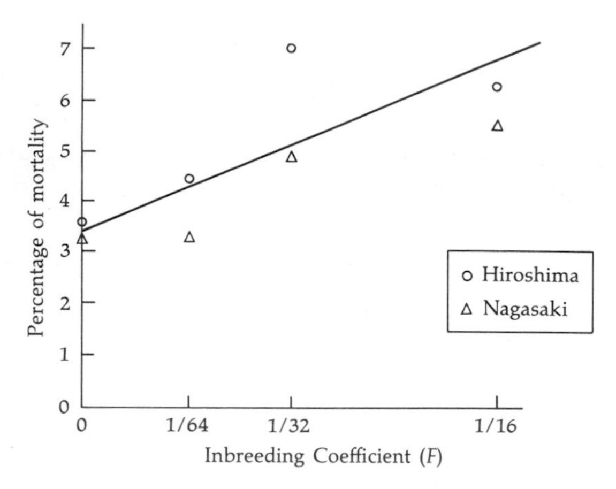

Figure 11.8
Percentage of mortality among children as a function of inbreeding coefficient (F) in two Japanese cities. The samples indicated are for progeny of second cousins ($F = \frac{1}{64}$), of $1\frac{1}{2}$ cousins ($F = \frac{1}{32}$), and of first cousins ($F = \frac{1}{16}$). (Data by W. J. Schull and J. V. Neel, *The Effects of Inbreeding on Japanese Children,* Harper and Row, 1965.)

were also examined. It should be noted that the recessives that contribute to the increased mortality reported in Figure 11.8 must have originated at least three generations before the events of World War II, and therefore have nothing to do with the atom bombs.

Mortality appears to increase almost linearly with the inbreeding coefficient. The data deviate somewhat from a straight line, mostly because of the error inherent in estimates based on a limited number of observations.

The slope of the increase in mortality with increasing F provides some information on the number of recessive lethal and detrimental genes.

This quantity (the slope) is related to the concept of *genetic load,* which originated in a paper by the famous geneticist H. J. Muller, who discussed what he called "our load of mutations." Mutations are often detrimental, if not actually lethal (see Chapter 8), and a population inevitably carries a certain load of detrimental genes originating through mutations. Dominants are rapidly eliminated (Chapter 8), but recessive genes can be hidden and so can accumulate. Estimation of the size of the genetic load is therefore difficult, but consanguinity studies help in this estimation.

Attempts at an exact treatment of the problem have on the whole been frustrated by the complexity of the phenomenon. Inbreeding exposes not only recessive defects, but also those in which both homozygotes are to some extent affected (where the heterozygote has a selective advantage). This is the case, for example, with sickle-cell anemia and thalassemia in malarial areas, and more generally with any balanced polymorphism due to heterozygote advantage (see Chapter 9).

At first, it was hoped that it would be possible to distinguish the effects of the two types of detrimental genes—those that are straightforward detrimental recessives (mutational load), and those whose presence is favored because of heterozygote advantage (segregational load). Such separation of the two types of load has not really been possible. A joint estimate of the two loads can be made, but this estimate must be considered very approximate because of a further complication. There is no proof that the total mortality (or detrimental effect) measured is due to single genes; the genes involved may interact in complex ways that would make the analysis much more difficult, if not impossible.

With these reservations, therefore, we now proceed to estimate how many recessive detrimental or lethal genes are carried, on the average, by each individual. Clearly, every individual can carry many recessive lethal genes, because he or she will not be affected by any of them, so long as each is heterozygous.

Let us consider mortality, and assume at first that all genes are simple recessives that are fully lethal when homozygous. In the general (randomly mating) population, one lethal gene with a frequency q kills a proportion of the population equal to q^2. But inbred individuals (with an inbreeding coefficient of F) are killed with probability $q^2 + Fpq$. In the appendix, we demonstrate that—with many genes, each with a small frequency q—the slope of the curve shown in Figure 11.8 is equal to Σq (that is, to the sum of the gene frequencies of all the lethal genes involved). For the Japanese data summarized in Figure 11.8, the slope is 0.4. Other studies give somewhat higher values, depending to some extent on the precise definition of "mortality" used. Including all deaths between birth and reproduction and averaging over various studies, we obtain a value of about 1 for Σq.

The quantity Σq is an estimate of the total number of lethal genes expected to be present in a gamete.

Because each zygote is formed from two gametes, this result indicates that each individual carries, on the average, two lethal genes in the heterozygous state. It should be noted that this result must be an underestimate. If a lethal gene does not kill all homozygotes, but only a fraction of them (say s, as defined by the selection coefficient), the slope of the mortality curve plotted against the inbreeding coefficient is actually equal to Σsq. This quantity has been called the number of *lethal equivalents* (Fig. 11.9). In this terminology, a gene that kills only 10 percent of the individuals homozygous for that gene is equivalent to 10 percent of a single lethal gene. The estimate given earlier is actually an estimate of about two lethal equivalents per person. Therefore, an individual may on the average carry more than two detrimental genes, some of which kill less than 100 percent of their homozygous carriers.

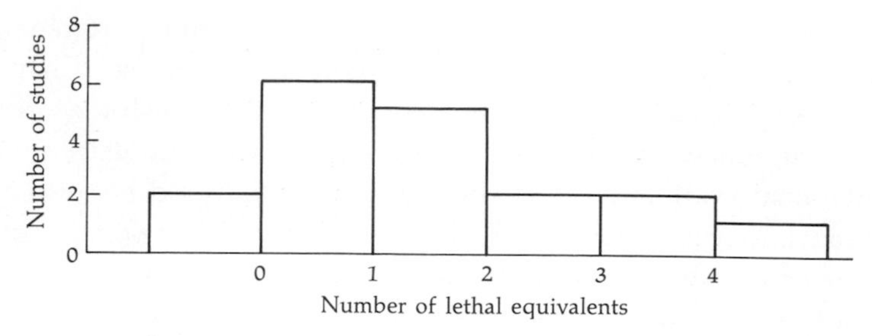

Figure 11.9
Numbers of lethal equivalents. A total of eighteen studies in a number of different populations is summarized in this graph, showing the number of studies that reported each possible number of lethal equivalents. Two studies reported negative values, which cannot actually exist and may indicate biases in the observations. The median number of lethal equivalents is slightly above 1. This should be the load of lethal equivalents per gamete; an individual will carry twice as many. Lethal equivalents are an underestimate of the actual number of detrimental genes.

It is of interest to know that each of us is likely to carry at least two lethals—which, as already noted, do not normally manifest themselves because they are recessive. In addition, however, we may each carry several detrimental genes, again hidden because of recessiveness, but which determine a handicap that is not sufficiently serious to be associated with a significantly increased mortality. An analysis of detrimental recessives found among inbred individuals can be carried out following exactly the same approach used for mortality. Some estimates obtained in this way from different studies are shown in Table 11.2. They vary considerably according to the criterion of severity adopted, but they do show that each individual carries several detrimental genes.

The same table gives another figure of interest: the overall increase in risk of disease in the progeny of first cousins as compared to marriages between nonrelatives. Again, the figure varies from study to study, but the average is not far from a factor of two. Because of the approximately linear relationship between the increased risk of detrimental disease and F, it is possible to compute from this result the increase in risk for other degrees of consanguinity. With uncle–niece or aunt–nephew matings, the increased risk is expected to be about three times the level for the general population, while for sib–sib incest it will be of the order of five times as high. The early Romans already knew that blind, deaf-mute, and malformed offspring came from brother–sister matings.

Of course, it is important to realize the difference between retrospective and prospective frequencies. For the rarest recessive defects, a majority of afflicted individuals may be offspring of consanguineous marriages. However, the incidence of a given affliction among such offspring is still very low as a percentage of all offspring of consanguineous marriages.

Table 11.2
Estimated numbers of detrimental equivalents, and increased risk of disease in progeny of first-cousin matings

Country	Type of defect	Estimated number of detrimental equivalent genes	Increase in risk of disease among progeny of first-cousin matings (see note)
France	Conspicuous abnormalities	2.2	3.8
Italy	Severe defects	0.6	1.9
	Various less severe diseases and defects (see note)	4.6	—
Japan	Major morbid conditions	0.7	1.4
	Minor defects	0.4	1.3
Sweden	Morbidity	2.0	2.3
U.S.A.	Abnormality	1.2	2.3

NOTE: The differences between data from different countries are probably due to the variety of criteria used in defining defects. The increase in risk of disease among progeny of first-cousin matings is given as the ratio of the incidence among progeny of such matings to the incidence in the general population. The "various less severe diseases and defects" reported in the Italian study included the following (with the associated estimates for detrimental equivalent genes): mental retardation (1.4), eye diseases (0.8), ear diseases (0.6), tuberculosis (0.2), malformations (0.5), and smaller defects (1.1).

Summary

1. Mating between relatives, or inbreeding, causes an increase in the frequency of homozygotes among the offspring. The inbreeding coefficient F helps to estimate the expected level of this increase.
2. Recessive phenotypes will appear with greater frequency among the progeny of inbred matings than in the general population.
3. Conversely, consanguinous or related couples will be represented more frequently among the parents of carriers of rare recessive traits. This phenomenon is more pronounced for rarer traits.
4. As most recessives are detrimental, inbreeding causes a loss of fitness called "inbreeding depression." Conversely, crosses between some inbred lines in animals and plants show "hybrid vigor."
5. One can estimate, from the rate of increase of death and defects with increasing F, that each of us carries at least some recessive lethals in the heterozygous state and altogether probably more than two recessive detrimental genes.

Exercises

1. Why does mating between relatives tend to increase the frequency of homozygotes?

2. What are the usual phenotypic consequences of increased inbreeding?

3. Why are average inbreeding levels in most human populations quite low?

4. Why are consanguineous parents likely to be overrepresented among the parents of individuals with deleterious recessive traits?

5. How have attempts been made to use the overall effects of inbreeding to provide information on the distribution of the number of recessive (or deleterious) genes?

6. There is considerable social and geographic stratification of consanguineous matings. If you planned a survey of the effects of consanguinity on mortality, morbidity, or other physical characteristics that may be influenced by socioeconomic conditions, which of the four following possibilities would you prefer as nonconsanguineous controls and why: (a) matching for ABO blood groups; (b) matching by socioeconomic status within a fairly large area, such as California or England; (c) matching by geographic area, such as within county; (d) taking the sibs of one or the other of the consanguineous mates?

7. The average size of individuals with inbreeding coefficients close to 1 may be reduced, in some species, as much as 50 percent with respect to "outbreds." Assuming that this extreme value is also valid in man, and given that the reduction may be approximately proportional to the average inbreeding coefficient, what average reduction of general size (stature) would you expect (a) in the most highly inbred human populations? (b) in the progeny of first-cousin matings?

8. In a family of five children, two have shown an unusual sensitivity to a common drug such as aspirin, while the others can take the drug with no adverse effects. The parents are first cousins. What is a possible explanation for this observation?

9. Diabetes usually appears in older people, but it tends to be more serious when it appears in younger people. There may be a variety of genetic forms of diabetes, although this variety is not clearly documented. In a country in which the average frequency of first-cousin marriages is 1 percent, the parents of diabetics with the juvenile form are first cousins in about 5 percent of cases (a statistically significant difference). (a) What is a likely explanation for these observations? (b) Does this result prove that juvenile diabetes is inherited?

10. Only twelve married couples live on the hypothetical island of Santa Maria. Two of them are first cousins, one an uncle and his niece, three are half second cousins once removed, and two are third cousins. The others are unrelated. What is the average degree of inbreeding in this isolate?

11. Show that a mating between half sibs has an inbreeding coefficient $F = \frac{1}{8}$ using the method of Table 11.1. HINT: The parents of two half sibs should be represented as having genotypes *ab* and *cd* for one half sib and genotypes *cd* and *ef* for the other half sib (the common parent having genotype *cd*). Note that the first sib can be *ac, ad, bc,* or *bd,* and the second sib can be *ce, cf, de,* or *df.* Set up a table of matings between the two sibs and count homozygotes. These can be only *cc* or *dd,* because there is only one common ancestor, whose genotype is *cd.*

12. If a recessive gene (say for microcephaly) has a frequency $q = 0.001$, compute the frequency of first-cousin matings expected among the parents of microcephalics, given that in the general population the frequency of first-cousin matings is $\frac{2}{100}$. HINT: The frequency of microcephaly among the progeny of first-cousin matings is expected to be $q^2 + Fpq$, where $p = 1 - q$.

References

Cavalli-Sforza, L. L.
1969. "Genetic drift in an Italian population," *Scientific American*, vol. 221, no. 2, pp. 30–37.
Cavalli-Sforza, L. L., and W. F. Bodmer
1971. *The Genetics of Human Populations.* San Francisco: W. H. Freeman and Company. [Chapter 7 gives a more extensive treatment of the subject of this chapter.]
Morton, N. E., J. F. Crow, and H. J. Muller
1956. "An estimate of the mutational damage in man from data on consanguineous marriages," *Proceedings of the National Academy of Sciences*, vol. 421, pp. 855–863.
Muller, H. J.
1950. "Our load of mutations," *American Journal of Human Genetics*, vol. 2, no. 2, pp. 111–176.
Schull, W. J., and J. V. Neel
1965. *The Effects of Inbreeding on Japanese Children.* New York: Harper and Row.

12

Drift and Migration

Random events play a role in evolution; the smaller the size of the population involved, the greater their effect. Variation due to chance fluctuations is called genetic drift. This chance variation can account for part of the variation that is observed between populations. Subdivision into many small local populations creates greater opportunities for drift; intermigration, however, tends to cancel out differences thus arising. Migration, in addition to counterbalancing drift, may play an important role of its own in evolutionary events.

12.1 Some Examples of Drift

The Pingelap atoll in the Pacific Ocean (at latitude 6°N, longitude 160°E) was devastated by several typhoons. The famine that followed one typhoon (Lengyeky) around 1775 took very many lives, leaving perhaps 30 survivors, one of whom was the nanmwark (chief), Mwahuele. Today, the island has some 1,600 people. Among them, as many as 5 percent have turned out to be affected by a recessive disease, achromatopsia, a form of color blindness that is not X-linked and, in the case of the Pingelapese, is accompanied by severe short-sightedness and other ocular disturbances.

The local explanation for the disease is that the god Isoahpahu fell in love with several women of Pingelap. Disguising himself as the husband, and using other tricks, he fathered several affected children. Isoahpahu has good night vision (like cats) and, as the night vision of people affected with achromatopsia is less affected than is their day vision, this lends credence to the explanation.

The genetic interpretation obtained from reconstructing the genealogies is that at least one person in the past, namely nanmwark Mwahuele, was heterozygous for the disease. He had, as chiefs often do, several wives and children. At the time of the population-size bottleneck after the typhoon, he was probably the only heterozygous carrier among the 30 survivors. If this is true, there was only one copy of the gene out of 60, or a gene frequency of $\frac{1}{60} = 1.4$ percent. But today, with 5 percent homozygous recessives, the gene frequency is much higher, of the order of 23 percent. How could this happen? Mwahuele had a somewhat higher number of children than the rest, but this is probably not a sufficient explanation.

Figure 12.1
The Pingelap Islands in the Pacific Ocean, some 1,000 miles south of the Hawaiian Islands. Population is about 1,600 persons. (Aerial photo courtesy of N. E. Morton.)

Small, isolated populations frequently have exceptional genetic constitutions.

This example is far from being a unique occurrence. There are isolated alpine villages, for instance, in which there is a high frequency of albino individuals. This is a recessive trait, usually with a frequency of less than $\frac{1}{10,000}$ in the general population, but in a few of these villages its frequency may be as high as several percent. In other villages, the frequency of deaf-mutes, of blind people, of those with one or another type of mental deficiency, or of individuals affected by some other less deleterious trait may be remarkably high.

In general, the concentration of some peculiar genetic defect in a population is a characteristic typical of an "isolate"—a community that receives very few immigrants and is thus relatively separate genetically from the general population. The smaller the population and the longer it has been isolated, the more likely it is that one will find such anomalies. However, similar effects are also found in groups that today are relatively large—for example, in some religious "isolates" such as the Amish or Hutterites. These are rural communities following certain religious beliefs; they immigrated to the United States as small groups, and since then have grown considerably in population size without much interbreeding with other groups in the general United States population.

The phenomenon is more likely to be observed if a population has gone through a "size bottleneck"—a period during which population size was greatly reduced.

For many isolates, such a "bottleneck" occurred at the time that the group became genetically separated from the main population. The island of Tristan da Cunha in the South Atlantic was originally settled in 1817 by one Scotsman and his family. (Earlier settlers and military men stationed there left the island in that year.) Over the following decades, the original family was joined by shipwrecked sailors, some women from St. Helena island, and a few European settlers. However, the population of islanders has remained relatively isolated from genetic exchange with other populations. In 1961, when a volcanic eruption forced evacuation of the island, there were 264 native islanders (Fig. 12.2). After two years in England, 248 islanders returned to Tristan da Cunha, and the population has grown slightly since then.

It seems intuitively clear that the gene distribution in such an isolate will reflect the distribution of genes that happened to be present among the small group of founders. Geneticists often discuss this "founder effect," but it is important to remember that the unusual distribution is not only established at the first generation; it is renewed in each generation. Still, it is useful to focus on the founder effect, as it may sharpen our understanding of the isolate phenomenon.

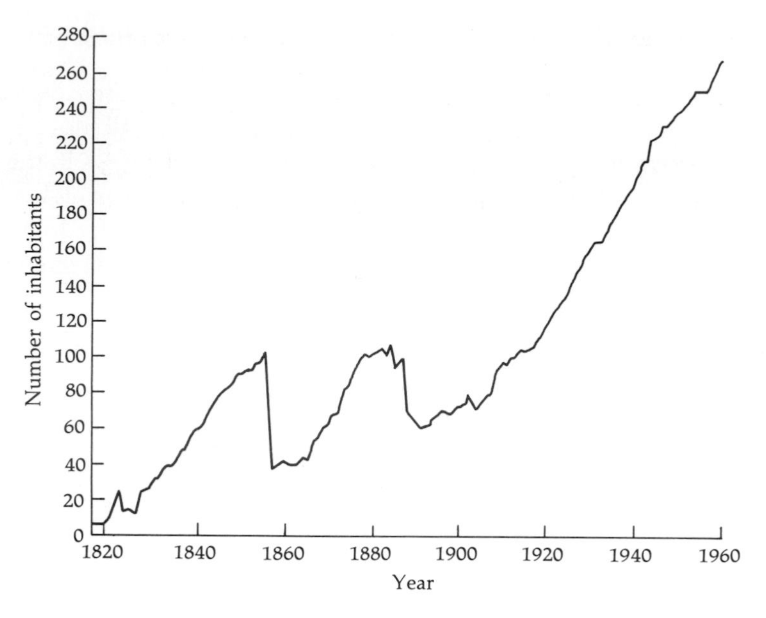

Figure 12.2
Tristan da Cunha population growth. The island was originally settled by about two dozen sailors, fishermen, and women early in the nineteenth century. There have been relatively few immigrants from outside populations since then. (Military men and other outsiders who live on the island for short periods of time rarely interbreed with the population of islanders.) The two population-size "bottlenecks" in the second half of the nineteenth century were associated with major disasters. A volcanic eruption forced evacuation of the island in 1961, but the population survived as an isolate in England and was returned to the island in 1963. (From D. F. Roberts, *Nature,* vol. 220, p. 1084, 1968.)

In the case of Tristan da Cunha, the current population of islanders can all trace their ancestry to about two dozen individuals who came to the island in the early nineteenth century. In other cases, the number of founders may have been even smaller. This was perhaps true of the Pitcairn Island population, which was started by the mutineers of the *Bounty* and also prospered.

It is useful to think of the founder generation as a small sample selected at random from some much larger population. Suppose, for example, that there are five individuals in the founder group. If we select five individuals at random from a European population, we can expect that about one in 100 such randomly selected groups will happen to consist of five individuals all of blood type O. This is a low probability, but it is certainly not inconceivable that a founder group of five individuals might all happen to be of blood type O (genotype *OO*). In such a case, all of their children must also be of type O. The alleles *A* and *B* have disappeared from this population. We say that allele *O* has been fixed and that alleles *A* and *B* are extinct. Of course, it is also possible that the founder group might happen to carry only allele *A* (in which case *A* is fixed and *O* and *B* are extinct), and so on through other possibilities.

Thus, the chance event that determined the composition of the genotypes of the founders has inevitable consequences for all successive generations. If allele *O* is fixed in the founder generation, the other alleles *A* or *B* can reappear only through mutation or through immigration by carriers of these alleles from outside populations. Barring mutation or immigration, the fixation and extinction are irreversible. However, it is important to note that this remarkable effect is not limited to the original founder generation.

Each generation is the founder of all successive generations.

Peculiar events in any one generation have consequences for all future ones. This fact may be appreciated more clearly through a further "thought experiment." Suppose that four of the five founders of an isolated population are of blood type O, and the fifth is an *AO* heterozygote. In this population, *B* is extinct but *A* is not—although it is present at a lower frequency than in other populations. Neither *A* nor *O* is fixed. Now consider the second generation, made up of the children of the founders. (In most human populations, generations are not too clearly separated. Marriages may occur between people of different ages, and in practice generations "overlap." There is no clearcut separation of generations as there is, for example, in annual plants that are born, reproduce, and die each year. However, for our purposes, it is convenient to imagine that generations are separate.) The single founder carrying an *A* gene might happen not to produce any offspring. Even if he or she does reproduce, the mate must be an *OO* homozygote, so each offspring has only a $\frac{1}{2}$ chance of receiving an *A* gene. It would not be too unlikely for all children of the *AO* founder to happen to receive only *O* alleles. Thus, the fixation of *A* that did not occur in the founders' generation may well occur in the second or a later generation. Whenever it happens (if it does), the *O* allele becomes fixed, the other alleles become extinct, and polymorphism at this locus disappears for all future generations—except for new mutation or immigration.

Figure 12.3 shows how a population of five individuals (ten genes) that starts with frequencies of 50 percent for one allele and 50 percent for another could, by chance alone, lose one of the two alleles in the course of a few generations. Note that the frequency of the allele that will eventually be fixed does not necessarily increase all the time (Figure 12.4). Note also that, for simplicity, we have not represented the mating phase in the reproduction of individuals, but we have replaced the population directly by the pool of gametes produced by all individuals. We know that random mating of gametes gives the same result as random mating of pairs of individuals, so this simplification is justified.

We know that the outcome of chance events can be anticipated on a probability basis, and we shall now see what useful predictions can be made.

12.2 Simple Expectations of Genetic Drift

The events that are depicted in the example of Figure 12.3 can be anticipated by probability calculations. To avoid using mathematics at this point, we resort to chance experiments, of which coin tossing is perhaps the simplest example.

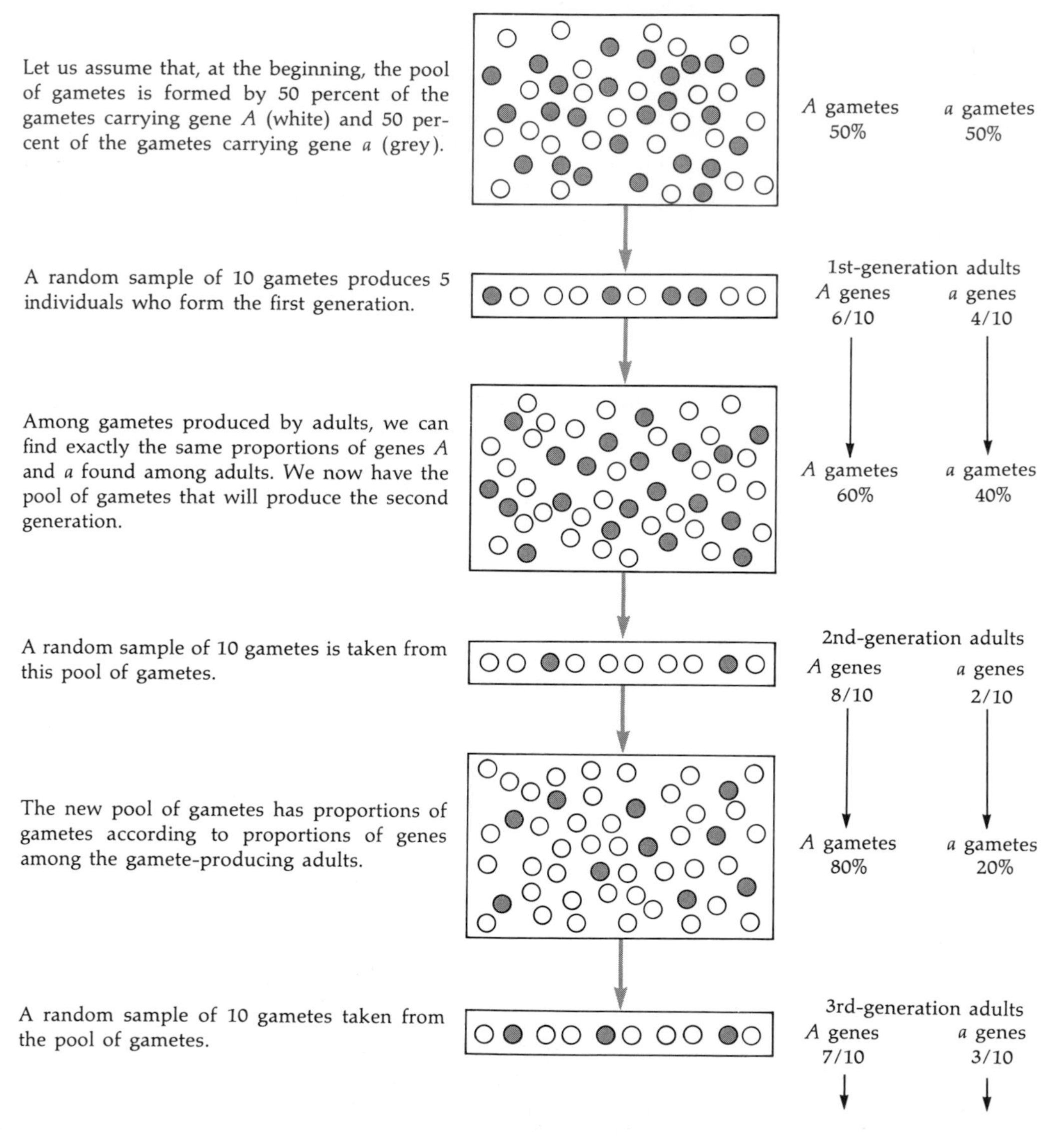

Figure 12.3
Chance extinction of an allele in a small population.

Chance phenomena are easily simulated.

We can easily imitate with coin tossing the first generation of Figure 12.3: take ten coins, toss them, and record the number of heads and tails. Suppose we consider heads as the shaded allele in Figure 12.3, and tails as the white allele. Then a

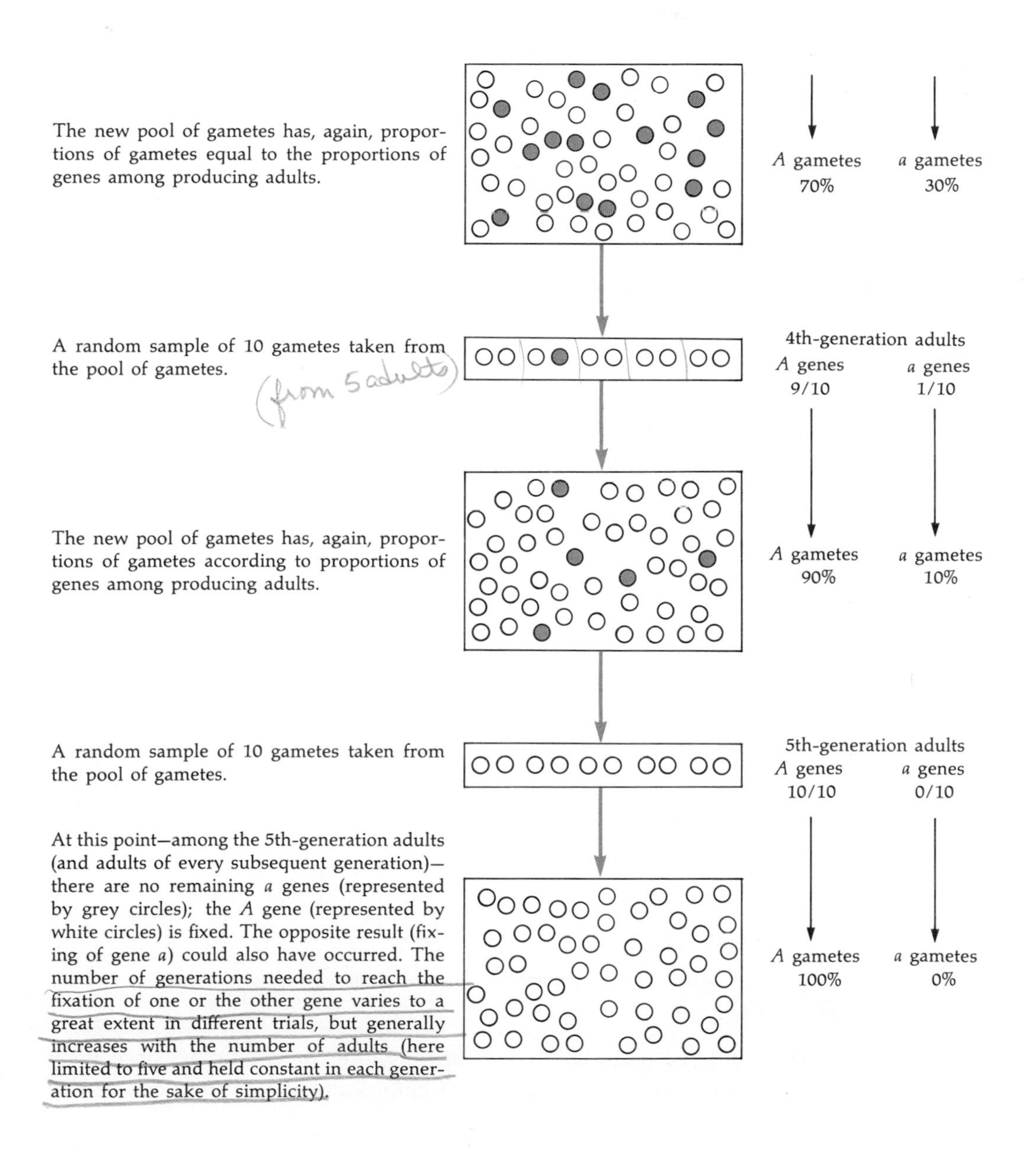

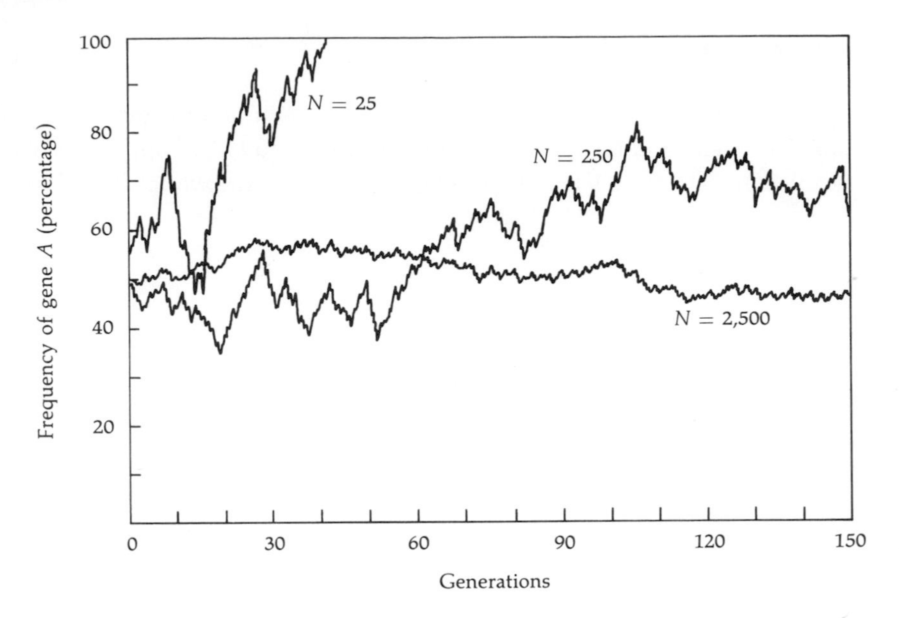

Figure 12.4
Chance and population size. The graph shows the results of computer experiments to simulate chance effects in three populations of different size, each starting with a gene frequency of 50 percent. The smallest population ($N = 25$ individuals) shows fixation of allele A after 42 generations. The medium-size population ($N = 250$) shows less important fluctuations, and has not reached fixation of either allele after 150 generations. Note that the frequency of allele A did become greater than 80 percent shortly before generation 110, but that chance events in succeeding generations happened to carry the frequency back toward 50 percent. In the largest population ($N = 2,500$), fluctuations are quite small, and fixation is very unlikely in any particular generation. However, fixation will eventually occur if the experiment is continued long enough.

result such as that in Figure 12.5—namely, four heads, six tails—would correspond to having, in the next generation, four shaded and six white alleles. In every such trial, the result is likely to be different. There are regularities, nevertheless, that will be apparent by repeating the tosses. Figure 12.5 shows the results of doing the experiment 1,000 times, throwing ten coins each time. The most frequent outcome was the even one, of five heads and five tails; but all other possible outcomes, from 0 heads 10 tails to 10 heads and 0 tails may be observed. The farther the outcome is from the 5:5 result, the less frequently it occurs.

A particular deviation from expectation is less likely, the greater the deviation and the larger the sample.

This result could be predicted mathematically (see Appendix), but once again we shall make use of an experiment to test empirically another property of this

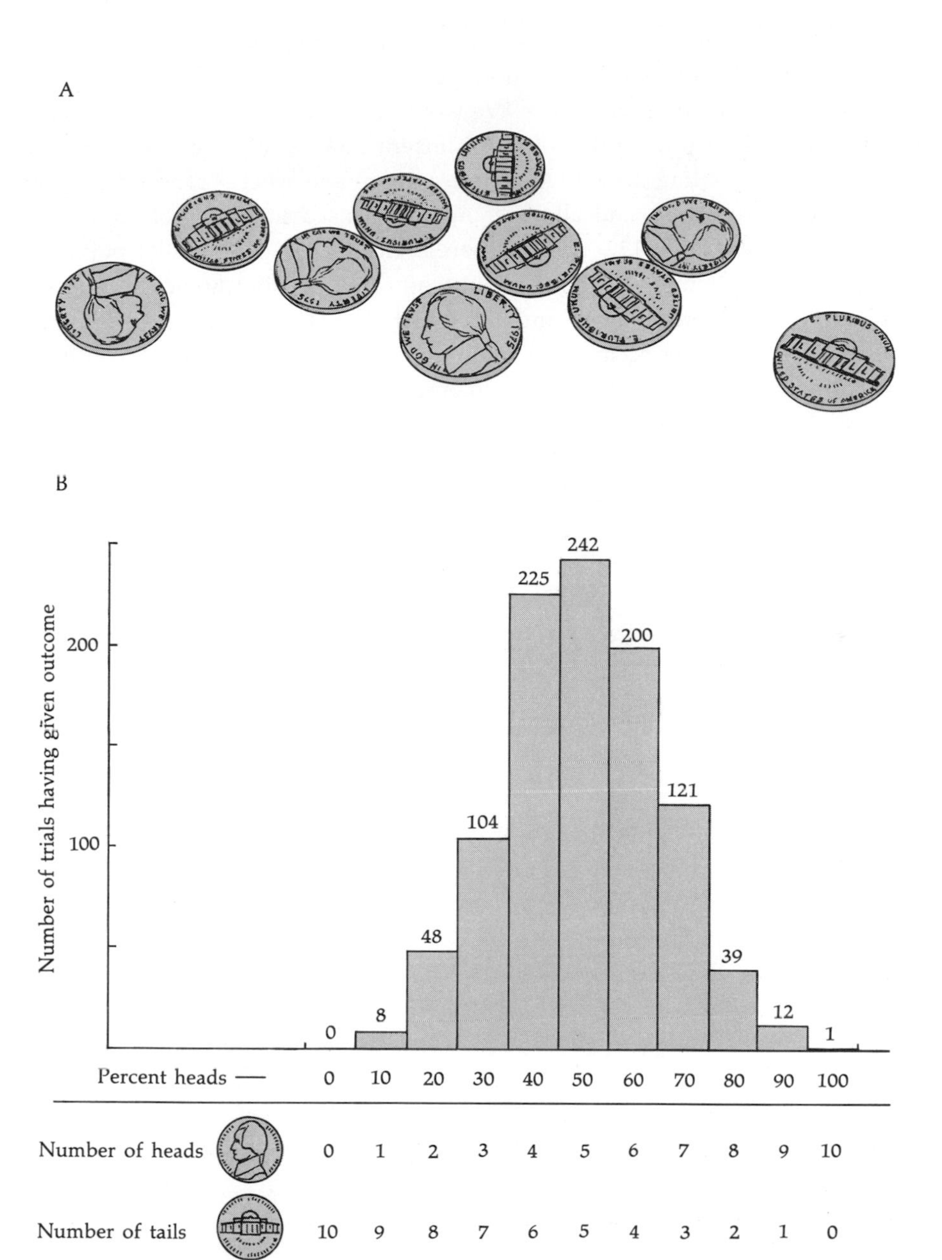

Figure 12.5
A trial consisting of a single throw of ten coins. (A) The first trial gives an outcome of four heads and six tails. (B) After 1,000 trials, it is clear that the most frequent result for a trial is five heads and five tails, but every possible result except one was obtained on at least one trial.

game of chance. Suppose we have a larger population, say of 100 individuals instead of 10. We now have many more possible outcomes, from 0/100 to 100/0, but the general picture is the same. We would have to toss 100 coins (and repeat that whole work of tossing them for a sufficient number of times). We can avoid the effort by simulating the coin tossing in a computer, where "random numbers" replace coins, roulettes, and all other machines used in games of chance. The result is given in Figure 12.6, and compared with that for the smaller population. Again, the odds are in favor of an even outcome (50:50); now, however, deviations from 50:50 are less pronounced, and 0/100 and 100/0 (fixation of one, with extinction of the other) never occurred in just one generation. Only rarely does one find a deviation from the most frequent outcome as large as ±10 percent—that is, an outcome less than 40 percent, or larger than 60 percent.

If we wish to use similar "coin-tossing" experiments to study the second and later generations in Figure 12.3, we must modify the design of our experiment. In generation 2, the gene frequency is 40 percent, not 50 percent. To simulate the results of gamete formation in this case, we need a coin that falls on one face 40 percent of the time and on the other face 60 percent of the time. Obviously it is not easy to design such a coin, but we can resort to the computer to simulate

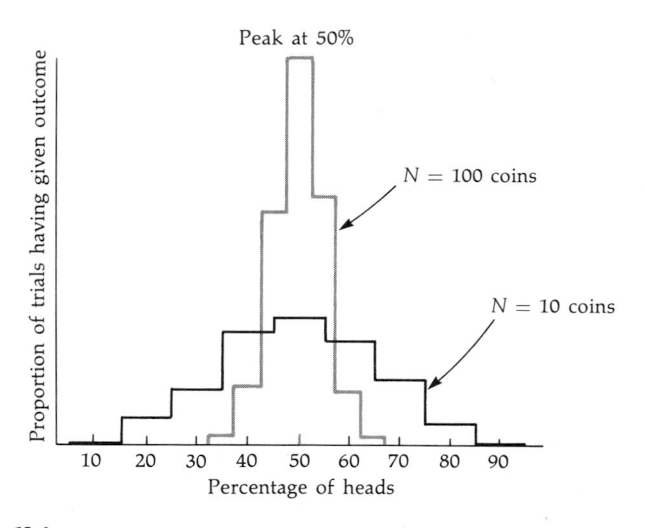

Figure 12.6
Experiments in coin tossing. This graph repeats the results of the experiment shown in Figure 12.5 ($N = 10$ coins tossed at each trial), and shows for comparison a computer simulation of a similar experiment with $N = 100$ coins tossed at each trial. Note that a larger sample on each trial (equivalent to a larger population at each generation for a genetics experiment) yields smaller deviations from the most frequent outcome. This graph shows the relative proportion of trials resulting in each possible outcome, with the outcome of a trial expressed as the percentage of coins showing heads. Each experiment involved 1,000 trials.

such a badly biased coin. We can write a computer program that will make a semi-random selection of head or tail on each "toss," but in such a way that the overall average will be 40 percent heads. Figure 12.7 shows the results of such a simulation. Note that the most frequent outcome is indeed 40 percent heads, and that again the larger population (larger sample on each trial) yields a smaller spread of outcomes. In general, regardless of population size, the most likely outcome of a generation's reproduction is a generation of offspring with the same gene frequency as that in the parental generation. As population size increases, the chances of a large deviation from this most likely outcome become much smaller.

We can summarize as follows some predictions about the effects of chance over a single generation.

1. The most frequent (most likely) event is the production of a generation of offspring with a gene frequency close to that of the parental generation. More precisely, the expected gene frequency in the offspring generation is equal to that in the parental generation.
2. Deviations from the expected frequency can occur in either plus or minus directions, with equal probability.

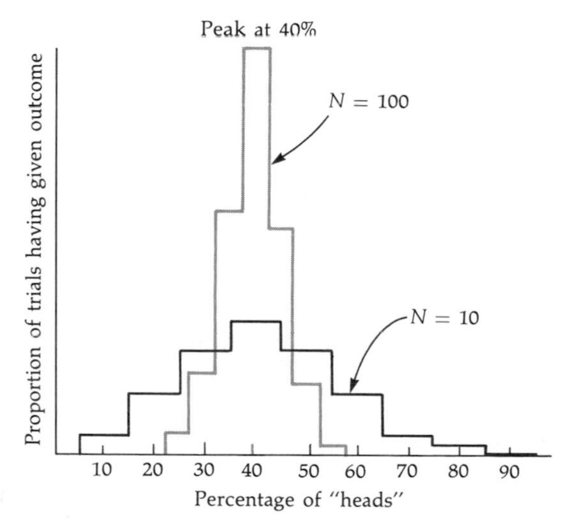

Figure 12.7
Experiments in tossing "biased" coins. This graph shows the results of computer simulations of two coin-tossing experiments, each using a "biased coin" that gives heads only 40 percent of the time. In one experiment, the sample population was $N = 10$ "coins tossed" at each trial; in the other experiment, $N = 100$. In each experiment, 1,000 trials were made. In the smaller population, the spread of outcomes is greater than in the larger population. In both experiments, the outcomes cluster around the expected proportion of 40 percent heads.

3. On the average, a larger population will experience smaller deviations from the parental gene frequencies.

The preceding statements summarize the effects of chance over one generation. But each generation acts as "founder" for the following one, and thus the effects of chance accumulate with the passing generations. An event that is very unlikely in a single generation becomes quite likely to occur at least once over a great many generations. (Similarly, it is very unlikely that a bridge game dealt from a randomly shuffled pack will give each player all the cards of a single suit. However, considering the number of bridge games dealt in the world each year, it is quite likely that such a game will be dealt somewhere at least once in every few years. The fact that it occurs in a particular game is surprising; the fact that it occurs somewhere is not surprising at all.)

Suppose that we repeat the experiment of Figure 12.3 a number of times. We find that, if we continue for a sufficient number of generations, the experiment always ends in the fixation of one or the other allele (Fig. 12.8). Although it is unlikely that one of the alleles will become fixed over a single generation, it is very likely that one will become fixed over 25 generations. Here then are some other important predictions dealing with the effects of chance over many generations.

4. The accumulation of chance events will always lead, in the long run, to the fixation of one allele and the extinction of all other alleles. (Remember that we are assuming the absence of any effects such as selection, mutation, or immigration that might counteract the effects of chance.)
5. The probability that a particular allele will become fixed (rather than extinct) depends on its initial frequency. In Figure 12.8, we began with 50 *A* alleles and 50 *a* alleles, and we find in the end that *A* is fixed about half the time and *a* about half the time. Had we started with different frequencies—say, 10 percent *A* and 90 percent *a* alleles—we would expect corresponding differences in the proportions of fixation—in this case, *A* being fixed about 10 percent of the time and *a* about 90 percent of the time.
6. The time needed for fixation of an allele varies from one experiment to another, but the average time is a function of population size. The larger the population size, the greater the average number of generations needed to reach fixation (Fig. 12.9).

The process of chance fluctuations of gene frequencies is called random genetic drift *(often abbreviated simply as "drift"). The smaller a population, the more significant the effects of drift over a finite number of generations.*

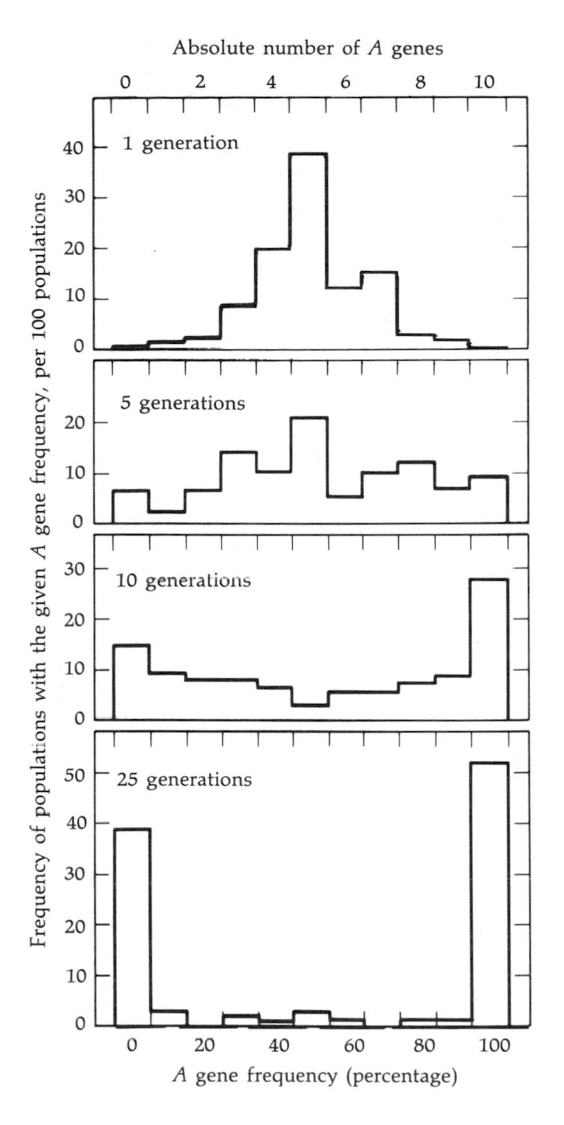

Figure 12.8
Time needed to reach fixation. The graphs show the results of 100 repetitions of the experiment carried out once in Figure 12.3. In each experiment, there was a population of 10 genes (that is, five diploid individuals), and an initial gene frequency of 50 percent gene *A*. The upper graph shows the distribution of results after a single generation in each experiment—for example, slightly less than 40 of the 100 experiments showed a frequency of 5 (50 percent) *A* genes after the first generation. Results are also shown after 5, 10, and 25 generations in each experiment. Note that fixation of one allele or the other has been reached in about 90 of the 100 experiments after 25 generations.

The term drift is to some extent misleading, for it may suggest a direction (as in everyday use of the word to describe the path of an object positively transported, say by currents). In fact, the effect of drift has no direction whatsoever, and the population will finally fix on a randomly chosen allele. Drift is thus one cause of variation of gene frequencies over time (or over space—that is, over different populations studied at the same time). It is clearly not the only force affecting gene frequencies; we have already seen the effects of mutation and selection.

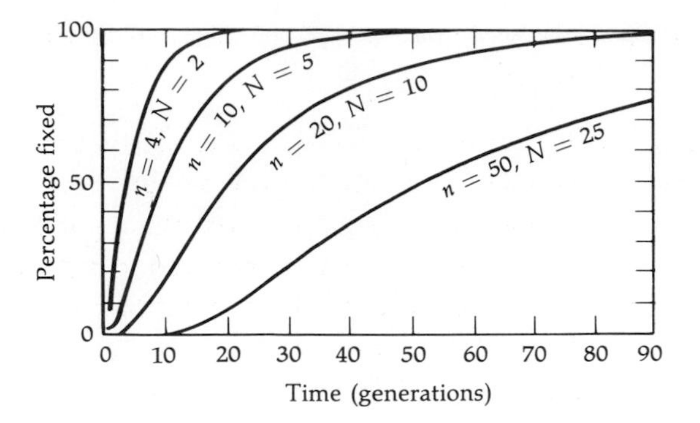

Figure 12.9
Population size and time needed to reach fixation. This graph summarizes the results of a very large number of experiments similar to those described in Figure 12.8, using populations of different sizes. A large number of experiments were carried out at each of four population sizes (*n* is the number of genes, *N* the number of diploid individuals in the population). The curves show the proportion of experiments in which fixation had been reached as a function of the number of generations passed. All experiments began with a gene frequency of 50 percent. Note that the median fixation time (the time required for fixation to occur in half of the experiments) increases with increasing population size.

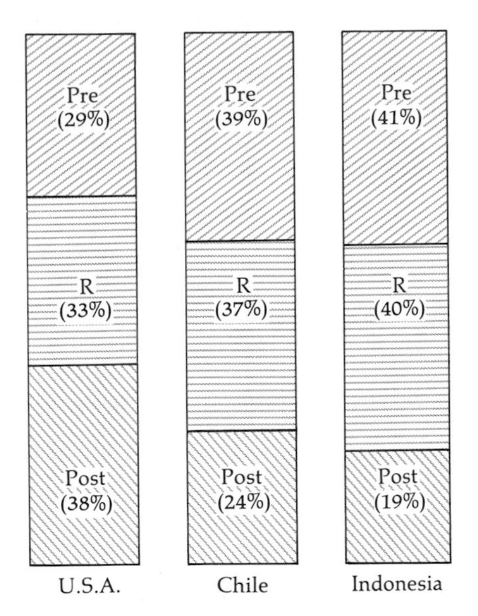

Figure 12.10
Effective population size. The charts show the proportion of the female population that is in prereproductive (Pre), reproductive (R), and postreproductive (Post) ages in three different countries. (The limits of the reproductive period are here taken as 15 to 40 years of age.) Only individuals in the reproductive ages contribute to the "effective population size," which is what matters from the point of view of genetic drift. This group makes up approximately one-third of the total population. (The group in the reproductive period tends to make up somewhat more than one-third of the total population in less developed countries.) A more precise estimate of effective population size can be made by taking account of the actual contribution each age group makes to reproduction, using age-specific birth rates.

Population size is the major determinant of drift effects. But the size that matters for drift effects is less than the census size.

Population size, by which is usually meant the total number of individuals of all ages in a population, is generally available from census-type records. However, the relevant number of people in a population from the point of view of genetic drift is only that part of the population that is involved in reproduction, because individuals in the pre- and postreproductive age groups are not contributing genes to the immediately next generation.

The relevant population size for drift purposes is often called the "effective" population size, N_e. Roughly speaking, a population can be divided into three broad age groups: prereproductive, reproductive, and postreproductive. The age distributions of many human populations are now such that approximately one-third of the population is in the reproductive age group (Fig. 12.10). This means that a rough approximation to the effective population size is provided by taking one-third of the census population size. Many more sophisticated methods have been devised for this estimation. It is obviously important to make appropriate allowances for the difference between the census and effective population sizes in order to obtain a proper assessment of the effect of population size on genetic drift.

12.3 Migration and Population Structure

Migration is an important factor in the history and geography of a species. With the discovery of the Americas, an enormous flow of people, almost all from over-crowded Europe, came to the new continents. In a few hundred years, the ethnic picture of the Americas was radically changed. This process has happened a great number of times in the history of man, though of course, more slowly when means of transportation were less efficient. But even in very early times, long-range migrations must have occurred, as when America was occupied by people from Northeast Asia, probably crossing a then-existing land bridge across the Bering Strait. These long-range migrations have populated and repopulated the earth.

The outcome of many such migrations must have been that of forming splinter groups that occupied new areas.

These regions were perhaps previously uninhabited or were inhabited more sparsely, for instance, by people having a less efficient technology of food production. Many such events must have occurred under some pressure—population growth or natural events causing food shortage, or other sources of stress. The flow of people from the old place to a new area may have continued for some time. Then the flow may have ceased, especially if events made further migration difficult or undesirable. If the new area was large enough, the displacement in it may have continued over the generations, bringing remote descendants of the original emigrants very far from the place where the latter were born. By such mechanisms, splinter groups are established; processes of *fission* take place. Populations thus separated will evolve independently and differentiate one from the other.

But migration can also be responsible for the opposite phenomenon—namely, increases in the homogeneity of populations. There is a great deal of short-range migration, or migration without definite or long-lasting trends, which causes a

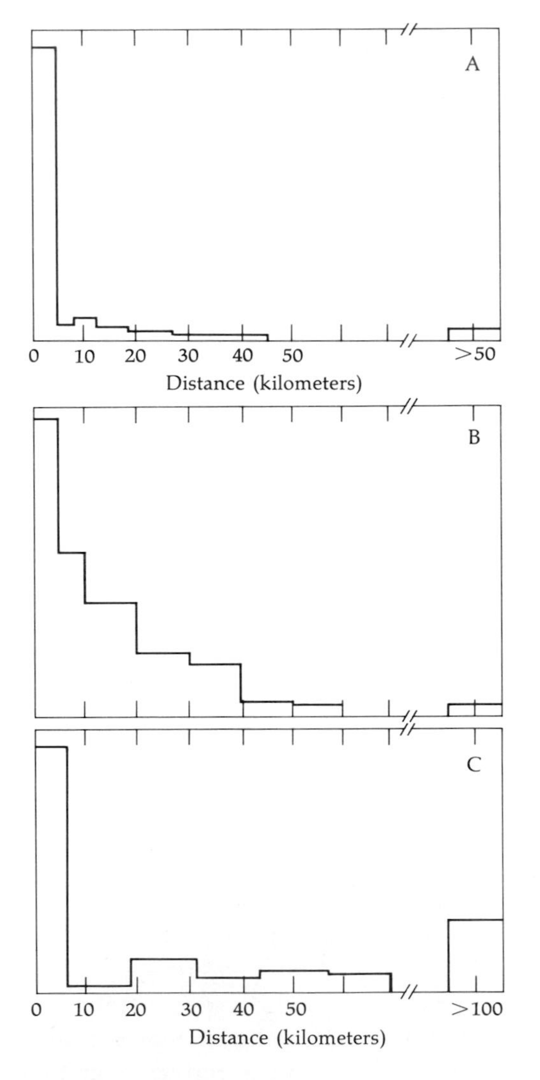

Figure 12.11
Distances between birthplaces of husbands and wives. The three graphs show the distributions of distances between birthplaces of husbands and wives in samples taken from three populations. (A) The upper Parma Valley of northern Italy, where the population density is 50 persons per square kilometer, endogamy (the frequency of marriages in which mates are from the same village) is 55 percent, and the average village size is 200 persons. (B) An African rural population (the Issongos of the Central African Republic), with a population density of 1 to 2 persons per square kilometer, endogamy of 20 percent, and average village size of 100 persons. (C) African pygmies (Babingas of the Central African Republic), with population density of 0.2 person per square kilometer, required exogamy (marriage outside the camp unit), and an average camp size of 25 persons.

continuous internal flow of people. In a country like the United States, very few people live most of their life where they were born. Even in more conservative societies such as rural ones, where most people spend most of their lives in their birthplace, some migration inevitably takes place—some of it, for instance, at the time of marriage. When the bride is born in a village different from that of the groom, one of the two must move away from his or her birth place. Similarly, even in the most isolated groups, some gene exchange goes on (Figs. 12.11 and 12.12).

An effect of this "creeping" type of migration is the partial cancellation of the influence of drift.

Two or more populations that are fully isolated—that is, have zero migratory exchange—evolve independently and can freely differentiate one from the other. Those that are of small size will be particularly subject to the effect of drift, and will differentiate more, the smaller they are and the longer the time of isolation. But intermigration will tend to cancel differences that would otherwise arise. The more intensive the migration, the less will be the differences due to drift (or other causes of differentiation). With time, a balance will be reached between drift and

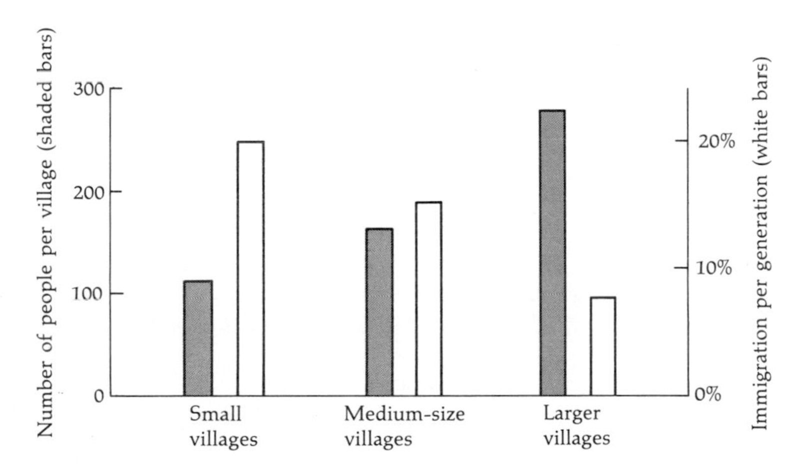

Figure 12.12
Village sizes and intermigration. The graph shows an example from the upper Parma Valley of northern Italy, a mountainous region where villages have been and still are of small size. Villages are grouped in three size classes. The average population size for each size class is given (based on actual census data), and the average percentage of the population made up of persons who have immigrated to the village during their lifetimes. Most of these immigrants are women who were born in nearby villages and married residents. (Data from M. Skolnick et al., unpublished.)

migration, differences between populations being (on average) greater for smaller populations or for smaller amounts of intermigration. With adequate demographic information on population sizes and migration patterns, we can use mathematical models or computer simulations (Fig. 12.13) to predict what genetic differences to expect (on average) between real populations (such as those between villages). We can evaluate the actual genetic differences between villages by studying the gene frequencies of polymorphic genes in the different villages. If drift is the only source of genetic variation among the villages (even though counterbalanced by migration), we should find that there is a close correspondence between the genetic differences predicted (on the basis of demographic data such as population size and intermigration) and the genetic differences observed (by study of genetic markers). Figure 12.14 shows the results of one such study.

Drift can have extreme consequences (such as fixation of an allele) only if the groups are very small or if the times over which differentiation has proceeded are very long. If migration is substantial and village sizes are not very small, genetic variation between villages is too small to be detected. However, in relatively rugged mountains away from cities (and, in general, where population density is very low), villages tend to be small—perhaps each containing just a few hundred persons. Under these conditions, we do find some perceptible variation in gene frequencies, even between neighboring villages (Fig. 12.15). The curve in Figure 12.15 shows the expected genetic variation, predicted on the basis of demographic data from the populations of some 20 villages. The plotted points show actual measurements of genetic variation made in the corresponding real villages. The observed variation found here (and in other similar studies) is well within the range expected on the basis of the demographic study and simulation, thus suggesting that the balance of drift and migration may indeed be responsible for the observed differences between villages.

Drift may thus be a cause of intergeographic differentiation; its effects may be expected to be prominent in areas of low population density and low rates of intermigration.

We may also reason that (on the average) geographically closer populations should be more similar genetically than are more distant ones. Observations confirm this expectation (Fig. 12.16).

However, if we extend our consideration to a wider area, it becomes more likely that other complications will have to be taken into account. Thus, selective conditions may change from one area to another. Then we may find that the frequencies of certain genes differ in the two areas because of adaptations to local conditions (differing patterns of selective advantage for the alleles). Rarely, but occasionally, this can happen over short distances, if the environmental conditions leading to variations also change over short distances. That very strong selective

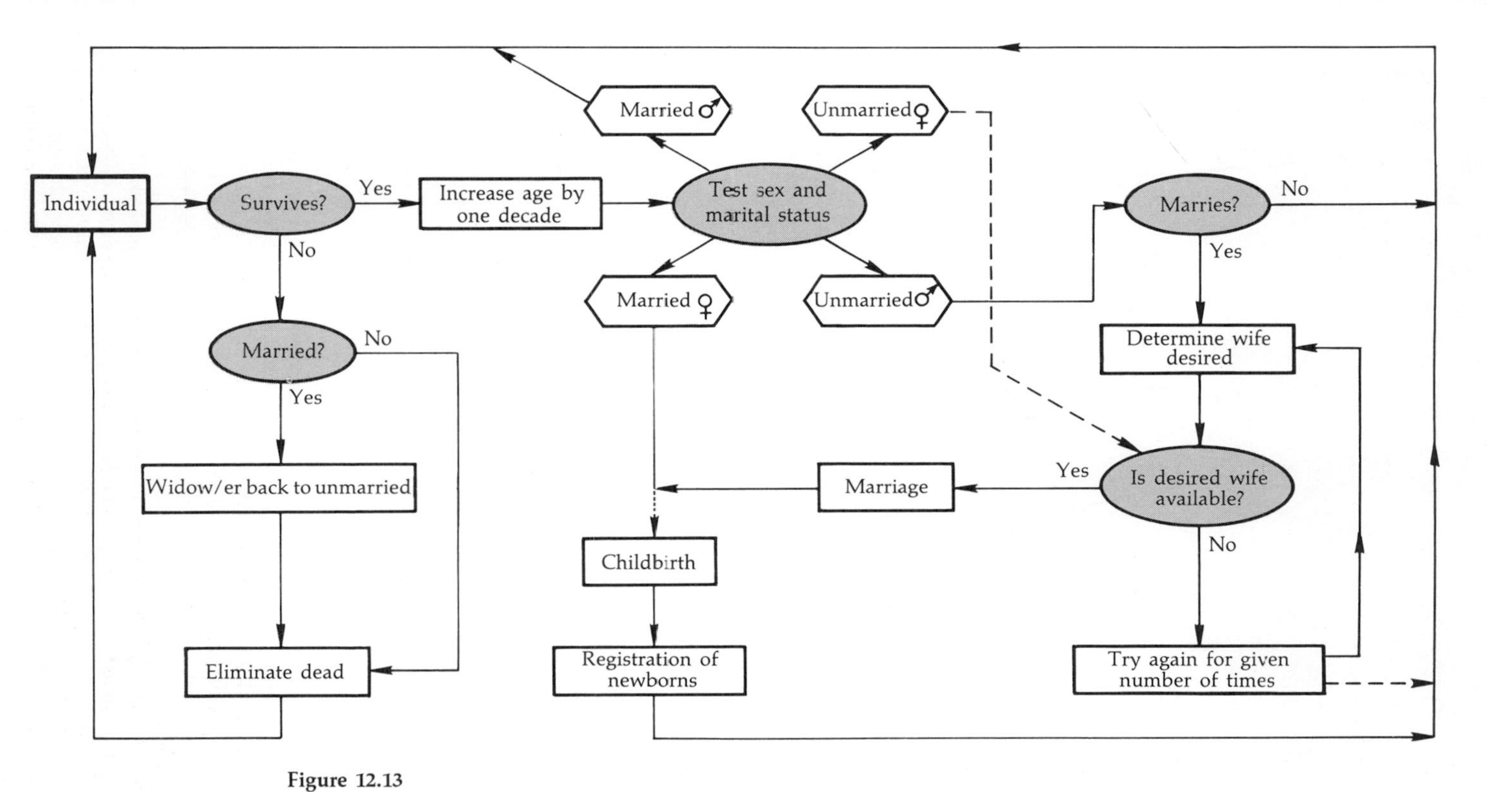

Figure 12.13
Flowchart of artificial population used to simulate effects of inbreeding and drift. Dashed arrows indicate portions of the chart that have been simplified. In the computer simulation based on this flowchart, questions are answered on the basis of probability tables that are constructed to correspond to the distributions of age at marriage, death, and so forth in some real population under study or in a hypothetical population. (From L. L. Cavalli-Sforza and G. Zei, *Proceedings of the Third International Congress on Human Genetics,* Johns Hopkins Press, pp. 473–478, 1967.)

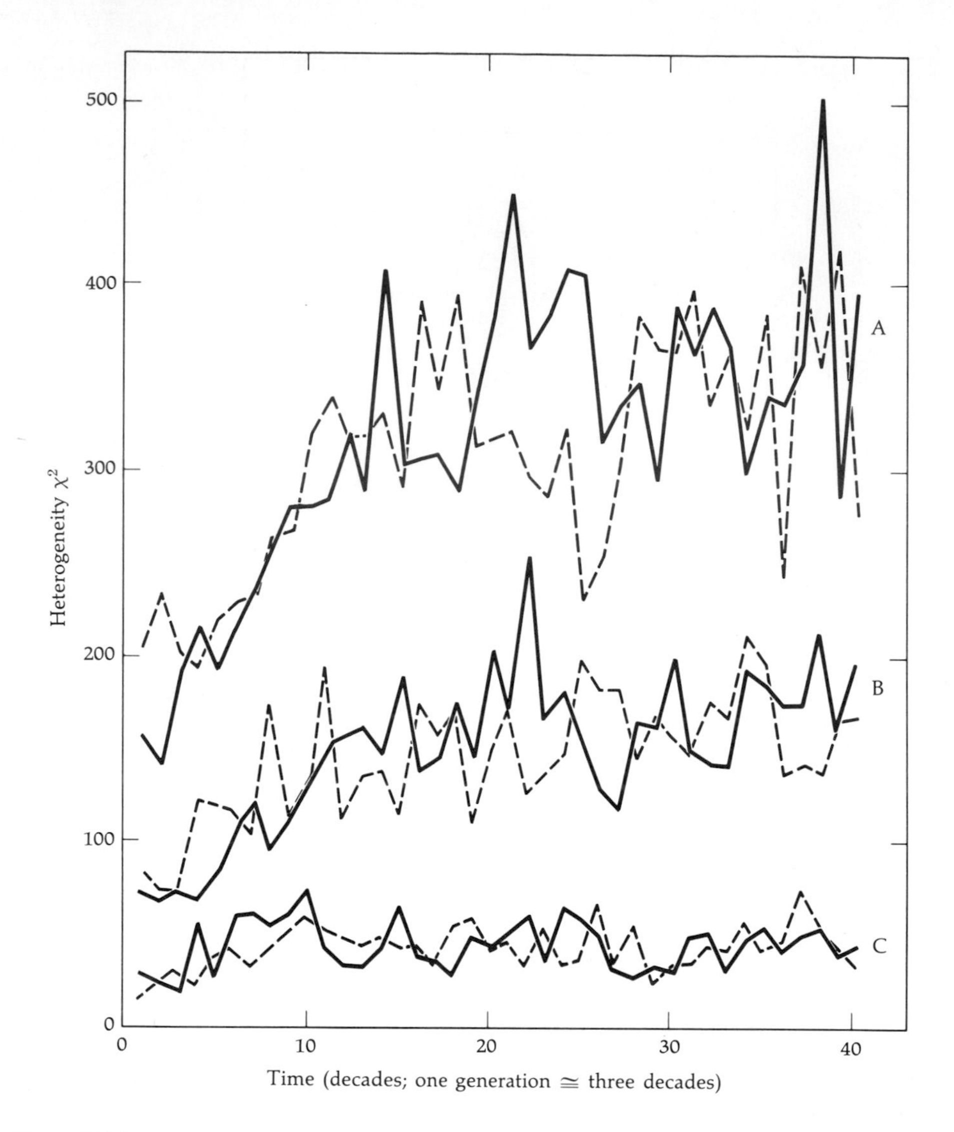

Figure 12.14
Computer simulation of drift. The flowchart shown in Figure 12.13 summarizes part of the computer program used to obtain these results. This simulation attempts to evaluate the effects of drift in 22 villages of the upper Parma Valley, using population sizes and intermigration patterns based on those observed in reality. The heterogeneity (χ^2) is a measure of the genetic differences between villages. In this simulation, the gene frequencies of three blood groups were assumed to be initially homogeneous for all the villages. (Assumption of high initial heterogeneity leads to the same results for final heterogeneity.) The dashed and solid lines show the results of duplicate experiments for each blood group. (A) The Rh system, with seven alleles considered. (B) The ABO system, with four alleles considered. (C) The MN system, with two alleles considered. The final heterogeneity obtained in these simulations compares favorably with that observed in the actual populations, showing that drift and intermigration are the major forces determining the level of genetic variation between villages. (From L. L. Cavalli-Sforza, "Human populations," in *Heritage from Mendel,* ed. A. Brink, Univ. of Wisconsin Press, 1967.)

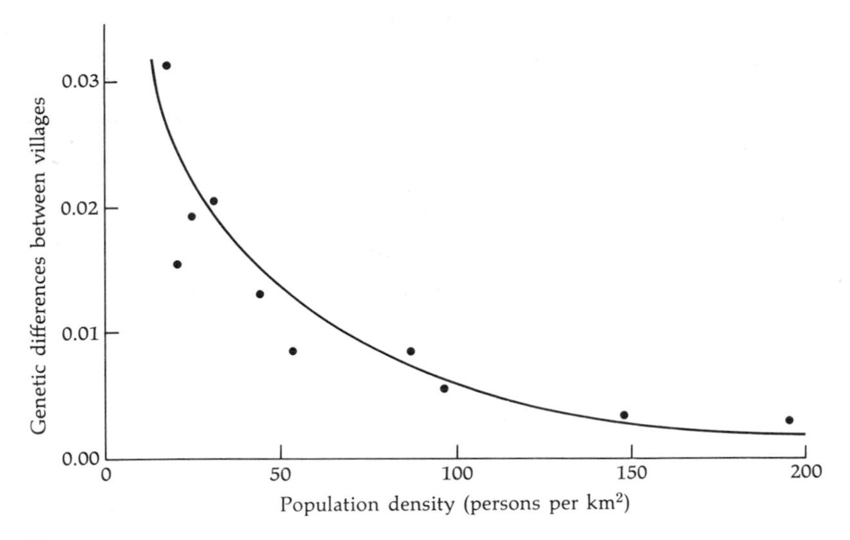

Figure 12.15
Population density and genetic variation. This graph summarizes the results of a study in which the Italian province of Parma was divided into subareas, and the genetic variation between villages measured in each subarea. The genetic variation between villages is plotted as a function of the average population density of the subarea. (From L. L. Cavalli-Sforza, "Genetic drift in an Italian population," *Scientific American,* August 1967. Copyright © 1967 by Scientific American, Inc. All rights reserved.)

agent, malaria, provides particularly striking examples of such short-range selective variation (Fig. 12.17).

Other complicating factors are simply consequences of the history of a particular population. Suppose that a group immigrates into a region and tends to settle in a particular area of the region. If the new group has different gene frequencies (at least for some loci) from those of the older population in the region, a gradient of gene frequencies may ensue (Fig. 12.18). Such a gradient is often called a *cline.* The effects of migration may eventually cancel out a cline, but the "evening-out" may take some time, especially if the initial difference was relatively large.

An observed cline is difficult to interpret unambiguously, however. Mongolian invasions from the East in the twelfth century may be responsible for the observed cline of the *B* gene (Fig. 12.18), given that this allele has an especially high frequency in the East. But historical data are, here as elsewhere, usually insufficient to evaluate the plausibility of explanations based on migrations. Alternative hypotheses—for example, that the cline is or was maintained on the basis of local selective differences (as is probably true of G6PD in Fig. 12.17)—are not so easily excluded.

The study of surnames provides an alternative approach to the study of populations—an approach that is less demanding than the analysis of genetic markers.

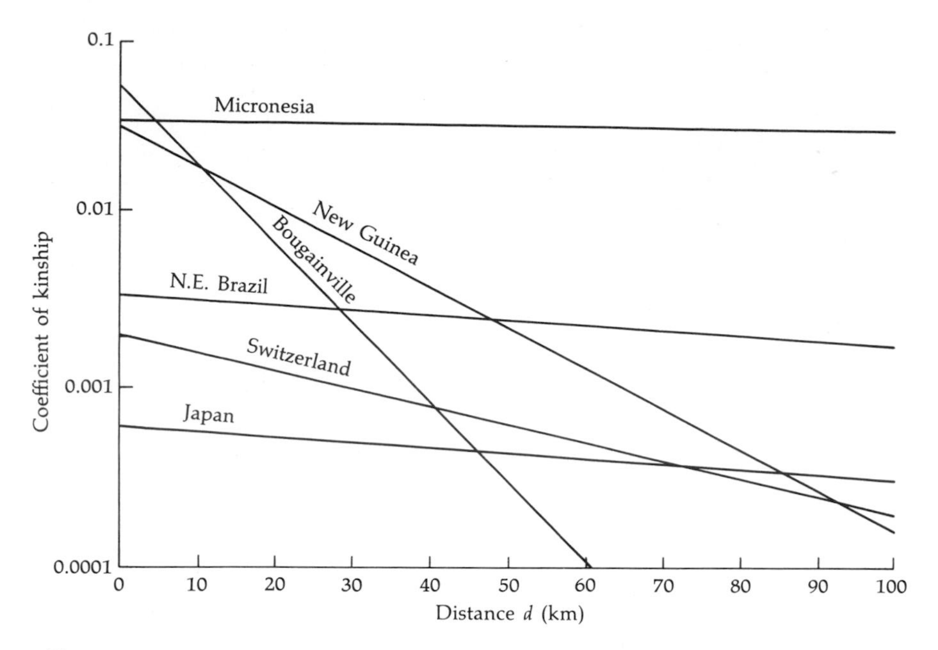

Figure 12.16
Genetic relationship and geographic distance. Each of the straight lines is interpolated from "kinship" values between pairs of villages (or higher administrative divisions) for a particular population. The kinship coefficient is a measure of genetic relationship (computed on the basis of gene frequencies) and is closely connected with the inbreeding coefficient F discussed in Chapter 11. (From J. S. Friedlaender, *Proceedings of the National Academy of Sciences,* vol. 68, pp. 704–707, 1971.)

In most societies, surnames are transmitted through the male line, and thus are inherited in much the same way as Y chromosomes (ignoring the fact that women also have surnames but no Y chromosomes). Under the conditions that lead to a high level of genetic drift, many surnames will be eliminated and only a few will remain. Thus, in the smaller and more isolated parishes of the Parma Mountains (Fig. 12.15), there is an average of only 12 surnames for a village of 200 to 300 people. In some villages, more than 50 percent of the inhabitants have the same surname. More extreme cases have been reported, in which only one surname is found in a village.

Counting the number of surnames in a given place can be a quick way to obtain information on the major quantities that determine drift—namely, population size and migration. We have already mentioned that surnames can be considered as Y-linked; more precisely, a surname can be considered formally as a Y-linked gene with many alleles (each different surname corresponding to an allele).

This approach has one important limitation, however. Because the transmission of surnames is limited to the male line, the information given by surnames reflects

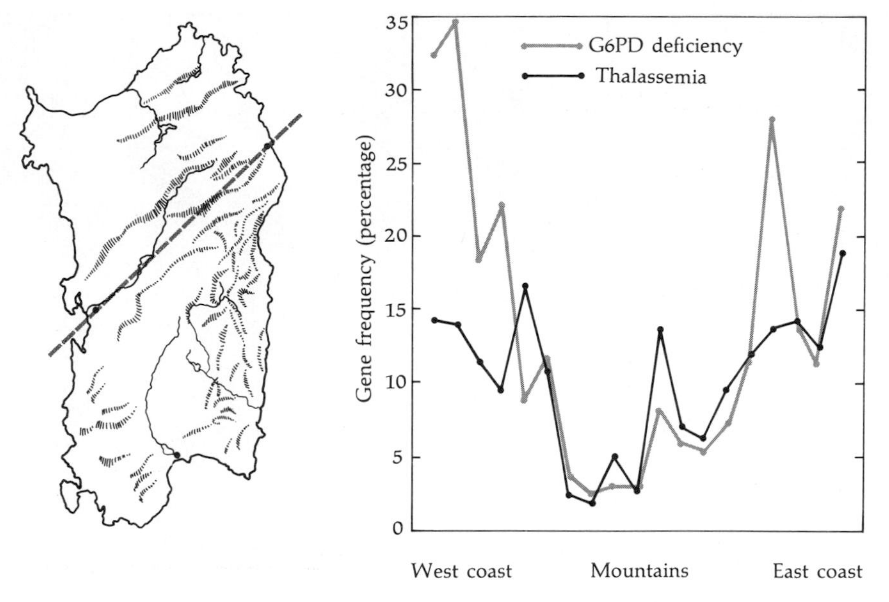

Figure 12.17
Variations in gene frequency over short distances, related to changes in selective advantages. The graph shows frequencies of genes for thalassemia and for glucose-6-phosphate dehydrogenase (G6PD) deficiency in central Sardinia. The populations are arranged in an approximately linear sequence from near Oristano on the west coast to Posada on the east coast, as shown by the dashed gray line on the map. Malaria is almost absent in the central mountainous regions, but is very frequent on both coasts. The genes may be in part maintained in the central regions by migration. (After F. B. Livingstone, *Abnormal Hemoglobins in Human Populations,* Aldine, 1967; data from M. Siniscalco et al., *Nature,* vol. 190, pp. 1179–1180, 1961.)

only the demography of males. In many places the social customs involve patrilocal marriage—that is, a new couple most frequently takes up residence in the place where the husband (and the husband's father) was born. This custom is often connected with inheritance of land. In situations of this kind, surnames tell only half the story—the half that refers to males. In patrilocal conditions, males migrate less than females do, so surnames are more strongly affected by random drift than is the average autosomal gene, for which both sexes count equally.

12.4 Gene Flow

A special type of gene diffusion goes by the name of *gene flow.* Two initially different groups may exchange genes at a low rate, sometimes only (or mostly) in one direction. Because the rate of exchange is low, the two groups may remain distinct,

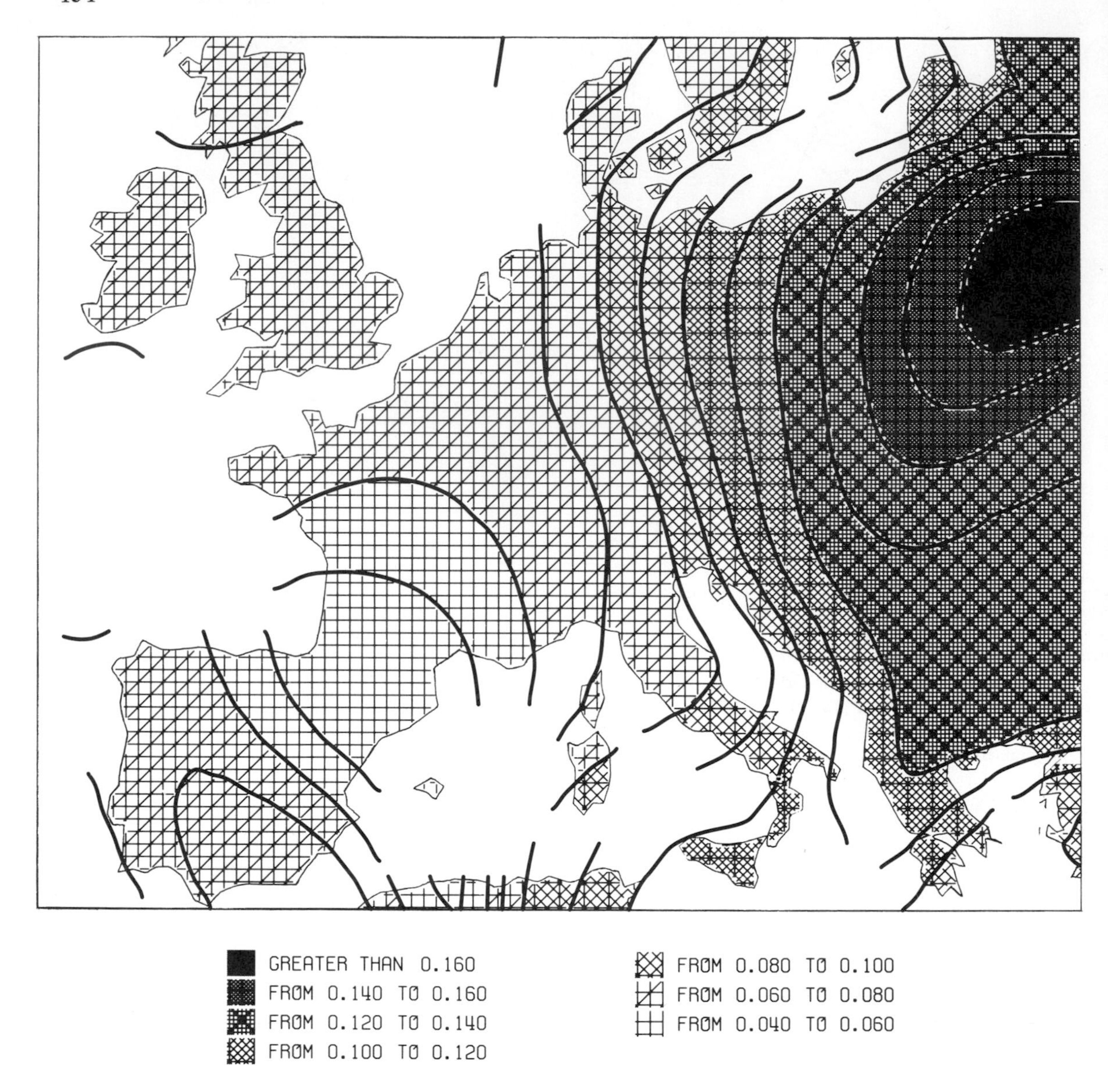

Figure 12.18
A cline, or gradient of gene frequency. The computer-generated map shows the frequency of the *B* allele of the ABO blood-group system in Europe. Note the gradual change from high frequencies near central Asia toward low frequencies in western Europe. The average "slope" of the cline is about 1 percent per 400 kilometers. The cline is probably a remnant of early population migrations and may be gradually disappearing, but it might also be maintained by selective pressures (as is the case in Fig. 12.17). (Courtesy of D. E. Schreiber, IBM Research Laboratory, San Jose, and of R. Matessi.)

despite the continued intermingling. Social barriers may help to maintain the distinction between the groups. In the long run, the process must result in elimination of the genetic differences between the groups but, if the rate of gene flow is small, it may take a long time.

The case of black and white Americans provides an interesting example of gene flow.

Blacks were brought to America as slaves during a period starting about ten and ending about five generations ago. The gene frequencies of black Americans are somewhat different from those of their probable African ancestors, deviations being toward the gene frequencies of white Americans. This observation indicates that there has been significant gene flow between the populations of white and black Americans. Because marriage across racial barriers was effectively forbidden in America until a few years ago, the simplest explanation is that gene flow has taken place through extramarital matings.

Gene systems that most clearly differentiate black and white populations—such as Rh, Fy, and Gm—are the most informative for studies of gene flow.

For example, among Africans the Duffy blood-group allele Fy^0 (phenotype a$^-$b$^-$) is almost the only allele present; its frequency is close to 100 percent. Among Caucasians, the Fy^0 allele is rare; almost all Fy genes are Fy^a or Fy^b. For simplicity, assume that the frequency of Fy^0 is 0 percent in Caucasians and 100 percent in Africans. Then the frequency of Fy^a and Fy^b among black Americans indicates the total percentage of white ancestry accumulated over the generations. A simple formula can be used to take account of the fact that the frequency of Fy^0 among whites is not exactly 0 percent. Averaging over the most informative genes for which data are available, we find that the percentage of white ancestry varies greatly according to the city of origin (Fig. 12.19).

The percentage of white ancestry is highest for blacks from northern cities. Immigration of blacks from the South to the North started after 1865, but it is difficult to determine whether the greater gene flow in this group occurred before or after migration to the North. For example, the likelihood of a black individual migrating to the North might have been influenced by his or her degree of white ancestry. Over the entire American black population, the admixture (percentage of white ancestry) is estimated to be about 30 percent. If this admixture occurred equally at each generation, then the "gene flow" per generation may have been on the order of 5 percent.

These estimates are, of course, average values for the entire black population. In the case of an individual, pedigree information (which is only rarely available)

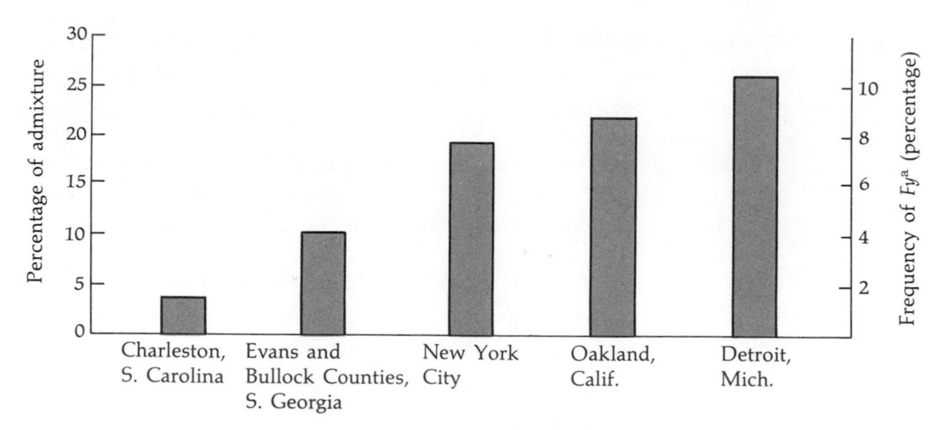

Figure 12.19
Racial admixture in U.S. blacks. The scale at the left indicates admixture with the white population (percentage), derived from the frequency of Fy^a genes shown on the scale at right. The admixture values are computed on the assumption that the ancestral African populations had zero frequency for the Fy^a gene, while the frequency of this allele is about 40 percent in the American white population. (Based on an analysis by T. E. Reed, *Science,* vol. 165, pp. 762–768, 1969.)

would be needed to determine the extent of admixture. In the absence of a pedigree, the degree of individual admixture can be assessed only very approximately, for very few genes are known that show racial differences large enough to be informative for small samples.

An interesting spinoff of these investigations is the finding, not wholly unexpected, that hemoglobin-S genes do not conform to the other genes in this type of analysis. Black Americans seem to have less hemoglobin S than they would be expected to have from considerations of admixture alone—that is, the frequency of the *S* allele among American blacks is lower than that among Africans by a factor greater than the changes in other gene frequencies that are explained by gene flow with the white population. The simplest explanation is that American blacks have lost *S* alleles because, in the absence of malaria, the allele is not maintained by any selective advantage for the heterozygote. Thus the *S* allele is being eliminated slowly but surely by selection against the homozygous recessives (the sickle-cell anemia patients).

Summary

1. Every real population is finite in size. The genes that are passed from one generation to the next are limited in number, and are equal to twice the number of individuals that actually reproduce. This number is the "effective population size" and is approximately one-third of the population census size.

2. Gene frequencies are therefore subject to an effect of statistical sampling at every generation, and this drift effect causes random fluctuations. The magnitude of the fluctuations is greater, on average, the smaller the population size.

3. The effect is especially noticeable and easy to appreciate when a population is started by a small group of founders. It is however a general effect, because every generation is the founder of all successive generations. Thus the effects of drift accumulate over time.

4. The final result of any drift process is the elimination of all alleles except one, which is then "fixed." This process may take a very long time if the population is large, the average number of generations required being approximately proportional to the effective population size.

5. Intermigration between populations limits the effects of drift. A balance will finally be established between migration and drift. Independent demographic and genetic studies of the phenomenon agree in showing that genetic variation in relatively isolated human populations is largely due to drift.

6. One-way migration (gene flow) may change the genetic constitution of populations. Expected change can be estimated when the genetic composition of the initial populations is known.

Exercises

1. Why does the effect of genetic drift increase as population size decreases?

2. What is the significance of population-size bottlenecks for generating genetic variation?

3. Why is fixation the inevitable outcome of genetic drift in the absence of new mutations?

4. In what way does migration counter the effect of random genetic drift?

5. How can estimates of population admixture, using genetic markers, serve as a basis for detecting the effects of natural selection?

6. It has been hypothesized that the virtual absence of the *A* and *B* alleles of the ABO bloodgroup system from most American Indians could be the effect of drift. Tests on pre-Columbian mummies have suggested the presence of A and B antigens. These conclusions are not entirely acceptable because bacterial contamination may have altered the results. If they could be accepted at face value, what would be the most likely explanations for the absence of A and B antigens among present American Indians?

7. Ashkenazim Jews, who lived in Russia, Poland, and Germany for at least 1,000 years, have much

lighter skin, hair, and eye color than do Sephardim Jews, who lived in the Mediterranean area. These traits are mostly inherited, though not in a simple way. At least two explanations are possible for these observations. (a) What are they? (b) What investigations would you do to help distinguish between them?

8. After the mutiny on the *Bounty* (in 1789), six Englishmen took eight Polynesian women with them and established themselves on Pitcairn Island. Their descendants multiplied at a rate that led to a doubling of the population every 20 years, almost until the present. How likely is it that this population, assuming there were no later immigrants, contains the genes for (a) albinism, (b) PKU, (c) the *O* allele of the ABO blood groups, (d) the gene for color blindness? Assume for simplicity that drift played a role *only* at the founding generation. (e) What consequence has this assumption on the probability thus calculated? (Give a qualitative answer.)

References

Cavalli-Sforza, L. L.

1969. "Genetic drift in an Italian population," *Scientific American*, vol. 221, no. 2, pp. 30–37.

1973. "Some current problems of human population genetics," *American Journal of Human Genetics*, vol. 25, no. 1, pp. 82–104.

Cavalli-Sforza, L. L., and W. F. Bodmer

1971. *The Genetics of Human Populations.* San Francisco: W. H. Freeman and Company. [Chapter 8 gives a more extensive treatment of drift.]

Crow, J. F., and C. Denniston

1974. (Eds.) *Genetic Distance*, New York: Plenum Press.

Crow, J. F., and M. Kimura

1970. *An Introduction to Population Genetic Theory.* New York: Harper and Row. [Chapter 7 gives a detailed derivation of drift, and Chapters 6 and 9 include migration problems.]

Harrison, G. A., and A. J. Boyce

1972. (Eds.) *The Structure of Human Populations.* Oxford: Clarendon Press.

Jacquard, A.

1974. *The Genetic Structure of Populations.* New York: Springer-Verlag. [Recently translated to English by D. and B. Charlesworth from the 1970 Paris edition.]

Kimura, M., and T. Ohta

1971. *Theoretical Aspects of Population Genetics.* Princeton, N.J.: Princeton University Press.

Malecot, G.

1969. *The Mathematics of Heredity.* San Francisco: W. H. Freeman and Company. [Recently translated by D. M. Yermanos from the 1948 Paris edition.]

Reed, T. E.

1969. "Caucasian genes in American Negroes," *Science*, vol. 165, pp. 762–768.

13

Molecular Evolution

The analysis of amino-acid sequences in proteins has opened up new vistas for the understanding of biology. A given protein may show very little difference when analyzed in species that are closely related; but in very distant species, a large number of changes is likely to have occurred. Even so, it is abundantly clear that homologous proteins have evolved from a common ancestor. Study of molecular evolution supplies new and remarkable insight: now one can begin to explore the relative roles of the major evolutionary factors—mutation, selection, and drift—from a very general point of view.

13.1 Evolutionary Trees from Proteins

Protein chemistry can now provide the sequence of amino acids that form a polypeptide chain (Chapter 3). A considerable amount of knowledge has thus accumulated on the chemical composition, in particular the amino-acid sequence, of several proteins—and some particular proteins have been studied in many species.

It has become clear that the same protein—for instance, cytochrome C (a respiratory enzyme made up of about 100 amino acids)—is extremely similar in closely related species; but the less related the species are, the more different the proteins.

Differences are usually estimated in terms of "amino-acid substitutions."

Consider the following sample of sequence of the end of the cytochrome chain (the last eight amino acids) from three different species.

Man	. . .	Tyr	Leu	Lys	Lys	Ala	Thr	Asn	Glu
Dog	. . .	Tyr	Leu	Lys	Lys	Ala	Thr	Lys	Glu
Chicken	. . .	Tyr	Leu	Lys	Asp	Ala	Thr	Ser	Lys

(Recall the three-letter codes for amino acids given in Table 3.1.) Among these eight amino acids, there is one amino-acid difference between man and dog, and three between either man and chicken or dog and chicken. These differences are the amino-acid substitutions.

The number of differences for the α chain of hemoglobin (which in man has 146 amino acids) is given for some species of mammals in Table 13.1. The evolutionary closeness of man and gorilla shows up very clearly, as there is only one amino-acid difference between these two species. The data indicate in addition that (based on this chain) we are closer to pigs than to rabbits! In fact, we have diverged from rabbits by 26 amino-acid substitutions, and from pigs by only 19. Approximately the same is true for gorillas.

Data on amino-acid substitutions in different species provide the basis for reconstructing a tree of origin of the species.

The simplest way to build such a tree is to start by joining the two closest species. These are, in the case of Table 13.1, man and gorilla, which have a difference of just one amino acid, and so are placed nearest each other in the tree shown in Figure 13.1. Among the other species, the one nearer to man and gorilla (which will now be considered as combined) is pig, because it shows 19 to 20 differences as compared to 26 to 29 for the rabbit. By similar (but usually more sophisticated) criteria, one can reconstruct more complex evolutionary trees. Figure 13.2 is such a tree built on the basis of cytochrome-C data only. This protein is thus far unique

Table 13.1
Number of amino-acid differences between four mammals in the α chain of hemoglobin

	Man	Gorilla	Pig	Rabbit
Man	0	1	19	26
Gorilla		0	20	27
Pig			0	27
Rabbit				0

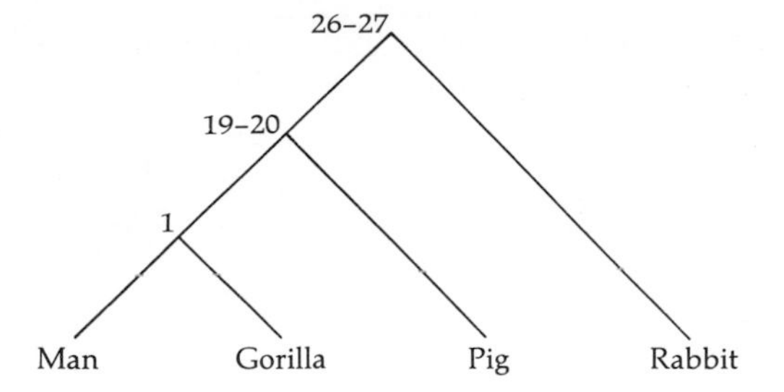

Figure 13.1
An evolutionary tree. This tree is constructed from the data in Table 13.1 (for the hemoglobin α chain). The numbers indicate the number of amino-acid differences found between pairs of species whose lines join at that point in the tree.

among those analyzed, in that it is found in a great variety of organisms, connecting man to both the vegetable and animal kingdoms. The picture obtained clearly fits well with our ideas of the unitary origin of life on earth.

The α and β chains of hemoglobin, and other proteins, or parts of them, have been similarly analyzed. Figure 13.3 shows a tree reconstructed on the basis of the α chain of hemoglobin, and Figure 13.4 a tree using the combined information from four different proteins.

Paleontologists have inferred from the fossil record the approximate time at which the common ancestors of the extant species lived. These data are indicated in millions of years in Figures 13.3 and 13.5.

The availability of paleontological information makes it possible to study the rates at which molecular evolution has proceeded.

In fact we can estimate, for instance, that the last common ancestor to pig and man lived some 80 million years ago. We know that there are, in the α chain of hemoglobin, 16 amino acids (out of 146) that differ between these two species, or about 11 percent. Dividing this difference by the time elapsed (0.11/80), we have 0.0014, the (approximate) chance that a given amino acid of the α chain of hemoglobin will change over a million years. We should actually take half this amount, or 0.0007, because evolution proceeded in *both* branches from the common ancestor to pig *and* to man. The study of rates of molecular evolution has given some interesting results that are summarized in the next section.

13.2 Rates of Protein Evolution

The sequencing of a protein is time demanding, and therefore the full amino-acid sequence is known only for a relatively limited number of proteins. Results have, however, accumulated on a certain number of proteins, and some generalizations are possible.

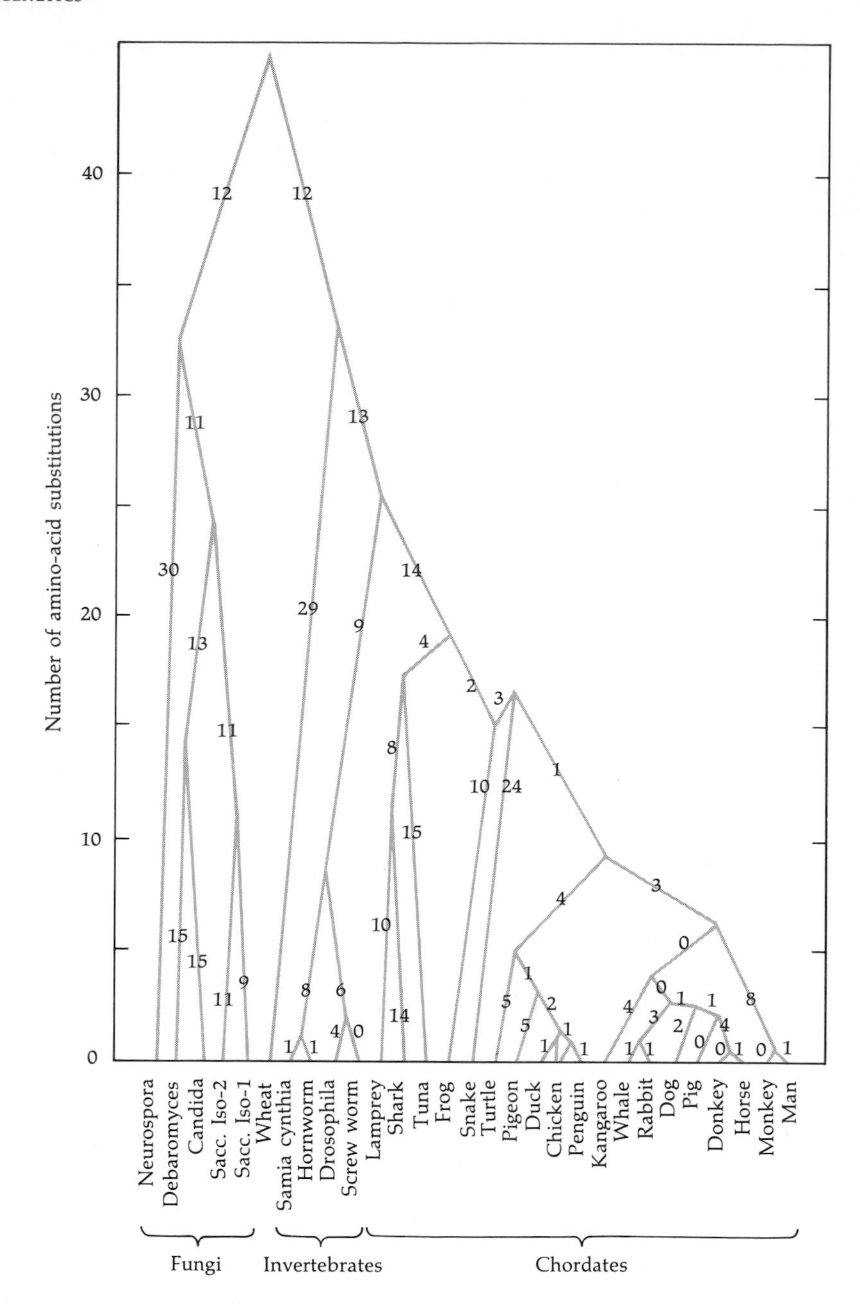

Figure 13.2
Evolution of cytochrome C. Each line segment in the tree is marked with the estimated number of amino-acid substitutions represented by the corresponding evolutionary changes. (After a tree constructed by E. Margoliash.)

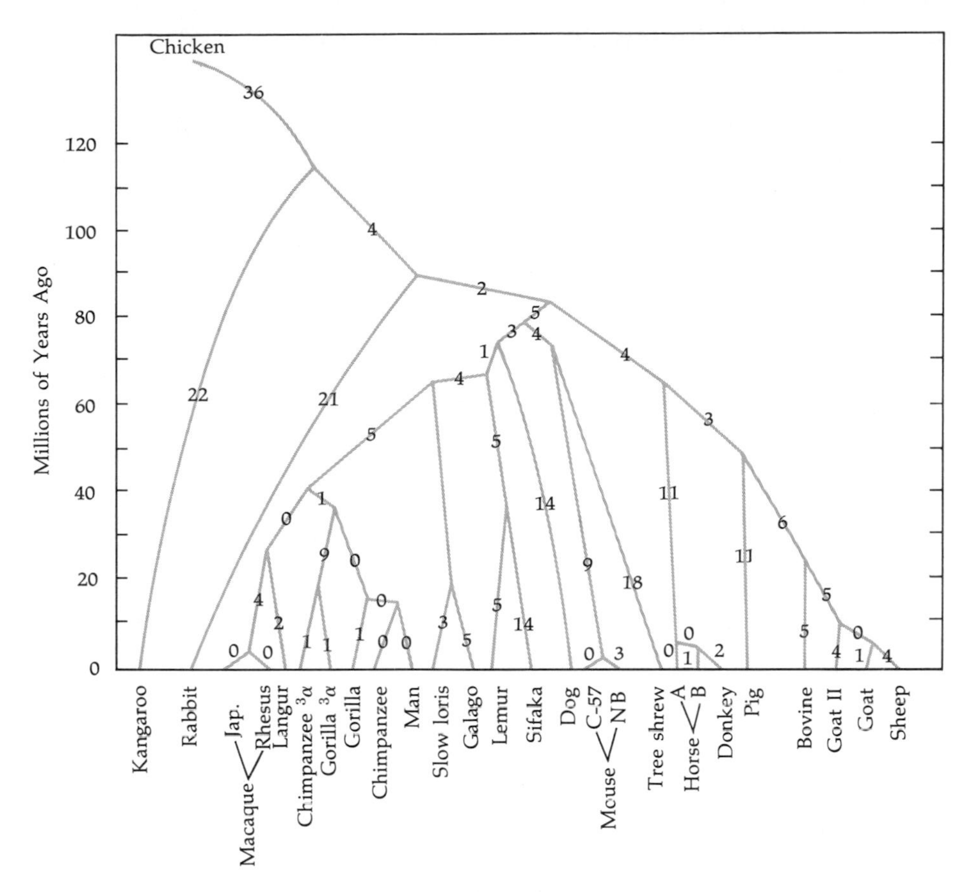

Figure 13.3
Vertebrate evolutionary tree based on α globin sequences. The vertical scale represents time in millions of years; numbers in the tree are numbers of amino-acid substitutions. (From M. Goodman et al., *Journal of Molecular Evolution,* vol. 3, pp. 1–68, 1974.)

For a given protein the rate of change is fairly constant.

Figure 13.5 shows the rates of evolution of cytochrome C, hemoglobin, and fibrinopeptides (some short polypeptides that are excised from fibrinogen, a plasma protein, at the time of blood coagulation, when fibrinogen is transformed into fibrin). There is for each protein a nice proportionality between the differences among the organisms and the time since their common ancestors lived. But the rate of change is clearly different for the three proteins.

Different proteins have different overall evolutionary rates.

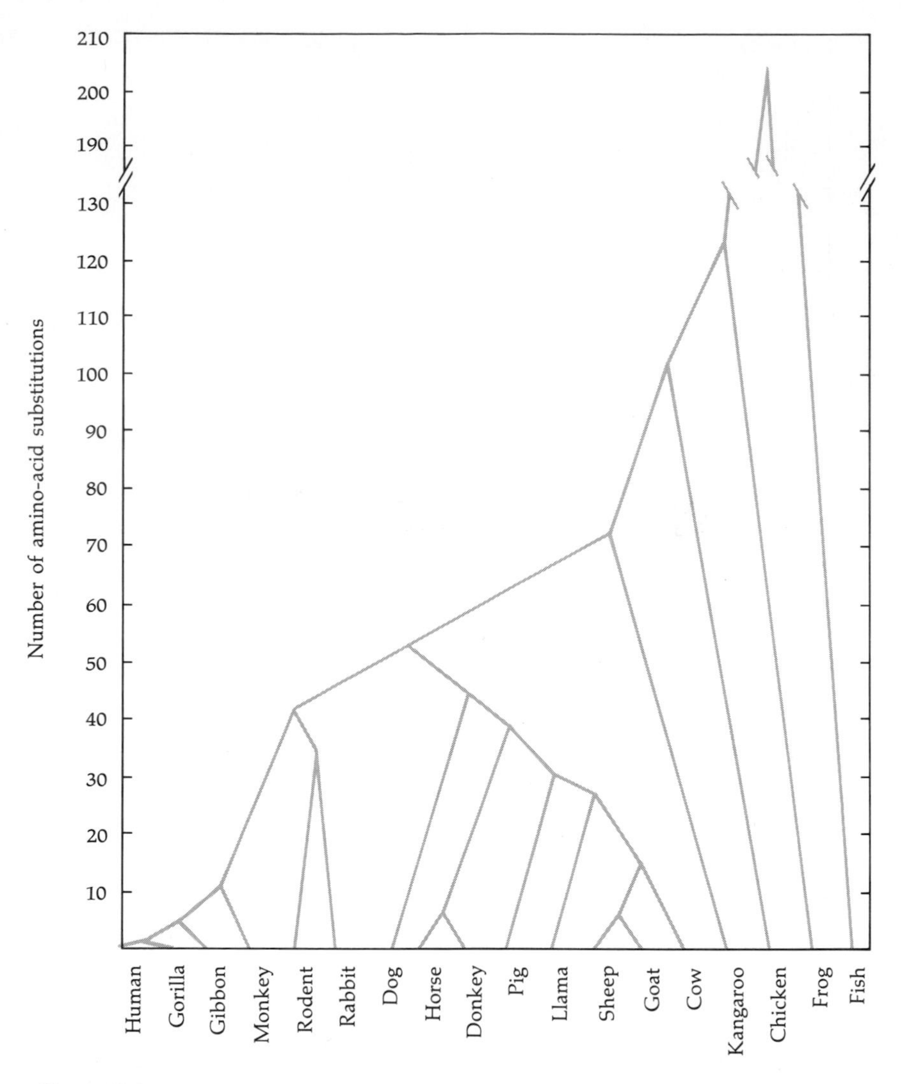

Figure 13.4
Composite evolutionary tree, based on evolution of hemoglobins α and β, cytochrome C, and fibrinopeptide A. The nodes representing common ancestors are placed on a scale proportional to the estimated number of amino-acid substitutions (in all four proteins) that occurred during the descent from the common ancestor to modern descendants. (After E. H. Langley and W. M. Fitch, *Journal of Molecular Evolution,* vol. 3, pp. 161–177, 1974.)

Figure 13.6 gives average evolutionary rates estimated for a number of proteins and polypeptides. Among those listed, cytochrome C is the slowest and fibrinopeptide A the fastest evolving substance. What is the source of the difference in rates? There probably are many different reasons, which still are not fully under-

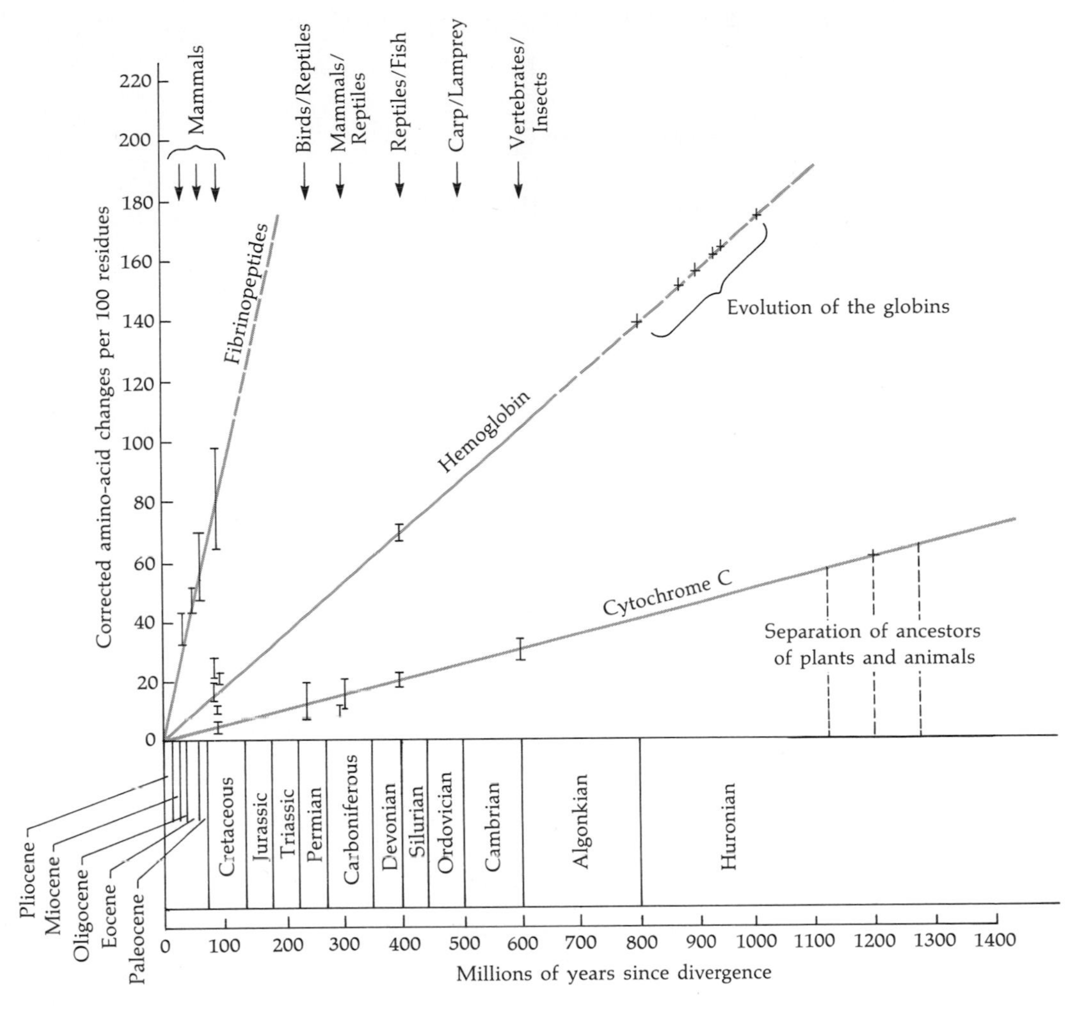

Figure 13.5
Rates of molecular evolution. The graph shows rates of macromolecular evolution in the fibrinopeptides, hemoglobin, and cytochrome C. Mean errors in amino-acid differences are indicated by the vertical bars. (After R. E. Dickerson, *Journal of Molecular Evolution,* vol. 1, pp. 26–45, 1971.)

stood. But some simple examples can show at least a few of the constraints on the evolution of amino-acid sequences.

Some amino acids, or amino-acid sequences, vary less than others.

A detailed study of the evolutionary fate of particular amino acids of protein segments reveals important differences. Cytochrome C, for example, has some interesting features. Because it is found in a great variety of organisms, this protein

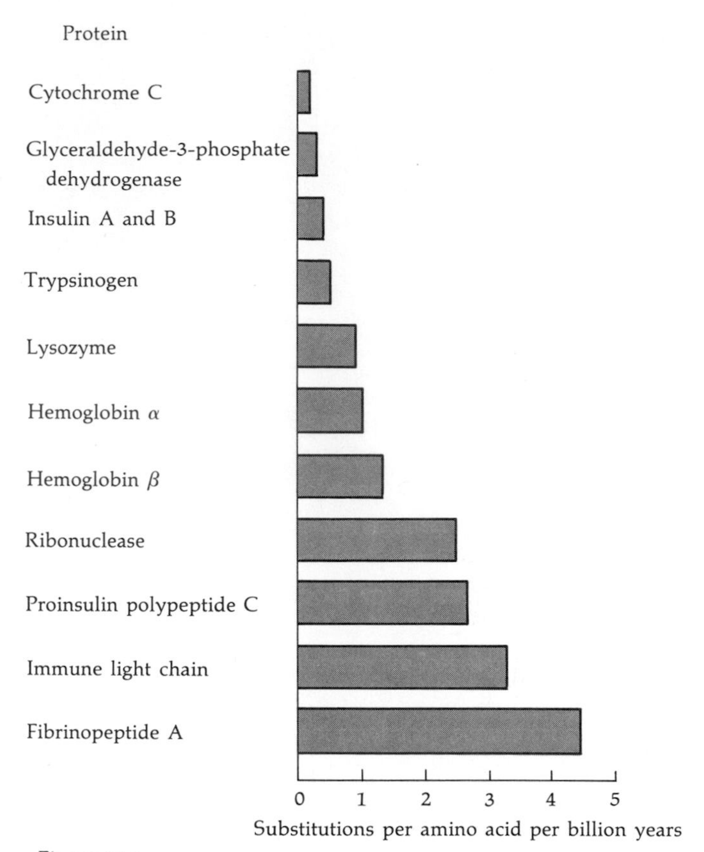

Figure 13.6
Rates of amino-acid substitution in various proteins and polypeptides. (For sources of data, see R. C. Lewontin, *The Genetic Basis of Evolutionary Change,* Columbia Univ. Press, 1974.)

can tell us about a longer time period of evolution than can many other proteins. It also has been one of the most slowly evolving proteins, at least among higher organisms. Perhaps it already (by the time higher organisms evolved) had a sufficiently long evolutionary history that many possible variations had been tested, and the fittest variants fixed, so that little further improvement was possible. There is one region of cytochrome C, formed by a sequence of eleven amino acids (positions 70 through 80) that is completely invariant in all modern organisms that have been tested:

70 75 80
Asn–Pro–Lys–Lys–Tyr–Ile–Pro–Gly–Thr–Lys–Met

Any change in this sequence would probably be incompatible with function, and hence with life: such mutations, if they occur, are therefore lethal and are not propagated.

Natural selection stabilizes those parts of the molecule which are important for function.

The 70 to 80 region of cytochrome C is in close proximity to one surface of the heme group, which is the iron-containing molecule that is attached to the polypeptide portion of cytochrome C. This region also includes a binding site for the enzyme cytochrome oxidase, which acts on cytochrome C. Most probably, any change in this 11-amino-acid sequence would be detrimental to the interactions with these molecules and hence to the functioning of the protein. There are no other sequences of amino acids more than two amino acids long that are completely conserved in cytochrome evolution. Conservation may be especially marked at sites that are involved in interactions with other molecules (Figure 13.7).

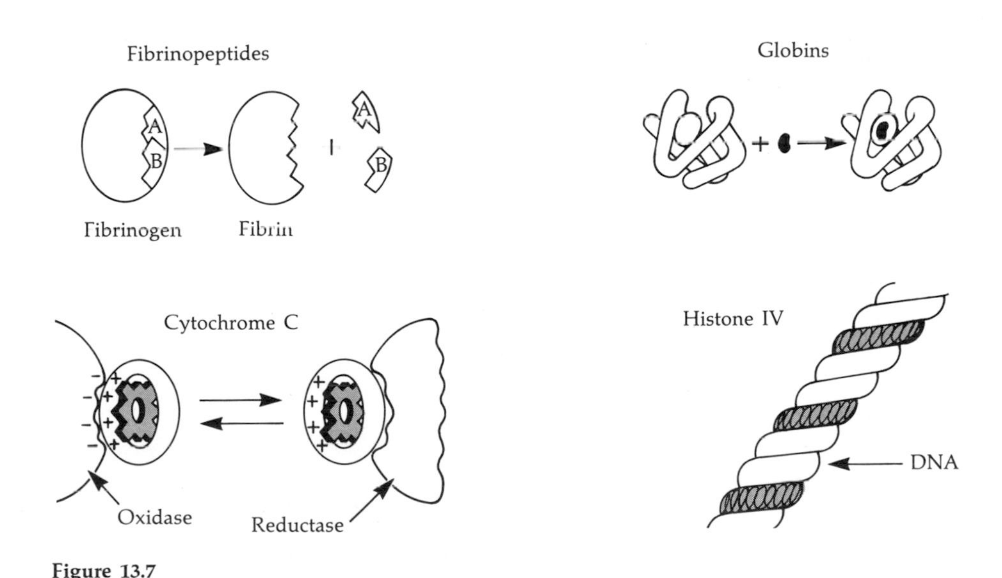

Figure 13.7
Rates of evolution and complexity of protein interactions. The more complex the interactions of a protein with other molecules or macromolecules, the longer the time period needed for a single change to occur in the protein. The discarded fibrinopeptides (A and B in the fibrinogen molecule) have undergone an average of one amino-acid substitution every 200,000 years, but the rate of molecular evolution in the histone IV bound to DNA within the nucleus has been 500 times slower. The rate for cytochrome C is lower than that for the globins, primarily because cytochrome C interacts with other macromolecular complexes, whereas hemoglobin binds only to O_2 and CO_2 in solution. (After R. E. Dickerson, *Journal of Molecular Evolution,* vol. 1, pp. 26–45, 1971.)

Internal parts of a protein molecule are also likely to be conserved. As a rule, these regions are formed by hydrophobic amino acids (those that "shy away" from water), while hydrophilic ones (that "like" water, or tend to associate with water molecules) are at the outside. This leads to further constraints: hydrophobic amino acids tend to be replaced by hydrophobic ones, and hydrophilic by hydrophilic. One interesting exception to this rule is the substitution that transforms hemoglobin A into S (leading to sickle-cell anemia). This substitution is at position 6 of the β chain, which is an "outside" site. Here, hydrophilic glutamic acid is replaced by hydrophobic valine. As we know, the properties of the S molecule are altered. In particular, its solubility is decreased, at least under conditions of low oxygen tension, and pathological consequences follow.

Amino-acid substitutions are not the only changes observed in molecular evolution.

Deletions and additions of single amino acids (or short sequences of them) are also observed, but these sorts of changes tend to be rarer than simple substitutions. Rarely, more radical changes that may affect a more extensive sequence of amino acids are also observed. Duplications, often of whole genes, have occurred and are of special importance, as we see in the next section.

13.3 Duplication in Evolution

Drosophila melanogaster, the fruitfly that was for several decades the favorite organism of geneticists, has giant chromosomes in the salivary glands of its larvae. These chromosomes have characteristic patterns of bands to which genes can be assigned, and so have permitted considerable improvement in the resolution of certain types of genetic analyses (see Fig. 2.27).

One phenomenon, first noted in Drosophila, is that certain chromosome segments are duplicated.

This duplication is often in tandem (the duplicated segment being located immediately near the original one in the same chromosome), or sometimes also in other parts of the chromosome or chromosomes. Gene duplication has long been thought to be a basic mechanism by which the genome can increase in size. By providing alternate sites of synthesis of similar molecules, it can also allow the evolution of specialization, for particular purposes, of molecules that were initially similar or identical because of their origin by duplication from a single common gene.

Molecular evolution provides many examples of duplication.

Among the largest series of proteins showing evidence of duplication are the globins and the immunoglobulins. The globins include the α, β, γ, δ and ϵ chains of hemoglobin (see Chapter 3) as well as myoglobin, a protein that performs the same oxygen transport function as hemoglobin, but is located in the muscles. These globin molecules all appear to have a common origin. Myoglobin, however, does not form tetramers like hemoglobin. Their tree of origin, based on their presence in various groups of animals, is given in Figure 13.8. The latest duplication known to us is that which led to the δ chain. This chain associates with the α to produce hemoglobin A_2 (composition $\alpha_2\delta_2$; see Section 3.5 for the explanation of this symbolism). Hemoglobin A_2 is a minor component making up only 2 percent of all the hemoglobin in the adult. Its absence would therefore probably have no practical consequence for the carrier. The presence of A_2 in some mammals (primates) directly on the human line of evolution sets its time of origin at

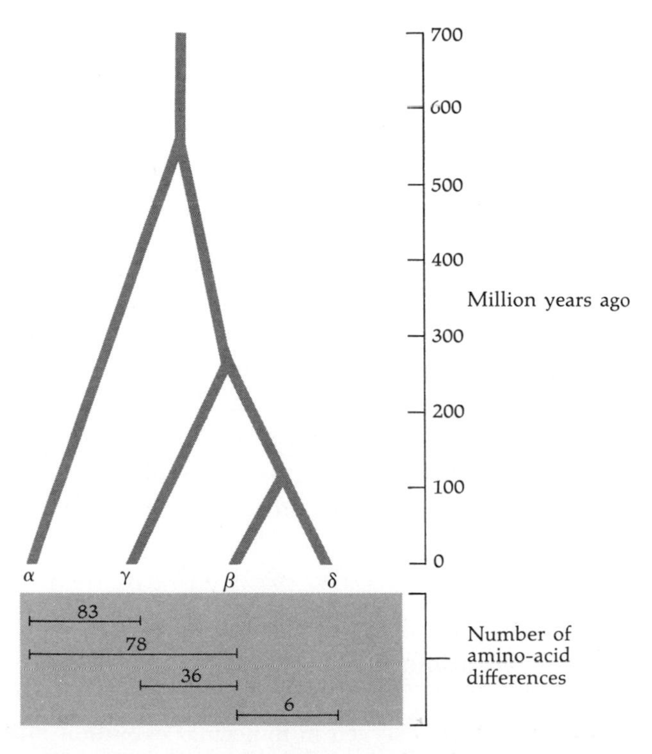

Figure 13.8
Molecular evolution of hemoglobin chains. The probable evolutionary tree of the α, β, γ, and δ hemoglobin chains is shown, with estimated times of evolutionary separation.

some 40 to 50 million years ago. The δ chain has apparently had little time to diverge from its progenitor, the β chain (from which it differs by only a few amino acids), and no chance at all to become important enough in the economy of the organism to take on some special function. The δ-chain gene is still closely linked to the β gene, from which it probably originated as a tandem duplication. This is shown by pedigrees in which variants of β and δ are always transmitted in complete linkage (Chapter 3). As discussed in Chapter 3, unequal crossing over between the β and δ genes must have occurred repeatedly, generating the Lepore-type hemoglobins, in which a β and a δ chain are fused. This suggests that the two genes may actually be next to each other on the chromosome.

Duplication favors specialization of function.

The earliest duplication known in the globin genes—that leading on the one hand to myoglobin and on the other to the hemoglobins—has generated two molecules, which are by now fully differentiated. Each has adapted to its own niche in the organism, one in the muscles and the other in the red cells. Similarly, the γ chain has specialized for fetal function and the β for adult function.

Duplication favors unequal crossing-over.

The Lepore hemoglobin is also an example of how a duplication can, by unequal crossing-over, generate further duplications. Again, there are other examples. *Haptoglobins* are proteins found in serum; they have the property of binding hemoglobin. They may serve the function of transporting hemoglobin from broken-down red cells to sites where the products are metabolized and reutilized. There are several haptoglobin alleles that are frequent in all human populations, of which one (Hp^2) probably originated by unequal crossing-over in a heterozygote for two previously existing alleles (Hp^{1S} and Hp^{1F}; see Fig. 13.9). This is not a complete duplication, however, for the Hp^2 protein has less than the total expected length for a duplication. There is another allele that looks like a triplicate of the Hp^1 type; it probably arose through unequal crossing-over between an Hp^1 and an Hp^2 allele.

13.4 The Fate of Mutant Genes in Evolution, and Overall Evolutionary Rates

The amino-acid substitutions observed in molecular evolution must be the outcome of nucleotide substitutions in DNA. These "point" mutations keep arising, even though at the low rate typical of mutation. What is their fate, once they have arisen? Why do some spread to the whole population and eventually become

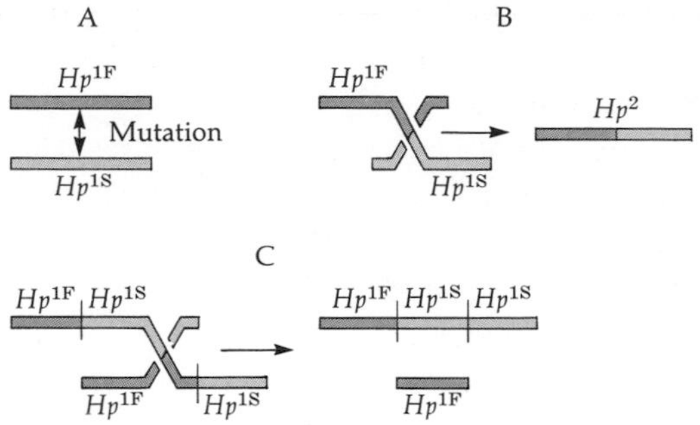

A B

Hp^{1F} Mutation Hp^{1S} Hp^{1F} Hp^{1S} Hp^{2}

C

Hp^{1F} Hp^{1S} Hp^{1F} Hp^{1S} Hp^{1F} Hp^{1S} Hp^{1S} Hp^{1F}

Figure 13.9
Duplication and triplication in haptoglobin alleles. (A) There exist two haptoglobin Hp^1 alleles, which presumably originated by mutation one from the other; it is not clear which of the two was ancestral. (B) Unequal crossing-over between the two alleles probably generated the Hp^2 allele. (C) Unequal crossing-over in an Hp^2 homozygote generated another allele that probably is a triplication.

fixed? There are, in principle, two possible answers: (1) the new mutant is at a selective advantage; (2) it is not at an advantage, but fixation is determined by chance. In other words, the two explanations invoke either natural selection or random genetic drift.

The expected fate of one new mutation introduced into a population has been analyzed theoretically, originally by R. A. Fisher in 1922. Any mutant, even if it arises only once in the history of a population, can eventually be fixed, but the probability that it will be fixed is very low. This probability depends on whether the mutant has a selective advantage or disadvantage, or whether it is selectively neutral.

We discuss in Chapter 12 the probability of fixation by random genetic drift of an allele present in a population with a given frequency. The probability of fixation is equal to its frequency. The frequency of a single mutant at the time it arises is one divided by the total number of genes, which is $2N$ (twice the total number of individuals in a diploid species). Thus the probability of fixation of a new mutant by random genetic drift is just $\frac{1}{2N}$. If, however, the gene has a selective advantage, its probability of fixation will be more than $\frac{1}{2N}$, and will be higher for higher advantages. The probability will be lower than $\frac{1}{2N}$ if it is at a disadvantage. The graph in Figure 13.10 indicates that a gene with a large advantage will not necessarily be fixed; it may be eliminated by drift. This is most likely to happen, if it does, in the early stages of a new mutant's history, when there are still only a few copies of the gene in the population. On the other hand, it is not impossible that a moderately disadvantageous mutant can be fixed, especially if the population is very small. Population bottlenecks may therefore be especially important in leading to the fixation of such genes.

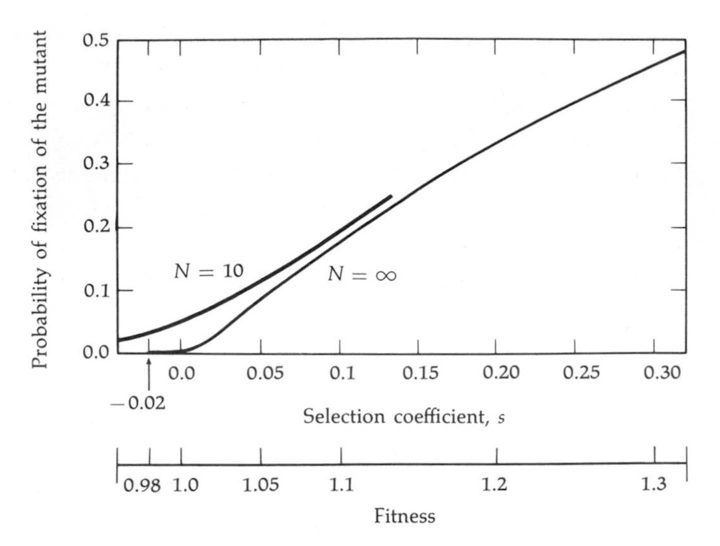

Figure 13.10
Probability that a mutant, arising only once, will eventually be fixed. The horizontal axis is the selection coefficient s of the mutant; a perfectly neutral mutant has $s = 0$. Note, however, that in a small population ($N = 10$) the probability of fixation, shown on the vertical axis, does not vary markedly for small variations of s around zero. For a neutral mutant, the probability of fixation is $\frac{1}{2N}$. Even with a marked selective advantage, there is no certainty of fixation. (This curve is based on a formula given by M. Kimura.)

The hypothesis has recently been put forward that molecular evolution is mostly due to "neutral" mutations.

This has started a scientific controversy between supporters of this hypothesis ("neutralists") and those who believe that practically all molecular evolution is the result of natural selection ("selectionists"). It is, of course, impossible to deny *entirely* the role of natural selection: without it, no adaptation could occur, and there is no other thinkable mechanism for coping with environmental challenges. Not even the most ardent neutralists would deny that many mutations are likely to destroy or impair considerably the physiological properties of the molecule they affect, and that such mutants never, or extremely rarely, will get fixed. More than 90 percent of all mutations that occur may perhaps be of this nature. The discussion really centers around the other mutations, the residual 10 percent.

The question is really how often a mutation is advantageous, and how often it is neutral.

The curve of Figure 13.10 shows that the probability of fixation does not change much if the fitness of a mutant is slightly below or slightly above unity (taking the fitness of the nonmutated, parental, genotype equal to one). Mutations whose fitness is close to one, on either side, are "effectively" neutral. Moreover, if the population is small, it can tolerate deleterious mutants whose fitness deviates more from unity, more so than if it is large. The concept of "neutral" itself thus depends on population size. We have very little information as yet on how the fitness of mutations varies from one mutant to another. We can only say that a large fraction of mutations must be deleterious enough that they have no chance of ever being fixed. Their fitness is very close to zero and they constitute the lefthand peak in the fitness distribution (Fig. 13.11).

At the other tail of the fitness distribution, there can hardly be a discontinuity between advantageous and neutral mutations. We have already pointed out that the definition of "neutral" must embrace a range of mutants around the neutrality point (which corresponds to a fitness value of one), and that this range is larger for smaller populations. There are further complications when one considers that fitnesses change with different environments, and that even members of the same species are exposed to a considerable range of environments. Furthermore, in diploid individuals such as man, three genotypes can be formed with two alleles at a locus. Whenever a mutant is introduced into an otherwise homogeneous population, one should therefore take into account three different fitnesses for the three corresponding genotypes.

It has, up to now, been very difficult to determine what fraction of all mutations revealed by studies of molecular evolution are "neutral" and what fraction are advantageous. But clearly, some mutants that were fixed must have been advantageous or else there would have been no adaptation. It is unlikely, for example,

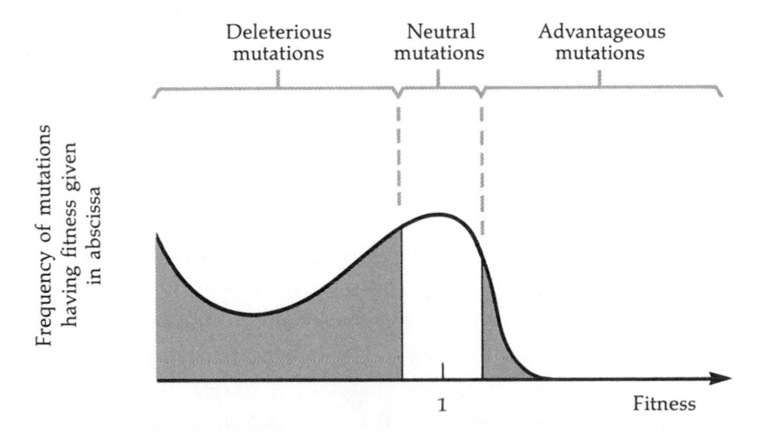

Figure 13.11
Hypothetical distribution, illustrating the variation in fitness of new alleles. In a small range (smaller than this graph may indicate) around fitness 1, mutations may be considered as selectively neutral.

that if our cytochrome were substituted with that, say, of snails, we would function in exactly the same way.

Molecular evolution has supplied an estimate of overall evolutionary rates.

The fossil record shows that there have been living organisms on earth for at least some three billion years. The age of the earth itself cannot be much more than five billion years. This is the time during which biological evolution must have occurred, starting with very simple organisms, and leading to the very complex ones now in existence, unless we assume that living organisms of some complexity arrived from other parts of the universe.

It is very difficult to get an estimate of evolutionary rates from these considerations alone. But molecular evolution can provide such estimates. Averaging over the known proteins (Figure 13.5), it appears that one amino acid will on average be substituted by another about every billion years. This may seem a very slow rate, but it is not.

The observed rate of amino-acid substitution is not far from that expected if all mutations were neutral, but this by no means proves neutrality.

Suppose for a moment that all mutations were neutral. The probability of fixation of a neutral mutant, at the time it arises, is $\frac{1}{2N}$. Let μ be the average rate at which mutations occur per generation. This is the average rate for one individual gene; but there are $2N$ genes in a population of size N and hence, every generation, $2N\mu$ mutations are produced. Of this number, only the fraction $\frac{1}{2N}$ will eventually be fixed. Thus the fraction of mutations arising per generation that will eventually be fixed is the product of the two quantities $2N\mu$ and $\frac{1}{2N}$, which is just μ. Under the condition that all mutants are neutral, therefore, μ (the mutation rate per gene per generation) is equal to the substitution rate (computed as the number of mutational substitutions occurring in a given time interval, divided by that time).

Assume that μ, the average mutation rate for one gene, is of the order of one in a million. This rate refers to a generation, and to a gene that may code for a protein containing, on the average, 300 amino acids. The mutation rate per amino-acid site will thus be one in 300 million per generation. A generation—in vertebrates, for example—may last on the average 3 years, making the mutation rate per amino-acid site per year equal to one in about a billion (10^9, or one thousand million) years. The agreement between this rate and the observed rate of amino-acid substitution has been used as an argument in favor of the neutralist hypothesis. In reality, however, these figures—in particular the mutation rate—are known only very approximately and may vary up or down by an order of magnitude or more. Moreover, there is an indeterminacy inherent in the problem, which depends on the totally unknown distribution of fitness values. The estimate above would agree just as well with the hypothesis that perhaps only 5 percent or less of all nondeleterious mutations are truly neutral and 95 percent or more have a selective advantage (which would probably make the selectionists happy). At the other extreme, the observed rates are compatible with an opposite hypothesis—namely, that 95 percent of all nondeleterious mutations are neutral and 5 percent advantageous (which neutralists would probably concede). Thus, neither theory nor observation have yet resolved the neutralist-versus-selectionist controversy, and other data are needed to throw further light on the problem.

13.5 The Prevalence of Polymorphisms, and Average Levels of Heterozygosity

In studies of molecular evolution, one amino-acid sequence from a protein is taken to represent the species. In fact, this is tantamount to taking one individual allele as representative of a gene. But we know that there are polymorphisms—that is, that often more than one allele per locus is present at substantial frequencies in a species. This complication is, in fact, expected. Fixation of one allele or another can hardly occur in a very short time. On the contrary, from the time of their first appearance, fixation of alleles may take a large number of generations. The mean time required for fixation is of the order of $2N$ generations (where N is the population size) for a neutral gene and, of course, usually shorter for an advantageous gene (Fig. 13.12 and 13.13). Meanwhile, the allele that will eventually be fixed coexists with any other allele(s) still present.

As discussed earlier, electrophoretic techniques have revealed that about one out of three enzymes shows sufficient genetic variation to be classified as a polymorphism. This is, moreover, an underestimate of the amount of genetic variation, because it is known that electrophoresis only reveals a certain proportion of enzyme differences.

High genetic variation in protein-coding genes is a general rule.

There are no major differences, in this respect, between various animals (including man) and plants. The amount of heterogeneity at a locus can be described by the number of alleles present, but this does not take account of the frequency of each allele. It would, for example, not distinguish between a locus in which there is essentially only one allele and a great number of rare variants (as for hemoglobin in Caucasians), and one like HLA, for which there exist many alleles, all at polymorphic frequencies.

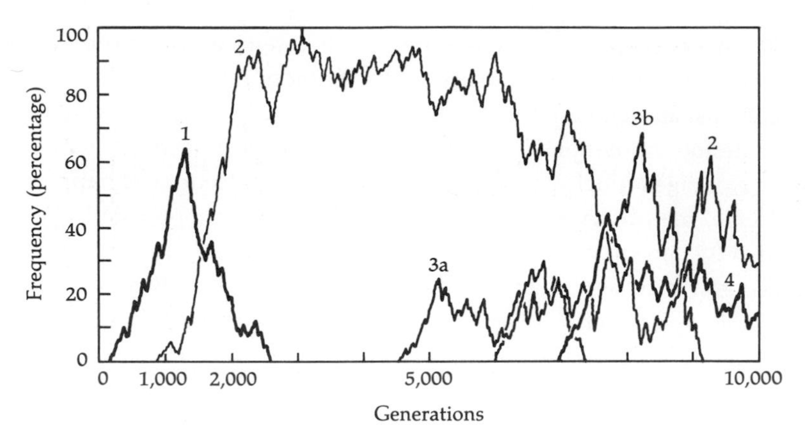

Figure 13.12
An experiment in computer simulation of evolution under neutral mutations. The numbers 1, 2, 3a, 3b, and 4 indicate mutations that reached substantial frequencies during the experiment, and whose frequencies are traced throughout all generations after their appearances. One mutation (2) was fixed for a short time (until other mutations appeared to make the locus polymorphic again), and this was the only mutation to reach fixation during the experiment. The simulation involved a population of $N = 2{,}000$ haploid individuals, with a mutation rate of 1 in 10,000.

One way to evaluate the degree of polymorphism at a locus is by the degree of heterozygosity.

Heterozygosity is measured by the percentage of individuals who are heterozygous at that locus for any pair of known alleles. Figure 13.14 gives data on heterozygosity per locus for 20 enzymes known to be polymorphic in man. The degree of heterozygosity would be close to zero for the hemoglobin genes, even though altogether over 100 alleles are known. But the number of known alleles depends on the number of individuals examined, which is in the hundreds of thousands for hemoglobin, because of its clinical importance. The degree of heterozygosity (d.h.) is on the whole independent of the number of individuals examined, and gives a simple quantitative idea of the overall level of polymorphism. Averaging over all genes (including those for which there is no polymorphism and therefore d.h. is zero or close to zero), one finds an average d.h. of 6 percent in man, and similar values have been obtained for other animals. Loci studied by immunological techniques as in the case of the blood groups may give a higher value for d.h., probably because this technique, even though more restrictive in some ways, may pick up more mutants at a locus than do the electrophoretic methods.

At least a part of this genetic variation may be transient.

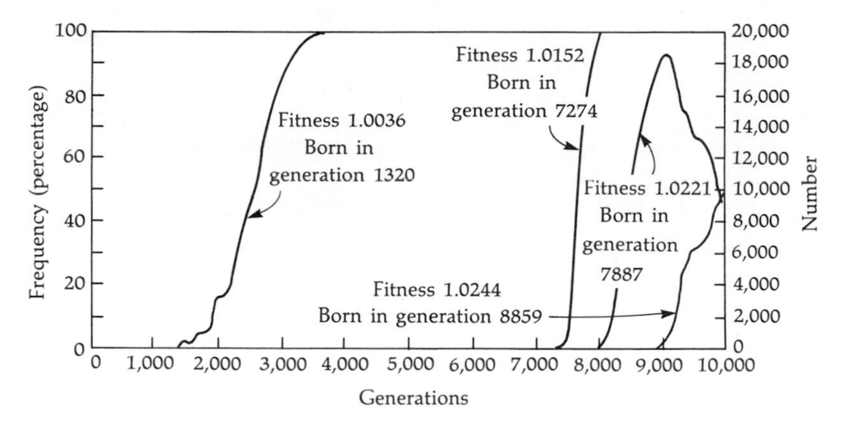

Figure 13.13
Computer simulation of drift and selection. In this simulation, new mutants have different fitnesses. The overall distribution of fitnesses is that shown in Figure 13.11 (a normal distribution with mean fitness 0.99 and standard deviation of ±0.01). The simulated
Computer simulation of drift and selection. In this simulation, new mutants have different fitnesses. The overall distribution of fitnesses is normal with mean fitness 0.99 and standard deviation of ±0.01 delete parenthesis to read ±0.01. The simulated

The number of alleles and the degree of heterozygosity at a locus represent a cross-section of a population at a given time, and may differ for different populations. Thus, hemoglobin is essentially monomorphic (that is, not polymorphic) in most Caucasians and Orientals, but polymorphic in Africans. One problem that arises is how would the picture shown by such a cross-section change from one time to another? To answer this question, we would clearly need to sample the same population at different times but, because of the slowness of biological evolution, this is not possible in practice. Sampling over very long time intervals would be needed to be of any value, at least for human populations. So far, it has also been more or less impossible to use the fossil record in this way. Attempts to use human skeletal material have not given very satisfactory results, for instance, in determining blood groups. There is a claim that Egyptian mummies have frequencies of ABO similar to present frequencies in Egypt, and that American Indian mummies do not; but the ABO substances are easily destroyed by bacteria, and substances similar to them are produced by some bacteria. The possibility of such contaminations, therefore, makes these results uncertain. Moreover, changes of gene frequencies that might have occurred in a few thousand years are trivial, unless very strong selection is acting.

Some polymorphisms are less transient than others.

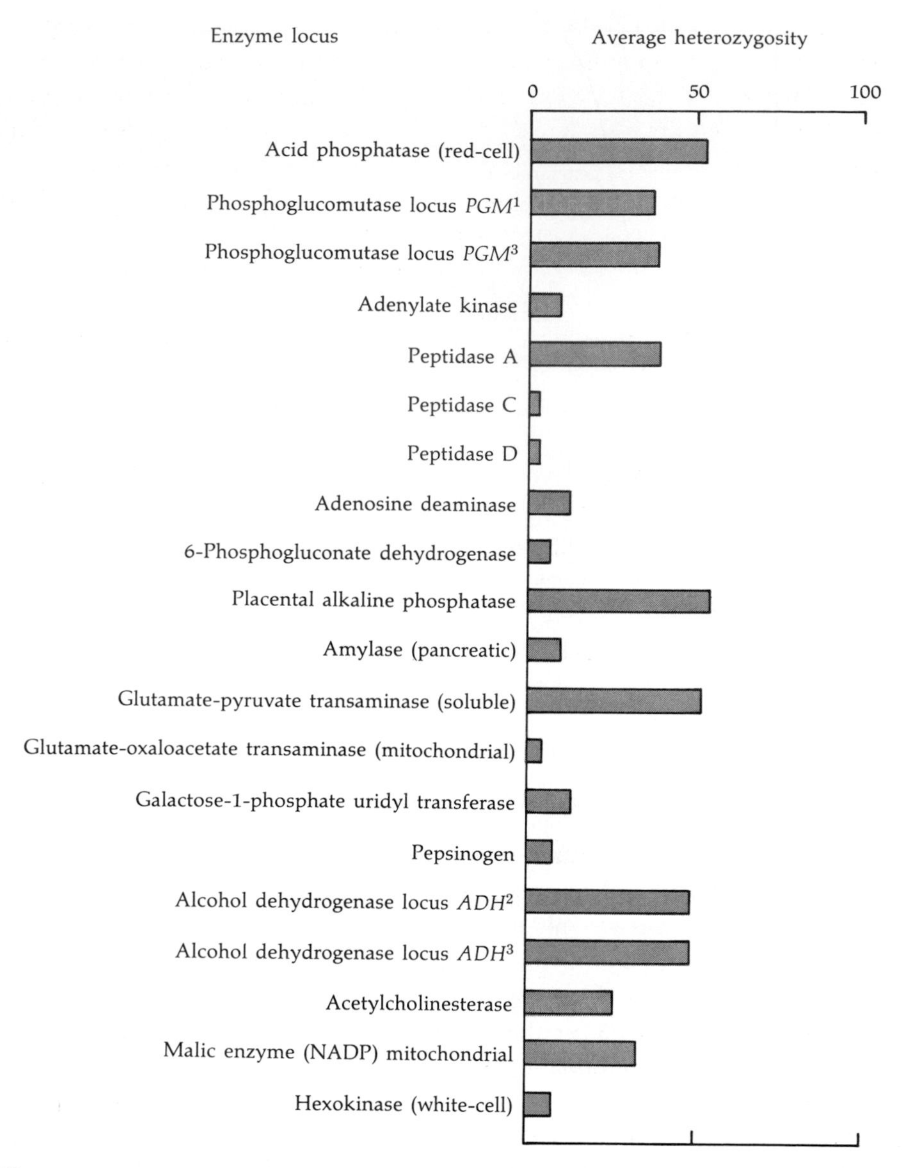

Figure 13.14
Heterozygosity of polymorphic human loci. This chart shows the average heterozygosity (proportion of individuals who are heterozygous) evaluated by electrophoretic methods for 20 enzyme loci found to be polymorphic in man. (For sources of data, see H. Harris and D. A. Hopkinson, *Annals of Human Genetics,* vol. 36, pp. 9–20, 1972.)

In the absence of analyses over a sufficiently long period of time, it is impossible—except in special cases—to say whether the frequency of an allele is on the increase or decrease because of selection, or fluctuating randomly because of drift, or at a stable equilibrium. The last case is that of a *balanced polymorphism,* and we have seen (Chapter 9) that this is most probably true for sickle-cell hemoglobin. There may be many other cases of a stable polymorphism due to advantages of the heterozygote, but this was the first found, and it is still in practice the only one for which adequate evidence is available. What fraction of all polymorphisms is stable for similar reasons we cannot say, but there are theoretical grounds for thinking that it may not be the majority. Some at least, as already pointed out, must be *transient polymorphisms:* the gene is polymorphic because it is caught at a time when one allele is in the process of replacing another, because of either selection or random genetic drift. Molecular evolution shows that substitutions do occur. In that part of the time between the first appearance of a mutant and its fixation, especially in the latter part of the process when the new mutant is relatively frequent but not yet fixed, the corresponding locus will by definition be polymorphic.

On the other hand, even stable or balanced polymorphisms like sickle-cell anemia, as pointed out in Chapter 9, are by no means permanent. The selective agent, malaria, may disappear because of changes in the environment; the parasite may evolve, or other more efficient genetic modes of resistance may develop. It is possible that many polymorphisms at one time expressed an adaptation to external conditions, such as diseases, that no longer exist. Such polymorphisms may now be neutral (or may even have reversed their adaptive value). These are, then, "relic polymorphisms."

In the long run, therefore, even stable polymorphisms are transient, and everything is in a state of flux.

Formulas developed on the basis of events in molecular evolution show that the observed overall amount of polymorphism is not far from the amount that would, in any case, be expected. This approach also, however, does not seem to be able to discriminate properly between natural selection and drift. Probably the best approach is that of studying each polymorphism on its own merits and so trying to determine possible causes of selection. We already have evidence, as in sickle-cell anemia, of the selective forces involved in some cases. One might attempt direct estimation of fitness values; however, this can usually only be successful if large selective differences are involved. To demonstrate that one allele shows no selective advantage or disadvantage over another to a degree that would be significant from an evolutionary point of view, observations would be required on an enormous number of individuals, presently well beyond our means. Other approaches may prove more fruitful in answering the problem of the relative roles of selection and drift in evolution.

We are certain that natural selection plays the fundamental role in shaping adaptation. We have become aware, however, that chance phenomena may also play an important part, but we are still far from being able to tell their relative importance, except in special circumstances. Moreover, studies of molecular evolution have so far largely been based on protein-coding genes. The evolution of DNA that does not code for protein is another subject for study. It may be that regulatory genes—which need not code for proteins, or may produce proteins at concentrations that are too low for study by present techniques—are subject to evolutionary change at rates different from those for the genes that have so far formed the basis for analysis of evolution at the molecular level.

One important complication is that our study of evolution, limited to one gene at a time, is likely to be, in some respects, an oversimplification.

Natural selection acts on individuals, not directly on genes. Genes, moreover, are not transmitted independently, but are organized in chromosomes. A gene subject to strong selection will affect, indirectly, all genes in the neighboring chromosome region. This creates considerable complications: theoretical models that take this effect into account are complex and, so far, their analysis is limited. But consideration of linkage relationships is important, as shown by the linkage disequilibrium observed for the HLA region (Chapter 10).

Summary

1. Proteins evolve mostly by amino-acid substitutions. The comparison of the same protein in different organisms shows that divergence, evaluated by the number of amino-acid differences, increases with increasing evolutionary separation.
2. Different proteins evolve at different rates, and some parts of a protein molecule are conserved more than others.
3. The average rate of substitution per amino acid is of the order of one per billion years. This rate corresponds approximately with expectations from theoretical considerations.
4. Substitutions that are deleterious are always eliminated by natural selection. Some substitutions may confer no selective advantage or disadvantage on carriers—that is, they may be selectively neutral.
5. The proportions of substitutions that have occurred by natural selection and those that behave as effectively neutral are difficult to ascertain; undoubtedly, however, natural selection must play a role in molecular evolution.
6. In every population, at least one gene out of three is polymorphic. Many of these polymorphisms may be transient, and may reflect the fact that gene

substitution is taking place. Some polymorphisms will be balanced by heterozygote advantage.

7. The amount of genetic variation in any population as measured by the extent of polymorphism, or average level of heterozygosity, is very large.

Exercises

1. What are the major contributions of molecular studies to the understanding of evolution?
2. Why do different proteins exhibit different evolutionary rates of amino-acid substitution?
3. Why do some polypeptide sequences within a given protein show different rates of amino-acid substitution from other sequences within the same protein?
4. What is the significance of duplication in molecular evolution?
5. Outline the arguments for and against the hypothesis that most amino-acid substitutions observed in molecular evolution are neutral.
6. What are the arguments for and against most observed polymorphisms being selectively neutral?
7. When the extent of chemical differences between pairs of amino acids was indicated on an arbitrary scale, it was found that the amino-acid substitutions observed in molecular evolution are relatively rarer, the greater the difference between amino acids. (a) Is this evidence in favor of the neutralist or the selectionist hypothesis? (b) Is it a decisive argument in favor of either?
8. There are some differences between the patterns of amino-acid substitutions observed in molecular evolution, and those observed in rare hemoglobin variants in man. For instance among the latter, several mutations have been found affecting one of two histidine sites believed to be important for the heme attachment. Substitution of these amino acids leads to methemoglobin, which has an altered affinity for oxygen, causing a pathological syndrome. These substitutions are not observed in molecular evolution. Can you explain why?
9. Nonpolar amino acids (which are hydrophobic—that is, "abhor" water) are usually found inside a protein molecule. Polar (hydrophylic, or water-compatible) amino acids are usually external. Can you give one reason why polar amino acids are usually substituted by other polar ones, and nonpolar by nonpolar ones in molecular evolution?
10. A polypeptide chain is usually folded where the amino acid proline is found. The functions of a protein are largely determined by its molecular shape. Would you expect substitutions involving proline to be more, less, or equally frequent when compared to other substitutions?
11. Differences in shape and size of organisms are often determined by genes affecting growth rates of groups of cells in the developing organism. What category or categories of genes are most likely to be responsible for such differences? Are they necessarily coding for protein genes? Are they necessarily *not* protein-coding genes?
12. Growth hormone is produced in the pituitary gland. Its absence (associated with recessive mutants observed in both mice and man) causes dwarfism. The growth hormone of some species (such as man) is inactive in the mouse. How could this evolution have occurred?

References

Cavalli-Sforza, L. L., and W. F. Bodmer
1971. *The Genetics of Human Populations.* San Francisco: W. H. Freeman and Company. [Chapter 11 gives a more extensive treatment of molecular evolution.]

Dickerson, R. E.
1971. "The structure of cytochrome *c* and the rates of molecular evolution," *Journal of Molecular Evolution,* vol. 1, pp. 26–45.
1972. "The structure and history of an ancient protein," *Scientific American,* vol. 226, no. 4, pp. 58–73.

Harris, H., and D. A. Hopkinson
1972. "Average heterozygosity per locus in man: An estimate based on the incidence of enzyme polymorphisms," *Annals of Human Genetics,* vol. 36, pp. 9–20.

Kimura, M.
1968. "Evolutionary rate at the molecular level," *Nature,* vol. 217, pp. 624–626.

Kimura, M., and T. Ohta
1971. *Theoretical Aspects of Population Genetics.* Princeton, N.J.: Princeton Univ. Press.

Lewontin, R. C.
1974. *The Genetic Basis of Evolutionary Change.* New York: Columbia Univ. Press.

Neyman, J., and E. L. Scott
1971. (Eds.) *Berkeley Symposium on Mathematics, Statistics and Probability, Vol. V, Proceedings of the Sixth Symposium: Evolution.* Berkeley: Univ. of California Press.

PART III

COMPLEX TRAITS AND THE NATURE–NURTURE PROBLEM

In Parts I and II we describe the analysis of a category of traits that share an important feature: they can be explained by a simple genetic scheme. In such cases, given the phenotype of an individual, we can easily infer his genotypic constitution. Dominance effects may cause some uncertainty about the exact genotype of an individual, but very little else is uncertain in these cases. There are many traits, however, that are not so easily explained by such simple models.

There are two basic reasons for complexity. The first is the existence of multiple loci that affect the phenotypic trait being studied. It is easy to see that a multiplicity of loci will generate potential confusion. When we study genetic differences due to two alleles at one locus (the simplest situation), we have only three genotypes to distinguish. If there are more than two alleles, or more than one locus, or both, there is a great increase in the number of genotypes that must be studied. In the models discussed thus far, theories were tested through experiments that involved counting the number of individuals with one or another genotype. With very many genotypes to be counted, and to be accounted for by any theory, we have a much more difficult task before us.

The other reason for complexity is even more formidable. In some conditions, genotypes cannot be sharply distinguished from one another; they overlap phenotypically. In some cases, a different technique of phenotypic analysis may help, but such an approach is not always possible. An obese person may have a genetic tendency to overeat, but he will not become obese if his food intake is limited. Studies of this trait are hampered by the difficulty of separating genetic tendencies from environmental and behavioral factors that may "mask" the expression of the genotypes. How can we identify "potentially obese" individuals who have not had the opportunity to eat enough to become obese? If we knew more about the physiology of obesity, we might be able to detect such a person—perhaps from his hormonal condition—but we cannot do this yet.

Food is only a part of the environment in which we live. Environmental conditions may be powerful factors altering the expression of a genetic trait, and they control the development of every individual. The organism is well buffered against the vagaries of environmental conditions, but there are inevitable environmental effects—at least for many traits. These traits will often be subject to a combination of genetic and environmental influences—of nature and nurture, as it is often put. The resolution of these two kinds of influences requires special techniques of analysis, to which this part of the book is dedicated.

The "environment," as we use the term here, is very broadly defined. It includes every external influence on development, including conditions in the mother's womb. In short, it includes everything that is not accounted for by genetic instructions encoded in the chromosomal DNA. It is clear that behavioral traits may show especially complex interactions between nature and nurture. Behavioral traits are, in addition, among those of greatest practical interest. Therefore, we devote special attention to them in Chapter 17.

14

Basic Models of Polygenic Inheritance

Traits may be inherited in ways that are too complex to be described by the simple Mendelian schemes we have studied so far. But those models can be extended to cover more complicated situations. Very frequently, confusion is also generated by the existence of sources of variation other than genetic differences. Quantities known as "heritabilities" have been designed to estimate the relative importance of genetic factors of variation. But there are many obstacles that often make it difficult to accept such estimates or to use them for practical purposes.

14.1 Continuous and Discontinuous Variation

Mendel arrived at his discovery by carefully selecting traits that were easy to study. Differences investigated were sharply defined, with no intermediates. By reducing the analysis to essentials, he could predict successfully the behavior of a genetic mechanism without having any idea of its physical basis. What we call Mendelian inheritance refers to traits that follow the scheme Mendel devised (see Chapter 2). When we look at traits that satisfy uncomplicated Mendelian inheritance, we are in the ideal condition of applying a simple model that gives clearcut expectations and, because of its simplicity, lends itself very well to mathematical analysis. But for this to be true, we must select traits that satisfy the requirements.

The basic condition for simple Mendelian analysis is that genotypes of individuals can be clearly identified. This is easier, the smaller the number of gene differences involved—possibly only one locus, with few alleles (two or three)—and the smaller the range of environmental differences that affect the manifestation of the trait. Ideally, Mendelian analysis aims at the analysis of differences between individuals at the level of DNA. The more removed the trait studied is from the DNA, the more complex is likely to be the situation.

Let us consider a trait that is on the borderline of being a good Mendelian character—namely, taste sensitivity to phenylthiocarbamide or phenylthiourea (PTC). It was found many years ago that some individuals (nontasters) can hardly taste this substance, while it has a very bitter taste for others (tasters). Taste sensitivity to PTC can be tested by a refined method—namely, by determining the individual taste-sensitivity threshold. This is done by asking an individual to differentiate between eight cups, four of which contain a suitable concentration of PTC, while the other four contain no PTC (but are otherwise indistinguishable). If the individual can correctly tell which cups contain PTC and which do not, he is declared able to taste PTC at the concentration added to the cups. The test is repeated with a series of dilutions, ranging from the most dilute (solution 13) up to the most concentrated (solution 1). Solution 1 contains twice as much PTC as does solution 2, solution 2 twice as much as solution 3, and so on. The taste threshold is then the lowest concentration at which an individual can correctly differentiate the cups (with and without PTC), and is a characteristic of that particular individual.

The results of determining the taste threshold on a large number of individuals can be plotted as a bar diagram: their "frequency distribution" (Fig. 14.1; in the Appendix, the meaning of a frequency distribution is discussed in detail). The distribution given in Figure 14.1, for the English population, clearly has two peaks, and is therefore called bimodal, the *mode* being a frequency peak.

Among the English, solution 5 seems to provide a useful standard for distinguishing between tasters and nontasters. Any person who cannot taste solution 5 will, of course, be unable to taste higher dilutions (solutions 6 through 13). However, most of these "nontasters" are able to taste solution 1, and many can taste solutions 2 and 3 as well. On the other hand, most of the individuals who can taste solution 5 will also be able to taste solutions as weak as 8, 9, or even 10. Use of solution 5 as a test for tasting ability gives the sharpest possible distinction between tasters and nontasters. However, the shape of the distribution indicates immediately that the separation between tasters and nontasters is imperfect. There is an overlap between the two classes (tasters and nontasters) that is an inevitable source of error in genetic studies of this trait.

In the Chinese population, the distribution is discontinuous. The frequency of individuals with thresholds of 2, 3, or 4 is negligibly small. Therefore, solution 3 can be used to provide an unambiguous division of the population into tasters and nontasters. In the English population, PTC sensitivity is not strictly discon-

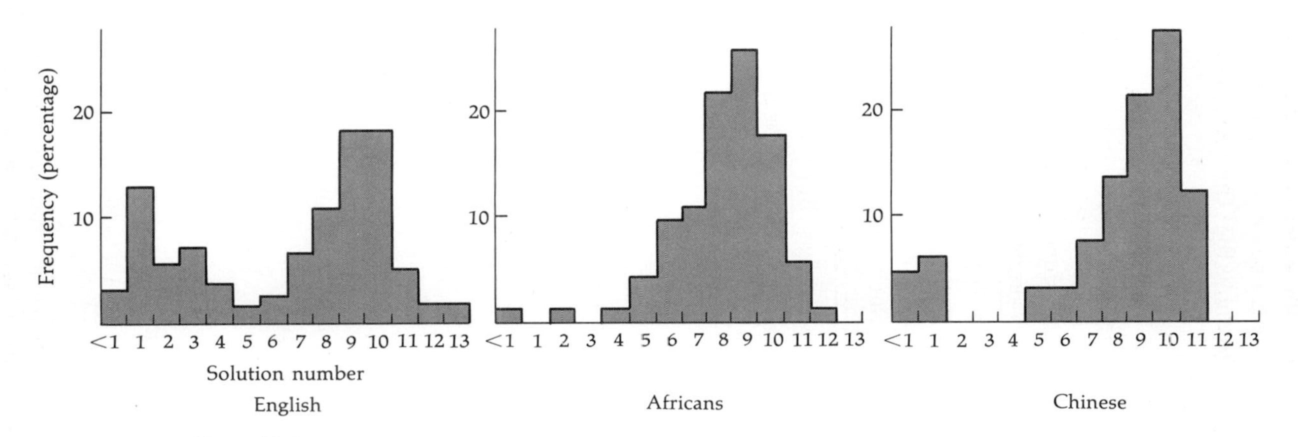

Figure 14.1
Distributions of taste thresholds for phenylthiocarbamide (PTC) in English, African, and Chinese populations. The strongest solution used (solution 1) has a concentration of 0.13 percent PTC in water; solution 2 has half this strength, solution 3 half that of solution 2, and so on. Solution 5 discriminates best between nontasters (at the left) and tasters (at the right) in the English population. For Africans and Chinese, who have a lower proportion of tasters, solution 3 probably provides the best discrimination. Clearly, classification of a sample of English individuals into tasters and nontasters may give rise to significant errors of classification. (From N. A. Barnicot, *Annals of Eugenics,* vol. 15, 1950.)

tinuous, but the bimodality is strong enough that most individuals can be placed unambiguously in one category or the other. The inheritance of this trait is sufficiently close to Mendelian that it is usually considered as due to a single-gene difference. In fact, family studies show that the incapacity to taste PTC tends to behave as a Mendelian recessive trait.

Many traits show a continuous, homogeneous variation.

The distributions of many traits show no signs of bimodality; unlike PTC among Caucasians they have only a single peak. Figures 14.2, 14.3, and 14.4 show examples of such unimodal distributions for height, weight, and intelligence quotient (IQ). These distributions are regular and "bell-shaped," the height and IQ distributions being essentially symmetrical around the peak, while the weight distribution is somewhat skewed. Such distributions give no hints about how to form phenotype classes that might correspond to different genotypes. Any possible subdivision may seem to be fruitless. But even here, rare extreme variants (not sufficiently numerous to give rise to a separate peak) are found that are due to

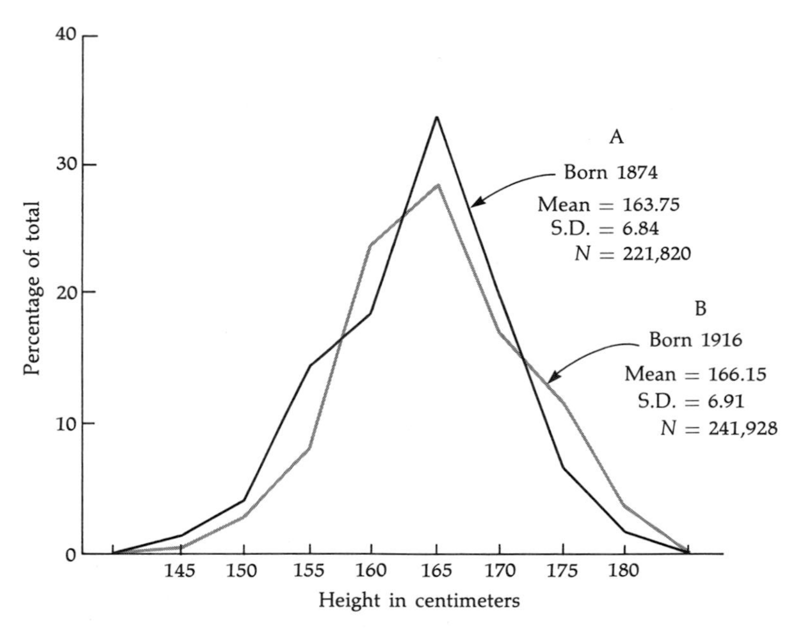

Figure 14.2
Distribution of stature (height) of 20-year-old Italian males. Class intervals of 5 cm are used. Curve A shows individuals born in 1874; not shown are individuals measuring less than 142.5 cm (0.7 percent of total) or over 182.5 cm (0.3 percent). Curve B shows individuals born in 1916; not shown are individuals measuring less than 142.5 cm (0.7 percent of total) or over 182.5 cm (0.7 percent). (Data from A. Costanzo, *Annali di Statistica,* 1948.)

a single-gene difference. This is true, for instance, of some types of dwarfism that are simple recessives, and which hormonal analysis shows to result from a lack of pituitary growth hormone. It is also true for another type of dwarfism, achondroplasia, a dominant condition usually discernible from other forms in a variety of ways. Similarly, some cases of mental deficiency at the extreme lower end of the IQ curve can be shown to be due to single-gene differences. Several mental deficiencies are caused by particular biochemical changes, as in the case of PKU. There is thus evidence for a single-gene difference underlying many of these extreme variants for height, weight, or IQ.

Many extreme variants in otherwise continuously distributed traits are thus the outcome of single genes. Because these have major effects, they are called *major genes.* But is there room for effect of genes in the minor differences that are observed among the less extreme variants—that is, in the middle portion of the distribution? It is clearly more difficult to ascertain genes that have minor effects on these traits. One may be tempted to call them *minor genes;* and, in fact, they are often so called.

Direct evidence for minor-gene effects can be obtained in many ways. One is to look at major genes, which have themselves been ascertained through clearcut manifestations, but which may affect other traits in a minor way. Phenylketonuria (PKU) is a good example. Phenylketonurics are ascertained biochemically by an assay of phenylalanine levels in blood (see Chapter 3), and affected homozygotes

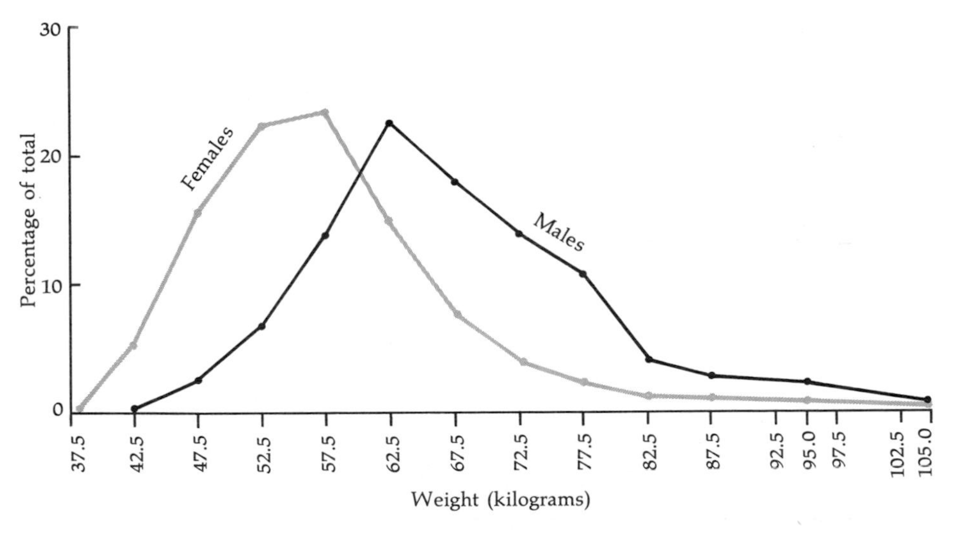

Figure 14.3
Distribution of weights of 17-year-old U.S.A. youths, 1966–1970. The dark line shows the distribution for males; the gray line shows that for females. Average ages for the samples were 17.51 years for males and 17.50 years for females. Not shown are those weighing more than 110 kg (0.5 percent of males and 0.5 percent of females). (Data from *U.S. National Health Survey, Series 11,* National Center for Health Statistics.)

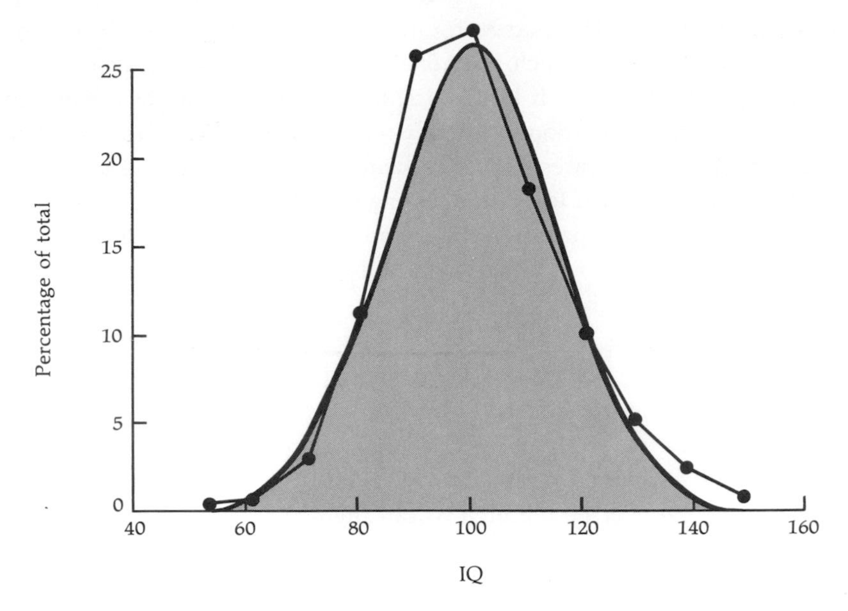

Figure 14.4
Distribution of IQ among the 14,963 children born in Scotland on February 1, May 1, August 1, and November 1, 1926. The smooth curve is the theoretical normal distribution with mean of 100 and standard deviation of 15. (Data from MacMeekan.)

show a clear difference in this characteristic from homozygous normals or from heterozygotes (Fig. 14.5). IQ tests on homozygotes also show a clearcut difference, though now there is a very slight overlap with the normal IQ distribution, due to other factors affecting the difference and giving rise to some PKU individuals who have higher IQ than the lower "normals." If one used IQ to discriminate PKUs from normals, one would not have a perfect classification—just as in the case of PTC taste sensitivity.

Other traits also show a difference between the two classes, but one that is much less marked. Head shape differs, for instance, but only in a minor way, as does hair color. Incidentally, the latter difference is expected, considering that hair pigment is formed from tyrosine, and PKU individuals, who cannot synthesize tyrosine from phenylalanine, therefore have a lower concentration of this amino acid. The four traits—percentage of phenylalanine in blood, IQ, head shape, and hair color—are progressively less affected by the gene difference, as indicated in Figure 14.5. On head size and hair color, PKU has minor effects only, while its effect is large on the traits through which it was discovered.

There must be many other genes which, like that for PKU, have only a modest effect on head shape or hair color. When, therefore, we select a particular trait

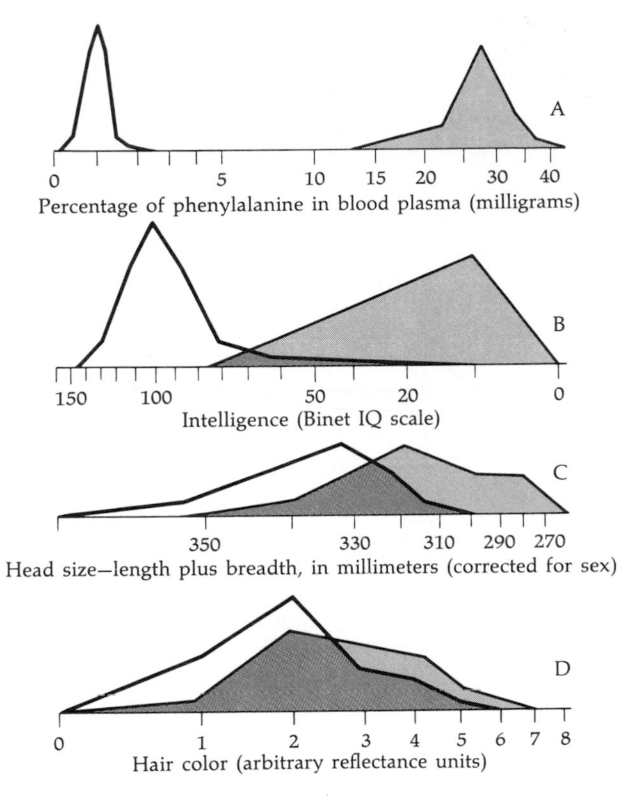

Figure 14.5
Frequency distributions for various traits among phenylketonurics and control populations.
In each graph, the curve for phenylketonurics is the one that extends farther toward the right. (A) Blood phenylalanine levels. (B) IQ. (C) Head size. (D) Hair color. (After L. S. Penrose, *Annals of Eugenics,* vol. 16, pp. 134–141, 1952.)

for study, we may expect that a variety of genes, whose direct action may be difficult to establish, affect the trait to some small extent. In particular, it is likely that continuously varying traits such as stature are affected by many minor genes.

In some cases, a perfectly continuous distribution may obscure a regular Mendelian trait.

Electrophoresis of the red-cell enzyme, acid phosphatase, shows that six genotypes determined by three alleles (*A, B, C*) can be clearly distinguished (Fig. 14.6). The alleles give rise to enzymes that have some differences in enzymatic activity—*A* being the least active, *B* intermediate, and *C* the most active. Heterozygotes have activities that are almost exactly intermediate between the respective homozygotes. The *CC* homozygote is not shown in the figure, being rare (the *C* gene frequency is the lowest of the three). The enzyme activity alone would not be very

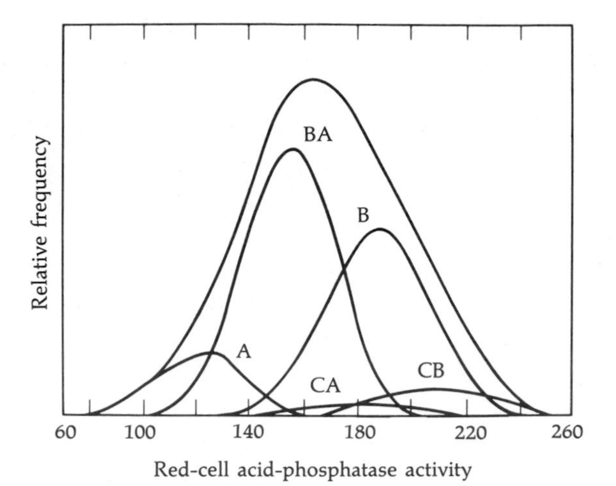

Figure 14.6
Distributions of red-cell acid-phosphatase activities. The upper curve shows the distribution in the general population; the other curves show distributions for the separate phenotypes (A, BA, B, CA, and CB) that are distinguished by electrophoretic analysis of the enzyme. The curves are constructed from values of the enzyme activity and from the relative frequencies of the phenotypes observed in a randomly selected population. Note that the distribution in the general population (upper curve) shows a perfectly continuous variation. The phenotypes cannot be distinguished on the basis of enzyme activity alone, because of their overlapping distributions. (From H. Harris, *Proceedings of the Royal Society,* ser. B, vol. 164, pp. 298–310, 1966.)

useful for distinguishing the six genotypes. The genotypes overlap (some more and some less completely) using the criterion of enzymatic activity.

Many sources may contribute to the variation between individuals of the same genotype: (1) error of the assay; (2) other genes that may influence the trait; and (3) "environmental" variation that may cause differences between individuals with the same genotype. It is theoretically possible to distinguish between the three sources of variation. The first is most easy to evaluate, by repeating the assay on the same samples. The distinction between the next two sources of variation is less easy. Problems of this kind will, in fact, occupy us throughout most of this part of the book.

14.2 Environmental and Polygenic Variation

Experimental breeders can produce "pure lines," made up of individuals all of which are almost identical genotypically, by inbreeding repeatedly for many generations (Chapter 11). They can cross these pure lines and make observations on the hybrids, cross the hybrids among themselves, and obtain the so-called F_2 generation. (In breeding experiments, the parental or P generation is made up of individuals from distinct, pure-breeding lines. The F_1 or first filial generation is made up of the offspring obtained when these lines are crossed. If the members of the F_1 generation are crossed among themselves, their offspring are called the F_2 generation. The F_3 generation is obtained using members of the F_2 generation as parents, and so on.) Breeders can carry out other crosses at will, and so they can obtain information about the inheritance of "complex" traits. By setting up conditions of growth that are as homogeneous as possible, they can "control" the environment—that is, reduce the effect of environmental variation from one individual to another, thus ensuring that all individuals are raised in conditions that are as homogeneous as possible. The breeder cannot eliminate environmental variation entirely, but at least he can ensure that individuals from different lines or crosses are grown in what is, *on average*, the same environment.

In Figure 14.7, we represent three hypothetical experiments carried out by a mouse breeder with two pure lines of mice of different weight. The weight difference is assumed to be due to a single gene. We use A to represent the allele for which all heavy mice are homozygous, and a the allele for which all light mice are homozygous. If the breeder is completely successful in standardizing the environment in which each mouse of either line is grown, there will be no variation within

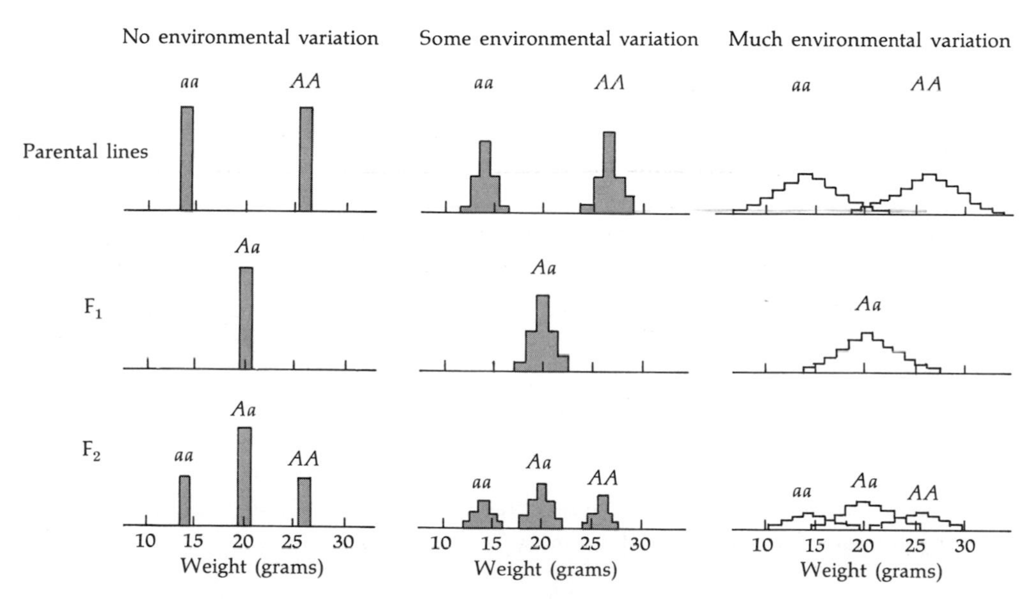

Figure 14.7
Hypothetical crossing experiments. The experiments involve crosses between pure lines of mice, differing only for a single gene determining weight. With no environmental variation, strictly Mendelian distributions would be expected (lefthand column). With moderate (center) to strong (right) environmental variation, the genetic effects are "blurred" by environmental effects on the trait, particularly in the case at the right.

the line. Let us assume that in such a case, all heavy mice weigh exactly 26 grams, and all light mice weigh exactly 14 grams. Let us further suppose that the heterozygote *Aa* is exactly intermediate between the two homozygotes, *Aa* and *aa*. Then the F_1 generation will be made up of mice all weighing 20 grams each (Fig. 14.7, left). Crossing F_1 mice together will produce again the two parental types of 14 and 26 grams (one-quarter of the mice being of each of these two genotypes) and the F_1 intermediate types (one-half of the mice).

No mouse breeder is likely to be that successful in standardizing environmental conditions. Some mice may find more food, others less. Many factors—including, for example, infections—that the breeder is unable to control may add to the variation. However, so long as, *on average,* conditions remain the same for the various lines and crosses, the means will remain the same as before. But now there is, in addition, individual variation within the parental lines and the crosses. The picture now looks like those in the middle or the righthand portion of Figure 14.7, which differ in the degree of environmental variation hypothesized. In the middle figure, this variation is not sufficient to prevent the distinction between the three genotypes. In the righthand figure, however, the variation does blur this distinction, and the overall distribution of the F_2 shows hardly any signs of the underlying genetic heterogeneity, almost as in the case of the acid-phosphatase activity shown in Figure 14.6. Here, spurious homogeneity is created only by environmental variation.

Continuous variation can also be generated by polygenic systems.

We call the determination of a trait *polygenic,* or multifactorial, when many genes affect the same trait and their effects are cumulative. Let us start, for simplicity, with two genes (loci) and assume that our breeder crosses lines with the same weights, as before, but now the difference is due to two genes with alleles *A,a* and *B,b.* These two genes account for the divergence in weight of the two lines in a very simple manner: every time one substitutes a capital (*A* or *B*) for a lower-case (*a* or *b*) letter, 3 grams are added to the weight of the mouse. The heavy parental line (genotype *AABB*) then weighs 12 grams more than the lighter (14-grams) one (genotype *aabb*), and therefore weighs 26 grams; the F_1 (genotype *AaBb*) weighs 20 grams, exactly as in Figure 14.7. If the two genes are on different chromosomes, they will segregate independently (Section 2.5), and all nine possible combinations will be formed, as in the following diagram.

	AA	*Aa*	*aa*
BB	*AABB*	*AaBB*	*aaBB*
Bb	*AABb*	*AaBb*	*aaBb*
bb	*AAbb*	*Aabb*	*aabb*

The proportions for gene *A,a* will be $\frac{1}{4}$ *AA*, $\frac{1}{2}$ *Aa*, $\frac{1}{4}$ *aa*, and the same for *B,b*; those for genotypes *AABB*, *AaBB*, . . . , and so on will be obtained as the products of the frequencies of genotypes for the single genes, as follows.

	AA $\frac{1}{4}$	*Aa* $\frac{2}{4}$	*aa* $\frac{1}{4}$
BB $\frac{1}{4}$	$\frac{1}{16}$	$\frac{2}{16}$	$\frac{1}{16}$
Bb $\frac{2}{4}$	$\frac{2}{16}$	$\frac{4}{16}$	$\frac{2}{16}$
bb $\frac{1}{4}$	$\frac{1}{16}$	$\frac{2}{16}$	$\frac{1}{16}$

The phenotype of each genotype can be computed by the rule that every capital letter adds 3 grams to the basic weight 14, of the *aabb* genotype, giving the following weights for the nine genotypes.

	AA	*Aa*	*aa*
BB	26	23	20
Bb	23	20	17
bb	20	17	14

There are only $\frac{1}{16}$ of all mice with phenotype 26; but there are two genotypes with phenotype 23 (*AaBB* and *AABb*), accounting for $\frac{4}{16}$ of all mice. Continuing for all phenotypes, we can give the results as a bar diagram of the F_2, as in the lefthand panel of Figure 14.8.

This reasoning can easily be extended to any number of genes, and the middle and right portions of Figure 14.8 show the results for three and six genes. We have kept the differences between parental strains at 12 grams, and so the effect of each gene inevitably decreases. In Figure 14.7, with a one-gene model, an allele with a capital letter contributed 6 grams; in Figure 14.8 (I), the contribution of such an allele was 3 grams; in (II) 2 grams; and in (III) 1 gram. In these crosses we have not introduced any environmental variation. But we can see that by simply increasing the number of genes (as in Fig. 14.8, part II), the gaps between the genotypes become progressively smaller. Qualitatively, the result of variation due to many genes is similar to—and in practice, indistinguishable from—that depicted on the right of Figure 14.7, where we had only one gene, but with environmental variation.

We cannot determine, just from the observation of a bell-shaped distribution, whether the variation is environmental or whether there are one, two, or many genes affecting the trait. But a comparison of the F_2 distributions of Figure 14.8 indicates a way to estimate the number of genes, which we shall use in Section 14.6. Looking at the spread of individual values in the F_2, we can see that the increase in the number of genes concentrates the individual values around the mode (or

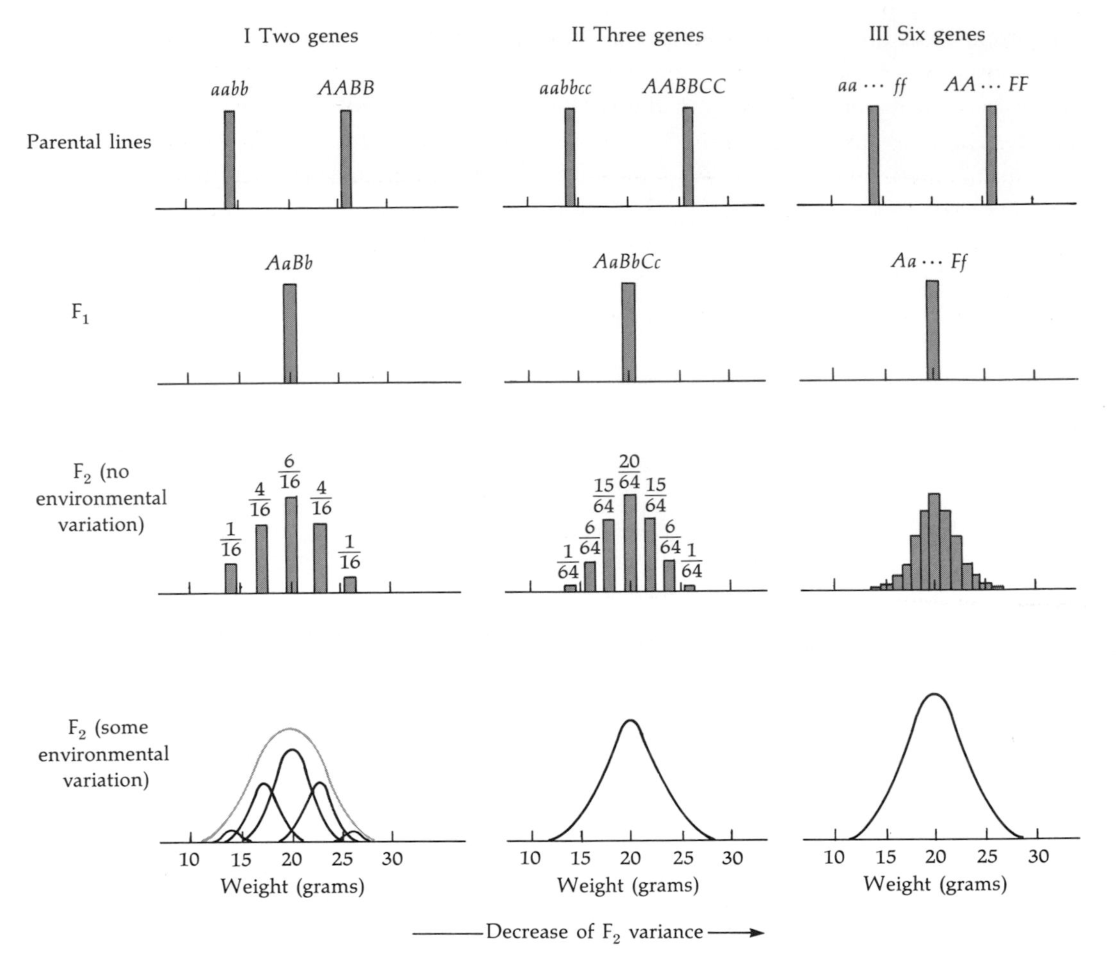

Figure 14.8

Crosses between two lines differing by two genes, three genes, or six genes. Distributions are shown for P, F_1, and F_2 generations in the absence of environmental variation. The bottom line shows the F_2 distributions as they would appear with some environmental variation. In each case, the homozygous "lightweight" parental line has a weight of 14 grams. In case I, alleles *A* and *B* each cause an increase of 3 grams in weight each time they occur in an individual. In case II, alleles *A*, *B*, and *C* each add 2 grams; and in case III, alleles *A*, *B*, . . . , *F* each add 1 gram. Note that an increase in the number of genes determining the same difference between the two parental strains leads to a decrease in the variance of the F_2 distribution.

most commonly observed value). For this and for many other purposes, we need a measurement of the spread—that is, of the variation of individual values around the mode.

The almost universal measure of variation is a quantity called the variance.

Its definition and mode of computation from data are discussed in detail in the Appendix. The variance is very simply related to another quantity, also used by statisticians: the standard deviation (SD). The variance is simply the square of the SD. The SD is expressed in the same units in which the original variates and their means are expressed, and so its meaning is easier to grasp intuitively than that of a variance (Fig. 14.9). But variance and SD are easily and unambiguously transformed one into the other. R. A. Fisher demonstrated an important property of the variance, as we see in the next section—namely that it can be partitioned into fractions, each of which has a specific meaning, of interest for the geneticist.

14.3 Partitioning of the Variance and Computing Heritability

The variance is a fraction whose denominator need not concern us for the time being. Its numerator is "the sum of squares of deviations from the arithmetic mean." The computation of the variance is a relatively simple operation. Suppose we have three observations, say the height measurements of three subjects (in inches):

$$70, \quad 65, \quad 69.$$

We first compute their arithmetic mean by summing $70 + 65 + 69 = 204$, and dividing by the number of observations, which is 3, giving the mean as $204/3 = 68$. The deviations (or differences) from the mean of the three observations are

$$70 - 68 = 2, \quad 65 - 68 = -3, \quad 69 - 68 = 1.$$

Continuing with the use of our recipe, we must square the deviations,

$$2^2 = 4, \quad (-3)^2 = 9, \quad 1^2 = 1,$$

and sum them:

$$4 + 9 + 1 = 14.$$

This result is the numerator of the variance. The denominator is simply the number of observations minus one. The numerator is often called *SS* (short for Sum of Squares).

The reason for introducing this quantity is that we want to compute it for a population of individuals of different genotypes raised in different environments, in order to see how to separate environmental from genetic variation. It is important to emphasize that most of the concepts that we develop here require experimental control of the environment. This can be done in man only rarely, and is much easier to do with experimental animals and plants. Therefore, the concepts

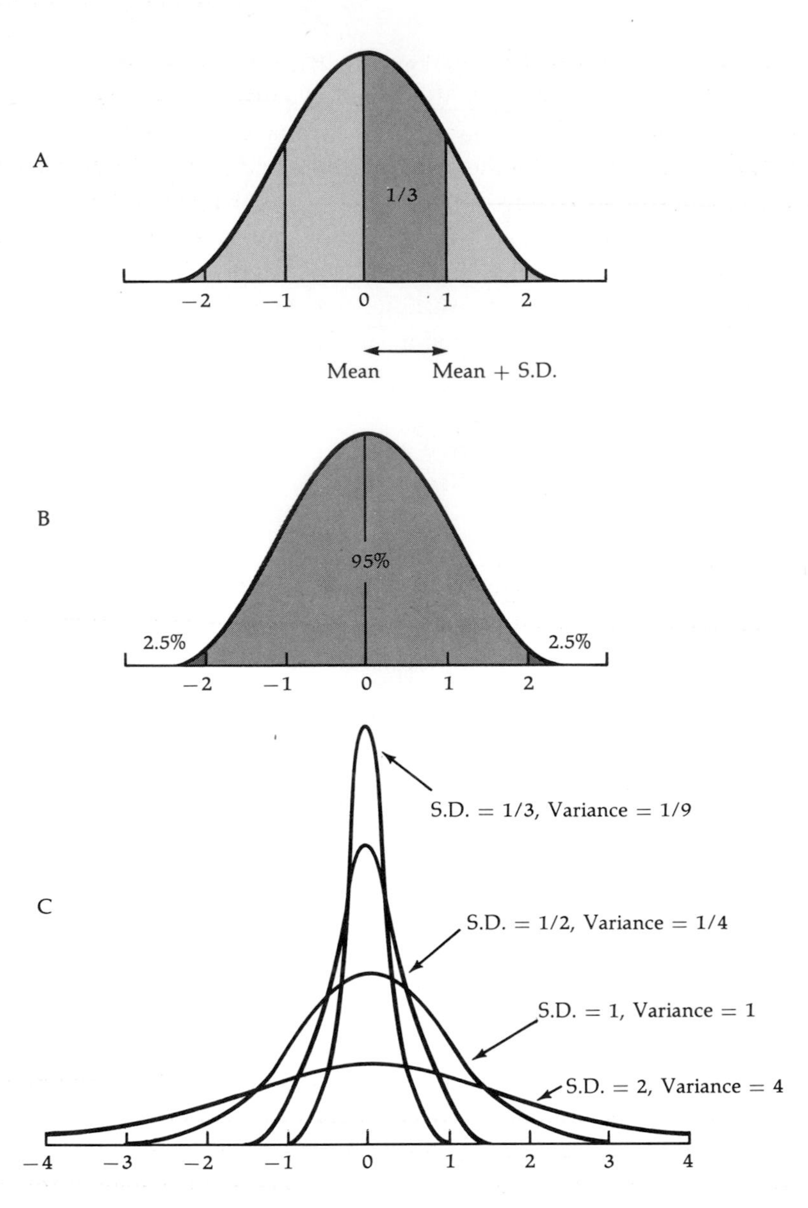

Figure 14.9
Meaning of standard deviation and variance. The distributions are gaussian (or normal) in shape (see the Appendix for theoretical definition). The scale of the horizontal axis in graphs A and B is constructed so that the mean is at zero and the standard deviation (SD) is equal to one (and so is its square, the variance). About one-third of all individuals in the distribution fall between the mean and one SD above the mean (dark area in graph A). The distribution is symmetrical around the peak (mean), so two-thirds of the individuals fall between one SD below the mean and one SD above the mean. As shown in graph B, 95 percent of the individuals in the distribution fall between two SDs below the mean and two SDs above the mean. Graph C shows four normal distributions, all with the same mean but with different SDs. The smaller the SD, the smaller the spread or dispersion of the individual values.

that we expound are more clearly visualized in terms of experiments on organisms other than man. But we need to clarify them in order to avoid some important pitfalls in the analysis of human data.

To simplify things to the extreme, let us imagine that there exist only two genotypes (A, B) and two environments (E1, E2); and the population consists of only four individuals, one of type A living in E1, one of type A in E2, one of type B in E1, and one of type B in E2. The trait we study is the amount of hemoglobin in the blood, which we will assume to be normally around 14 grams per 100 milliliters. The hemoglobin in the four individuals is found to have the following values.

Individual		Genotype		Environment		Hemoglobin		
Individual	1	Genotype	A	Environment	1	Hemoglobin	16	g/100 ml
"	2	"	A	"	2	"	14	"
"	3	"	B	"	1	"	12	"
"	4	"	B	"	2	"	10	"

There is some variation among the four individuals. Some are anemic, especially those of genotype B. The mean for all four individuals is 13 g/100 ml. It is easy to check that

$$SS = (16 - 13)^2 + (14 - 13)^2 + (12 - 13)^2 + (10 - 13)^2 = 20.$$

This value is called the phenotypic or total SS.

Suppose that we were able to eliminate environmental differences, obtaining the following results.

Individual		Genotype		Environment		Hemoglobin		
Individual	1	Genotype	A	Environment	1	Hemoglobin	16	g/100 ml
"	2	"	A	"	1	"	16	"
"	3	"	B	"	1	"	12	"
"	4	"	B	"	1	"	12	"

In this case, the mean amount of hemoglobin is 14 g/100 ml, and $SS = 16$. This SS refers to *genetic* variation only, because the environment has been equalized. (Note that choice of environment 2 would yield hemoglobin values of 14, 14, 10, and 10, with a mean of 12, but the SS would remain the same at 16. E2 is not as good as E1 for production of hemoglobin, but there is no change in the genetic variation.) Note that the genetic SS ($SS_G = 16$) is less than the total SS ($SS_T = 20$).

Now let us do another experiment: equalize the genotype, while letting the environments differ as in the original experiment. In this case we take four individuals, all of genotype A, and place two of them in E1 and two in E2.

Individual		Genotype		Environment		Hemoglobin		
Individual	1	Genotype	A	Environment	1	Hemoglobin	16	g/100 ml
"	2	"	A	"	1	"	16	"
"	3	"	A	"	2	"	14	"
"	4	"	A	"	2	"	14	"

In this case, the mean is 15 and $SS = 4$. (Had we made all four individuals of genotype B, we would have obtained a mean of 11 and $SS = 4$. Genotype A is better than genotype B at making hemoglobin in these environments, but the value of SS for uniform genotype is independent of the particular genotype.) The estimate of SS obtained in this case measures *environmental* differences only, so we call it SS_E.

We have now measured the total variation ($SS_T = 20$) and two "partial" contributions—one due only to genotypic differences ($SS_G = 16$) and the other due only to environmental differences ($SS_E = 4$). Graphically, we can represent this result as follows.

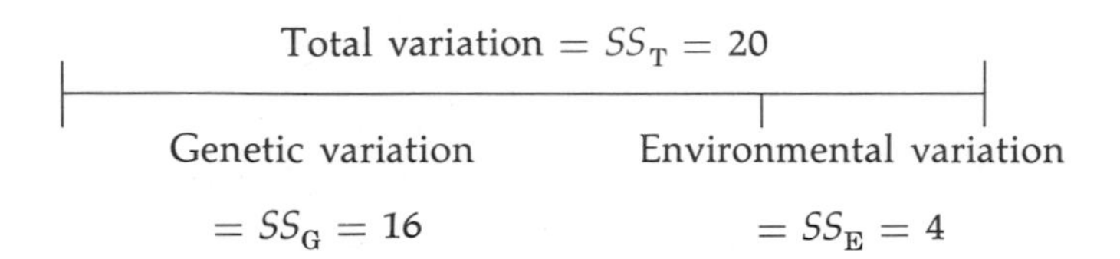

The message is that the phenotypic (total) variation can be partitioned into two fractions—genetic and environmental—which sum to the total.

There are, however, specific conditions for the validity of partitioning the total variation into genetic and environmental fractions. The message just given is correct only in part. The example was constructed in a special way: we assumed that genotype A is "better" (more exactly, makes more hemoglobin) than B in *all* the environments tested (E1 and E2) *by the same amount* (4 grams). This is not always necessarily true. In fact, whenever adequate testing is done, it is rarely *strictly* true. But, if breeders are careful in controlling sources of environmental variation, they may be able to minimize this potential source of confusion.

We explore the possible sources of confusion further in a later section. Here we just consider the simpler situation, in which genetic and environmental variation add up to the total variation. We have demonstrated additivity for SS. Each SS value is turned into a variance by dividing it by the same denominator—namely, the number of observations less one. Therefore, the additivity shown for SS values ($SS_G + SS_E = SS_T$) is also valid for the respective variances. Let V_T represent the phenotypic or total variance, V_G represent the genetic or genotypic variance (due to differences between genotypes), and V_E represent the environmental variance (due to effects of differences in the environmental conditions to which the individuals examined were exposed). Then, we have

$$V_T = V_G + V_E.$$

Or, in words, the total variance can be partitioned into a variance due to effects of environmental differences (in short, the environmental variance) and a variance due to effects of different genotypes (the genetic variance). The genetic variance divided by the total variance (V_G/V_T), expressed as a fraction or a percentage, is called the *degree of genetic determination*, meaning the proportion of variance due to genetic effects. This value is also sometimes called the *heritability* in the broad sense. In the example we have been discussing, the heritability is $\frac{16}{20} = 0.8$, or 80 percent. In later chapters, we discuss the determination of this value in actual genetic studies.

Neither the environmental nor the genetic variance is in any sense a "constant" quantity.

The two variances measure the differences that have been experienced by the individuals forming the population. Two genetically similar populations grown by breeder X and breeder Y may differ in their heritabilities. If breeder Y is less punctilious, the environmental conditions for his animals may show greater fluctuations, and therefore V_E and V_T may be larger for his populations than for those of breeder X, while V_G may remain the same. Heritability will then be smaller in the breed raised by Y. The effect of genes may also change in different environments: genotypes *A* and *B* of our hemoglobin example may, in two environments unlike E1 and E2, differ by less (or more) than 4 grams. In such a case, V_G will change, and the estimate of heritability may be affected. Thus, a heritability is strictly valid only for a given population under the particular conditions of measurement.

14.4 Crosses Between Pure Lines

Crosses between "pure lines," developed as discussed in Section 14.2, can now be analyzed using the concepts described in the two preceding sections. We use first an example from plants: the corolla length in a species similar to the tobacco plant (Fig. 14.10). The first generation of hybrids (F_1) between the inbred parental lines (P) is formed of individuals all genotypically identical. The variance in the F_1 can therefore be taken as an estimate of V_E. The same is true for each of the parental lines; in fact, their variances are quite similar, judging by the spread of the curve, and are also comparable to that of the F_1 (the parental line on the left has a slightly smaller variance, but we shall ignore this difference). The F_2, however, has a larger variance; in the F_2, genotypes segregate. The variance of the F_2 must therefore be the sum of $V_E + V_G$. The F_2 total variance is computed from the original data to be about 43; the estimate of V_E from averaging the variances of the two Ps and the F_1 is about 7. Thus, the genetic variation V_G can be obtained by subtracting $V_E = 7$ from $V_T = V_G + V_E = 43$, obtaining $V_G = 43 - 7 = 36$. The heritability in the F_2 is thus $\frac{36}{43} = 84$ percent. Note that this is a simplified

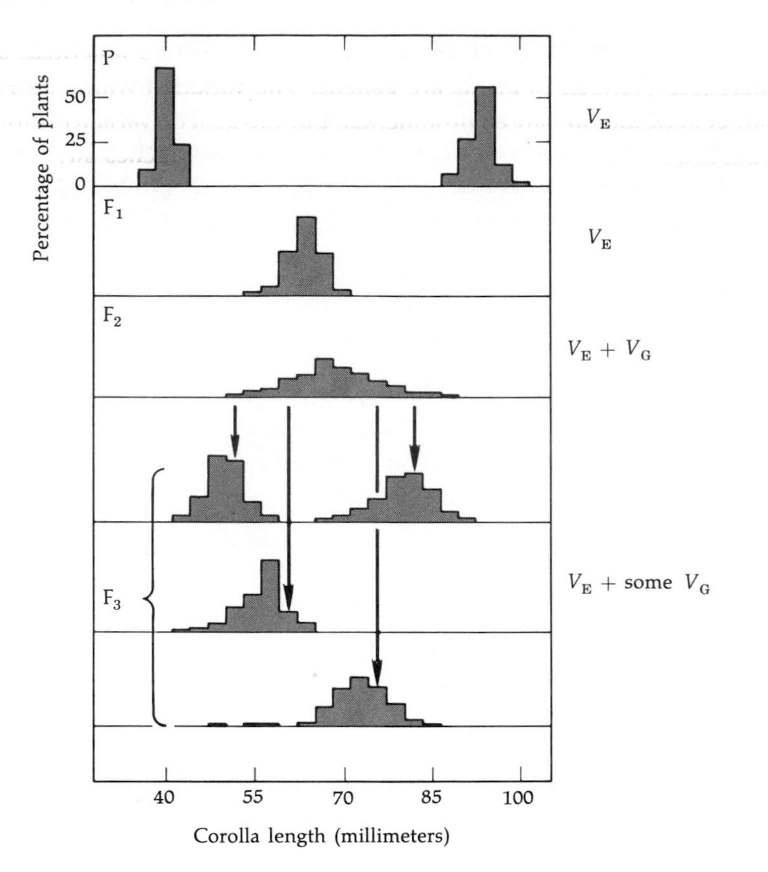

Figure 14.10
Inheritance of corolla length in Nicotiana longiflora. The results are shown as the percentage frequencies with which individuals fall into classes, each class covering a range of 3 mm in corolla length and centered on 34, 37, 40, . . . mm. This grouping is quite artificial, and the apparent discontinuities in the distributions are artifacts of the class groupings; corolla length actually varies continuously. The means of F_1 and F_2 are approximately intermediate between those of the parents. The means of the four F_3 families are correlated with the corolla length of the F_2 plants from which they were obtained by selfing, as indicated by the arrows. Variation in P and F_1 is all nonheritable (environmental), and hence is less than the variation in F_2, which includes additional variation arising from the segregation of the genes involved in the cross. Variation in F_3 is, on the average, less than that of F_2 but greater than that of P and F_1. The amount of variation among the different F_3 families differs according to the number of genes that are segregating. (After K. Mather, *Biometrical Genetics,* Methuen, 1949; and E. M. East, *Genetics,* vol. 1, pp. 164–176, 1916.)

treatment. A more exact one would take account of the fact that the variances of the two Ps and F_1 are not quite equal, but tend to increase with increasing mean.

The F_3 distributions arise from "selfing" (many plants can be self-fertilized) individuals of the F_2, whose corolla lengths are indicated by the positions of the arrows. The peak of every F_3 corresponds closely to the value of the trait in the

F_2 individual that was selfed. This confirms that the trait is inherited, and that the differences between F_3 plants are genetic. The variation within an F_3 family is in part genetic and in part environmental, but the genetic variance within an F_3 distribution is less than that in the F_2. If there are many genes affecting corolla length, it is probable that an individual of the F_2 does not have *all* the gene differences that existed between the two parental lines. In fact, that individual is likely to be homozygous for some of those genes, and heterozygous for others. It is only the latter that contribute to the genetic variance in the F_3. Correspondingly, the heritabilities in the four F_3 families are less than that in the F_2.

How many genes affect the corolla length in this cross? In other words, for how many genes affecting the corolla length did the two parental lines that were crossed differ? The three examples of Figure 14.8 suggest a way to answer this problem: the variation in the F_2 decreases with increasing number of genes when the difference between the two parental lines is kept (as in that figure) constant. In fact, a simple formula can be used to compute the minimum number of genes from P, F_1, and F_2 data. It is found to be about 9 in the case of the corolla length. It is, however, beyond our scope to give such formulas, or to point out their pitfalls.

There is perhaps only one trait (skin color) for which this approach to determining the number of genes is possible in man.

Man, as we know, is genetically very variable, and even apparently inbred populations are actually very little inbred and still very variable. Thus, there are no such things as pure lines in man, and it would be unthinkable to use this approach in human genetics. Yet, there is one *trait* to which this approach could be applied: skin color. For this characteristic (and almost for no other), there are populations that have been naturally selected to be almost extreme, black or white. This does not at all guarantee "purity" of the lines, but at least makes the genetic differences between the lines greater than that within, so that, to a first approximation, the latter can be neglected. There has since occurred some crossing between the "lines" (Fig. 14.11). The F_1 and F_2 peaks are intermediate between the two parental lines, indicating no dominance of the genes involved. The results of the F_2 agree with the hypothesis that there are perhaps four genes determining skin color in man. Each of these four genes may be responsible for about one-fourth of the skin color difference between blacks and whites; the full difference is the outcome of their cumulative action.

14.5 Artificial Selection

All animal and plant breeding has been largely dependent on "artificial" selection; namely, the choice as reproducers of individuals that show some desirable quality. Whether it was yield, their resistance to disease or to climatic condition, in cereals;

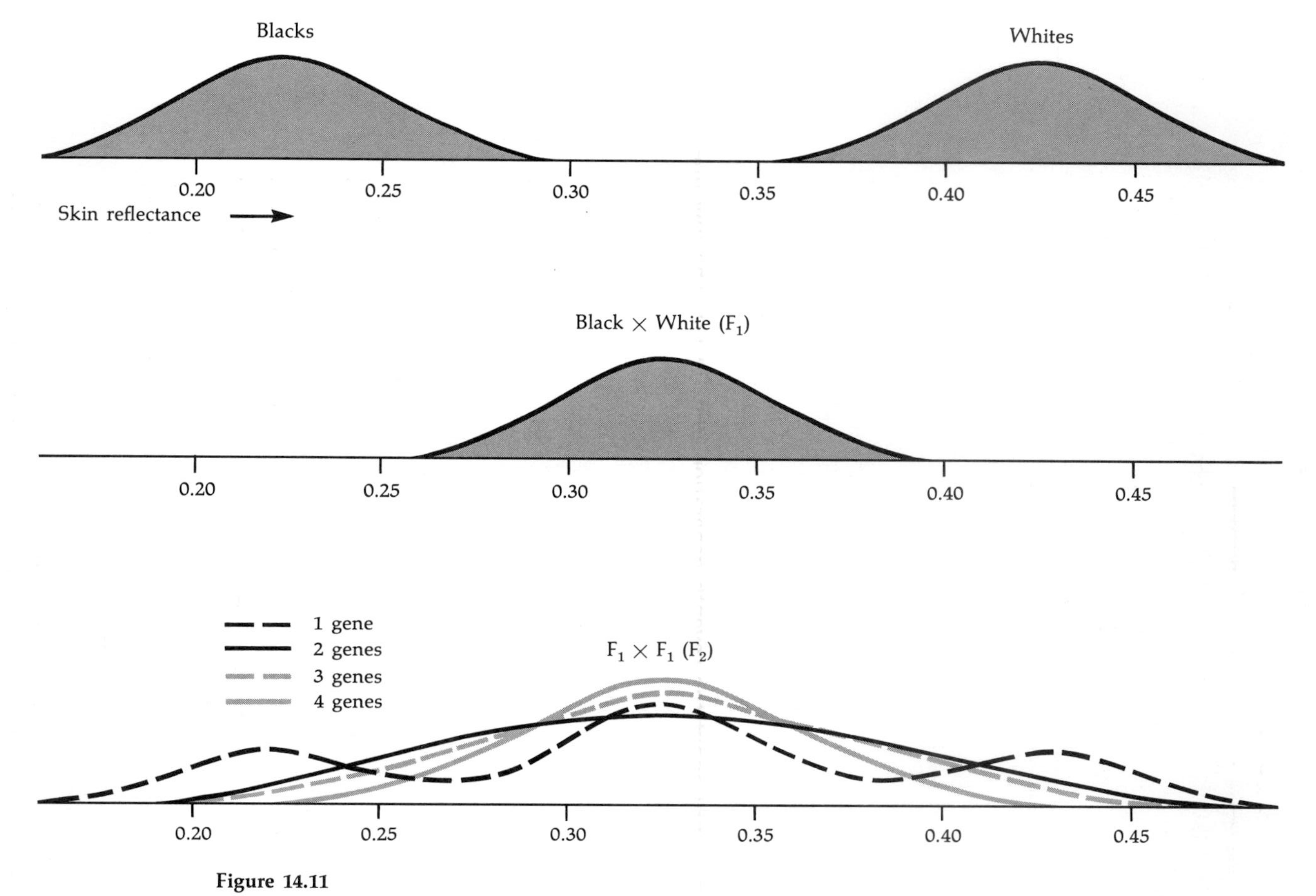

Figure 14.11
Skin-color distributions in blacks and whites. Skin color is measured by the skin reflectance for light of 685 mμ wavelength. For the F_2 generation, distributions shown are those expected under various hypotheses about the number of genes involved. Observations on F_2 and backcrosses tend to resemble those expected if the trait is determined by three or four genes. (Based on research by G. A. Harrison and J. J. T. Owen, *Annals of Human Genetics,* vol. 28, pp. 27–37. An independent analysis on American blacks by C. Stern yields essentially the same results.)

speed or endurance in horses; coat color or odd shapes in pets—when the trait was inherited, selection has permitted the attainment of some degree of fixation of the desired traits in the progeny.

Most desirable traits (like cereal yield) are continuous and depend on many genes. Selection permits only small advances to be made each generation, but repetition over many generations can eventually bring about substantial improvement. Response to selection for a quantitative trait is a powerful way of showing the existence of genetic variation. Only if there is genetic variation can we expect a trait to change under artificial selection. If, for example, the variation in stature were fully environmental, the offspring of extremely tall individuals would show a stature distribution identical to that for the offspring of normal or short individuals.

An example of a selection experiment on rats for an interesting behavioral trait was started by E. C. Tolman in 1924, and repeated by R. C. Tryon, who carried it on for 22 generations. The character selected for was the capacity to run a maze. Both a "high" and a "low" selection line were maintained, choosing as reproducers at every generation those that performed best (for the high line) and worst (for the low line). Using as a score for ability the mean number of errors made while running the maze, there was a decrease in the number of errors made in the high line, until about the seventh generation (Fig. 14.12). The low line,

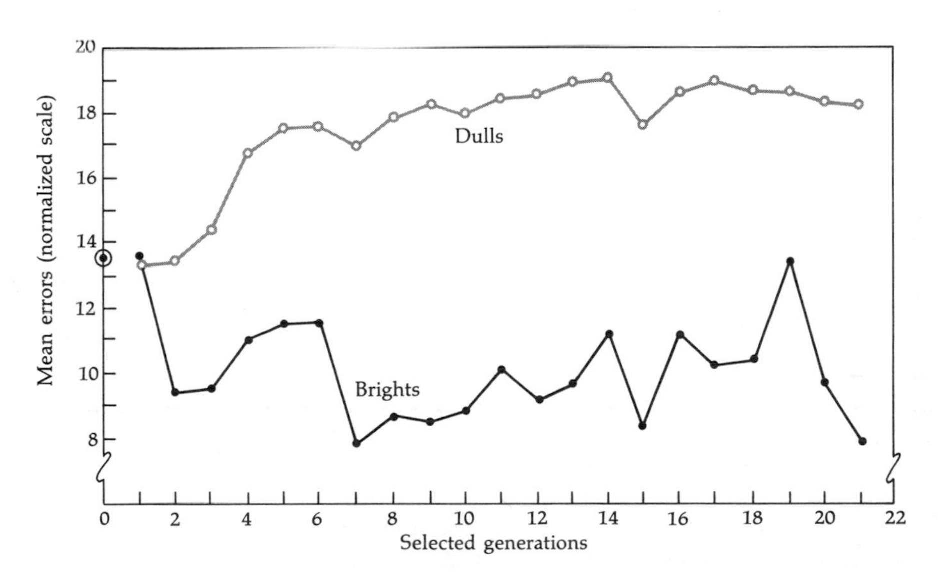

Figure 14.12
Results of selective breeding for maze brightness and maze dullness in rats. These are the results obtained by R. C. Tryon in his classic experiment. (From G. E. McClearn, in *Psychology in the Making,* ed. L. J. Postman, Knopf, 1963.)

selected for a large number of errors, apparently continued to change until about the thirteenth or fourteenth generations. The observed means are inevitably subject to large oscillations, as the populations involved were small (see Appendix). It is hard to say whether the experiment really measured "brightness" and "dullness," or rather some other related characteristic. But it did show that there is some inheritance for a trait that is at least vaguely connected with IQ in man.

By a selection experiment, one can measure the existing genetic variation, but of a specific kind.

A simple selection experiment can tell us how much genetic variation there is, relative to that of environmental origin. Let us select the "best" individuals of a population for some measurable, continuous trait. We compute (1) the mean of *original* population (M_{or}) from which the individuals were selected, and (2) the mean of the individuals *selected* as reproducers (M_{sel}). These latter are allowed to reproduce and their progeny measured for the same trait. We then compute also (3) the mean of their *offspring* (M_{off}). Figure 14.13 shows three possible cases. In the first at the left, the offspring of selected individuals have almost the same mean (22.7) as their parents (23.0). In Figure 14.13 at the right, the progeny of those selected has about the same mean as the original population. Selection has had almost no effect. The example in the middle of Figure 14.13 is intermediate; the mean of the offspring of those selected ($M_{off} = 21.5$) is midway between the original population ($M_{or} = 20$) and that of the chosen reproducers ($M_{sel} = 23$).

The efficiency of selection is measured by heritability, H_N, (also called, for reasons that we shall see, heritability in the narrow sense). This is a ratio of the actual gain to that expected under full effect of selection. If no selection is practiced, one would expect the mean of the offspring to remain equal to that of the original population, $M_{or} = 20$. If the trait were inherited in full, and by the simplest mechanism, the progeny of the selected parents would be expected to have mean equal to their parents, $M_{sel} = 23$, implying an expected gain due to selection equal to $M_{sel} - M_{or} = 23 - 20 = 3$ units. This is the gain expected if the offspring are phenotypically identical on average to their parents. Such a result can occur only if there is no environmental variation and if, in addition, the individual phenotype reflects in a simple way the genotype. That is, any reshuffling of genes under the process of gene assortment under sexual reproduction must cause no change, on average, of the phenotype. As we shall see later this condition also is not necessarily met. In the hypothetical example shown at the left of Fig. 14.13, however, it is assumed that $M_{off} = 22.7$; the average trait in the children of the selected parents is almost as high as the parents' mean (23). The actual gain over the mean expected in the absence of selection, $M_{or} = 20$, is thus $M_{off} - M_{or}$

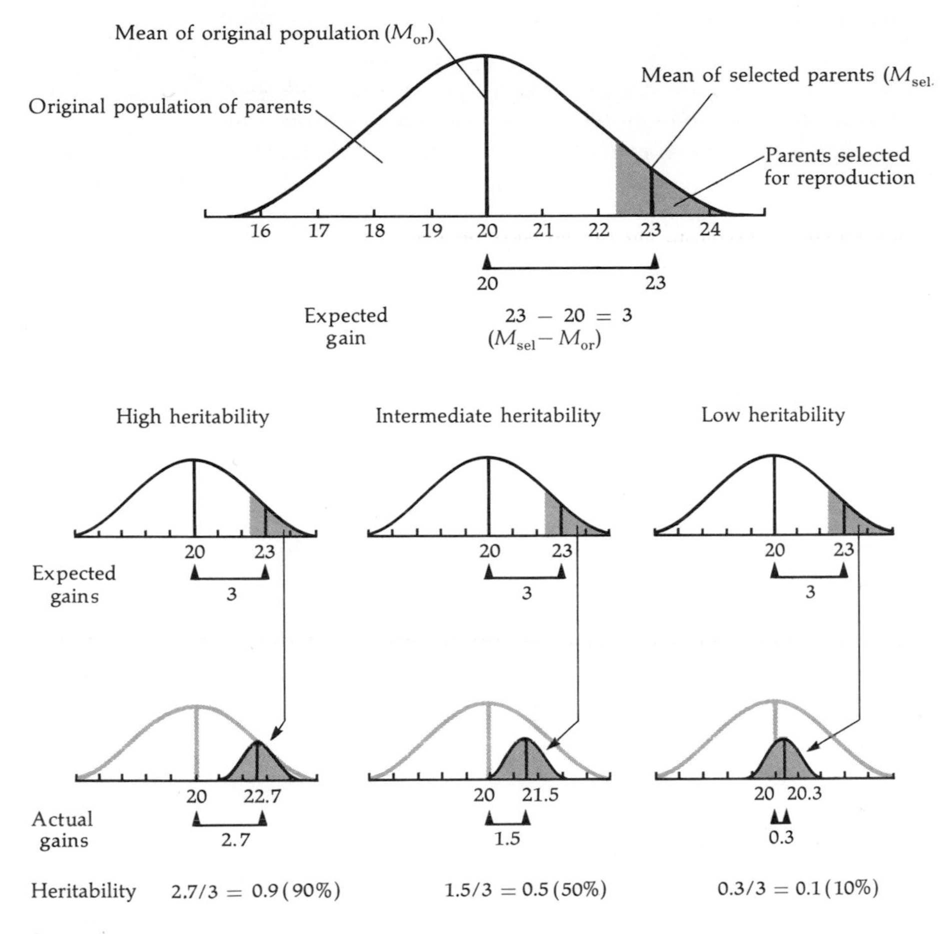

Figure 14.13
Heritability (in the narrow sense) computed from the outcome of experiments in artificial selection. The shaded area in the original population shows individuals selected for reproduction; they all have high values of the trait (averaging M_{sel}) because it is desired to raise the average value of the trait by use of selected reproducers that show high values of the trait. The average value of the trait in the original population is M_{or}, and that in the offspring of selected individuals is M_{off}. The higher the value of M_{off}, the more successful the experiment. Narrow heritability is the ratio between ($M_{off} - M_{or}$) and ($M_{sel} - M_{or}$); this value thus measures the relative degree of success of artificial selection. The distribution of offspring from the selected group is shown in the shaded portion of the progeny distributions; the dashed distribution given for comparison is that of the original population. In the experiment on the left, the progeny of the selected group has a mean similar to that of their parents, so the heritability is high. In that on the right, selection has almost no effect; heritability is low. In the center experiment, the situation is intermediate.

= 22.7 − 20 = 2.7 units, almost the same as the expected gain, 3 units. The narrow heritability is

$$H_N = \frac{M_{off} - M_{or}}{M_{sel} - M_{or}} = \frac{22.7 - 20}{23 - 20} = \frac{2.7}{3} = 0.9, \text{ or } 90\%.$$

Such a high heritability is rarely if ever observed. In practice values intermediate between those shown at the center and right of Fig. 14.13 (which have heritabilities of 50 percent and 10 percent, respectively) are more frequently encountered. In an extreme situation in which all the variation is of environmental origin, selection of parents with a high (or a low) value of the trait has no effect on the progeny. When the trait is not inherited the heritability will therefore be zero; but an H_N value of zero does not necessarily mean that the variation is entirely of environmental origin. One other possible complication is, for instance, dominance.

Dominance limits the efficacy of selection.

We now come to consider a special case: assume for simplicity only one gene difference, and also assume that the heterozygote, far from being intermediate between the two homozygotes, has an extreme manifestation of the trait (Fig. 14.14). One then speaks of *overdominance.* When the "best" are selected for reproduction, no effect can be achieved, because the *aa, AA,* and *Aa* types will all reappear in the progeny in the proportions expected for an F_2. Clearly, dominance can confuse and even annul the results of artificial selection.

It can be shown that the measure of narrow heritability used above refers, in fact, to the difference between the two homozygotes, which is called the "additive" component, and ignores dominance effects. A more elaborate treatment than we give here is needed to show why the fraction of genetic variance revealed by artificial selection is called "additive" (see the Appendix). Simply stated, the additive portion of the genetic variance (V_A) is the genetic variance that would exist if the heretozygote were exactly intermediate between the two homozygotes.

Another fraction of the genetic variance, that due to dominance (V_D), depends on heterozygotes not being exactly intermediate between the respective homozygotes, for any of the genes contributing to the trait. The genetic variance can be split into the two fractions (additive and due to dominance)

$$V_G = V_A + V_D.$$

The total variance, earlier analyzed into an environmental and a genetic portion can thus be subdivided into three fractions:

$$V_T = V_A + V_D + V_E.$$

The broad heritability, or degree of genetic determination is

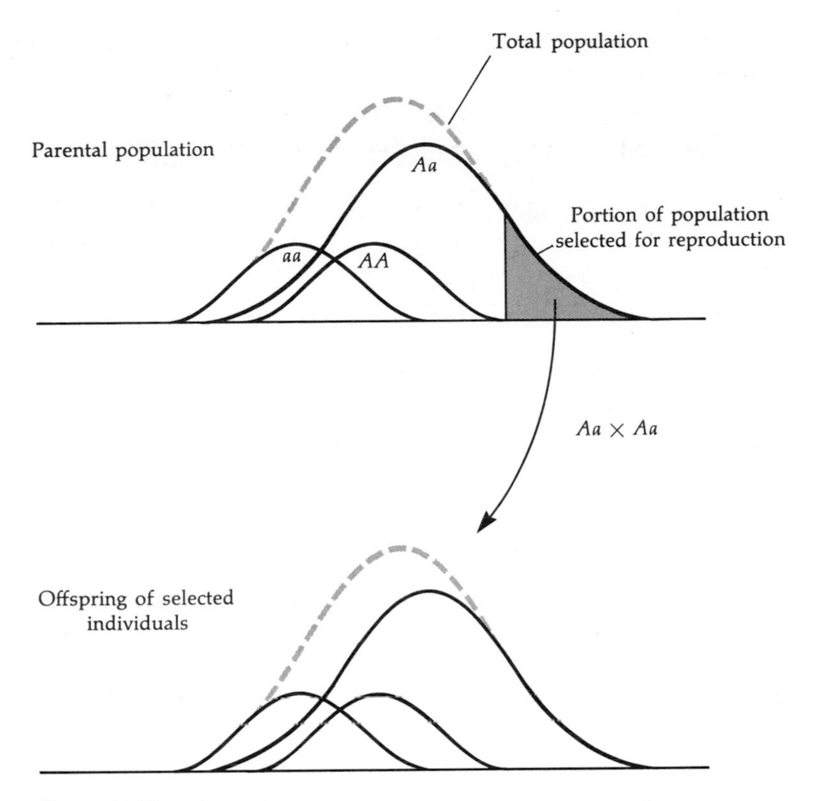

Figure 14.14
Inefficiency of artificial selection when there is overdominance. In this example, individuals are assumed to differ only in a single gene. The dashed line indicates the distribution of the total population.

$$H_{\mathrm{B}} = \frac{V_{\mathrm{G}}}{V_{\mathrm{T}}}$$

and measures cumulatively all genetic effects. The narrow heritability, as determined from measures of actual relative to expected gains in artificial selection, is

$$H_{\mathrm{N}} = \frac{V_{\mathrm{A}}}{V_{\mathrm{T}}}.$$

Because $V_{\mathrm{G}} = V_{\mathrm{A}} + V_{\mathrm{D}}$, the narrow heritability is in general smaller than the broad one.

This analysis of the total variation into the three fractions (additive, due to dominance and environmental) is valid only under a series of simplifying assumptions. We will indicate here two of them, which are especially important in applications to human populations. With nonexperimental organisms like man, it is unfortunately very difficult to test if these assumptions are met. If they are not, the estimates of the fractions of total variance called environmental, genetic, or additive may be severely biased.

A potential serious limitation is due to genotype–environment interaction.

In the example of hemoglobin used in Section 14.3, we had two genotypes *A* and *B* for which we gave the phenotype values in two environments. They are given again in Table 14.1, together with two new genotypes, *C* and *D*.

Note that genotype *C* is insensitive to the difference between the two environments: the hemoglobin level is unaffected. In genotype *D*, however, the situation of *A* and *B* is reversed—namely, environment E2 is now better than E1. When there are such combinations of genotypes and environments—that is, the reaction to various environments (as calculated in the last column of the table) is different for different genotypes—then V_G and V_E do *not* add to the total variance. Rather, we have

$$V_T = V_G + V_E + V_I,$$

where V_I is called the *genotype–environment interaction variance.*

An interesting example of this interaction was shown with the rats selected for maze-dullness and -brightness. These rats were, throughout the experiment, bred in ordinary laboratory conditions. The difference in capacity to cope with the maze problem was developed with artificial selection under those conditions. Does it remain unaltered when the rats are grown under other conditions? An experiment was done in which rats from the two lines were raised in enriched, exceptionally favorable conditions. The opposite conditions (severe deprivation) were also tested. In both cases, the difference between the two lines disappeared: both lines did equally well on the maze test when raised in an excellent environment, and equally poorly when raised in an unfavorable environment (Fig. 14.15). Thus, there is a genotypic difference between the two lines, but it can be revealed phenotypically only in some environmental conditions; in others, there is no, or almost no, phenotypic difference.

Table 14.1
An example of genotype–environment interaction

Genotype	Phenotypes in two environments		Change due to environment
	E1	E2	
A	16	14	+2
B	12	10	+2
C	13	13	0
D	11	14	−3

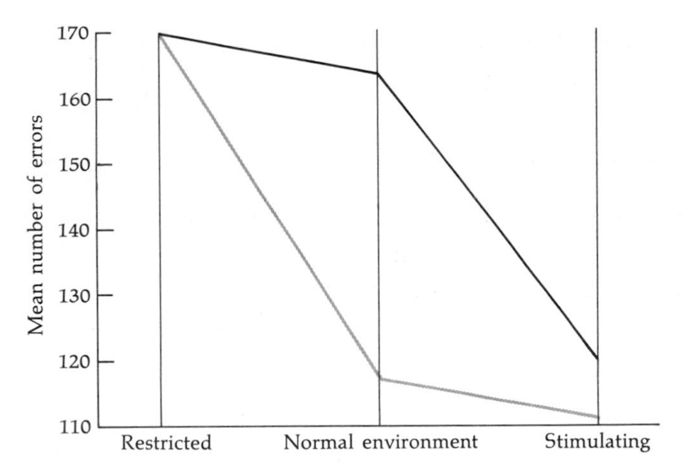

Figure 14.15
Genotype–environment interaction. The graph shows results of an experiment carried out by R. Cooper and J. P. Zubek of the University of Manitoba. The experiment involved two strains of rats: one selected for "brightness" (cleverness at finding their way through a maze) and the other selected for "dullness." In a normal environment, bright rats (lower curve) made only 120 errors, whereas dull rats made about 165 errors. When both strains were raised in a restricted environment, however, both made about 170 errors. When raised in a stimulating environment, both strains of rats did almost equally well. (From W. F. Bodmer and L. L. Cavalli-Sforza, "Intelligence and race," *Scientific American,* October 1970.)

In the rat example we have a controlled experiment in which the environment can be manipulated almost at will. When we examine a human population, we cannot control the environment. We can still, however, impose some *statistical* control; for instance, we can evaluate socioeconomic conditions on the basis of income, education, and many other quantities that can be observed, and correct the data for the possible effects of these variables. But it is misleading to think that such control always completely eliminates complex environmental effects. Moreover, the factors of the environment that are significant in determining a phenotype may escape our analysis, if they are—as they often are—unknown to us (see also Section 17.2).

Nonrandom distribution of genotypes among environments is another source of complication. It is called genotype–environment correlation.

The partition of the variance can be affected in yet another way that may invalidate estimates of V_G and V_E, namely, if there is *genotype–environment correlation.* The assortment of genotypes among the various environments, under the simple hypotheses on which the variance of a trait is usually partitioned, is assumed to be unbiased. For instance, in our example of hemoglobin in Section

14.3, one individual of genotype *A* was placed in E1 and one in E2; and the same was done for individuals of genotype *B*. If in a large population, different genotypes occupied the various possible environments at random, no bias would arise. But often this is not the case. Inequalities in the distribution of genotypes among environments gives rise to a genotype–environment correlation. The correlation can be in one of two opposite directions. There may be a tendency to compensate for a genotypic defect; this is the case for many diseases. A potential diabetic who is on a diet containing a smaller than normal amount of carbohydrates may, in part at least, compensate for his defect. By creating a more favorable environment for himself, he decreases the phenotypic difference between himself and a genotypically normal individual. The opposite may happen in highly competitive conditions where weaker individuals, under limited resources, may be unable to obtain, for instance, as much food as is necessary, and therefore become weaker. Phenotypic differences are then amplified.

The existence of these potential complications poses important caveats to the interpretation of the relative roles of nature and nurture, especially in human populations.

Summary

1. Environmental differences between individuals, or the presence of several genes affecting a trait, make the analysis of some traits very complex. A trait affected by many genes, each with small effect, is called polygenic.

2. In principle, one can partition the total variation (V_T) into a portion due to genotypic differences (V_G) and another due to environmental differences (V_E). Heritability in the broad sense is the ratio of V_G to V_T.

3. Response to selection reveals a fraction of the genetic variance, called "additive," V_A, that is due to differences between homozygotes.

4. Narrow heritability is the ratio of the additive V_A to the total variance, V_T. It is less than the broad heritability whenever there is dominance.

5. Crosses between "pure lines" can give some information on the minimum number of genes at play in a polygenic trait. In man, this approach has very limited potential, but has been used in the analysis of skin color.

6. Genotype–environment interaction limits the utility of heritability measurements. It is the consequence of different reactions of genotypes subjected to various environments. Genotype–environment correlation, the unequal representation of genotypes among various possible environments, is also a source of bias in estimates of heritability.

Exercises

1. What are the major characteristics of complex quantitative traits that distinguish them from simply inherited Mendelian traits?

2. How would you define the distinction between major genes and minor genes?

3. Does the fact that a trait has a unimodal continuous distribution imply that it must be influenced by the environment? Justify your answer.

4. What has been the contribution of the study of pure lines to understanding the inheritance of quantitative characters?

5. What information can artificial-selection experiments provide about the variance components of a quantitative trait?

6. Define heritability and justify your definition.

7. What is genotype–environment interaction? How does its existence affect the variance analysis of a quantitative trait?

8. From the data of Figure 14.5, the approximate values shown in Table 14.2 can be computed. In order to evaluate the difference between normals and PKU for each of the four traits on a scale that permits comparison of the various traits, divide the difference between the means of the two groups by the standard deviation. This measure estimates the discrimination between the two groups for each trait. List the four traits in descending order of their discriminating power.

Table 14.2
Data for Exercise 8

	Means		Standard deviation*
	Normals	PKUs	
Phenylalanine level in blood†	−0.05	1.35	0.11
IQ	100	18	15
Head size	335	311	12
Hair color	1.9	2.6	1.0

*Average of the SDs of the normal and PKU distributions.
†To equalize approximately the variation of the two groups, values for phenylalanine concentration in the blood are given on a logarithmic scale.

9. Compute the expected distribution in the F_2 of a cross, as in Figure 14.8, between two lines of 14 and 26 grams, respectively, if the difference between the two parental lines is due to 4 genes.

10. a. Compute the broad heritability from the following data on mouse adult weight (grams).

	Mean	*Standard deviation*
Strain A	25.2	2.1
Strain B	18.1	1.9
F_1 (A × B)	19.2	1.8
F_2	19.6	3.3

b. Can you say whether the genetic variance is likely to contain a dominance component greater than zero? Assume that differences between means of more than 1 g are significant.

11. Selecting from the F_2 of the cross given in Exercise 10 the 10 percent of heaviest mice, it was found that they weighed on average 26 grams. Mating these mice among each other, the progeny was found to weigh on average 22.8 grams. (a) Compute the narrow heritability. (b) Compare it with the broad one estimated in Exercise 10.

12. Selecting from the F_2 of the cross given in Exercise 10 the 10 percent of lightest mice, they were found to weigh on average 16.1 grams. The progeny of these mice weighed 18.8 grams. Compute the narrow heritability. This is lower than that found in the experiment of Exercise 11; can you give a possible explanation for this?

13. The following is a selection of the data given in Section 14.5.

Genotype	Environment	Phenotype
A	E1	16
A	E2	14
C	E1	13
C	E2	13

Compute by the methods discussed in Section 14.3 the following *SS*s: SS_T, SS_G, SS_E. Check that the last two sum to a quantity that is less than the first; compute the interaction sum of squares (SS_I), given that $SS_T = SS_G + SS_E + SS_I$. NOTE: To assume that *A*, or *C*, individuals are grown in the same environment, you can average between the phenotypic values of a genotype in different environments. To assume that there are no genotypic differences, average over the genotype values for each environment.

14. Repeat the calculation of Exercise 13 for the following set of data.

Genotype	Environment	Phenotype
A	E1	16
A	E2	14
F	E1	12
F	E2	14

15. The genetic variance in the F_3 is expected to be half that of the F_2. Given that the total variance of the F_2 is 43 in the *Nicotiana* cross of Figure 14.10, and the environmental variance V_E is 7, what is the heritability in the F_3? How does it compare with that in the F_2?

16. Can you explain why the expected decrease in the genetic variance in the F_3 variance is one half that of the F_2? HINT: Assume there is only one gene segregating in the cross.

References

Cavalli-Sforza, L. L., and W. F. Bodmer
1971. *The Genetics of Human Populations.* San Francisco: W. H. Freeman and Company. [Chapter 9 expands considerably upon this topic.]

Falconer, D. S.
1960. *Introduction to Quantitative Genetics.* Edinburgh: Oliver and Boyd.

Harrison, G. A., and J. J. T. Owen
1964. "Studies on the inheritance of human skin color," *Annals of Human Genetics,* vol. 28, pp. 27–37.

Mather, K., and J. L. Jinks
1971. *Biometrical Genetics.* 2nd ed. New York: Cornell Univ. Press.

Stern, C.
1973. *Principles of Human Genetics.* 3rd ed. San Francisco: W. H. Freeman and Company. [Chapter 18 gives an extensive treatment of the basic models of polygenic inheritance.]

15

Similarity Between Relatives

The models designed to study the inheritance of traits that are not transmitted in a simple Mendelian fashion can also be applied to man. Special methods are needed that make use of a phenomenon with which we are all familiar: the similarity between relatives. We proceed by measuring this similarity, and then comparing it with what it should be under specific hypotheses. We can thus obtain an interpretation in genetic terms. Although this interpretation does not go very deep, it does at least give us a general picture. The possibility that the similarity between relatives may be due to phenomena other than biological inheritance should, however, always be kept in mind.

15.1 The Measurement of Similarity

It is a foregone conclusion that children resemble their parents. All fathers and mothers have the experience of discovering a number of close similarities between their offspring and themselves. Some of these are so striking as to make the power of inheritance immediately clear, whether the resemblance involves body structure (in general or in detail), facial attitudes, voice inflections, and so on. But dissimilarities are also noticeable, and they are sometimes quite marked.

One way to investigate the "power of inheritance" with respect to a given trait—or, as it is often called, the heritability of the trait—is to measure the similarity between relatives. This approach was fostered in the last century by Francis Galton, who developed statistical methods with this aim in mind. Modern methodology is based on correlation and regression coefficients—techniques of which we will give only the essentials, without the detailed formulas and computations (some of which can be found in the Appendix). Modern methods differ somewhat (mostly in detail) from those introduced by Galton. In fact, one of the phenomena he encountered gave rise, as we shall see, to the name "regression coefficient" that is still widely used in modern statistics.

To study similarity between relatives, one must collect data on an adequate number of relatives. It will be intuitively clear that the closer the relationship, the greater the expected similarity, and the easier it is to study. Attention is therefore concentrated on the closest types of relationship easily observed in man. Of special importance are relationships between twins, which will be investigated in more detail later in this chapter. We shall, for now, limit our attention to the closest relationships most commonly observed, those between parent and offspring and between sib and sib.

It is convenient to present the data on similarity between relatives as a correlation diagram.

Assume that we have measurements of a trait, say stature, in a number of parents and their children. In the case of stature, the sexes will be separated, because it is known that sex has an effect on stature. Taking, therefore, the stature of fathers and sons, we can plot every pair of father/son data as a dot in a diagram, as in Figure 15.1. The cluster of points so obtained has a considerable scatter, but it shows a tendency to aggregate around a straight line at about a 45° angle.

The coefficient of correlation, r, measures the tendency of points to aggregate around a straight line.

In an extreme case, all the points would actually fall on a straight line. The correlation coefficient (r) is then exactly equal to one (Fig. 15.2A). At the other extreme, the points would show no tendency to aggregate around a straight line (Fig. 15.2E) if r is zero. There exist between these two extremes all degrees of tendency to collect around a straight line, and the other parts of Figure 15.2 illustrate some examples of this. (Further details and methods of computation of r are given in the Appendix.)

Suppose, for example, that we want to predict the IQ of a child, given that of one of its parents. Parents are first grouped into intervals of, for example, ten IQ points, as in Figure 15.3. We then compute the average IQ of the children of parents whose IQ is 70. This value is plotted as an open circle in Figure 15.3. The same is done for parents of IQ 80, and so on. The regression line is a straight line drawn as close as possible to all these open circles. Note that this regression line is flatter than a line drawn to show what the IQ of a child would be if it were identical to the parent. This shows that the IQ of the child is closer to the general mean than that of the parent, the phenomenon discovered by Galton, to which he gave the name "*regression* to the general mean."

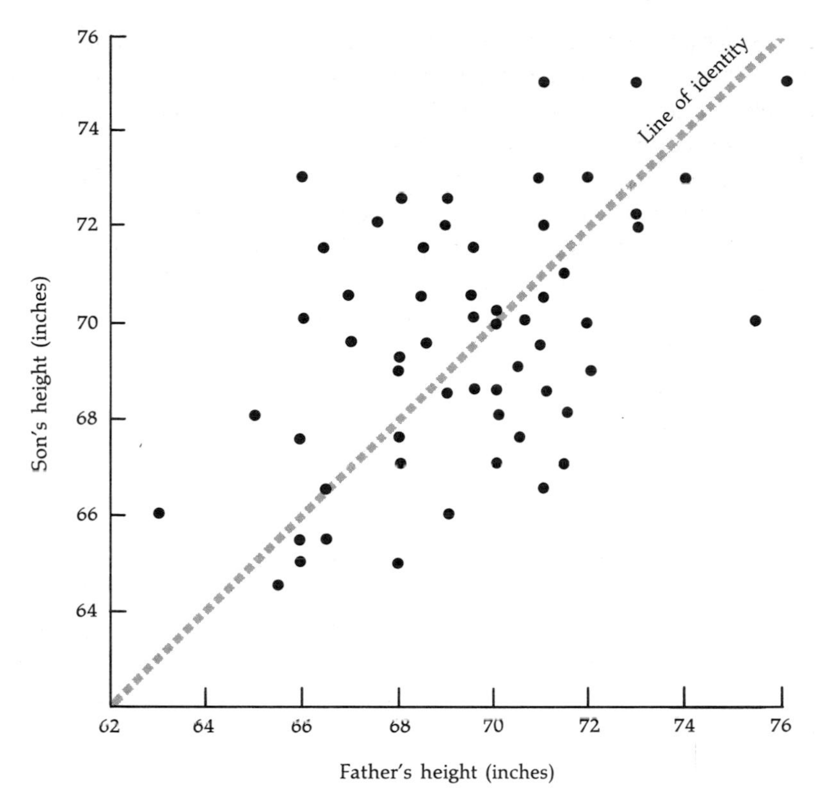

Figure 15.1
A correlation diagram. Each point plotted in the diagram represents a pair of data for the stature of a father and his son. If the stature of the son were identical to that of the father in every case, all observations would fall on the dashed line (line of identity). It is easy to see that this 45° line includes all possible pairs where the son's height is identical to the father's height. At the other extreme, if the son's height were completely unrelated to the father's height, the points would be scattered randomly around the diagram, with no tendency to cluster along the line of identity. The correlation coefficient (r) measures the extent to which the points tend to deviate from a random pattern toward positions near a straight line. A rough estimate of the value of r in this case can be made by comparison with the diagrams of Figure 15.2.

The regression coefficient is the slope of the regression line.

The slope of a straight line in a graph, as in Figure 15.3, can be given a numerical value; Figure 15.4 shows some examples. Thus, a line at a 45° angle has a slope of 1; a slope of 0.5 corresponds to an angle of about $26\frac{1}{2}°$. That of the IQ of the sons versus IQ of the fathers has a slope of about 0.5, and this is therefore the value of its regression coefficient. (Methods for computing regression coefficients are given in the Appendix.)

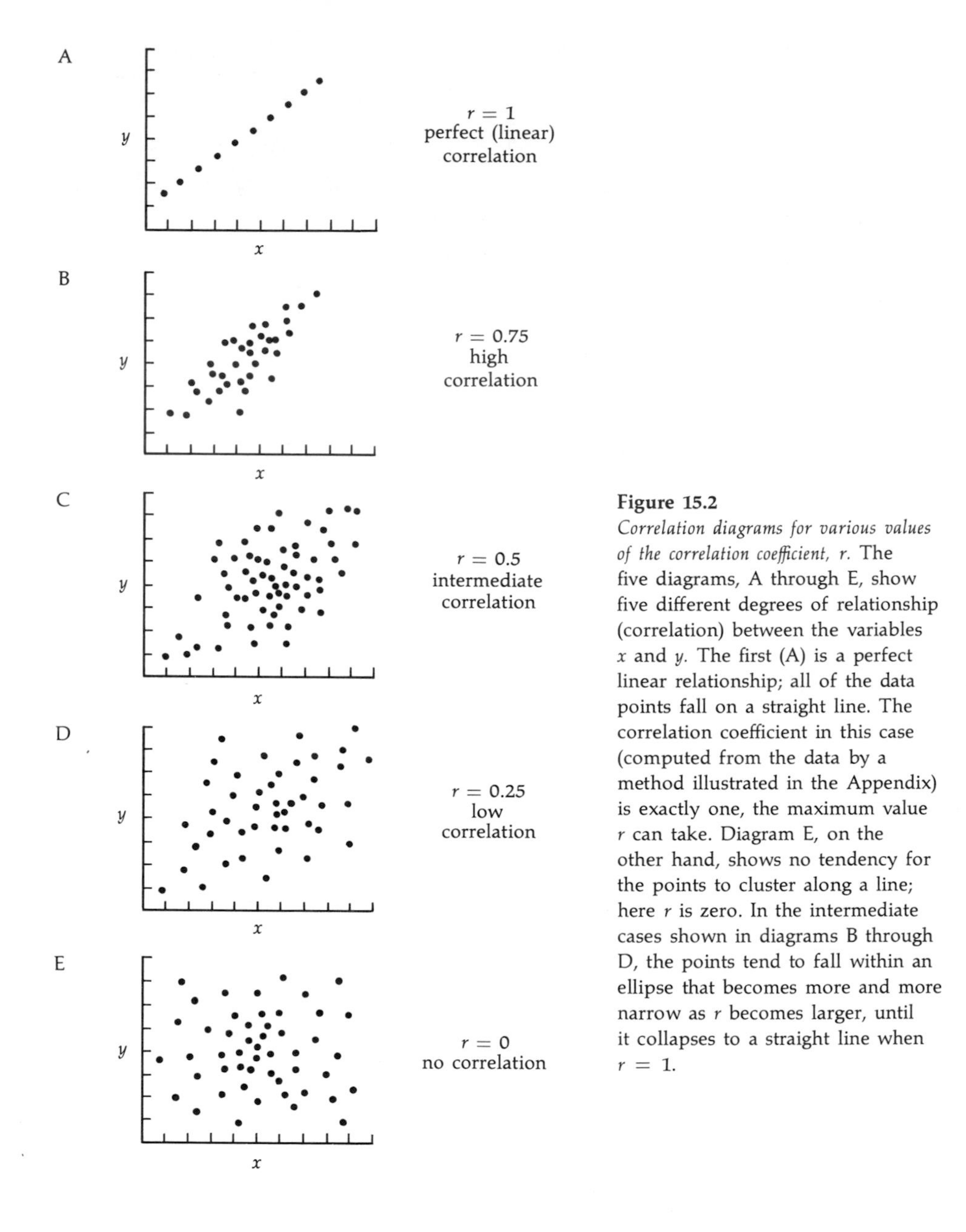

Figure 15.2
Correlation diagrams for various values of the correlation coefficient, r. The five diagrams, A through E, show five different degrees of relationship (correlation) between the variables x and y. The first (A) is a perfect linear relationship; all of the data points fall on a straight line. The correlation coefficient in this case (computed from the data by a method illustrated in the Appendix) is exactly one, the maximum value r can take. Diagram E, on the other hand, shows no tendency for the points to cluster along a line; here r is zero. In the intermediate cases shown in diagrams B through D, the points tend to fall within an ellipse that becomes more and more narrow as r becomes larger, until it collapses to a straight line when $r = 1$.

There is a precise relationship between regression and correlation coefficients.

This relationship is considerably simplified in the case of correlations between relatives, where the two coefficients are very similar to each other and, under certain conditions, tend to be identical.

In Table 15.1, we give some observed correlation coefficients between relatives

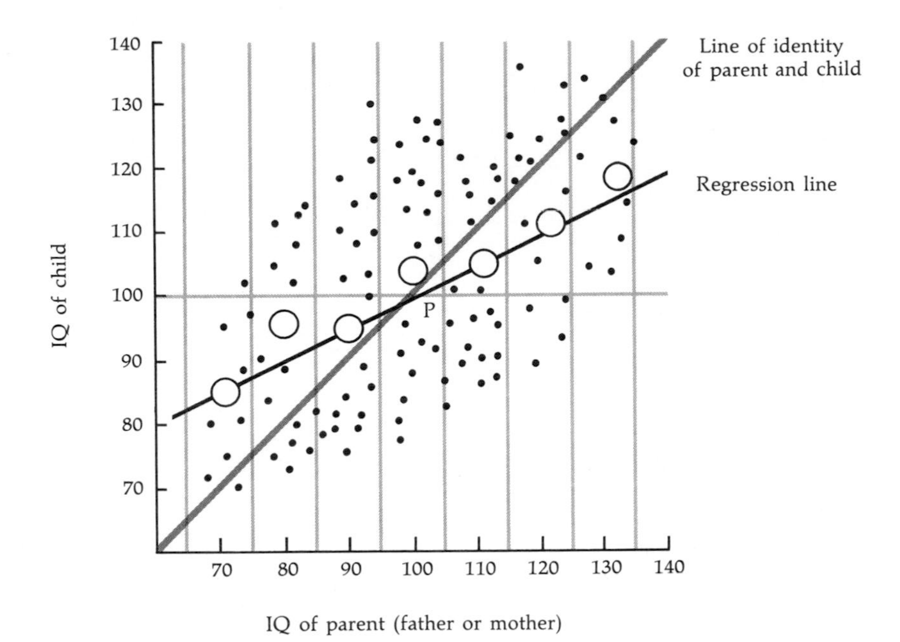

Figure 15.3
Regression of children on parents for IQ. Each data point in this diagram represents a pairing of a child's IQ with the IQ of its father or mother. The vertical lines separate the parents into classes with a range of ten IQ points, centering on IQs of 70, 80, and so on. Each open circle represents the mean IQ of all children born to parents within that IQ class. The regression line is a straight line drawn to come as near as possible to all of the open circles, and also to pass through the point P (which corresponds to the mean IQ of children plotted against the mean IQ of parents). Note that the slope of the regression line is less than that of the line of identity. If the regression line were horizontal, it would mean that the mean IQ of each group of children was the same (and equal to their general mean), regardless of the IQ range of the corresponding parents. In this example, as in most actual studies of human traits, the position of the regression line indicates some tendency for children to differ from their parents in a direction that brings them closer to the general mean (a regression toward the mean).

for a few traits of interest. Parent–offspring correlations are the averages of father–son, mother–son, . . . correlations. These are all expected to be the same, except for traits influenced by X-linked genes. For their interpretation, we must turn to consider first what correlations are expected between relatives for a complex trait, under the supposition of Mendelian inheritance.

15.2 Correlations Between Relatives Under Mendelian Inheritance

In a paper published in 1918, with more or less the same title as this section, R. A. Fisher showed how to predict the correlations to be expected between relatives, on

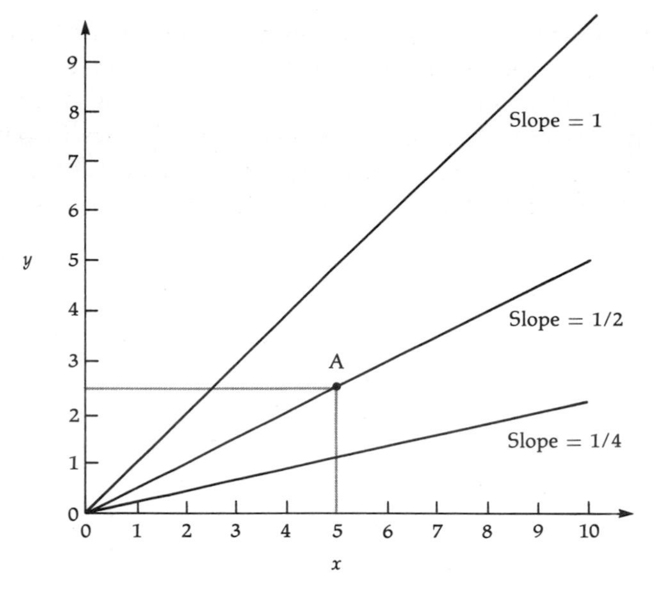

Figure 15.4
Slopes of straight lines passing through the origin of coordinates x and y. If a straight line passes through the origin, its slope can be measured simply by taking the ratio of ordinate to abscissa at any point along the line. For example, at point A we measure the slope of the line passing through that point as $\frac{y}{x} = \frac{2.5}{5.0} = 0.5 = \frac{1}{2}$. Verify that the same value for slope will be obtained using any point along the line.

the assumption that biological inheritance is Mendelian. Mendelism had been rediscovered in 1900; but it had been applied only to discontinuous traits. The inheritance of continuous traits, such as stature and weight, had been interpreted on a completely different basis by Francis Galton in the nineteenth century. While Mendel postulated a "particulate" inheritance involving genes that keep their individuality in hybrids, Galton (who did not know of Mendel's work) hypothesized that the hereditary material is a continuum, so that, in the hybrid, the properties of the parents are irreversibly mixed together to give what was called

Table 15.1
Similarity between relatives, measured by correlation coefficients, for certain traits in man

	Correlation coefficient	
Trait	Parent–Offspring	Sib–Sib
Stature	0.51	0.56
IQ	0.48	0.58
Fingerprints (ridge counts)	0.48	0.50
Blood pressure (systolic)	0.24	0.33

"blending" inheritance. Blending inheritance actually predicts fairly well the observed correlations between relatives. Fisher, however, proved that Mendelian inheritance can predict them even better. Today we know that practically all biological inheritance must follow the Mendelian, not the blending, pattern. It so happens, however, that many predictions (like the regression of children on parents) are identical or at least quite similar in the two theories.

Under the simplest assumptions, the expected correlation between parents and offspring is $\frac{1}{2}$.

To predict the correlations expected between parents and offspring, we may proceed as follows—and this line of reasoning happens to be the same for a Mendelian (polygenic) or a blending theory. If a father has an IQ of say 140, what will the IQ of his children be? This will depend, of course, also on the mother, and if mating is random, the mother will have, on average, an IQ equal to that of the general population, which is conventionally set at 100. If IQ were completely inherited, a child of a father with IQ 140 and an average mother with IQ 100 would be intermediate—that is, have an IQ of 120 (in the absence of dominance). Under the same hypotheses, a father of IQ 80, with the average mother of IQ 100, will have children on average with an IQ of 90. We can plot these two points and draw a straight line through them (Fig. 15.5). All other points similarly computed for

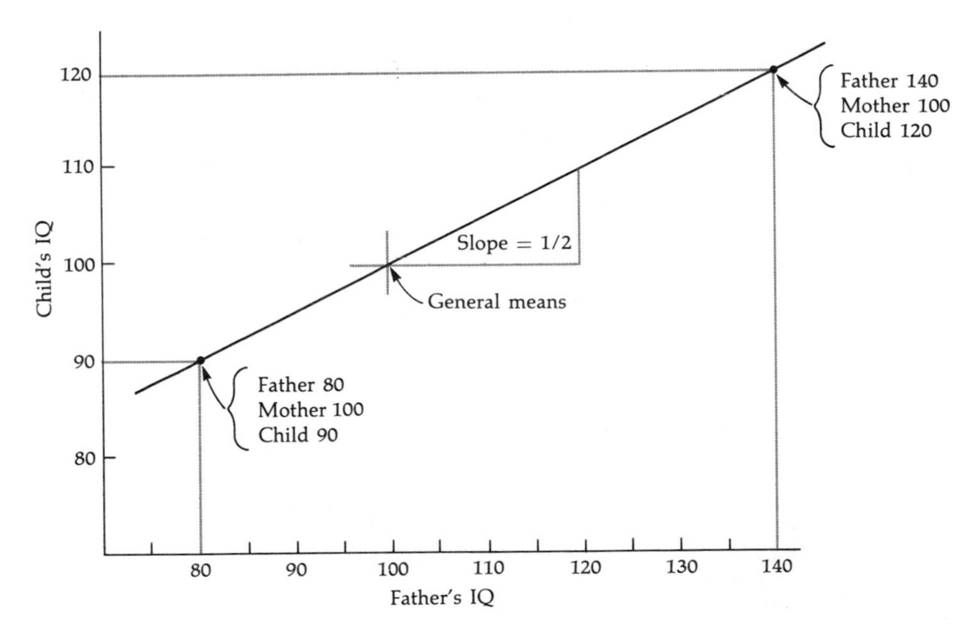

Figure 15.5
Predicted correlation of $\frac{1}{2}$ between parents and offspring, for perfect Mendelian inheritance with no dominance.

other fathers will fall on the same straight line. The line has a slope of $\frac{1}{2}$, and this is the regression (also the correlation) expected between parent and child, assuming perfect Mendelian inheritance with no dominance. Exactly the same value is expected for the correlation and regression between two sibs.

Dominance lowers the correlation between relatives, and causes a difference between the parent–offspring (P/O) correlation and that between sibs (S/S).

In Table 15.1, note that each sib–sib correlation is a little higher than the corresponding parent–offspring correlation. Observed correlation coefficients are actually subject to a considerable degree of statistical error, due to the small number of observations on which they are usually based. Therefore, small differences in such coefficients should not be taken too seriously. However, there is a good reason why we expect the S/S correlation often to be higher than the corresponding P/O correlation. The P/O correlation is very much like the narrow heritability discussed in Chapter 14; it is determined only by the additive variance. If there is dominance, it can be shown that the P/O correlation will be less than the S/S correlation. In fact, the difference between the S/S and P/O correlation coefficients should be equal to exactly one-quarter of the proportion of the total variance due to dominance. If there is dominance, both correlations will be decreased to values below 0.5, with the P/O correlation decreased more than the S/S correlation.

Other effects are due to environmental variation and nonrandom mating.

Individual variation in developmental conditions can reduce the similarity between sib and sib, or between parent and offspring. In fact, one way of measuring such variation and of obtaining estimates of heritability involves comparison of expected and observed correlations. For example, the ratio of an observed to an expected correlation can be used as a simple measure of heritability. However, suppose that we make such a computation for the S/S correlation for IQ from

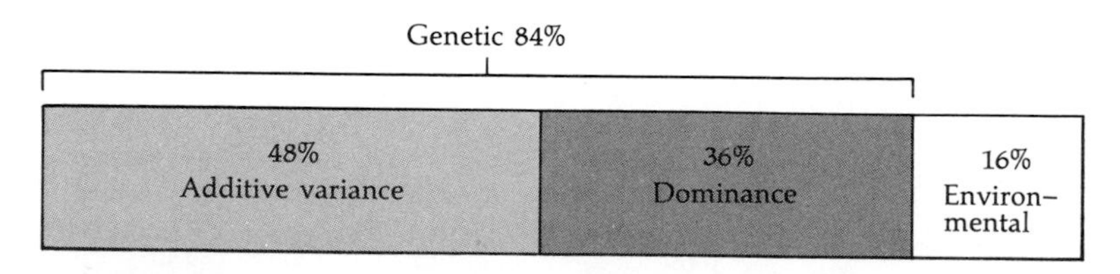

Figure 15.6
Partition of the variation for blood pressure, using data from Table 15.1. Table 15.2 outlines the computation of the values shown here.

Table 15.1. Dividing the observed correlation by the expected correlation, we obtain $\frac{0.51}{0.50} = 1.02$, or an estimate of heritability as 102 percent. This result is absurd, because heritability can at most be 100 percent. Error due to statistical fluctuations must be tolerated in such computations—and such error can be much larger than that observed here. In fact, a heritability of 100 percent would be suspiciously high for IQ because we have other evidence—as we see later—that there are important environmental sources of variation in IQ. The correlation coefficients for blood pressure, however, seem to give more reasonable estimates of heritability (Fig. 15.6).

From data on P/O and S/S correlations, we can make a preliminary analysis of the total variation into three parts: that due to dominance, that due to additive gene effects, and that due to environment.

The three components of the variation are given as percentages of the total variation in Figure 15.6 for the blood-pressure data from Table 15.1. The dominance fraction is computed from the difference between the S/S and the P/O correlations, multiplied by four. The broad heritability is the sum of the dominance and additive fractions, or 0.84 in this case. Table 15.2 provides details of these calculations.

There is another possible complication that may explain the tendency for some observed correlation coefficients to be near or above 0.5, even though environmental effects should keep them below 0.5. We know that, for many traits, mating is not random. Most of the traits studied in preceding chapters—such as blood types or enzyme levels—cannot be observed directly by the people around us. Such traits follow the Hardy-Weinberg rule, indicating that mating is essentially random with respect to the phenotypes. However, some of the traits we are discussing now—such as stature, weight, and IQ—are readily perceived by other individuals. In fact, the phenotypes of these traits form important parts of our everyday descriptions of people, and often influence our judgments about their desirability as friends or mates. A tall person may more often than not be attracted by a tall mate, and similarly at the other end of the scale. Computing the correlation between husband and wife for such a character, we are measuring the extent of nonrandom or *assortative mating.*

A positive correlation coefficient indicates a tendency for similar values to be paired—like with like. A negative correlation coefficient indicates a tendency for opposite values to be paired—like with unlike. (A correlation coefficient of -1.00 would mean that all points on the correlation diagram are located on a straight line sloping downward from left to right.) The husband-wife correlation is nearly always positive for readily detectable traits, indicating that individuals tend to seek mates similar to themselves. For height, the husband-wife correlation is

Table 15.2
Analysis of the variance for blood pressure, using data from Table 15.1

Step	Relationship	Source
1	V_T = total variance.	Definitions.
2	V_A = additive variance.	
3	V_D = variance due to dominance.	
4	V_E = environmental variance.	
5	$V_T = V_A + V_D + V_E$.	
6	$r_{P/O}$ = parent–offspring (P/O) correlation.	
7	$r_{S/S}$ = sib–sib (S/S) correlation.	
8	$r_{S/S} = \frac{1}{2}(V_A/V_T) + \frac{1}{4}(V_D/V_T)$.	Relationships derived from mathematical expressions for these quantities. See Note.
9	$r_{P/O} = \frac{1}{2}(V_A/V_T)$.	
10	$r_{S/S} - r_{P/O} = \frac{1}{4}(V_D/V_T)$.	Step 9 subtracted from Step 8.
11	$V_D/V_T = 4(r_{S/S} - r_{P/O})$.	Rearrangement of Step 10.
12	$V_A/V_T = 2r_{P/O}$	Rearrangement of Step 9.
13	$V_E/V_T = 1 - (V_D/V_T) - (V_A/V_T)$.	Rearrangement of Step 5.
14	$V_D/V_T = 4(0.33 - 0.24) = 0.36$.	Inserting data from Table 15.1 in Steps 11, 12, and 13.
15	$V_A/V_T = 2 \times 0.24 = 0.48$.	
16	$V_E/V_T = 1 - 0.36 - 0.48 = 0.16$.	

NOTE: This analysis indicates that the variance for blood pressure can be partitioned as follows: 36 percent of the total variance is due to dominance effects, 48 percent to additive variance, and 16 percent to environmental effects—as shown in Figure 15.6. For more complete derivation of these expressions, see pp. 531–537 of L. L. Cavalli-Sforza and W. F. Bodmer, *The Genetics of Human Populations*, San Francisco: W. H. Freeman and Company, 1971.

about +0.3. For IQ, the correlation is even higher—near +0.4. In part, these high correlations derive from social stratification. There is an increase in average stature and average IQ with increasing socioeconomic status. The tendency to choose a mate within the same social class would therefore lead to a positive correlation for these traits, even if there were no direct individual tendency to try to choose a mate of similar height or IQ.

The husband–wife correlation coefficient for a trait provides a measure of the extent of assortative mating (AM).

A positive value of the correlation coefficient for AM has an effect similar to that of inbreeding: it increases the frequency of homozygotes. It also inflates the correlation values between relatives above those expected for random mating.

Because AM can be measured, its effect on these correlations can be predicted (Fig. 15.7). The increase in correlations between relatives depends on the intensity of the assortment. Assume, for example, that mating is completely assortative

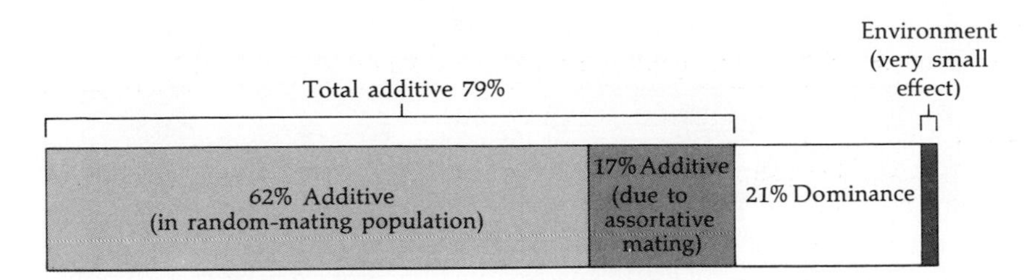

Figure 15-7
Partition of the variance for stature, with allowance for assortative mating. In Table 15.2, the total additive variance is computed as $2r_{P/O}$. Taking account of assortative mating, we now compute the total additive variance as $[2/(1 + AM)]r_{P/O}$, where AM is the correlation between mates. Using the known value of AM = 0.28, and $r_{P/O} = 0.51$ from Table 15.1, we obtain a total additive variance of $[2/(1 + 0.28)] \times 0.51$, or about 79 percent. The portion due to the assortment (17 percent) and the dominance contribution (21 percent) are calculated using slightly more complicated formulas (see L. L. Cavalli-Sforza and W. F. Bodmer, *The Genetics of Human Populations,* W. H. Freeman and Company, pp. 543–548, 1971). The analysis outlined here follows the pattern set forth by R. A. Fisher in 1918. The estimates of the genetic components given here, however, may be too large, for reasons that will be discussed more fully in relation to IQ.

(r between husband and wife = +1.00). This means that an individual with IQ of 140 always marries another individual with IQ 140, an individual with IQ 80 marries another with IQ 80, and so forth. With V_D and V_E equal to zero, the children of a man with IQ 140 married to a woman of IQ 140 will also have an IQ of 140, and so on for other possible matings. Plotting these values on a graph, we obtain a line with a slope of 1 (see Fig. 15.4) as the regression line for the child versus the parent. The parent–offspring correlation is +1.00, and there is no tendency for regression toward the population mean.

If positive assortative mating is incomplete, the effect will not be so drastic, but there will be an increase in the correlations between first-degree relatives (P/O and S/S), which can thus reach values above +0.5.

On the basis of these considerations (and using not-too-complicated formulas), it is possible to partition the variance for traits such as stature and IQ, taking into account the tendency for assortative mating (Figs. 15.7 and 15.8). Note in these figures that there are two components of additive variance. One is the fraction that would be found if the population were mating at random. The other, which is smaller, is due to assortative mating and the consequent increase in homozygotes. We explore this problem further in a later section, but first we must study a very special category of relatives: twins.

15.3 Twins

One member of a twin pair had a heart attack in a city on the East Coast; ten days after, so did the other twin, who lived in the Midwest. The first twin turned

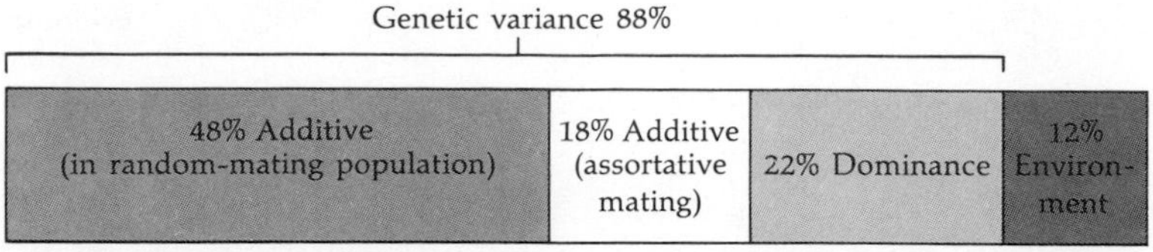

Genetic variance 88%

48% Additive (in random-mating population)	18% Additive (assortative mating)	22% Dominance	12% Environment

Figure 15.8
Partitioning the variance of IQ. This analysis, done in 1965, did not give adequate consideration to data on adoptions, and overemphasized results on MZ twins reared apart—results later shown to be unsatisfactory. As a result, the genetic variance (the three portions at the left) is considerably inflated. For comparison, see the more satisfactory analysis in Figure 17.4. (Data from C. Burt, *British Journal of Psychology,* vol. 57, pp. 137–153, 1966.)

out to be resistant to a drug widely used in such cases, which is a very rare event. But it was no strange coincidence that the second twin was also found to be resistant to the same drug by the doctors in an entirely different city. The twins had identical genotypes!

There are two categories of twins: identical (MZ) and fraternal (DZ).

The more interesting group is that of "identical" or monozygous (MZ) twins. They develop from a single zygote (a single egg fertilized from a single sperm), which divides into two cells, each capable of giving rise to a complete embryo. The second category, that of "fraternal" or dizygous (DZ) twins, arises from two independent zygotes formed at the same time—that is, two eggs fertilized by two independent sperm cells. Whereas MZ twins have the same genotype (and differences between them can only be of environmental origin), DZ twins have genotypes that are only as related as those of two ordinary sibs.

Twin births are not a particularly rare event. But, whereas MZ twins are born with a frequency of about 1 every 200 gestations that is fairly independent of race and mother's age, DZ twin incidences vary much more. They increase in frequency with increasing age of the mother until about 38 years, after which the frequency drops sharply. The frequency of DZ twin births also differs considerably between races, being highest for Africans (about 6 times the frequency of MZ), intermediate for Caucasians (slightly more than twice as frequent as MZ), and lowest among Orientals (where the frequency is about $\frac{1}{2}$ that of MZ). Finally, DZ twin births are influenced by treatment with gonadotropic hormones. Therapy with such hormones, which is a treatment for some types of female infertility, quite often results in the birth of twins or triplets. In contrast with MZ twinning, the tendency to have DZ twins shows a trace of inheritance. Triplets, and even more so quadruplets, are exceedingly rare, and therefore only exceptionally of use in the study of inheritance.

It is usually easy to distinguish between identical and fraternal twins on close inspection.

For research purposes, the best method is to test the twins for as many strictly inherited traits as possible. Identical twins must be identical for every one of these—including, of course, sex. If twins are discordant for even a single Mendelian trait out of many, they must be DZ. Using ten or more common polymorphic loci—such as blood groups ABO, MN, and so on—the chance that a DZ twin pair will be identical for all the loci tested (though not for others), and thus will be misclassified as MZ, is of the order of one in a hundred or less.

Because MZ twins have identical genotypes, the comparison of MZ–MZ correlations with sib–sib correlations should provide a good measure of the extent to which a trait is genetically determined. For a trait that is almost completely determined by heredity, we would expect to find a very high correlation between the members of an MZ twin pair, a smaller correlation between DZ twins or between nontwin sibs, and the smallest correlation between nonsibs. On the other hand, for a trait with very low heritability, we would expect to find low and very similar correlations for all of these pairings. However, in practice, the study of twins involves many complications that require very careful interpretation of results.

Very occasionally, the members of a DZ twin pair have "mixed" genotypes, which must have resulted from an exchange of cells, probably due to vascular connections in the placenta, allowing cross-circulation of blood to occur between them. During the fetal period, immunological tolerance (see Glossary) is the rule, allowing the survival and implantation of the foreign cells in the other twin. These exchanges are often reciprocal. The carriers of such tissue-graft exchanges are called chimeras.

Chimeras in man are rare, and one can consider DZ twins to be, in practice, just like two ordinary sibs from a genetic point of view.

Unlike two sibs, however, they have all the peculiarities of twins: they have grown together in the same uterus, usually later in the same house; they have the same age; they usually establish a close relationship with each other and often live much of their lives together, at least during childhood. They usually go to the same schools and have the same friends. It is clearly reasonable to suppose that some of the similarity between twins derives from such circumstances rather than from their genotypes. A possible test of this hypothesis is to compare the resemblance of DZ twins with that of sibs; if the former is higher, then commonality of the environment of the twins may have played a role. But even if DZ twins are not more similar than ordinary sibs, it would still be dangerous to consider that the data on MZ twins are entirely free from this source of bias. For these and related reasons, studies of twins can often only suggest, but not prove, biological inheritance.

The time coincidence of a heart attack in the two twins mentioned at the beginning of this section is extremely close. In general, identical twins are very similar in their rates of development. Figure 15.9, for example, shows that the age of onset of menstruation is extremely close for MZ twins, while DZ twins behave almost like ordinary sibs. Naturally, sibs are more similar than two unrelated individuals (the last column on the right).

Even superficial observations of some traits in monozygous twins gives an immediate feeling that the power of biological inheritance may be strong.

The meaning of many features of electroencephalograms is still largely unknown, but extensive individual variation exists, and some of it may be subject to relatively simple inheritance mechanisms. Figure 15.10, showing the electroencephalograms of three pairs of MZ twins, offers a sample both of the difference between genotypically different individuals, and of the similarity between MZ twins of the same pair.

The expected correlation between members of MZ twin pairs for a trait that is determined only genetically is one. The observed correlation between MZ twins is a somewhat inflated measure of broad heritability.

We have already mentioned that the twin-study approach to measurement of heritability tends to underestimate the extent of environmental variation, simply because twins are usually raised in very similar environments. The well-known human geneticist L. S. Penrose once commented that uncritical use of twin data would lead to the conclusion that clothes are inherited! As a matter of fact, many

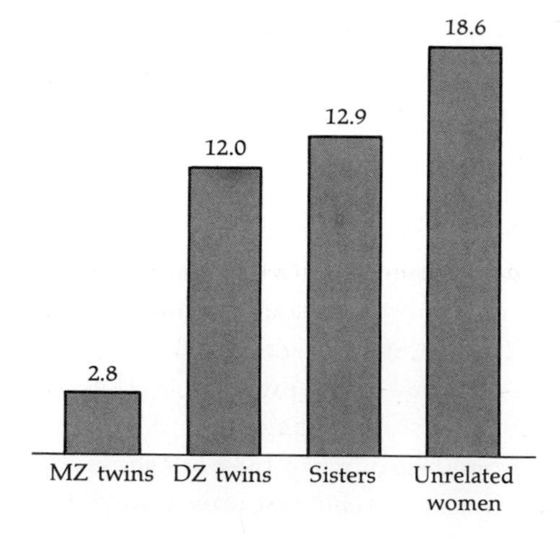

Figure 15.9
Mean difference (in months) in age at menarche, for various categories of relationship. (Data from E. Petri, *Zeitschrift für Morphologie und Anthropologie,* vol. 33, pp. 43–48, 1935.)

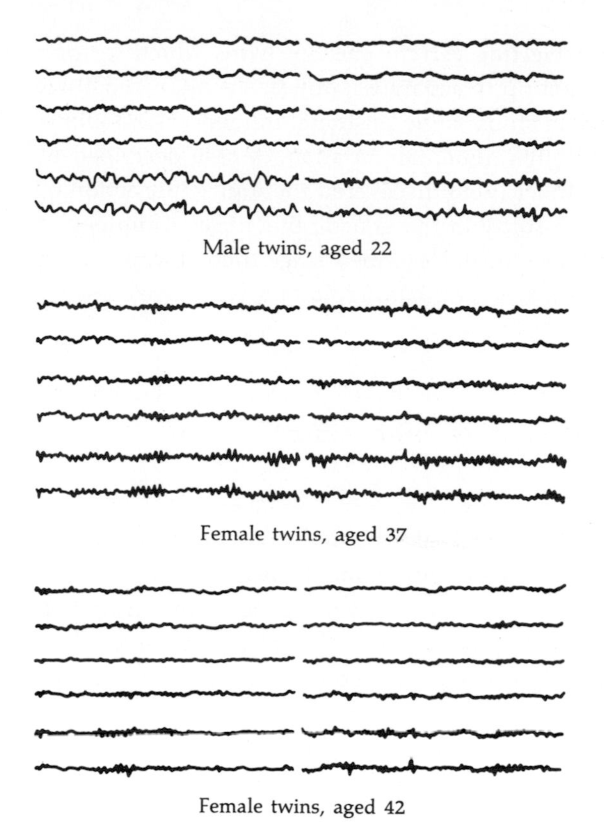

Figure 15.10
Electroencephalograms of three pairs of monozygous twins brought up apart. The left and right sides of the figure compare the tracings of the pairs of twins; they show practically complete concordance for all features of the tracings. The six tracings shown for each individual are recordings from different parts of the head. (From N. Juel-Nielsen and B. Harvald, *Acta Genetica,* vol. 8, pp. 57–64, 1958.)

members of twin pairs, especially when young, do tend to wear identical clothes—and the correlation in clothing is higher for MZ twins than for DZ twins. Today, however, psychologists recommend that twins be encouraged to develop independent personalities, and so it is likely that the custom of dressing twins identically will gradually disappear. Meanwhile, twins—especially MZ twins—do tend to spend most of their time together and certainly influence each other to a considerable extent. They also appear to develop a special relationship, with one twin becoming socially dominant over the other.

Especially for behavioral traits, it is difficult to accept that the environmental difference between two members of a twin pair is representative of that between two random individuals in the population. Therefore, estimates of heritability using only MZ twins, or contrasting MZ with DZ twins, are especially likely to be biased. In addition, many of the formulas that are used for estimating heritabilities from twin data contain approximations that may often not be valid.

It has been said that twins are a perfect experiment of nature to test the power of inheritance. But this experiment of nature is far from ideal. The possibility remains of improving on nature, selecting certain pairs of twins which for one reason or another have been brought up in separate families. Twins reared apart provide an important opportunity to study some postnatal influences and eliminate part of the uncertainty as to environmental variation. A case described in 1925 by the U.S. geneticist H. J. Muller (who discovered the mutagenic action of X rays) appears to be the first such study in the genetic literature; a number of other cases of twins reared apart have been described since then. Twins reared apart are a special case of an important class of data, adoptions, which we consider next.

15.4 Biological Versus Other Types of Inheritance

The analysis of stature given in Section 15.2 assigns such a major role to genetics that the estimate of environmental variation is almost nil. Yet we have many other indications that stature is subject to strong environmental effects. Stature has, for example, increased on average by more than an inch per generation during this century, at least in Western countries. This increase is usually attributed to improvement in diet, and perhaps to decrease of infectious diseases. Evidence that the increase might have a genetic component is quite unconvincing.

There are differences in average stature between socioeconomic classes. The number of children in a family is also a contributing factor; the larger the family size, the smaller the average stature. All of these observations are in agreement with the hypothesis that improvement in environmental conditions tends to increase size. The partition of variance made in Figure 15.7, leaving no room for environmental differences, is therefore somewhat strange.

To settle this problem, we can use data on MZ twins reared apart. They should, superficially at least, provide the perfect answer; individuals with identical genotypes, raised in different families, should provide a way to partition unequivocally the environmental and genetic sources of variation.

MZ twins truly reared apart are, unfortunately for the geneticist, rare.

Four studies that have been published to date have reported only a total of 122 such pairs: 53 from London; 38 from England at large; 19 from the U.S.; and 12 from Denmark. The largest study, on London children (by Sir Cyril Burt), has been severely criticized recently because of several deficiencies in the way the data were collected and reported. The other English study (by J. Shields) is much more accurately described. The detail of Shields' report makes one realize, however, one of the weaknesses of these studies. Ideally, one member of the twin pair should be raised by the "biological" family, and the other in a totally unrelated family.

In fact, in the Shields study, in over $\frac{2}{3}$ of the cases, the twin brought up in a family other than its own was raised in that of grandparents or uncles. Almost certainly, environmental conditions in such cases will be more similar for the twins than for random nonsibs. It is, unfortunately, the case that almost all studies of this kind are likely to be affected by biases of this sort.

Even though the conclusions do not have the desired rigor, nevertheless the analyses remain of interest. Results given in Table 15.3 show that separation of the twins somewhat decreases the correlation for height, as well as for weight. The heritability for stature remains high (90 percent), even when computed from separated MZ twins. That for weight is somewhat lower, but still reasonably high (77 percent). The most interesting trait measured in this series of observations was IQ; its results are discussed in the chapter on behavioral genetics.

The evaluation of biological inheritance (by which is meant that transmitted by genes and chromosomes) makes use of the similarity between biological relatives, including twins.

Measuring the strength of biological inheritance only on the basis of similarity between biological relatives may, however, lead to serious error. In so doing, an important fact is ignored: parents and children, sib and sib, usually share one common environment—namely, that within the family. Many properties in addition to chromosomes are shared by the family: food, customs, income, and so on. Some of these environmental parameters are inherited (socially, of course). Wealth certainly is, as also are many more subtle influences (Fig. 15.11). We have indicated this in the figure with the arrow labeled "sociocultural inheritance."

This term is used here quite loosely to indicate all aspects of the environment that are transmitted from one generation to the next. The arrow labeled "phenotypic inheritance" is not always easily distinguished from the sociocultural inheritance; with it, we indicate, for instance, the transmission of skills, learned directly by the children from their parents. This transmission was more important some time ago, and has become less and less significant in recent times. Still today,

Table 15.3
Correlations for height and weight between twins raised together and apart

	MZ twins brought up		DZ twins brought up
	Together	Apart	Together
Height	0.95	0.90	0.52
Weight	0.88	0.77	0.54

NOTE: The correlations given here are averages over the three largest studies.

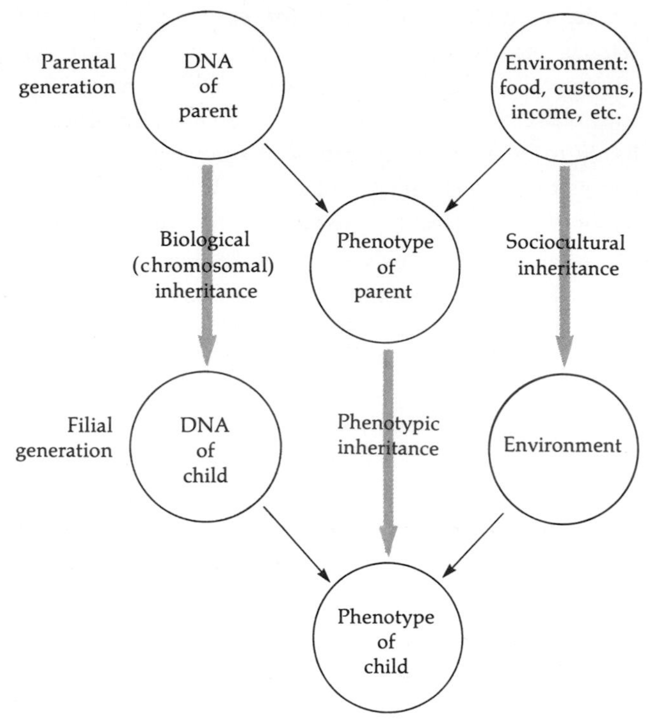

Figure 15.11
Types of inheritance. In a biological family, many properties are transmitted from parent to child by mechanisms other than the chromosomes. Vertical arrows represent modes of transmission from one generation to the next: biological, phenotypic, and sociocultural.

however, the family environment remains an important factor in development, and many of its elements are transmitted from generation to generation. In studying the similarity between relatives, one often overlooks the fact that the inheritance studied may not be that transmitted by the chromosomes, but may be due to some of these other nongenetic forms of transmission.

One approach that provides a safeguard against this problem is that of investigating adoptions.

In Figure 15.11, there are three vertical arrows indicating different modes of transmission. If they all affect the trait being studied, the correlations between biological relatives will express only the sum of the three types of inheritance. But the study of adoptees makes it possible to discriminate between the biological and the other types of inheritance (Fig. 15.12).

In principle, one can study the correlations between the adoptee and his biological relatives (parents, sibs) on the one hand, and those between him and his foster

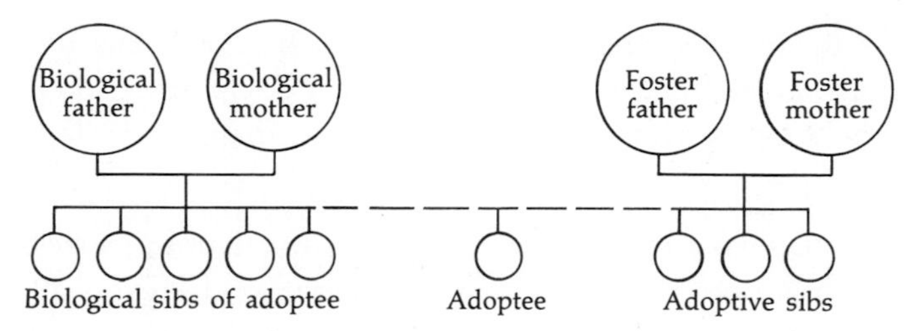

Figure 15.12
Adoption studies. The advantage of studying adoptions is the possibility of separating the effects of biological transmission from the effects of other types of transmission between generations.

relations on the other. The former measure the biological, the latter the other types of inheritance. In practice, the biological relatives are not always known, especially the father; an illegitimate maternity may, for example, be a cause for giving a child for adoption. There are likely to be no foster sibs because sterility of the foster parents is a common reason for adoption. A more serious complication is that adoptees often come from poorer families, thus forming a potentially biased sample; or that adoption agencies use poorly defined criteria for "matching" the biological and the adopting family. Some matching is done for external appearance, for example with respect to skin, eye, and hair color. Finally, adoptions do not always take place immediately after birth and, even if they did, they could not account for effects of the prenatal maternal environment.

In spite of all these limitations, adoptions make an essential contribution to testing for biological versus other types of inheritance. Adoptions are, however, also comparatively rare. They have been used especially for investigating behavioral traits, which are always strongly suspected to be subject to sociocultural or to phenotypic transmission. Examples of the application of adoption studies to behavioral traits are discussed in Chapter 17. "Physical" traits may also be influenced by the external environment, and it would be very desirable to have more adoption data for these as well. Even for physical traits like stature or blood pressure, estimates of heritability that do not include adoption data should be considered as suggestive (rather than rigorous or accurate) measures of biological inheritance.

Summary

1. The study of similarities between relatives provides a method for investigating biological inheritance in man.

2. Correlations measure similarities. Regression coefficients under these conditions are very much like correlation coefficients.

3. One can estimate the expected correlations and regressions for all types of relationships. In the absence of dominance and environmental effects, the expected correlation between first-degree relatives (parent–offspring or sib–sib) is 0.5.

4. Dominance decreases these correlations. Parent–offspring correlation reflects only the additive variance. Environmental variation decreases all correlations equally. Assortative mating, if positive, increases all correlations.

5. MZ twins have an expected correlation of one. The correlation between members of MZ pairs is an estimate of broad heritability—influenced, however, by the special circumstances under which twins grow up.

6. MZ twins separated at birth are potentially an excellent source of material for distinguishing nature from nurture. In practice, there are not many such pairs available for study, and the few studies that exist are not entirely satisfactory.

7. The major safeguard against confusing biological and sociocultural inheritance is given by the study of adoptions.

Exercises

1. What is the basis for using similarity between relatives to assess heritability?

2. Why are correlation and regression coefficients used in the study of similarity between relatives?

3. Why may sibs be more similar to each other with respect to a quantitative trait than are parents and offspring?

4. What is the effect of assortative mating on estimates of heritability?

5. How have data on twins been used to separate the effects of genotype and environment on quantitative traits?

6. What is a chimera?

7. Outline some of the major limitations of twin-data analysis.

8. What are the major features that distinguish cultural from biological transmission of inherited traits?

9. Why are adoptions important for the analysis of quantitative traits in human population? What are their major limitations?

10. S. Holt (1961) has studied the density of ridges in fingerprint patterns. The parent–offspring correlation was 0.48 and the sib–sib correlation 0.50 ± 0.04. (a) Estimate the heritability. (b) Is there likely to be dominance? (c) The correlation between husband and wife was 0.05 ± 0.07. What could be the fraction of variance due to assortative mating? Is it likely to be of importance? NOTE: for an understanding of standard errors (values preceded by the ± sign), see Appendix.

11. It is a well ascertained fact that human stature has increased continuously over the last 100 to 150 years in all Western countries. The change is frequently of the order of one standard deviation per generation. Assume that, in one study, the stature of fathers born over the 30-year period

1900 to 1930 was found to be directly correlated with the stature of their adult children. Is the computed correlation a valid estimate of the genetic relationship? If it is biased, in which direction is it biased?

12. What effects can the following factors have on the correlation for birth weight between MZ twins reared together? Indicate the direction of the effect on the correlation in each case as increase or decrease.
    a. In families of higher economic status, food is richer and in general more adequate.
    b. The cytoplasm of the single zygote forming the twins may divide unequally between the two twins.
    c. Some births tend to take place prematurely, while others occur after periods of gestation that are longer than average (more generally: there is individual variation in the length of gestation).
    d. One of the twins is usually situated in a less favorable position in the maternal uterus.

15. The mean differences between various pairs of relatives are given in Figure 15.9. It can be shown that the square of the mean difference is approximately 1.4 times the variance. On the basis of this and the data of Figure 15.9, compute the heritability of age at menarche in two different ways (exclude DZ twins). Compare the estimates. Does the comparison suggest anything?

16. Averaging over many studies of IQ in biological relatives, the following correlation coefficient values were obtained: parent–child 0.48; sib–sib 0.58; DZ twins reared together 0.63; MZ twins reared together 0.88. Assume that differences of 0.03 between these means of correlation coefficients may be significant.
    a. Give separate estimates of heritability from each of these 4 data.
    b. Compare them with each other and discuss the comparison.
    c. How would the estimates of heritability be affected by the fact that the correlation for IQ between husband and wife is 0.4?

17. The correlation between IQ of adopted child and adopting parent is on average 0.23 (see Table 17.3). Does this affect the conclusions obtained in Exercise 16 and, if so, in which direction?

References

Bulmer, M. G.
1970. *Biology of Twinning in Man.* London: Oxford Univ. Press.

Cavalli-Sforza, L. L., and W. F. Bodmer
1971. *The Genetics of Human Populations.* San Francisco: W. H. Freeman and Company. [Chapter 9 gives a more extensive treatment of similarity between relatives.]

Fisher, R. A.
1918. "The correlation between relatives on the supposition of Mendelian inheritance," *Transactions of the Royal Society of Edinburgh,* vol. 52, pp. 399–433. [Reprinted in *Collected Papers of R. A. Fisher,* ed. J. H. Bennett, South Australia: Univ. of Adelaide Press, 1971.]

Mittler, P.
1971. *The Study of Twins.* Baltimore, Md.: Penguin.

Shields, J.
1962. *Monozygotic Twins Brought Up Apart and Brought Up Together.* London: Oxford Univ. Press.

16

All-or-None Complex Traits

Traits that do not vary continuously, but are either present or absent in a given individual, are sometimes called "all-or-none traits." Some of them are, of course, determined by a single-gene difference. But many all-or-none traits are not simply inherited in this way, even if there is evidence for some genetic determination. If we assume, however, that there is a continuous variation in the liability to manifest such traits and that there is a liability "threshold" for their manifestation, we can extend to all-or-none traits the models that we have already used for continuous variation. The major applications are in the analysis of certain types of diseases.

16.1 The Threshold Model

Harelip is a relatively rare congenital defect that affects about one person per thousand. Some affected people have just a very slight fissure of the lip that is barely noticeable, while in others the whole height of the lip may be involved. Harelip is due to an incomplete fusion of two portions of the lip in embryological development. Surgery carried out early enough (in the first two months after birth) can repair it successfully.

When an MZ twin has harelip, the other also shows the defect about half of the time. This is evidence that there *may* be an inherited component to the defect, but the similarity between MZ twins is no rigorous proof of this. An infection, for example—or some trauma or toxic substance to which the mother was exposed during pregnancy—might also be hypothesized to be responsible. The process of lip fusion that is completed during normal development may be especially sensitive to some external disturbances; the developmental process, however, may be more sensitive in some genotypes than in others.

In other words, the liability to the defect, rather than the defect itself, may be inherited.

The liability to a defect, or to a disease, may in general be influenced by a variety of factors, both genetic and environmental. If we could measure the liability by criteria other than the occurrence of the disease, we could carry out an analysis of the trait "liability" rather than the trait "disease." But, in the absence of such a measurement, we find ourselves in a situation similar to being able, for example, only to distinguish "giants" and "nongiants," rather than to measure height. The threshold model hypothesizes that individual liability varies, but that only when the liability is greater than a certain value (the threshold) does the trait become apparent.

The liability may be affected by one or many genes and, in addition, by environmental factors.

The simplest hypothesis we can make is that the liability has a "normal" distribution, as in Figure 16.1A. Above the threshold, all individuals are affected. A normal distribution may, as discussed in previous chapters, arise from a truly polygenic situation, which is in practice always accompanied by environmental effects on individual variation. If, however, one gene difference has a disproportionately large effect compared to all others, then we may find the model of Figure 16.1B more appropriate. In this case the variation in liability is largely due to the single gene difference (A, a), but the genotypes overlap substantially because of variation due to environment, and perhaps also because of "residual genotypic" effects. The result is that there are affected individuals of all three genotypes, though they are relatively more frequent among the most liable genotype aa, less in the heterozygotes, and least frequent of all among the AA homozygotes. The relative incidence of the trait in a genotype is also called the "penetrance" in that genotype. It can be seen from the graph shown in Figure 16.1B that the penetrance of aa is about 70 percent, that of Aa about $\frac{1}{3}$, and that of AA very small. Given a high frequency of allele A (as in the figure), however, the number of AA affected will not be trivial compared with that of aa and Aa individuals. Almost half of all the affected are, in this example, heterozygotes.

When there is no way of measuring the liability directly, it is difficult in practice to choose between polygenic and monogenic models.

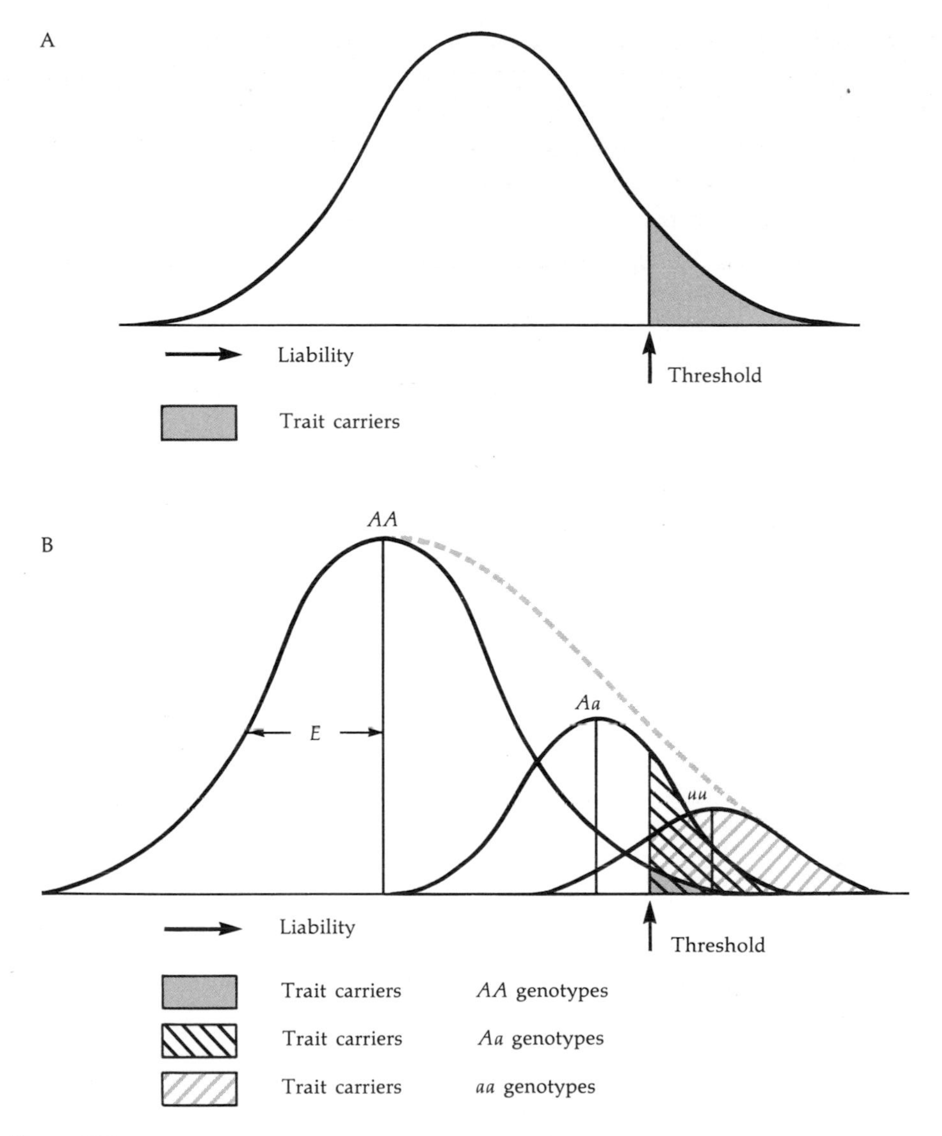

Figure 16.1
The threshold model for non-Mendelian traits. The liability varies between individuals according to the distributions represented. (A) A "normal," single-peaked distribution that might be due to polygenic determination. (B) A composite distribution made up from the sum of three distributions, due to three genotypes created by the presence of two alleles (A, a) at a single locus. In both cases, the variation around the peak or peaks may be in part environmental and in part genetic. In case B, genetic variation (if present) is due to genes other than the major gene difference, A or a. Individuals whose liability is above the threshold manifest the disease. The diseased individuals correspond to the shaded areas under the distribution curves. The dashed curve in B represents the sum of individuals from the various genotypes, in the part of the distribution where the same liability is found in different genotypes. E is the standard deviation of the liability among individuals with a given genotype. (After K. Kidd and L. L. Cavalli-Sforza, *Social biology*, vol. 20, pp. 254–265, 1973.)

The two models give rise to very similar expectations for all the correlations between relatives, if one must rely only on the incidence of the trait because no direct measurement of the liability is available. In principle, explanations based on single-gene differences are preferable because they correspond to situations that are simpler, in physiological terms. But if there is no way to choose, one must wait for other criteria of analysis to indicate which may be the correct model. The best hope is to find biochemical or physiological measurements that play a role in the determination of the trait.

The polygenic model is slightly simpler than the monogenic one when it comes to analysis of actual data. The major practical use of these models is for evaluating risks to future progeny in a family whose pedigree is known, a problem that arises frequently in genetic counseling (see Chapter 20).

Table 16.1 shows some examples of incidences among relatives of those affected for harelip and two other congenital malformations. Note that the percentage affected among the relatives decreases regularly with decreasing degree of relationship. Although this is no rigorous proof of biological inheritance of the trait, it is certainly highly suggestive. Naturally, the most informative relatives are MZ twins, but they are rare. Sibs are the next closest relatives of interest. For various reasons, they are preferable to parents for this kind of analysis if a choice must be made.

The incidence in sibs of the affected, and the incidence in the general population, are sufficient for an initial analysis of threshold traits.

Figure 16.2 shows in graphical form the use of these two quantities to classify diseases into three major groups: dominants, recessives, and those with a complex pattern of inheritance. For the latter, one can also evaluate approximately from the graph the heritability, assuming a polygenic model for the liability distribution.

Table 16.1
Percentage affected among the relatives of persons with a congenital malformation using, as examples, harelip, dislocation of the hip, and pyloric stenosis

	Harelip (with or without cleft palate)	Congenital dislocation of the hip	Pyloric stenosis
MZ twins	50%	50%	15%
Sibs	3.5%	4%	2%
First cousins	0.7%	0.4%	0.4%
Second cousins	0.3%	0.15%	0.15%
Incidence at birth in general population (approx.)	0.1%	0.1%	0.3%

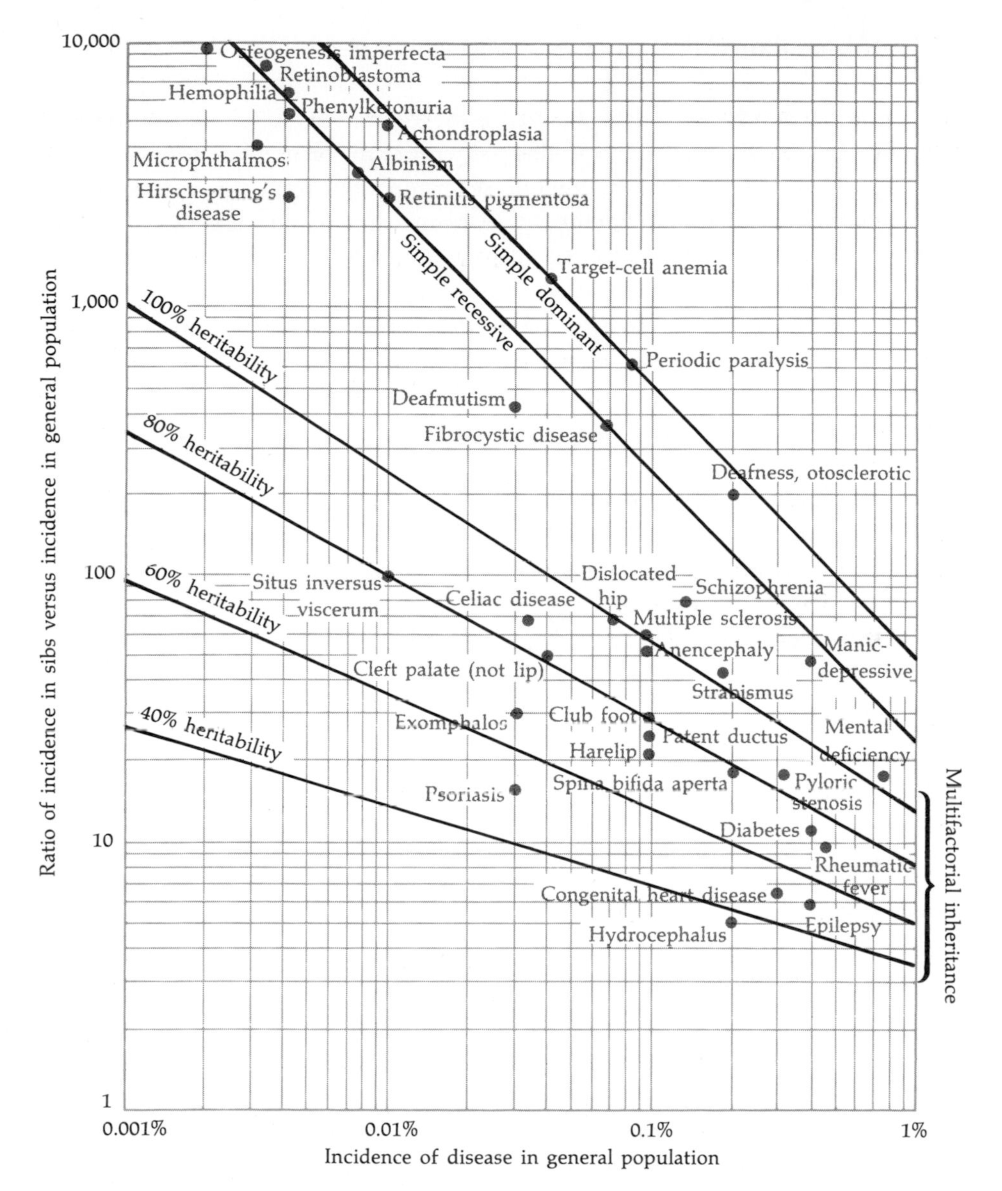

Figure 16.2
Heritability of diseases. Diseases are classified into single dominants, single recessives, and those with complex inheritance; classification is made on the basis of incidence in the general population and in the sibs of the affected. For the latter, heritability is given assuming a polygenic model for the determination of an underlying, continuously distributed liability. Formulas can then be derived that relate the heritability to the incidences among sibs of affected individuals and in the population at large. The lines for the expected relationships for heritabilities of 40, 60, 80, and 100 percent are based on these formulas. They indicate, for example, a heritability of about 40 percent for hydrocephalus (water in the brain), 80 percent for clubfoot, and close to 100 percent for dislocated hip. (Modified from H. B. Newcombe, in *Papers and Discussions of the Second International Conference on Congenital Malformations,* ed. M. Fishbein, International Medical Congress, 1964.)

16.2 Concordance in Twins

A time-honored procedure for analyzing the heritability of threshold traits is that of comparing the concordance (similarity for an all-or-none trait) of MZ twins and DZ twins. If there is some genetic determination, the concordance of MZ twins should be higher than that for DZ. Concordant twin pairs are those in which both twins have the trait, and discordant pairs are those in which one does and the other does not. In Table 16.2, we show an example of twin-concordance data for epilepsy, a relatively common disturbance, often due to trauma or other environmental effects, but sometimes not associated with any detectable external cause. The concordance for MZ twins (37 percent) is found to be higher than that for DZ twins (10 percent). This suggests genetic determination in a proportion of the cases. But the concordance of MZ twins is so far from 100 percent, the value expected for complete genetic determination, that environmental factors are clearly important in at least about $\frac{2}{3}$ of the pairs. In fact, the difference between the MZ and DZ concordance rates is relatively small, and it may reasonably be asked whether it could not be due to chance alone, given that the numbers of twins observed are small. We can test this hypothesis by using the statistical test known as χ^2 (see Appendix). The result of the test shows that it is very unlikely that the difference between MZ and DZ is due to chance, and so we can accept the difference as real.

Twin-concordance data can be used to obtain an estimate of the heritability.

There are many formulas in use, but few are satisfactory. Moreover, data limited to MZ and DZ twins do not tell us about dominance. The estimate of heritability changes, depending on whether dominance is present or absent. One can, however, give upper and lower limits to the degree of genetic determination (broad heritability) on two different assumptions: that dominance is present, and that it is absent. Data on twin-concordance rates for some diseases, derived from the Danish registry of twins, are presented in Figure 16.3, together with the corresponding heritability estimates. Death from acute infection has, not unexpectedly,

Table 16.2
Twin concordance for epilepsy

Twins	Concordant pairs (percentage)	Discordant pairs	Total pairs
MZ	10(37%)	17	27
DZ	10(10%)	90	100
Total	20	107	127

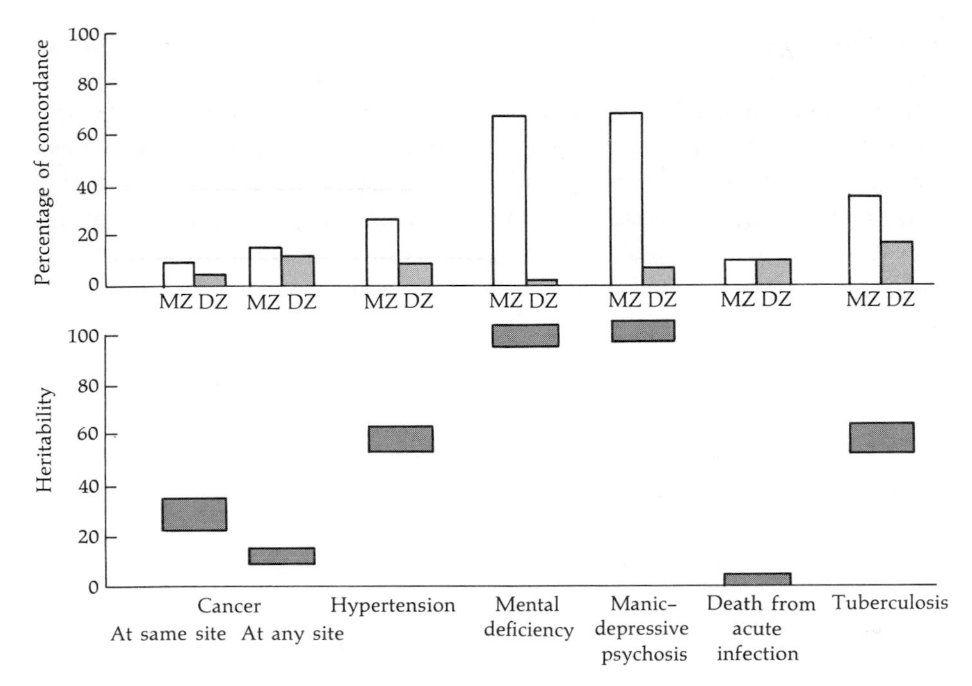

Figure 16.3

Degree of genetic determination for some diseases, based on twin studies. The upper graph shows concordances in MZ and DZ twins. The lower graph shows the range of heritability estimates. The extremes of the range are determined by assuming that dominance is present (upper limit) or absent (lower limit). (Based on data from the Danish Twin Registry; see L. L. Cavalli-Sforza and W. F. Bodmer, *The Genetics of Human Populations,* W. H. Freeman and Company, p. 583, 1971.)

a heritability of zero. Mental deficiency, and also manic-depressive psychosis (of which we shall learn more in the next chapter), have heritabilities around 100 percent, while the other values are intermediate. It is interesting that some infectious diseases like tuberculosis do show a relatively high heritability. Clinicians have for long had the impression that there is a predisposition to tuberculosis, and that this predisposition (which is, of course, just the same as a liability) may well be inherited. Without the germ causing the disease, of course, tuberculosis cannot develop. But under conditions in which the disease is ubiquitous, the heritability estimated from twin studies may be more or less valid. This is not entirely unexpected in view of the evidence, discussed in Chapter 10, for some genetic control of immune responses.

In principle, twin studies have some serious limitations.

If taken in isolation, they cannot inform us about the mode of inheritance (dominant, recessive, polygenic, and so on). For this, family studies are necessary.

Neither are they fully convincing as to the information they provide about the degree of genetic determination, because MZ and DZ twins cannot be expected to be subject to a range of environmental variation that is comparable to that relevant for the population as a whole. In extreme cases, there does remain, however, some useful information. A concordance of 100 percent in MZ twins (and lower than 100 percent in DZ twins) gives a fairly high presumption of genetic control, while equal concordances in MZ and DZ twins on the whole exclude genetic factors. The need to consider, in most cases, other sources of evidence than that obtained only from twins should, in any case, be kept in mind.

Discordance in MZ twins may provide useful information about environmental factors.

A concordance value for a trait in MZ twins that is less than 100 percent shows that genes are not the sole determinants of the trait and that other factors must be operating. A careful study of the discordant MZ twin pairs may reveal the nature of such factors. The Swedish Twin Registry, for example, was used to analyze the history of MZ twins who were discordant for a common acute problem—namely, ischemic heart attack. This has some genetic background, but the concordance rate for MZ twins is not very high, indicating the existence of environmental factors. In the search for these, no evidence was found that smoking, obesity, or high blood cholesterol were responsible for the difference between the discordant twins. But one finding of interest was that the twins who were subject to heart attacks had, unlike their luckier cotwins, severe problems and dissatisfactions with their jobs or with their home lives. The existence of psychosomatic disorders is of course no novelty, but further evidence of this kind from MZ twin studies may be especially useful.

Summary

1. The transmission of many all-or-none traits is incompatible with simple Mendelian models. It is still, however, amenable to genetic analysis, by assuming that there is an individual liability to a given disease, in part controlled by genes, and that only those individuals who are above a certain liability threshold manifest the disease.
2. All-or-none traits in twins are often studied by comparing concordance rates for MZ and DZ twins.
3. Twin studies have many limitations, but give interesting suggestions for further research, for the study not only of genetic, but also of environmental factors.

Exercises

1. What are the essential features that distinguish all-or-none complex traits from simply inherited Mendelian traits?

2. What is the threshold model? How does it help with the interpretation of the inheritance of all-or-none complex traits?

3. How are the incidences of an all-or-none trait in relatives of people with the trait and in the general population used to provide estimates of the trait's heritability?

4. What is twin concordance? How can it be used to obtain heritability estimates?

5. Certain virus infections during pregnancy cause fetal malformations (such as cataracts when the mother has German measles—rubella). Certain drugs administered to the mother also have effects on the fetus. When there are twin fetuses, (a) does a high concordance of the malformation in twins indicate genetic factors? (b) is it necessary to find significant differences between the MZ and DZ concordance rates to be sure of genetic effects? (c) what can these genetic factors be due to, when the causal factor is an infectious disease?

6. The following observations were made (Shields, 1962) on the smoking habits of female twins.

	Alike	*Unlike*
MZ, brought up apart (S)	23	4
MZ, brought up together (T)	21	5
DZ	9	9

Is there evidence of inheritance? HINT: First test MZ(S) versus MZ(T) by the appropriate test of significance (see Appendix). If no difference is found, one can pool the MZ data and test MZ versus DZ.

7. The incidence of spina bifida—a congenital malformation that, if not too severe, can be successfully operated on shortly after birth—is about $\frac{1}{200}$ births; sibs of affected are affected with a frequency of about 9 percent. What type of inheritance does this suggest?

8. If dominance is unimportant, and the incidence of spina bifida in sibs of patients affected by this disease (see Exercise 7) is 9 percent, what will be (to a first approximation) the frequency of affected offspring when a parent is affected?

References

Cavalli-Sforza, L. L., and W. F. Bodmer
1971. *The Genetics of Human Populations.* San Francisco: W. H. Freeman and Company. [Chapter 9 includes a discussion of all-or-none complex traits.]

Edwards, J. H.
1969. "Familial predisposition in man," *British Medical Bulletin,* vol. 25, no. 1, pp. 58–64.

Falconer, D. S.
1967. "The inheritance of liability to diseases with variable age of onset, with particular reference to diabetes mellitus," *Annals of Human Genetics,* vol. 31, pp. 1–20.

Liljefors, I., and R. H. Rahe
1970. "An identical twin study of psychosocial factors in coronary heart diseases in Sweden," *Psychosomatic Medicine,* vol. 32, pp. 523–542.
Shields, J.
1962. *Monozygotic Twins Brought Up Apart and Brought Up Together.* London: Oxford Univ. Press.

17

Behavioral Genetics

Behavior is, according to a dictionary definition, "the externally apparent activity of a whole organism." Human behavior thus encompasses the entire area of human activity—voluntary and involuntary. It is the outcome of many years of training of each individual by parents, teachers, and many other people. The legal system exists to compensate for imperfections or failures in individual behavior that adversely affect social life. Thus, the study of behavior also has outstanding importance from a social point of view, but it is not difficult to guess that our knowledge is most limited in this area.

This state of ignorance is especially apparent if we note that there coexist today two opposing schools of thought about behavior. The environmentalists tend to view behavior as almost exclusively controlled by the previous experience of the individual. The hereditarians, on the other hand, minimize the effects of experience and emphasize innate (genetic) factors to account for an observed behavior. Of course, neither school would accept such an extreme and inevitably naive summary as representative of its position. However, this overstatement of the conflict is useful to indicate the two opposite poles that can be readily recognized in most discussions of behavior—positions that emphasize either nature or nurture.

Experimental analysis of animals has begun to contribute to the resolution of this controversy, especially in recent years. In man, we do have some general evidence that behavior can be determined genetically. But serious problems of interpretation arise when a specific trait is investigated. In this chapter we deal particularly with the thorny question of IQ, and with the most common mental diseases and behavioral abnormalities.

17.1 Experimental Analysis of Behavior

Everyone is familiar with our most common pets, cats and dogs. Probably no one disputes the fact that the difference in behavior between a cat and a dog is largely innate—that is, genetic. The differences in skills between dog breeds also are probably—in part, at least—genetic. But the development of these skills—in herding, hunting, racing, watching, tracking, guarding, and so on—usually requires considerable training, so that the relative contributions of nature and nurture to interbreed differences are not so sharply defined.

Much experimental analysis of the genetics of behavior can be done with animals. Some of this research is now being done with the fruitfly, *Drosophila,* whose study contributed so much to our knowledge of general genetics. The animal species closest to man that is favorable for this kind of research is the mouse. Much work by mouse breeders and geneticists has led to the development of a number of independent inbred lines, each consisting of individuals who are very similar genetically to one another. As we have seen, most individuals of an inbred line are homozygous for the same allele at most loci. Because mice (like any other species) have a great deal of genetic polymorphism, any two inbred lines are likely to carry different alleles at many loci. When inbred lines are crossed, heterozygotes for these loci are formed. The first generation of hybrids can then be backcrossed to either of the parental lines; in the backcross progeny, gene differences will segregate according to Mendelian proportions. Thus, the effects of gene differences between the two original lines reappear in the backcross progeny.

Methods of assay exist for a great variety of specific behavioral activities in rats and mice.

Methods of assay for genetic and environmental effects have been highly developed in research with rodents. For example, pharmaceutical companies use tests on mice and rats to find out whether a particular drug has psychological activity. Purebred lines have been developed to provide genetically uniform populations for these tests. Individuals from such lines are more homogeneous in their responses than are randomly bred animals, permitting the attainment of greater precision in drug assays.

If two inbred lines, grown in well-controlled environmental conditions, have different mean values for a trait, it is very likely that the difference is determined genetically. (This conclusion is as good as the control of the environmental conditions.) Crossing of inbred lines and study of the segregation of the trait—for example, in the backcross progeny—can help to determine the number of gene differences involved. Such research shows that inbred strains may differ from each other by only one or a few genes that influence a given trait (although they usually also differ by many other genes influencing other traits).

Genetic analysis is, of course, much easier when single-gene differences are studied. Recent research has demonstrated some cases where a single gene affects differences in patterns of behavior between certain inbred mouse strains. In particular, such a single-gene difference has been demonstrated for avoidance behavior (the avoidance of adverse stimuli such as, for example, a small electric shock; Fig. 17.1). Under some experimental conditions, for instance, avoidance behavior may involve moving to a "safe zone" of the cage to escape an adverse stimulus. The success of these experiments opens up interesting possibilities for further analysis and characterization.

The validity of genetic analysis of behavior in experimental animals depends largely on accurate experimental control of environmental conditions. These include potentially also maternal effects, both prenatal and postnatal. Thus, it is possible to insert a fertilized egg into the womb of a mouse of a different line, or to have a newborn mouse brought up by a foster mother. Experimenters in this field are well aware of the necessity and the difficulties of experimental control of all the possible variables, and of the existence of many subtle environmental effects. For instance, the handling of mice in their early life has been shown to have an effect on their brain development.

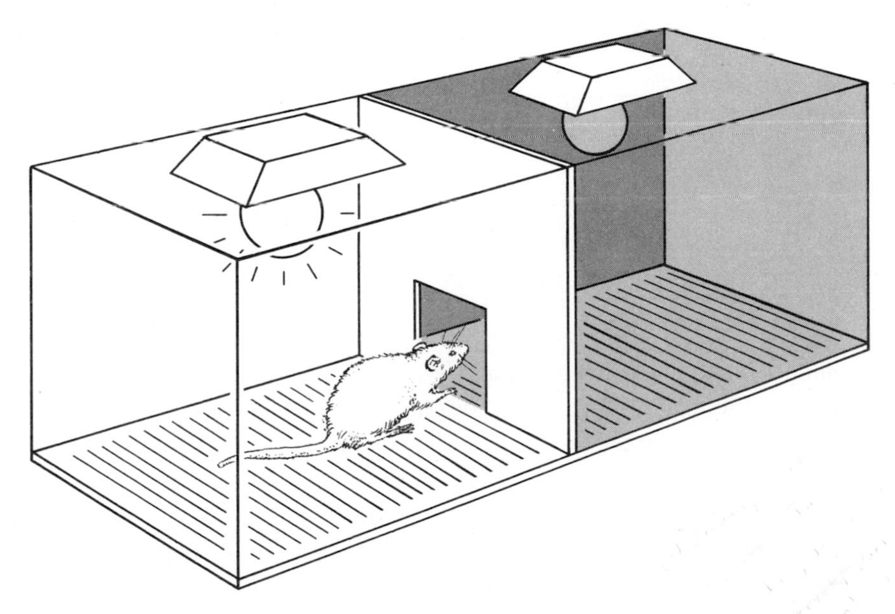

Figure 17.1
Testing avoidance behavior in mice: Warner's test for measuring the learning capacity of mice. To avoid an electric shock administered through the metallic floor, the mouse learns to enter the dark side of the cage as soon as a light appears. After 30 seconds, a new cycle starts and, to avoid the shock, the mouse must again pass to the other side of the cage. (After Oliverio, *Le Scienze*, 1974.)

There are other techniques of genetic analysis for behavioral genetics. We have already discussed an example of artificial selection for "brightness" and "dullness" in rats (Section 14.4). Selection experiments lead to the concentration in two or more different mouse breeds of the genetic differences that were present in the starting population used—those differences that responded to selection. We have also seen how a phenotypic difference accumulated in this way by selection may not be apparent in changed environmental conditions, giving substance to the concept of genotype-environment interaction. These findings make it imperative to define and carefully standardize environmental conditions for the reproducibility of experiments and reliability of conclusions.

17.2 Genetic Determination of Behavior in Man

There are likely to be many environmental factors (including social ones) that are of importance in shaping behavior in man. Animal breeders are familiar with the role that the environment can play, and can exercise some control over it by careful planning and execution of experiments. But in man we have no experimental control over these factors. Thus, when we turn to man, we are faced with a task which is in some respects much more difficult.

One possibility is the method of "statistical" control.

Suppose we want to know whether XYY individuals are really prone to criminal behavior, as has been suggested by the initial discovery of a disproportionately high number of them in a maximum security prison hospital (see Section 4.2). Among the first observations made on XYY individuals was their higher stature than XY males. Does the higher stature itself have direct effect? It would seem that it may, for prison inmates tend on average to be taller than people "outside," quite independently of their having an extra Y chromosome. One might hope to overcome this difficulty, whatever its origin, by comparing XYY and XY males who have approximately the same stature. This is ordinarily done by "matching"—namely by selecting XY "controls" as close as possible to the XYY that are available, with respect to stature and all other traits that may matter, such as age, education, and income. The XYY individuals will then be compared to the matched XY controls—say for their criminal record, or for appropriate related psychological measurements. The correction for the effects of collateral traits such as stature can also be done, not by actual selection of matched "controls," but in other ways, making use of statistical methods. But the effectiveness of statistical control is always inevitably limited, among other things, by the nature of the collateral traits that are taken into account. If those that are really important (and they may well be unknown to us) are ignored, the procedure will be quite ineffective in removing these sources of error.

Chromosomal aberrations show that behavior is controlled genetically in man, as it is in animals.

The XYY problem is at present still not completely solved. There seem to be some slight deleterious effects at the psychological level due to the extra Y chromosome, but they certainly do not seem to justify the stigma with which this aberration was initially associated.

There is less ambiguity concerning the behavioral consequences of other chromosomal aberrations (Section 4.2). The first one to be discovered, 21 trisomy, or Down's syndrome, has sufficiently typical characteristics that it was described as a syndrome long before the recognition of its chromosomal causation. The behavioral aspects of 21 trisomy are also typical. There is always moderate to severe mental retardation, which can only in part be altered by special training. These children are unusually affectionate and enjoy social contacts. A significant increase in the incidence of moderate mental retardation is also found in the other common chromosomal aberrations, namely Turner's (XO) and Klinefelter's (XXY), though the majority of affected individuals fall within normal IQ ranges.

These proofs of genetic control of behavior are important, but not so useful from the point of view of understanding the mechanisms involved. Differences due to a whole chromosome, and therefore very many genes, are usually too gross to permit further analysis. Moreover, we do not know enough about the mechanism of chromosome imbalance to make useful hypotheses for explaining such complex phenomena.

Further evidence and insight about the genetic determination of behavior comes from the study of certain single-gene differences.

The most important mental deficiency due to a single-gene difference in most Caucasian populations is phenylketonuria (PKU). We have seen in Chapter 3 that PKU is due to a genetic defect of a liver enzyme, phenylalanine hydroxylase, which transforms phenylalanine into tyrosine. PKU individuals retain a great excess of untransformed phenylalanine (see Sections 3.3 and 14.1), which is detrimental for brain development in the early years of life. The exact mechanism of the toxic effect is not yet known. Many other inborn errors of metabolism also lead to mental deficiency, and it is interesting that some of them have very peculiar behavioral consequences. The rare defect of the enzyme HGPRT (hypoxanthine guanine phosphoribosyl transferase), a sex-linked recessive determining the Lesch–Nyhan syndrome (see Section 5.2), induces a very peculiar behavioral aberration, compulsive biting, in addition to symptoms such as incoordination, mental retardation, and aggressiveness. Affected children literally destroy their fingers and even their lips by biting. The missing enzyme is known normally to be very active in the brain and is important in the metabolism of purines, but there is as yet no clue as to the way in which the biochemical defect determines this peculiar behavioral anomaly.

Twin studies indicate genetic control for many behavioral traits.

There have been attempts to quantify almost every aspect of man's psychological activities. There are thus tests both for "global" intelligence and for many of its various, separate aspects. There are tests for extroversion or introversion, for neuroticism, for masculinity or femininity, and for many other attributes. Reliability and usefulness varies with the particular tests. We shall not discuss this aspect of the subject, which is beyond the scope of this book, but shall give some results obtained by using the tests in studies on twins.

We have seen that evidence from twin studies is not as strong as might be desired, but that it certainly can be indicative of genetic, as well as of environmental, effects. This is true also in the determination of this very elusive but fascinating entity, the human personality (see Fig. 17.2). Much psychometric investigation has concentrated on measurements that go under the name of IQ, and to them we devote the next section.

17.3 IQ: The Roles of Genetics and Environment

Francis Galton, who was active in the second part of the last century, was probably the first to attempt to measure human intelligence. He was also responsible, as mentioned in previous chapters, for pointing out the potentialities of twins for genetic research, for introducing measurements of resemblance between relatives and, finally, for showing how specific talents—such as those for music or mathematics—may be concentrated in genealogies, suggesting their biological inheritance. Galton was himself a member of a family of unusual scientific talents, being the first cousin of Charles Darwin, in whose genealogy many other brilliant scientists were and are found.

Present IQ tests owe much to the procedures that were established— in part under the influence of Galton's work—by the French psychologist, Alfred Binet, for measuring potential achievement in schools. The tests were devised in answer to a request from the French Ministry of Education.

There are now an enormous number of procedures for testing IQ. They all depend on answers to a number of questions and problems. These are assigned

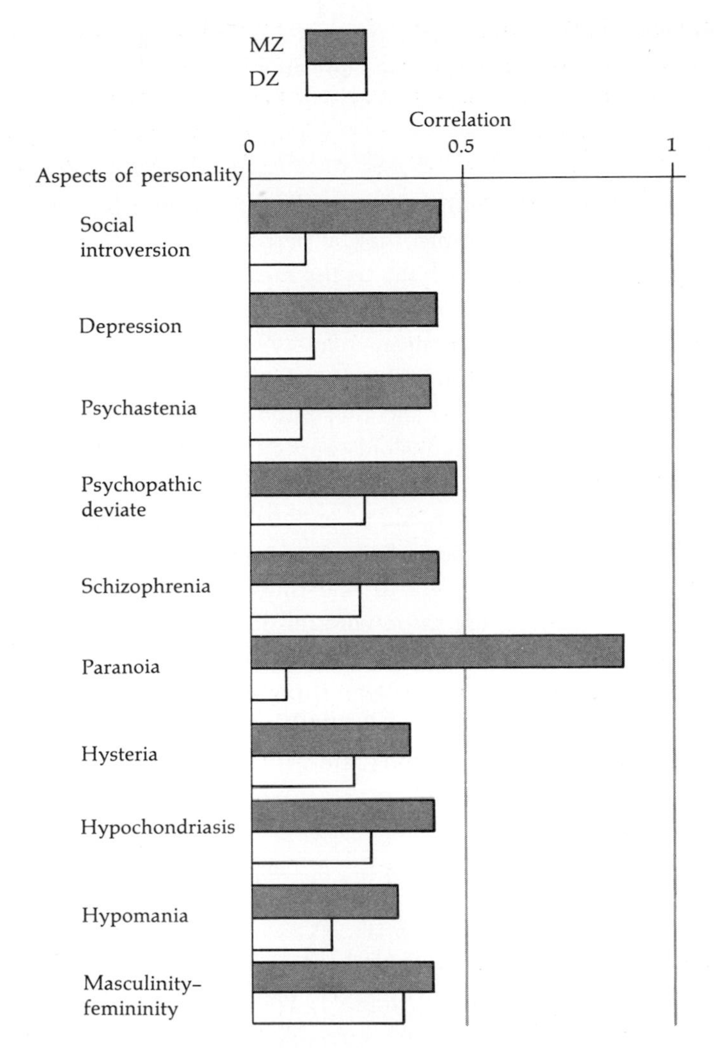

Figure 17.2
Correlations between normal MZ and DZ twins for performance on the Minnesota Multiphasic Personality Inventory. (Data from a survey by S. G. Vandenberg, in *Recent Advances in Biological Psychiatry*, vol. 9, ed. J. Wortis, Plenum, 1967.)

scores. The sum total of the scores obtained by a given individual is transformed into an IQ test score by a process of standardization. This is carried out by testing a large population sample, selected to be representative of a country or of a segment of it. The original scores are transformed so that the IQ value has a mean of 100 and a standard deviation (SD) of 15, independently of age and sex. Naturally, questionnaires differ for different ages. Tests have also been adapted for adults and for preschool children. But tests given at 2 years of age or earlier correlate very little with results at later ages, and presumably test different abilities.

IQ has a normal distribution.

Most IQ tests give a distribution that is close to normal (see Appendix and Fig. 17.3). This means, for example, that 50 percent of all individuals are above and 50 percent below the 100 mark that is the mean. Because the standard deviation is 15, about $\frac{1}{3}$ of the individuals have IQ scores between 85 and 100. The value of 85 is one SD below the mean—namely, $100 - 15 = 85$. Another $\frac{1}{3}$ lie between 100 and 115; and the remaining $\frac{1}{3}$ is about equally divided among those below 85 ($\frac{1}{6}$ of all the population) and those above 115 (the other $\frac{1}{6}$).

The normality of the distribution is only approximate. At the lower tail, there is an excess of mentally deficient individuals. In a strictly normal distribution, between 20 and 50 IQ points there should be only 0.05 percent of all the population, but in this range are actually found five times as many, or 0.25 percent of the population. These individuals were at one time classified in Great Britain as "imbeciles," while "idiots" were those who fell in the range below an IQ of 20. These are found with a frequency of 0.06 percent, whereas in a truly normal distribution, they should occur with a vanishingly small percentage. The classification now adopted in the U.S. is similar to that proposed by the World Health Organization. It distinguishes profound (below IQ of 25), severe (25–39), moderate

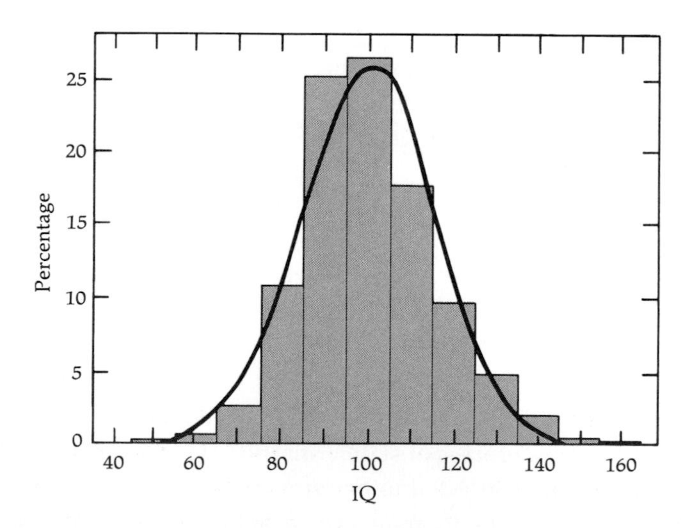

Figure 17.3
Distribution of IQ among Scottish children. This is the same distribution as that of Figure 14.4, here shown as a histogram. (From K. Mather, *Human Diversity*, Free Press, 1964; data from MacMeekan.)

(40–54), and mild (55–69) mental retardation. In general, one expects a normal distribution to hold when several factors, each of small magnitude, all contribute to determining the trait. Normality may then be the consequence either of a polygenic system or of a multitude of small environmental effects, as discussed in Chapter 14. The excess of cases in the lower tail of the distribution is largely due to single deleterious genes, or to environmentally induced severe traumas.

How stable is the IQ of an individual over time?

Retesting the same individual after a short time with a related test gives results very similar to those obtained the first time. The correlation between first and second tests declines as the time interval between them increases. There are, however, very few followup studies of individuals over long periods. One such study showed that some individuals' IQs remained approximately constant between 2 and 18 years of age. Others, however, showed changes, and all possible types of change were represented: increases, decreases, an initial increase with a final decrease, and vice versa. Among the environmental factors that were examined in this study, the only ones that showed some effect were extreme parental permissiveness or extreme severity, both of which seemed to be responsible for decreases of IQ with increasing age. Another study, made in Sweden, showed that a group of students whose IQ was measured at 12 years of age, and who then went through an especially demanding high-school curriculum, had an average increase of 11 points when they were retested after finishing high school. This may be an effect of a stringent scholastic discipline.

In summary, IQ is by no means perfectly stable with age, and in fact, its fluctuation over the lifetime of an individual may be far from negligible. But it is perhaps stable enough to justify the search for genetic factors that influence it. Possibly, long-term instability is itself a factor of importance, but a difficult one to study.

IQ measurements are controversial: What does IQ actually measure?

There is a great variety of opinions among psychologists as to the value and meaning of IQ tests. Some dismiss them entirely, or view them only as a very narrow facet of human personality, and consider IQ as a naive and even misleading measurement. At the other extreme, some psychologists (especially perhaps educators) view IQ testing as one of the most precise psychometric techniques. The latter group includes some who believe that there is a "general intelligence factor," *g*, which is largely innate and dominates the results of IQ measurements. The *g*-factor theory is, however, far from being generally accepted. Other psychologists prefer to analyze and measure, rather than a single factor, many different "abilities"—such as verbal, numerical, inductive, and so on. Yet others distinguish a large number of more specific abilities, perhaps up to 120 in all.

Definitions of intelligence can only be vague—such as "innate general cognitive ability," or "the ability to carry out abstract thinking"—and are not operationally too useful. In fact, in practice, intelligence often is defined operationally (if not verbally) as the ability to score well on a particular IQ test being used for the research. It is important to remember that many assumptions about intelligence are built into the IQ tests.

Binet's original IQ test (1905) was developed to aid in predicting the success or failure of French school children. During the development and standardization of the test, children's scores were compared with their teacher's estimates of their abilities and intelligence and with later reports on how well they had actually succeeded in subsequent years of school. Test items that correlated well with these checks on validity were retained; other items were dropped. Similar procedures have been followed in the preparation of subsequent IQ tests.

An American revision of the Binet test was prepared in 1916 at Stanford University by L. M. Terman and others. This test and its 1937 and 1960 revisions have long been regarded as the standard American IQ tests. Results obtained with the 1916 Stanford-Binet test indicated higher average IQs for boys than for girls. This might have reflected a tendency for some questions to favor boys (because of the nature of their experience), rather than an innate tendency toward higher intelligence in males. In preparation of the later revisions, test items were eliminated if one sex or the other seemed to do better on those items. In this way, the tests were adjusted to give almost identical IQ distributions for males and females.

Therefore, the statement that there is no significant difference between males and females in Stanford-Binet IQ tells nothing about possible genetic differences in intelligence between the sexes. It simply indicates that the test developers were successful in their aim of devising a test that gives identical means for the two sexes, while continuing to function well as a predictor of scholastic success. No similar attempt was made to eliminate test items that gave significantly different scores for children of varying cultural, racial, or socioeconomic backgrounds. Subsequent studies have pointed out a great many cultural biases in the Stanford-Binet test items—questions that we might expect a child from a background other than white-American-middle-class to find confusing. In many cases, in fact, we might expect an intelligent child from certain backgrounds to give a particular "wrong" answer.

To meet these objections, many researchers have tried to develop "culture-free" or even "culture-fair" tests of intelligence. The goal is a test that can be administered to persons of any cultural background; an individual's score should correlate well with his scholastic achievement, career accomplishments, teachers' evaluations of intelligence, and other independent validity tests. No test has yet been devised that achieves this goal. Attempts to make cross-cultural comparisons of IQ—either between individuals or between groups—are therefore likely, on the whole, to be unsuccessful, at least with present techniques.

As definitions are vague and culture-fairness not achieved, one wonders what scientific and practical use IQ tests may have. The major application may be for scholastic placement in a culturally homogeneous society: this was an important reason for their development and may remain the major, if not the only, rationale for their continuing use. While leaving discussion of social aspects for Chapter 21, we may add here that IQ tests in Caucasians do correlate fairly highly ($r = 0.6$ to 0.7) with scholastic achievement.

What are the chances of sorting out genetic from nongenetic factors in the determination of human behavior?

Let us first consider the correlations between relatives reared together (Table 17.1). Medians of the many studies are given here for convenience, rather than means (see Appendix) and, in addition, we give the observed range of r values (smallest and largest values observed). The extent of the range indicates considerable heterogeneity between studies, which is hardly surprising, if only because different tests were used in the different studies. The correlations for MZ twins reared together are extremely high; nobody, however, would accept these values as an estimate of the broad heritability. The other correlations are also too high for a simple theory of genetic determination to hold. For instance, DZ twins and sibs have values above the 0.5 expected for random mating, "perfect" inheritance (that is, no environmental effects), and no dominance. But as we have already noted, mating is far from random for IQ; the correlation between husband and wife is quite high—at least around 0.4. This raises all expected genetic correlations for sib–sib and parent–offspring comparisons from 0.5 to 0.71 (namely, by a factor of $2/[1 + 0.4]$; see Fig. 15.7), making the observed correlations appear much more reasonable. There is also an indication of dominance, because the parent–offspring correlation is lower than that between sibs.

Taken at face value, and even after corrections for assortative mating, these results indicate a very high heritability for IQ. But they cannot be trusted, if taken

Table 17.1
Correlations for IQ between relatives reared together

Relatives	Correlation coefficients: Median	Correlation coefficients: Range	Number of studies
MZ twins	0.88	0.76–0.98	26
DZ twins	0.63	0.44–0.92	26
Sibs	0.58	0.45–0.67	8
Parent–offspring	0.48	0.35–0.58	6

SOURCE: Based on data compiled by C. Jencks and others, *Inequality*, New York: Basic Books, 1972.

in isolation. Relatives reared together cannot provide data that distinguish between biological inheritance and transmission of environment from parent to child, or direct phenotypic transmission (see Chapter 15). A considerable effort has been made by several investigators to secure data from adopted children to counteract this problem.

Existing data on MZ twins reared apart unfortunately do not throw much light on this question.

We have already mentioned the hopes and disappointments generated by the study of MZ twins reared apart. The largest set of data (by C. Burt) is the one that has been most severely criticized, especially by psychologist Leon J. Kamin. Even Arthur Jensen, who has in the past derived much support for his hereditarian theories from this set of data, conceded recently that Cyril Burt's estimates were to be treated with caution. Table 17.2 summarizes the available data. The average of the four existing correlation values (including Burt's) is 0.81, and it is mainly from this estimate that the "80 percent heritability of IQ" widely quoted by some of the "hereditarians" has been derived. If one excludes Burt's study, the average correlation drops somewhat. The next highest value is J. Shields', which is based on a very well-reported set of observations. But here, as we have already noted, many (more than $\frac{2}{3}$) twins reared apart were raised by close relatives. The broad heritability estimate would drop further if we excluded Shields' data, but we should then be left with very few cases indeed (31), and the statistical error of a correlation coefficient computed on so few observations is very large, leaving us with a rather uncertain estimate.

The 80 percent estimate of broad heritability must include genotype–environment interactions and correlations (see later) that cannot be separated from the genetic variance by these data alone. Including these terms with the genetic part

Table 17.2
Correlations for IQ between MZ twins reared apart

Study	Number of twin pairs	Correlation coefficient (r)
C. Burt (Great Britain)	53	0.86
J. Shields (Great Britain)	38	0.77
H. H. Newman, F. N. Freeman, and K. J. Holzinger (U.S.A.)	19	0.69
N. Juel-Nielson (Denmark)	12	0.73
Total	122	

SOURCE: Based on data compiled by C. Jencks and others, *Inequality,* New York: Basic Books, 1972.

of the variance cannot give a result comparable to that which would be obtained if one could equalize environments, and certainly exaggerates estimates of the genetic component to the determination of IQ.

The study of adoptions helps to provide a clearer picture of the situation.

Because of the rarity of adoptions, only a few studies have been done on IQ of adopted children, and a summary of some of the available results is given in Table 17.3.

There is a nonzero correlation between adopted children and their adoptive parents or sibs, which demonstrates an influence of the environment. Therefore, the estimates of heritability derived from correlations between biological relatives must be too high because they tend, as already emphasized, to include in the genetic component other, nongenetic sources of correlation between individuals belonging to the same family. There are, unfortunately, numerous possible biases in most adoption data, somewhat limiting their reliability. Adopted children are not a random sample of the population, and the placement of these children in foster families by adoption agencies is not random. This generates some perplexity when one tries to make an exact partition of the total variance.

An attempt at a general analysis of all these data and at estimating "genotype–environment" correlations has been made by C. Jencks and coworkers, with the results summarized in Figure 17.4. Genotype alone explains a little less than half of the variance, but the calculated value (about 45 percent) is subject to considerable uncertainty, because of the paucity of the data. It could be as low as 25 percent or as high as 65 percent. The portion of the total variation due to genotype–environment correlations includes at least some of the variation due to nonbiological transmission (see Fig. 15.11).

Table 17.3
Correlations for IQ of adopted children

Pairing	Correlation coefficient		Number of studies
	Median	Range	
Between adopted child and adopting parents	0.23	0.07–0.37	6
Between adopted child and natural child	0.26	—	1
Pairs of independently adopted children reared together	0.42	—	1

SOURCE: Based on data compiled by C. Jencks and others, *Inequality*, New York: Basic Books, 1972.

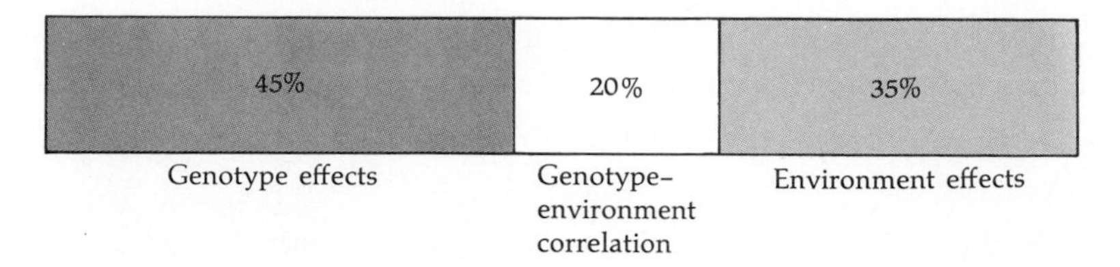

Figure 17.4
Estimated variation components in IQ, using correlations in adoptive and biological relatives from U.S. white and English populations. (Data from the analysis by C. Jencks and coworkers, *Inequality,* Basic Books, 1972.)

A more satisfactory way of representing genotype–environment relationships is to view them in terms of the "plasticity" of the phenotypic response to the environment.

The environment is a very complex entity: very many dimensions would be needed to express it adequately. A single measurement of environmental quality would in most cases be an oversimplification. But let us, for the sake of constructing a simple model, imagine that we can simplify the environment down to one single quantity. We use this as the abscissa in the graphs of Figure 17.5, while the ordinate is the phenotype. The hereditarians' view is that the phenotypic response is essentially independent of the environment: this theory generates the horizontal lines of Figure 17.5A. All (or most) of the individual differences are entirely predicted by the genotypes. The view of the environmentalists is that genotypic differences are trivial, so that all genotypes fall on the same curve and the individual variation is due solely to the different environments to which different individuals are exposed (Fig. 17.5B).

The correct view almost certainly is that the genotype determines only the mode of response to the environment—that is, the phenotypic response is "plastic." A given genotype may give rise to different phenotypic values, depending on the environment; but the curve relating the phenotypic response to the environmental conditions varies from one genotype to the other. Some genotypes may be more stable (for example, 2 in Fig. 17.5C), while others (such as genotypes 1 or 3) are more plastic. The rate and even the direction of change of phenotype with environmental change (contrast genotypes 1 and 3) may vary with the genotype. The picture may also vary with the range of environmental conditions that the individuals of a population experience. If, for example, the range of environmental conditions is limited to that indicated as "a" in Figure 17.5C, the environmentalists' hypothesis is almost correct. Under conditions "b," however, the hereditarians' view comes closest to being satisfactory—while both are unsatisfactory in the "c" range. Inevitably, the correct picture is a complex one, and we are far from having reached satisfactory conclusions. Scientific discussions tend to be

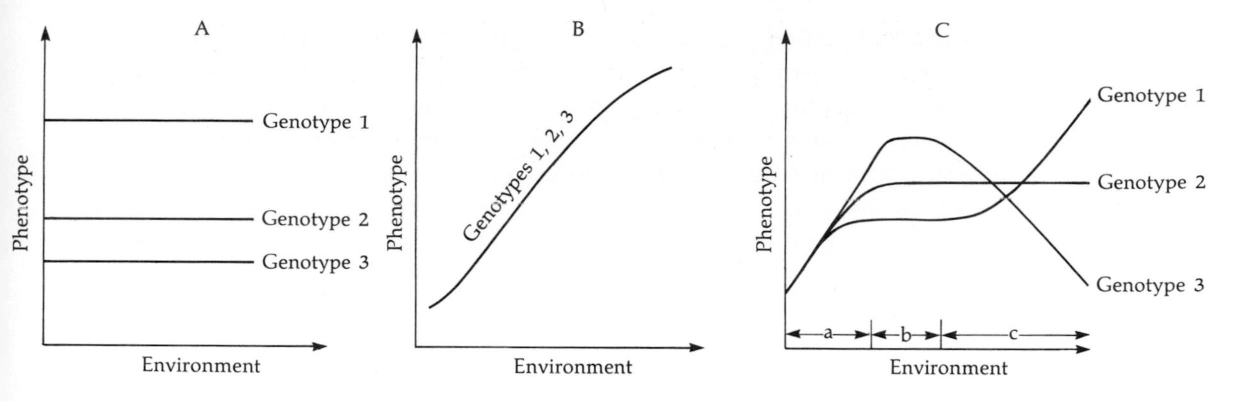

Figure 17.5
Schematic representation of the dependence of phenotype on genotype and on environmental conditions. (A) The hereditarians' view. (B) The environmentalists' view. (C) Genetic variation of individual response to environment. In environmental range a, environmentalists are approximately correct. In range b, hereditarians are approximately correct. In range c, neither view is right.

more heated, the greater the state of ignorance and the more important the subject, and there could hardly be a better illustration of this than the continuing debate on the inheritance of IQ.

17.4 Group Differences in IQ, and Environmental Effects

Considerable public interest has recently been focused on IQ differences between races and between social classes. We deal with these two problems fully in Chapter 21, after discussing racial differentiation. But there are some general concepts that can be introduced at this stage of our discussion. The main reasons for focusing our attention on IQ are public interest and the fact that IQ has been studied more extensively than any other psychological attribute. Criminality, another important and much studied social and psychological attribute, is discussed briefly in the following section. Only a very small fraction of criminal activity, however, can be adequately analyzed because much—and usually most—of it is either not reported or does not result in a conviction (leaving the identity of the criminal uncertain).

In the view of a number of psychologists, IQ tests are sufficiently culture-biased ("anglocentric" according to one) that much time spent on them is wasted. Today, IQ tests are criticized for being biased toward British (or American) middle-class culture—or perhaps, more generally, biased in favor of the values of Western industrialized civilization. It is worth noting that 2,000 years ago Cicero certainly expressed the consensus view of the most distinguished Romans of his day when he recommended avoiding taking a Briton into one's house, for you could not find a more stupid slave! As a general rule, members of any culture have tended to regard most members of other cultures as unintelligent.

IQ tests, however, are closely related to scholastic performance (within the culture for which the test was standardized), and scholastic achievement is closely allied with the procedures by which people are channeled into their professions and occupations. Though these procedures often may not be particularly efficient, yet improvements in some kind of IQ testing and better methods of encouraging students to enter appropriate careers seem, in general, desirable. Despite the inadequacies of educational and IQ testing procedures, it is likely that encouraging a person with an IQ of 70 to become a civil engineer or a lawyer would be about as wise as expecting a tone-deaf Italian to become an opera singer.

With the introduction of conscription during World War I, the opportunity arose for testing the IQs of a large number of conscripts drawn more-or-less randomly from the population of young U.S. males. The results showed differences between white and black Americans. In every state, blacks had *on the average* a lower IQ than that of whites from the same state. But some other interesting facts also emerged: for example, blacks from some northern states had average IQs that were *higher* than those of whites from some southern states. The test employed in World War I was especially "culture-loaded" and could almost be regarded as a direct test of scholastic achievement. These data, which foreshadowed much of the current discussion of IQ differences between the races, illustrate the complexity of the problem of deciding whether such a difference could possibly have a genetic component. As discussed in Chapter 21, there is also evidence for differences in mean IQ between the various social classes, again raising the question of genetic components to differences between population groups.

Thus far in this chapter, we have emphasized the problems of estimating the genetic component to the determination of IQ. It is, of course, the evidence for a significant genetic component, as measured by estimates of heritability, that leads to the question of genetic differences between population groups. However, in addition to considering the genetic contribution to IQ, it is important to ask about the evidence for demonstrable environmental effects on IQ.

The widespread use of correlations may obscure some of the evidence for the existence of environmental effects on IQ.

Children who come up for adoption are often from the lower classes. One hundred children born to young, white, unmarried mothers, adopted into middle-class families, were the object of a study by M. Skodak and H. M. Skeels. Correlations between the children's IQs and those of their biological mothers were higher than those with their adoptive parents (the latter being around the usual value of 0.25), though smaller than the correlations between biological parents and children reared in their biological family. This result confirms the existence of some genetic effect on IQ, and is in line with the estimates of Figure 17.4.

There was, however, another very interesting feature of the data from this study, which correlation studies alone would not ordinarily show. The biological mothers had an average IQ of 85. The adopted children, who had the good fortune to be brought up in homes that were on average richer than the homes of their biological parents, had an average IQ of 107—fully 22 points higher than that of their biological mothers. We do not know how much the biological fathers contributed to their children's IQ, because the fathers were not available for study. But it is reasonable to assume that the more favorable environment offered by the foster homes contributed, if not all the 22-point difference, at least more than half of it. Thus, the family environment may account for nearly 15 IQ points (or about one SD of the IQ distribution). This variation is certainly not enough to explain in purely environmental terms *all* the known variations for IQ. But even the staunchest hereditarians must concede that there exists some environmental variation between individuals within a population—variation that can affect IQ.

Other environmental effects on IQ have been demonstrated.

One clearcut effect that is believed to be environmental is the reduced IQ associated with being a twin. Twins, whether MZ or DZ, have an IQ that is on average 5 points lower than that of singletons. The possibility that the effect is genetic is remote; for one thing, twinning itself is under very loose genetic control. Triplets show an even greater reduction of IQ, averaging about 9 points.

Twins have a lower birth weight and a higher mortality than singletons in the perinatal interval (from shortly before to soon after delivery). Thus, one possible hypothesis for the twin effect is that it is due to prenatal damage. This hypothesis is, however, contradicted by the fact that twins who survive the perinatal death of a cotwin have a normal IQ. This observation seems to give a role in the environmental depression of IQ in twins to postnatal rather than to prenatal factors. Possibly, the effect is to be sought in the inevitably diminished parental care devoted to an individual twin, or to more complex psychological factors. Twins, especially if they are identical, have rather unusual lives. Their life together apparently creates a special environment usually involving very deep emotional and other ties.

It has long been known that there are differences between first and later born offspring in many respects. An analysis by an IQ-like test, conducted on a very large body of data collected on Dutch conscripts provides a clear example of this phenomenon (Fig. 17.6). IQ decreases steadily with increasing order of birth, for all family sizes, and both for rural and urban boys. Psychologists have not yet provided an explanation for this striking phenomenon, though here (as in the case of twins) the amount of parental attention per child may be an important factor.

By contrast, another study on Dutch conscripts has helped to show that the severe famine experienced in Holland during the winter of 1944 and 1945, because of World War II, had no apparent effect on the IQs among those who were exposed to it, either *in utero* or after birth. Evidence for detrimental effects of malnutrition is thus far confined to severe cases of qualitative rather than quantitative inadequacy of the diet.

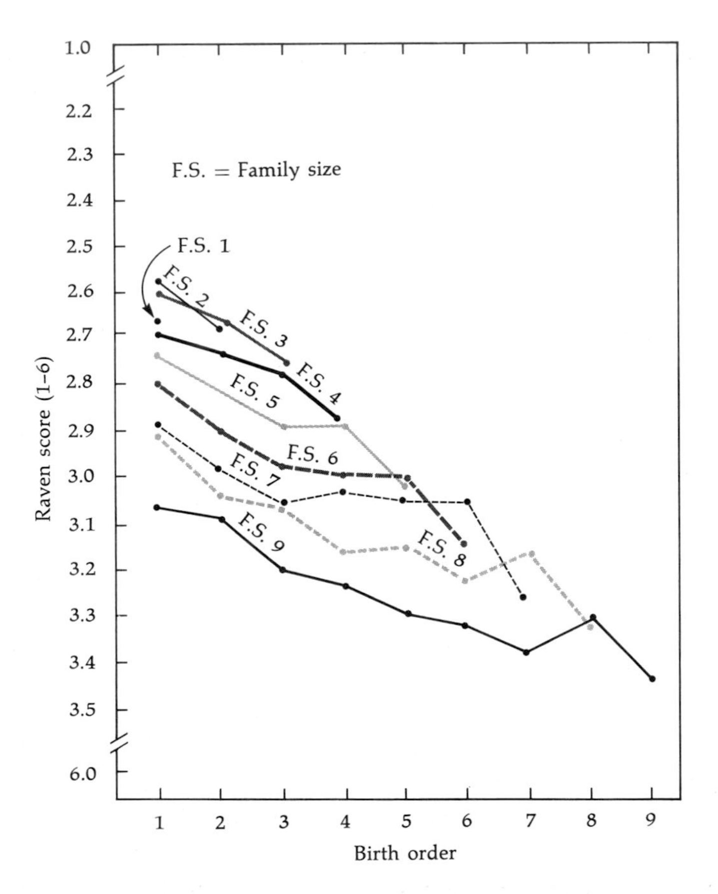

Figure 17.6
Relation between birth order and IQ (Raven score), based on data from Dutch conscripts. Curves are given for different family sizes varying from a total of one to nine children. (From L. Belmont and F. A. Marolla, *Science,* vol. 182, pp. 1096–1101, 1973. Copyright 1973 by the American Association for the Advancement of Science.)

Rather than dedicating much research to improving the validity of the figures that attempt to define the relative roles of nature and nurture in the determination of IQ, as heritability is supposed to do, it would surely be much more rewarding to increase our knowledge of *specific* genetic and environmental factors that may control IQ. We should also look at many other aspects of our mind, of which IQ is just one narrow facet. In this way we might hope to contribute to improving the conditions for our normal mental development.

17.5 Abnormal Behavior

The threshold beyond which a pattern of behavior becomes abnormal is of course arbitrary, and is inevitably a function of the culture to which one belongs. Some manifestations of abnormal behavior are, however, "objective," such as suicide, which may be an extreme and objective sign of a severe depressive condition. Many individuals, however, may suffer from severe depressions without ever being induced by external contingencies to a suicidal act. Suicide is not, in fact, a good measure of the intensity of a depressive condition.

If alcoholism were measured by the amount of alcohol consumed, one would not be taking into account the different effects which the same amount of alcohol ingested can have on different individuals. "Alcoholics," in a study done in Iceland, were identified by another objective criterion: an alcoholic was defined as a person who had to be referred at least once to a physician for treatment of an alcoholic condition. In this way, it was found that 11 percent of all males were alcoholics, as compared to only 1 percent of females. These figures, which superficially seem to indicate X-linked recessive inheritance (see Chapter 6), might on the other hand reflect only a sex difference in the opportunities for exposure to "acute intoxication," or even in the chances of being referred to a physician when found in such a condition.

Many mental diseases are further complicated by the fact that there may be no physical symptoms. The diagnosis depends entirely on what the doctor learns about previous history, and on the patient's outward behavior. A distinguished Stanford professor of psychology and some of his students managed to get themselves diagnosed as having severe mental disturbances, ranging from schizophrenia to other types of psychoses, in respectable psychiatric hospitals. They offered relatively simple descriptions of spurious symptoms. The psychiatrist is relatively powerless against such false pretensions, for there are no physical signs on which to rely. Most mental disorders are, moreover, quite heterogeneous, and the constellation of symptoms can vary greatly from one individual to another. Moreover, in the last analysis, there is little agreement as to the criterion for distinguishing between normal and abnormal.

Of 114 twins judged to be schizophrenics by I. Gottesmann and J. Shields, other well-known experts gave the same diagnosis in different numbers of the same subjects, ranging from 77 to 32 out of the original 114. This was not so much due to inconsistency of the diagnosis, but to differences in the willingness to consider a case sufficiently severe to merit the disease label. Given these difficulties, it is not too surprising that estimates of the incidence of schizophrenia, for instance, in the U.S. population vary from 0.5 to 2 percent.

Schizophrenia is the single most important mental disease.

In spite of the variation in the figures of incidence that we have mentioned, schizophrenia remains the most important mental disease. Today, drugs are available that greatly reduce the severity of the major symptoms and make it possible to keep many patients at home. Previously, more than half of the beds in psychiatric hospitals in the U.S. were occupied by schizophrenics. The disease involves a severe disturbance of feelings, of volition, and of the capacity to relate normally to people. After a usually insidious onset, the patient regresses to a life of his own, with less and less activity, ending frequently in complete immobility, sometimes in strange attitudes (catatonia) and with a total absence of voluntary actions. In one particular form of schizophrenia, called hebephrenia, the affective reactions are not absent but are inappropriate. In another form (paranoid), there are bouts of persecution mania or of delusions, and absurd thinking. Especially in these last two forms, auditory (the patient hears "voices") or visual hallucinations are common. The age of onset of schizophrenia is variable, but usually relatively early in adult life; the disease tends to appear earlier in males than in females. There is some social stratification of disease incidence, with lower classes being proportionately more affected. There are, however, no known significant ethnic differences in its incidence or manifestation.

The contribution of genetic factors to schizophrenia has been much debated. Familial correlations are relatively high, with the incidence in sibs of the affected being of the order of 12 percent, that in dizygous cotwins of the same magnitude, while monozygous cotwins of affected have an incidence of at least 40 to 50 percent. These data are summarized in Figure 17.7. The most perplexing finding is the enormous variation in twin concordance, which ranges from 15 to 90 percent in different studies. This variation is correlated with the degree of severity that is used as a threshold for defining the disease. One conclusion seems inescapable from these data. Even though they suggest a substantial degree of inheritance, the relatively low concordance of monozygous twins shows that environmental differences also play an important role. Attempts at fitting threshold models to the observations (see Chapter 16) have not made it possible to choose between polygenic and single-gene hypotheses.

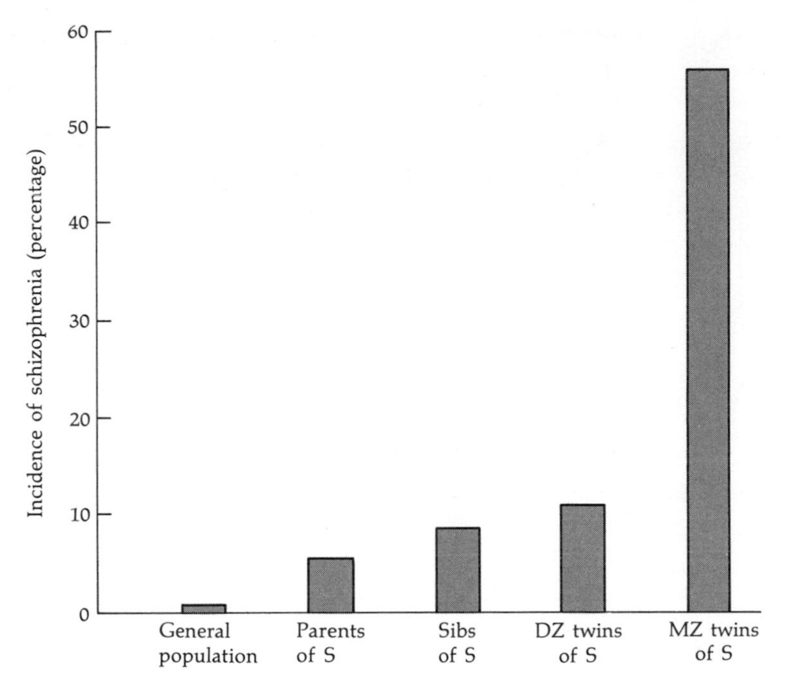

Figure 17.7
Incidence of schizophrenia among relatives of schizophrenics (S). These values are averages over available studies; the data, particularly those for twins, show considerable variation from study to study. (Data from compilation by D. Rosenthal, *Genetic Theory and Abnormal Behavior,* McGraw-Hill, 1970.)

It should be stressed once again that family data alone do not prove that the disease is inherited biologically. The contribution of environmental factors, which is shown by the comparatively low concordance in MZ twins, makes it important to obtain other independent evidence.

Adoption data support the hypothesis of partial genetic determination of schizophrenia.

Adoption data on schizophrenia have been collected (see Figure 17.8). In one study, the children of schizophrenic mothers who were adopted at birth into normal families ("probands") have been compared with "controls," who were adoptees from nonschizophrenic parents, matched for sex, age, and economic conditions with the probands. Mental diseases, mostly of the schizophrenic variety, have been found among the probands in greater variety and number than among the controls. These observations do not exclude the possibility that mothers having a schizophrenic attack during pregnancy may have toxic substances circulating in their blood, which may affect the fetus adversely. Other

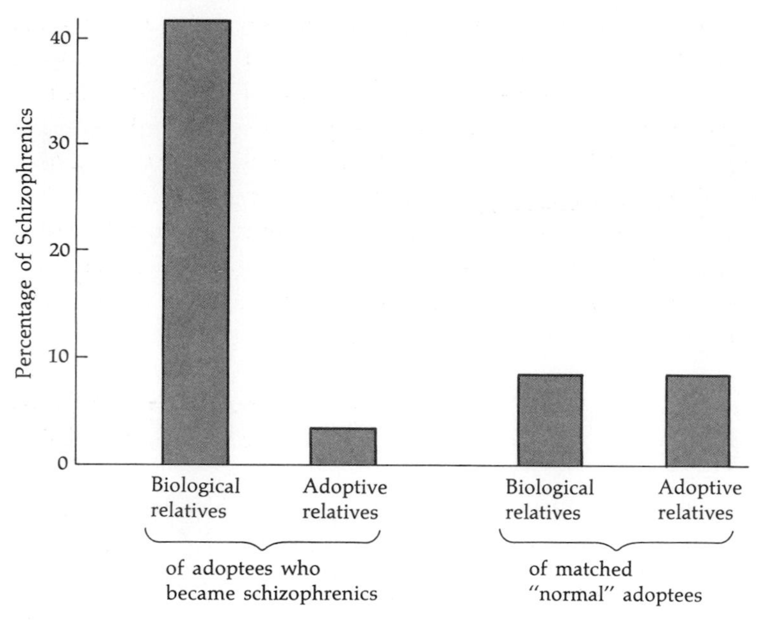

Figure 17.8
Incidence of schizophrenia among relatives of adoptees who became schizophrenic and of matched controls. In these data, the incidence is reported as the percentage of families in which at least one member became schizophrenic. (Data from S. S. Kety et al., in *The Transmission of Schizophrenia,* ed. D. Rosenthal and S. S. Kety, Pergamon Press, 1968.)

observations on adoptees, however, which were conducted in Denmark, excluded this hypothesis. They showed that an adopted individual who becomes schizophrenic has a much greater incidence of schizophrenia among his biological than among his adoptive relatives. Matched "control" adoptees (who never developed the disease) show no such increase. These observations thus demonstrate that there is an important genetic element to this disease.

Future progress is most likely to come from other sources. Association of the disease with biochemical abnormalities is actively being sought. Past biochemical experience with this disease has been rather unrewarding, and many claims of significant discoveries have later been disproved. Recent observations, however, have created new grounds for hoping that a biochemical abnormality will soon be identified, and perhaps will lead to a better understanding of how to treat this distressing and relatively common disease.

Manic–depressive psychosis also has a genetic component.

The second most frequent disease among the more serious mental disorders is manic–depressive psychosis. Here again there are no anatomical or objective signs of disease. In one somewhat rarer form, attacks of depression and despair alternate with manic phases, in which elation, self-assertive aggression, and grandiose delusion dominate. This disease is sometimes called *bipolar,* in contrast with the *unipolar* type, in which only attacks of depression occur. The bipolar disease has a lower age of onset (around 25 years) than the unipolar one, and has a higher familial incidence. The female sex is somewhat more often affected, and this alone provides a very vague indication that the disease might be an X-linked dominant. This suggestion is confirmed by important findings about the inheritance pattern. For the study of a trait thought to be controlled by an X-linked dominant gene, it is useful to contrast two types of matings:

affected father × healthy mother;
healthy father × affected mother.

In the first mating, only daughters should be affected, and all of them should be affected if penetrance is complete, but all sons should be healthy. In the second mating, on the other hand, there is no special reason for expecting different proportions of affected progeny in the two sexes. The observations are fairly close to the expectations; there are almost no affected sons in the first type of mating, in contrast with the second. As a few affected sons are found where none are expected, it may be that penetrance is not complete. Alternatively, the diagnosis may be unsatisfactory and, in particular, the distinction between this and other depressive conditions may be difficult to make. Finally, the disease may be determined by different genes, some of which are autosomal, and one or more of which is an X-linked dominant. Genetic heterogeneity of a disease is a frequent occurrence.

Further evidence in favor of the hypothesis of an X-linked dominant controlling bipolar manic–depressive psychosis comes from linkage studies.

If a *bona fide* gene is on the X chromosome, it should show linkage with other X-linked genes. A search for linkage with color blindness and with the Xg blood group has given positive results. The postulated gene seems to be about halfway between CB and Xg. Genetic heterogeneity (or any of the other causes of error) still, however, gives rise to some anomalous results that await clarification. Pending final confirmation, the case of an X-linked dominant gene for bipolar manic-depressive psychosis seems reasonably good. It will probably be the evidence from linkage studies, when this has been further clarified, that establishes the case conclusively. The same kind of evidence could be sought in many other situations. Autosomes are so far less well mapped genetically than is the X chromosome (see Chapter 5), but the gradual accumulation of linkage data will undoubtedly make this approach much more powerful in the future for all kinds of genetic disease.

The evidence for a genetic component for many other behavioral abnormalities is still much debated.

The range of behavioral abnormalities is very wide. It includes mental retardation, many forms of which, as we have seen, are well characterized genetically and biochemically. Research has been less rewarding in other forms of abnormality, ranging from criminal behavior to homosexuality—which, though once perhaps considered an abnormality, is now increasingly accepted as normal. Renewed interest in adoption studies and the strengthening of rigor in the design and analysis of these difficult observations may well help to improve our knowledge in this area.

The most elusive behavioral abnormalities are those that manifest certain rather transient mental disturbances. The oscillations in the incidence of these disturbances are comparable to those observed with infectious diseases. Heroin addiction 20 years ago was mostly a disease of blacks in the U.S., but is now mostly a "white" disease. There might indeed be individual differences in the liability to drug addiction, and they may well be genetic. But the chances of showing this by conventional genetic techniques are slim. There may well also, for example, be a genetic liability to smoking or to drinking, and there exist claims for this, mainly based on twin studies that are somewhat better substantiated in the case of alcoholism than for smoking.

There have been attempts to show that criminal behavior has a genetic basis. This is an especially difficult proposition, given the fleeting character of the phenomenon. Table 17.4, which contrasts the incidence of different types of crimes among Irish immigrants to the U.S., their offspring, and other "native whites," shows how strong are cultural effects in patterns of crime. This thought may seem reassuring, if one believed that environmental influences are easier to counteract

Table 17.4
Conviction rates in the New York Court of general Sessions (1908–1909)

	Relative incidence of convictions among		
Crime	Irish immigrants	Children of Irish immigrants	Native "whites" of native fathers
Homicide	2.3	1.0	0.5
Rape	0.0	0.3	0.7
Gambling	1.2	2.7	3.6

NOTE: These percentages are based on very small numbers of individuals. Because the differences are not statistically significant, they can be taken only as an indication of an effect. Data from studies of other ethnic groups show similar trends.
SOURCE: E. H. Sutherland and D. R. Creasey, *Principles of Criminology*, Philadelphia: Lippincott, 1960. Based on data collected by the U.S. Immigration Commission, 1911.

than genetic differences. This, however, is not necessarily true. Recent years have seen a considerable increase in criminality and such short-term changes are certainly never genetic. (Even population changes occurring over hundreds of years are unlikely to be predominantly genetic.) Nevertheless, attempts at understanding and preventing the increase in crime have largely failed. The control of environmentally determined phenomena may sometimes be as difficult as that of genetically determined ones. Among the major advances in the prevention and therapy of behavioral abnormalities have been dietary prophylaxis of PKU, and the drug treatments of schizophrenics and of depressive states, many of which are now believed to have a genetic basis.

The genetic analysis of behavior, as can be seen, is still in its infancy, though some results have been obtained. Clearly, genetics can achieve its greatest success when the traits studied turn out to be determined by one or few genes, and when the biochemical patterns can be understood. Knowledge at the biochemical level is likely to help considerably in the discovery of therapeutic treatment. Some avenues for the cure and prevention of mental problems have already been opened through genetics and through other methods of attack, and it can be hoped that serious behavioral disturbances may eventually become preventable—or, at least, therapeutically controllable—to a more satisfactory extent than is now possible.

Summary

1. Evidence for genetic control of behavior is now available from animal experiments. These take advantage of the possibility of precise control of the environment, thus making it possible to dissect even minute genetic differences.

2. In man, both chromosomal aberrations and single-gene differences, especially for inborn errors of metabolism, offer strong evidence for genetic control of behavior. This does not preclude the existence of large environmental effects, in particular with respect to "normal" variation.

3. IQ, because of its apparent ease of measurement and the interest of educators, has been studied more than any other behavioral trait. Perhaps as much as half of all the variation between individuals is genetic.

4. The question of the existence of a genetic component to IQ differences between races and social classes has been raised and will be answered in Chapter 21.

5. The differences in IQ between twins and singletons, and between adoptees and their biological mothers, provide examples of major environmental effects on IQ.

6. Abnormal behavior may also, in part, be inherited. The best evidence for genetic components has, in particular, been obtained for schizophrenia and for a form of manic–depressive psychosis.

Exercises

1. What aspect of the human way of life is a major limiting factor to obtaining solid heritability estimates from studies of similarities between relatives? Why is the problem especially difficult for behavioral traits?

2. What contribution, if any, has the study of individuals with chromosome abnormalities made to our understanding of the genetic determination of behavior?

3. In what ways can inborn errors of metabolism influence behavior?

4. What are the major features of the IQ distribution? What is IQ supposed to measure?

5. Give an overall assessment of the value and validity of currently available estimates of the heritability of IQ.

6. Describe some examples of data that illustrate definite effects of the environment on IQ.

7. How have adoption studies contributed to our understanding of the inheritance of schizophrenia?

8. What use can be made of polymorphic genetic markers (such as blood groups and enzymes) in the study of behavior genetics?

9. What would you consider to be the ultimate goal of behavior-genetic studies?

10. The following figures were given in a study on alcoholism: among 112 children of alcoholics adopted in normal families, there were 18 heavy drinkers; among 98 children of nonalcoholics adopted in normal families, there were 5 heavy drinkers. Do these data favor the hypothesis that alcoholism is, at least in part, genetically controlled? (NOTE: These figures are fictional, designed for easy testing, but recent actual observations agree with the conclusions you should draw from this hypothetical example.)

11. Assume that there is some reasonable assurance (from data in a utopian country where men and women have achieved complete social equality) that the higher incidence of alcoholism in males is genetically determined. Design observations to be collected that would allow you to test whether alcoholism is, at least in part, due to an X-linked gene. Describe the expected results (a) if the hypothesis is correct, and (b) if, instead, an autosomal gene with different expression in the sexes is involved.

12. Suppose that a group of Australian aborigines, living in a reservation, have an average IQ of 80. Australian whites in the general population have an average IQ of 100. The mean difference between the two groups, therefore, is 20 IQ points. Suppose that the broad heritability of IQ in Caucasians is estimated properly as 70 percent, and similarly that the narrow heritability is 50 percent. Is it correct to state that 70 percent of the 20-point difference between Australian aborigines and whites (or 14 points) is genetic? Should one rather say that 50 percent of the 20-point difference (or 10 points) is genetic? Discuss your reasoning. Are the answers to these two questions modified if the heritabilities within Australian aborigines are known to be essentially identical to those among Caucasians?

13. The following questions refer to the situation hypothesized in Exercise 12.
   a. Assume the simple model that the condition of having been brought up in a reservation lowers the average IQ by 20 points, if everything else—genetic and environmental variations between individuals, difference between homozygotes and heterozygotes, and so on—is identical in the two populations. Can one then explain the observed average difference on a purely environmental basis?
   b. Assume the simple model that the Australian aborigines are indeed genetically inferior in IQ to whites, on the average by 20 points. Are the observed data then consistent with the stated heritabilities?
   c. In the light of your answers to the first two parts of this exercise, is it correct to state that the observed difference in average IQ is perfectly compatible both with the hypothesis that the IQ difference between the two groups is entirely environmental and also with the hypothesis that it is entirely genetic?
   d. In the light of your answer to part c, is it correct to state that any within-group heritability is compatible with any explanation—genetic or environmental, in full or in part—of the mean difference between two groups?

References

Belmont, L., and F. A. Marolla
  1973. "Birth order, family size and intelligence," *Science,* vol. 182, pp. 1096–1101.

Cavalli-Sforza, L. L., and W. F. Bodmer
  1971. *The Genetics of Human Populations.* San Francisco: W. H. Freeman and Company. [Chapter 9 includes a discussion of behavioral genetics.]

Cavalli-Sforza, L. L., and M. W. Feldman
  1974. "Cultural versus biological inheritance: Phenotypic transmission from parent to children (A theory of the effect of parental phenotypes on children's phenotype)," *American Journal of Human Genetics,* vol. 25, pp. 618–637.

Jencks, C., et al.
  1972. *Inequality.* New York: Basic Books.

Kidd, K. K., and L. L. Cavalli-Sforza
  1973. "An analysis of the genetics of schizophrenia," (in Proceedings of IV International Congress of Human genetics, Paris), *Social Biology,* vol. 20, pp. 254–265.

McClearn, G. E., and J. C. DeFries
  1973. *Introduction to Behavioral Genetics.* San Francisco: W. H. Freeman and Company.

Rosenthal, D.
  1970. *Genetic Theory and Abnormal Behavior.* New York: McGraw-Hill.

PART IV

EVOLUTION, HUMAN WELFARE, AND SOCIETY

Clearly, an understanding of human evolution may help to answer many questions that we often pose to ourselves about our origins, our position in nature and, perhaps, our behavior and our aims. We observe biological and cultural diversity among humans–between races, for example. This diversity requires an explanation. We are probably unduly sensitive to the problems raised by the existence of diversity, so we should find it useful to explore its biological and cultural bases. Human genetics is relevant to social problems in other ways as well. What can be done to avoid genetic disease? How does genetic diversity between individuals affect individual and social life? Can predictions be made about the genetic future of man? What sensible measures, inspired by genetic knowledge, can be taken to improve human society? These and other problems have recently become the subjects of heated discussion; we devote this last part of our book to them.

18

Evolutionary Development of Modern Man

We are primates; our closest cousin is probably the chimpanzee. The fossil record shows some of the putative links in the chain leading to modern man. The picture tends to become progressively clearer, and inevitably of more direct relevance, the closer we come to the present. The major change that has affected us from a social and biological point of view in the last prehistorical epoch is the transition to an agricultural way of life. Not long after that transition, written history begins.

18.1 Evolutionary Origin of Man: The Fossil Record

The oldest rock that contains evidence of life existing when it formed is over 3 billion years old, but the fossil record is extremely sparse until about 600 million years ago (Table 18.1). Fish are the first vertebrates (backboned animals) in the fossil record; from them separated first amphibians and then reptiles. Mammals differentiated from the reptiles somewhat less than 200 million years ago. Primates, the order of mammals to which we belong, are about 70 million years old. In the modern world, our order includes prosimians (lemurs, lorises, and tarsiers) and anthropoids (monkeys, apes, and man).

Most students of the fossil record agree that the *dryopithecines* formed a stage in the line to man and the great apes. Dryopithecine fossils are common in Miocene rocks of Eurasia and Africa. They were arboreal, fruit-eating quadrupeds. Changes in their dentition indicate that they may, in temperate climates where fruits are seasonal, have changed their foraging habits and descended from trees to the ground. This change in habits may have been one of the stimuli to the acquisition of bipedal walking in the line leading to man.

At least one other line besides our own may have developed from these remote dryopithecine ancestors: the line leading to our nearest living cousins, the great apes (chimpanzees, gorillas, and orangutans).

Among our great-ape cousins, the nearest relative to us is perhaps the chimpanzee—but the chimpanzee line and ours *may* have diverged as much as 20 or 25 million years ago. The great apes have retained a quadrupedal posture like that of the dryopithecines; they have developed (or retained?) a special form of locomotion called knuckle walking. The brains of the great apes are about the same size (in proportion to body size) as were the brains of the dryopithecines. Unlike humans, great apes have 48 chromosomes in each diploid cell. Most human and chimpanzee chromosome pairs are sufficiently similar that their homology can be established (Fig. 18.1). After the pairs are matched up between the species, we are left with only two human and three chimpanzee chromosome pairs that are sufficiently dissimilar to prevent establishing their relationships with present techniques. In some pairs whose homology is clear, there are characteristic differences between chimpanzee and man. For instance, pericentric inversions (those that involve break points on opposite sides of the centromere) account for the different ratios of chromosome arms (see Chapter 4). It is not at all surprising that closely related species show evolutionary differences due to chromosome mutations, some of which may also determine changes in chromosome number.

Much excitement has been generated recently by new discoveries about the ancestors of man. Some recent finds in East Africa, in particular, have required major revisions of earlier theories, but it is probably too early to draw final conclusions, especially at a time when new observations are being reported at a relatively high rate. It is outside our scope here to analyze the fossil record in any detail. The interested student should consult one of several recent surveys (see References). We will, however, discuss very briefly the major aspects of the most recent evolutionary changes leading to modern man.

There are large gaps in the fossil record between dryopithecines and the appearance of the first men.

Some fossil bones (of about 10 million years ago) classified as *Ramapithecus* may be intermediate between dryopithecines and man, but the evolutionary position of *Ramapithecus* is still the subject of discussion (Fig. 18.2). *Australopithecus* is the name given to a genus (with various species) found in a variety of places in the Old World, dating from 5 to a little less than 1 million years ago. This animal

Table 18.1
The evolution of vertebrates in the direct line of descent to man

Era	Period	Epoch	Approximate time ago (million years)	Earliest appearance of
Cenozoic	Quarternary	Recent	0.01	
		Pleistocene	3	*Homo*
	Tertiary	Pliocene	12	
		Miocene	26	Dryopithecines
		Oligocene	37	
		Eocene	53	
		Paleocene	65	
Mesozoic	Cretaceous		136	Primates
	Jurassic		190	
	Triassic		225	Mammals
Paleozoic	Permian		280	
	Carboniferous		345	Reptiles
	Devonian		395	Amphibians
	Silurian		430	
	Ordovician		500	Fishes
	Cambrian		570	
Proterozoic or "Precambrian"				

was bipedal, with a brain size (400 to 600 cm^3) larger than that of modern gorillas and chimpanzees (350 to 400 cm^3). However, even considering *Australopithecus'* smaller body size (Table 18.2), his brain was much smaller than that of modern man (1,400 cm^3).

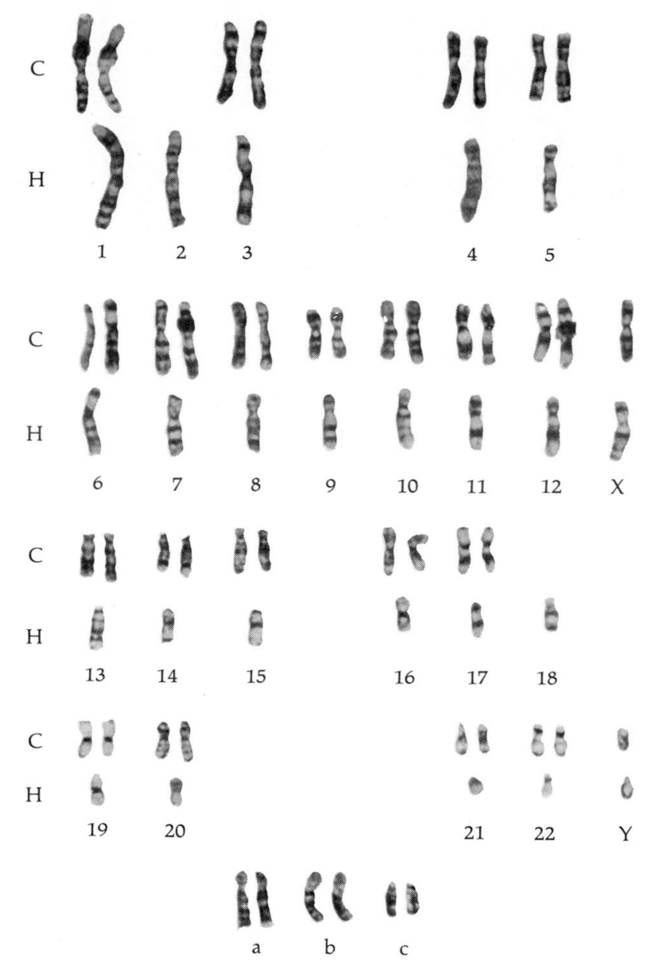

Figure 18.1
Comparison of human (H) and chimpanzee (C) chromosomes. Below each chimpanzee chromosome pair is shown one member of the human pair that is believed to be homologous. Pairs a, b, and c shown at the bottom are chimpanzee chromosomes for which no human homologs can be assigned. Similarly, for two human chromosomes (2 and 18), no homologs can be assigned in the chimp karyotype. Chromosomes are shown stained after trypsin banding. (Courtesy of M. Bobrow.)

The various species of *Australopithecus* include *A. africanus* (with smaller brain and body size), and *A. boisei* and *A. robustus* (larger brain and body). There is evidence that the australopithecines were tool makers. The fossil remains have been found in Africa, but they probably lived elsewhere as well. The latest finds seem to be relatively recent (*A. boisei* in East Africa about half a million years ago).

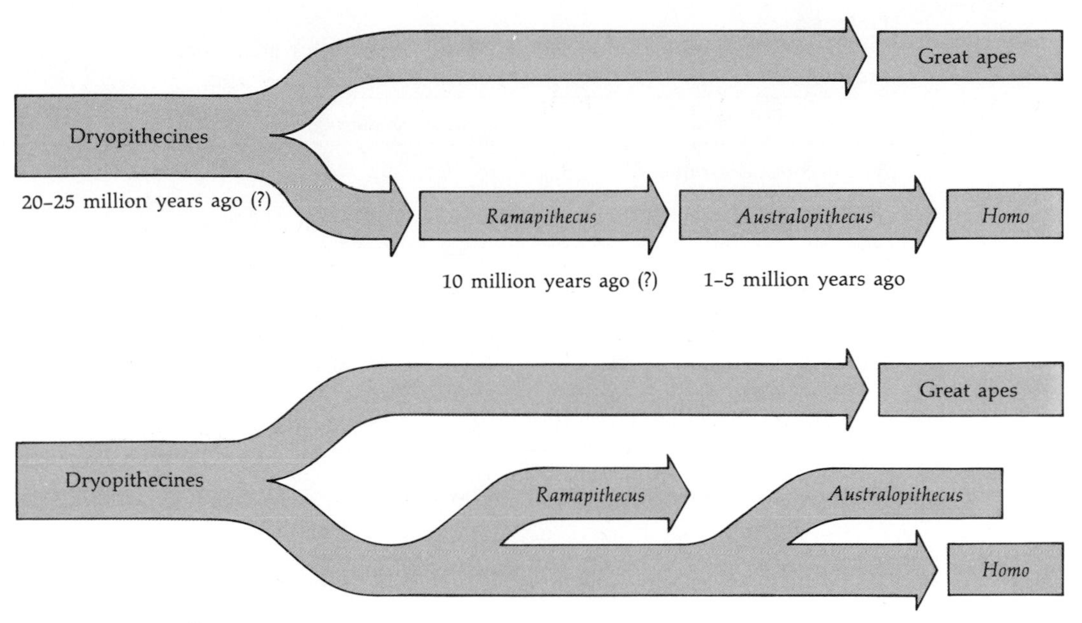

Figure 18.2
Some alternative interpretations of lines of descent from dryopithecines.

Recent discoveries indicate that the separation of Homo from australopithecines occurred earlier than had formerly been accepted.

Until some years ago, it was believed by many that *Homo* arose as a relatively late transformation of *Australopithecus,* but it is now becoming increasingly clear that the *Homo* line diverged from the australopithecines some millions of years ago. The discoveries that triggered this realization were made in East Africa, first at Olduvai (in Tanzania), and the most recent finds in the Lake Rudolph and Omo regions (of Kenya and Ethiopia). Here, finds were made in places where volcanic eruptions covered wide areas with lava flows and layers of ash, creating conditions satisfactory for preservation of fossils and also favorable for dating by a number of techniques—potassium–argon and others.

The finds made in East Africa have been assigned to the genus *Homo.* The most interesting and most fully investigated one to date seems to be the skull ER-1470, which has a brain capacity of about 800 cm^3, definitely larger than many later australopithecines, despite a relatively small stature (about 4 feet). The date is about 2.9 million years B.P. (before present)—certainly not less than 2 million—and there is clear evidence of hunting and stone-tool making. Although the discoverers have chosen not to give a specific name to this *Homo* specimen, many authorities tend to apply to it the formerly-used name of *Homo habilis* (habilines are tool

Table 18.2
Estimates of mean body weights and cranial capacity for great apes and hominids

Species	Body weight (kg)	Cranial capacity (cm^3)
Australopithecus africanus	32	450
Australopithecus robustus	40	500
Australopithecus boisei	47	510
Homo habilis	43	725
Homo erectus	53	1,050
Homo sapiens (Australian aborigines)	57	1,230
Chimpanzee	45	395
Gorilla (lowland)	105	505
Orangutan	53	393

NOTE: Values given are averages of the two sexes.
SOURCE: D. Pilbeam and S. J. Gould, *Science*, vol. 186, p. 892, 1974.

makers). Figure 18.3 shows one of several possible phylogenies of australopithecines and *Homo.*

There has been rapid evolution for increased brain volume.

There is no doubt that skull capacity is an important quantity, both when measuring evolutionary rates in hominid evolution, and for placing these fossils in proper perspective. Certainly, brain volume is not the only characteristic that distinguishes man from other animals, but it is a major one. It can often be measured (when enough skull fragments are available) and can supply an approximate, but significant, evaluation of intellectual abilities. It should be noted, however, that brain volume does vary among living humans. There is a correlation of brain size with body weight, but only a slight correlation of brain size with intellectual ability among living humans. Moreover, brain volume and skull capacity are not identical. They are, though, highly correlated, and the increases of skull size noted during hominid evolution are of such magnitude that they must imply a considerable increase in brain volume as well.

Figures 18.4 and 18.5 show the changes of hominid skull capacity with time. The data are approximate, but they indicate a period of fast growth with a particularly rapid phase until about 200,000 or 300,000 years ago; since that time, growth in brain size apparently has ceased. A trebling of size in a period of perhaps 10 million years is a very rapid evolutionary change. Clearly, many behavioral changes must have been associated with the increase in brain size. The areas of the brain connected with the use of the hand and of the mouth and tongue are much larger in man than in most other mammals (Fig. 18.6). Almost certainly,

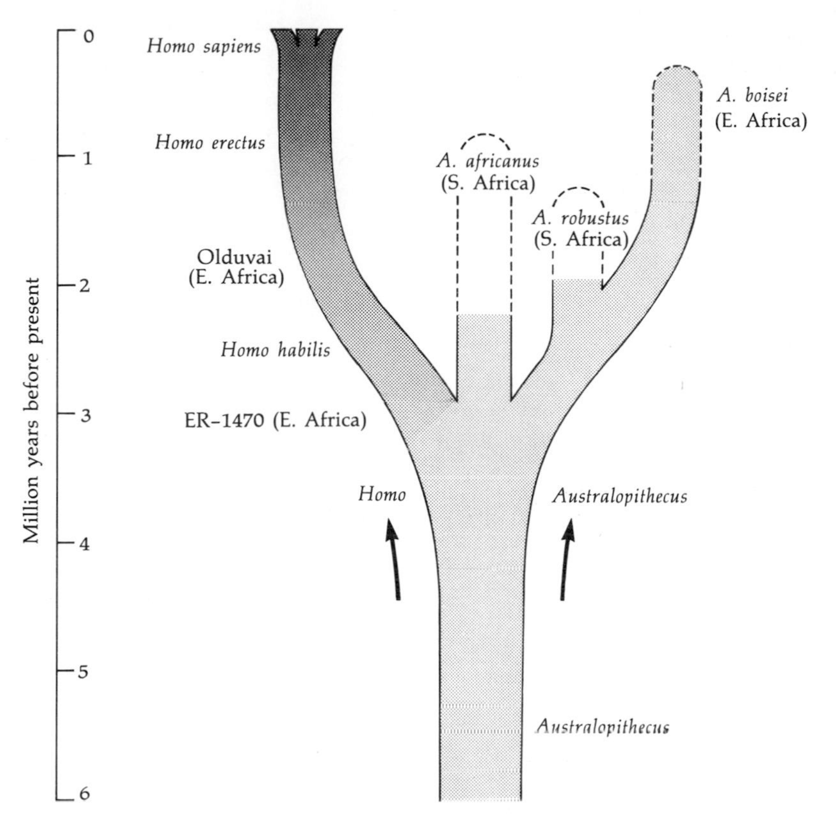

Figure 18.3
Provisional phylogenetic tree of hominids. Regions where major finds were made are indicated in parentheses. Dashed portions of the tree indicate relationships of greatest uncertainty. (After P. V. Tobias, *Annual Review of Anthropology,* vol. 2, pp. 311–334, 1973.)

the growth of these brain areas has been correlated with the development of abilities to make and use tools and to communicate through spoken language. There is very little, if any, evidence on the time of origin of language, but some students of the subject believe that it came late in human evolution. Figure 18.7 summarizes both biological and cultural developments during the transition from the earliest man (*Homo habilis* of East Africa) to modern man (*Homo sapiens*).

Men of the period from about 1,500,000 to about 300,000 years ago (during the transition from H. habilis to H. sapiens) are classified as Homo erectus, formerly called Pithecanthropus.

The name *Homo erectus* may be somewhat misleading, for australopithecines and the early *Homo habilis* already had erect postures. The brain of *Homo erectus* is larger than that of *H. habilis,* but smaller than modern man's. The volume is

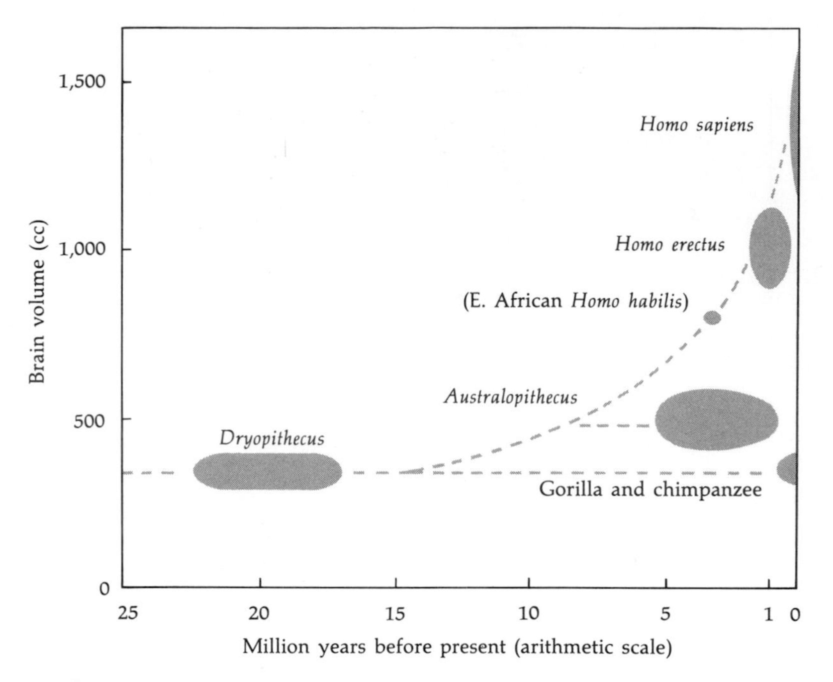

Figure 18.4
Brain-volume changes in hominid evolution. The time scale is arithmetic. Ovals indicate approximate ranges of variation.

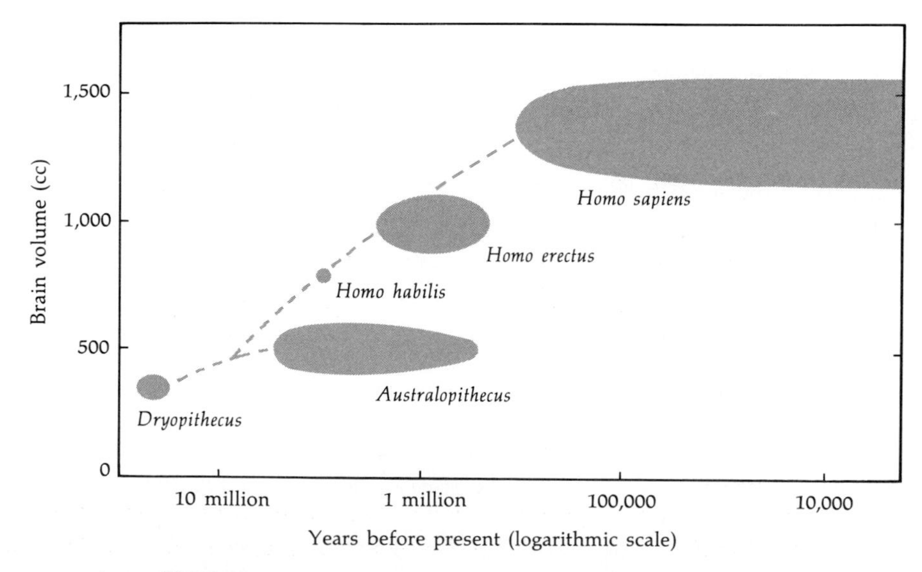

Figure 18.5
Brain-volume changes, as in Figure 18.4, but with logarithmic time scale.

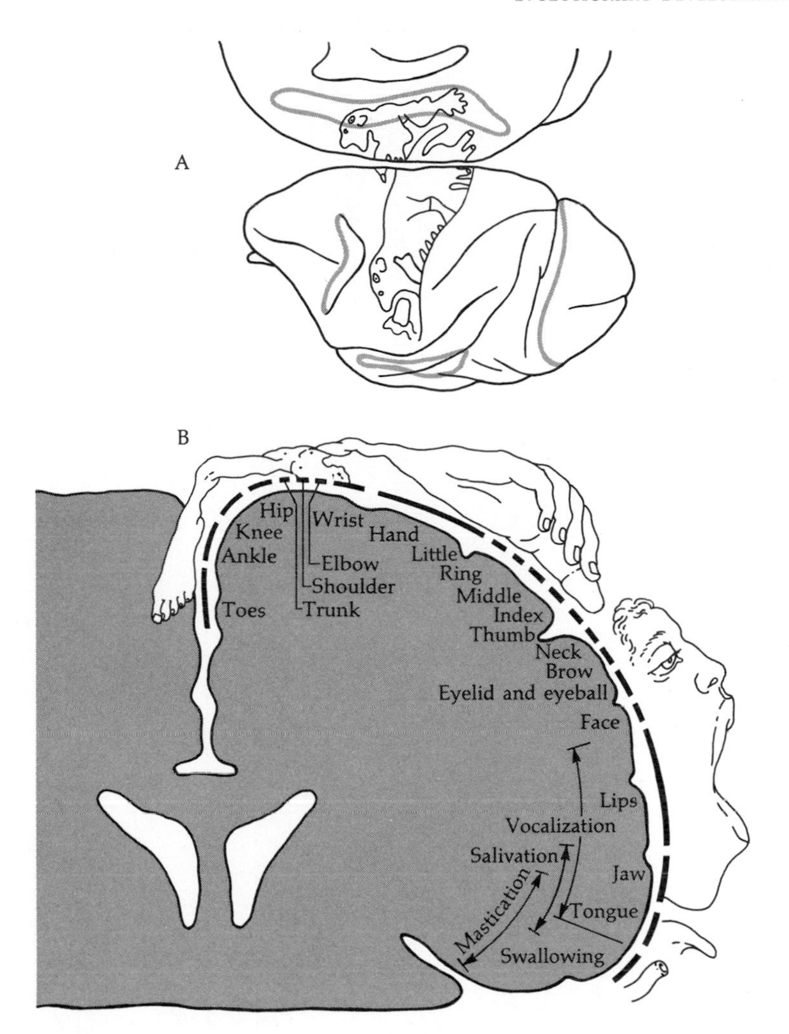

Figure 18.6
Functional maps of the motor cortex (A) in the monkey and (B) in man. Compared with that for the monkey, the map for the human motor cortex shows particularly high development in areas associated with the hand and face. (Part A from C. N. Woolsey, in *Biological and Biochemical Bases of Behavior*, ed. H. F. Harlow and C. N. Woolsey, Univ. of Wisconsin Press, 1958. Part B after W. Penfield and associates, Montreal Neurological Institute.)

around 1,000 cm^3 in some specimens. Some simian features are still present (Fig. 18.8). Skulls of *H. erectus* are found over a wide range in the Old World. Therefore, even if the earlier *H. habilis* was perhaps confined to East Africa, we are sure that the new species *H. erectus* was successful and spread. Naturally, the

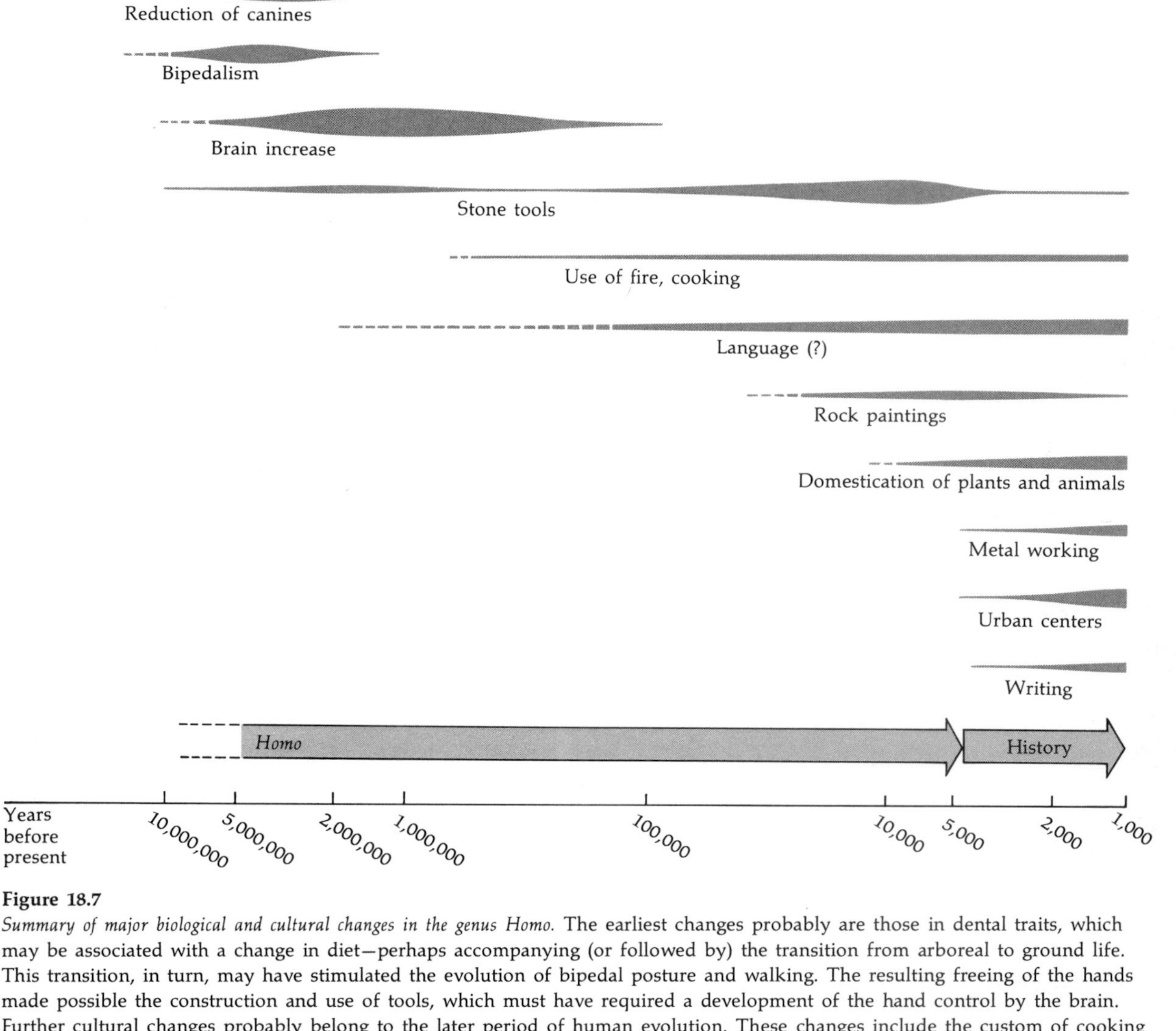

Figure 18.7
Summary of major biological and cultural changes in the genus Homo. The earliest changes probably are those in dental traits, which may be associated with a change in diet—perhaps accompanying (or followed by) the transition from arboreal to ground life. This transition, in turn, may have stimulated the evolution of bipedal posture and walking. The resulting freeing of the hands made possible the construction and use of tools, which must have required a development of the hand control by the brain. Further cultural changes probably belong to the later period of human evolution. These changes include the custom of cooking food and, therefore, of using fire. An even more important change was the acceleration and facilitation of social exchange made possible by improved vocalization and development of language.

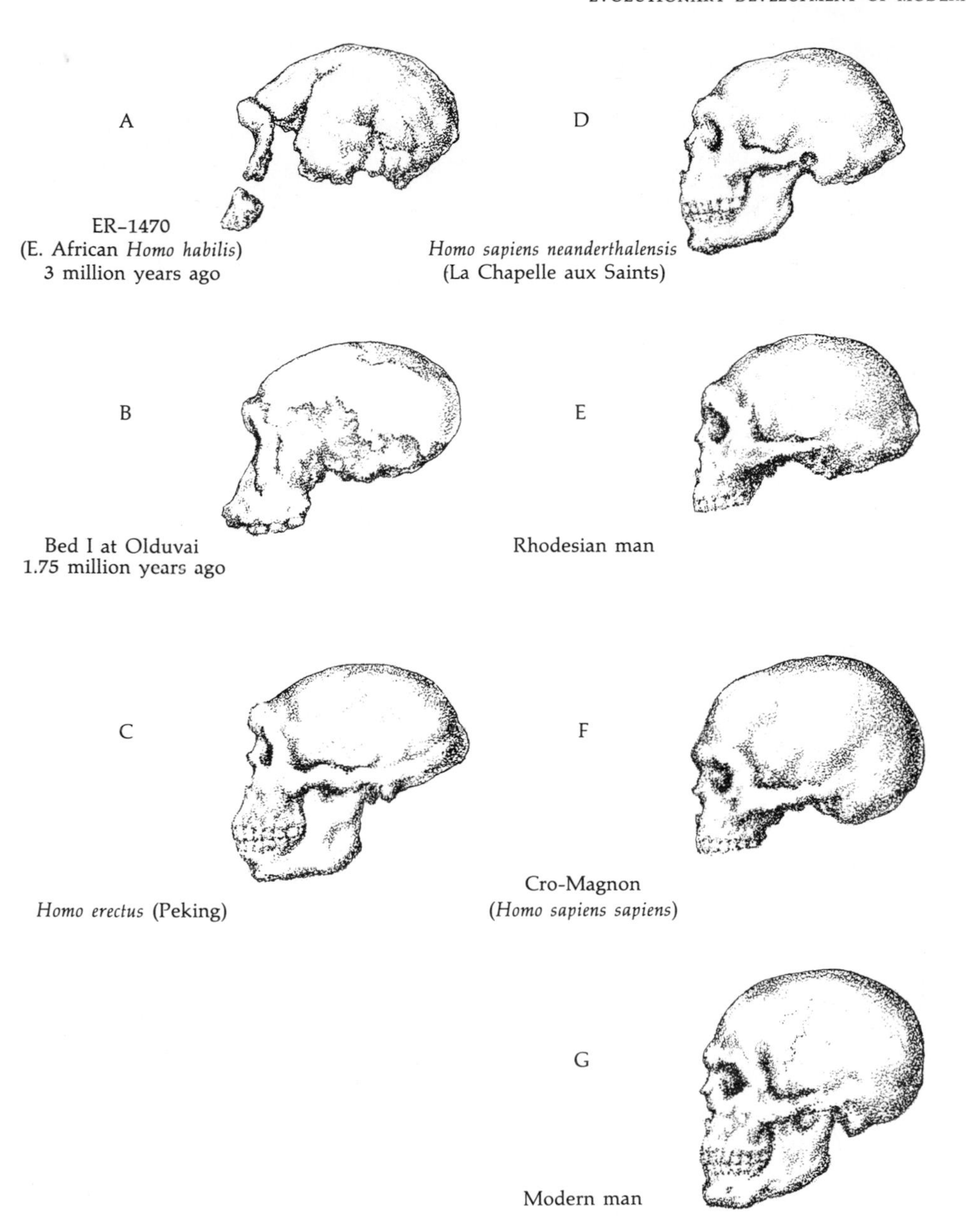

Figure 18.8
Skulls of modern man and some of his progenitors. (After W. Howells, *Evolution of the Genus Homo,* Addison-Wesley, 1973.)

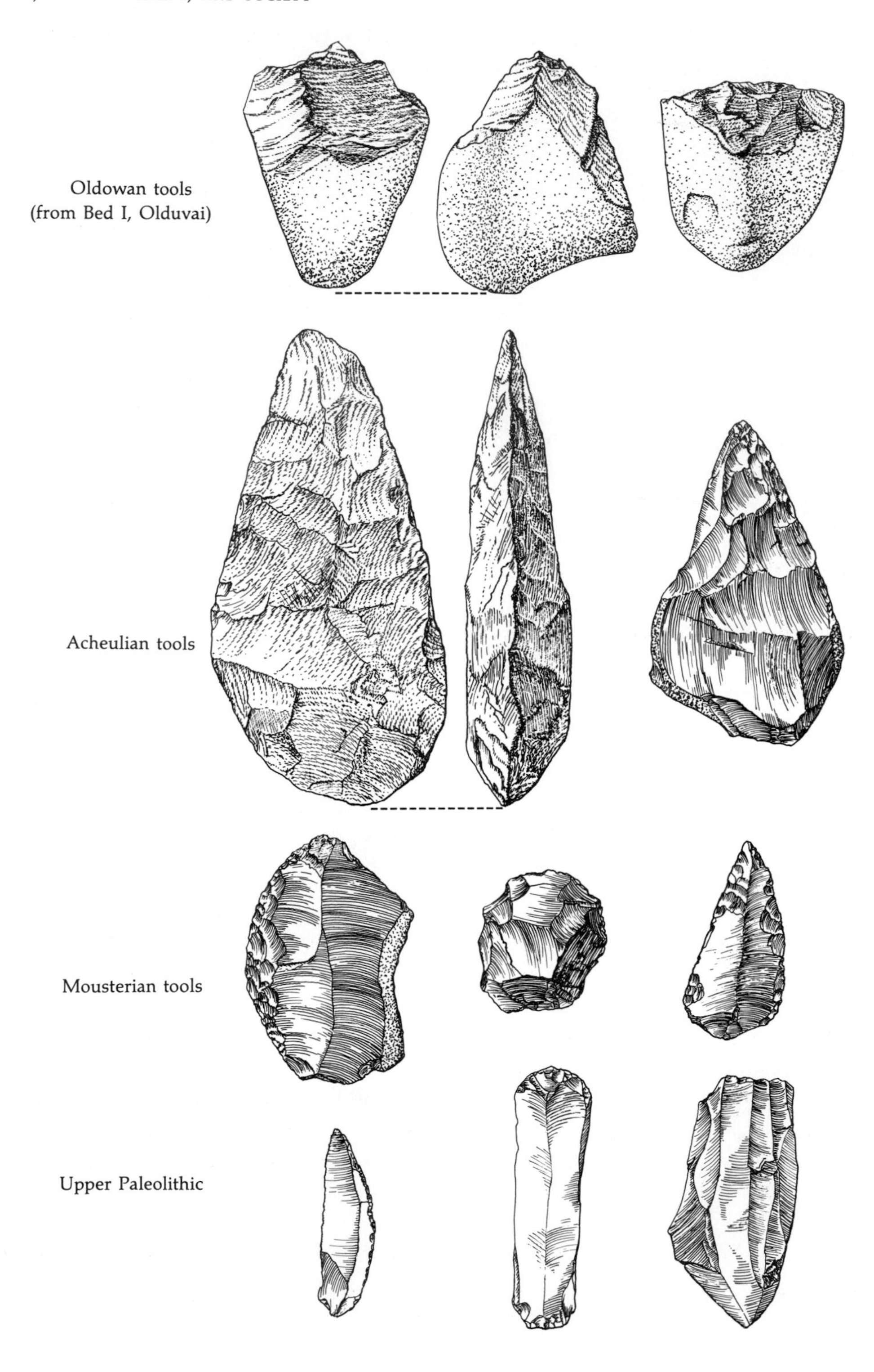
Oldowan tools
(from Bed I, Olduvai)
Acheulian tools
Mousterian tools
Upper Paleolithic

localization of *H. habilis* specimens in East Africa may well be simply a consequence of the better conditions for preservation and research that exist in that area.

Stone tools show slow changes throughout most of the period of human evolution.

The early habilines used stone tools that were essentially choppers and sharp knifelike flakes (Fig. 18.9). These are called Oldowan tools, after the Olduvai gorge in Tanzania. Careful excavation of the site at Olduvai had proceeded since 1931, yielding rich records of these obviously very ancient tools. In 1959, the years of research paid off with the discovery of human remains, clearly associated with tool-making activities, that could be dated to 1.75 million years ago—thus more than doubling the known range of human tool-using development. Later finds, like ER-1470, showed even greater antiquity for tool-using man; the types of tools, however, show very little change over this long period.

During the *Homo erectus* phase, a new tradition of stone-tool making, the Acheulean (bifacial hand axes), was added to the repertoire of Oldowan choppers—but only in Europe and Africa. In East Asia and Siberia, the older Oldowan tradition was continued with little change until rather late. Both Acheulean and Oldowan traditions show little tendency to change over time (remaining substantially unchanged for hundreds of thousands of years) or over space.

Beginning about 200,000 to 300,000 years ago, forms of humans appear that might be called Homo sapiens, with a skull size comparable to that of modern man. Some of these early large-brained men were of the type called Neanderthal men.

A number of these skulls found in Europe—and to some extent those of African origin (Rhodesian man)—show a peculiar morphology (see Fig. 18.8). The striking differences from modern man have prompted some paleontologists to classify these men as a separate species, *Homo neanderthalensis* (named after the place in Germany where one of the first finds was made). Others classify Neanderthal man as a subspecies of modern man: *Homo sapiens neanderthalensis* (with modern man being called *H. sapiens sapiens*).

Neanderthal man had a big brain—even somewhat bigger than that of modern man—but others of his features were more primitive and resembled those of *Homo erectus*. The face was at a different angle in relation to the skull, resulting in a lower cranium than that of modern man. Neanderthals developed a tool-making tradition of their own, the Mousterian. From the skull anatomy, researchers have

Figure 18.9
The evolution of stone tools. (After K. P. Oakley, *Man the Tool Maker,* British Museum, 1950.)

inferred the size of the pharyngeal and oral cavity (in which vowel sounds are produced). This research has led to the speculation that the Neanderthals could not articulate the principal vowels as fully as can modern man. They did, however, have a fairly well-developed culture, including what appears to be a cult of the dead.

About 40,000 years ago, at about the same time modern man spread out over the Old World, the Neanderthalers disappeared quite suddenly. Perhaps they were destroyed by modern man, arriving from the East. On the other hand, some remains found in Yugoslavia and in Israel seem to be intermediate between modern man and Neanderthal, suggesting that some form of hybridization may have taken place. It is not impossible that Neanderthal man disappeared simply through genetic absorption into a much more numerous group of invaders from the East—the modern men to whom we now turn our attention.

18.2 The Spread of Modern Man

The fossil record of man provides us with two kinds of materials: bones and artifacts. The former tell us more about biological and the latter more about cultural evolution. The oldest known skulls that are difficult to distinguish from those of modern man are about 30 to 40 thousand years old. We should be prepared for the possibility that newer finds will push this time back, but most anthropologists now believe that modern man made his appearance about then. The earliest known sites are in the Middle East but, again, this putative center of origin may be changed by future discoveries. It seems very reasonable to assume that this modern man was very successful in an evolutionary sense and spread rapidly over the whole earth. He soon replaced Neanderthals (with whom he may have partially mixed in Israel and in Europe). In Africa, he may have stumbled into Rhodesian man, the local counterpart of the Neanderthals. Until recently, Rhodesian man was believed to have disappeared somewhat later than the European Neanderthal, but this conclusion is now disputed. Expansion eastward is documented by finds in Australia, which are dated back to 32,000 years B.P.

In North America, there have been recent claims (using a new technique, the "racemization" of amino acids) that human bones found in California are 50,000 years old. There were earlier claims of evidence of human occupation some 30,000 years B.P. However, the evidence for substantial occupation of North America stands much later—no earlier than 15,000 years ago—by humans coming from Northeast Asia. Until about 10,000 years ago, North America was connected to Northeast Asia by a land bridge (Beringia) occupying the present Bering Strait. Similarly, Australia was connected to Southeast Asia. Rise of sea level at the close of the last glaciation covered the connecting lands about 10,000 years ago. With the disappearance of these land bridges, Australia and America were cut off from Asia, thus reducing considerably (if not ending) the flow of people between them.

One reasonable hypothesis holds that modern man (Homo sapiens sapiens) spread out from Western Asia around 30 or 40 thousand years ago, replacing or mixing with earlier human populations, and initiating the rapid cultural changes of the Late (or Upper) Paleolithic period.

Because of the large gaps in our knowledge, any interpretation of the data is bound to be speculative. One such speculation is that there was a "population wave" determining radiation—that is, a process of spread from a central area, accompanied by slow population growth. Thus, the subspecies *Homo sapiens sapiens* may have expanded from a nuclear area (perhaps in Western Asia) to all of the world during a period perhaps 30 to 40 thousand years ago. This new human group probably was able to replace preexisting ones, through it may have mixed with some of them. This expansion of modern man occurred at the beginning of the period called Late Paleolithic.

The Paleolithic period (or Old Stone Age) begins with the first tool makers more than 3 million years ago. It ends with the domestication of plants and animals about 10,000 years ago. We have mentioned that changes in tool shape and variety were very slow throughout most of this long period. However, towards the end of the Paleolithic, especially in the last 40,000 years or so of the period, such changes became quite dramatic. Previously, tools had been rather monotonous—few kinds, with little difference from place to place or from time to time—but now the number of different tools increased considerably (Fig. 18.10). Local variation of cultures became more and more pronounced.

The changes during the Late Paleolithic were still very slow, but the rate of change increased continuously and, in comparison to the very slow earlier pace, it is clear that there was an acceleration in the process of cultural evolution. Compared with the earlier cultural stability, the Late Paleolithic has almost the character of a cultural "explosion." Did this "explosion" accompany modern man's conquest of the world? Is it the consequence of the behavioral attributes of this invader? However we answer these questions, it is clear that—in this climate of more and more rapid change—a new technological development eventually appeared that greatly increased man's independence from nature.

18.3 Domestication of Plants and Animals

Throughout the Paleolithic, man was a hunter-gatherer. His food supply came from the animals he could catch and from vegetable products that he gathered. Some 10,000 years ago, a new development took place: domestication of plants

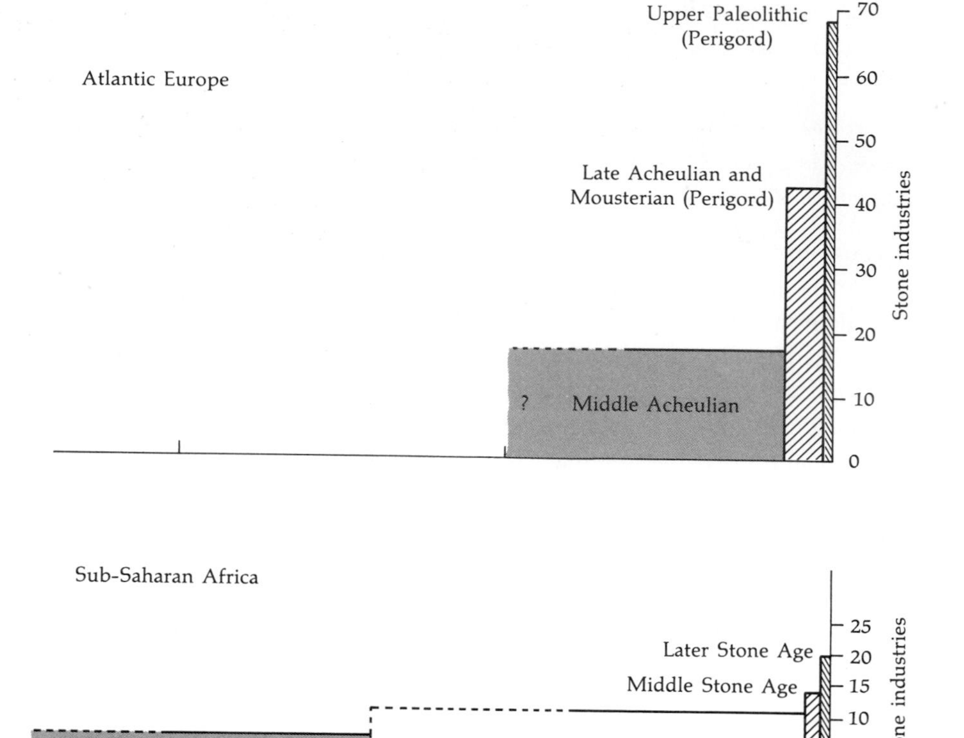

Figure 18.10
Cultural "explosion" of the Late Paleolithic. The graphs show the number of categories of different stone industries catalogued by archeologists, illustrating the sharp rise in variety of cultural activities in the late Stone Age. (From G. L. Isaac, in *Calibration of Hominoid Evolution,* ed. W. W. Bishop and J. A. Miller, Scottish Academic Press, 1972.)

and animals as sources of food. This development apparently occurred first in the Middle East, in a fairly wide area extending from eastern Turkey to Iraq and Israel (Fig. 18.11). Here, in different localities, one finds the earliest domesticated sheep, goats, pigs, and cattle—as well as wheat and barley. Other nuclear areas of agriculture probably developed independently with the domestication of other local wild plants (such as rice in China, corn and many other plants in Mexico).

Domestication represented a natural development toward making more easily available the sources of wild food that had been normally consumed.

Figure 18.11
Main centers of domestication of plants and animals. Dark-shaded areas represent centers more accurately known; light-toned areas indicate centers that appear "diffuse," probably because of insufficient evidence. (Modified from J. R. Harlan, *Science,* vol. 174, pp. 468–474, 1971.)

The Middle East is, even today, rich in the wild-growing cereals from which most modern cultivated cereals were developed. There is evidence of local harvesting and processing of cereal seeds that seems to antedate agriculture. In these temperate areas, the availability of most foods varies substantially with the seasons. Cereals are easy to store, but a bulky store of cereals is not easily transported by a wandering group of hunter–gatherers. Thus, the need to store food may have been the first stimulus to a sedentary or settled way of life, and hence to the attempt to provide more food locally by growing plants in a planned way, rather than relying on spontaneous growth in the wild. Other conditions—such as the depletion of wild resources, perhaps helped by population growth—may also have favored the development of agricultural techniques.

The period of development of domestication of plants and animals is sometimes called the Neolithic period (or New Stone Age). However, the Neolithic period is also sometimes defined in terms of another event: the introduction and use of pottery. The earliest pottery known is actually found outside the area of agriculture. One example is the development of pottery by the Jomon culture, a pre-agricultural economy in Japan, around 9000 B.P. (before present). However, the art of pottery was somehow learned by farmers of the Middle East by around 7,500 to 8,000 years ago. It quickly spread through the areas where agriculture had already spread, such as western Turkey and Greece. From these areas it was diffused mostly in association with the practice of domestication.

Domestication of plants and animals spread slowly.

The spread toward Europe of the early farming practices and their accompanying culture can be followed with some accuracy. Due to incomplete data, the spread in the other directions towards Asia and Africa is less clear. Figure 18.12 shows a map of the spread towards Europe and its timing. In spite of the irregular geography, the rate of spread is fairly constant at about 1 km per year. It thus took some 3,000 years for agriculture to spread from the Near East to the extreme north and west of Europe. The spread took place in other directions as well, with rates that are less well known.

Agriculture probably underwent a series of local adaptations during the spread. This is to be expected in view of the slow rate of spread and of the variety of the local environments. The Sahara was, at that time, not a desert. The wave of the farmers' culture could thus easily cross it, but must have stopped at the limit of the African forest. This limit corresponds with the "diffuse" center shown as a band across northern Africa in Figure 18.11. None of the domesticated plants and animals developed in the Near East was adapted for growth in the more humid conditions encountered there. Thus, new crops were domesticated there, which probably included sorghum and millet. The expansion eastward toward Asia seems to have stopped at the Indus River for reasons that are not well understood.

The spread of agriculture from the centers of origin to the rest of the world is not complete even today.

A number of people, located in places that are difficult to reach and little known even today, still make a living by hunting and gathering and have not adopted the agricultural style of life. A count of such people lists about 30 groups located in different parts of the world. The most numerous groups are the African Pygmies, who number from 100,000 to 200,000, and who occupy various areas of the African rain forest. In South Africa there are the Bushmen, totalling about 60,000, most of whom have recently begun to practice agriculture; only some 3,000 Bushmen are still hunter–gatherers. Other groups in Africa, America, Asia, and Australia are found, at various degrees of acculturation. Australian aborigines and Eskimos are prominent examples.

These groups may not necessarily give a faithful picture of the earlier stages through which humanity has passed. Many things may have changed since then. Moreover, contact with more advanced civilizations is inevitable and has effects

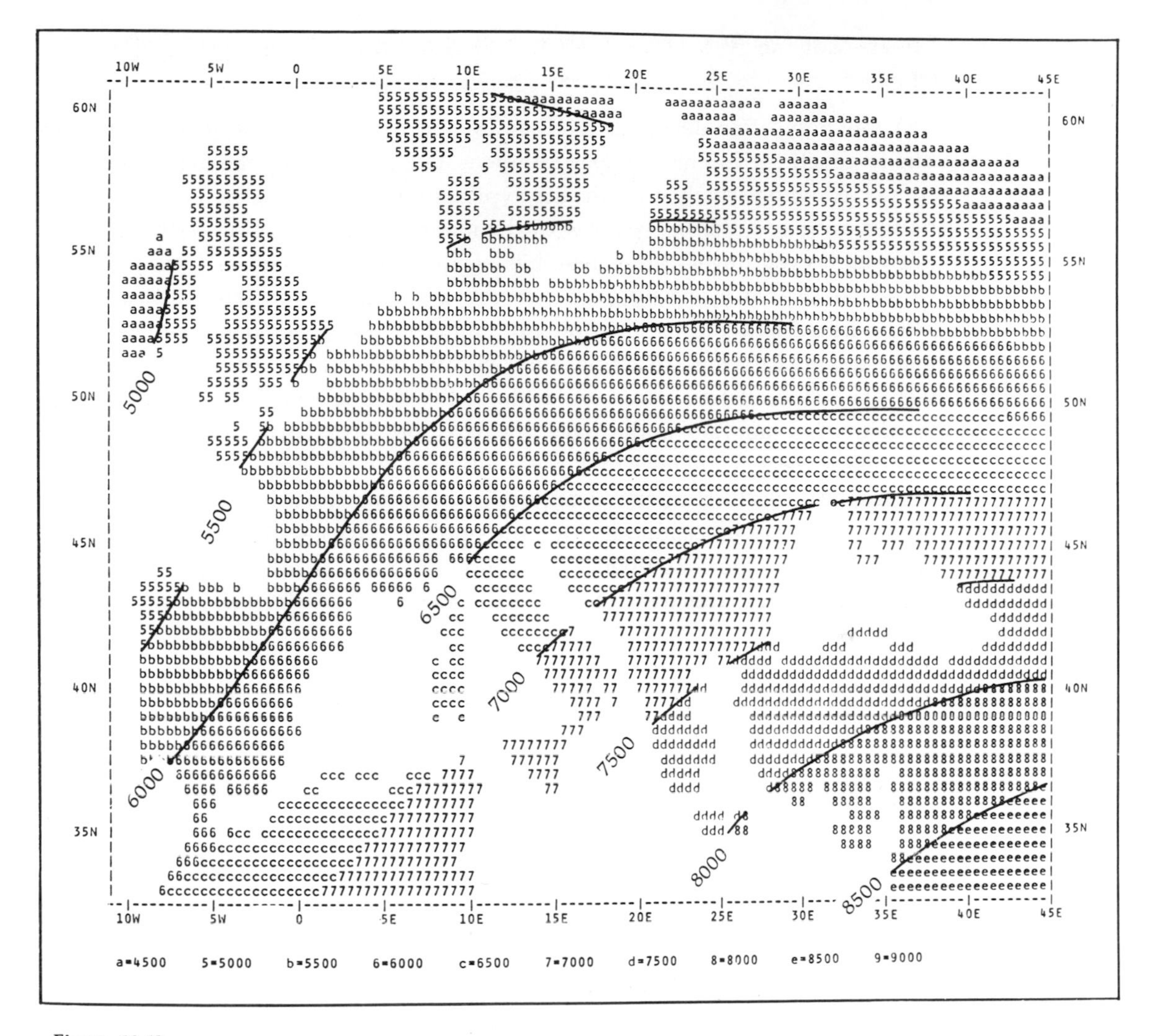

Figure 18.12
The spread of early farming from its center of origin in the Near East toward western Europe. The lines represent the time of first arrival in years before the present. The data summarized in this map are radiocarbon datings of the earliest farming-culture remains known in each region. Recently, a correction of the radiocarbon datings has been suggested. The application of this correction would add about 900 years to each of the ages shown here. (After A. Ammermann and L. L. Cavalli-Sforza, *Man,* vol. 6, pp. 674–688, 1971.)

even when isolation appears to be high. Nevertheless, hunter-gatherers and the primitive farmers that are still found today do give some clues to the condition of life in Paleolithic and Neolithic cultures, respectively. For the geneticist, some of the demographic characteristics are especially interesting.

Hunting and gathering involves a low population density, usually less than 0.2 inhabitants per km^2.

A low population density is likely to involve a higher rate of genetic variation due to genetic drift. This is in part compensated, however, by the higher mobility connected with partially nomadic life, typical of hunters and gatherers. Even if the hunter–gatherers' density was low, the total number of humans on earth may have been high even during the Paleolithic; as already mentioned in Chapter 7, the world population may have been between 1 and 10 million (Fig. 18.13).

The introduction of agriculture and animal breeding has one major effect: it increases the carrying capacity of the land and thus makes room for a potential increase in population size. Even without the refinement of modern industrial techniques, China and India have reached population densities of about 400 inhabitants per km^2, similar to densities observed in the Netherlands. These may be close to the saturation limits, at least by present standards. Early farming was not, of course, as efficient, but could provide for a density of perhaps 10 inhabitants per km^2. One consequence of adoption of early farming was the abolition of nomadic habits. This was not immediate, however. Early farmers, in the absence

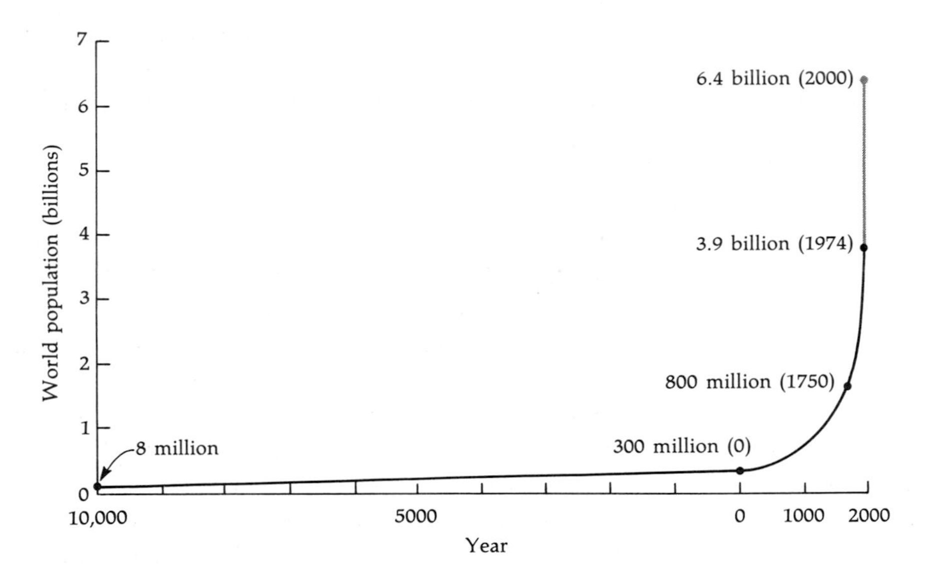

Figure 18.13
Growth of the human world population since the end of the Paleolithic period. In the preceding two or three million years, population growth must have been exceedingly slow. (See A. J. Coale, "The history of the human population," *Scientific American,* vol. 231, no. 3, pp. 40–51, 1974.)

of crop-rotation practices and ways of restoring soil fertility, had to shift to new fields every few years. It is only with the development of later agricultural techniques that maximum sedentariness of farmers was achieved.

The population growth accompanying the transition to agriculture was due to decreased death rates or to increased birth rates, or to both.

Contemporary hunter–gatherers and some primitive farmers have low birth rates, achieved mostly by long birth intervals. In some groups, birth control is also achieved through infanticide. Changes in social customs leading to increased birth rates may have accompanied or followed the transition to agriculture. Death rates may have decreased, initially at least, with improved food supply, because this increased the chances of surviving periods of famine thanks to food storage.

Undoubtedly the transition to agriculture has involved important changes in selective conditions.

Among Europeans and a few North African tribes that are pastoral nomads (and use milk as adults), one notices a high frequency of lactose tolerance among adults, which is rare or absent in other ethnic groups (see Chapter 7). Other changes of importance may have been due to the reduction in protein intake in all those groups which abandoned meat consumption in favor of vegetables. Many of these vegetables (such as corn or manioc) do not provide a balanced diet and thus are a cause of nutritional diseases such as pellagra or kwashiorkor, to which some individuals may be more sensitive than others.

It has been claimed that, in tropical and temperate climates, a consequence of agriculture was the increase in prevalence of malaria, especially of the *falciparum* type (see Chapter 9). The increased concentration of people (because of larger villages), their decreased mobility, the greater chances of reproduction of mosquitoes (the vectors of the malarial parasite) when the forest is cut for plantation purposes and small ponds form more easily—all may have contributed to spreading this disease and increasing its prevalence. Malaria alone may account for one-third of all mortality in a malarial environment. This can explain why malaria has had such important selective consequences, as we have seen in the cases of sickle-cell anemia, thalassemia, and G6PD. But, in addition to these direct selective effects, the transition to agriculture may have had other genetic consequences.

It is not clear, from the archaeological data, whether it is the farmers themselves, or the "idea" of farming, that spread from the centers where it all began.

The increase in population size of farmers and the mobility associated with the techniques of shifting agriculture may well have caused a diffusion of people ("demic" diffusion) rather than of the idea ("stimulus" or "cultural" diffusion). Probably both phenomena have occurred, but their relative importance remains a problem. Some computations indicate that "demic" diffusion may have played an important role. If so, the genes of the farmers must have spread with them, and the development of agriculture must have been accompanied by a very marked "demic" selection as defined in Chapter 7. This might explain why Europeans (and Caucasians in general) are fairly homogeneous genetically. If the population of Europe is largely composed of farmers who gradually immigrated from the Near East, the genes of the original Near Easterners were probably diluted out progressively with local genes as the farmers advanced westward. However, the density of hunter–gatherers was probably small, and the dilution would thus be relatively modest. In Africa, a progressive dilution of a Caucasian type diffusing slowly from the Near East could be the explanation for some of the gene gradients observed, especially in the northern and eastern parts of the continent (Fig. 18.14). The problem of the relative importance of demic and cultural diffusion of agriculture, however, is not yet settled and requires further investigation.

Another important consequence of the introduction of agriculture is the chances that it offered for urbanization.

The earliest town known is Çatal Hüyük, in Southwestern Turkey, which may be 9,000 years old and was inhabited continuously for several thousand years (Figs. 18.15 and 18.16). Houses were built very close to each other using mud bricks. When a house collapsed, a new house was built on the same spot and, in this way, large mounds were formed, called "tells" in the Middle East. The population size of Çatal Hüyük is difficult to assess, especially at the beginning, but it may have reached several thousand inhabitants even in Neolithic times. This must have been made possible by fairly intensive agriculture all around the town. Even if much of the rural population may still have lived in smaller villages, 100 to 300 persons in size, the potential for the origin of towns was created. One of the largest of such towns was that of Uruk, which at the date of 3000 B.C. had a population of perhaps 50,000 (Fig. 18.17).

With towns, a complex political organization must have become a necessity. Trade and transportation must also have been considerably stimulated. The introduction of metals (copper, bronze, iron) in lieu of stone is a late event that in several areas did not occur until historical time. The beginning of "history," as opposed to prehistory, is marked by the introduction of writing, which started at very different times in different parts of the world. The earliest writing observed dates to a little before 3000 B.C. in Sumeria and in Egypt.

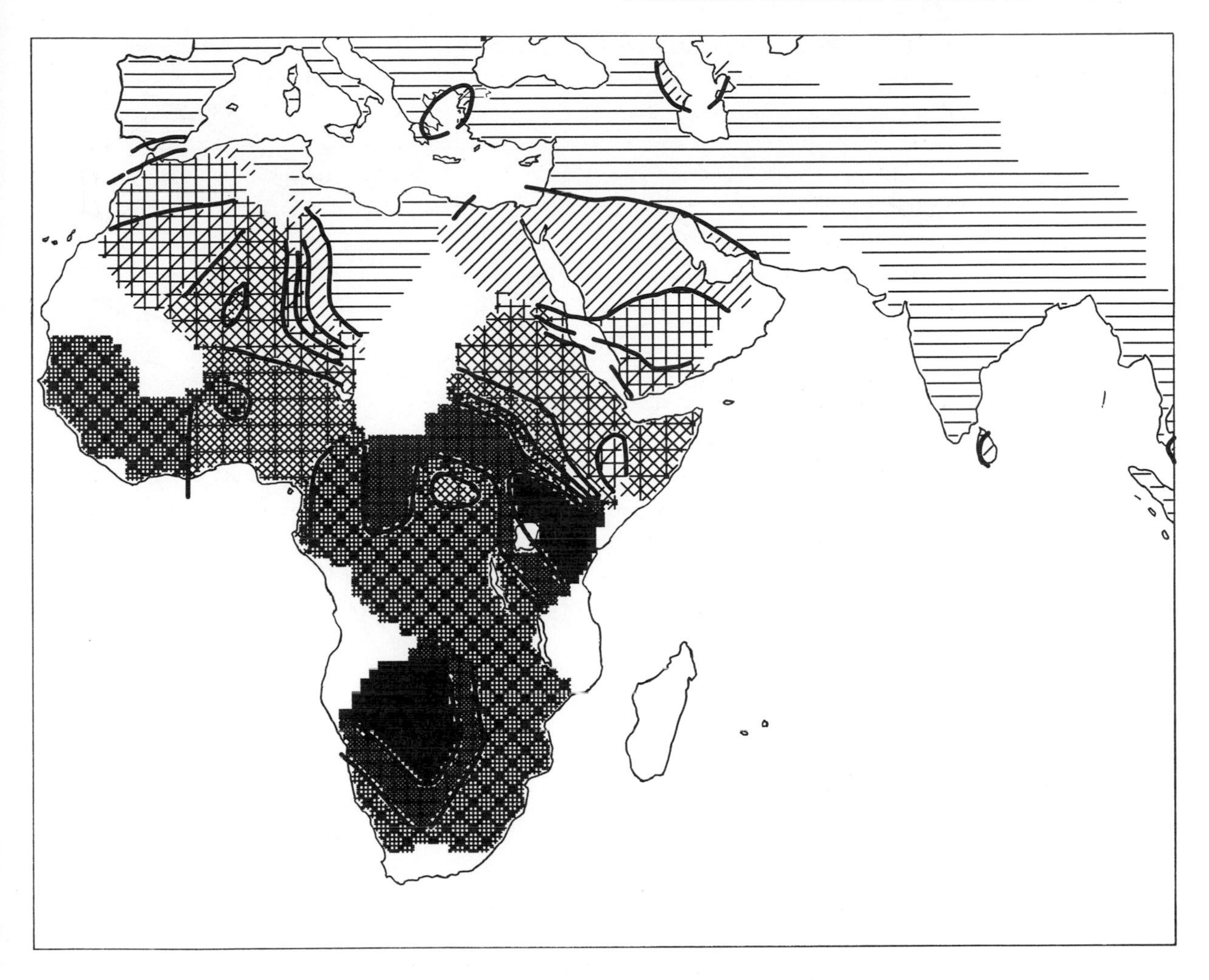

Figure 18.14
Gradients in gene frequencies of Rh^0, representing probable Caucasoid admixture across Africa. Only "proto-Africans" (Bushman and Pygmies) are shown in the areas with the highest frequencies in this computer-generated map. (Courtesy of D. E. Schreiber, IBM Research Laboratory, San Jose, California.)

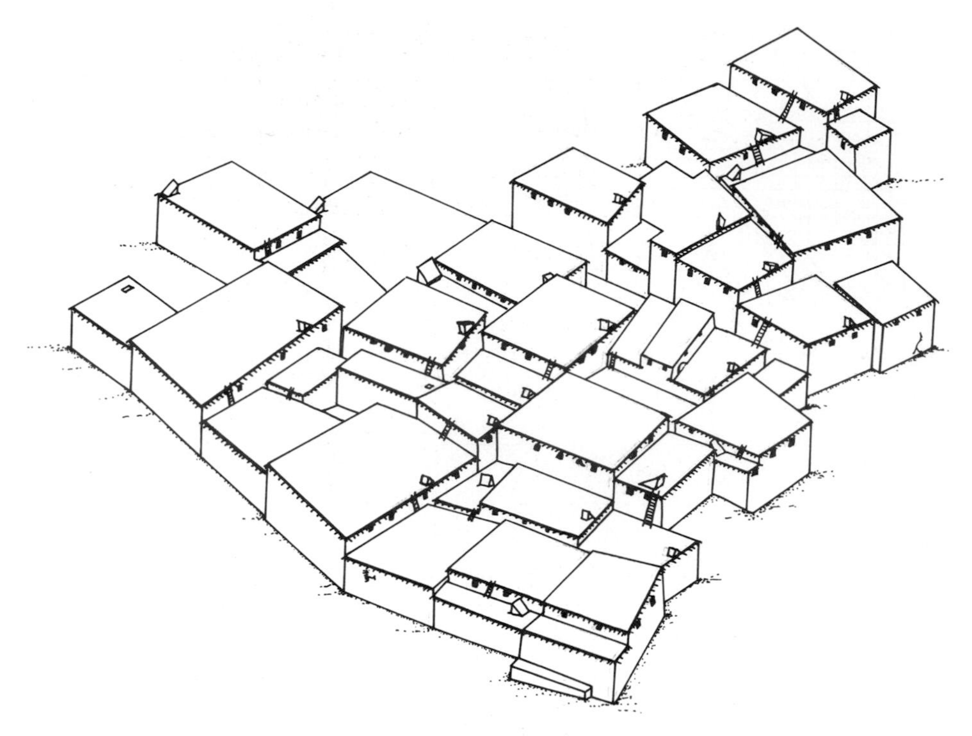

Figure 18.15
A reconstruction of the early Neolithic town of Çatal Hüyük in Turkey. Level VI is shown, with houses and shrines rising in terraces one above the other. Entrances to houses are on the roofs. (From J. Mellaart, *Çatal Hüyük,* Thames and Hudson, 1967.)

18.4 Some General Problems

There is one problem that arises again and again in the analysis of human evolution (as well as in that of other species which, like man, occupy a very large territory). How did it happen that the sequence of changes from australopithecines through *Homo erectus* to *Homo sapiens* apparently occurred *in parallel* over *most* of the Old World?

The first caveat to be kept in mind is that the real "facts" are few. Finds of fossil bones are rare, and few of the known finds can be dated with confidence. Artifacts are more common than bones, but again dating is seldom satisfactory. Because of their relative homogeneity over large distances and long times, the artifacts are relatively uninformative. Their homogeneity over space would seem to indicate a considerable cultural exchange—and hence contacts and migrations over long distances—even in remote periods. But the strength of this argument is attenuated by the remarkable stability of the artifacts over time—with relatively few changes through most of the Pleistocene. For example, exactly the observed

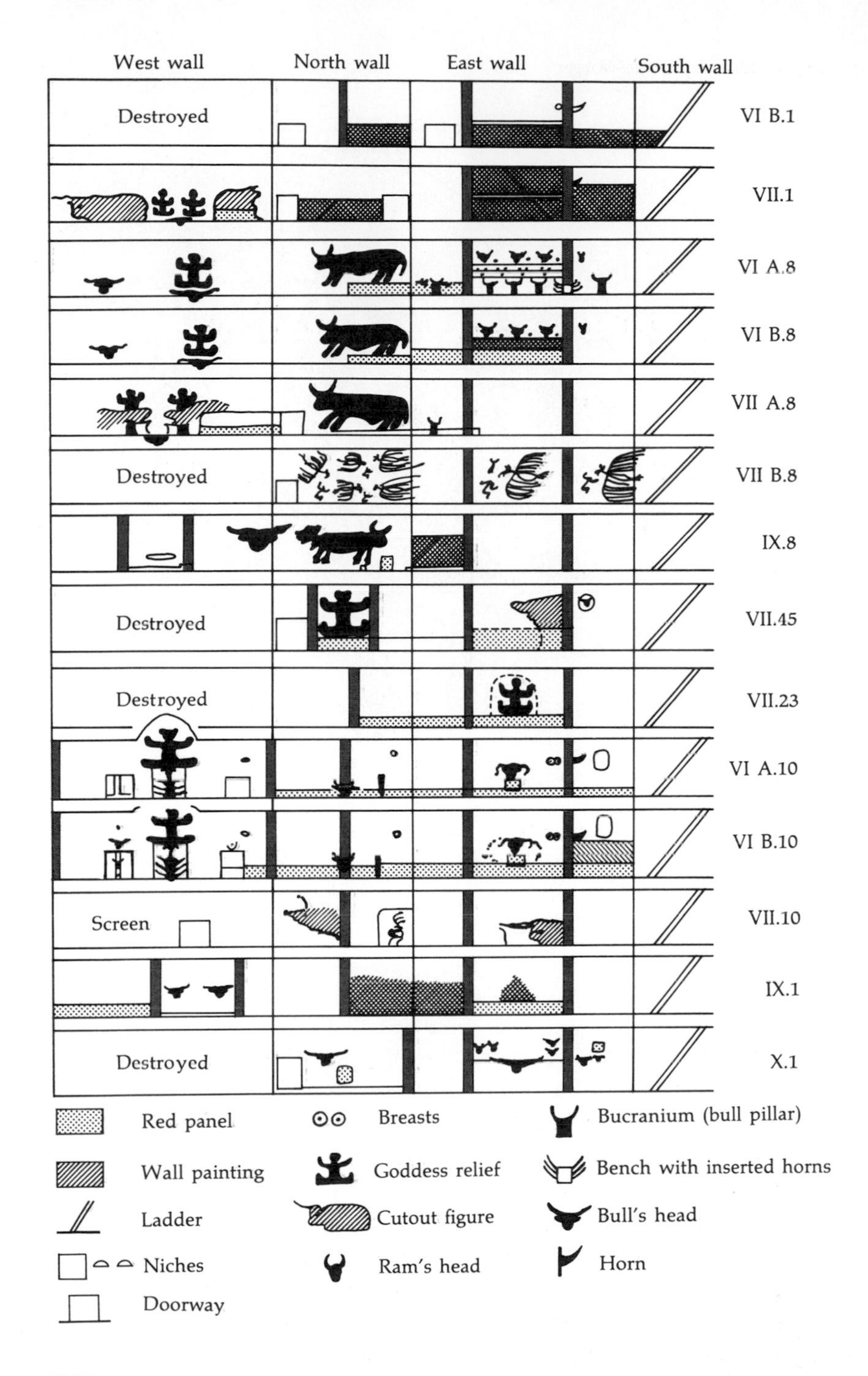

Figure 18.16
A section through the various levels of Çatal Hüyük. The lowest level is dated by radiocarbon to over 8,000 years ago. This section of the city was particularly rich in shrines, which were usually built one on top of the other with the passing of time. (From J. Mellaart, *Çatal Hüyük,* Thames and Hudson, 1967.)

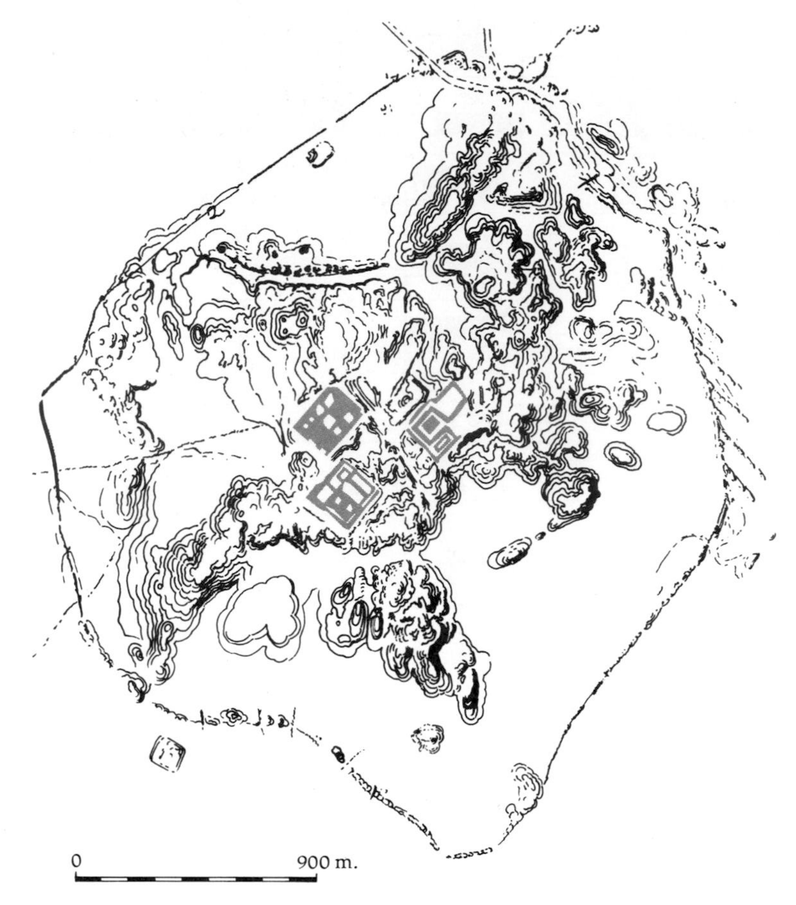

Figure 18.17
One of the earliest cities: Uruk (3000 B.C.). The perimeter indicates the walls of the town. Major temples were located in the central part. Uruk was one of the most important Sumerian cities.

cultural picture would be expected if there were *demic* spreads of one or a few populations to most of the inhabitable world at one or another time in the long history of human evolution, without much cultural exchange or contact during periods between spreads.

The picture can be clarified by more detailed discussion of some models one can reasonably assume about the evolution of the human species.

One model, which some anthropologists have cherished, supposes that the trend of evolution is to some extent predetermined by built-in properties of the evolving genome. Thus evolution would, more or less inevitably, take the same path in any evolving human population—even if it were isolated from contact with other populations. Based on this model, it has been postulated that the various human races have evolved independently toward a very similar final result in the last 500,000 years or so. "Independently" must mean with little or no genetic exchange, essentially in complete isolation. This model of *parallel independent evolution* is difficult to accept from a genetic point of view. Various environments to which the various populations adapt are so different, and possible genetic solutions to problems of adaptation so numerous, that it seems extremely unlikely that the same genetic results would be reached in many independent opportunities.

At the other extreme, one can postulate that exchange between people was extremely frequent, even in the early Pleistocene. Cultural homogeneity of stone tools over space may favor this view. Cultural exchange would inevitably be accompanied by genetic exchange. But the amount of genetic exchange would have to be exceptionally high to permit the evolution of the whole species in unison despite the very long distances between the areas in which evolution happened. This model of transformation of the whole species in unison over the whole inhabited Old World, with minor local differentiation, is, however, difficult to discard entirely. The scarcity of existing data makes a full analysis impossible. But the hypothesis demands that the amount of genetic exchange is shown to be sufficiently high to maintain substantial homogeneity over the very wide area of evolution of the genus *Homo.*

Irregular bursts of spread of people may be the explanation.

There is an alternative model, which we tend to favor, at least for specific periods. One of these periods, as we have seen, is that of the "agricultural revolution." This period is sufficiently close to us in time, and thus material is sufficiently abundant, that one can hope to accumulate satisfactory evidence to test it. Another such period may have been that of the spread of *Homo sapiens sapiens*—and earlier spreads such as those that accompanied the appearance of Neanderthals and earlier types.

The general idea is that major biological (or cultural, or mixed) changes may have given to a "local" population a potential to spread (radiate) from the area of origin. Each such spread may have taken a very long time but, in the vague picture we can reconstruct from few and uncertain data, the spreads themselves may be hard to prove or disprove. The occurrence of such demic spreads (radiation of people) is guaranteed if there is a positive population pressure and at least some migratory movement. The latter is practically always available. The former demands some advantageous novelty (biological or cultural). In human evolution there is much of both. If, however, a cultural innovation is easy to learn and accept and if communication is sufficiently easy, the innovation can spread very fast, and the spread will be mostly cultural and not demic. This may have been the case, for instance, for the spread of pottery, but perhaps not (or not exclusively) for that of agriculture, or for the earlier spreads.

The structure of human society is another important facet of human evolution.

An economy of hunting and gathering has accompanied most of human evolution. Since the very beginnings known to us, hunting was a communal affair for humans. The hunting of large mammals may have been an important factor in shaping social customs towards greater cooperation. Social life is by no means unique to man; it is common, for instance, among most primates. Everyday human life suggests that "social" attitudes are perhaps not as common or innate as might be desirable. Yet man may be more innately altruistic (on average) than, say, baboons or chimps. Needless to say, it is not easy to obtain estimates of such quantities. It may be added that, for a while, it was thought that "altruism"—conceived as the tendency to activities advantageous to the group, rather than to the individual—was unlikely to be successful in natural selection. Darwinian fitness is a highly individual property. If altruism is inherited, and the altruist sacrifices himself, altruistic genes are lost with their carrier. Thus it would seem that the Darwinian fitness of "genes" for altruism must be low and natural selection would disfavor them. But that loss can be compensated in many ways—for instance, the altruist may happen to save preferentially by his action his own relatives. The genes lost with the altruist will prosper with his relatives. Perhaps the most interesting conclusion from theoretical considerations is that strictly Darwinian fitness, in some cases, may be an oversimplification. We have already seen how in other cases one should introduce the idea of "intergroup" selection, as an addition to that of selection between individuals.

We are, of course, unaware of how much of the social behavior of man is innate (genetic) or learned. The fact remains that, when hunting large mammals, it is efficient to be in a group, both at the time of taking the prey and of eating it. Surviving hunters and gatherers show social structures that conform to the idea of small hunting groups ("bands"), of size, mobility, and social customs well adapted to group hunting. There are few, if any, social stratifications. Hierarchies are mostly nonexistent. It is only with larger group sizes and population densities, as made possible by agriculture, that more complex social organization, hierarchies, and leadership become common and necessary. Perhaps the instability, incoherences, and convulsions of our sociopolitical systems are an expression of our biological and cultural immaturity for life in large communities; evolution may have adapted us only to life in small groups. Given the rates of biological evolution, it would be somewhat hopeless to wait for biological adaptation to our new living conditions. Millions of years of hunting may have slanted us toward a behavior that is not necessarily adapted to the life we presently live. But we clearly must cope with the genotypes we have. It is, of course, encouraging to think that part at least of this adaptation may have been toward cooperation. But probably the demands for cooperative tendencies in the modern world are much greater than those that were sufficient for the smaller world of our most ancient ancestors. We can only hope that our genotypes provide us with sufficient plasticity to meet these new and challenging environments we have created for ourselves.

Summary

1. We are primates. The separation from our nearest cousins among primates, the chimpanzees, may date back 20 or 25 million years. Chimpanzees and other great apes have 48 chromosomes. Karyotype differences between humans and chimpanzees are due to various chromosome mutations, especially pericentric inversions. The difference in chromosome number must also be due to a chromosome mutation. The common ancestor of the great apes and man was probably the dryopithecine.

2. Recent finds indicate that human-looking ancestors lived in East Africa as early as 3 million years ago—perhaps even earlier.

3. From these early humans (*Homo habilis*) developed *Homo erectus* and then *Homo sapiens.* The increase in brain size that marked human evolution was more-or-less complete by about 200,000 or 300,000 years ago. The transition to the bipedal posture, which helped to free hands for tool making and using, occurred before the change in brain size was complete. The perfecting of language, on the other hand, may have come later in human evolution.

4. Modern man (essentially physically identical to living humans) may have begun to spread over the world some 30 to 40 thousand years ago, perhaps from an area of origin in Western Asia.

5. A cultural "explosion," as revealed by increases in number and local varieties of artifacts, accompanied the spread of modern man.

6. Domestication of plants and animals is a late event, marking the end of the Paleolithic and beginning of the Neolithic period, about 10,000 years ago. Agriculture arose, probably independently, in several nuclear areas—of which the best known and probably oldest is in the Middle East, where cereals (in particular, wheats and barley) and many animals were first domesticated.

7. From nuclear areas of origin, agriculture diffused radially at a relatively slow rate. The spread of agriculture may have been accomplished largely through the spread of farmers themselves.

8. The spread of farmers and the alterations in living conditions brought about by agriculture may have been responsible for several genetic changes in the human population.

9. There have been drastic and rapid changes in human cultural patterns during the past 10,000 years—leading in turn to drastic changes in the physical and social environment of the individual. This period has probably been too short for selection to have accomplished much genetic adaptation to the altered conditions in which modern man must live.

Exercises

1. Review the evidence for the fact that man is a primate whose nearest cousin is the chimpanzee.

2. Outline the final steps in the evolution of *Homo sapiens* from his more apelike ancestors.

2. What is the evidence indicating that agriculture spread from a center in the Middle East?

4. Give some examples of genetic changes that might be a consequence of the spread of farming and the development of agriculture.

5. The increase of brain size in the evolutionary line leading to man was about 3- or 4-fold, over a period of 10 to 25 million years. Compare this evolutionary rate with one of the fastest known, namely the increase in overall size of horses. *Eohippus* in the Eocene was about the same size as a fox terrier, and is the ancestor of the modern horse.

6. A heterozygote for an inversion, or a heterozygote for a translocation, is likely to have greatly decreased fitness. (Recall that in an inversion heterozygote, single crossovers may give rise to gametes containing duplications and deletions of large chromosome segments that are likely to be inviable; for events in a translocation heterozygote, see the discussion on Down's syndrome in Chapter 4.) Given the chromosomal differences between chimpanzee and man, discuss the relevance of the following facts. (a) Sometimes species, or at least populations, may go through extremely small population-size bottlenecks (see Chapter 11). (b) In wild populations of several *Drosophila* species, there are balanced polymorphisms for chromosome inversions; in some mammals there appear to be balanced polymorphisms for translocations.

7. Why should pottery spread faster than agriculture? Compare the factors that may be important in the spread of a relatively simple technological innovation with that of a radically new mode of life involving many aspects of everyday life.

8. What factors may contribute to the persistance of hunting–gathering societies in the twentieth century?

9. By what mechanisms could the use of clothes have contributed—if it did—to the evolution of hairlessness in man?

References

Ammerman, A. J., and L. L. Cavalli-Sforza
1971. "Measuring the rate of spread of early farming in Europe," *Man*, vol. 6, pp. 674–688.
Flannery, K. V.
1973. "The origins of agriculture," *Annual Review of Anthropology*, vol. 2, pp. 271–310.
Howells, W.
1973. *Evolution of the Genus Homo.* Menlo Park, Calif.: Addison-Wesley.
Isaac, G.
1972. "Chronology and the tempo of cultural change during the Pleistocene," in *Calibration of Hominoid Evolution*, ed. W. W. Bishop and J. A. Miller, Toronto: Scottish Academic Press and Univ. of Toronto Press.
Jorgensen, J. G.
1972. (Ed.) *Biology and Culture in Modern Perspective: Readings from Scientific American.* San Francisco: W. H. Freeman and Company.
Lamberg-Karlovsky, C. C.
1972. (Ed.) *New World Archeology, Foundations of Civilization: Readings from Scientific American.* San Francisco: W. H. Freeman and Company.
Laughlin, W. S., and R. H. Osborne
1967. (Eds.) *Human Variation and Origins: Readings from Scientific American.* San Francisco: W. H. Freeman and Company.
Mellaart, J.
1967. *Çatal Hüyük: A Neolithic Town in Anatolia.* London: Thames and Hudson.
Pilbeam, D.
1972. *The Ascent of Man: An Introduction to Human Evolution.* New York: Macmillan.
Tobias, P. V.
1973. "New developments in hominid paleontology in South and East Africa," *Annual Review of Anthropology*, vol. 2, pp. 311–334.
Ucko, P. J., and G. W. Dimbleby
1969. (Eds.) *Domestication and Exploitation of Plants and Animals.* London: Aldine.

19

Racial Differentiation

In many cases, biologists over the years have found it useful to divide a species into two or more subspecies, or races. The criteria for the definition of races—based on geographic distribution and various features of the body—yield classifications similar to those obtained using genetic markers. Use of genetic markers also shows very clearly that there are no "pure" races. Races are, in fact, generally very far from pure and, as a result, any classification of races is arbitrary, imperfect, and difficult. Yet anyone can see that there are certain relatively clear differences between a typical Caucasoid and a typical Mongoloid or a typical Negroid. What are the causes of racial differentiation? Which differences are cultural and which are biological?

19.1 The Human Species

A species is a set of individuals who can interbreed and have fertile progeny.

When Carolus Linnaeus originally defined the concept of a species in 1735, he believed that each species had been separately created by God and had remained unchanged ever since. Therefore, he viewed the species as a natural, unchanging, and precisely definable unit. Already in that century it was clear that the classification into species should bring together individuals capable of interbreeding with each other, but incapable of interbreeding with members of any other species. As taxonomists worked at the task of classifying all known organisms into species, they soon discovered that nature could not be fit perfectly into such rigid pigeonholes. In a classic work on the nature of species written in 1942, the zoologist Ernst Mayr put it this way: "no system of nomenclature and no hierarchy of systematic categories is able to represent adequately the complicated set of interrelationships and divergences found in nature." It is now clear that existing species have changed and that new species have been formed continuously throughout the history of life on earth. Therefore, there exist many situations that are intermediate between the ideal cases of a single species or of two distinct species.

The members of a species have, potentially at least, a common gene pool. The formation of a new species is a long and complex process. A very important factor in this process of speciation is *isolation* between two or more populations of the original species. Isolation creates two distinct gene pools, in which evolution may follow different patterns. Isolation may have many origins, the most important one being geographical. Geographical barriers, especially mountains and water, tend to reduce migration and so to create isolation.

In the absence of intermigration, genetic differentiation between populations may be expected as a result of (1) selective adaptation and (2) genetic drift. For two isolated populations, initially identical in genetic composition, it would not be surprising if selective adaptation leads to two different patterns of genetic change. This would occur if the two populations occupy differing habitats, so that they are adapting to different environmental conditions—thus producing different selective advantages for various traits in the two populations. Even if the two populations occupy similar habitats, adaptation might occur through different mechanisms in the two populations. Finally, genetic drift and random accumulations of different mutations might be expected to lead to differing evolutionary patterns in the two isolated populations.

Any gene flow between incompletely isolated populations will tend to reduce or prevent differentiation. Given sufficient isolation, however, differences between the populations will accumulate with time, and the genetic gap between the populations will gradually increase. Formally, the biologist tends to classify the two populations as different species when the capacity to interbreed has effectively disappeared. The effective incapacity to interbreed may be a secondary (and usually late) result of the differences accumulated in this process of genetic differentiation. It is more likely to occur, the longer the separation and the more significant the biological hiatus that has been created. Thus, the formation of different species does not (necessarily at least) occur abruptly. In zoological examples, one finds that interfertility often is reduced gradually as the two populations differentiate before it eventually disappears completely.

Because the process of differentiation that eventually leads to different species is usually a continuing one, the differences between isolated groups may be trivial or difficult to detect in the initial phases of the process. In such a case, taxonomists are likely to group these populations into a single classification. At the stage when some differences between the isolated populations have become detectable, the taxonomist is likely to classify the recognizably different populations as different "races" (sometimes called "geographic races" when they are geographically isolated) of the single species. When the process is more advanced and the differentiation between the two populations is quite noticeable, the taxonomist is likely to classify the populations as "subspecies" of the single species. (However, some taxonomists regard the terms "race" and "subspecies" as equivalent.) In any case, the decision about whether to classify the populations as different races or subspecies, and the decision about what characteristics to emphasize in describing the differences between them, are relatively arbitrary ones.

In general, among mammals, it may take on the order of a million years for two isolated populations to evolve from genetic identity to a situation of intersterility. It has apparently been about 20 or 25 million years since the human populations became genetically isolated from the great ape populations. Therefore, we should certainly expect the human species to be totally separate from the species of our closest cousins, the chimpanzees and gorillas. In fact, the modern great apes are even considered to belong to a different genus from our own. (The genus is a higher category of genetic differentiation than the species.) For obvious reasons, the experiment of interbreeding man with higher primates has not been carried out under controlled conditions. If nothing else, the possible embarrassment of how to bring up any progeny that might result from such a cross should deter any sensible person from yielding to this kind of curiosity. The differences in appearance, the differences in the number and morphology of chromosomes, and the presumably very long isolation of the populations are all indications—but not guarantees—that the differentiation between man and the great apes has gone well beyond the specific level.

By definition, different races are still potentially interfertile, and they often exchange individuals, especially at the boundaries of their geographic distributions.

Sometimes, movements of large groups may lead to partial fusion of groups that had formerly split. This process of fusion after fission cannot occur between species, whose gene pools are separated irreversibly because interbreeding is no longer possible.

However human races are defined, there is no evidence of any decrease in fertility in crosses between even the most distant human races (for instance, between Africans and native American Indians). It is clear that human racial differentiation has not had the time or the opportunity to reach a point of incipient speciation. A much longer period of geographic isolation would probably have been necessary for intersterility mechanisms to evolve. If it is accepted that *Homo sapiens s.* started racial differentiation at the time the species began to expand and spread throughout the world (about 40,000 years ago), then clearly this time has been too short for any significant reduction of interfertility between the human races to have occurred. It is quite possible that increases in geographic mobility during the past few centuries have reduced isolation of various human populations sufficiently to reverse the trend toward racial differentiation in the human species.

19.2 Human Races

Race is a more elusive concept than that of species.

A species may be divided into races if the differences between the populations so defined are of some significance, but the level of differences used as a threshold is entirely arbitrary. When differences are striking, the classification is easy; but even then the taxonomical work may be made difficult by the existence of gradual transitions between the groups so defined. In most cases, separate races will be defined only if the groups differ in several significant traits. However, in most cases, these traits vary independently of one another in the transitional populations between the races. Thus, boundaries between the races drawn on the basis of one trait will not coincide with boundaries drawn on the basis of another trait. These are precisely the problems faced by the taxonomist who attempts to classify the human species into races. It is not difficult to see why there is nearly complete continuity in the distribution of almost every single trait, as revealed by maps of their geographical distributions (Figs. 19.1 through 19.4).

Human races probably evolved at the time of spread of *Homo sapiens s.* The occupation of different habitats in this worldwide expansion created the basis for local differentiation. Three possible complications must be considered. One is that there probably were, in almost all areas, many successive waves of expansion, so that the process was complex, with later waves contributing to a reduction of the initial genetic differentiation. A second conceivable complication is that, even in his first expansion, *H. sapiens* may have found earlier local groups with which he mixed. The relative importance of this hypothetical mixture will depend on the relative size of the groups with which admixture occurred, if it did so at all. The only serious claim of this nature is with respect to admixture of *H. sapiens* with Neanderthal man in Europe. Another weaker claim of this sort has been put forward with respect to an analog of Neanderthal in Africa, Rhodesian man.

The recent increase in geographic mobility makes the task of distinguishing races even more difficult.

The third complication is perhaps the most real and formidable of all. In his only expansion, man may have already shown considerable mobility, but the times and rates of this movement are very poorly known. Most migrations, probably including those to Australia and America, could take place only by land, and made use of natural land bridges that were then in existence. Later on, however, man considerably increased his mobility, and much technological progress has been specifically directed at the aim of improving transportation on water and land (and, of course, more recently by air). Five centuries ago, transportation technology reached the point where it became possible to explore the whole earth during a fraction of the lifetime of an individual. But in the first exploration and occupation of the earth, it probably took several millennia for the same job.

The process of geographic exploration that started five centuries ago has affected the geographic distribution of man to such an extent that now the Americas and Australia are populated mostly by people of European descent. Today, all continents of the world are inhabited by representatives of the three major human races: African, Caucasian, and Oriental. The proportions of the three groups still differ considerably in the various countries, and the migrations are too recent for social barriers between racial groups to have disappeared. The trend, however, seems to be in the direction of greater admixture.

Had the incipient differentiation been allowed to continue for a million years or so, different human species *would* have formed on earth. The dropping of geographic barriers that has followed the recent cultural evolution may in the next centuries erode most of the differentiation that had previously developed. It is for these historical reasons that, when speaking of "aboriginal" groups, one generally refers to the habitat of people before the great geographic discoveries that began in the fifteenth century.

It should not be assumed that human movements were insignificant before the fifteenth century A.D.

Early Neolithic man used boats in the Mediterranean, as shown by the utilization of obsidian sources in islands, and the occupation of Crete around 8,000 years before the present. We have seen that the expansion of early farming in the Near East probably involved a creeping occupation of most of Europe, Arabia, Iran, India, and northern and eastern Africa. With the later development of organized trade, sea transportation, and political units of some size, mass colonization became possible. Examples in historical times include the colonization of the Mediterranean by Greeks and by Phoenicians.

All these movements of people are hardly compatible with clearcut, well-isolated geographical races—which in fact are not found. As we shall see, much

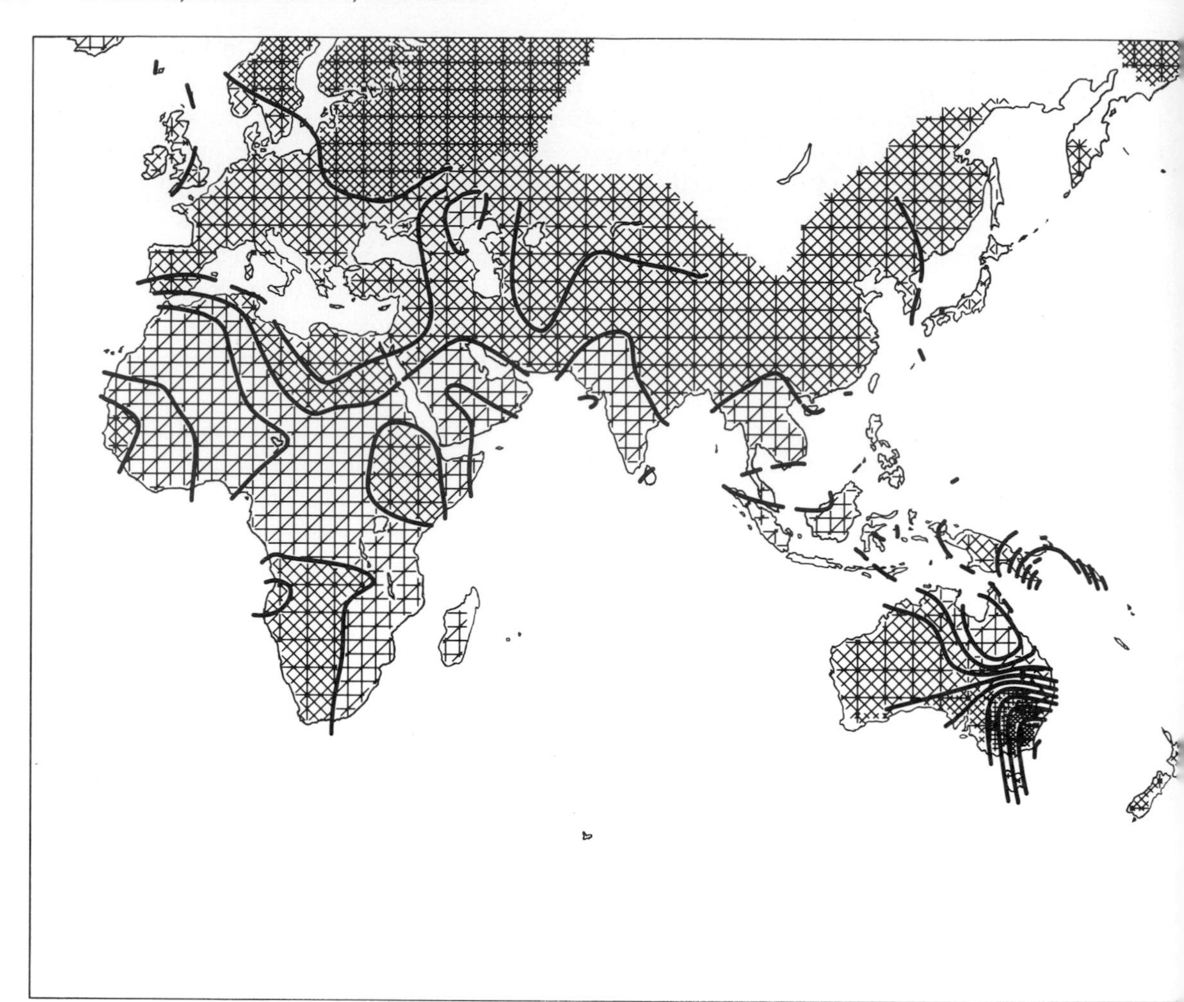

Figure 19.1
Approximate distribution of the blood-group gene A in the aboriginal populations of the world: a computer-generated map. (Courtesy of D. E. Schreiber, IBM Research Laboratory, San Jose, California.)

variation, observed at the level of single gene frequencies, or single anthropometric traits, is *clinal*—that is, shows a gradual change from one geographical region to another. A *cline* is a relatively regular change in a biological trait over a stretch of territory.

This background can explain why the job of classifying races is so difficult—the more so, the more detailed we try to be.

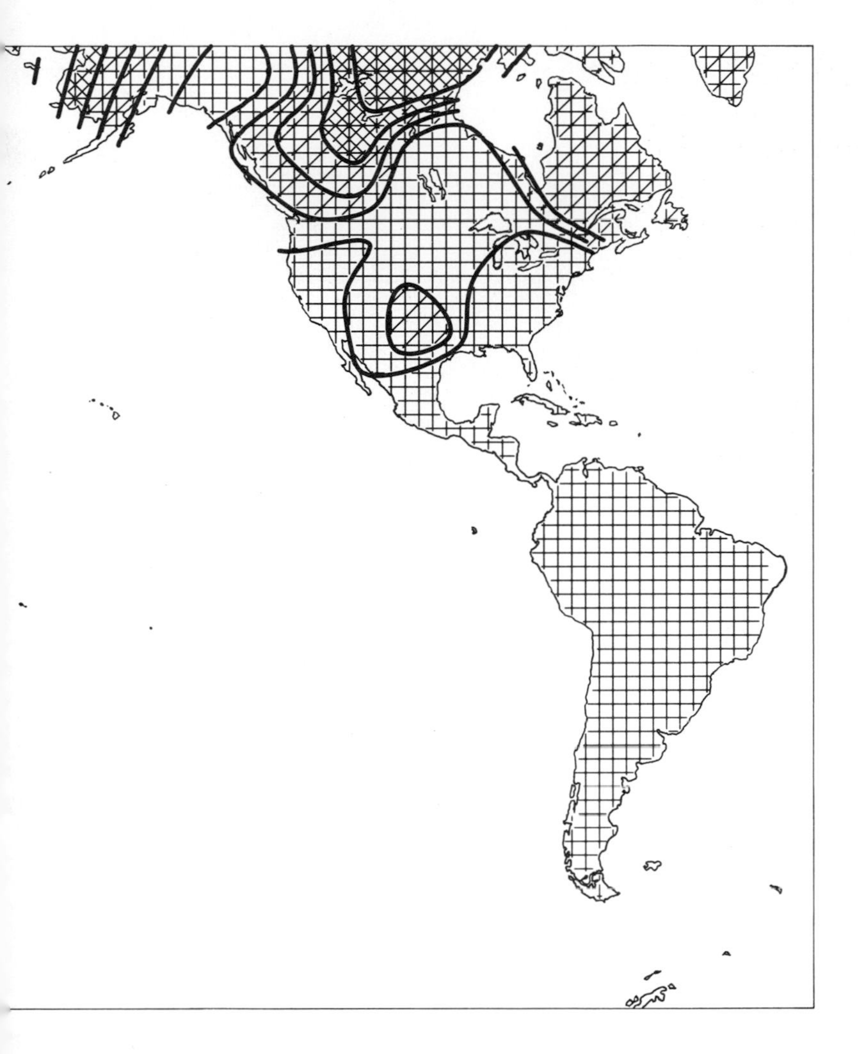

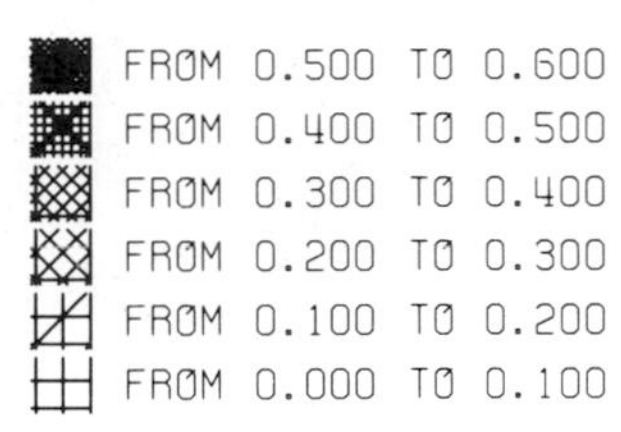

On the most general level, however, geographical and ecological boundaries (which acted as partial barriers to expansion and migration) help to distinguish three major racial groups: Africans, Caucasians, and a highly heterogeneous group that we may call "Easterners" (Fig. 19.5). The Easterners include subgroups that were separated in various older classifications, such as American Natives (American Indians) and Orientals (Chinese, Japanese, Koreans). Some regard Australian aborigines as a separate race, but they do not differ much from Melanesians. From

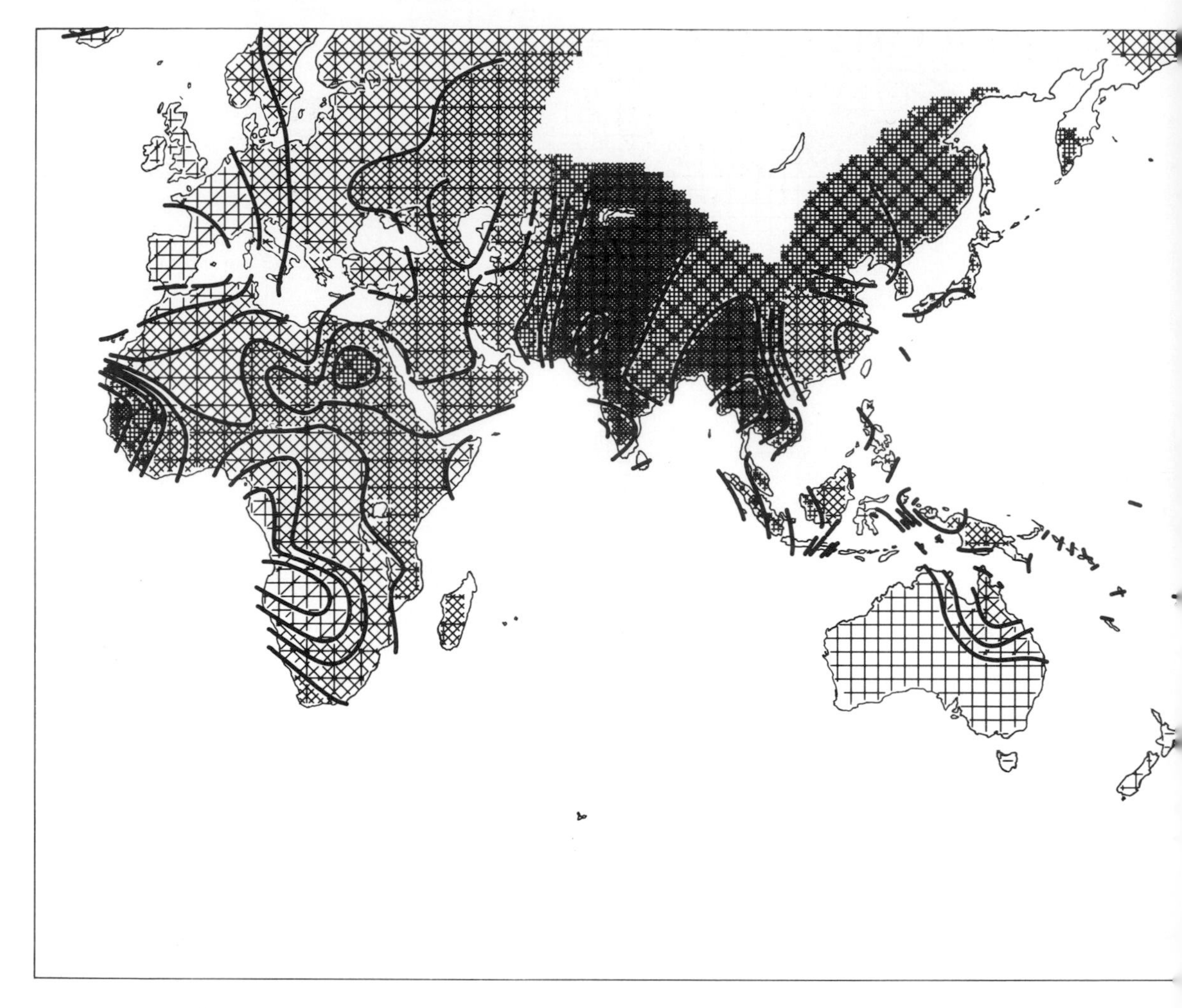

Figure 19.2
Approximate distribution of the blood-group gene B in the aboriginal populations of the world: a computer-generated map. (Courtesy of D. E. Schreiber, IBM Research Laboratory, San Jose, California.)

the Melanesians, we can trace a sequence of relatively gradual changes through the transition to Indonesians, then to Southeast Asians, and on to East Asians. American Natives and Eskimos probably both came from a related Northeast Asian stock from (or through) Siberia into North America. Eskimos, however, came much later than American Indians, and they subsequently expanded further eastward to Greenland.

The African continent contains, in the north and east, populations that have various degrees of admixture with Caucasians by all criteria of analysis. In the western, central, and southern parts of the continent, Africans are relatively

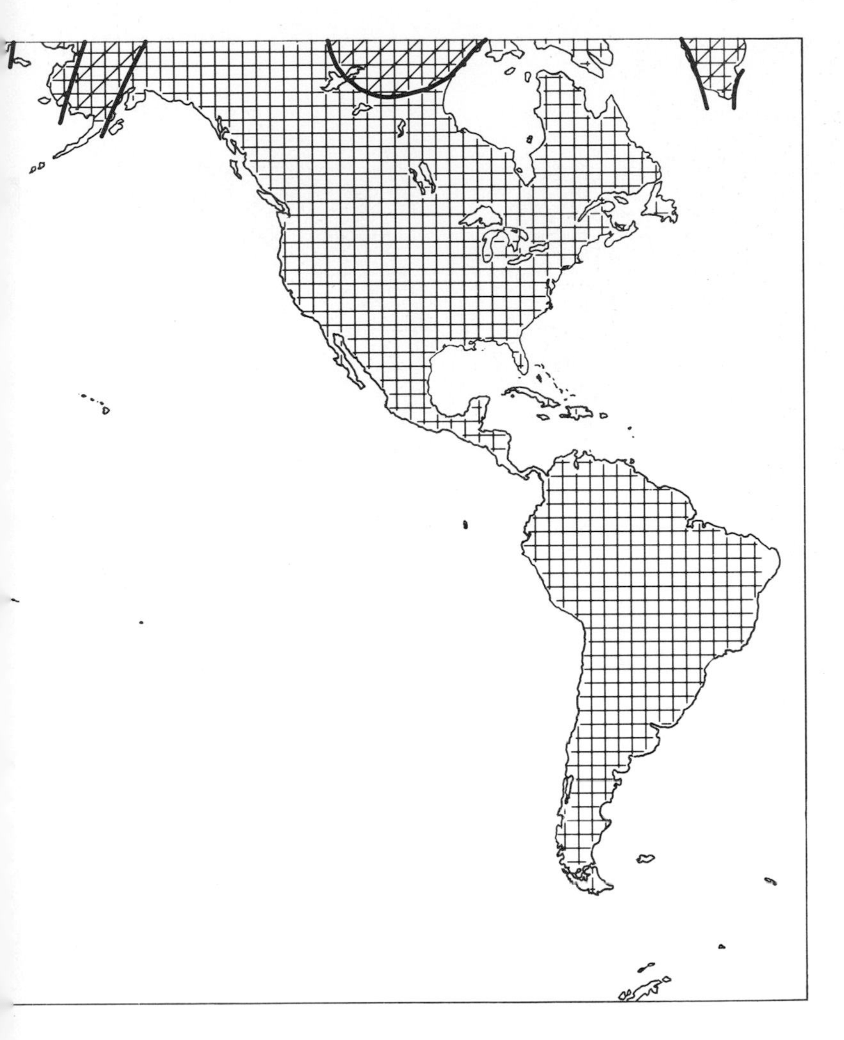

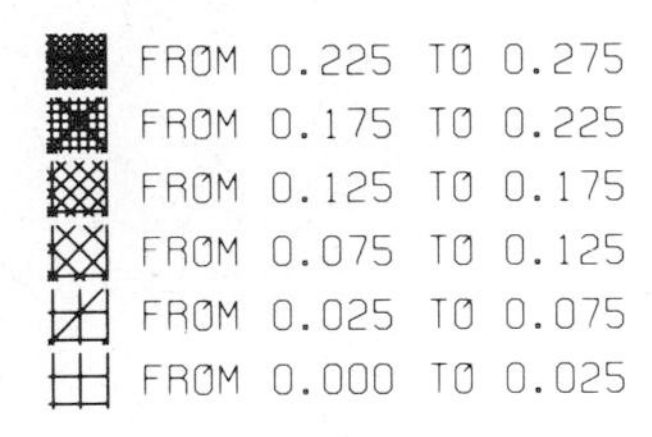

homogeneous—although some isolated groups of hunter–gatherers (like Pygmies and Bushmen) show cultural and physical peculiarities that suggest they should be considered somewhat separately. In fact, the Pygmies at least have attributes that indicate they may be "proto-African" groups—populations that have been the least altered by more recent events.

Figure 19.6 outlines a tentative history of African people that is in reasonable agreement with archeological and genetic data currently available. Caucasians are spread over Europe and in Southwest Asia as far as India, where there is a relatively gradual transition with Easterners. From Arabia toward East Africa and in

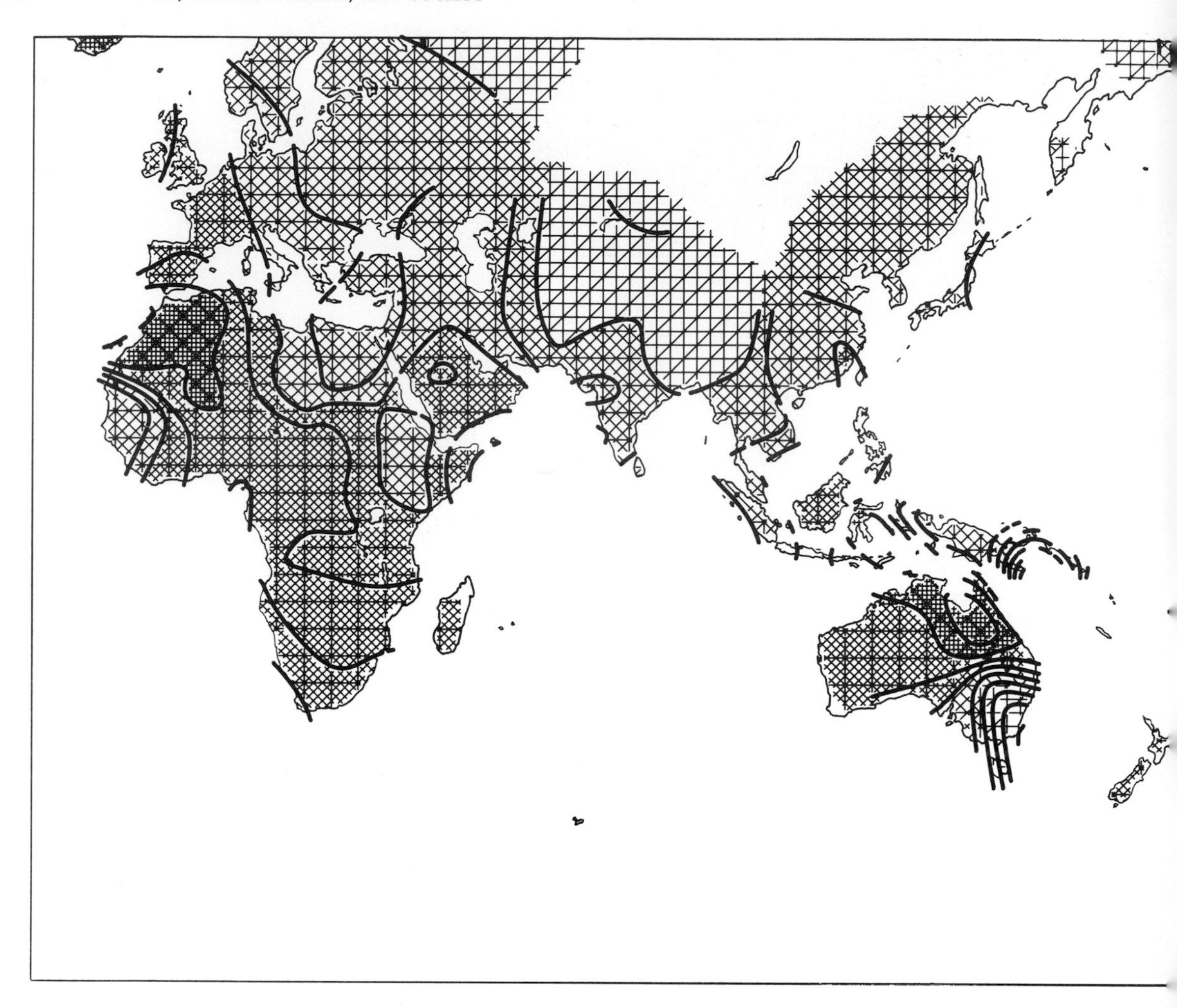

Figure 19.3
Approximate distribution of the blood-group gene O in the aboriginal populations of the world: a computer-generated map. (Courtesy of D. E. Schreiber, IBM Research Laboratory, San Jose, California.)

North Africa, there is an almost continuous transition toward the African type as one proceeds toward Central Africa.

Linguistic classifications are only moderately useful in defining human races.

Among other possible classifications of human populations, languages prove to be only relatively useful. Languages evolve much faster than genes do. In a few thousands of years, the languages of isolated populations change enough seriously to compromise intelligibility between the groups. Moreover, social and political

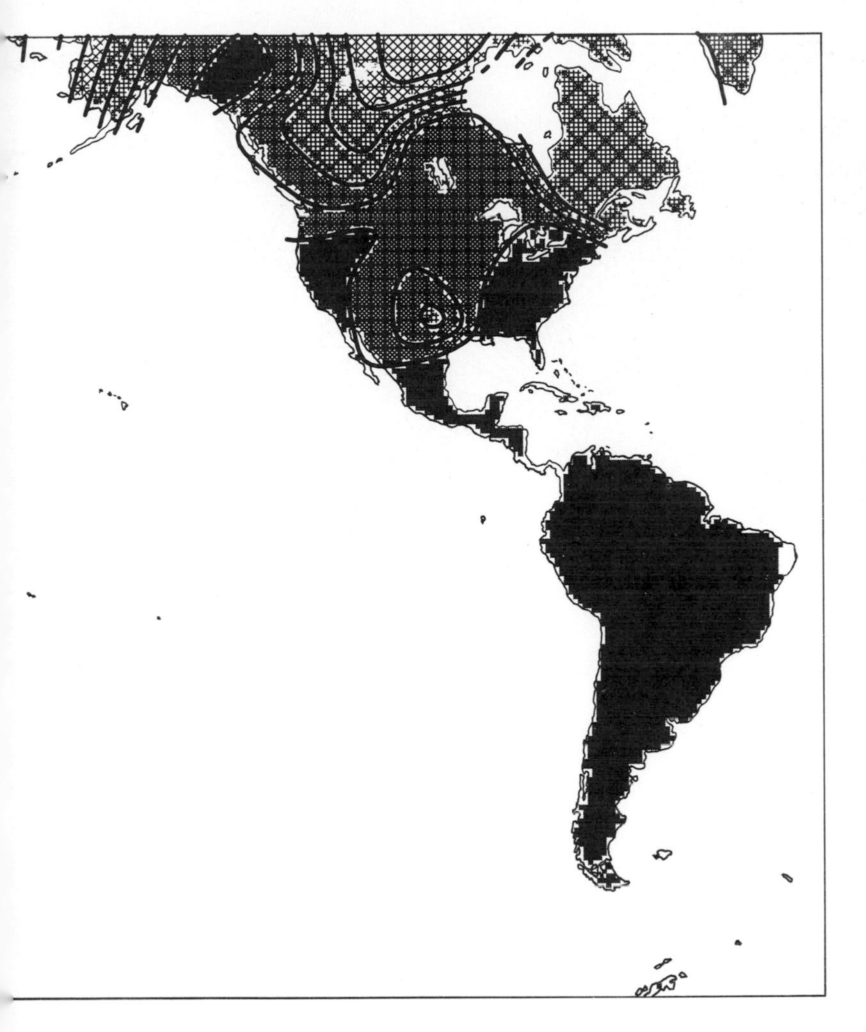

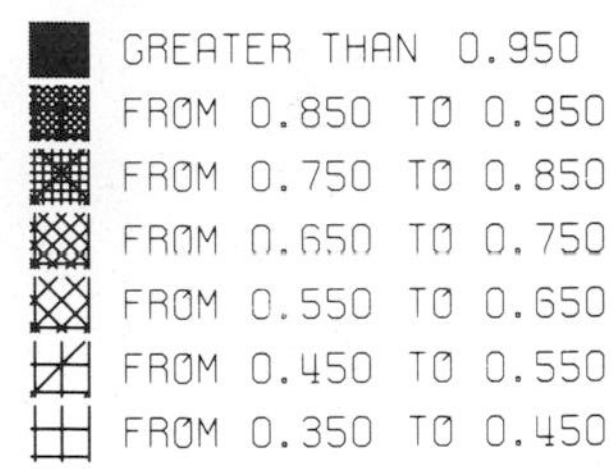

events may lead to the substitution of one language for a totally unrelated one in the space of a few generations—or at least such events may cause substantial transformations of an original language. Because of factors like these, linguistic distributions seldom correspond to biological or genetic distributions. Even so, some correlations between the linguistic and biological picture may remain, when analyzed over relatively small distances, so that in some areas populations more similar genetically also prove to be more similar linguistically. This is not a universal finding, however—nor should it be construed as due to any direct interaction between genes and language. (There is no evidence of any innate tendency to

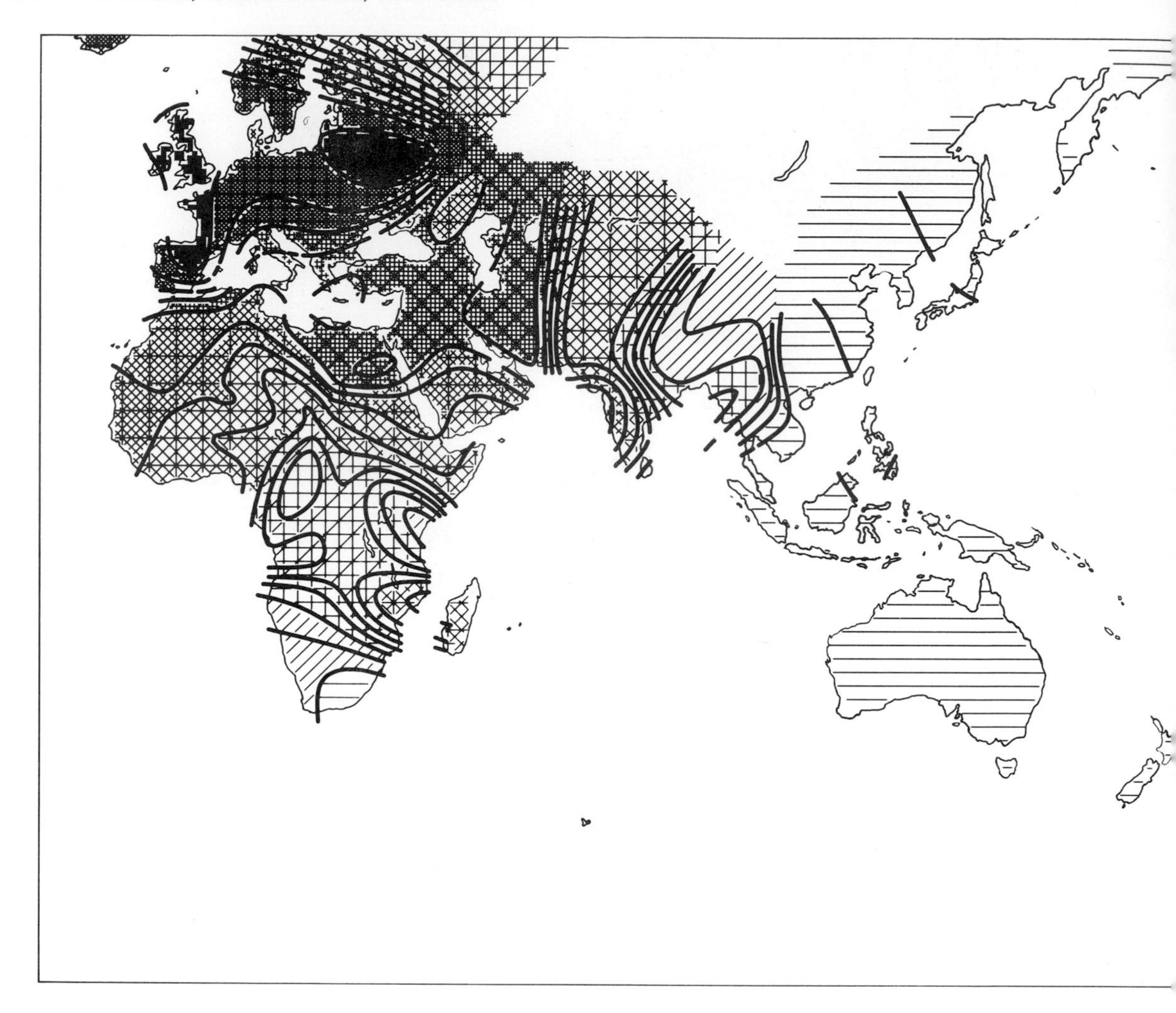

Figure 19.4
Approximate distribution of the blood-group gene Rh⁻ (haplotype cde) in the aboriginal populations of the world: a computer-generated map. (Courtesy of D. E. Schreiber, IBM Research Laboratory, San Jose, California.)

speak any particular human language rather than another.) When similarities are found between linguistic and genetic traits, they must be simply reflections of historical events that have had similar influences on both linguistic and genetic evolution.

The correlation between linguistic and biological traits does tend to hold, for instance, among Caucasians. Most Caucasians who inhabit Europe and parts of Asia today speak languages of the Indoeuropean group. The area where Indo-

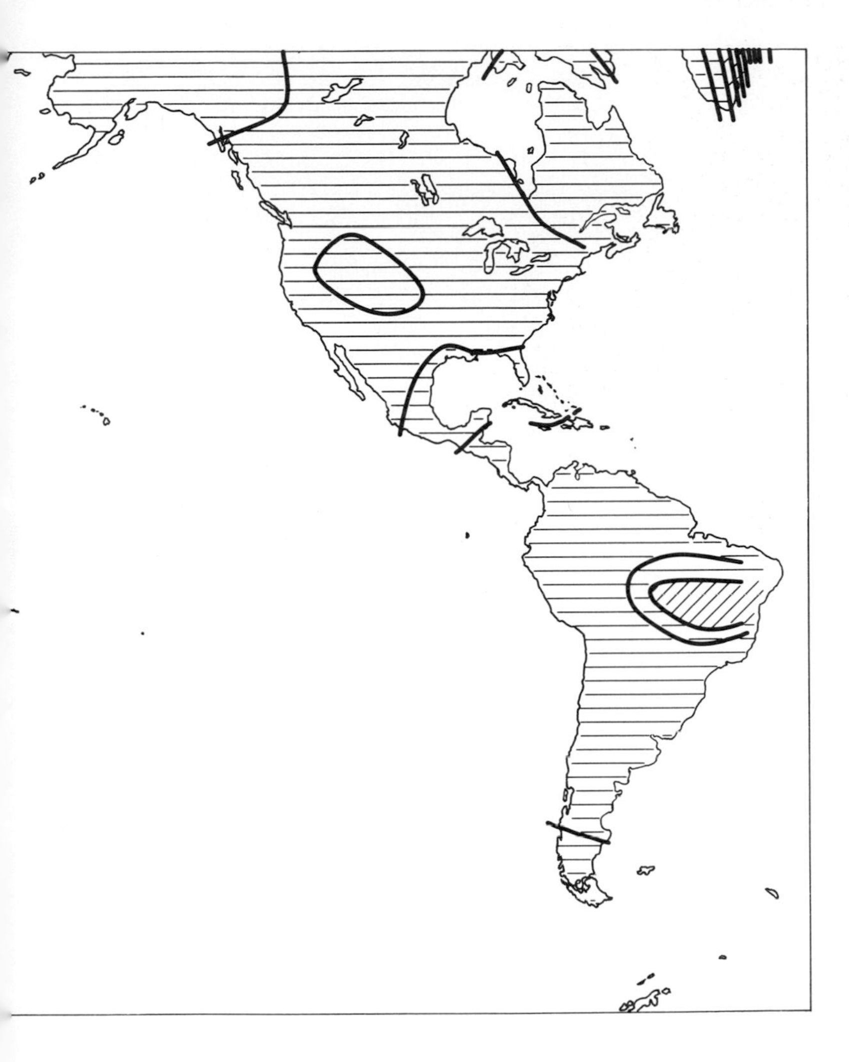

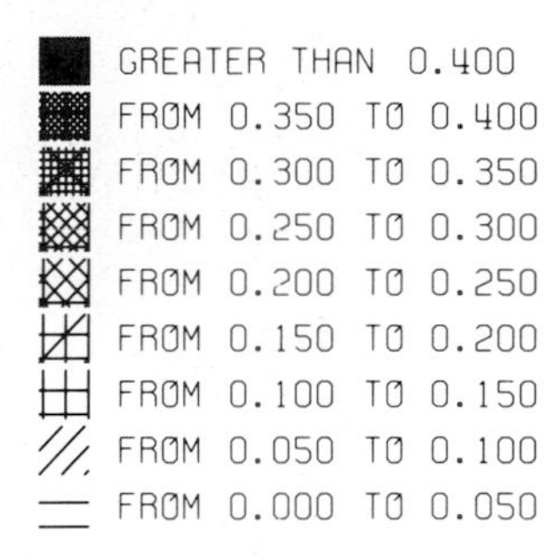

european languages are spoken overlaps almost exactly with the northern part of the expansion of early farming. In the southern area of the Neolithic expansion, semitic languages are spoken. This is the present situation, but the languages used when expansion occurred in the Neolithic are totally unknown. Linguists do attempt to recreate the evolution of languages by noting the kinds and degrees of similarities and differences between languages, but the theories set forth must remain largely speculative. Only with the development of writing, some 4,000

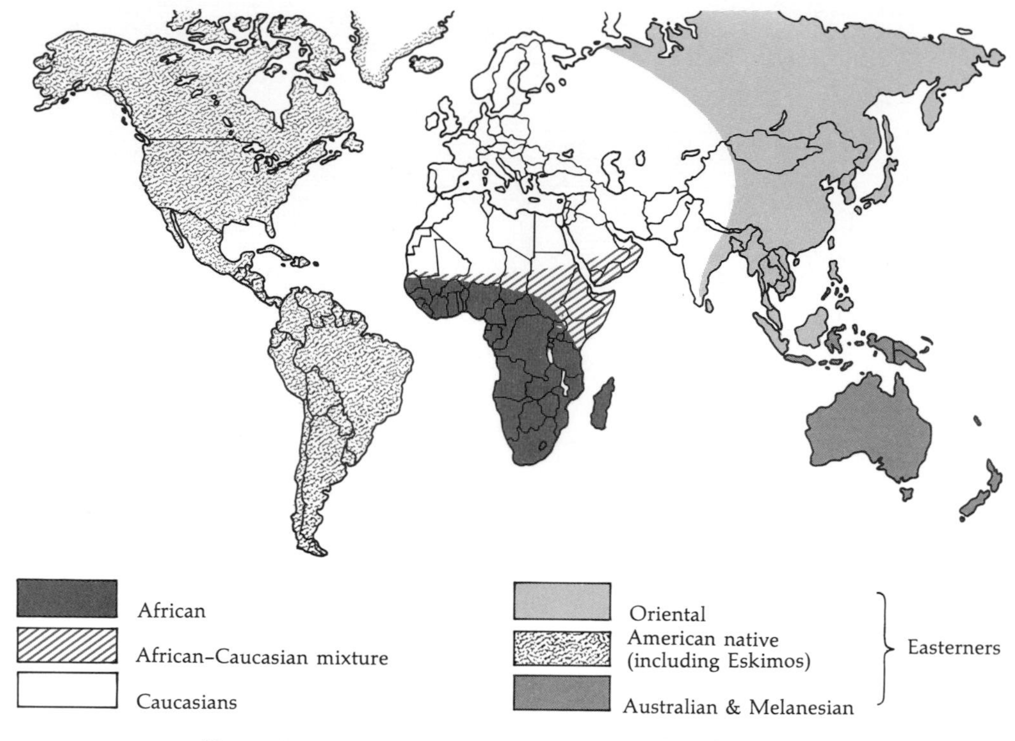

Figure 19.5
Geographic distribution of major ethnic groups. Boundaries are approximate.

years after the initial onset of farming, do we obtain a record of the languages spoken by various human populations.

Apart from the linguistic aspects, there is considerable genetic similarity (as we see in detail later) among all Caucasians, ranging from Europe to western Russia and to India and Arabia. As we have mentioned, populations around the fringes of this area show admixture of Caucasian traits with African or Easterner traits. However, the Caucasian population is not entirely homogeneous even in the areas of little admixture. We tend to side with those taxonomists who prefer to group the human species into a few large racial groups (such taxonomists have been called "lumpers"). Others ("splitters") prefer to distinguish a large number of groups differing in relatively subtler ways. The splitters have proceeded to subdivide the Caucasian race into smaller groups.

A fairly natural classification, which follows geographic boundaries, is that between European and extra-European Caucasians. Further splitting—for example, of the European group into Northern, Alpine, and Mediterranean subgroups—becomes more and more ambiguous and uncertain. It is true that the man on the

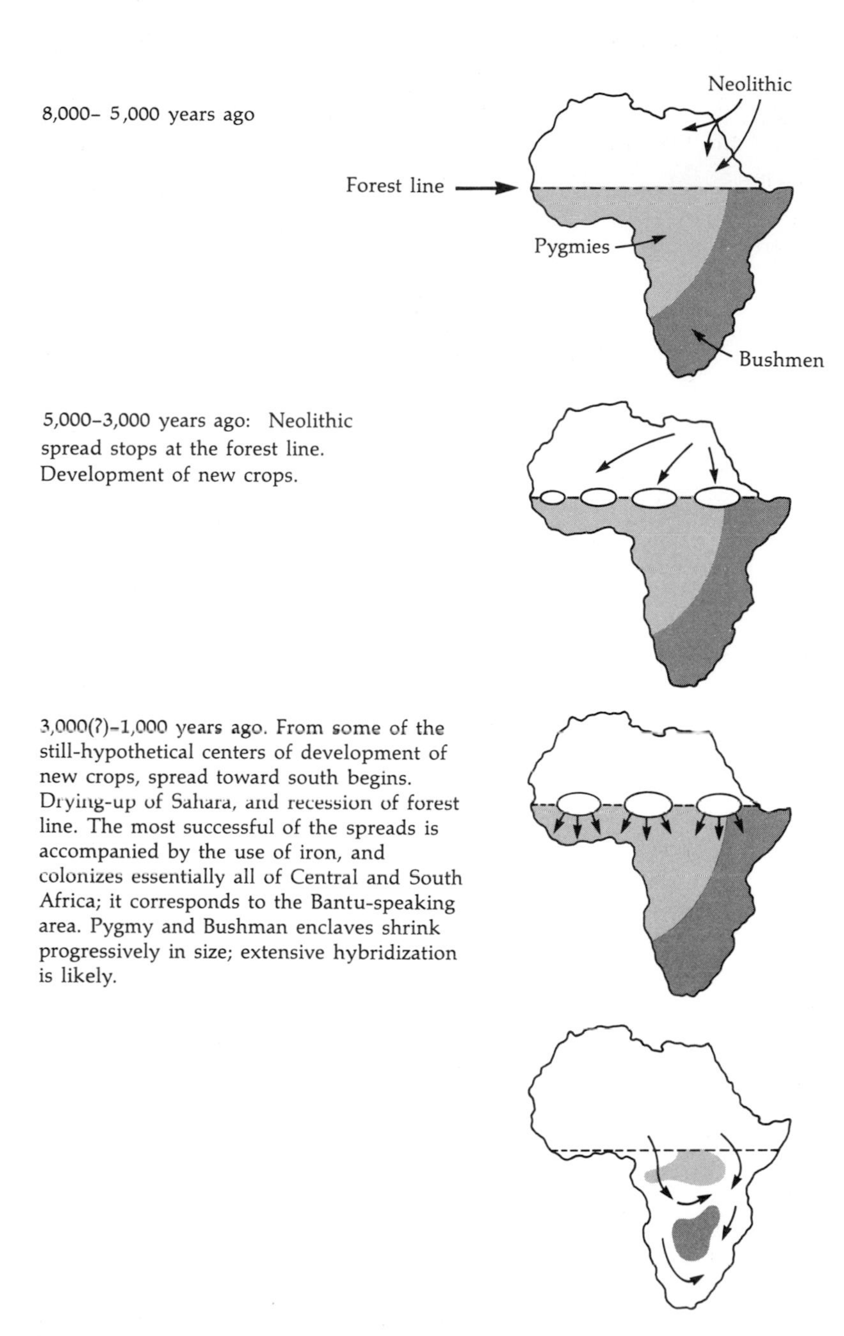

Figure 19.6
Ethnic history of the African continent: a speculative reconstruction.

street can usually guess quite accurately whether an individual's ancestors came from north or south Europe—the typical Scandinavian looks very different from the typical Italian or Greek. Yet the validity and usefulness of this breakdown into smaller and smaller groups is doubtful, at least on the basis of present data.

The bases of the groupings we have been discussing are largely geographic. But, as we shall see, the broadest geographic groupings do correspond to a large extent with the available genetic data.

19.3 Variation of Gene Frequency

The extent of ethnic variation in gene frequencies can be evaluated synthetically, as in Figure 19.7, which shows the variation in gene frequency between ethnic groups for a number of alleles of different polymorphic systems. The greatest variations are found at three loci: those for blood groups Duffy (Fy) and rhesus (Rh), and that for immunoglobulin G markers (Gm).

As pointed out in Chapter 12, Duffy is one of the best markers for distinguishing between people of African origin and others. Africans have exclusively (or almost so) the allele Fy^0, which is rare (3 percent) among Caucasians and even

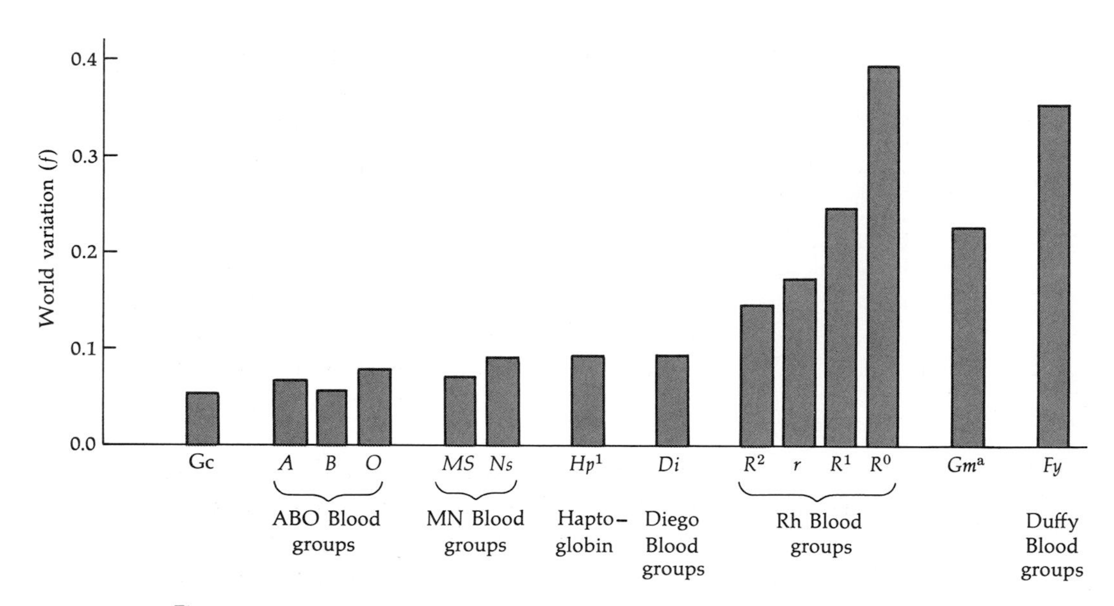

Figure 19.7
World variation in gene frequencies for some alleles. The quantity f used to measure variation is the variance of gene frequencies for a given allele, divided by $\bar{p}(1 - \bar{p})$, where $\bar{p}$ is the mean of the gene frequencies for that allele (see Appendix A.10). Among the alleles shown, Gc^1 is the least variable and R^0 the most variable. Gm^a is a conglomerate of alleles at the Gm locus; individual alleles show much greater variation if considered separately.

rarer among Easterners (see also Section 10.5). The Rh alleles also show conspicuous differences among groups. The Rh^- type (*r* allele) is most frequent among Caucasians (more than 50 percent in Basques, 40 percent in the rest of Europe, and about 30 percent elsewhere), while it is rare among Africans and among Easterners. Easterners have effectively only the R^1 and R^2 alleles, which are present at lower frequencies in the other groups.

The Gm locus shows even greater differences. In fact, these differences are so large that one must assume that this locus has been subject to very rapid evolution. Because Gm determines properties of antibody molecules, it seems reasonable to assume that it has been under diversifying (disruptive) selection due perhaps to association with a variety of local infectious diseases (see Chapter 10). However, direct evidence to confirm this theory is very meager. The variations between ethnic groups for all of these genetic markers are very probably too large to be explained by drift alone. We have, however, no good evidence of exactly what selective factors have been involved in the differentiation between groups.

Almost all other genetic markers show differences in gene frequencies among ethnic groups, though of smaller magnitude. HLA alleles have on the whole a smaller variation than those of the Fy, Gm, and Rh systems. The lower the variation between human groups, the less information is provided for distinguishing races. (If the variation were zero, polymorphisms would be identical in all races, thus permitting no distinction at all.) However, the HLA alleles do give substantial information on ethnic origins because of the great number of alleles.

Various HLA alleles show significant correlations with various diseases; these correlations are much more impressive than similar correlations found between diseases and other markers (see Chapter 10). We are not yet able to say whether these diseases are the selective agents that determined local variation in HLA allele frequencies. In some cases, however, the evidence points in the opposite direction. The allele *HLA-B8,* which is especially frequent among Caucasians, is found to be correlated with the incidence of coeliac disease, an intestinal disorder that is triggered by ingestion of gluten (a component of cereals). Caucasians introduced cereals in their diet before any other ethnic group, so they have been exposed to gluten over a longer period than any other group. Therefore, on the simplest hypothesis, we would expect selection against the *HLA-B8* allele to have produced the lowest frequency of this allele among Caucasians. The explanation of the apparent paradox is to be found in the fact that coeliac disease is rare and affects a very small proportion of the population, even among those carrying the *HLA-B8* allele. (It does, however, affect a higher proportion of those with the *HLA-B8* allele than of those without the allele.) If the frequency of *HLA-B8* has increased among Caucasians, the increase must have been caused by factors other than wheat consumption, which should have a weak tendency to decrease the *HLA-B8* frequency.

The Diego blood group is another interesting marker. The allele Di^a is relatively frequent in American Natives (where it was first discovered) and somewhat less frequent in other Easterner groups (although absent in Eskimos). It is not found at all in Caucasian or African populations.

The ABO blood system, the first genetic polymorphism discovered in man, has been the subject of many studies. The suggestion that this system could be used for analysis of ethnic origins was made (by L. Hirschfeld and H. Hirschfeld in 1918) soon after its discovery. However, it has turned out that the variations in ABO frequencies are not as great as the variations in the Rh, Fy, Gm, Di, HLA, and other systems—so ABO frequencies are not as informative on ethnic origins as are the other systems. Yet, the relatively low discriminating power of ABO is compensated by the large body of ABO data available from all over the world (Fig. 19.8). This data has been collected because of the importance of ABO type-matching for transfusions. With this data, maps of ABO allele frequencies were constructed quite early in the history of racial studies. One sharp variation in ABO frequencies does exist: alleles *A* and *B* are almost totally absent from Amer-

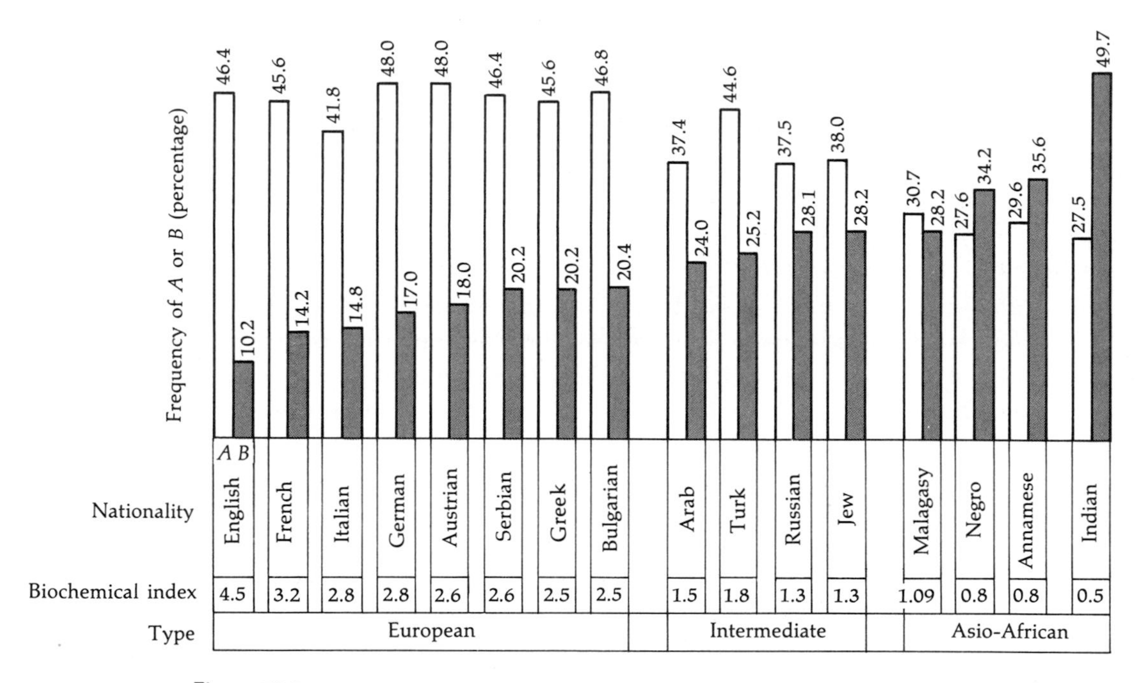

Figure 19.8

A classic study in human population genetics. This graph showing ethnic differences in ABO gene frequencies was the first use of genetic markers to study racial differences. The figures given are percentages of positive reactions with anti-A and anti-B reagents. The "Biochemical Index" is the ratio of A to B. (From L. Hirschfeld and H. Hirschfeld, *Anthropologie,* vol. 29, pp. 505–537, 1919.)

ican Indian populations and when present usually (but not always) indicate Caucasian admixture.

We have already mentioned that the genetic variation for almost any single locus tends to be clinal in many parts of the world. Thus, the *B* gene shows a decreasing gradient as one moves away from Central Asia in almost any direction (see Fig. 19.2). This gradient can be clearly distinguished even in some relatively small areas, for example along the length of Japan.

In many gene maps, there are peaks or troughs of gene frequency.

The variation observed between a peak and a trough is, by definition, clinal. Local selective advantages and disadvantages for the respective alleles are often hypothesized as explanations for the peaks and troughs of gene frequencies. However, other factors such as drift, demic selection, or historical accidents (mass migrations, for example) may be involved as well. To distinguish among these hypotheses is not easy.

The only cases where the selective agent has been identified are those polymorphisms that are known to represent adaptations to malaria. The interpretation of the selective advantage of sickle-cell anemia and thalassemia heterozygotes in the presence of malaria was originally advanced by J. B. S. Haldane, on the basis of the similarity between the geographic distributions of the malarial parasite and of the alleles in question. However, as we have seen, a geographic correlation is not sufficient evidence, and further tests are necessary.

Many other suggestions have been advanced on the basis of correlations between geographic distribution of some genes and selective conditions. For instance, it has been suggested that different ABO alleles confer differing resistances to smallpox, plague, syphilis, and other infectious diseases (see Chapter 10). But studies of the correlation of susceptibility to disease and ABO type in single individuals have given contradictory results.

The case of malarial adaptations is particularly clear.

The selective coefficients involved are very high. The disease is well known and widespread. Practically everybody in malarial areas is subject to attack by the parasites, and mortality due to malaria, directly or indirectly, is very high. Therefore, malaria is a very powerful selective agent. The disease is still active now in many areas—or was until a short time ago.

Diseases that once had selective consequences may now be extinct, because a parasite either died out or decreased its virulence so that it is no longer a pathogen. For such diseases, there is very little chance of being able to infer, retrospectively, what might have been the selective associations. Thus, as discussed in Chapter 13, some of the peaks and troughs of gene frequencies—and, in general, some polymorphisms—may be "relics" of selective effects due to an infectious or parasitic disease now extinct or modified.

19.4 A Tree of Origins

Very few, if any, of the genetic markers that we know (except perhaps for some Gm markers), taken in isolation, are sufficient to tell us about the ethnic origin of an individual. But it was realized very early in such studies that a combination of different markers might help in understanding more about evolutionary origins of human races. For example, William C. Boyd showed more than 30 years ago that a combination of blood groups ABO, MN, and Rh yields a classification of races very similar to that based on considerations of geography and classical anthropology. A synthesis of all differences observed between groups for all available genetic markers might give a better perspective than that obtained through analysis of the markers independently.

If the principle that differentiation tends to increase with the time of separation is correct, then it should be possible—on the basis of the total genetic differences observed between groups—to reconstruct their evolutionary history. This history can well take the form of a tree of descent similar to those constructed to depict the phylogenetic origins of species. An early attempt at such a reconstruction was based on the variations in five blood groups: ABO, MNS, Rh, Diego, and Duffy. Later analyses, using larger numbers of markers, have on the whole confirmed the original results. The latest results, using 58 independent markers (including many additions to the markers that we have already discussed), are shown in Figure 19.9.

Historical inference of this kind is inevitably weak, and confirmation from other sources is clearly necessary.

One problem is the existence of some disagreement between this tree and trees obtained using anthropometric characters (discussed in the next section). Another problem arises when we consider the amounts of genetic differentiation that have occurred along the various branches of the tree. In Figure 19.9, the length of each branch is represented, for simplicity, as constant. Actually, looking at the real branch length, it is found that Caucasians and Orientals (in particular the Chinese) show very short branches, much shorter than the other populations. What is the cause of this phenomenon? The length of a branch indicates the amount of evolutionary differentiation that has taken place. It is possible that larger or smaller evolutionary rates have operated in different branches of the tree, causing greater or smaller evolutionary differentiations.

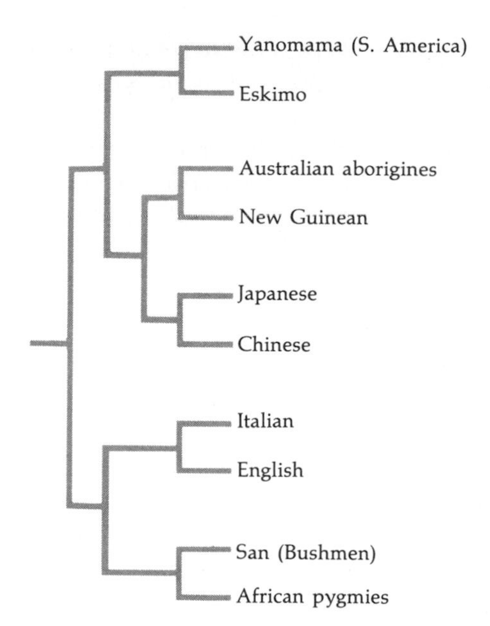

Figure 19.9
A tree of descent of selected human ethnic groups, based upon differentiation in genetic markers. A total of 58 independent alleles from 15 loci were used in preparing this tree, including ABO, MN, Rh, Gm, Diego, Fy, and HLA (A, B). (After L. L. Cavalli-Sforza and A. Piazza, *Theoretical Population Biology,* in press, 1975.)

On the other hand, the short length of some branches may be misleading in this respect because it can be explained by factors other than decreased evolutionary rate. Considerable hybridization with neighboring groups may also produce a shorter branch. Those groups that first developed agriculture and spread with it (such as the Caucasians and the Chinese) are most likely to have extensively hybridized with neighbors during the spread. The effect of hybridization will be to form a group whose characteristics represent a weighted average of the groups that contributed genes to the pool. In fact, hybridization tends to create a population similar to the hypothetical ancestral population of the two hybridizing groups. The effect of hybridization, then, will be to decrease the amount of differentiation, shortening the corresponding branch, and giving the appearance of a smaller amount of evolution.

An extreme example of the effects of hybridization is provided by "triracial isolates," which have in fact been studied in America. These are isolated populations whose gene pools include contributions from African, Caucasian, and Easterner (American Indian) ancestors. The gene frequencies in such groups are very close to those postulated for the original ancestral population in Figure 19.9. Therefore, if one attempted to place these isolates in such a tree, it would be necessary to attach them by a short branch very close to the origin—indicating very little genetic differentiation from the ancestral population. Such extreme examples indicate the caution that must be used in interpreting genetic differentiation as a direct measure of evolutionary descent.

The center of origin of Caucasians is believed to lie in an area geographically intermediate between Africans and Easterners. Therefore, it is reasonable to suppose that the Caucasian gene pool would contain a significant proportion of genes obtained from the other races by admixture, particularly because we believe that the Caucasian population spread outward into these other areas along with the spread of early farming techniques. This hybridization offers one possible (at least partial) explanation of the shorter branches leading to Caucasian groups. A similar explanation can be advanced for the Chinese.

19.5 Conspicuous Racial Traits

Our everyday impressions of racial differences do not come from genes, which cannot be directly detected. They come from visible traits—some of which are, indeed, extremely conspicuous. Among them, the most prominent is skin color. We have already seen that skin color varies in response to environmental conditions: everyone (including blacks) tans when exposed to the sun. The tanning reaction clearly has survival value. Those who cannot tan (albinos, for example) have serious problems when they are exposed to strong sunlight. However, skin-color differences between ethnic groups are, on the average, larger than what can be explained by environmental effects. It is clear that there is a large genetic component to the trait of skin color.

It is also apparent that natural selection must have been involved in determining the worldwide distribution of skin color (Fig. 19.10). Almost without exception, the skin is innately darker where sunlight is stronger (at lower latitudes). Exactly how did selection act? Perhaps sunburn acted as a selective agent, giving a selective advantage to darker skin in the tropics. Skin cancer is another factor that may have favored darker skin where sunlight is stronger; skin cancer is seven or eight times less frequent among blacks than among whites. (It has been confirmed that exposure to ultraviolet light does increase the frequency of skin cancer.) However, skin cancer is so rare (even among light-skinned individuals near the equator) that it is hard to believe this could have been the only selective factor involved.

One interesting hypothesis on the origin of skin-color differences involves the possible role of vitamin D.

The formation of vitamin D takes place in the skin under the action of ultraviolet light, using precursors available in the diet (see Chapter 7). A dark skin reduces the amount of ultraviolet reaching the skin capillaries where the transformation occurs, and thus reduces the amount of vitamin D that is formed. An insufficiency of vitamin D leads to the childhood disease rickets which, if severe,

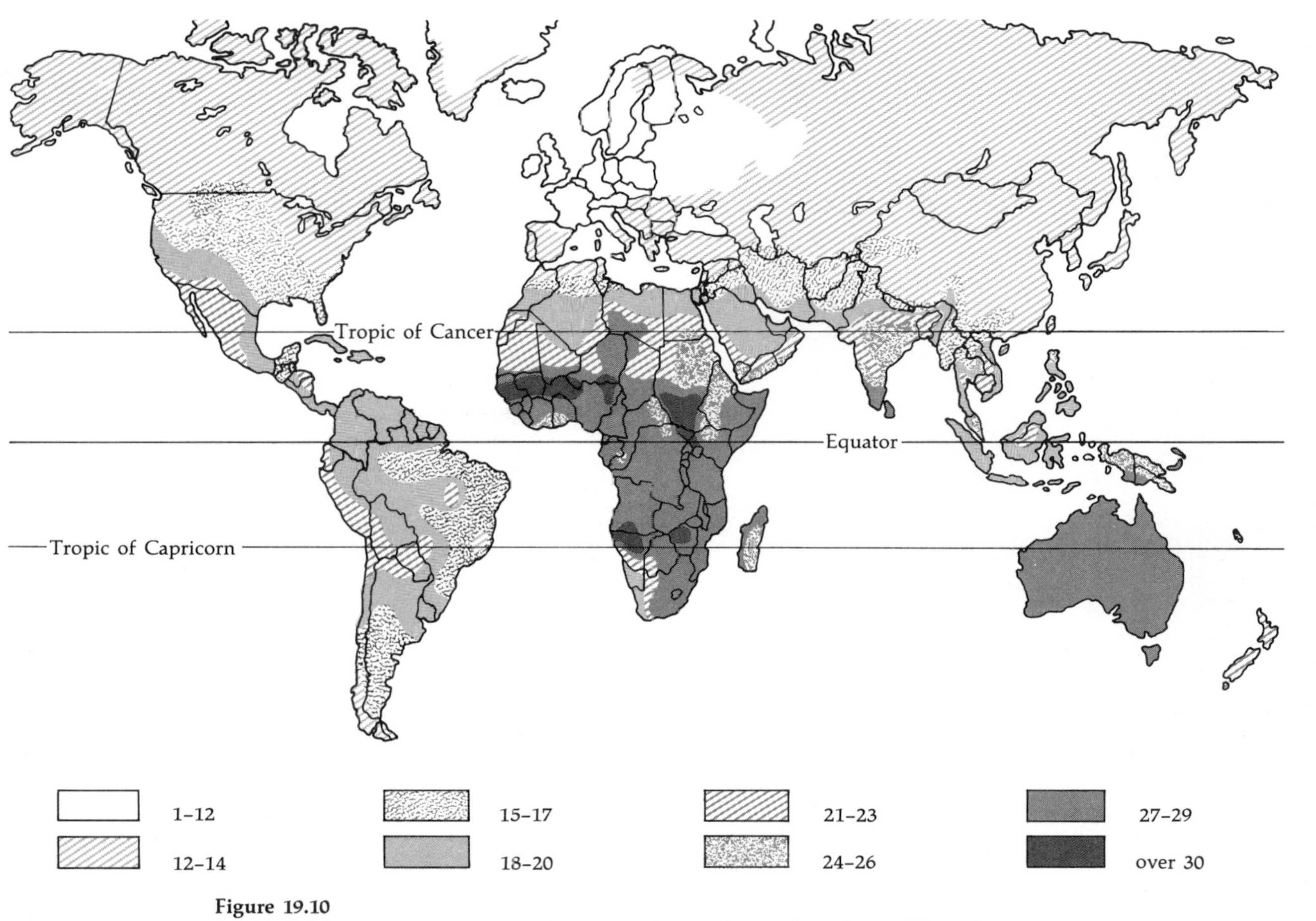

Figure 19.10
Approximate geographic distribution of skin color in human aboriginal populations. The values increase with darker skin color. (After R. Biasutti, *Razze e Popoli della Terra,* Torino: UTET, 1951.)

has serious consequences for both survival and reproduction. Therefore it seems quite reasonable to suppose that the need for vitamin D may have given an important selective advantage to lighter skin at high latitudes, where the sun is weak. The lightest skin color is found among Northern Caucasians. History provides an impressive confirmation of the importance of sunlight for vitamin-D production. When the industrial revolution in nineteenth-century England darkened the skies with smoke, rickets reached epidemic proportions. Hygienists of the time knew nothing about vitamin D, but they soon discovered that exposure to sunlight helped to prevent rickets.

F. Loomis, who proposed the vitamin-D hypothesis of skin-color origin, has further suggested that a darker skin would possess a selective advantage in the tropics. The selective mechanism in this case would be the prevention of excessive production of vitamin D, which may be toxic.

Impressive support for the vitamin-D hypothesis comes from the study of groups that form an exception to the general rule that whiter skins are found at higher latitudes. The Eskimos and Lapps have relatively dark skins, although they live at extremely high latitudes. However, unlike those Caucasians whose diets have been cereal-oriented for the past 5 to 8 thousand years, the Eskimos and Lapps have diets much richer in natural vitamin D; they have much less need to produce the vitamin with the help of the sun.

Many other conspicuous human traits have correlations with climate.

Hair and eye pigmentation shows a distribution similar to that of skin color (Fig. 19.11). This pigmentation may be affected by selective factors similar to those controlling skin color, or perhaps some of the genes that control skin color also control eye and hair pigments.

The peppercorn hair of Africans may be advantageous for preventing the rapid evaporation of sweat, thus protecting the head from excessive heat and the individual from sunstroke. The absence of facial hair among Easterners of northern origin (including American natives) may have the advantage of preventing frostbite. The epicanthic fold and the narrow eye slit of these same people may protect the eye from extreme cold and grit in wind or from the glare of the sun on snow. Smaller nostrils and longer noses among peoples living in high latitudes may aid in warming and humidification of inhaled air. Thinner or thicker lips have also been suggested as further examples of such adaptations to environmental conditions. All these suggestions are largely speculative and difficult to prove, but they certainly seem reasonable.

The hypothesis that body build shows adaptation to climate rests upon somewhat more solid evidence. Body size is inevitably greatly influenced by diet, but at least a fraction of the differences observed between groups is likely to be genetic

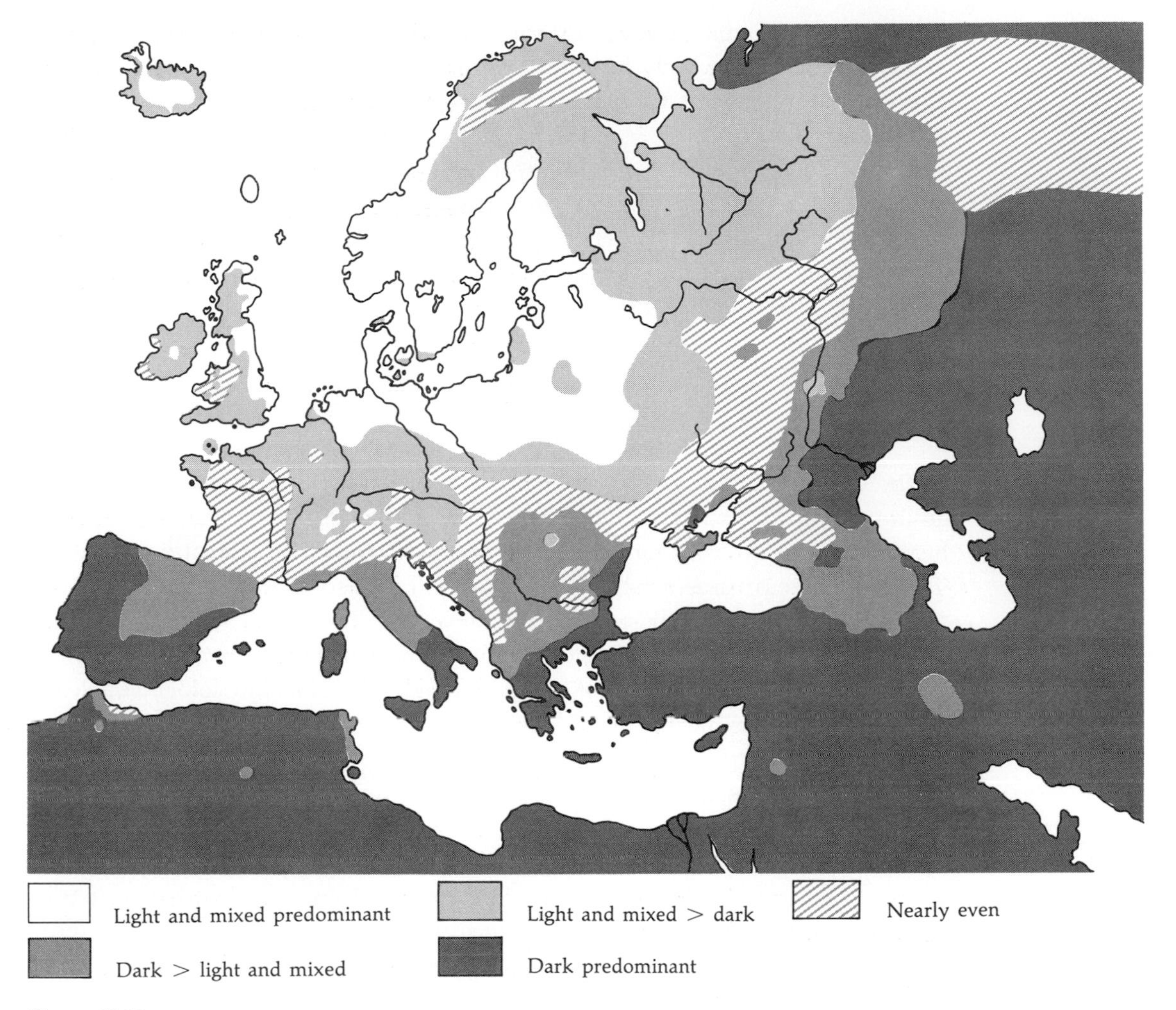

Figure 19.11
Pigmentation of hair and eyes among European populations. (From C. S. Coon, *The Races of Europe*, Greenwood Press, 1956.)

in origin. Moreover, there is abundant evidence from studies of nonhuman animals that body build is markedly affected by climate. In fact, the major rules of climatic variation were originally established through studies of variations between populations of the same or similar animal species living in different climatic conditions.

Bergmann's rule states that general size increases with latitude. The explanation of this correlation is that a larger body size tends to give a smaller ratio of surface to volume. Heat production is roughly proportional to body volume, and heat loss roughly proportional to body surface. Therefore, a larger body size will result in more retention of body heat. In the tropics, increased loss of body heat has survival advantage; in colder climates, there is a selective advantage to increased retention of body heat.

Similar considerations, based on heat loss, also explain Allen's rule: in colder climates, limbs are shorter in relation to trunk length. Again, the change observed in colder climates is one that tends to produce a smaller ratio of surface to volume. To some extent at least, both Bergmann's rule and Allen's rule apply to human traits (Figs. 19.12 and 19.13).

It is not at all surprising that adaptations to climatic conditions produce readily observable differences in characteristics of the body surface.

The body surface is in fact the interface between the internal and the external environment. As such, the body surface is of great importance for regulation of heat exchange, and therefore must play an important role in climatic adaptation. This tendency for superficial traits to reflect climatic adaptations has very important consequences for our study of human racial differences.

First, when evaluating differences between human races, we are likely to be unduly impressed by the superficial differences (such as skin color), which are not only readily visible, but are also conspicuous. Because these superficial differences are likely to be chiefly the result of major climatic adaptations, they are probably a biased sample which overemphasizes the genetic differences that exist between races.

A second consequence is that superficial differences may be misleading when we try to reconstruct the evolutionary history of human races. If these traits tend to measure adaptation to climate, it is not surprising that they tend to tell us more about a population's climatic history than about its evolutionary origins. In other words, we would expect all populations that have lived in cold climates for a number of generations to show similar superficial traits, even though the superficial traits of the populations from which they came may have been very different. Study of superficial traits alone would suggest that these cold-climate populations are closely related. Study of other traits less strongly affected by climatic adaption might reveal quite a different pattern of evolutionary history.

Figure 19.14A shows part of an evolutionary tree based upon differentiation in anthropological and anthropometric traits—measurements of superficial differences. Contrast it with the corresponding portion (Fig. 19.14B) of the evolutionary tree based upon differentiation in genetic markers (the tree discussed in the preceding section). The anthropometric tree corresponds to climatic exposure. Most populations of Australian aborigines and Africans have adapted to tropical environments; most American Indian and Caucasian populations have adapted to temperate or cold climates. As expected, the anthropometric tree indicates close

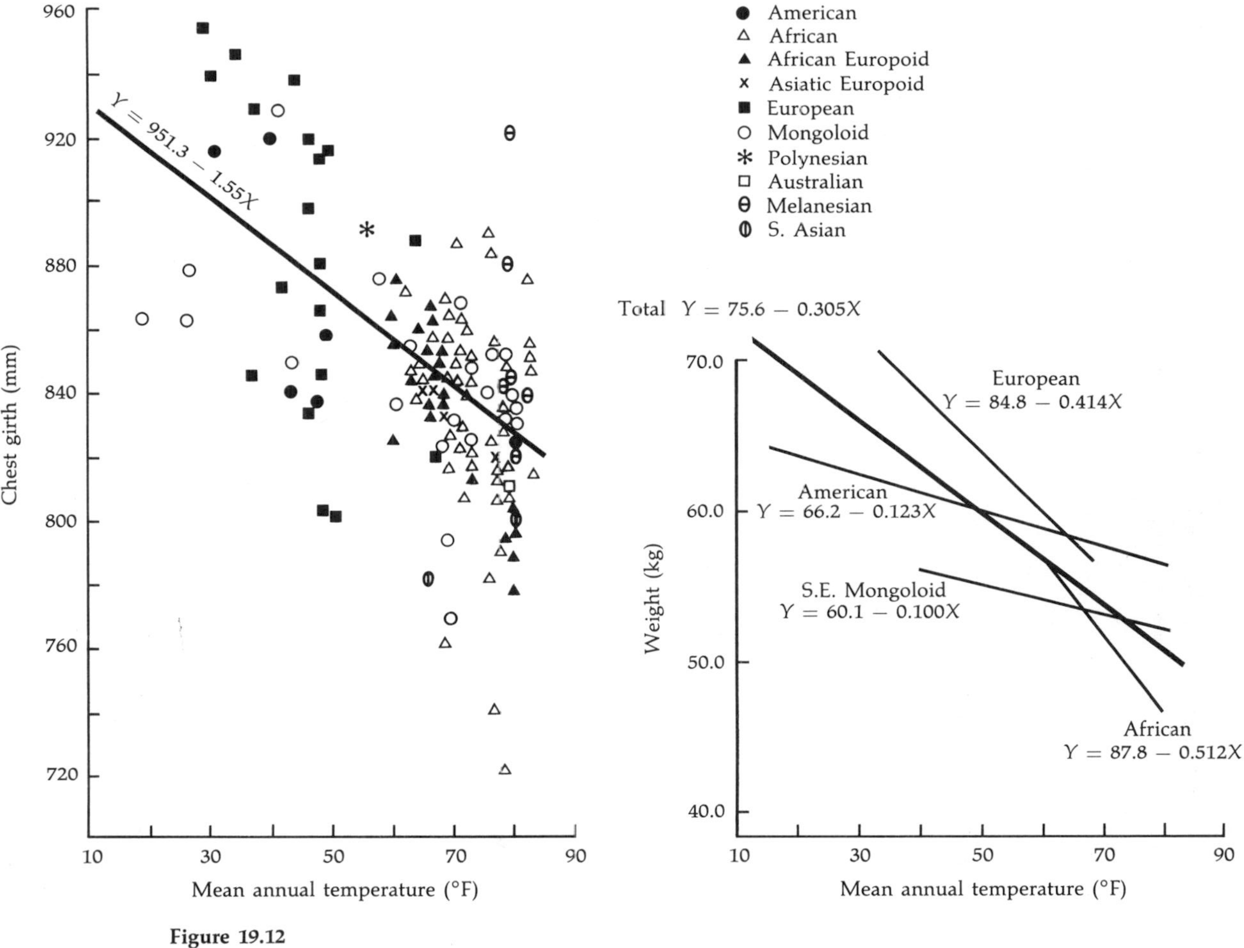

Figure 19.12

Some correlates of Bergmann's rule in human populations. Chest girth and weight, both correlated with the ratio of body surface to volume, decrease with increasing average temperature. Only aboriginal males are included in the samples used for this figure and for Figure 19.13. (From D. F. Roberts, *Climate and Human Variability*, copyright © 1973, Cummings Publishing Company, Inc., Menlo Park, California.)

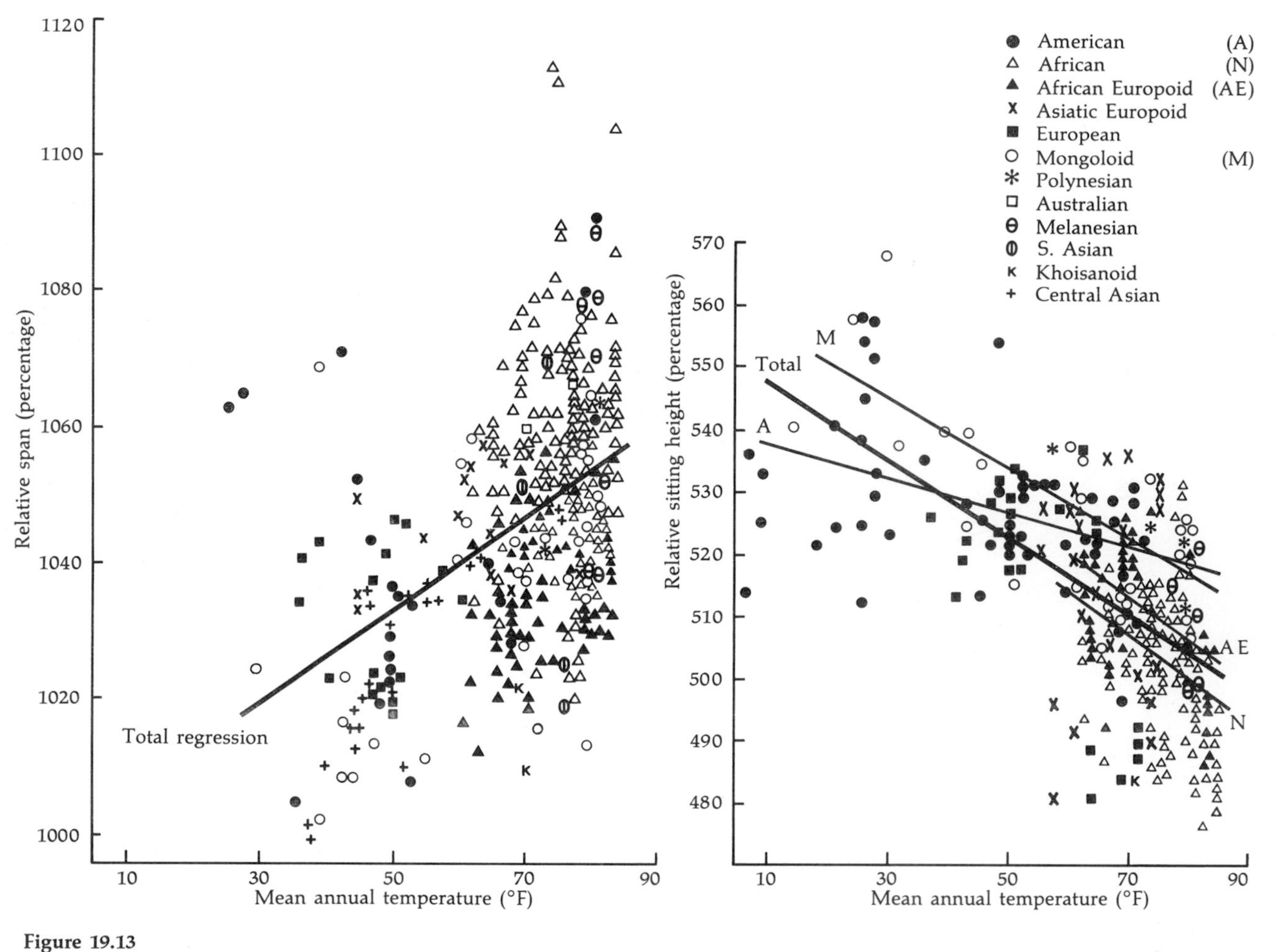

Figure 19.13
Some correlates of Allen's rule in human populations. Relative span is the distance between fingertips with arms extended, expressed as a percentage of stature. A larger relative span indicates that the arms are longer in relation to the overall height. Relative sitting height is a measure of the length of trunk plus head and neck, expressed as a percentage of stature. A smaller relative sitting height indicates that the legs are longer in relation to the overall height. Both measurements show a tendency toward decreasing relative limb length in colder climates. (From D. F. Roberts, *Climate and Human Variability,* copyright © 1973, Cummings Publishing Company, Inc., Menlo Park, California.)

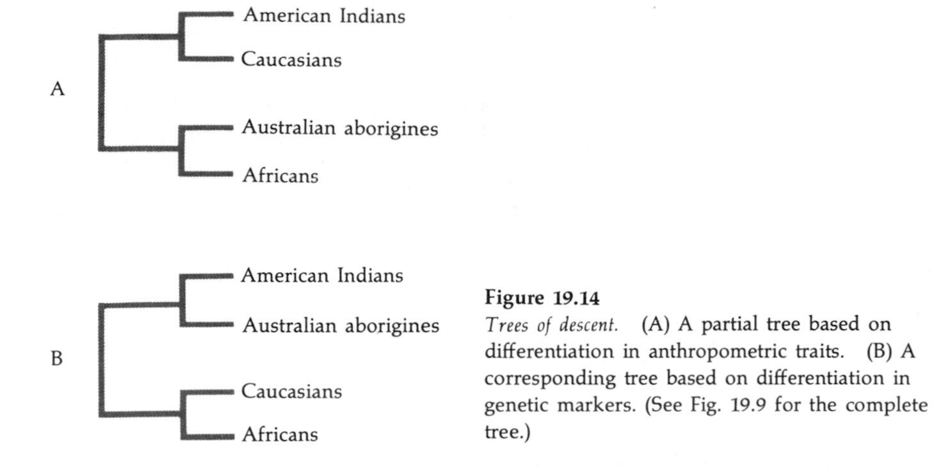

Figure 19.14
Trees of descent. (A) A partial tree based on differentiation in anthropometric traits. (B) A corresponding tree based on differentiation in genetic markers. (See Fig. 19.9 for the complete tree.)

relationships between populations now living in similar environments. The tree based upon a wide range of genetic markers (unrelated to superficial characteristics) suggests a quite different pattern of evolutionary relationships.

A further complication with anthropometric traits is that they show short-term responses to a wide variety of changes in environmental conditions.

We have already discussed variations in response to climatic conditions. However, the most remarkable variation that we have witnessed in recent years is a very rapid increase in average stature (Fig. 19.15). This change has been observed in all parts of the world in which economic development has taken place and is, in some way, a measure of this development. In part, the change has been one of acceleration in growth rates; but the final result, adult size, has also increased markedly. The change has been so rapid that genetic factors, if any, must be relatively unimportant. Environmental factors such as nutrition and disease are likely to be involved. The change in averages is of the order of two standard deviations, and a similar change is observed in all traits related to overall size.

The observed rapid change in height suggests strongly that stature (and related anthropometric traits) has a high percentage of nongenetic variation. Therefore, the variations with climate shown in Figures 19.12 and 19.13 must be, in part, physiological variations—that is, variations that take place during individual development (in direct response to climate) within a wide range allowed by genetic limitations. In order to study genetic variation between groups, it may be more appropriate to study variations in the slope of such regression lines—that is, to study the extent to which individuals of the group are able to adapt to different

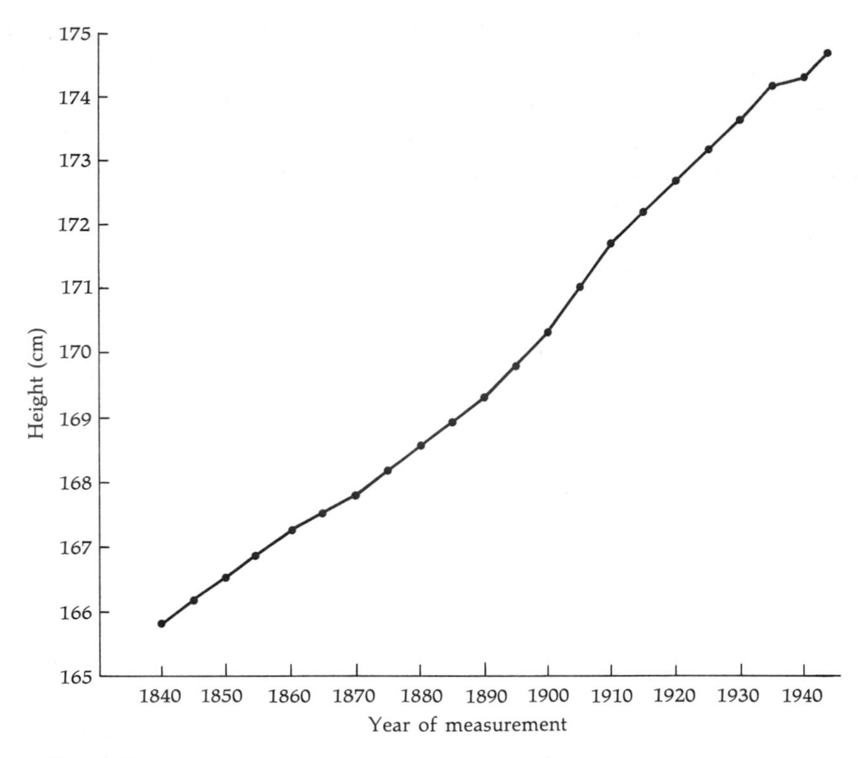

Figure 19.15
Increase in mean height of Swedish males aged $20\frac{2}{3}$ years. (From data published by B. J. Lundman, 1939, as cited by A. Costanzo, *Annali di Statìstica,* vol. 2, 1948.)

environments. However, this conclusion is only speculative, and direct investigations of this possibility have not yet been reported.

The capacity of anthropometric traits to show short-term changes in response to environmental variation generates further perplexity in the interpretation of anthropometric differences between races. Genes do not show such rapid change in response to environment, and therefore should be better indicators of the long-term evolutionary history of human races.

19.6 Differences Between and Within Groups

One tenet of old-fashioned racism—sometimes surviving even today—is that one should worry about keeping a race "pure." Hybridization, it is argued, will ruin a race! A simple calculation shows how fallacious is the idea of racial purity. Even if one takes a sample of people from a single small, relatively isolated village, one finds that no two individuals are genetically identical (except, of course, for identical twins). Furthermore, one finds that all (or almost all) polymorphisms present in the human species are also represented in the sample, through perhaps with somewhat different gene frequencies for the various alleles. The genetic heterogeneity between individuals, even those chosen from the same village, is so large that the chances of finding two genetically identical individuals who are not MZ twins are practically zero.

Today we have information on very many different genetic loci, and we know that about one-third of them are polymorphic. For each of these polymorphic loci, we can measure the frequency of heterozygotes in the human population. Averaging over all the polymorphic loci, we find that the frequency of heterozygotes is about 30 percent, a value that is called the *average degree of heterozygosity*. Since about one-third of all loci are polymorphic, we know that the chances of an individual being heterozygous for a randomly selected locus are about $\frac{1}{3} \times 0.30 = 0.10$, or 10 percent. Thus it is rather ludicrous to argue that an individual (much less a race) has a "pure" stock of genes that must be protected against admixture.

There is an enormous amount of genetic variability in store.

At equilibrium, all the variability lost each generation (by drift or by selection) must be resupplied by mutation. Figure 19.16 indicates the relative amount of variation produced by mutation every generation, and the amount present at a given time, which is several orders of magnitude higher. We still do not know the relative importance of the forces that maintain this variation. Its magnitude, curiously enough, is relatively independent of the size of the group that we examine. Whether we look at a village, at all the English, at the population of all Caucasians (or Africans or Easterners), or at the whole human species—the average degree of heterozygosity shows only a modest increase with the increase in size of the group.

This measurement can help us to understand the magnitude of racial differences by comparing them with individual differences. The difference between individuals within a group can be measured by the average degree of heterozygosity in that group. This quantity can also be considered as a difference observed between a gene taken from an individual at random, and the same gene taken from another random individual.

The probability that the two genes are different is estimated by the average degree of heterozygosity.

We can repeat this computation taking a gene from an individual of one race, and the same gene from an individual of another race. We then expect to see a greater difference because races do show some genetic variation. In fact, the

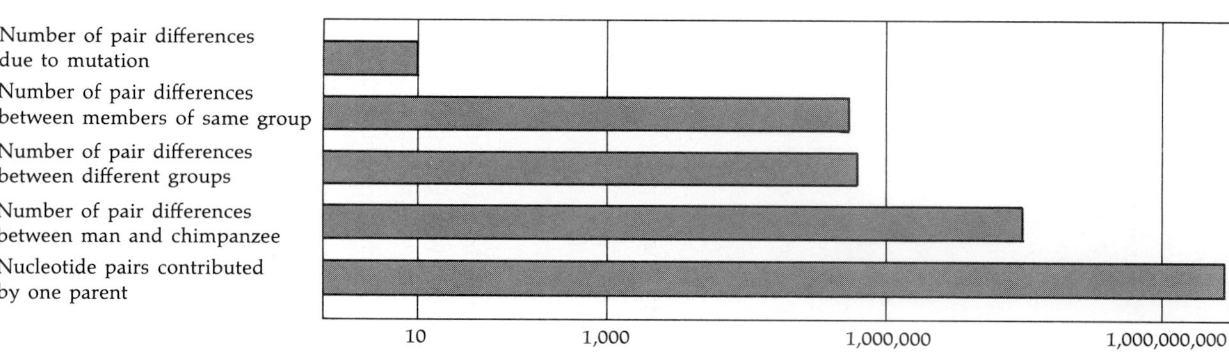

Figure 19.16
Numbers of genetic differences between individuals. This graph shows on a logarithmic scale the numbers of genetic differences estimated to exist between individuals of varying degrees of relatedness. The data used are the numbers of differences in the nucleotide pairs of the DNA that encodes proteins; these differences are observable as amino-acid substitutions in the protein. The observed number is extrapolated to estimate the total number of nucleotide-pair differences. For comparison purposes, the bottom line of the graph (e) shows the size of the haploid genome —that is, the number of nucleotide pairs contributed to an individual by one parent. This number is about 3 billion. The top line (a) indicates the number of differences that may arise in the haploid genome because of random mutation in a single generation; this number may be as small as ten. A much larger number of differences (half a million or more, line b) may exist between the haploid genomes of two randomly chosen members of a single ethnic group. This number is not greatly increased (c) if the randomly chosen individuals belong to different major divisions of the human population, such as Caucasian and Easterner. Differences between man and his closest relation among other species, the chimpanzee, number in the many millions (d). The figures are approximations only. (From L. L. Cavalli-Sforza, "The genetics of human populations," *Scientific American,* September 1974.)

increase in this difference is a measure of the genetic difference between races. This corresponds to computing the average degree of heterozygosity in an interracial cross. The result—whether we compare Africans and Caucasians, or Caucasians and Easterners, or Africans and Easterners—is about 35 to 40 percent, and so is somewhat higher than the comparison between two individuals from the same group (28 to 30 percent). (These data use 25 loci, each polymorphic in at least one racial group.) Clearly, the increase is small; this is tantamount to saying that the genetic difference *between* groups is small in comparison to that *within* groups. The difference between Africans and Easterners, incidentally, is a little larger than that between Africans and Caucasians or between Caucasians and Easterners, because Caucasians are somewhat intermediate between the other two groups. These differences can be approximately translated into DNA nucleotide differences with the results shown in Figure 19.16 (also see Exercise 6).

The conclusion is that overall genetic differences between races are in fact small.

If we had the impression that they were large, this must have been dictated by the differences we are used to seeing. But we have already indicated how this impression may be biased by the conspicuousness of these differences, and the reasons behind this bias—namely, climatic and other environmental adaptations.

19.7 Cultural and Historical Correlates of Racial Differentiation

If we turn our attention to cultural differences, we find that, by and large, there are important cultural correlates of racial differentiation. Today, rapid advances in technology have maximized economic, social, and political differences between people. Until a few thousand years ago, there were no political units of large size, and economic differences between groups were far fewer.

Linguistic differentiation was probably, until quite recently, even greater than it is today. If hunter-gatherers of modern times (such as the Australian aborigines) are representative of earlier periods, a group speaking a common language (usually corresponding to a tribe) must have had a size that was on the average even smaller than 1,000 individuals. Language is an important criterion for social unity, as a common language favors communication. Different languages, though of course not a complete barrier, are statistically important as a barrier to exchange of all sorts, including that of genes. With agriculture, the average size of linguistic units very probably increased above that of hunter-gatherers. It is now of the order of 5,000 in New Guinea, where a primitive type of agriculture is still practiced. With the transition from the tribal to the national organization, the size of linguistic units has increased further and is now over a million. But we have seen that the emergence of modern man has been accompanied by increasing cultural differentiation. Given the great increase in world population size, it may be that the number of different languages spoken was not greatly decreased in the last few thousand years. But certainly the average number of people speaking the same language has increased by some three orders of magnitude.

Even smaller linguistic units, however, cannot be considered as fully random-mating communities. One can find subtle genetic differences (in terms of gene frequencies) even between neighboring villages, or parts of a tribe, not to speak of different parts of one country. The ABO map of the British Isles (Fig. 19.17) shows significant differences in gene frequencies, with both north–south and east–west gradients. The source of these clines is probably historical in nature. It is

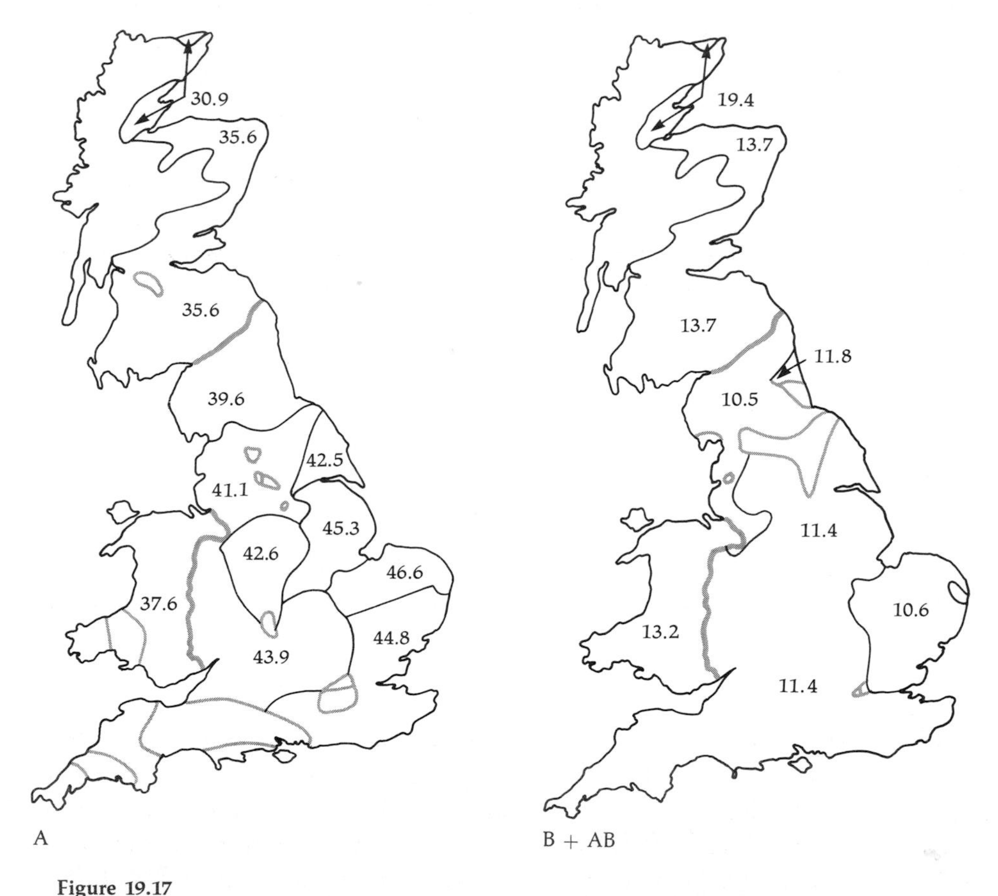

Figure 19.17
Percentage frequencies of A individuals and B or AB individuals in Britain. (Modified from A. Kopeć, *The Distribution of the Blood Groups in the United Kingdom,* Oxford Univ. Press, 1970.)

interesting to follow the history of occupation of the British Isles, as an example of the complexity of the origins of a population and of how genetic differences may reflect historical origins.

Before 5,500 years ago, the British Isles may have been peopled by a very small number of hunter–gatherers (on the order of 10,000 to 50,000, based on ecological and archeological information). These people may have descended from, or somehow been related to, the Late Paleolithic hunters who occupied northwestern Europe. The arrival of farmers from Europe around 5,500 years ago may have brought much new blood to the British Isles. Before the Roman conquest, other immigrants from Europe may have settled there in the Bronze and Iron Ages. Among the archaeological remains of the pre-Roman era are megalithic monuments, whose origin is obscure, but whose builders may also have contributed some genes. The Roman conquest, however, may have had no great influence in terms of genes. At that time the aboriginal population was already of the order of one million people. However, the Saxon conquest in the early Middle Ages probably did contribute quite a few genes and may have helped push earlier cultures, possibly also *genes*, farther toward the north and west (Figure 19.18).

The Norman conquest (again from the southeast) and the Viking settlements (directly from Scandinavia on the northern coasts and elsewhere about a thousand years ago) added to the genetic heterogeneity. Since that time and until quite recently, further external contributions of ancestry have been unimportant, but homogeneity could hardly have been achieved in the intervening time interval. The only force that could have given rise to homogeneity is internal migration, but this has since become increasingly centripetal—namely, toward the cities rather than toward the peripheral rural areas.

Most cities represent admixtures of people of all origins. The larger the city, the more remote the origins of many of its inhabitants.

City populations tend to under-reproduce. Only in very recent times has the birth rate of a city population equaled or exceeded its death rate. Thus, cities have been maintained, and increased, by immigration from the rural hinterland. A city population represents a sampling of genes of disparate, predominantly rural origin.

Even within a single city, there may be geographic variations in gene frequency—patterns of internal genetic heterogeneity. Recent immigrants may not have had time to admix with earlier arrivals. Typically, the most recent immigrants must begin at the bottom of the social ladder. Thus immigration of a new group to cities may create a stratification of gene frequencies in social classes; this stratification may sometimes take several generations to disappear. Under certain social systems, stratification in socioeconomic classes, if accompanied by barriers to

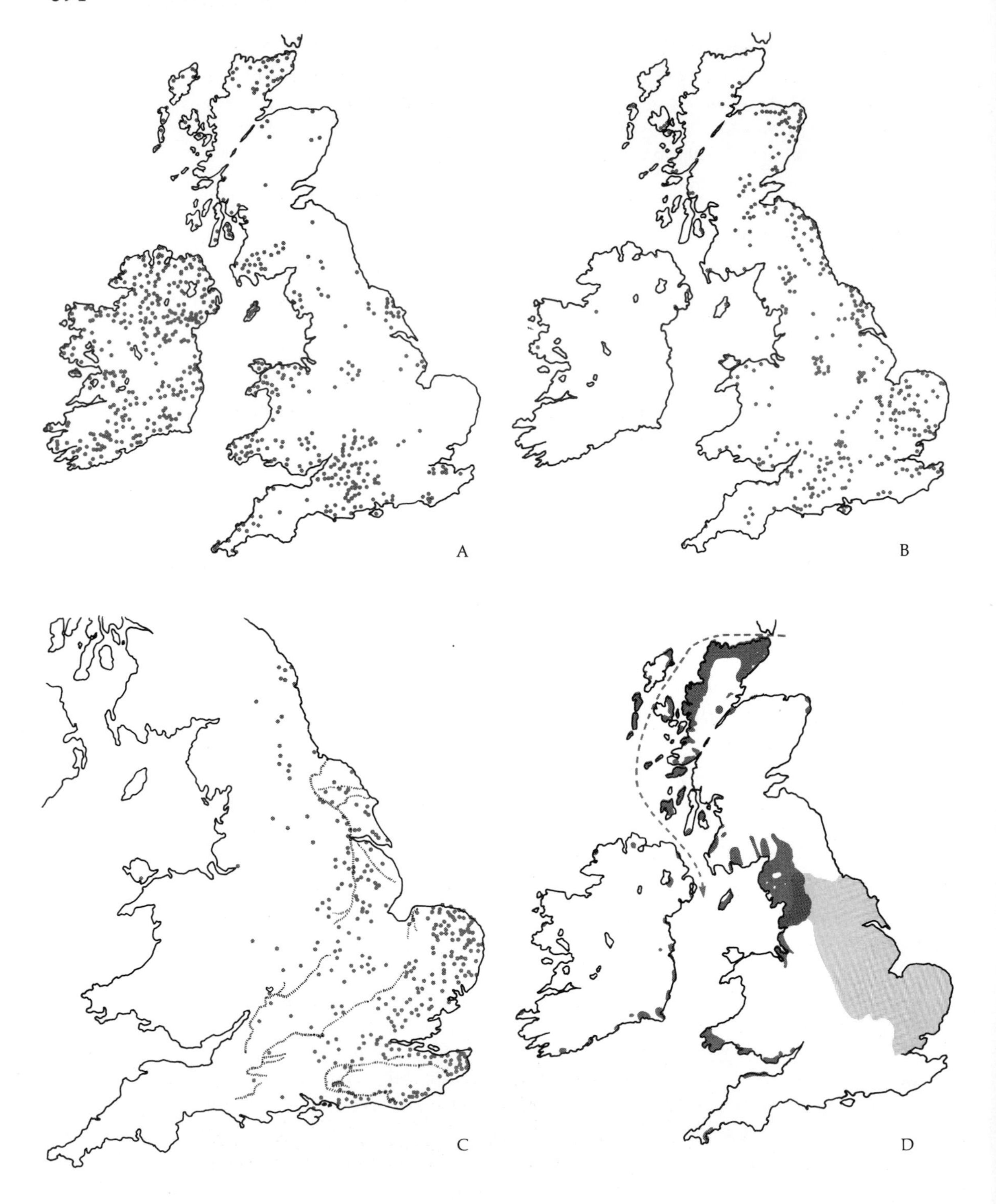
A
B
C
D

marriage between classes, may even promote the creation or magnification of gene-frequency differences between social strata. In India, where classes became *castes* (social strata between which genetic exchange is minimal or zero), there is considerable gene-frequency variation from caste to caste.

Not all genetic variation necessarily represents true adaptation to local environmental conditions.

If there is no further immigration for a long time, then random variation and/or adaptation to local selective conditions may proceed unimpeded, adding to or subtracting from the genetic differences of historical origins. With low densities of local population, or with a high degree of isolation, random genetic drift may further complicate the picture. In the case of Britain, local ABO gene frequencies remain similar to those of groups that immigrated in historical periods—for example, the Vikings in the north and west. Apparently, neither admixture nor drift due to local selective adaptation was powerful enough, given the time and circumstances, to blur the historical picture.

The existence of visible differences, which tend to accompany the invisible differences described by genes, certainly helps to create racial prejudice.

People of different origins usually show some physical external differences. Where there are also important cultural differences, the seeds for racism are more easily germinated. Racism is the belief in an inherent superiority of some "races," however defined. With very few exceptions, the race believed superior is, of course, one's own! According to racist beliefs, it is especially behavioral traits for which racial superiority is claimed, and it is further claimed that differences for such traits have a genetic origin.

Many analyses have shown and confirmed that social and political conditions are of great importance in determining the intensity and direction of racist feelings. But there are also other important factors favoring the development of racism, such as the constant confusion between biological and cultural attributes. Other factors are psychological, such as the inevitable preference for the culture

---

Figure 19.18
Distribution of archeological finds in Great Britain. (A) Megalithic monuments. (B) Beakers (Early Bronze Age). (C) Primary areas of early Anglo-Saxon colonization. (D) Areas of colonization by Daneslaw in the ninth century (light-toned) and Norse (dark-toned). Maps C and D are drawn mostly on the basis of place-name distributions. (Parts A through C from D. F. Roberts in *Genetic Variation in Britain,* ed. D. F. Roberts and E. Sunderland, Taylor and Francis, 1973; sources are given by Roberts. Part D after H. J. Fleure, *A Natural History of Man in Britain,* Collins, 1951.)

Figure 19.19
African Pygmies. There may be 200,000 of these "Forest People" forming clusters scattered in various parts of Central Africa. Their average stature is the smallest of any human group (see also Fig. 1.4), and they still live largely by hunting and gathering, wherever the forest has not been destroyed. Genetically, they differ somewhat (though not greatly) from other Africans, and also from the Bushmen or San. Probably, like the San, they have had less admixture than most other Africans. Pygmies have considerable interest in dancing and singing; they were known to the ancient Egyptians for their ability in dancing. Note the size of the breast-feeding child. Hunter-gatherers and other groups with primitive economies tend to have long intervals between successive births, a custom that has helped keep their numbers small. Children may be breastfed until they are 3 or 4 years old, but they are also given other types of food very early.

in which one is raised, the drive for identity, and the insecurity which, while it promotes social rapport, also tends to make us distinguish between "us"—the people on whom one can rely—and "them"—those who cannot be trusted.

The fundamental common element underlying the belief that there are biological grounds for racist ideas is the inevitable tendency to correlate different facts, and then to express these correlations uncritically as causal relationships. The same conditions that create genetic isolation also create cultural isolation. Therefore, if genetic differentiation has been favored, then cultural differences probably have arisen as well. It is all too easy to jump to the conclusion that cultural differences (usually regarded as a cultural "inferiority" of the other race) are caused by genetic differences.

Figure 19.20
Australian aborigines. When Australia was rediscovered in 1770, perhaps 300,000 aborigines were still in a Late Paleolithic culture, living as hunter-gatherers and making stone tools. The men shown here are from near Mt. Liebig (Northern territory); they are burning acacia leaves for ashes to add to native tabacco quid. The women are from Warupuju (in the eastern desert of Western Australia). These photographs were taken in the early 1930s. (Courtesy of Dr. N. B. Tindale.)

Figure 19.21
Melanesians. Melanesians live in New Guinea, New Caledonia, Fiji, Solomon, and other islands of the southwest Pacific. They subsist on agriculture, usually of a fairly primitive type. The women shown here are from the Highlands of East New Guinea; they belong to a tribe which, like many other New Guinean tribes, abandoned the use of stone tools only a few years ago. Australian aborigines are living relics of Late Paleolithic culture, while Melanesians exemplify Neolithic culture. Genetically, Australian aborigines and Melanesians are fairly closely related.

For example, there are differences in skin color and differences in socioeconomic position (income, job status, social status, and so on). There is a correlation between skin color and socioeconomic position. In various countries, this correlation is in the same direction—whiter skin correlates with higher socioeconomic position. From these facts, many people conclude that whiter skin (and associated genetic traits) *causes* socioeconomic superiority. They do not consider or remember the historical events that may have led to the correlations. The assumption that biological differences cause the cultural ones is likely to present itself as an apparently rational explanation.

Once accepted, the explanation soon spreads. Racist ideas are often taught to children—in very subtle ways as well as in explicit teachings. It is difficult to free oneself of prejudice, especially if acquired early and subsequently confirmed by the experience of the differences that led to the "explanation" in the first place. In the subsequent generations, as in the first, the usual preference for one's own culture and the mistrust of unfamiliar people reinforces racist tendencies. As the

Figure 19.22
South American natives. Some South American tribes, in particular those living in the tropical forest or in its immediate neighborhood, have had very little contact with outsiders and are still largely unacculturated to the modern world. In particular, the Xavantes, Makiritare, and Yanomama tribes have recently been subjects of much genetic research. The Yanomama tribe is recently increasing in numbers, despite internal and external warfare. Yanomamas live in small villages, in considerable isolation one from the other, or from other tribes. It is therefore not too surprising that they show a high heterogeneity between villages for genetic markers and also show considerable differences from other tribes of American natives. The men shown here are Xavantes preparing for dances; the girls are Yanomamas. (Courtesy of J. V. Neel.)

Figure 19.23
Central American natives. A Mayan woman from the northern Yucatán peninsula (Mexico) near the door of her home. Like many other people living in the tropical forest, Mayans are small. Mayans had a flourishing civilization until the European conquest of America. Today they number in excess of 2 million.

racist ideas become common in the culture, they lead to maintenance and perhaps even magnification of the socioeconomic or cultural differences, thus in turn further stimulating and reinforcing racist beliefs.

Even if there were an innate tendency, a real drive to indulge in racism (for instance, for the psychological reasons mentioned above), it is clear that modern society must benefit from an education leading to a rational analysis of the sources of the differences. We must learn to live with the difference without being prey to the emotions that such racial differences may elicit and that can cause such disruption of our social life.

We are unable, by present techniques, to detect behavioral differences between races that we can seriously consider genetic.

Genetic differences between groups are, as far as we can tell, small on the average in comparison with differences between individuals. The well-known problem of IQ and race is considered in Chapter 21.

Figure 19.24
Tuaregs. A Tuareg camel trader from an oasis near Djanet in the central Sahara. Tuaregs have an unknown origin; their genes are mostly Caucasian with some African admixture. They have considerable social stratification by skin color, as they used to have slaves from sub-Saharan Africa. There are perhaps 300,000 Tuaregs spread over most of the Sahara desert. They are pastoralists and also operate most of the trade on camels still going on in the Sahara.

Genetic differences between races for behavioral traits may exist, though there is so far no evidence for their existence. If any were shown, we should have to learn to live with them. We are learning to live with much more profound genetic differences *between individuals,* which are, after all, real. Until the last century, the deaf and the blind were considered insane and were often institutionalized. We have, fortunately, learned that the burden of these people, whether it has a genetic origin or not, can be considerably alleviated and that even such very severe handicaps are compatible with normal function in society.

It can hardly be denied that different groups show behavioral differences. The simplest presumption is that these are basically cultural in origin. It will be a difficult task to show convincingly that they have a genetic component. Many of these differences (which are clearly not genetic) are surely a distinct bonus, and the maintenance of a cultural pluralism with respect to such variation can only benefit society as a whole. The existence of so many national arts, crafts, and cuisines is an enrichment of our lives. Other cultural differences may not be so beneficial insofar as they prevent communication and agreement. But cultural evolution is much faster than biological evolution. Differences may arise quickly

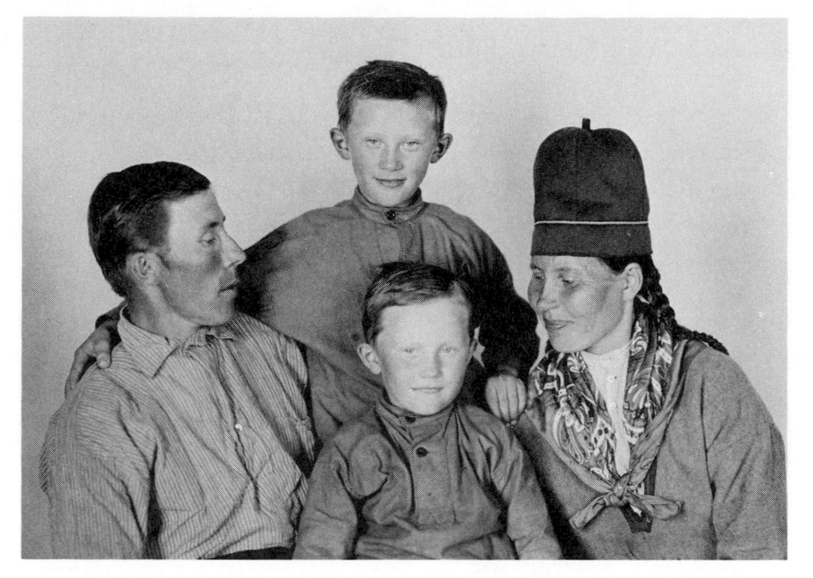

Figure 19.25
Lapps. Lapps live in the northern part of Scandinavia (northern parts of Sweden, Norway, and Finland). Depending on areas in which they live, they are (or were) hunters, fishers, or reindeer herders. They number about 40,000. Their origin is obscure, but they have been in the area for at least 2,000 years. Genetically, they are closer to Europeans than to Asians, though they may have some admixture. (Courtesy of Lars-G. Lundin.)

and also disappear quickly. "Quickly," however, is relative to a historical time scale; sometimes it may not be as fast as one might, in some cases, wish. All those behaviors that are imprinted on us early in our life are likely to accompany us until death. So long as the family has a part in the upbringing of the child, such behavior may be transmitted culturally to the progeny and may survive longer than the individual. The resistance to change of many cultural traits in immigrants to the U.S. is an indication of the power of cultural inheritance. This happens even under conditions that would appear, superficially at least, to favor rapid dilution and disappearance of customs differing from those that prevail.

Summary

1. We belong to one species. There are no known reductions of interfertility in interracial crosses.
2. Races are more difficult to define than species. The degree of differentiation at which two groups are considered different races is arbitrary.

3. Because of human history, it is especially difficult to define races in man. Much variation is continuous for practically every trait.
4. A reasonable partition is into Africans, Caucasians, and a heterogeneous group of Easterners, which includes all aboriginal populations who live in the Pacific area.
5. Gene differences are in good agreement with this classification. Cumulating single gene differences, one can reconstruct a tree that approximately describes human racial differentiation.
6. Superficial traits like skin color and body build are fairly clearly correlated, for good reasons, with climate.
7. Genetic differences between races are small as compared with differences within races.
8. There are many cultural correlates with genetic differentiation, but there is no good reason to assume that there is any direct control of one by the other.
9. Racism arises in part from a confusion between biological and cultural evolution.

Exercises

1. What is the evidence for the fact that *Homo sapiens* is a single species?
2. Outline the basic methods by which genetic markers can be used to delineate human races and construct their phylogeny.
3. Why do superficial traits—such as skin color, body build, and facial features in general—seem relatively unsatisfactory as markers of racial differences?
4. Compare and contrast linguistic and biological evolution.
5. There are few exceptions to the rule that, once an allele is found in one major racial group, it is also found in the other two (leaving aside subgroups for which there is evidence of recent hybridizations). Is this evidence in favor of, or against, the idea that the formation of human races occurred relatively recently? What conclusion does it suggest with respect to the major exception to this rule—namely Gm, which shows some of the most extensive differences even between populations as closely related as the Northern and Southern Chinese?
6. In Chapter 13 we state that the average degree of heterozygosity at a locus in a typical human population is 6 percent, while in Section 19.6 we put it at 28 to 30 percent. Reconcile, on a semiquantitative basis, these two statements, noting that the second statement is based only on informative loci and includes immunologically as well as electrophoretically detected polymorphisms.

7. The correlation between body size and temperature is also found for white Northern Americans, where it is about the same as that shown for all human populations in Figure 19.12. Does this reinforce or weaken the hypothesis that the correlation is the result of natural selection?

8. Indicate reasons for and against the idea that, from an evolutionary point of view, morphometric measurements are less informative than gene-frequency data.

9. Look for possible flaws in the following reasoning. (a) The most striking ecological differences between habitats of human racial groups are climatic. (b) Climatic adaptation especially affects surface characteristics. (c) Surface characteristics are particularly conspicuous. (d) Therefore, the differences that we see between human racial groups tend to exaggerate the true differences.

10. C. S. Coon, S. M. Garn, and J. B. Birdsell (prominent investigators of human racial differentiation) have speculated that the Mongoloid (Oriental) face is the result of engineering through natural selection to resist the very cold climate of Northeast Asia in which the Mongoloid group probably evolved. They suggest selective advantages for the following traits: (a) breathing passages arranged to deliver to the lungs air of maximal warmth and humidity; (b) insulation by fat paddings of the most sensitive parts of the face; and (c) reduction of all appendages through which heat may be more easily lost and which may be more liable to damage from frostbite. Describe and discuss experiments and observations that might help test these hypotheses. Consider also the problem of whether the differences are genetic, or might (at least in part) express individual acclimatization.

References

Cavalli-Sforza, L. L.
1974. "Genetics of human populations," *Scientific American,* vol. 231, no. 3, pp. 81–89.

Clark, G.
1971. *World Prehistory: A New Outline.* Cambridge: Cambridge Univ. Press.

Coon, C. S.
1965. *The Living Races of Man.* New York: Knopf.

Giblett, E. R.
1969. *Genetic Markers in Human Blood.* Oxford: Blackwell Scientific Pub.

Katz, S.
1974. (Ed.) *Biological Anthropology: Readings from Scientific American.* San Francisco: W. H. Freeman and Company.

Mourant, A. E.
1954. *The Distribution of the Human Blood Groups.* Oxford: Blackwell Scientific Pub.

Roberts, D. F.
1973. *Climate and Human Variability.* Menlo Park, Calif.: Addison-Wesley.

Roberts, D. F., and E. Sunderland
1973. (Eds.) *Genetic Variation in Britain.* London: Taylor and Francis.

20

Genetics and Medicine

Of all the technological conquests of man, it is perhaps the medical advances that we would be least willing to forego. Though genetic diseases have on the whole been among the last to yield to therapy, many examples now show that this picture is changing. The overall impact of genetics on health is reviewed in this chapter. Present possibilities for therapy, as well as future prospects and developments, are assessed. Differences in the incidence of certain genetic diseases among ethnic groups, inherited differences between individuals in response to drugs, and some of the newer possibilities for approaches to genetic manipulation are considered. The problems of medical genetics that concern the whole of society are discussed in Chapter 21.

20.1 An Overall Picture of Genetic Disease

At the beginning of this century, the British physician Archibald Garrod had already pointed out that some inherited recessive disorders can be explained on the assumption that they are due to genetic blocks in specific metabolic pathways—or, as he called them, "inborn errors of metabolism" (see Chapter 3). Albinism, alkaptonuria (the inability to metabolize homogentisic acid, see Fig. 3.11), and two other conditions known at that time, were discussed by Garrod as examples of such inborn errors.

Today, V. A. McKusick's catalog, Mendelian Inheritance in Man *(1975 edition), lists 2,336 conditions known or thought to be determined each by a single mutant gene or pair of mutant alleles (Fig. 20.1).*

*20010 ABETALIPOPROTEINEMIA (ACANTHOCYTOSIS) 331

Features are celiac syndrome, pigmentary degeneration of the retina, progressive ataxic neuropathy, and a peculiar 'burr-cell' malformation of the red cells called acanthocytosis (sometimes incorrectly written 'acanthrocytosis'). Intestinal absorption of lipids is defective, serum cholesterol very low and serum beta lipoprotein absent. Few cases have to date been discovered and almost all have been Jews. Autopsy (Sobrevilla et al., 1964) and biopsy of peripheral nerves show extensive central and peripheral demyelination. Lees (1967) demonstrated that the lipid-free apoprotein of beta-lipoprotein is present in abetalipoproteinemia. The defect must concern formation of the complete macromolecule. See LIPID TRANSPORT DEFECT OF INTESTINE for a disorder with some of the same features as abetalipoproteinemia.

Dische, M. R. and Porro, R. S.: The cardiac lesions in Bassen-Kornzweig syndrome. Report of a case, with autopsy findings. Am. J. Med. 49: 568-571, 1970.

Dodge, J. T., Cohen, G., Kayden, H. J. and Phillips, G. B.: Peroxidative hemolysis of red blood cells from patients with abetalipoproteinemia (acanthocytosis). J. Clin. Invest. 46: 357-368, 1967.

Fredrickson, D. S., Gotto, A. M. and Levy, R. I.: Familial lipoprotein deficiency. In, Stanbury, J. B., Wyngaarden, J. B. and Fredrickson, D. S. (eds.): The Metabolic Basis of Inherited Disease. New York: McGraw-Hill, 1972 (3rd Ed.). Pp. 493-530.

Isselbacher, K. J., Scheig, R., Plotkin, G. R. and Caulfield, J. B.: Congenital beta-lipoprotein deficiency: an hereditary disorder involving a defect in the absorption and transport of lipids. Medicine 43: 347-361, 1964.

Lees, R. S.: Immunological evidence for the presence of Beta protein (apoprotein of beta-lipoprotein) in normal and abetalipoproteinemia plasma. J. Lipid. Res. 8: 396-405, 1967.

Mier, M., Schwartz, S. O. and Boshes, B.: Acanthocytosis, pigmentary degeneration of the retina and ataxic neuropathy: a genetically determined syndrome with associated metabolic disorder. Blood 16: 1586-1608, 1960.

Salt, H. B., Wolff, O. H., Lloyd, J. K., Fosbrooke, A. S., Cameron, A. H. and Hubble, D. V.: On having no beta-lipoprotein. A syndrome comprising a-beta-lipoproteinaemia, acanthocytosis and steatorrhoea. Lancet II: 325-329, 1960.

Schwartz, J. F., Rowland, L. P., Eder, H., Marks, P. A., Osserman, E. F., Hirschberg, E. and Anderson, H.: Bassen-Kornzweig syndrome. Deficiency of serum beta-lipoprotein. Arch. Neurol. 8: 438-454, 1963.

Sobrevilla, L. A., Goodman, M. L. and Kane, C. A.: Demyelinating central nervous system disease, macular atrophy and acanthocytosis (Bassen-Kornzweig syndrome). Am. J. Med. 37: 821-832, 1964.

RECESSIVE

20013 ABSENT EYEBROWS AND EYELASHES WITH MENTAL RETARDATION (PSEUDOPROGERIA SYNDROME)

Hall et al. (1974) reported two brothers with mental retardation, absence of eyebrows and eyelashes, progressive spastic quadroplegia, micrcephaly, glaucoma, small and beaked nose. One had had a 'cervical spinal cyst' removed at age 1 year and the second had occipital cranium bifidum occulatum. The parents were unrelated. They and three brothers were normal.

Hall, B. D., Berg, B. O., Rudolph, R. S. and Epstein, C. J.: Pseudoprogeria — Hallermann-Streiff (PHS) syndrome. Birth Defects Orig. Art. Ser. 10 (7): 137-146, 1974.

*20015 ACANTHOCYTOSIS

Cederbaum et al. (1971) observed a family in which three sibs had developed progressive chorea and dementia and had acanthocytes in the peripheral blood. A brother and sister had died at ages 32 and 39 and the proband was a 41-year-old-male. Both parents were healthy but consanguineous. Two children of the proband were healthy. They suggested that the same disorder may have been present in the family of Critchley et al. (1967, 1970). Although in that family the disorder was thought to be dominant (10050), the inheritance could be recessive.

Cederbaum, S., Heywood, D., Aigner, R. and Motulsky, A.: Progressive chorea, dementia and acanthocytosis: a genocopy of Huntington's chorea. Clin. Res. 19: 177 only, 1971.

*20020 ACATALASEMIA

Acatalasia was first discovered in Japan by Takahara, an otolaryngologist, who, in cases of progressive oral gangrene, found that peroxide applied to the ulcerated areas did not froth in the usual manner. Heterozygotes have an intermediate level of catalase in the blood. The frequency of the gene, although relatively high in Japan, is variable. The frequency of heterozygotes is 0.09 percent in Hiroshima and Nagasaki but is of the order of 1.4 percent in other parts of Japan (Hamilton et al., 1961). Acatalasia has been detected in Switzerland (Aebi et al., 1962), and in Israel (Szeinberg et al., 1963). In both of the latter situations the homozygotes showed some residual catalase activity suggesting that this may be a different mutation than that responsible for the Japanese disease in which catalase activity is zero and no cross-reacting material has been identified. Hamilton and Neel (1963) presented evidence that at least two forms of acatalasia exist in Japan. In an extensive kindred with acatalasia in two sibships, heterozygotes showed catalase values overlapping with the normal. Hypocatalasia has also been found in the guinea pig, dog and domestic fowl (see review by Lush, 1966). Electrophoretic variants of red cell catalase have been described (see dominant catalog). These may be determined by genes allelic with that for acatalasemia. Shibata et al. (1967) found that an immunologically reactive but enzymatically inactive protein about one-sixth the size of active catalase is present in red cells of acatalasemics.

Figure 20.1

Part of a catalog of genetic diseases. This is a single page from V. A. McKusick's *Mendelian Inheritance in Man: Catalogue of Autosomal Dominants, Autosomal Recessives, and X-Linked Phenotypes,* 4th ed. (Baltimore: Johns Hopkins Press, 1975.)

Their modes of inheritance have been classified as 90 percent autosomal (about half dominant and half recessive) and 10 percent X-linked (Fig. 20.2). The great majority of these "Mendelian" traits are diseases. Often, diseases that appear superficially to be similar turn out to be due to mutations at different gene loci. This "genetic heterogeneity" can be established in at least three ways.

1. By pedigree analysis—for instance, there are several types of muscular dystrophy (a disturbance of muscular development and function), some autosomal recessive, some autosomal dominant, and at least two X-linked.
2. The clinical pattern of the disease—for example, the X-linked forms of muscular dystrophy are distinguished from an autosomal dominant type by early versus late age of onset; clinical features other than age of onset further help to distinguish the various types; and sometimes individual pedigrees exhibit a unique disease pattern.
3. Biochemical data can also help to distinguish various forms of a disease. This applies particularly to the large number of types of mental deficiencies, many of which can be distinguished by a biochemical analysis. Phenylketonuria is of course an especially important example of a form of mental deficiency that can be diagnosed biochemically. There are, in addition, many other types of *amino-acidurias* (a name for diseases due to an inability to metabolize an amino acid)—and also, for example, disturbances of blood, fat, and sugar metabolism—that can be a cause of mental deficiency.

Recessive traits, whether determined by autosomal or X-linked genes, are each likely to be due to a defective enzyme with a profound deficiency in activity.

The specific enzyme involved has been identified in almost 150 recessive disorders (141 are listed by McKusick in his 1975 edition). The heterozygote carriers

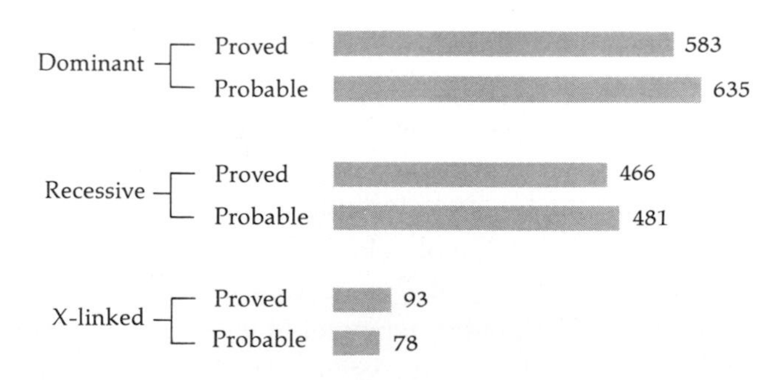

Figure 20.2
Modes of transmission of traits (mostly diseases) known to be inherited in a Mendelian fashion in man. Numbers are those of known traits ("proved") and of less well-known ones ("probable"). (Data from V. A. McKusick.)

of an allele determining an inactive enzyme usually have about half the amount of enzyme that is found in the normal homozygote. But, as discussed in Chapter 3, the activity of an enzyme metabolizing its substrate is not necessarily proportional to the amount of enzyme, unless the enzyme is present in very small amounts (see Fig. 3.33). Thus, the carrier heterozygote, although he has half the normal amount of enzyme, usually metabolizes the substrate almost as efficiently as the normal homozygote. Natural selection appears to have adapted diploid organisms to produce enough enzyme per gene for one copy of an active gene in an individual to be sufficient for ordinary needs. It is for this reason that enzymatic defects, which constitute the majority of inborn errors of metabolism, are recessive traits.

Dominant conditions are more frequent in man, relative to recessives, than is the case in laboratory animals such as Drosophila and mice.

The reason is fairly simple: in laboratory animals, it is quite easy to inbreed and so reveal recessive defects more readily. In man, high levels of inbreeding are exceptional, as discussed in Chapter 11, and so the majority of recessives remain hidden. Support for this explanation is indicated by the facts that (1) a disproportionately high frequency of traits determined by the X chromosome has been described (93 out of 1,142, or 8.1 percent of the total, while the X chromosome contains only about 5 percent of the total DNA); (2) most known X-linked traits, unlike the autosomal ones, are recessive. This must be a consequence of the greater ease with which genes on the X chromosomes, particularly those determining recessive traits, can be identified by pedigree analysis.

The biochemistry of most dominant defects is poorly understood.

Dominant defects are, however, less likely to be due to changes in the enzyme molecule itself. One example in which a protein defect has been identified is methemoglobinemia, which is a form of hemoglobin abnormality. It arises from the substitution of one of two amino acids, in either the alpha or the beta hemoglobin chains, which link the heme group (to which oxygen is attached) to the protein chain (Fig. 20.3). The abnormal amino acid present in the mutants does not allow the proper functioning of the heme–oxygen relationship. Thus, the mutant hemoglobin tends to remain in an irreversible oxidized state and cannot carry oxygen properly. It also has a different color from normal hemoglobin, and as a result, individuals with this hemoglobin have a peculiar bluish discoloration of their skin. In heterozygotes, half of the hemoglobin molecules are malfunctioning, and this is enough to cause disease, and so give rise to a dominant genetic defect.

Homozygotes for most genes determining dominant defects are almost never seen. This is, in part at least, because matings that produce homozygotes are

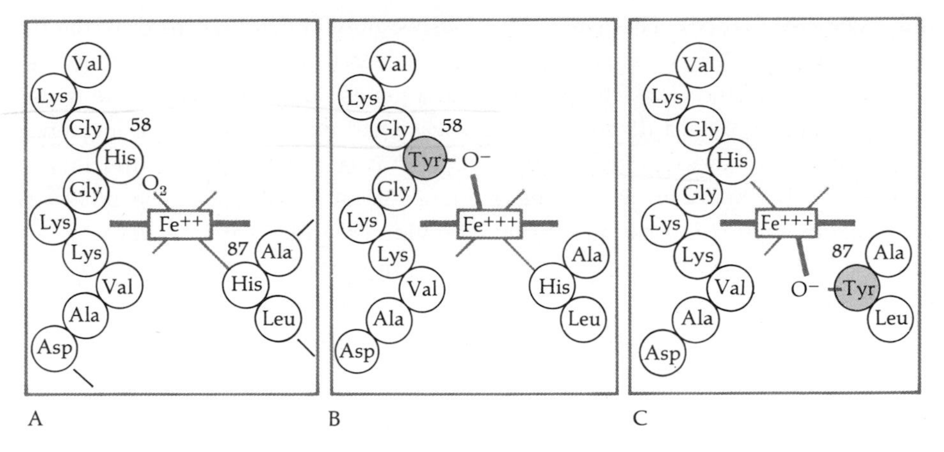

Figure 20.3
Amino-acid positions in α hemoglobin chains to which the heme group is attached. (A) Normal oxygenated hemoglobin. (B) A hemoglobin variant (methemoglobinemia M Boston) in which tyrosine has substituted for the histidine normally at position 58. The tyrosine structure differs from that of histidine in such a way as to allow formation of a complex between the heme iron (now in a ferric state), a single oxygen atom, and the side of the chain (R group) of the tyrosine. The resultant complex interferes with normal function. (C) Another hemoglobin variant, similar to meth. M Boston, except that the tyrosine is substituted for the histidine at position 87. (After H. Harris, *The Principles of Human Biochemical Genetics,* North Holland Pub. Co., 1970.)

rarely, if ever, formed; both parents must be heterozygotes for the same rare gene. It is also likely, in addition, that homozygosity for genes that determine a dominant defect would be a lethal condition. This is very probably the case, for instance, with achondroplastic dwarfism, in which putative homozygous children of two affected parents die *in utero* or in the neonatal period with an extremely severe form of the disease. Dominant disorders may be associated with alterations in regulatory mechanisms operating at the level of the gene or at the metabolic level (see Chapters 3 and 9).

Most genetic diseases are rare, and many may have been observed only once—which is usually not enough to establish their mode of inheritance.

Many genes can probably mutate in such a way that a recognizable defect is eventually produced; that is, they can mutate to give rise to a deleterious allele, whether dominant or recessive. There must be at least hundreds of thousands of different genes in a human individual, so that the variety of possible genetic diseases is almost unlimited. Moreover, different alleles of a gene may produce different diseases. Hemoglobin-gene mutations can give rise to a variety of diseases, ranging from hemolytic anemias (in which there is instability of hemoglobin and therefore extensive destruction of red cells) to methemoglobinemias and other disturbances. The genetic diseases we now know must be only the tip of the iceberg (Fig. 20.4). Each mutant gene, however, usually tends to remain rare.

As we saw in Chapter 8, the incidence of a genetic disease is twice the mutation rate, for a dominant disorder, if the disease has a zero fitness. The incidence can be higher if fitness is greater than zero—that is, if lethality or infertility is incomplete. The factor by which the incidence is increased above the mutation rate depends on the fitness of the affected individuals (and of carrier heterozygotes in the case of a recessive). In general, however, most diseases determined by one mutant allele tend to remain rare. Only under special circumstances—for example, if the disease is recessive and heterozygotes show increased fitness compared to either type of homozygote—will it be likely to become more frequent. This is, of course, the case for sickle-cell anemia and thalassemia (see Chapter 9). Another reason for a relatively high incidence of a genetic disease is multiplicity of genetic causes. What seems superficially to be a single phenotype may be the sum of many

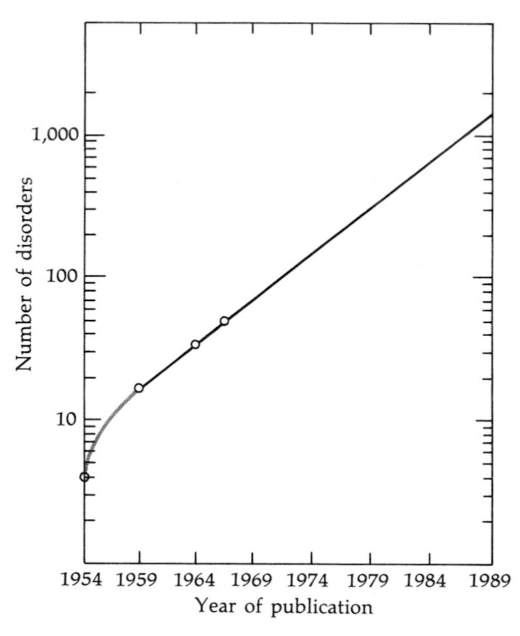

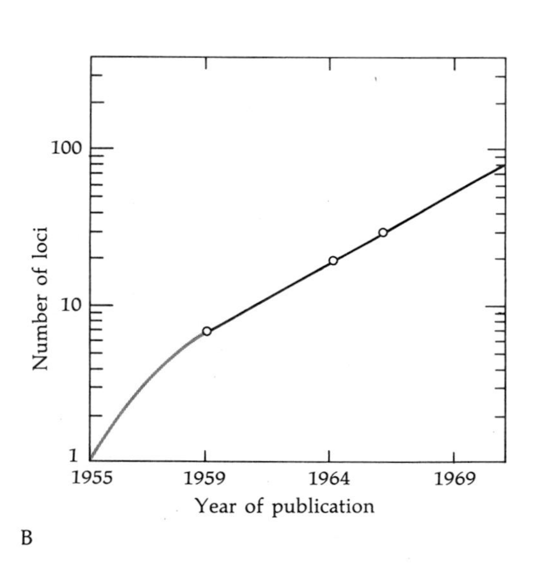

Figure 20.4
Rate of discovery of new genetic diseases. (A) The number of enzyme-deficiency disorders discovered (year of publication is used). Note that the vertical scale is logarithmic. Knowledge continues to rise exponentially; data for recent years show that the growth continues to follow the line drawn here. (B) The number of known loci specifying human protein variants. Again, the increase is exponential. (From B. Childs, *American Journal of Diseases of Childhood,* vol. 114, p. 465, 1967. For data from recent years, see B. Childs, *Yale Journal of Biology and Medicine,* vol. 46, pp. 297–313, 1973.)

different genetic, perhaps clinically indistinguishable, diseases. This is the case for blindness, congenital muscular dystrophy, deafness, and unclassified mental deficiency.

Apart from "gene mutations," which account for the Mendelian catalog of disease in man, two other important categories of genetic diseases should be remembered: chromosomal aberrations and complex multifactorial genetic diseases.

Chromosomal aberrations are described in detail in Chapter 4, and here we should only remember that many chromosome abnormalities are sufficiently serious that their carriers usually do not reproduce, so that almost all cases are due to new "mutations." The syndromes they generate are complex, representing the outcome of imbalance for a relatively large number of genes, because most chromosomes may contain at least tens of thousands of genes. Some chromosome aberrations tend to appear more frequently, such as trisomy 21 and the sex-chromosome abnormalities, and these form a large fraction of all the observed aberrations at birth. As pointed out in Chapter 8, about 0.31 percent of births are affected by autosomal aberrations, of which more than one-third are Down's syndrome, and about 0.20 percent are affected by X-chromosome abnormalities. Many other aberrations are very rare, and may have only ever appeared once.

The other category of genetic diseases, those with a complex inheritance pattern, includes a wide variety of relatively frequent and severe diseases, such as schizophrenia, diabetes mellitus, congenital dislocation of the hip, cleft lip and/or palate, and other congenital malformations and chronic illnesses. In general, it is reasonable to expect that many of these diseases will be heterogeneous genetically. If genetic heterogeneity can be resolved by pedigree analysis, and studies of the clinical and biochemical patterns of the disease reveal several genetic entities, then the problem of understanding the nature of the inheritance of a complex genetic disease may be considerably clarified. Biochemical analysis is, in particular, most useful when incomplete penetrance of the gene is the explanation for an apparently complex pattern of inheritance, because the biochemical defect may not always be accompanied by clinical manifestations. Even in phenylketonuria, which has a very high penetrance, there are a few homozygotes who, though they manifest the biochemical abnormality, develop normal intelligence. Levels of phosphate in the blood can be used to help in the diagnosis of vitamin D-resistant rickets, a dominant disorder determined by an X-linked gene. The presence of this gene, detected by low serum phosphate concentrations (Fig. 20.5), does not always lead to overt signs of rickets.

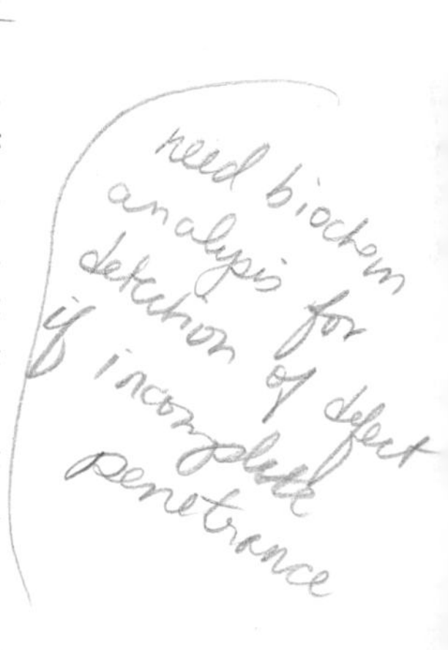

An even more striking case involves the protein, alpha-1-antitrypsin. This is a normal blood protein that is identified by its capacity to neutralize the activity of the enzyme trypsin. The function of trypsin, which is normally produced in

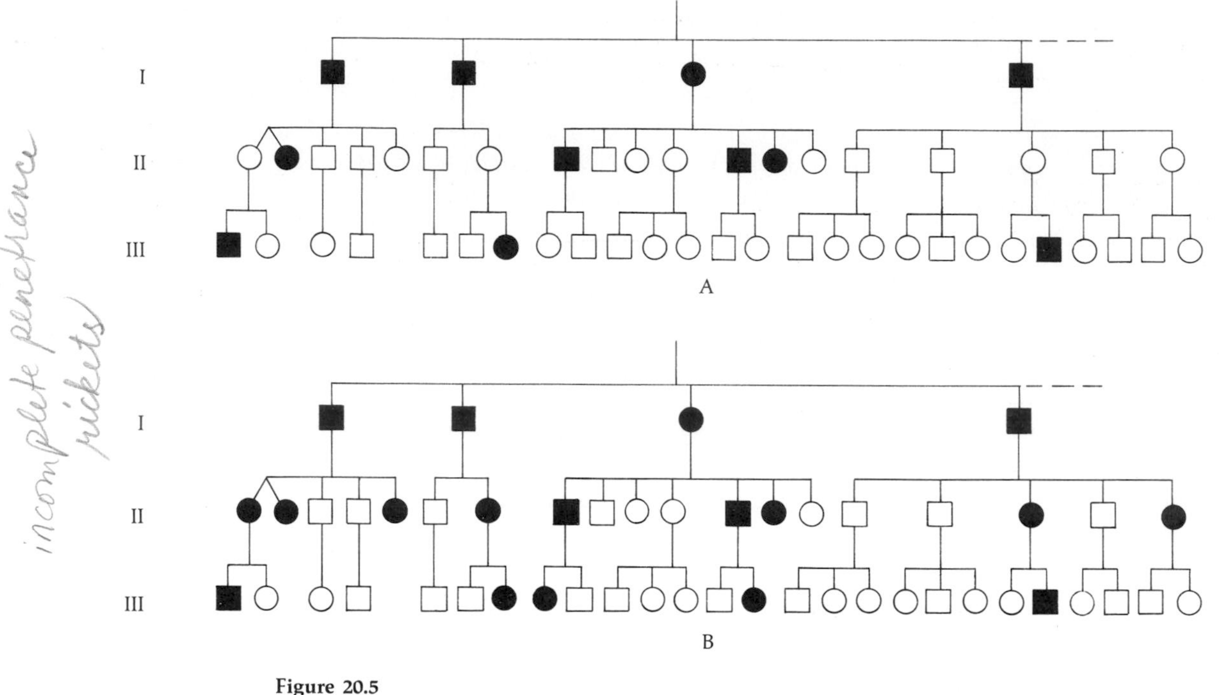

Figure 20.5
Pedigree of hypophosphatemic (vitamin D–resistant) rickets. Identification of the biochemical lesion involved in a disease often allows clarification of the pattern of transmission. (A) Pedigree prepared using skeletal deformity as the disease phenotype; the mode of inheritance is unclear. (B) Pedigree prepared using low serum phosphate as the phenotype; an X-linked dominant trait is clearly indicated. The skeletal deformity is a secondary effect, not always present. (For sources of data, see V. A. McKusick, *Human Genetics,* Prentice-Hall, 1969.)

the pancreas and secreted into the small intestine, is to digest dietary proteins. There exist several alleles of the alpha-1-antitrypsin gene, some of which give rise to considerably reduced activity. A high frequency of patients who have pulmonary emphysema (a loss of elasticity of the lungs that results in progressively more severe pulmonary disability) show antitrypsin deficiency. Not all the affected are, however, homozygotes for the allele that is associated with antitrypsin deficiency; some heterozygotes also appear to be affected. Furthermore, this is certainly not the only cause of emphysema, but only one out of many. It is not difficult to see why there can be many causes of emphysema; but why are some individuals genetically susceptible, while others are not? There must be other factors involved, some of which are likely to be environmental (such as smoking), which interact with genetic factors to cause emphysema.

What is the overall impact of genetic factors on disease?

Somewhat less than 1 percent of all individuals born are suffering from some serious genetic disease that is probably determined by a strictly Mendelian trait. Tables 20.1, 20.2, and 20.3 give approximate incidences of major genetic diseases among Caucasian populations (also see Fig. 20.6). Chromosomal aberrations also are responsible for an additional incidence of somewhat less than 1 percent of serious diseases among newborn individuals. However, as discussed in Chapter 8, these aberrations contribute a great deal to spontaneous abortions and may, in fact, be responsible for up to one-third of these. The number of recognizable spontaneous abortions is as high as 15 to 20 percent of the number of births, so at least 4 to 5 percent of all conceptions probably involve chromosomal abnormalities. It has been estimated that as many as 25 or 30 percent of all fertilized zygotes may be lost through spontaneous abortions (with most of those occurring very soon after fertilization being undetected). Spontaneous abortion is an important factor in natural selection, eliminating many grossly unfit zygotes.

Stillbirths (deaths of a fetus that occur in the last two or three months of gestation) and neonatal and early infant deaths (those occurring in the first few months of life) are also often due to genetic defects, but they are much less significant numerically than spontaneous abortions. Although many congenital malformations have a major genetic component, others do not, resulting from a virus or other infection, or from an environmental insult affecting the fetus during pregnancy. "Congenital" should not be taken to be synonymous with "genetic"; it only

Table 20.1
Approximate incidences at birth of the most common dominant diseases

Disease	Incidence per million births
Deafness (many different forms)	100
Blindness (many different forms)	100
Huntington's chorea (late-onset disorders of motion and speech, and destruction of brain)	100
Epiloia (tendency to form certain tumors in skin and other tissues)	50
Marfan's syndrome (abnormalities of connective tissue)	50
Achondroplasia (or chondrodystrophy; form of dwarfism)	50
Neurofibromatosis (skin tumors)	50
Dystrophia myotonica (muscle wasting)	50
All other dominant diseases	150
Total	800

NOTE: Because of the low frequencies of most genetic diseases, it is very difficult to obtain reliable estimates of their incidences. The estimates given in this and the following tables are based on a variety of sources relating to the Caucasian population and must be taken as only approximate. As discussed in the text, some of the diseases may have incidences that vary considerably from one population to another. We are grateful to J. H. Edwards for providing the information on which many of these frequency estimates are based.
SOURCE: After A. Jones and W. F. Bodmer, *Our Future Inheritance*, Oxford Univ. Press, 1974.

Table 20.2
Approximate incidences at birth of the most common recessive diseases

Disease	Incidence per million births
*Severe mental defects (exluding aminoacidurias)	800
*Deafness (severe)	500
Cystic fibrosis	400
*Blindness	200
*Adrenogenital syndromes	100
Albinism	100
Phenylketonuria (PKU)	100
*†Treatable aminoacidurias (excluding PKU)	50
*†Untreatable aminoacidurias (excluding PKU)	50
‡Mucopolysaccharidoses (all forms)	50
Tay–Sachs disease	10
Galactosemia	5
Total	2,365

*These diseases exist in multiple forms.

†Aminoacidurias are diseases like PKU that are caused by accumulation of an amino acid because of an enzyme block due to an inborn error of metabolism. Some 13 different diseases are included under these two headings.

‡Mucopolysaccharidoses are diseases like Hurler's syndrome that are caused by a defect in the metabolism of certain types of complex sugars (the mucopolysaccharides), whose accumulation in various tissues causes a variety of gross developmental abnormalities.

NOTE: See footnote to Table 20.1.

SOURCE: After A. Jones and W. F. Bodmer, *Our Future Inheritance*, Oxford Univ. Press, 1974.

Table 20.3
Approximate incidences at birth of the most common X-linked diseases

Disease	Incidence per million births
Duchenne's muscular dystrophy	200
Hemophilia (A and B)	100
All other X-linked diseases*	200
Total	500

*The Hunter and the Lesch–Nyhan syndromes, both of which can be detected *in utero*, occur with frequencies of a few births each per million.

NOTE: See footnote to Table 20.1.

SOURCE: A. Jones and W. F. Bodmer, *Our Future Inheritance*, Oxford Univ. Press, 1974.

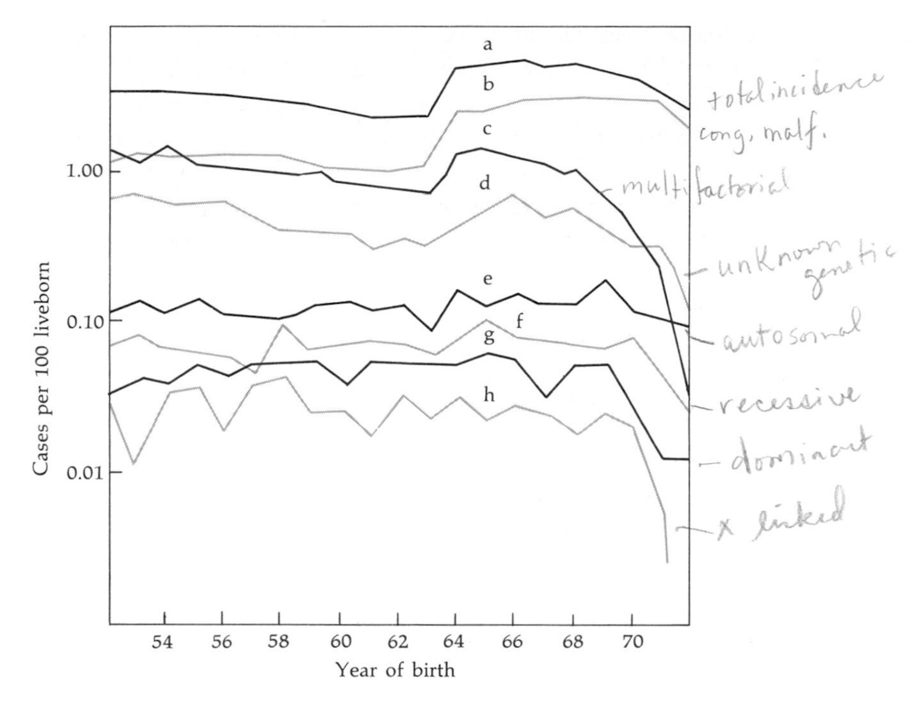

Figure 20.6
Frequency of genetic diseases among liveborn in British Columbia. Frequencies are given as the number of affected individuals per 100 liveborn. Line a shows the total incidence of genetic disease; b shows congenital malformations; c shows other multifactorial diseases; d shows diseases of unknown genetic mode; e shows diseases determined by autosomal chromosomes; f shows recessive diseases; g shows dominant diseases; and h shows X-linked diseases. Data for births from 1964 through 1972 include records from both the congenital anomalies surveillance system and the handicap registry; data for earlier years are obtained only from the handicap registry. Note that these data differ somewhat from estimates given in Tables 20.1, 20.2, and 20.3. For possible sources of bias in these and other data, see the original publication. This study is a first attempt at a systematic, ongoing, statistical collection of data on genetic diseases. (From B. K. Trimble and J. H. Doughty, *Annals of Human Genetics*, vol. 38, p. 199, 1974.)

means observable at birth. Many genetic defects are not detectable at birth, while many congenital defects, as already pointed out, are not genetic.

Apart from these genetic causes of death or severe disease, there are many other diseases in which genetic factors are likely to be involved.

Even for infectious diseases, there is likely to be some inherited liability (or, conversely, resistance), as discussed in Chapter 10. Heart and vascular diseases, which are nowadays the leading group of causes of death in developed countries, are (in part at least) genetically determined. The most important single such cause of death is ischemic heart disease, which accounts for about one-quarter of all

Heart disease

deaths in developed countries. Among survivors of myocardial infarctions occurring before 60 years of age, about 20 percent are heterozygous for a gene causing hyperlipidemia—that is, an increase in the lipo (fat-containing) proteins of the blood and other tissues—and a majority of those who die at an early age are homozygous for this same gene.

Cancer is the second leading group of causes of death in developed countries. Some tumors, such as retinoblastoma, are simply inherited as single Mendelian traits, but these are very rare. Some cancers may be due to viruses, but even then, genetic differences in susceptibility may play an important role. Allergic diseases—in particular ragweed hay fever, one of the common pollen allergies—are probably associated with particular HLA factors, so that they too must be to a significant extent genetically determined. A number of autoimmune-type diseases (in which antibodies are made against one's own tissues or organs), and also some diseases that are known or presumed to be the consequence of slow virus infections, also show close associations with HLA or HLA-linked factors, as discussed in Chapter 10. These also, therefore, must be to some extent genetically determined.

It has, for example, been estimated that between 10 and 20 percent of admissions to pediatric hospitals in North America are connected with overt genetic disease. Thus, one way or the other, genetic causes may account for a very substantial fraction of our disease load, as indicated in Table 20.4, which represents an attempted synthesis based on approximate Caucasian incidence data. Future research is certain to alter these figures substantially, undoubtedly in an upward direction.

Table 20.4
An attempted synthesis of the overall impact of genetic disease

Type of genetic disease		Incidence (percentage of all conceptions or births)
Deleterious (single-gene) effects:	Dominant	0.08
	Recessive	0.2
	X-linked	0.05
	Total	0.33
Chromosome abnormalities		0.54
Complex genetic traits*		(2.0)
Congenital malformations†		(1.2)
Contribution to other diseases (heart disease, cancer, etc.)		?‡

*Including schizophrenia, manic–depressive psychosis (bipolar), diabetes, etc.
†This incidence includes one-half the total of all such malformations, assuming that there is an average genetic contribution of about 50 percent to such cases.
‡This value is certainly more than several percent.

20.2 Treatment of Genetic Disease

Medicine has had its most outstanding success in fighting infectious diseases. These were the major causes of mortality until a few decades ago in the Western countries, and still are in developing countries. Their control through hygiene, chemotherapy, immunization, and antibiotics has decreased prereproductive mortality from about 50 percent almost to 2 percent, a 25-fold decrease. Among the remaining diseases, those with a genetic basis have formed—and still form—a hard core of afflictions whose therapy is much more difficult to handle. But, nevertheless, much progress has been made.

One of the first inherited diseases to be treated was *diabetes mellitus.* This disorder is, as we have seen in Chapter 16, not very clearly defined genetically, but the inherited element is sufficient to consider diabetes as a predominantly genetic disease. Some forms of the disease have recently been shown to be associated with the HLA system. It is likely that multifactorial inheritance—involving genes at a number of loci, which interact with specific environmental factors—accounts for most, if not all, cases of diabetes mellitus. A diet rich in carbohydrates might, for example, precipitate the occurrence of the disease only in certain genetically susceptible individuals. Many of the clinical manifestations of the disease result from inadequate production (or other reasons of functional insufficiency) of the hormone *insulin* (Fig. 20.7), which is made by specialized cells of the pancreas. One of the important medical discoveries of the early twenties was the isolation of insulin and the demonstration that its use could usually permit diabetic patients to live in a relative state of health.

The treatment of diabetes mellitus by insulin is the first, and perhaps most important, example of substitutional therapy, in which a gene product that is not present in sufficient amounts is administered from the outside.

There are numerous therapeutic examples in which a deficient gene product can be replaced. In hemophilia A, a specific protein factor needed for normal blood clotting is defective or is produced in insufficient amounts. Hemophilic patients must receive injections of the factor to treat bleeding episodes, and prophylactic administration permits these individuals to undergo surgery, and may even prevent hemorrhages. Whereas insulin is a relatively simple molecule, which has actually been experimentally synthesized and can be obtained from animals for human use, the antihemophilic factor is a large protein, still well beyond modern synthetic capabilities. Only the human protein can be used if adverse reactions are to be avoided. The supply of this factor, obtained from blood of normal individuals, is inevitably limited, and the treatment costly. In severe hereditary endo-

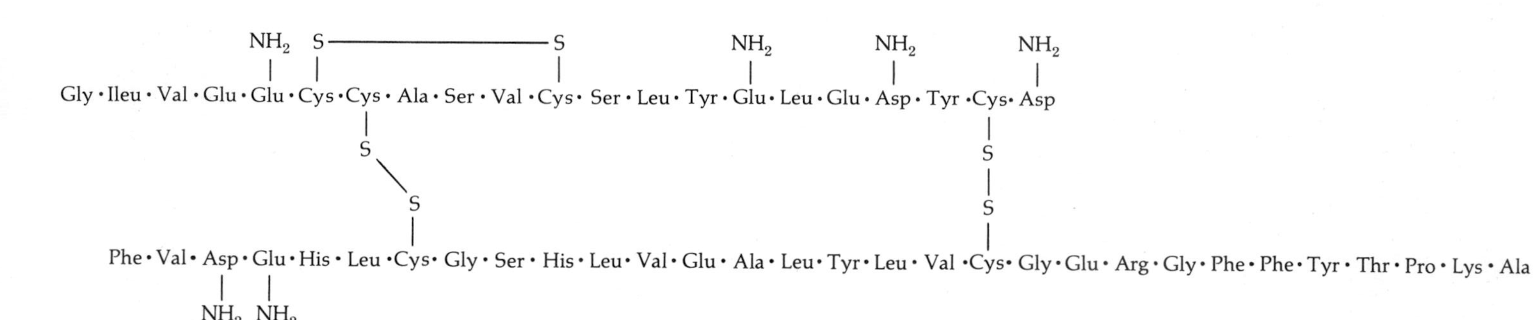

Figure 20.7
The insulin molecule. Insulin is composed of 51 amino acids. The molecule consists of two chains, joined together by sulfur atoms (S–S, or disulfide, bridges). The chain shown at top here is called the glycol chain because it begins with glycine; the other chain is called the phenylalanyl chain.

crine disorders which are due to an insufficient production of a hormone, administration of the hormone itself may be sufficient to control the disease. Often, as in the case of growth hormone (which has, however, recently been synthesized in the laboratory), it is once again necessary to use a human rather than an animal source.

Substitutional therapy is not successful in every case in which the lack of an identified gene product is responsible for the disease. Thus, in phenylketonuria (PKU), simple administration of the missing enzyme would probably not be sufficient. The enzyme is produced and localized in the liver, and so far there is no known way of introducing it from the outside into the liver cells. After simple injection, it would probably be rapidly lost from the blood. Moreover, enzyme provided from the outside, if obtained from animals, might soon give rise to antibodies that could cause a severe immunological reaction and would, in any case, rapidly eliminate the injected enzyme. PKU has, of course, been successfully treated by a dietary treatment involving a reduction in the intake of phenylalanine (see Chapter 3). Dietary treatments, and administration of specific substances designed to alleviate the consequences of disease, have been tried successfully in a variety of genetic diseases. For example, avoidance of galactose, lactose, and fructose, respectively, alleviates the problems associated with galactosemia, lactose intolerance, fructosemia. Restriction of copper and iron intake helps for Wilson's disease and hemochromatosis, respectively, these being diseases in which the control of utilization of these metals is impaired.

A knowledge of the biochemical basis of a defect is very important for devising a reasonably effective therapy. In some of these diseases, such as diabetes and hemophilia, the treatment can be given after the development of the symptoms of the disease. In others, such as PKU, it cannot; to be effective, the treatment for PKU must be started as early in life as possible, because the developing brain is sensitive to the toxic effect of excessive amounts of phenylalanine derivatives. This requires that children at risk be tested long before development of severe symptoms, and in practice, during the first weeks after birth. Genetic screening of all newborns for PKU has now been initiated in many countries in order to deal with this problem (Fig. 20.8). The social issues raised by such a program are discussed in Chapter 21.

Prophylactic treatment also plays a major role in the treatment of rhesus incompatibility, as discussed in Chapter 10.

The treatment is very effective (Fig. 20.9). If treated with anti-Rh immunoglobulin within three days of their first delivery of an Rh-positive child, only 2 percent of Rh-negative women develop antibodies, as compared to 17 percent of untreated

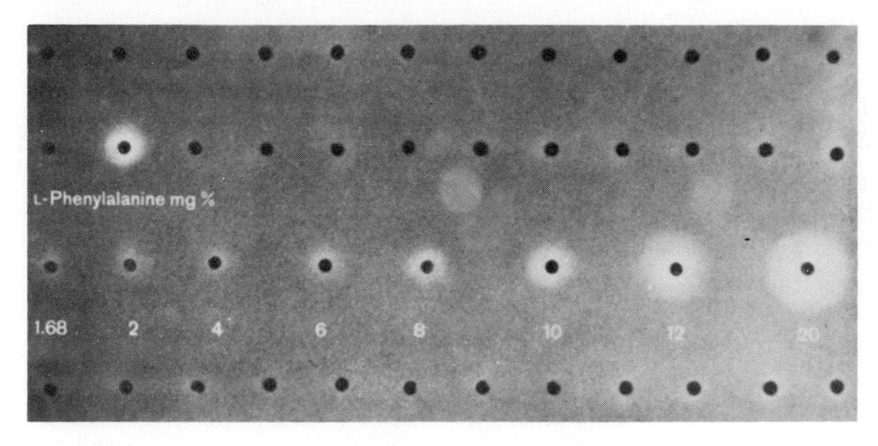

Figure 20.8
The Guthrie PKU-screening test. Each black spot is a disk impregnated with a blood sample. The third row is a series of controls impregnated with various known concentrations of phenylalanine (as indicated below each disk). A positive reaction consists of a halo, which represents inhibition of bacterial growth in the neighborhood of the disk. A positive reaction is seen in the second row. (From A. Jones and W. F. Bodmer, *Our Future Inheritance,* Oxford Univ. Press, 1974.)

women. The anti-Rh immunoglobulin is prepared using carefully selected Rh-negative *male* blood donors, who are injected with Rh-positive red cells. This treatment is important only in Caucasian populations, because the frequency of Rh-negatives is very low in other groups. It has been computed that one volunteer Rh-negative male donor can satisfy the needs of one million Caucasians. When prophylaxis fails, or if a woman is already immunized against the Rh factor by

---

Figure 20.9
The Rh prophylactic procedure. (A) Maternal and fetal red blood cells (RBC) in an Rh-incompatible but ABO-compatible pregnancy. (B) At the time of delivery, there is a leakage of fetal RBC into the maternal bloodstream. The Rh group is a protein foreign to the maternal lymphocytes, and its presence stimulates stem cells, leading to production of an anti-Rh antibody. (C) In subsequent pregnancies, the anti-Rh antibody crosses the placenta, where it can complex with Rh^+ fetal RBC, causing destruction of the fetal RBC. (D) Maternal and fetal RBC in an Rh-incompatible *and* ABO-incompatible pregnancy. (E) Leakage of fetal RBC does not stimulate production of anti-Rh antibodies, because of the extremely rapid neutralization of the fetal cells by the anti-B antibodies normally present in the bloodstream of a B^- individual. (Similarly, people who are B have anti-A antibodies, and people who are O have both anti-A and anti-B antibodies.) (F) Subsequent pregnancies continue uninterrupted because anti-Rh antibodies were never produced. (For unknown reasons, anti-A and anti-B antibodies do not usually cross the placenta.) (G) Rh prophylaxis is thought to create a situation similar to that shown in parts D through F. An injection of anti-Rh antibodies just before delivery neutralizes any fetal cells before production of anti-Rh antibody is stimulated. The injected antibodies are soon destroyed in the mother's body, and no further production of such antibodies has been triggered. (After C. A. Clarke, in *Medical Genetics,* ed. V. A. McKusick and R. Claiborne, H.P. Pub. Co., 1973.)

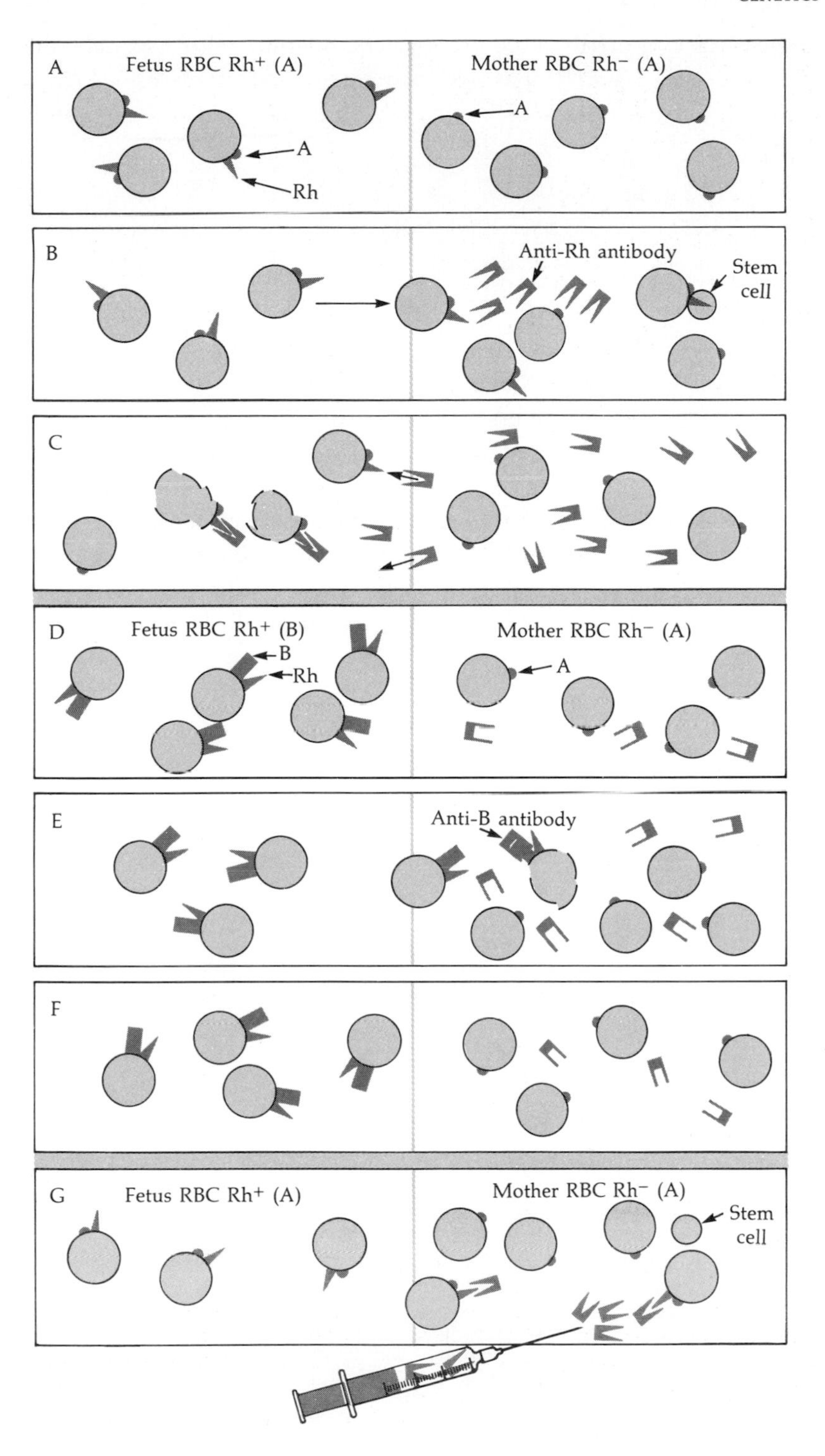
A
Fetus RBC Rh+ (A)
Mother RBC Rh− (A)
A
Rh
A
B
Anti-Rh antibody
Stem cell
C
D
Fetus RBC Rh+ (B)
B
Rh
Mother RBC Rh− (A)
A
E
Anti-B antibody
F
G
Fetus RBC Rh+ (A)
Mother RBC Rh− (A)
Stem cell

pregnancies that occurred before prophylactic treatment was available, there is a good chance of saving an Rh-positive baby by exchange transfusion, a procedure in which the newborn's blood is replaced with donor's blood shortly after birth.

In the case of a number of genetic diseases for which there is, so far, no satisfactory treatment, there is still another major approach available: the disease may be detected by suitable testing of the fetus at a stage of development early enough to allow the possibility of carrying out an induced abortion.

In this "treatment," the affected individual is not cured, but is removed from the population well before birth. The procedure that has made this possible is known as amniocentesis; a sample of up to 10 or 15 milliliters of the amniotic fluid that surrounds the fetus and that contains cells of fetal origin is collected using a needle and syringe. The fluid and the cells so obtained can be examined in various ways and the cells cultured for further tests (Fig. 20.10). The procedure can be carried out at 14 to 16 weeks of pregnancy, and so far appears to carry, at most, a low risk for mother and fetus. This leaves, on the whole, enough time for tests to be carried out within the period during which a pregnancy can legally be terminated (in those countries that allow induced abortions to be carried out), even if the cells first must be cultured before chromosome or biochemical analysis can be done. If the fetus is found to be afflicted with a serious genetic disease, then the mother can be offered the possibility of an abortion.

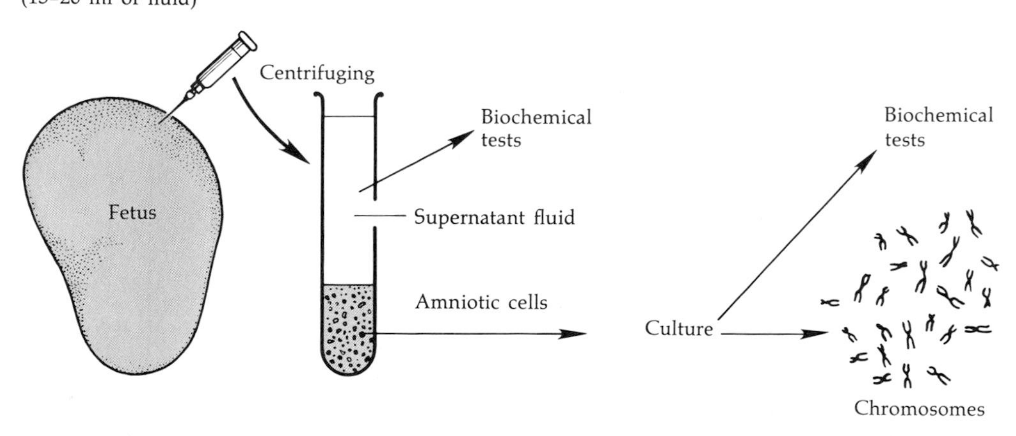

Figure 20.10
Amniocentesis procedure. Cells of the amniotic fluid can be cultured and subjected to karyotype analysis and to biochemical tests. This procedure permits prenatal detection of enzymatic and chromosomal inherited disorders. (From A. Jones and W. F. Bodmer, *Our Future Inheritance*, Oxford Univ. Press, 1974.)

What diseases can be detected following amniocentesis, and so can be treated by a "prophylactic" abortion?

The major group of relevant genetic diseases are the chromosomal aberrations—many of which, of course, give rise to very serious defects, including Down's syndrome, which cannot be cured. A second category that can be detected biochemically following amniocentesis includes those inborn errors of metabolism that are not yet treatable (see Table 20.5 for a list of detectable diseases). The availability of the technique of amniocentesis has thus led to considerable development of new methods for the examination of fetuses.

There is a third group of diseases that can now be detected *in utero* by amniocentesis. These are the neurological developmental abnormalities, anencephaly and spina bifida, which form a substantial fraction of the congenital malformations, having a total birth incidence of about 3 per 1,000. These diseases are not simply inherited, but come into the category of complex genetic diseases that have a significant genetic component, as well as being subject to major environmental influences. It was found recently that, when the fetus is affected by spina bifida or anencephaly, a protein called α-fetoprotein, which is not usually present except at very low levels, is released into the amniotic fluid in relatively large amounts. This protein can readily be detected in samples of the amniotic fluid using a simple immunodiffusion technique (see Fig. 5.21), in time for an abortion to be possible. More recent data have even suggested that a test for α-fetoprotein carried out on a maternal blood sample might work for the early detection of these severe congenital malformations. Unfortunately, these tests seem to be most reliable with the more severe forms of the malformations, which are, in any case, often stillborn. The "milder" forms, which may be missed even with the test on amniotic fluid, are the ones that cause the parents most anguish, because they are likely to survive but may be severely disabled, both physically and mentally.

A major problem is posed by the fact that the majority of pregnancies are normal, and that only a relatively small fraction really need undergo amniocentesis.

This raises considerable practical problems of social as well as medical concern. Even though amniocentesis carries, at most, a low risk, and is relatively simple, there are not (so far, at least) sufficient grounds nor facilities for carrying it out on every pregnancy. A decision has to be made as to which pregnancies should be picked out for testing. This decision depends on a variety of factors, including the risk of occurrence, incidence, and severity of the disease. These and other social issues connected with genetic screening and amniocentesis programs are discussed in Chapter 21.

Table 20.5
List of some recessive diseases that have been (or in principle could be) recognized before birth

*Adrenogenital syndrome
Arginosuccinic aciduria (AA)
Citrullinemia (AA)
Congenital erythropoietic porphyria (abnormality in formation of an essential nonprotein part of the hemoglobin molecule)
Cystinosis (AA)
Fabry's disease (disease of glycolipid fat storage, due to inability to break down certain types of fat)
*Galactosemia (inability to use the sugar galactose; can be countered by a galactose-free diet)
Gaucher's disease (fat-storage disease)
Generalized gangliosidosis (fat-storage disease)
Glycogen-storage diseases II, III, and IV (Pompe's disease; inability to break down the glycogen–sugar complex that is used for the storage of glucose)
Homocystinuria (AA)
Hunter's syndrome (MP)
Hurler's syndrome (MP)
Hyperammonemia (defect in ability to use protein-breakdown products as an energy source; patients dislike foods containing protein)
Hyperlysinemia (AA)
I-cell disease (MP; fat-storage disease)
Juvenile GM_1 gangliosidosis (fat-storage disease)
Ketotic hyperglycinemia (AA)
Lesch–Nyhan syndrome
Lysosomal acid phosphatase (abnormality of nucleic-acid breakdown)
Mannosidosis (inability to use the sugar mannose; only one case described)
Maple-syrup urine disease (AA)
Metachromatic leukodystrophy (MP)
Methylmalonic aciduria (defect in ability to use protein-breakdown products)
Morquio syndrome (MP)
Niemann–Pick disease (fat-storage disease)
Ornithine-alpha-ketoacid transaminase deficiency (inability to use protein-breakdown products)
Orotic aciduria (defect in nucleic-acid metabolism)
Pyruvate decarboxylase (defect in energy metabolism)
Refsum's disease (fat-storage disease)
Sandhoff's disease (similar to Tay–Sachs disease)
San Filippo syndrome (MP)
Scheie syndrome (MP)
*Tay–Sachs disease
Xeroderma pigmentosum (inability to repair damage caused by ultraviolet light)

See notes at bottom of opposite page.

20.3 Genetic Counseling

Increased knowledge about genetic diseases and increased sophistication in the standards for their diagnosis and treatment or prevention has led to the development of a new medical speciality: medical genetics. As a result, specialized units have been formed, which are now found in many large medical centers, to deal with the particular problems of genetic diseases and to provide a counseling service for parents who seek advice about potential genetic risks to their future offspring. These are often called genetic counseling centers.

According to F. C. Fraser and colleagues, "genetic counseling is a communication process which deals with the human problems associated with the occurrence, or the risk of occurrence, of a genetic disorder in a family.

"This process involves an attempt by one or more appropriately trained persons to help the individual or family to (1) comprehend the medical facts, including the diagnosis, the probable course of the disorder, and the available management (the term 'burden' is sometimes used to describe these clinical aspects of the disease relevant to genetic counseling); (2) appreciate the way heredity contributes to the disorder and the risk of recurrence in specified relatives; (3) understand the options for dealing with the risk of recurrence; (4) choose the course of action which seems appropriate to them in view of their risk and their family goals, and act in accordance with that decision; and (5) make the best possible adjustment to the disorder in an affected family member and/or the risk of recurrence of that disorder."

Much activity revolves around informing prospective parents about the risks of genetic disease among their future offspring. When an affected child is born in a family, medical advice is often sought. Treatment of the child may fall within the province of the primary physician or general practitioner, but the parents may be worried about the possibility that another child will also be affected. This is generally the major reason for referral to a genetic counseling center (Fig. 20.11). Another possibility is that there may be a defect in a close relative of a potential

---

*Incidences of these diseases are given in Table 20.2. For some other diseases, such as the aminoacidurias (AA) and mucopolysaccharidoses (MP), incidences of the complete class of diseases are given in Table 20.2. The remaining diseases occur so infrequently (often just a few isolated families have been described) that no reliable estimates of their incidences can be given.

NOTE: A very brief description of the nature of each disease has been given, except for those discussed in the text. (AA denotes aminoaciduria; MP denotes mucopolysaccharidoses.) The fact that many of these diseases have similar biochemical bases is mainly a reflection of the available techniques for detection before birth. Only those diseases whose bases can be unravelled using available techniques are included in this list; this number is steadily increasing.

SOURCE: A. Jones and W. F. Bodmer, *Our Future Inheritance,* Oxford Univ. Press, 1974.

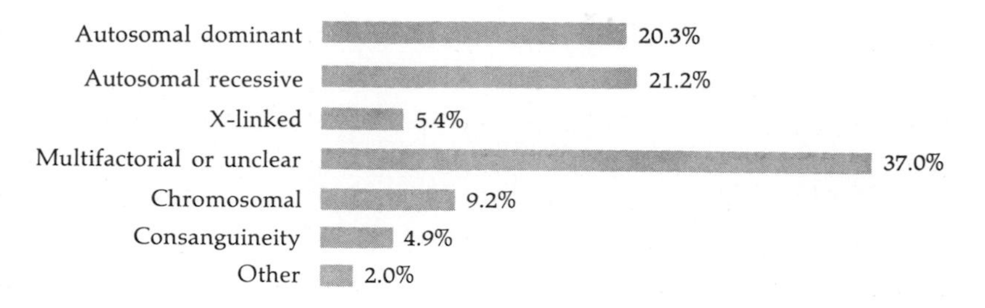

Figure 20.11
Reasons for seeking genetic counseling. Diagnoses for 349 families referred for genetic counseling. (Data from F. C. Fraser in *Ethical Issues in Human Genetics,* ed. B. Hilton et al., Plenum Press, 1973.)

father and mother, and sometimes genetic counseling is sought before marriage. This may happen, for instance, when two close relatives want to marry, or when there is a record of recurring disease in one of the families. Facilities for genetic counseling may be inadequate in areas in which there is a high prevalence of certain diseases—such as is the case, for instance, for thalassemia and for sickle-cell anemia in parts of southern Europe, Asia, and Africa. As we discuss in the next section, certain diseases in addition to the hemoglobin abnormalities are almost entirely confined to particular ethnic groups.

In most cases, people referred to counseling centers want to know the chances of conceiving another affected child.

They can usually be given an estimate of the risk of recurrence. But much further guidance may be needed, including psychological help. Parents of handicapped children often suffer from a painful sense of guilt. The availability of the new techniques for the therapy of genetic diseases can sometimes create undue hopes. Genetic counseling thus requires not only the availability of a suitable laboratory where appropriate diagnostic tests can be carried out to help define risk, but also experts trained in the technical and in the social and psychological aspects of a field that requires considerable communication with those seeking advice. As much of the information given is to help a couple at risk in making a decision about further reproduction, it is essential that the information given should be well understood. One condition for such communication is that the couple should have requested genetic counseling voluntarily.

The information provided by the genetic counselor includes estimates of risk, the extent of availability and effectiveness of treatment (whether prophylactic or not) and the possibility of prophylactic abortion. Opinions vary as to whether the counselor should advise the couple directly to have another child or not, or whether he should simply provide the "facts" and offer no further advice. It is important, of course, to present and explain the facts in a completely unbiased way. In practice, it was found in one follow-up study that, when the verdict was a high risk of another affected child (say, greater than 10 percent), couples refrained from taking the chance in two-thirds of the cases, while when the risk was low (10 percent or less), only one-quarter were deterred.

The estimate of risk is easily made in cases of traits determined by a single mutant gene or pair of mutant alleles, or usually for chromosomal aberrations. For dominant, fully penetrant diseases, one parent having the disease clearly means a 50 percent chance of a diseased child at every birth, which is among the highest recurrence risks. For recessive diseases, two normal parents known to be heterozygotes for the same allele have a chance of $\frac{1}{4}$ of having an affected child with each pregnancy. If, however, normal parents have had no previously affected child, but are concerned because of the existence of diseased relatives, their genotypes may not be clear. For a number of conditions, however, there exist biochemical tests that can detect heterozygotes—for example, by showing a reduced enzyme level, as can be done in the case of heterozygous carriers of the galactosemia gene. X-linked traits, of course, have their own clearcut rules of transmission, the most important of which is that father does not transmit to son.

For genes with low penetrance, and diseases for which the genetic component is ill-defined, it is possible to provide empirical risks based simply on observed incidences of affected offspring under various conditions. It is generally possible nowadays to provide satisfactory risk estimates. All risk estimates are, however, unfortunately just probabilities. It is for this reason that amniocentesis can be so helpful, because in appropriate circumstances it can turn a vague probability into a practical certainty. For example, two parents who are known to be heterozygotes for a gene causing a recessive disease that can be detected *in utero* can now be assured of having only normal children, provided the mother is prepared to undergo amniocentesis and electively to abort abnormal fetuses.

It is clear that more consumers would take advantage of genetic counseling if the service were more widely available and its existence more widely known.

In addition, it is helpful for potential consumers (any and all of us) to have received a suitable education concerning genetic principles in secondary school. The overall decrease in mortality and the disappearance of many infectious diseases has been made possible in part by spreading knowledge about hygiene through schools and other forms of education. Some instruction in genetics is increasingly given to students at school, but knowledge about "genetic hygiene" is not yet widespread. If people were, for example, aware that late childbirth (namely, above 40 years of age for the woman) greatly increases the chances of having a child with Down's syndrome, the necessary precautions (namely, amniocentesis and chromosome examination) could be carried out, and many unnecessary births of affected individuals would be prevented. A list of the available counseling centers in the United States is distributed by the National Foundation—March of Dimes, 800 Second Avenue, New York, New York 10017.

20.4 Ethnic Variation in Genetic Diseases

Some *bona fide* genetic diseases show widespread ethnic variation. In some cases, the reason is clear, such as for all the polymorphisms that are associated with malaria, including sickle-cell anemia, thalassemia, and G6PD deficiency. These diseases are found only in locales where malaria is or has been endemic, or in populations with migrants from malarial areas. The cause of this ethnic variation is the geographic distribution of the selective agent—namely, the malarial parasite. Ethnic variation has, however, also been found for other much rarer diseases. Whenever a genetic disease is rare everywhere, we assume that it is maintained by mutation–selection balance, and not by an advantage of the heterozygote, which is the classical cause of balanced polymorphisms. If, however, we find that a genetic disease is relatively frequent in some ethnic groups, and rare or absent in others, where should we look for an interpretation? Before we discuss the possible hypotheses, let us consider some actual examples.

There are at least three recessive diseases that show widespread ethnic variation (Fig. 20.12).

Cystic fibrosis (CF), discussed in Chapter 2, is a generalized disorder affecting many, and perhaps all, exocrine glands (glands with ducts)—mucus-producing and others. Thick secretions block the duct of the pancreas, with a resultant inability to deliver into the small intestine enzymes that are essential for digestion of various nutrients. Thick mucus secretions in the bronchial tree of the lung cause obstruction, which predisposes to recurrent pneumonia and chronic pulmonary disease. The sweat of patients with CF contains abnormally high concentrations of electrolytes, making them relatively sensitive to extremely high environmental temperatures. The biochemical basis of this disease is still obscure, and treatment is still of limited value, mainly helping to counteract recurring chest infections and improving digestion of food. Among Caucasians, especially Northern and Central Europeans, the incidence of CF at birth is approximately one in 2,000 to 3,000 births, corresponding to a gene frequency of about one in fifty, and so a frequency of heterozygous carriers of $\frac{1}{25}$. Among Orientals and Africans, the disease is much rarer.

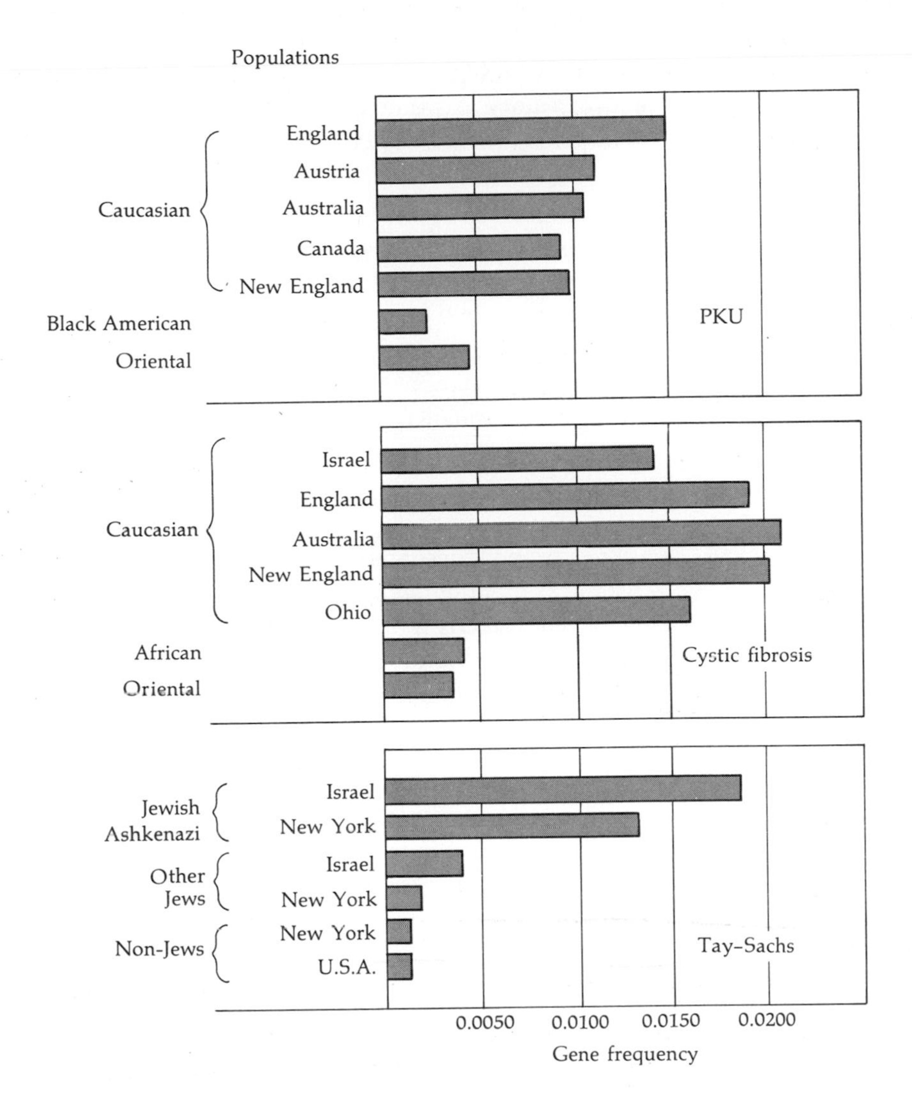

Figure 20.12
Ethnic distribution of genes causing recessive lethal diseases. The frequency of genes causing PKU and cystic fibrosis is higher in Caucasian populations than in either Negroid or Oriental populations. Although the gene frequency varies among different Caucasian populations, in all cases it is much higher than in the other racial groups. Tay–Sachs disease is much more frequent in a particular group of Jewish people, the Ashkenazi, than either in other Jewish groups or among non-Jews. (Data from D. K. Wagener and L. L. Cavalli-Sforza, *American Journal of Human Genetics,* vol. 27, pp. 348–364, 1975.)

The frequency of PKU similarly varies widely among ethnic groups, being most frequent among Caucasians, especially Northern Europeans and particularly the Irish, where its frequency can be as high as one in 5,000 births. It, however, is about 100 times less frequent among Africans and Orientals.

Even more marked differences are found between Caucasian groups in the incidence of Tay–Sachs disease (TSD). This degenerative brain disorder of infancy is due to an autosomal recessive mutation in the gene controlling the glycolipid-degradative enzyme, hexosaminidase A. The age of onset in TSD is about four to six months, and affected children show progressive mental deterioration, paralysis, deafness, blindness, and convulsions, with death finally occurring between three and five years of age. The incidence of TSD among Ashkenazi Jews (whose ancestors originated in Eastern Europe) is about 1 in 4,000, while that among other Jews or non-Jews is some 100 times smaller.

What is responsible for this ethnic variation?

It is unlikely that rates of mutation to these specific gene changes differ in different populations. One possibility is random genetic drift. The mutant allele may simply be one that has happened to increase in frequency, by chance, in a given population, or was present in a small group of emigrants that founded a new large population. Another possible explanation is a moderate advantage of the heterozygote over the normal homozygote in some ethnic groups, but not others—perhaps reflecting differences in environmental conditions. It has been suggested that heterozygotes for Tay–Sachs disease are more resistant to tuberculosis, which was an especially frequent disease in the crowded ghettos of Northeast Europe, but this is highly speculative. A direct search for heterozygote advantage in Tay–Sachs disease has so far given inconclusive results.

Some quantitative considerations may help at this point in understanding the difficulty in answering this problem. Consider a genetic disease that is lethal or highly deleterious in the homozygote, such as one of the three discussed above (CF, PKU, TSD), and which has a frequency at birth of one in ten thousand. If this frequency is maintained by mutation alone, the mutation rate must be one in 10,000 to keep the frequency unchanged over generations. But if there is a selective advantage favoring the heterozygote, which may be as little as 0.1 percent, the mutation rate for maintaining equilibrium at the same incidence at birth is ten times smaller ($\frac{1}{100,000}$)—while if the heterozygote advantage is 1 percent, the mutation rate would be 100 times smaller (namely, $\frac{1}{1,000,000}$). Such small selective advantages are extremely difficult to measure directly in man. We have seen that with hemoglobin S, one can give a reasonably precise estimate of a selective advantage of the heterozygote, which is about 10 percent, using data on tens of thousands of individuals. The number of observations needed to estimate a selection coefficient with a given precision increases with the reciprocal of the square of the selection coefficient. Thus, with a selection coefficient that is ten times smaller (say, 1 rather than 10 percent), 100 times more individuals would be needed.

Another possible mechanism that has been considered for these ethnic variations—but for which again the evidence is inconclusive—is that of linkage to an advantageous gene. Whenever a given allele has increased in frequency because of a selective advantage, a small chromosome segment in the neighborhood of the relevant locus will also have increased in frequency to some extent, passively, because of the linkage. The segment will just be carried along for a while with the advantageous allele. If such a chromosome segment happened to include a recessive lethal, the frequency of the latter would be increased; this is sometimes called the "hitch-hiking" effect.

It is difficult at present to conclude which mechanism is responsible for the high incidence of genetic diseases such as PKU, cystic fibrosis, and Tay–Sachs in some ethnic groups but not in others. The issue has important implications for choosing the appropriate strategy to prevent genetic disease. Thus, the possible benefits of a genetic screening program depend on the number of cases found and successfully treated, and this will depend on the relative frequency of the disease. It would be a waste of time and money to search for a disease in a population where it does not exist, or where it is extremely rare; at the opposite extreme, the more frequent the disease, the more beneficial the program. Where should one draw the line? These questions are considered in more depth in Chapter 21.

Many diseases with a complex genetic basis, or in which genetic factors play a less significant role than in those just discussed, also vary in frequency between ethnic groups.

In some cases, notably involving infectious diseases, the origin of the differences may be simply related to differences in standards of hygiene and in the prevalance of infectious agents. In other cases, social habits of one form or another may be responsible—as, for instance, in the connection between smoking and lung cancer. A number of cases remain, however, that are poorly understood and where the cause of the variation may be differences in frequency of genes conferring partial resistance or susceptibility to the disease involved. Multiple sclerosis, for example, has a higher frequency in Caucasians than in most other groups. This degenerative disease of the nervous system is one of the diseases that, as described in Chapter 10, is significantly associated with factors of the HLA system. One possible explanation, therefore, for its frequency distribution in different ethnic groups (which is consistent with the observed data) is that this is due to differences in the frequencies of the relevant HLA haplotypes. Certainly, ankylosing spondylitis (the disease that is highly associated with HLA type B27), appears to show a population distribution consistent with that of B27. The large number of factors, both genetic and environmental, that distinguish one ethnic group from another, however, may often make it very difficult to determine the specific cause of a difference in disease incidence between two populations.

20.5 Pharmacogenetics

The hereditary defect of glucose-6-phosphate dehydrogenase (G6PD), which occurs in about 10 percent of black American males, was discovered because it is the cause of serious hemolytic crises following treatment with the antimalarial drug, primaquine. It was later shown that people with this enzyme deficiency are also highly sensitive to the toxic effects of a number of other drugs, varying from the antibacterial sulfonamide drugs to various analgesics. The mechanism responsible for the hemolysis is not fully understood, but it is probably related to the level of a substance called reduced glutathione in the red blood cell. When G6PD is low or absent, as in G6PD-deficient individuals, the level of reduced glutathione in red cells is low. In those patients, drug-induced red-cell destruction is preceded by the rapid fall in intracellular levels of reduced glutathione. The role of reduced glutathione in protecting the red cell from destruction is not clear, but it may be related to the maintenance of the cell membrane.

A number of other cases of inherited differential sensitivity to drugs, in addition to G6PD deficiency, are now known.

A well-known example results from deficiency of the enzyme pseudocholinesterase found in serum. An allele determining low activity of this enzyme has a frequency of about 2 percent among Caucasians, and much lower or zero frequencies in other ethnic groups. Homozygotes for this gene, which occur with a frequency of about 1 in 2,000 among Caucasians, show unusual sensitivity to muscle-relaxant drugs such as succinylcholine or suxamethonium, drugs used during anesthesia that are normally hydrolyzed by this enzyme. The homozygote treated with such a drug cannot inactivate it quickly enough, because he lacks the necessary enzyme and so may have prolonged paralysis of respiratory muscles during and after anesthesia, requiring artificial respiration for long periods.

Most drugs introduced into the organism are further metabolized by a variety of enzymatic reactions, such as acetylation (which is the addition of the acetyl group, CH_3CO–), conjugation with other substances, and oxidation. Usually, the products of drug metabolism are nontoxic or less toxic than the drug, and also are inactive. A common process of drug detoxification by the organism is acetylation. Consider, for example, the drug isoniazid, which was introduced for the treatment of tuberculosis. When acetylated by the appropriate enzyme (an acetyl transferase), this drug is much less active therapeutically, and is also less toxic. About 50 percent of Caucasians and black Americans are slow inactivators of this drug (Fig. 20.13). They are homozygous for a recessive gene that determines an inactive acetylating enzyme found in the liver. Rapid inactivators are homozygous or heterozygous for the gene that determines the active hepatic acetyl-transferase. There appears to be no significant effect of this difference in drug inactivation on the response to treatment of tuberculosis. In principle, slow inactivators could be treated with a smaller amount of drug. They are also slow inactivators of certain other drugs, including an antidepressant and an antihypertensive agent, which are

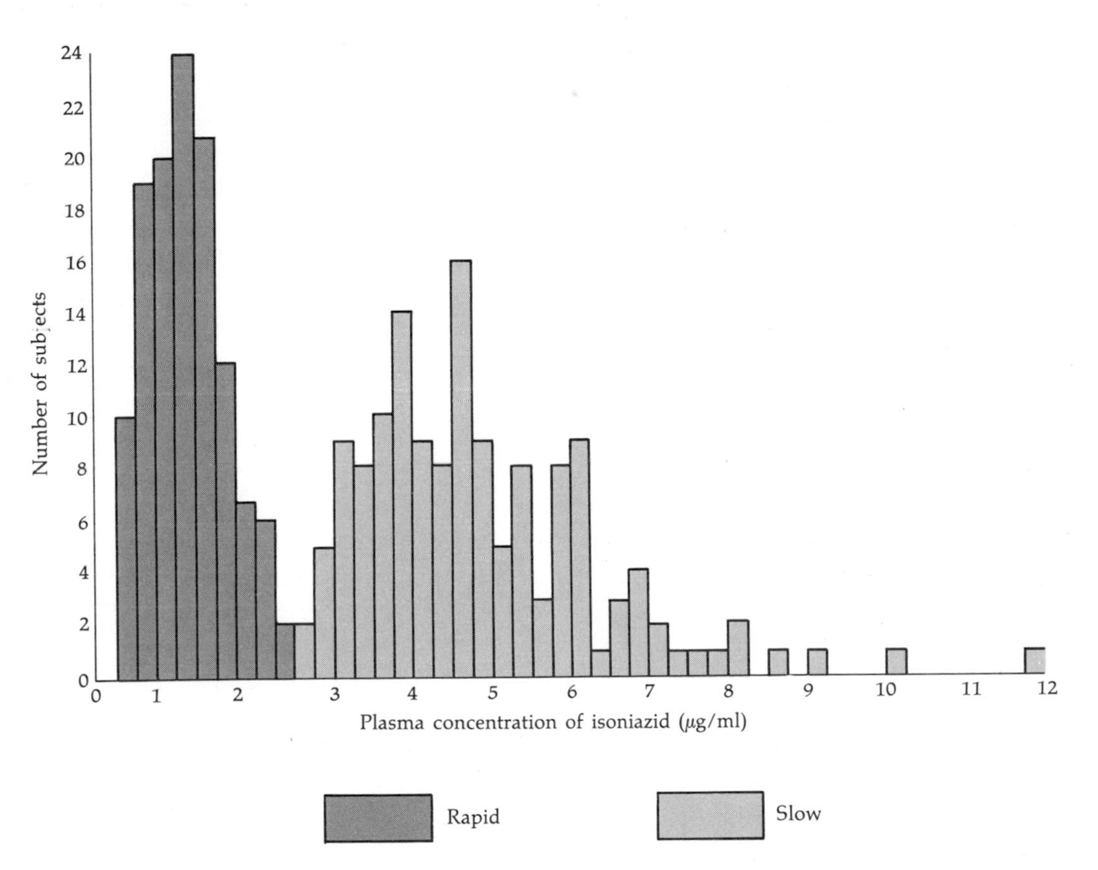

Figure 20.13
Distribution of plasma isoniazid levels, illustrating rapid and slow inactivation. The graph shows the distribution of plasma levels of the drug in 267 individuals six hours after an oral dose of the drug. Rapid and slow excretors can be clearly separated, as indicated by the distinct bimodal distribution. (From B. N. La Du, in *Medical Genetics,* ed. V. A. McKusick and R. Claiborne. Copyright 1973 by H.P. Publishing Co., Inc., New York.)

chemically related to isoniazid. When treated with any of these drugs, slow inactivators are more likely to develop side effects than are fast inactivators. This example is of special interest because it is connected with a polymorphic variant that shows considerable differences in gene frequency between ethnic groups.

Other rare but severe examples of drug idiosyncrasies due to peculiar genetic conditions are known. One example comes from the group of genetic diseases known as porphyrias. Porphyrias are disturbances in the metabolism of heme. Acute intermittent porphyria, an autosomal dominant disorder, has been hypothesized to have affected, among others, King George III of England. Patients with this disease (which, though very rare, has been found especially in Sweden) have a variety of clinical manifestations, from seizures to paralysis, and severe intestinal pains simulating an acute abdominal condition. They usually also have severe psychiatric disturbances. Severe, even fatal attacks of porphyria, can be precipitated in these patients by administration of barbiturates and sulfonamides. Idiosyncrasies to specific drugs are also not uncommon, though their genetic basis is often not known.

The reverse of drug hypersensitivity also occurs: some individuals may be abnormally resistant to certain drugs. For succinylcholine, for example, "drug resistors" have been described (all members of one family) who require perhaps 20 times the usual concentration of the drug to achieve the same effect. These individuals are heterozygotes for a gene that determines a pseudocholinesterase molecule with increased activity. A case has been described of identical twins who happened to have heart attacks within a few days of each other in different cities; both turned out to be resistant to the effects of an oral anticoagulant (sodium warfarin) commonly administered in such cases. Warfarin is actually commonly used as a rat poison, and its use often leads to selection for resistant rats, perhaps by a genetic mechanism similar to that operating in the resistant twins.

Some drugs are inactive, unless transformed into the active compound by enzymes produced by the organism. Individuals who have a genetic defect in such enzymes may prove insensitive to the drug. Unquestionably, there is a great variety of genetically controlled differences in the sensitivity of individuals to drugs, which parallels other described genetic variations, both rare and polymorphic.

20.6 Genetic Engineering

Recent advances in molecular and cell biology have raised the hope that a new technology may soon be created to correct and improve the genotype by acting directly at the DNA level.

This approach to curing genetic diseases is sometimes called "gene therapy." Practically, all these developments are still, at the time of this writing, difficult and to some extent speculative, but the rate of progress is fast, and a few innovations along these lines may be perhaps expected in the next decade.

In bacteria, as discussed in Chapter 3, DNA from a donor can be integrated into a recipient's chromosome by a crossing-over–like mechanism. In some bacterial species, as we have seen, this is possible simply by adding chemically purified DNA to recipient bacteria (transformation). Another mechanism of DNA transfer in bacteria is transduction, in which a bacterial virus (a phage) may carry in some of its particles DNA from the bacterial host in which it grew, and inject this DNA into another bacterium. In certain cases, this new host is not killed and the DNA can be integrated into the bacterial chromosomes.

The experiment of inserting genes into mammalian cells by mechanisms similar to those known for bacteria has been attempted a number of times. The use of purified DNA as a vector for gene transfer had equivocal success. Other approaches, possibly, offer more hope. For instance, a mutant line of mouse cells, incapable of synthesizing the enzyme thymidine kinase, has been exposed to the herpes simplex virus, which contains a gene coding for this enzyme. The mouse cells thus treated appear to have acquired (in a seemingly permanent way) the capacity to synthesize this enzyme. Recently, more success in such experiments has been achieved by adding preparations of whole chromosomes rather than purified DNA. Other even more convincing experiments have also been done along related lines. It is now known that the DNA of several oncogenic (cancer-causing) viruses, such as polyoma (a mouse virus) and SV40 (a monkey virus), is capable of integrating into human chromosomes. It has been widely realized that this property may be used for "genetic engineering." If, for example, an individual carrying a genetic defect were treated with a virus, somehow made harmless and carrying in its genome a segment of normal human DNA, corresponding to the defective one in the patient, one could hope to "cure" the patient of his disease by substituting the "good" DNA for the "bad" copy in his cells.

There are innumerable difficulties in such a genetic-engineering technology.

It will usually be necessary not only to get the "good" DNA to its appropriate place in the cell, but also to "cure" the right type of cell. Many genetic defects are restricted to cells of particular tissues, such as the liver, and it is these cells that must be "cured." The problem of getting the right gene to the right place in the right cell is still a very formidable one. Great care would also be needed in evaluating the risks of treatment with a virus potentially oncogenic for man. What criteria could be used to assure that the virus really is harmless? Human experiments clearly are not possible and animal experiments may be an unreliable guide. However, there may be safe ways of neutralizing the oncogenic (or other undesirable) activities of a virus.

There now exist ways of joining DNA segments of entirely different origin, using enzymes (called "restriction" enzymes) that break DNA in such a way that the ends remain "sticky" and can easily be joined to other similarly treated DNA fragments.

This procedure allows, in principle, the joining of any two DNA fragments, even if of remotely different origins. It has become possible, for instance, to join DNA from toads to bacterial DNA (Fig. 20.14). The hope is to be able to use the bacteria to grow small defined sequences of, for example, human DNA, and also to make the appropriate gene product in the bacterial environment. The major application of these techniques, however, will lie not so much in curing more genetic diseases, as in the opportunity they provide for understanding basic cellular processes on the one hand and, on the other, for harnessing bacteria to make useful therapeutic products and, especially, for novel developments in plant breeding. Novel combinations of characteristics in plants can be envisaged that could eventually revolutionize agricultural practice and greatly help to alleviate the world's food-supply

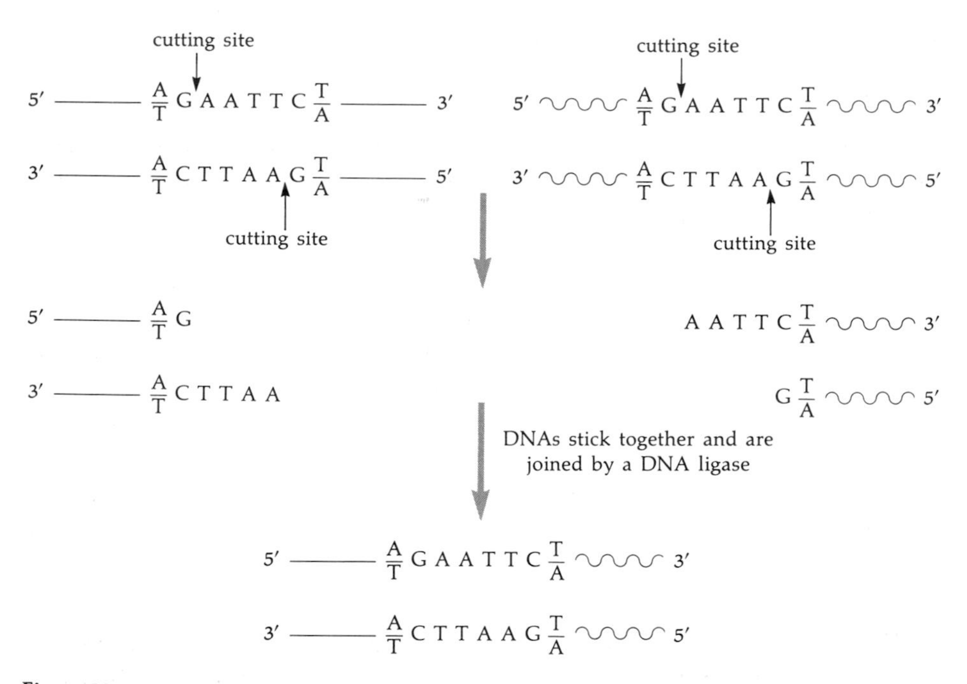

Figure 20.14
Restriction-enzyme joining of DNAs of diverse origins. Restriction enzymes nick the DNA at specific locations; a particular enzyme will cleave only one particular nucleotide sequence. Some endonucleases stagger their cuts in such a way as to generate "sticky" ends, where the ends of the DNA sequences are complementary. The fragments may then reassociate under appropriate conditions, and can be rejoined with a DNA ligase. Any DNA fragments—even if from different organisms—that have been cleaved by the same endonuclease (of the right sort) can be joined with each other. This procedure allows construction of hybrid DNAs.

problem. One possibility often mentioned is the introduction of the capacity to fix nitrogen into plant species, such as the cereal crops, that do not now have it. This could create enormous savings in the use of nitrogeneous fertilizers. Useful human gene products that might be "mass produced" in bacteria include hormones, such as insulin and the pituitary growth hormone, as well as specific antibodies to be used for passive immunization, and parts of, for example, viral proteins that would be nontoxic but could be used for vaccination.

Questions have been raised about the safety of some of these procedures because of the possibility of producing quite novel dangerous organisms (see the report of the Working Party chaired by Lord Ashby, listed in the References.) On the whole it seems, however, that most of these experiments can be done quite safely if carried out with proper precautions.

Other potential developments come from techniques of cell biology, including somatic cell genetics, as discussed in Chapter 5.

Cell-fusion techniques might be an alternative way of genetically curing a deficient cell removed from a patient. The difficulty with this approach is that unwanted genetic material may be transferred, and so ways must be found to select those cells that have taken in just the requisite amount of DNA needed for their genetic cure. These cells could then be transplanted back into the same patient. In animals (including mammals for some experiments) it has, in addition, been possible to obtain parthenogenesis (the development of unfertilized eggs, which would give rise to females only), to transplant nuclei from one cell to another (and, in some of these cases, to obtain the development of a full embryo with the DNA of an adult cell), to fuse embryos, obtaining "chimeras," so that one individual is composed of cells of two or more different genotypes, and also thus to make artificially monozygous twins. All these techniques of experimental cell biology may one day—still rather a long time away—prove to be of interest from a medical point of view.

It is still at this moment difficult to foresee how such complicated and expensive procedures could be widely used to cure some of the very rare inborn errors of metabolism.

Less ambitious than genetic engineering based on "microsurgery of DNA," but also perhaps closer at hand, is the correction of genetic defects by making "artificial chimeras"—that is, by transplantation of normal tissue into a defective individual.

This approach has, for example, worked in a few cases for bone-marrow or fetal-thymus transplantation in individuals affected by rare genetic immune-deficiency diseases, disorders in which patients have an inability to produce antibodies or have defective cellular immunity; however, the inborn incapacity of these individuals to respond to transplantation makes this example atypical. One could, for example, try to use a normal liver transplant to cure diseases such as those of glycogen storage. These are due to deficiencies in liver enzymes that control the synthesis and breakdown of glycogen, the complex molecule that is the main form in which carbohydrate is stored. The problem, however, with the more wide-spread application of this approach, is that of graft rejection (see Chapter 10). One answer to avoiding transplant rejection, and to permit tolerance of the graft by the host and vice versa, might be transplantation of material from a normal fetus into an affected fetus. This is technically possible, and model experiments in animals have already been conducted to verify it, but formidable obstacles both technical and social still must be overcome before application of such an approach could be envisaged.

Even if the techniques of genetic engineering and related approaches could work, and were tested beforehand in animal experiments to show their safety, many of them would pose difficult ethical and social problems. These problems are the subject of the next chapter.

Summary

1. Genetic diseases constitute a large part of all disease; thus far, we have probably seen only the tip of the iceberg. Defects with a complex determination are the major contributors, and their importance is largely unknown. Known single-gene diseases and chromosomal aberrations affect a few percent of all births.
2. The capacity to treat diseases—by substitutional therapy, dietary treatment, or other methods—is continually increasing.
3. If a disease is untreatable—or causes a considerable burden to the affected individual and his or her family—there is now the possibility, in a number of cases, of detecting the disease *in utero* early enough in pregnancy to permit an elective abortion of an abnormal fetus. Detection is accomplished through the technique of amniocentesis, in which fetal cells are withdrawn and examined by biochemical and cytological techniques.
4. Genetic counseling is a communication process that attempts to help patients with genetic diseases and their relatives to make appropriate decisions about conceiving future offspring. The risks of having affected offspring in future pregnancies can be estimated accurately in many cases; this information constitutes one important element in such decisions.
5. There are considerable variations in the incidence of genetic diseases among various ethnic groups. Among the causes of such variation are geographical

distributions of infectious agents or other factors that give rise to differential natural selection.

6. There is much individual variation in response to drugs. A particular individual may show increased or decreased sensitivity to a specific drug or a group of drugs. These differences usually reflect differences in individual enzymatic constitution, which in turn are usually genetically determined.

7. Genetic engineering through DNA "transplants" is a dream still far in the future that could perhaps eventually revolutionize the approach to the treatment of genetic diseases. It is, however, fraught with technical difficulties and ethical problems.

Exercises

1. There is a very large number of different genetic diseases. Each disease occurs with a relatively low frequency. Discuss the reasons for and the consequences of each of these two facts.

2. Discuss (with examples) possible reasons why some genetic diseases are more frequent than others.

3. Why are there fewer recessives (relative to dominants) in man than in most experimental animals?

4. Why is analysis at the biochemical level more likely to give a Mendelian pattern of inheritance than is analysis at the clinical level?

5. What avenues may in the future make substitutional therapy possible in an inborn error of metabolism?

6. List criteria that make it possible to use amniocentesis followed by elective abortion successfully for a given disease.

7. Discuss the purposes and limitations of genetic counseling.

8. List and briefly describe the most important cases of differences in ethnic incidence of genetic diseases.

9. Give some examples of pharmacogenetic differences that are of practical concern in medicine.

10. What approaches may genetic engineering take for the therapy of genetic disease? What major obstacles are likely to be encountered?

11. Two unmarried individuals, both carriers for a recessive deleterious disease, come in independently for genetic counseling. One is heterozygous for Tay–Sachs disease; the other is heterozygous for sickle-cell anemia. Discuss how the advice given to these two individuals might differ.

12. A program of abortion of all carrier and affected individuals is suggested as a way of eliminating an autosomal recessive lethal gene. At the beginning of the program, the frequency of this gene is 0.01.
    a. How many deaths per 10,000 births prior to the program could be attributed to this gene?

b. How many abortions per 10,000 fetuses will be required in the first generation of the program?
c. Suppose that fifty such recessive genes, each with a gene frequency of 0.01, can be detected. All genes are sufficiently spaced that they can be considered as genetically independent. What proportion of all pregnancies must be aborted if the program is extended to include all fifty genes?
d. What is your opinion on the feasibility of such a program?

13. A recessive lethal is maintained by mutation-selection balance at an incidence of $I = q^2 = \mu$, where q is the gene frequency of the lethal and μ is the mutation rate. If successful therapy is now introduced, restoring the fitness of the recessive homozygote to one, the incidence will double in approximately $t = 1/2q$ generations.
    a. Compute the time in years for doubling the incidence of the recessive lethal if the initial incidence is (1) $I = 1/10{,}000$; (2) $I = 1/100{,}000$; (3) $I = 1/1{,}000{,}000$. Assume that a generation lasts 30 years.
    b. Give examples of diseases to which each of the times in part a might apply, and discuss each case.
    c. If you are enthusiastic enough to seek extra credit, try explaining the following statement: the formula $t = 1/2q$ was obtained as an approximate solution of the equation $(t\mu + q)^2 = 2q^2$.
    d. List the major assumptions that must be made if the formula $t = 1/2q$ is to be valid.

14. A dominant deleterious gene is maintained by mutation–selection balance at an incidence of $I = 2\mu/s$, where μ is the mutation rate and s is the selective disadvantage of the heterozygote. If successful therapy is introduced, reducing the disadvantage s to zero, the time in generations for doubling the initial incidence is $t = 1/s$.
    a. Following the lead of Exercise 13, explain the formula.
    b. List the major hypotheses under which it is valid.
    c. Compute the mutation rate and the time for doubling the incidence, if at the time of introducing the therapy, $I = 1/100{,}000$ and s is (1) 0.1; (2) 0.5; (3) 1.
    d. Are there genetic disorders in man for which this situation applies?
    e. What would be the doubling time for Huntington's chorea (with $I = 1/20{,}000$; $\mu = 1/400{,}000$) if it could be treated successfully?

References

Bergsma, D.
1973. (Ed.) "Contemporary genetic counseling," *Birth Defects: Original Article Series*, vol. 9, no. 4.

Cohen, S. N.
1975. "The manipulation of genes," *Scientific American*, vol. 233, no. 1, pp. 24–33.

Emery, A. E.
1973. (Ed.) *Antenatal Diagnosis of Genetic Disease.* Edinburgh: Churchill-Livingstone.

Fraser, F. C.
1974. "Current issues in medical genetics: Genetic counseling," *American Journal of Human Genetics*, vol. 6, pp. 636–659.

Harris, H.
1974. *Prenatal Diagnosis and Selective Abortion.* London: Nuffield Provincial Hospitals Trust.
Jones, A., and W. F. Bodmer
1974. *Our Future Inheritance: Choice or Chance.* Oxford: Oxford Univ. Press.
Leonard, C. O., G. A. Chase, and B. Childs
1972. "Genetic counseling: A consumer's view," *New England Journal of Medicine,* vol. 287, pp. 433–439.
McKusick, V. A.
1971. "Ethnic distribution of disease in non-Jews," *Israeli Journal of Medical Science,* vol. 9, pp. 1375–1382.
1975. *Mendelian Inheritance in Man.* 4th ed. Baltimore, Md.: Johns Hopkins Press. vol. 9, pp. 1375–1382.
McKusick, V. A., and R. Claiborne
1973. (Eds.) *Medical Genetics.* New York: H. P. Publishing Co.
Ramot, B.
1973. (Ed.) "Genetic polymorphism and diseases in man—Sheba International Symposium," *Israeli Journal of Medical Science,* vol. 9, nos. 9/10.
Stanbury, J. B., J. B. Wyngaarden, and D. S. Fredrickson
1972. *The Metabolic Basis of Inherited Disease.* 3rd ed. New York: McGraw-Hill.
Stevenson, A. C., and B. C. Davison
1970. *Genetic Counseling.* Philadelphia: Lippincott.
Working Party on the Experimental Manipulation of the Genetic Composition of Micro-Organisms (Lord Ashby, Chairman)
1975. *Report.* London: Her Majesty's Stationery Office.
World Health Organization Expert Committee on Human Genetics
1964. *Second Report: Human Genetics and Public Health.* Geneva: WHO.
1969. *Third Report: Genetic Counseling.* Geneva: WHO.

21

Genetics and Society

The extent of genetic variability in the human population is emphasized and documented in many places throughout this book. This genetic variability affects all aspects of the individual—from behavior to hair and eye color, from blood types to serious organic diseases. Almost every aspect of genetic variability has some social implications, and to these we dedicate this final chapter.

21.1 Social Aspects of Medical Genetics

The approaches to the treatment of genetic disease that are reviewed in Chapter 20 raise a number of social, legal, and ethical problems.

Some of these problems arise from the conflict between the rights of individuals and those of society—a conflict that is in no way unique to genetic disorders.

Identification and isolation of patients who carry a contagious disease is a clear example of how such a conflict may arise in another field of medicine. Problems arise also because legal, religious, and moral codes differ between countries, cultures and individuals. Legal codes, of course, change to some extent with the demands posed by changing moral attitudes. Technological progress has also created a number of new legal problems to which the law tries to respond.

Abortion was recently legalized in a number of countries, among which the five largest are China, India, the U.S.S.R., the U.S.A., and Japan. Twenty other countries also have legalized abortion; a total of 58 percent of the world's population now lives under laws that permit abortions under at least some situations. In most cases, the decision to legalize abortion was made largely in response to problems of family planning and women's liberation, but was also an important step toward coping with genetic disease. Religious codes, however, may change more slowly. Thus, while Protestants take a much more liberal attitude towards abortion, Roman Catholics and Orthodox Moslems remain staunchly opposed. The Koran prohibits abortion after 80 days of pregnancy, which unfortunately precludes amniocentesis. Religions do not change rapidly, but individual moral views, observance of religious norms, and even affiliation to religious groups are often more flexible.

There are several unique aspects of genetic diseases that distinguish them markedly from other types of disease.

First, genetic disease is predictable. This is true at the population level, because the incidences of genetic disease do not change appreciably within the span of a generation or a lifetime. It is also true for families in which one or more persons are affected by a given disease, as the chance of recurrence in other members is often predictable.

Second, genetic diseases are transmitted vertically (between generations), rather than horizontally (between contemporary individuals) as most other diseases are. Genetic diseases follow the DNA transmitted from parent to offspring, and they do not respond to hygienic norms set up to avoid contagion. There are some important nongenetic diseases that show vertical transmission; these are the so-called "slow" virus infections, such as breast cancer in mice. Kuru is a virus disease in man (discovered in a New Guinea tribe) that causes mental deterioration (Fig. 21.1). For a time after its discovery, kuru was thought to be a genetic disease because of its pattern of vertical recurrence. It now seems, however, that kuru only mimicked a genetic disease because of local social customs, which involved contact between women and dead relatives, thus favoring contagion between generations (Fig. 21.2).

Third, many genetic diseases are hardly influenced by the usual environment. However, once their biochemical basis is known, they do become amenable to treatment—as, for example, in the case of PKU.

Fourth, because recurrence of a genetic disease can be predicted, its presence in a family (or in a population) may become a social stigma. Even minor misunderstandings of the scientific situation by the public (and, worse, by legislators) may help turn a minor or nonexistant handicap into a serious problem. For example, sickle-cell anemia (homozygosity for the gene for hemoglobin S) is a serious physical handicap, but the much more frequent heterozygous state is not.

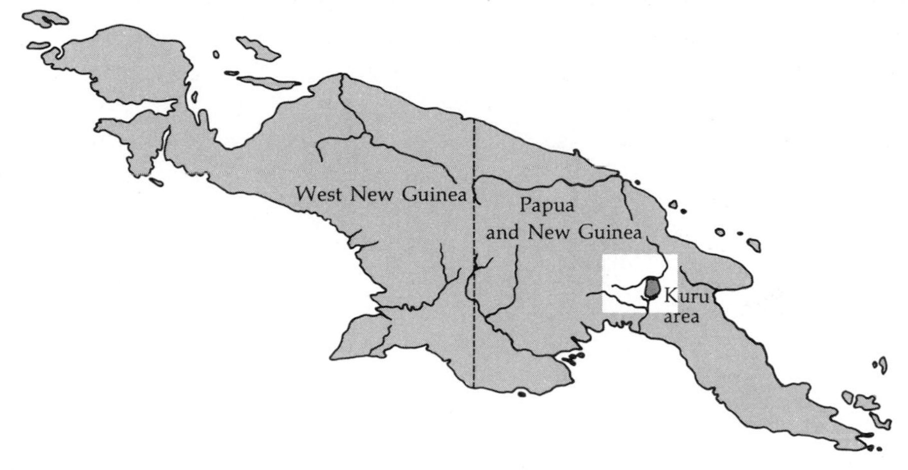

Figure 21.1
Geographic distribution of kuru. The small dark gray area (within the rectangle) on this map of New Guinea indicates the group of adjacent valleys from which all known cases of kuru have come. Kuru is a severe degenerative disease of the central nervous system that results in death within three to nine months after appearance of the first symptoms. Over 80 percent of the cases occur in the Fore cultural and linguistic group; the incidence rates and prevalence ratios are both about 1 percent. (From D. C. Gajdusek, in *Tropical Neurology,* ed. J. D. Spillane, Oxford Univ. Press, 1973.)

Nevertheless, a stigma has in some cases been inadvertently imposed on heterozygotes by ill-advised legislation that has raised confusion about the two conditions. Of course, a stigma may also be associated with some nongenetic diseases such as syphilis and tuberculosis—and, even more clearly, leprosy. The chance of contagion in the late stages of syphilis and in many cases of leprosy is so minimal that this fear cannot justify the stigma associated with the disease.

Fifth, a particular type of guilt is often experienced by the parents of individuals affected by genetic disease, particularly congenital malformations.

The apparent increase in the population incidence of disease attributed to genetic causes is due (1) to the decrease in the incidence of diseases (notably infections) that have an obvious environmental component; (2) to increases and more widespread knowledge of and interest in genetics; and (3) to the recognition that many syndromes of unknown origin have a significant genetic component. The "autoimmune" diseases which have been associated with HLA (including various types of arthritis and skin ailments, and other serious diseases whose causes remain to be elucidated, such as multiple sclerosis) are an example of this last point.

One of the great difficulties facing the progress of medical genetics is the great multitude, but individual variety, of genetic diseases.

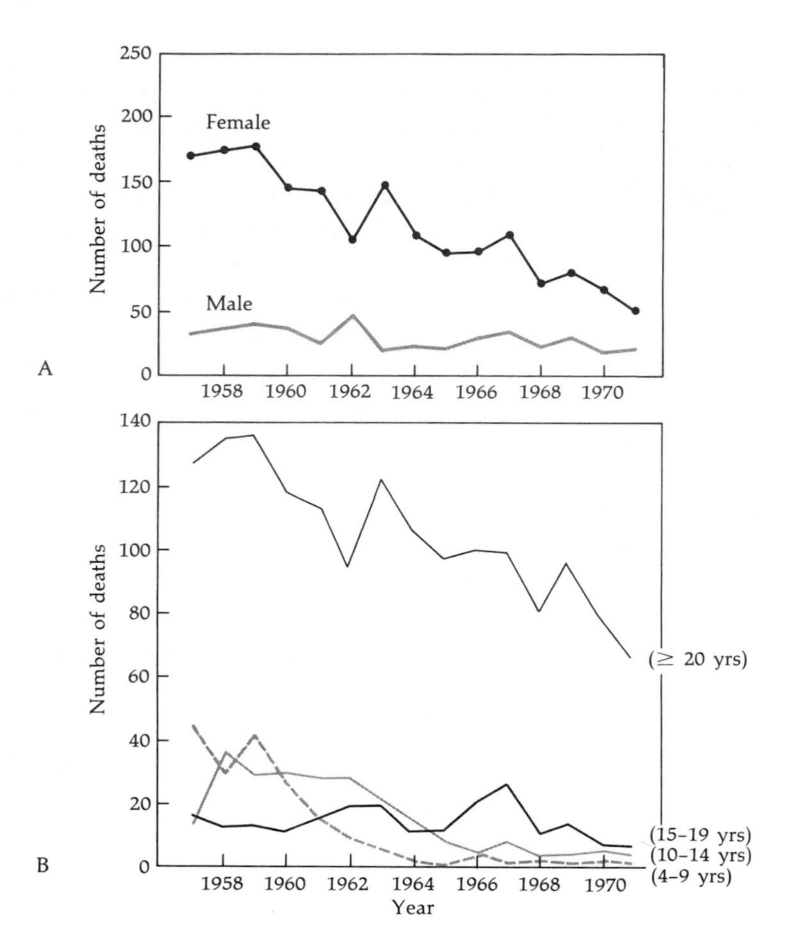

Figure 21.2

Number of deaths caused by kuru. (A) Sexual distribution of kuru deaths. (B) Age distribution of kuru deaths. Prior to the early 1960s, all ages beyond infants were affected in the New Guinea population at risk (Fig. 21.1). Although common in children of both sexes and in adult females, the disease was rare in adult males. Since the early 1960s, there has been a reduced frequency of the disease, and preadolescent children with kuru are no longer seen. For a long time, kuru was an enigma to medical researchers. Consider the following facts. (1) No persons from outside the kuru region have ever developed kuru after residing in the region, but some individuals from that region have developed kuru after years of absence from the region; all such cases occurred in families wherein kuru had occurred in the past. (2) No cases of kuru developed among people who ate and lived with the natives of the region who developed kuru while away from home. Non-Fore residents in the kuru region never developed kuru. (3) The culture and diet of the kuru region resemble those of surrounding kuru-free peoples. Adult men, the group least afflicted by kuru, often eat with the women and children and share a similar diet. (4) Among the Fore group, kuru was the major cause of death. Surrounding groups showed disease patterns similar to those of the Fore for all diseases except kuru.

The preceding facts might be explained by a simple genetic hypothesis: a kuru gene is dominant in the adult female and rarely penetrant in the adult male; homozygotes, whether male or female, develop the disease at an early age. This genetic hypothesis, however, does not

Each of the clinical entities is in general relatively rare; thus, while a large number of different problems is posed to the medical scientist, the number of useful cases available for the study of each type is small, and these are generally widely scattered and often relatively inaccessible. The rarity is of course the consequence of selective elimination, while the multitude and variety of different diseases is due to the great number of different genes whose mutations can lead to pathology. Thus, having on the whole successfully coped with many of the more common diseases, medical research is now faced by a multitude of rare and difficult ones whose total incidence is considerable. It is possible that the total incidence of genetic disease may increase in the future, because of selective relaxation due to medical treatment (discussed in Section 21.3), but it also might decrease as a result of voluntary selective breeding techniques (discussed in Sections 21.2 and 21.4). Both directions of change raise substantial social problems.

The availability of dietetic, surgical, and other types of treatments that can in large measure relieve patients of symptoms and consequences of certain genetic defects, has created the justified hope that a new era has come for the therapy of genetic diseases. But the cost of the treatment may be high both for the individual and for society. Who should bear the cost? Is it feasible to support a large number of possibly very expensive approaches to the treatment of many very rare diseases? These questions are discussed in the Section 21.2.

Another important social problem arising from progress in medical genetics is that a potentially important technique of prevention—namely, abortion of an abnormal fetus following amniocentesis (selective abortion)—clashes with the attitudes of certain religious groups, and with the prevailing laws in a number of

---

explain the maintenance of such an enormously high gene frequency in the face of such a high rate of gene loss. Thus the nature of kuru remained a puzzle until certain other facts were brought to light.

Until 1957, the Fore peoples practiced ritual cannibalism of dead kinsmen. The women prepared the corpses for the cannibalistic ritual, with infants present and children crowding around. The women worked bare-handed, and the ritual including squeezing brain tissue into a pulp with the hands. The flesh of the dead person was eaten bare-handed. Throughout this procedure there were no attempts at any sort of hygiene. Men rarely ate the flesh of dead kuru victims. The final key to the puzzle fell into place with the discovery that kuru is caused by a virus that is infectious by peripheral inoculation routes (such as open sores and the respiratory tract) or by ingestion. It is very heat-resistant. The kuru virus is present in its highest concentration in the brain of a victim—a concentration of over 1 million infectious doses per gram. Self-inoculation of women, infants, and young children (who were kinsmen of a dead kuru victim) was a certainty. Symptoms of the disease might not appear until many years after infection, because the kuru virus is a "slow" virus that can remain dormant in the body for a long time before being "triggered" into active replication. With the decline in ritual cannibalism among the Fore after 1957, there has been a steady decline in the incidence of this disease. (From D. C. Gajdusek, in *Tropical Neurology,* ed. J. D. Spillane, Oxford Univ. Press, 1973.)

countries. However, even in countries in which abortion is—or was until a short time ago—completely illegal and severely punished, a large number of abortions have always taken place for birth-control reasons. This suggests that a substantial proportion of people do not disapprove morally of abortion, even when the law and dominant religion are against it. However, it is important to realize that abortion may not always be entirely without psychological consequences for the woman and always has attendant medical risks, however small. In addition, we believe it to be important that abortion should never be imposed, but should be the informed choice of the particular couple.

The decision in favor of abortion may be relatively easy for a couple that has already experienced the birth, say, of a child with Tay–Sachs disease, but in the absence of such a prior experience, a couple may find it more difficult to decide.

The difficulty will be even greater when the anticipated disease is mild. Factors pertaining to the burden imposed by the disease affecting the decision to abort include the duration of disease, its effect in shortening life, the suffering of the affected individual, and that of the family. Even for a genetic disease that is relatively mild but chronic and requires continuous treatment and extreme caution throughout life, as is the case for hemophilia, the odds may still be in favor of abortion. Hemophilia cannot yet be diagnosed by amniocentesis, but the sex of the child can; female heterozygotes for this disorder now have the option to abort all male fetuses, which would of course include, on average, fifty percent normal unaffected males.

It is difficult to draw the line at which to decide for selective abortion.

The problem is aggravated by the existence of a wide range of defects, from those that are very serious to others with relatively trivial consequences. Where should one draw the line? For example, consider the case of allele *B27* of the HLA system, which is known to carry a risk of up to a few percent of getting the disease ankylosing spondylitis (see Chapter 10). Antigen *B27* can be recognized in the fetus following amniocentesis. Should such fetuses be aborted, especially perhaps in families in which the disease appears to be concentrated? Suppose amniocentesis reveals that the fetus has an extra Y chromosome. Should abortion follow? The criminal tendencies of XYY individuals are far from proven, and if they do exist, can be manifested in at most a small proportion of males with this karyotype; so that there is certainly no obvious moral compulsion to abort an XYY fetus. In neither of these cases would there seem to be a case for abortion. But conditions considered mild or tolerable by one family will be intolerable for another. If parents wish it because they are afraid of the potential consequences, they may be seriously harmed by the legal bar against such abortions in those countries in which the bar still exists. However, we do stress the clearcut need for expert genetic counseling, especially where amniocentesis and selective abortion are potentially available.

Human reproduction is a sensitive subject, and this naturally accounts for much of the concern surrounding the prevention of genetic disease by selective abortion.

In several primitive societies, infanticide is, or was, practiced as a means of birth control, for eugenic purposes, or simply for sex selection. This is not acceptable in our modern society which recognizes the right to live of the newly born infant. The capacity to live outside the womb seems a natural definition of "being born" and underlies many laws. This definition limits the performing of elective abortions usually up until 24 to 28 weeks of gestation.

Even so, the correction of certain birth defects may require decisions that could be interpreted to be on the verge of infanticide. An acute example of this is spina bifida. Children born with this congenital defect can, if it is relatively mild, be at least partially treated by appropriate surgery (Fig. 21.3). However, if the defect is more serious, the surgery, while keeping the child alive, may not be able to prevent severe mental and physical handicap. The practice has been established in a number of medical centers in Great Britain and in the United States that the

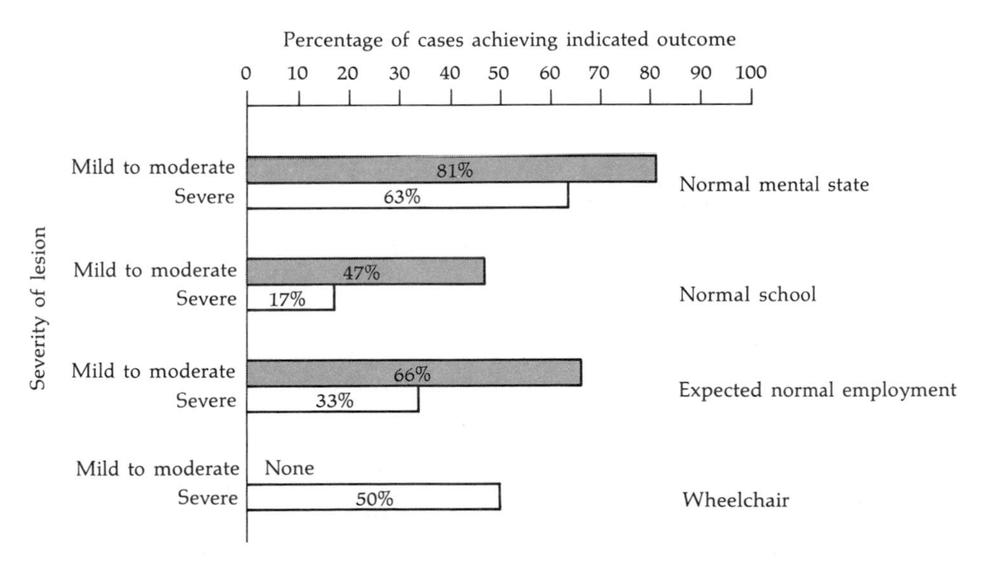

Figure 21.3
Outcome of treatment of spina bifida as a function of the severity of the disease. These figures are for the period 1961 through 1965.

milder cases which are recognized as curable should be treated vigorously, and that severe cases should receive minimal treatment. These untreated children will undoubtedly usually die during their first year of life. So far, this decision has been left to the doctor, based on his expert knowledge of the disease, and thus, his ability to distinguish mild from severe cases. Clearly, this is a case where it may be very difficult for the parents to make an objective decision for themselves.

There are clearly many delicate questions raised by these issues for the whole of society—not simply for the scientists and doctors. They also require the collective wisdom of lawyers, legislators, clergymen, philosophers, and others for their solution. An informed public and extensive discussion of these and numerous other questions that are continuously raised by technological progress is essential. Experience shows that the law often has difficulty in keeping up with technological progress. There is inevitably a lag between recognition of the need for a law and its approval, and this lag should be kept as short as possible. Wide dissemination of the relevant information to the public coupled with discussion of the issues in a democratic society are the best catalysts for a smooth and rapid adaptation of our laws and customs to this rapidly changing world.

21.2 Genetic Screening

Genetic screening is the systematic testing of newborns, or individuals of any age, for the purpose of ascertaining potential genetic handicaps in them or in their progeny that may require treatment or prophylaxis.

There are now at least a few genetic diseases (such as PKU and Tay–Sachs disease) for which screening may be desirable under appropriate conditions. For any given genetic disease, there are two major questions that must be asked in turn to determine the strategy to be followed in dealing with the disease.

The first question is: Is corrective treatment available?

The answer now is yes for a variety of genetic diseases, which still constitute a small fraction of the total, however. Treatment is limited to the individual's phenotype and does not remove the gene. Thus, some descendants will be potential gene carriers and may perhaps even be affected by the disease. This may create some misapprehension with regard to what will happen in future generations, a problem we consider in Section 21.3. When corrective treatment is available, however, there is a further question of importance.

How early should the treatment be administered?

Usually, of course, the earlier the better. However, the treatment may still be effective if applied at the appearance of the first overt signs of the disease, so that a screening program to detect affected individuals may not be needed. On the other hand, if (as in the case of PKU) the treatment is ineffective or less effective if not applied earlier than the appearance of clinical symptoms, an early mass-screening program is essential for the treatment to be successful. Diseases in addition to PKU that can be very readily and easily diagnosed at birth include galactosemia (an inborn error in the metabolism of the sugar, galactose), maple-syrup-urine disease (the hereditary inability to metabolize the amino acids leucine, isoleucine, and valine), and other metabolic disturbances. For the treatment of PKU to be effective, it is necessary to monitor all births and test blood of newborns within the first weeks of life by suitable laboratory methods, as discussed in the previous chapter. The outcome of treatment by a phenylalanine-poor diet, if started as soon as possible after birth and maintained into childhood, appears to be very good. In the case of rhesus hemolytic disease of the newborn, it is necessary to monitor every woman at the time of her first pregnancy by testing for Rh type. Only the 15 percent (in Caucasian populations) of women who are Rh^- need then be followed up initially by Rh typing the husband and child. Immediately after the first Rh^+ child is born, prophylactic treatment of the mother with anti-Rh antibodies, as described in Chapters 10 and 20, is given in order to prevent her immunization and thus protect the next infant from the danger of Rh hemolytic disease.

The advisability and practicality of screening pregnant women for their Rh blood type has never been questioned, at least for Caucasians, for whom the frequency of hemolytic disease of the newborn—without prophylaxis—is of the order of one in every 200 births. However, questions of cost must be taken into account when considering rarer diseases, such as PKU. Ideally one would like to provide for the complete treatment, however expensive, of any disease, however rare. But when resources are limited, the question of cost must be considered in order to establish priorities among the diseases to be treated. In some areas of the United States with a predominantly black population, PKU screening was dropped after a few years, during which not a single case was discovered. This was, of course, a consequence of the fact that, among blacks (and also Orientals), PKU is about ten times less common than among Caucasians. This emphasizes the importance of the incidence of the disease in any consideration of the cost-effectiveness of screening programs.

What is the cost-effectiveness of the PKU screening program?

The cost of the actual PKU screening test in the United States (in 1974) is about 50 cents per sample, but to this must be added the extra costs for collection and delivery, as well as overheads to allow for salaries of technicians and other personnel. The total cost per sample varies considerably with place and time. Estimating it to be about $1.25, and multiplying by the approximately 3,000,000 births per year to be screened, one obtains a total annual cost for the screening program of about $4,000,000. The number of PKU births per year, given an approximate incidence of 1 in 15,000 (the current best overall available estimate for the United States), will be about 200. The estimated annual cost of the dietary treatment is about $1,000, and this treatment must be maintained for six years. Thus the total cost of treatment for the 200 new PKU births per year is 6 × 200 × $1,000, or $1,200,000. When added to the screening costs, this gives an annual cost for the whole program of about $5,200,000. The alternative cost in the absence of screening and dietary treatment would be that of maintaining the 200 new PKU births per year in an institution. A conservative estimate of the annual cost of institutionalization per individual is $5,000, and this must be maintained for the whole lifetime of the PKU individual from about ten years of age—say an average of 30 years, allowing for a somewhat reduced averaged chance of survival. Thus, the cost of institutionalizing the 200 affected individuals produced per year is around 200 × 30 × $5,000 = $30,000,000, which is about 6 times as high as the cost of the screening and treatment program. Thus, even ignoring the obvious personal benefits of the treatment for the individual and his family, the cost-benefit analysis is heavily in favor of the screening program in this case. This analysis is not likely to be much affected by unavoidable inaccuracies in some of the figures used, and applies equally to the United Kingdom, where PKU screening is also carried out on a very high proportion of newborns (Fig. 21.4).

The PKU program satisfies all the criteria needed to justify a screening program.

The tests are technically reliable with a low proportion of false negatives or positives, the cost is low, and the treatment is both beneficial and safe. Occasional cases are found in which the phenylalanine blood level is high, but which are not classical PKU. However, these cases can be sorted out by careful retesting, so avoiding the risk of giving the dietary treatment to someone who does not need it. The major subsequent problem that can arise with the PKU screening program is if affected women have children. It has been found, in the few cases known, that such children of PKU mothers will always, in the absence of any preventive treatment, be severely mentally retarded, whatever their genotype. Apparently, the increased phenylalanine blood level in the PKU mother affects her developing fetus, creating a "phenocopy" of the genetic defect. The preliminary evidence

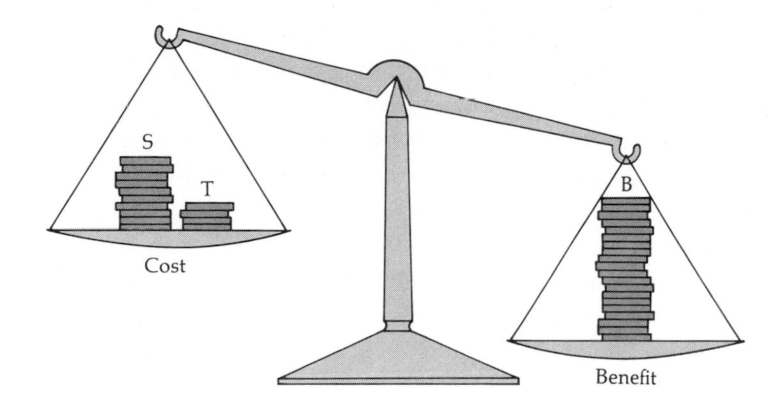

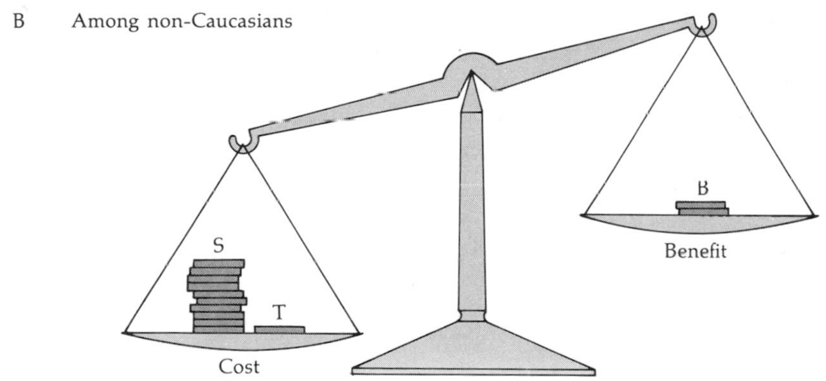

Figure 21.4
Cost–benefit analysis of genetic screening for prophylaxis of phenylketonuria (PKU). Elements of direct cost that can be more easily estimated include S (the cost of screening each newborn, multiplied by the number of births) and T (the cost of treating identified PKUs). The benefit (B) is the cost of the treatment of diseased individuals if screening were not carried out (and prophylaxis therefore were impossible). The magnitude of B depends on the incidence of the disease, while the costs are largely independent of incidence. (T, which is proportional to incidence, is much smaller than S, which is independent of incidence.) (A) In Caucasian populations, the incidence of PKU is sufficiently high that the cost–benefit ratio is favorable to screening. (B) In other ethnic groups, where incidence of PKU is much lower, the ratio is probably unfavorable to screening. These calculations do not include other less direct costs, more difficult to ascertain, nor do they include humanitarian considerations, which would always weigh in favor of the screening procedure.

suggests that this may be prevented if the mother is put on a phenylalanine-deficient diet during pregnancy. It is then essential to keep track of any woman who has been identified as having PKU in order to prevent her from having PKU "phenocopy" offspring.

Clearly, the favorable cost–benefit ratio for PKU is a function of its relatively high population incidence. Suppose, for example, the disease were ten times less frequent, while the unit costs remained the same. Then the costs of screening alone, which of course do not alter with the decreased incidence, would be greater than the cost of institutionalization ($4,000,000 vs. $3,000,000, respectively). The economic efficiency of screening can, however, be significantly increased if it turns out to be possible and convenient to screen for several diseases with the same blood sample. Each of these diseases may be less frequent than PKU and so might not by itself justify the screening; but if they are combined, and if automated screening techniques are used, the operation may become efficient enough to be worthwhile. These analyses of cost of screening for genetic diseases have neglected a number of factors whose cost is difficult to estimate. The medical expenses incurred in care of the patient, the loss to the family and society of an earning and contributing individual—these are some of the factors that are not part of the cost equation. It is even more difficult to put into the bill those considerations that are most important from the human point of view, like those derived from the amount of suffering endured by the untreated patient and his or her family. Clearly, all these factors can only help offset the balance in favor of screening programs.

The second question concerning the strategy of screening for a genetic disease arises if corrective treatment is not available: Can the condition be detected by amniocentesis in time for selective abortion?

The answer, as discussed in the previous chapter, is positive for all chromosomal aberrations, and for a number of metabolic disturbances. Amongst these, one of the most striking, which we shall use as a model, is Tay–Sachs disease. Recall that this autosomal recessive disease—which is due to a defect in the enzyme hexoseaminidase A (Fig. 21.5)—is associated with death (usually by 3 to 5 years of age) and is much more common among Jews of Central and Eastern European origin (Ashkenazi Jews) than in all other groups. Although there is no known treatment for the disease, heterozygotes can be recognized by a relatively simple blood test, and homozygous fetuses can be ascertained following amniocentesis. The suggested strategy is therefore to screen all married couples to ascertain heterozygotes for the Tay–Sachs gene. When both members of a couple are found to be heterozygotes, they are warned that they have a chance of one in four at each conception of having an affected child. Each pregnancy from such a couple is monitored by

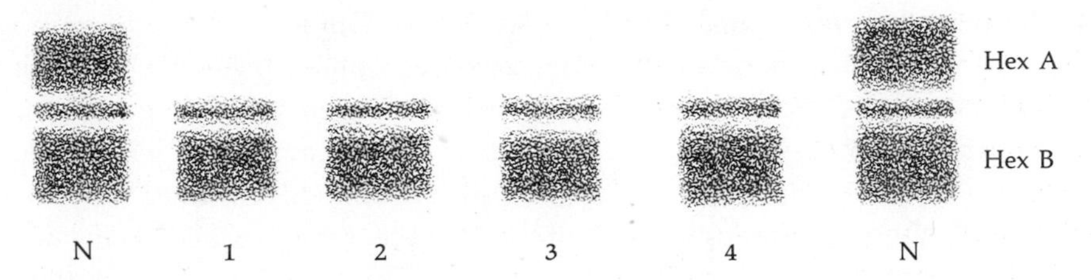

Figure 21.5
Demonstration of hexoseaminidase A, used in screening for Tay–Sachs disease. Starch-gel electrophoresis of liver tissues obtained from normal subjects (N) and from aborted fetuses with Tay–Sachs disease (1 through 4). Although hex B is present in both groups, hex-A activity is absent from the liver of affected fetuses. (After M. M. Kaback and J. S. O'Brien, in *Medical Genetics,* ed. V. A. McKusick and R. Claiborne. Copyright 1973 by H.P. Publishing Co., Inc., New York.)

amniocentesis for the presence of an affected fetus, and this information provides the couple with the option to terminate the pregnancy and try again for a normal child. This strategy has a favorable cost–benefit ratio for Ashkenazi Jews, but not in other populations, and it is now practiced in several Jewish communities. The main benefit of the program is to relieve the suffering of parents and child during the few years an affected child would remain alive. In this way, the consequences are perhaps less dramatic than for PKU, where untreated homozygotes live till about 40 or 50 years of age. One important countereffect of such a screening program is that it may raise unnecessary anxieties in normal, heterozygous, carrier parents. Once again, the only satisfactory way to counter this problem is by the provision of an adequate and intensive education before, during, and after screening programs, for the public to understand the issues involved.

Amniocentesis for chromosomal aberrations raises similar issues to those considered for Tay–Sachs disease.

It does not yet seem either justified or realistic to do an amniocentesis on every pregnancy. Thus, somehow, cases involving a high *a priori* risk must be identified. Obviously, parents who have already had an offspring with either a dominant or a recessive deleterious trait, or who carry a balanced translocation chromosome, represent the highest risk situation. In these cases, if the techniques for diagnosis following amniocentesis are available, the procedure should be carried out. A mating between two heterozygotes each carrying one copy of the gene for Tay–Sachs disease is an example of a high-risk couple for whom prenatal detection is available.

In the case of chromosomal abnormalities, the very much increased incidence of Down's syndrome among the offspring of older women (over age 40 the risk is $\frac{1}{100}$ or higher; see Chapter 4) identifies these as a relatively high risk group. More than a third of Down's-syndrome births occur to women over forty, who produce at most a few percent of all births. Screening all conceptions of women over 40 for chromosome abnormalities by amniocentesis could therefore reduce the incidence of Down's syndrome by some 30 to 40 percent. The overall cost of an amniocentesis with a chromosome analysis is about $200. Thus, the cost of screening the (on average) 100 births to women over 40 years of age per detected Down's syndrome is about 10 $\times$ $200 = $20,000. The cost of institutionalizing one affected individual for 30 years at $5,000 per year, on the other hand, is about 30 $\times$ $5,000 = $150,000, giving a very advantageous cost–benefit ratio of about $150,000/$20,000 = 7.5. On this basis, there would seem to be no doubt of the advantage of screening births to older women for chromosome abnormalities. In fact, the cost of screening 750 conceptions by amniocentesis is the same as for institutional care of one patient with Down's syndrome—a fact which suggests that, on purely economic grounds, it might even be worth screening all pregnancies for this condition (the reader will remember that the incidence of Down's syndrome is close to 1 per 700 births).

There are at least three questions about the advisability of screening all or a high proportion of pregnancies by amniocentesis.

The first, as already mentioned, is the possible risk of the procedure of amniocentesis itself. The indications so far are that the frequency of abortions and still births following amniocentesis is low, but it is still difficult to state that the procedure is entirely free of risk. Other major complications affecting the mother and/or fetus seem to occur infrequently—provided, of course, the procedure is carried out by a competent practitioner. On the other hand, to detect effects at a frequency level comparable to the incidence of Down's syndrome among births to women of all ages (about 1 in 700) would require far more data than are presently available.

The second question concerns the practice of institutionalizing children with Down's syndrome. Not everyone agrees that this should be done, though we believe that this is probably the best way to avoid the various problems that a family may face if it must look after such an affected individual for thirty or more years.

The third and extremely important problem, of course, is that the personnel and clinical and laboratory facilities are simply not yet available to undertake very widespread amniocentesis. It may take many years and dollars to achieve the necessary facilities, and they are in competition with other requirements for medical care. This emphasizes the fact that all the costs, even on the strictly economic side, are not easily built into the cost–benefit equation. The benefits of *in utero* detection followed by abortion, however, seem to be very clearcut in the case of Down's syndrome. A difficult question is to decide at what (maternal) age, or by what other criterion, to draw the line in providing the service.

Some typical data on recent experience of prenatal diagnosis by amniocentesis are shown in Figure 21.6. These data do not, however, include α-feto protein tests for the detection of spina bifida and anencephaly, which have recently become a major cause for amniocentesis following the birth of an affected child, because the recurrence risk is then up to 3 or 4 percent. The major indications for amniocentesis given in the table are a previous child with Down's syndrome, advanced maternal age, or segregating factors in the family, including autosomal and sex-linked recessives and translocations.

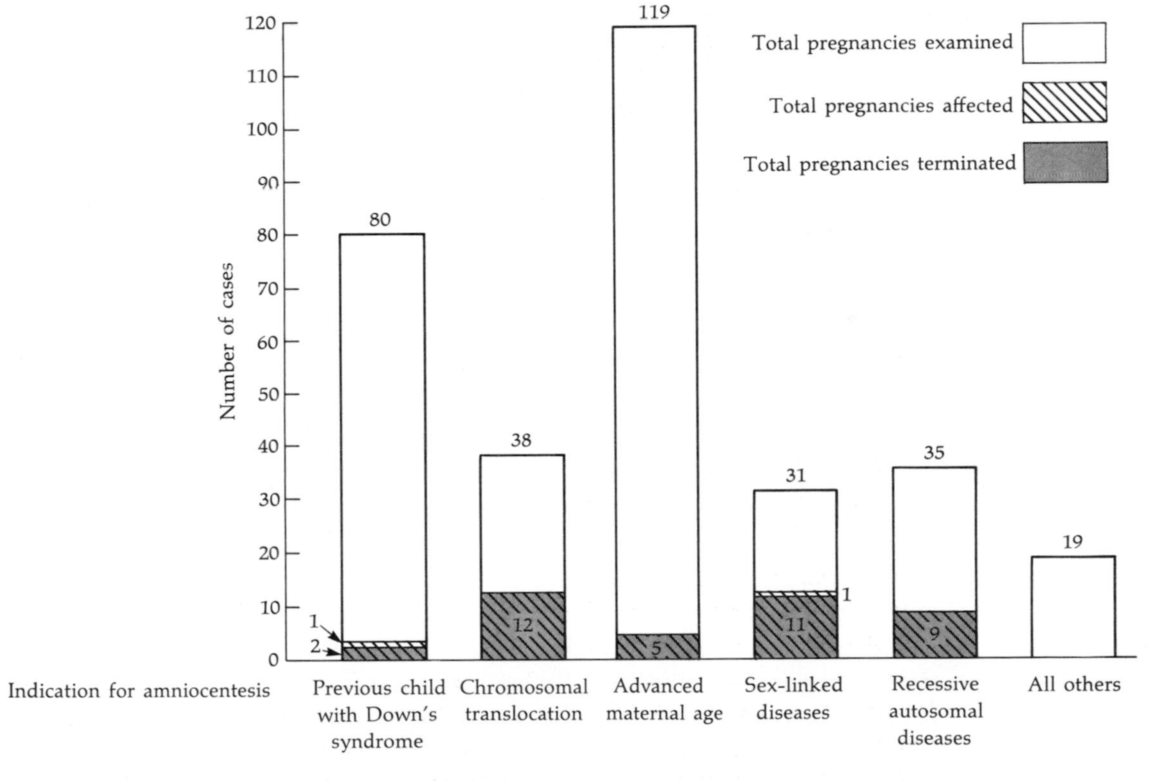

Figure 21.6
Experience of prenatal diagnosis by amniocentesis. The data are collected from ten independent studies reported from North America and Europe during the years 1967 through 1972. The height of the bars indicates the number of pregnancies examined. (Based on data collected by M. Ferguson-Smith.)

New techniques will undoubtedly increase the range of abnormalities detectable *in utero.* Ultrasound techniques that accurately detect the position of the placenta and identify twins are becoming a standard part of the amniocentesis procedure. Especially important would be the development of approaches for detecting some of the commoner congenital malformations other than spina bifida and anencephaly. One possibility is limited visual inspection of the fetus with a small fiber-optic telescopelike instrument (fetoscope), passed through a hollow needle into the uterus. Another possibility for diagnosing fetal malformations is the detection of substances (other than α-feto protein) that are released in the amniotic fluid and may be markers of an abnormality.

There are, unfortunately, a number of genetic diseases—including some of those with the highest incidence—for which it is not possible to provide treatment, nor can they be detected in utero *following amniocentesis.*

The most notable cases are cystic fibrosis (which is the autosomal recessive disease with the highest frequency in Caucasians) and the hemoglobin diseases, including sickle-cell anemia and thalassemia—which, of course, may occur with quite high frequencies in areas where malaria is or was endemic. In each of these disorders, affected individuals can be detected soon after birth and, with respect to hemoglobin diseases, heterozygous carriers of the recessive gene can also be identified. Is genetic screening justified for these diseases?

One approach that has been suggested for sickle-cell anemia, for example, is to screen young adults for the presence of the trait and then encourage gene carriers to avoid marrying other gene carriers.

Carriers of the gene include here both homo- and heterozygotes for S hemoglobin. Because all individuals with sickle-cell anemia must be offspring of matings between gene carriers, a program discouraging such matings could, if fully effective, essentially eliminate at a stroke the disease from the population without putting serious limitations on the reproduction of gene carriers. It might be hoped that an appropriate education program would be enough to prevent gene carriers from marrying each other, without any other forms of persuasion. However, attempts at screening for sickle-cell hemoglobin have, at least in the United States, encountered several difficulties. As already mentioned, laws for screening blacks have been passed that do not distinguish between the heterozygote and the homozygote, and (partly as a result of this) heterozygotes have been stigmatized. The worst problem, however, arises from the misunderstandings created by the racial distribution of the disease, which have even led to accusations that screening programs are directed at genocide of the black population. If it ever became possible to detect sickle-cell anemia *in utero,* the situation might well change, because then a program such as the Tay–Sachs screening program might help to solve the problem. There would, in that case, be no need to engage in a program of discouraging matings between certain types of "genetically marked" individuals, which has its disadvantages. One approach to *in utero* detection that may become possible in the not-too-distant future will follow from the development of techniques for sampling fetal blood at an early stage in pregnancy. Fetal blood can then be tested for the hemoglobin genotype. The relevant hemoglobins are of the "adult" type, but they are found in small but sufficient amounts also during fetal life, at a stage early enough for selective abortion.

Though cystic fibrosis can be detected at a very early stage of the disease, this seems to make little difference to the prognosis of affected individuals, so that there is clearly less of a case as yet for screening newborns for this disease, as is done for PKU.

There is, however, a rather peculiar and somewhat cumbersome test available for the detection of heterozygous carriers. This test is based on the fact that the serum of patients contains a factor that inhibits the beats of the cilia (or hairs) of freshwater mussel or oyster gills. If therefore an *in utero* test on amniotic cells for cystic fibrosis were developed, it would once again be possible to institute a program such as that being carried out for Tay–Sachs disease in Ashkenazi Jewish populations. So far, it is not possible to distinguish the behavior of cells cultured from heterozygotes and from affected homozygotes with the "cilia" test, which for this and other reasons is not suitable for prenatal detection on a large scale. Clearly, it would be most helpful to know what is the primary biochemical defect in cystic fibrosis, but so far there are only tantalizing suggestive results and no firm conclusions. There is little doubt that, because of its high incidence, cystic fibrosis would give a favorable cost–benefit ratio if suitable screening tests and *in utero* detection techniques became available. At the moment, the only practical recourse for a pair of heterozygous parents is to avoid further reproduction after the birth of an affected child, if they wish to be sure of avoiding the anguish of having further affected children.

Genetic screening programs and approaches to the treatment of genetic diseases have certainly advanced considerably in recent years. They are, however, still somewhat limited in their applicability and are often coupled with the need for abortion, which is never desirable and is still against the moral principles of a substantial number of people. An additional factor to bear in mind is the long-term effects of such programs on attitudes both toward normal heterozygous carriers of deleterious recessive genes and toward affected offspring born despite the availability of "prophylactic" abortion. Hopefully, an adequate education of the general public in these areas of genetics will prevent the generation of unnecessary anxieties and will allow unimpeded further progress to be made in the increasingly important area of the treatment of genetic disease.

21.3 Selection Relaxation and Dysgenic Effects of Medicine

Modern medicine permits the survival of patients afflicted by diseases that were once incurable. For those diseases that are inherited, it is often claimed that such survival will increase the incidence of the disease in the future. This result is called the *dysgenic* effect of medicine. (Dysgenic is the opposite of "eugenic," which refers to improvements in the human gene pool. The practice of "eugenics," the deliberate effort to eliminate deleterious genes from the human gene pool, is discussed in Section 21.4.)

The history of modern medicine is essentially very short, extending only over the last century or so.

The most dramatic progress in medicine has occurred within the past fifty years or less. Surgery is an ancient art, but its effectiveness was certainly rather limited before antisepsis and anesthetics were introduced in the second half of the nineteenth century.

The "demographic transition" from high birth and death rates to low birth and death rates began not more than 150 years ago, and then only in some countries (see Chapter 7). The decline in death rates was initiated, not by new therapies (which came later), but by improved hygienic, economic, and social conditions. A much earlier change in survival conditions, however, was initiated by the agricultural revolution. The development of farming must have greatly improved the selective advantage of many types other than those most fit for a life of hunting and gathering. The urbanization that soon followed led to still further changes in the style of life (see Chapter 19). This did not lead to a total relaxation of natural selection, but rather to a change in the traits selected for. Undoubtedly for some traits there must have been relaxation but, for others, different and even more stringent selection occurred. The "dysgenic" effects of medicine are therefore just examples of selection relaxation, which is a more general phenomenon.

There are a few suggestive examples of selection relaxation, though in no case is the evidence really compelling. The high incidence of color blindness among Caucasians (8 percent in males and lower in other races) as compared with the lower incidence among hunters and gatherers or other groups that made the transition to agriculture rather late, is a possible example of selection relaxation.

When selection against a gene is relaxed (in the true sense of the gene becoming selectively neutral), its gene frequency can increase by two mechanisms: either mutation pressure or random genetic drift.

The mutation rate that would be needed, however, to increase the gene frequency of color blindness from the hunter–gatherers' level to the Caucasian level in the time since Caucasians adopted agriculture is much too high. The difference in gene frequency between Caucasians and hunter–gatherers is of the order of 5 percent. When this is divided by the number of generations since the transition to agriculture, namely 400 (about 10,000 years at 25 years per generation), we get an approximate estimate of 1 in 8,000 for the mutation rate to color blindness needed to alter the gene frequency in 400 generations. A mutation rate of 1 in 8,000 is very high compared with other known rates and would actually be the highest known for any single gene (see Chapters 4 and 8). More than one gene is known to be involved in color blindness, so that the sum of their mutation rates must be considered. Even so, it seems likely that genetic drift or other factors contributed to raising the gene frequency of color blindness among modern Caucasians. Perhaps there is some other (as yet unknown) reason why environmental conditions may have favored the color-blindness gene during the time since transition to agriculture and so allowed it to increase by positive selection. Only a slight selective advantage, perhaps too small to be detectable, would be needed to effect such an increase by selection.

A more convincing case of selection relaxation than color blindness is that for eye refraction.

Most people require spectacles sooner or later. The spherical refraction (Fig. 21.7) of Caucasian eyes shows a much higher individual variation than that of other ethnic groups. The more this index diverges from zero, either in the positive (long-sight, or hyperopia) or negative (short-sight, or myopia) direction, the more people require spectacles. A higher proportion of Caucasians require spectacles, because the distribution of the spherical refraction index has longer tails than in other population groups. There is a fairly large inherited component to this trait, and so one must consider the possibility of relaxed selection for refraction among Caucasians. Caucasians also seem to have "worse" genes, which would normally be selected against, for a variety of other traits, including abnormal nasal septa (which increase the chances of infections of the nose and neighboring systems), and the capacity to produce milk. Will the life-saving action of obstetricians in difficult deliveries increase the proportion of difficult births in future generations? The use of the rhythm method for birth control could perhaps increase irregularities in menstrual periods, as more children will be born from women whose cycle is irregular, and therefore cause failure of birth control. Will

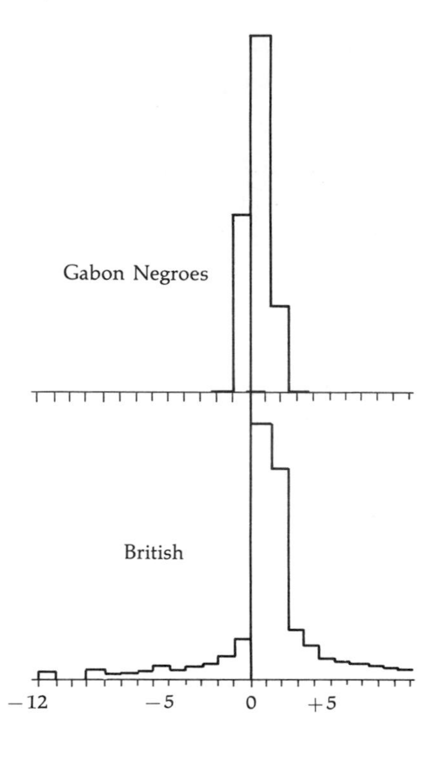

Figure 21.7
Spherical refraction in different populations. Frequency distributions of refractions are shown in samples from an African and a Caucasian population. Note the spread of the tails of the Caucasian distribution. The greater the deviation from zero, the greater the dependence on glasses for normal vision. (After R. H. Post, *Humangenetik,* vol. 13, p. 253, 1971.)

harelip and spina bifida increase because selection against them has been relaxed by surgical treatment? And so the list continues.

What is the magnitude of the expected dysgenic effect of selection relaxation due to medical advances?

Even though there are few or no data, reasonably precise computations can be carried out, at least in specific cases, because of the predictability of genetic disease. Consider a recessive lethal such as PKU with a birth incidence of 1 in 15,000. If we assume the heterozygote has the same fitness as the normal homozygote, it will take about 50 generations for the disease to double in frequency (that is, rise to 1 in 7,500) if the affected PKU individuals are completely cured. A rarer recessive starting at 1 in 100,000 (such as galactosemia) will take even longer (some 140 generations) to double its frequency. Recessives thus will increase extremely slowly under fully relaxed selection (Fig. 21.8).

Dominant defects will, however, increase in frequency more rapidly. If they are maintained entirely by mutation, as is almost the case for achondroplastic dwarfism, they will double in frequency in the first generation and keep increasing by the same amount in the later generations. Thus, in 10 generations, or about 250 years, they will be at about 10 times their initial frequency. In the case of

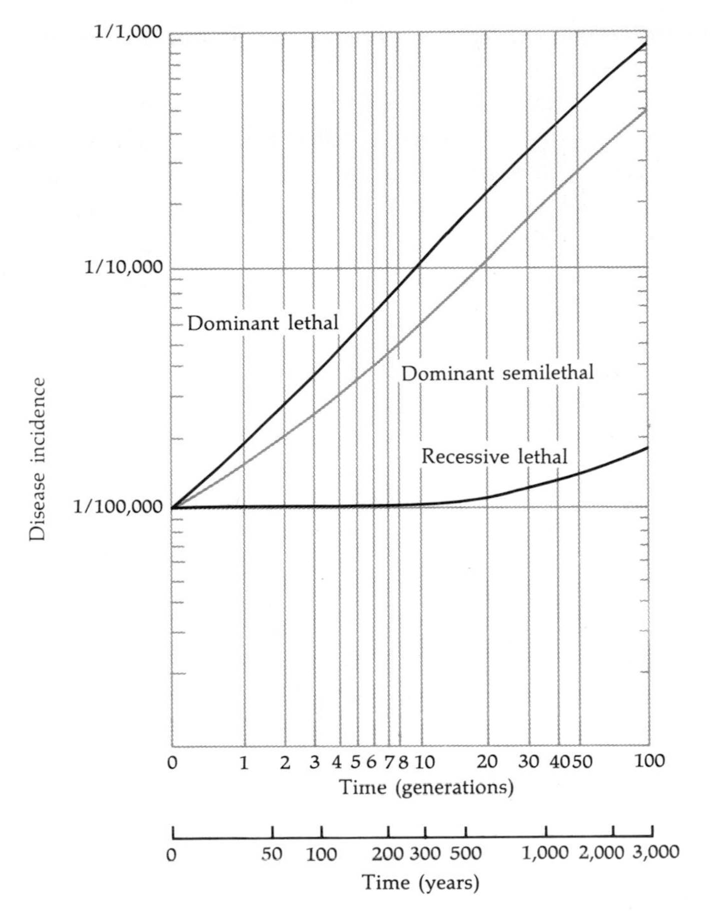

Figure 21.8
Dysgenic effect of medicine. A dominant lethal with an incidence at birth of 1 in 100,000 would increase in incidence in later generations if a treatment became available that would normalize the fitness of carriers. A dominant semilethal (for which the heterozygote had a fitness 50 percent of that of normal homozygotes before the treatment became available) would increase in incidence somewhat more slowly. A recessive lethal would show a perceptible increase only after some ten generations. The shape or slope of the curves in this logarithmic presentation would not be altered substantially (at least in the first several generations) if the initial incidence were some value other than 1 in 100,000.

dominant defects for which heterozygotes have a relatively high fitness, such as for Huntington's chorea, the increase in frequency will be much slower, requiring ten generations or so to double in frequency.

The effect of selection relaxation on traits with a complex pattern of inheritance will depend on their heritability and initial frequencies. In general, the increase under relaxed selection will be slow, the maximum possible rate (rarely attained) being a doubling in one generation, while the minimum is a totally negligible increase in one generation.

The dangers of other hazards that we face over the next thirty years or so are certainly much higher than these effects of relaxed selection. The average increase in frequency of genetic diseases will be so slow that very little change will be perceived in the next 50 years, even with the help of excellent vital and health statistics. It would seem therefore that the dysgenic effect of medicine is not a real threat. By the time it may have a clearly perceptible global effect, 200 or 300 years from now—when the incidence of severe genetic disease may have doubled on average—our descendants almost certainly will have discovered simple methods of therapy, unless our civilization will have been destroyed by its own lack of wisdom!

Genetic screening and prophylaxis may sometimes have delayed, and also definite, dysgenic effects.

As already mentioned, the cure for PKU creates a problem when PKU treated women have pregnancies. Hopefully, use of the appropriate diet during their pregnancy will prevent their offspring from becoming PKU "phenocopies."

In the case of hemophilia, the strategy of aborting hemophilic males may have a very small dysgenic effect, for if these are replaced by female children, the frequency of heterozygous females carrying the hemophilia gene will increase. The best strategy would be that of aborting *only affected* males, but at the moment these cannot be recognized *in utero* so that, in order to avoid the birth of diseased children, all males must be aborted. In either case, the expected increase in gene frequency due to the birth of an increased number of heterozygous females has been shown to be extremely small. In some other cases of serious X-linked diseases, such as the Lesch–Nyhan syndrome (see Section 5.2), affected males can be recognized *in utero* and aborted.

Similarly, for Tay–Sachs disease, a very slight increase in gene frequency is expected if all homozygotes are aborted. On the assumption that the total number of live children born to the at-risk families, which involve matings between heterozygotes, is the same as in other families, more heterozygous children will be produced than in the absence of selective abortion. But here also the resulting increase in gene frequency is so small that it can be considered as practically negligible.

The effects of medicine are not, however, entirely dysgenic.

One consequence of malaria eradication is that of starting the elimination of genes such as those that cause sickle-cell anemia and thalassemia. With the disappearance of the heterozygote advantage, which exists only in the presence of malaria, these diseases become single deleterious recessives, and in populations in which they are at high frequency because of previous malarial history, they will therefore decrease in frequency. This will happen at the relatively slow pace characteristic of selection against recessives (Fig. 21.9). For example, where thalassemia now has an incidence of 1 percent, its incidence will be halved in five generations, and halved again in five more generations. Sickle-cell anemia (which is not quite as detrimental as thalassemia) will decrease a little more slowly, its incidence going down by a factor of three (rather than four) by ten generations from now. Ten generations is approximately the period spent in America by the

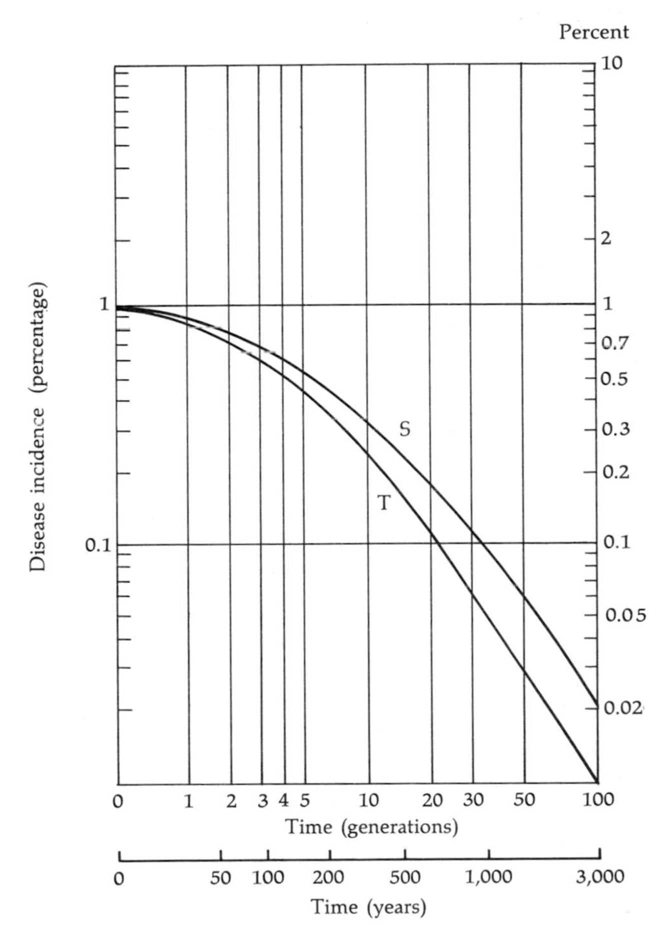

Figure 21.9
Eugenic effect of medicine for diseases maintained by heterozygote advantage. Thalassemia and sickle-cell anemia are prominent examples of such diseases; elimination of the diseases that determine the heterozygote advantage has a eugenic effect. The curves show the fall in disease incidence to be expected for sickle-cell anemia (S) and for thalassemia (T) as a result of establishment of malarial prophylaxis.

Africans who were first brought over as slaves. It should be expected, therefore, that the incidence of sickle-cell anemia among black Americans is lower than it was originally and this is found, even when the admixture of Caucasian ancestry is taken into account. For genes that have a heterozygote advantage maintained by some environmentally induced or contagious disease, the eradication of the disease may thus have *eugenic* effects.

On the whole, however, it seems that the overall effect of medicine is more likely to be dysgenic than eugenic, and each new type of intervention—medical, hygienic, or surgical—has a chance of adding to the future burden. Nevertheless, the total burden accumulates very slowly, so that even the detection of a change may require a very large body of data. It should be remembered that recently observed increases in the incidence of certain diseases (such as diabetes or cancer) have quite other causes, such as improved diagnosis, changing age distribution of the population, changing environment, and changing customs.

Many much more serious problems than the dysgenic effects of medicine are likely to afflict the world in the nearer future. Even if this were not true, we are not likely to give up technological progress, especially in medicine, only because of a fear of increasing the incidence of disease in the rather remote future. Technological progress in general is irreversible, except perhaps following catastrophes, and the same is true of its acceptance. We are not likely to give up eye glasses to avoid an increase in the frequency of refraction defects amongst our descendants. One cannot easily turn the clock back on technological development.

21.4 Eugenics

Can one improve the human stock by selective breeding in favor of more desirable, or against less desirable, human types?

Eugenics is the term coined by Francis Galton in the last century for the science of improving the future qualities of mankind by selective breeding. Though the term is relatively new, the idea is quite old—at least as old as the classical Greek philosopher, Plato. In early Roman times, the Tarpeian rock was said to be the place for the disposal of handicapped children, and more recently, Frederick II of Prussia is said to have given the "best" girls to his "best" soldiers to promote the formation of an "elite." The success of plant and animal breeders in improving their stock by artificial selection suggests that one may be able to obtain, by similar techniques, considerable improvement in our presently so highly fallible human stock.

There is usually a distinction drawn between *positive* and *negative* eugenics. The first indicates the enrichment of desirable qualities, while the second refers to the elimination of undesirable qualities. It is certainly on the whole easier to decide which traits are "bad" and should perhaps be bred out of the population, and so we shall start our discussion with negative eugenics. Only inherited characters can respond to selection and so genetic diseases are obvious targets for negative eugenics.

Few people would perhaps argue against the benefits of removing, say, the PKU or the Tay–Sachs or the cystic-fibrosis gene from the population, if this could be managed at a "reasonable" cost, but it is not always easy to decide whether a gene is "good" or "bad."

Even when considering an apparently deleterious, disease-causing gene, it may be difficult to decide whether it is desirable to increase or decrease its frequency.

In the absence of malaria, sickle-cell anemia has nothing obvious to recommend itself, as far as we know. Nevertheless, the *S* gene is not universally "bad." In malarial environments, it is "good," and only very recently, in those few areas where malaria has been eradicated, has it been turned from a "good" to a "bad" gene. There would be no advantage at present to eradicating the *S* gene in Africa, even if this were feasible. Even in the case of, say, the cystic-fibrosis gene, there might conceivably be a slight hidden advantage of the heterozygotes.

When we consider other relatively frequent diseases, such as schizophrenia, we may wonder whether the eradication of the genes that are involved—were this at all possible—might not be as undesirable as the eradication of the sickle-cell gene in Africa. Schizophrenia is only partly inherited, and it is not clear how many genes are involved. But assuming for a moment that it is a simple recessive, it would be difficult to explain the high frequency of the disease unless either (1) heterozygotes for the gene have an advantage and this maintains the high frequency, or (2) the gene was previously neutral or advantageous, but the environment has changed, so that it is now deleterious. If the first explanation is true, eradicating schizophrenia might well be a mistake. It has even been jokingly suggested that, as so many artists appear to have a schizoid temperament, the same general genotype is responsible for both schizophrenic and artistic tendencies. Would, then, doing away with the gene also throw away the theater, movies, paintings, entertainment, and art, all in one shot, leaving us in a rather dull world? The problem obviously is that we often even do not know what genes are involved, nor do we really know what maintains gene frequencies at their present level, especially in the case of some of the most common, and so most important, diseases. Rarer diseases are more likely to be maintained by a balance between mutation and adverse selection so, in these cases, elimination of the gene is almost certain to be beneficial.

This means that a limited policy of negative eugenics might be desirable if feasible, and might counteract the possible eventual increase in the burden of genetic disease brought about by modern medicine.

The same reasons that (fortunately) make the dysgenic effects of medicine very slow, also (unfortunately) make the response to negative eugenics slow.

One can predict reasonably well the improvement expected under various negative eugenic programs, ranging from the discouragement of reproduction to the forced sterilization of the "unfit" (which is rarely practiced, but is in principle encouraged by the laws of a few countries or states).

Genetic counseling itself is bound, on average, to have a slight eugenic effect, as couples at risk are sometimes discouraged from reproducing—the more so the higher the risk. The effect can only be small, for perhaps only 1 percent of couples seek genetic advice. As we have already discussed, there would hardly be much improvement from making genetic counseling mandatory, because those who seek advice do so on a strictly voluntary basis, and so are unlikely to be a random sample of couples at risk. It must be emphasized that the primary purpose of genetic counseling is not eugenic. It does not aim at improving the population level of a gene, but is (and should be regarded as) a practice designed to benefit the persons seeking advice on an individual or family basis. Making the service mandatory for eugenic reasons might easily discredit the whole operation of counseling, and would at best have a marginal effect.

Some idea of the total potential impact of negative eugenics can be obtained from an estimate of the effect of a campaign that would effectively avoid completely the reproduction of the "unfit" through sterilization, abortion, or other means.

The results obtained would depend upon the operational definition of the "unfit." Three different definitions could be used.

First, the "unfit" might be defined as homozygous carriers of deleterious recessives.

This approach would essentially involve a total reversal of present medical trends. For example, it would call for the elimination (at least from the reproducing population) of PKU patients, rather than their cure. Because these individuals have (or had until recently) a fitness close to zero, gene frequencies and incidences would probably remain much the same as present levels, despite the eugenics program. The only effect of the program would be the prevention of a future ("dysgenic") increase, which (as discussed in Section 21.3) is very slow.

In a second approach to eugenics, the "unfit" might be defined to include apparently healthy heterozygous carriers of deleterious recessives, as well as the recessive homozygotes.

In a program based on this definition, heterozygotes for recessive defects (such as sickle-cell anemia, cystic fibrosis, and many others) would be sterilized or at least discouraged from breeding. (Presumably they would be allowed to live out their own lives.) At first glance, such a program might seem desirable, for it offers the opportunity to remove in a single generation many deleterious genes from the gene pool. (The program would have to be continued in future generations to remove those deleterious genes that reappear through mutation.) However, only about 10 percent of the recessive defects known today can now be detected in the heterozygotes—moreover, the deleterious recessives known to us must be only the tip of the iceberg. As shown in Chapter 11, each individual probably carries (on average) about two or more deleterious recessives. Thus, only a very small fraction of the population (perhaps 1 percent or less) is totally, or relatively, free from seriously deleterious recessives. An effective eugenics program then would permit only this tiny fraction of the population to reproduce! Clearly, such a strategy would be absurd. Because there are so many recessive genetic diseases, each with a relatively small incidence, the application of such a program only to certain defects would have relatively insignificant effects on the overall problem.

A third approach to such a eugenics campaign might define as "unfit" only homozygotes for recessive defects and both homozygotes and (sick) heterozygotes for deleterious dominants.

The elimination of heterozygotes for deleterious dominants would be effective only for those diseases (such as Huntington's chorea) whose Darwinian fitness is relatively high. However, in most such cases (as with Huntington's chorea), the symptoms of the disease appear only *after* reproduction. There is, so far, no way to detect the presence of the gene before the age of reproduction. Only half of the offspring of affected individuals carry the deleterious gene (on average), but at present the only way to apply such a program would be to prevent *all* children of affected individuals from reproducing. (Symptoms usually appear in an affected individual after his own reproduction, but before his offspring reach the age of reproduction.) Such an approach would require the effective sterilization of a number of entirely healthy people, though it would essentially eradicate the disease (except for mutational reappearances).

For dominant diseases with a low Darwinian fitness, such as achondroplastic dwarfism (which has a fitness of about 20 percent), such an approach would reduce by about 20 percent the number of affected individuals in the next generation. The balance of the incidence in each generation is due to new mutations. If the fitness of the heterozygotes is zero (as for Down's syndrome), the eugenics program could have no effect on the incidence of the disease in the next generation, because this incidence is maintained entirely by new mutations. Similar computations can be made for X-linked or polygenic traits. In general, these computations show either that the eugenic effects would be small or nonexistent, or that a program of mass sterilization would be required, which certainly cannot be justified on the basis of present knowledge.

There is, however, one approach to negative eugenics that may (potentially) be slightly more rewarding.

This approach involves encouraging certain pairs of individuals to avoid matings—an approach discussed in Section 21.3 in connection with the sickle-cell trait. If matings between heterozygotes for the *same* recessive gene were successfully discouraged, the disease could be substantially reduced in incidence or even eliminated. Of course, the responsible genes would remain in the population's gene pool, but only in heterozygotes. The problem is that this approach requires that every individual be screened for all possible genes in time to avoid marrying other carriers, and this would be a very costly exercise indeed. Because most of the genes involved are very rare, the restrictions on mating patterns that this would impose are, on the other hand, quite minimal. Another possibility, pointed out many years ago by J. B. S. Haldane, is to avoid consanguineous matings. As discussed in Chapter 11, homozygotes for rare recessives are often the offspring of consanguineous unions. Avoidance of close inbreeding would, therefore, clearly reduce the incidence of rare recessive diseases. Modern demographic trends seem to be achieving this aim without much need for further encouragement.

Positive eugenics—namely, breeding in favor of "desirable traits"—has been discussed from time to time by scientists, but for obvious reasons has never gained much public support.

There are many problems with eugenically motivated selective-breeding programs. Not the least of these are deciding what traits to breed for and the implied restriction on the liberty of individuals to reproduce as and when they wish to. One approach to positive eugenics was strongly advocated by the distinguished U.S. geneticist, H. J. Muller, who was the discoverer of the mutagenic action of X rays, for which he received the Nobel prize. Muller supported the notion, which he called "eutelegenesis," that semen from certain carefully chosen, distinguished and important men should be collected and stored for the purpose of artificially inseminating as many women as possible. In this way, at least, the genes from "desirable" males could be selectively propagated. Technically this is quite feasible, and in fact, AID (Artificial Insemination by Donor) is turning out to be a valuable therapy for male infertility when a couple wishes to have children. It is a reasonable alternative to adoption, which (because of the decline in the birthrate, and so of the number of children available for adoption) is inevitably becoming more and more rare. The possibilities of AID arose mainly out of studies with economically important farm animals. Most cattle breeders, for example, now make use of artificial insemination, using carefully selected bulls for genetic improvement of the stock. Semen can be stored more or less indefinitely in liquid nitrogen at a temperature of $-196°C$.

In spite of its apparent advantages as a therapy for infertility, AID raises a number of major legal and ethical questions.

A number of these problems center on the question of the legitimacy of the AID child. In most countries so far, the AID child is in a sort of legal limbo, while in others it is frankly illegitimate and may even be a ground for suing for divorce. Should the sperm donor be identified, and what are his responsibilities if the husband dies prematurely? Does the child have a right to know who his or her biological father is, and should he in fact know whether or not he is a child of AID? Then, there is the question, of course, of who should be chosen as donors, and whether they should be paid. Commercial sperm banks have already been established in the United States; in Britain, regulation of sperm banks and moves to prevent their commercialization have been called for. This is a formidable list of problems, although it should be possible to deal with them in a way that does not violate most people's moral and ethical principles.

AID could certainly be used for eutelegenesis as advocated by Muller; but most, if not all, societies would find repugnant any restrictions on the right to choose husband or wife and to have children.

Thus, eutelegenesis could be considered only on a voluntary basis, and it is unlikely to appeal to a significant proportion of the population. The most difficult question is always the choice of donors. What are the desirable traits to be selected for? IQ would seem to be favored by some academics. In fairness to Muller, it should be said that his list of desirable traits did not give much prominence to IQ, but rather emphasized qualities such as altruism, moral values, and usefulness to society. Muller himself demonstrated how difficult it is to be consistent in one's choice of desirable people. In an early list of desirable sperm donors, he included Marx and Lenin, but after a sojourn in Russia, which provoked his disenchantment with the Russian system of the time, he struck these two names off his list. It is virtually impossible to evaluate the potential usefulness of an individual to society, except perhaps in a fairly narrow sphere of activity. Moreover, almost all of the qualities considered as desirable, especially in the social sphere, are not known to have a substantial inherited component. In any case, simply establishing the existence of a genetic component is not sufficient, because it is only the additive component (which contributes to the "narrow" heritability; see Chapter 15) that matters in artificial selection, and this may be more difficult to estimate than the overall genetic component. It is not likely to be an easy matter to estimate the relative effect of inheritance on traits such as altruism or selfishness, love, hate, or jealousy. These problems, of course, apply to all eugenic programs, whether carried out by eutelegenesis or in other ways.

In the desire to improve society by eugenic means, it is easy to forget that variety is an essential ingredient of our society.

We need an enormous variety of talents and skills. Selection for just a few traits, however desirable, carries with it the risk of decreasing genetic variation. This decrease is an inevitable consequence of all artificial-selection breeding programs, and it creates problems of which animal and plant breeders are well aware. The selection of a few, well-chosen pure lines to generate "miracle crops" may prove disastrous when a new environmental situation is faced, such as a new pathogen or unusual climatic conditions. In animal breeding, the hidden defects of selected sires have been responsible for a number of serious problems generations later—as in the case of two Holstein Friesian bulls which were widely used as sperm donors by the Swedish dairy industry, but each of which carried a recessive lethal gene. The resulting defects were only discovered among their descendants after the gene frequencies had already reached a high level and the defects were causing serious financial losses. Many similar stories can be found in the annals of animal breeding.

However well intentioned, it seems likely that eugenic programs are, so far at least, doomed to failure.

Even if there were to be an expansion of eugenic programs, the eugenist—or rather, his descendants—would have to wait a long time, measured in generations, before seeing any effect of the advocated measures. There are, after all, many more urgent problems to solve that can yield returns in a time that is more compatible with the lifetime of an individual and that of his progeny and his grandchildren.

The eugenic movement was, of course, greatly discredited by some of the monstrous abuses that were perpetrated under its guise. The worst—not soon to be forgotten—was the massacre of 6 million Jews (and many other "unfit" individuals) on Hitler's orders. Biases in immigration laws in the United States and elsewhere are relatively minor, but still are significant examples of the political abuse of the eugenic movement. These examples illustrate the fact that eugenic principles are often coupled with the notion that there are good and bad races, a notion undoubtedly cherished by Francis Galton himself. Racism and eugenics frequently seem to go hand in hand.

Selection of the sex of one's children, though not strictly eugenics, is certainly closely related, and is already technically possible to a limited extent.

Thus, following amniocentesis, it is possible to abort a fetus of the unwanted sex. This is, however, certainly not a practice which should be readily condoned. Experiments have been reported more than once that claim the separation of X and Y sperm, with a view to using the separated sperm for artificial insemination to produce children of the desired sex, but none have so far been substantiated. Though these techniques are not yet available, it may not take too long before they are developed. However, even then one might wonder how many mothers would be willing to have an abortion, or a child by artificial insemination with their husband's sperm, in order to have a child of the desired sex. It seems likely that simpler methods would have to become available before sex selection would become at all widespread.

Widely divergent opinions have been expressed on the possible social consequences of "sex selection." Some have predicted apocalyptic consequences, such as the creation of large changes in the sex ratio, followed perhaps by massive reversals, creating strong and unpleasant competition between people of the same sex. Others consider that there may be some advantage, for example, to eliminating children of the unwanted sex who may be the victims of parentally induced maladjustment. Another important accompanying factor could be a contribution to birth control. Sex selection would make it easier to achieve the desired sex composition of the family with, on average, a smaller number of children. The results of recent questionnaire studies have actually suggested that widespread use of sex selection might not, in fact, have much effect on the sex ratio. Clearly, however, the provision of a choice of the sex of children is not something to be entered into lightly, especially if a reliable and cheap method ever becomes available.

Concern is often expressed about the future consequences of present genetic manipulations.

How worried should we be about man taking his future evolution into his own hands? Undoubtedly, all such developments must be watched carefully and their future impact carefully evaluated. However, as emphasized in our discussion of genetic engineering, many of the more dramatic possibilities—paraphrased in the popular press by such terms as the "test-tube baby"—are still a long way off. As the U.S. geneticist Joshua Lederberg pointed out, by analogy, the fact that we can build a bridge across the San Francisco Bay does not mean that we can build a bridge across the Pacific Ocean—or that, if we could, we would be willing to devote the necessary resources to doing so.

21.5 IQ, Race, and Social Class

Much public discussion has centered in recent years (as already mentioned in Chapter 17) on problems of IQ and society, especially in the United States. There are two major issues: (1) Does the lower IQ measured in some races (essentially American blacks) as compared to others (usually Caucasian) have a genetic or an environmental basis? (2) Is the lower IQ measured in the lower socioeconomic classes (within a racial group) due to genetic or environmental factors?

Whenever a subject raises strong emotional feelings, scientific discussion inevitably becomes less easy and less constructive.

There is an obvious emotional background to the IQ question that is revealed by a consideration of the history and the social aspects of these issues. There has been for many years considerable discrimination against blacks in the United States. Only recently has this discrimination in part been removed, while at the same time attempts to compensate for past discrimination by so-called "affirmative action" programs have been initiated. The consequences of this history of oppression cannot be obliterated at a stroke, and may take several generations to sort out. The mere posing of the IQ question suggests, and sometimes appears even to be derived from, the desire to revive oppression and to create a rationale for its existence. Similarly, when the IQ question is posed at the level of social classes, it immediately evokes the suspicion of an attempt to justify and perpetuate social injustice. It is clearly very difficult to discuss these issues, however scientifically and objectively one may try, without raising emotions and prejudices of one sort or another.

The statement that the observed differences in IQ may have a genetic rather than environmental basis often appears to be coupled with the belief that nothing can be done to correct the difference if it is genetic. Remarkably, however, one of the few genetic diseases that can be treated successfully, namely PKU, has as its main symptom a very substantial lowering of IQ. This was not just a lucky coincidence, because there are several other genetically determined aminoacidurias that also lead to severe mental deficiency and that have been similarly treated with success. There is really no justification for "genetic fatalism."

The magnitude of the difference in IQ between U.S. blacks and whites is of the order of 15 IQ points, or one standard deviation of the IQ distribution (Fig. 21.10).

In other words, the IQ average for blacks is 85 as compared to 100 for whites, on whom the test is standardized. An immediate problem is the possible bias introduced by the fact that the test is standardized on Caucasians and so is biased in favor of their culture. This is almost certainly true to some extent: there exists no truly culture-free IQ test. Any intelligence test is a test of performance and can hardly be a test of innate ability, because IQ (as discussed in Chapter 17), like any other behavioral trait, inevitably depends on development having taken place in a specific cultural background. Sufficient differences between the cultural backgrounds of blacks and whites remain even after desegregation for an effect of

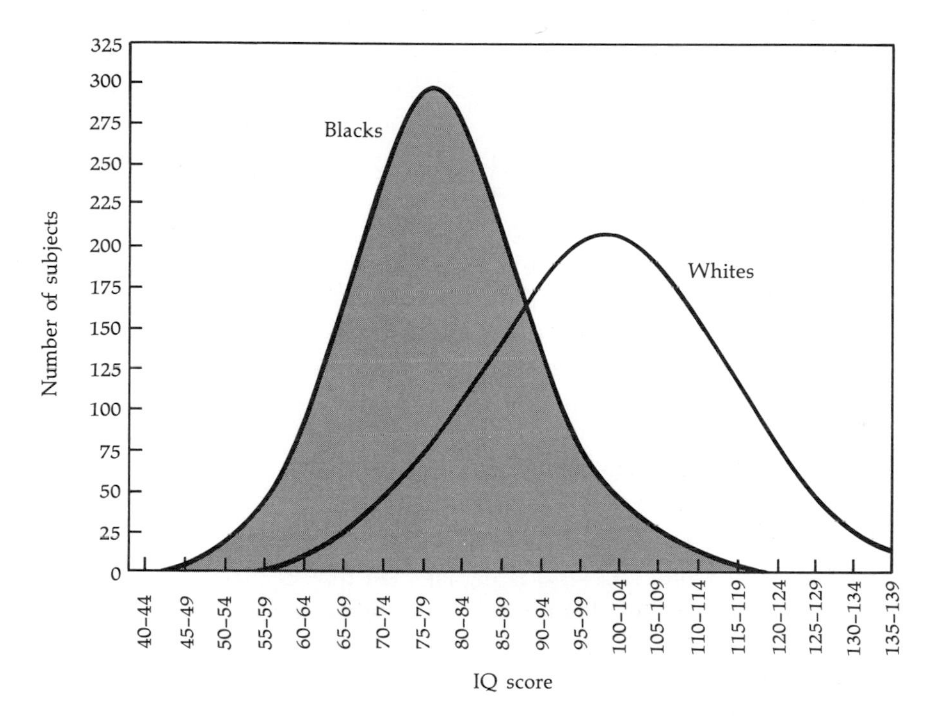

Figure 21.10
IQ difference between blacks and whites in U.S. population. The graph compares IQ distributions in a representative sample of whites and among 1,800 black children in the schools of Alabama, Florida, Georgia, Tennessee, and South Carolina. W. A. Kennedy of Florida State University, who surveyed the students' IQ, found that the mean IQ of this group was 80.7. The mean IQ of the white sample was 101.8, a difference of 21.1 points. The two samples overlap distinctly, but the difference in the two means is significant. Kennedy's result is one of the most extreme reported; other studies indicate a black-white IQ difference of 10 to 20 points. (From W. F. Bodmer and L. L. Cavalli-Sforza, "Intelligence and race," *Scientific American,* October 1970. Copyright © 1970 by Scientific American, Inc. All rights reserved.)

such differences to be expected. However, the magnitude of the effect is difficult, if not impossible, to assess using presently available data and approaches.

In his widely quoted article in the Harvard Educational Review *in 1969, Arthur Jensen argued that the black–white difference is likely to be mostly genetic in origin. The following major points outline his argument.*

1. IQ has a high heritability in Caucasians. Jensen, following Burt, estimated this heritability to be 80 percent.
2. The black–white difference is found within each socioeconomic group, and therefore is unlikely to be due to environmental causes.
3. The parent–child regression for IQ is the same in blacks as in whites, but is shifted downward, indicating the same level of genetic determination of IQ in blacks as in whites.

Since 1969, Jensen has added the following major considerations to his argument.

4. IQ tests that show a high heritability among Caucasians also show a greater difference between blacks and whites.
5. No environmental component that has been investigated—from nutrition to schooling—can account for the black–white difference. In his more recent book on educability and group differences, Jensen comes to the conclusion that between $\frac{1}{2}$ and $\frac{2}{3}$ of the average IQ difference between blacks and whites is genetic, but he does not properly explain how he arrived at this conclusion.

William Shockley, well-known physicist and Nobel-prize-winning coinventor of the transistor, has also actively embroiled himself in discussions of race and IQ. Shockley has added the following argument to the hereditarian position on the matter.

6. The mean IQ of blacks from Northern states (who have, on average, a greater proportion of white ancestry) is higher than that of blacks from Southern states. Shockley claims that the data indicate a one-point increase in IQ for each added percent of Caucasian ancestry in a black individual.

We believe that none of these findings can answer the question of the extent to which the black–white IQ difference has a genetic component.

Let us consider the six points raised by Jensen and Shockley, in turn, bearing in mind the principles of quantitative genetics covered in Part III of this book.

1. The heritability of IQ, both broad and narrow, is very probably lower than the value of 80 percent estimated by Jensen. However, even if the heritability were 100 percent (which would make the trait completely innate), this does not mean that the trait could not be modified by a change of environmental

conditions. For example, there could easily be *environmental factors that do not vary so much within either group as they do between the groups.* Such factors do not contribute to heritability estimates obtained within a group. These estimates only measure the importance of genetic and environmental variation in the group, and they can say nothing at all about differences between groups. Cultural differences between races—connected, for example, with family structure or patterns of child rearing—may simply have no parallels within racial groups. Note that stature has a heritability at least as great as that of IQ; nevertheless, mean stature has changed over the past 100 years as a result of environmental changes. PKU has close to 100 percent heritability, but an environmental change has effectively abolished its effect on IQ.

2. If all of the IQ difference were due to environmental causes, then the black–white difference would disappear when blacks and whites were compared in identical environments. However, disappearance of the difference within socioeconomic groups would be expected *only* if socioeconomic conditions as such are the only source of environmental effect. But the cultural segregation of blacks and whites does not disappear when blacks and whites of the "same" socioeconomic groups are compared, even assuming that those rather loose categories called "socioeconomic groups" are strictly comparable between blacks and whites. *Socioeconomic level must be very far from a complete description of the environment* that matters for the development of IQ.
3. Even with a purely cultural (rather than biological) theory of parent-to-offspring transmission of IQ, we would expect to observe the same regression to the mean in blacks and whites, but shifted downward in the case of blacks. Furthermore, a purely environmental component stimulating a phenotype in environment A and/or depressing it in environment B would cause exactly the same phenomenon. All that can be concluded from Jensen's observation about the regression to the mean is that the heritability (without distinguishing biological from cultural transmission) is about the same within blacks as it is within whites. This finding is of no help in distinguishing genetic from environmental factors.
4. The correlation between heritability and black–white differences for various types of IQ tests is subject to the same objection as that raised for the third point, because *the heritability estimates that are commonly given do not distinguish clearly between biological and cultural transmission.* Again, this finding does not distinguish genetic from environmental factors.
5. As discussed in Chapter 17, there are many environmental effects on IQ whose precise mechanisms are not yet understood. The well-documented decreases in IQs of twins (by an average of 5 points) and of triplets (by 9 points), in addition to the decrease in IQ with order of birth (see Fig. 17.6), are striking examples of subtle environmental effects, somehow dependent

on family structure. Until we know how to explain such environmental changes, we must always wonder whether we have not missed important environmental components in seeking an environmental basis for differences between groups. Another striking effect of the environment on IQ is shown by the increase in mean IQ of adopted children, as compared with their biological parents. The magnitude of the gain is about one standard deviation over the values that might otherwise have been expected, and so is comparable to the black–white difference (see Chapter 17). This effect can, of course, be ascribed to the "good" environment in the adopted home, but what exactly is it in the environment that makes it good? The attention paid to children's curiosity at the time they mature, and stimulation (or repression) by parents of this curiosity may be important elements that are subject to cultural influences. Many other emotional components that are difficult to evaluate also are likely to be influential, such as age-peer effects, for example. It seems likely that there is a *sensitive period of development* during which many factors can influence a child's future IQ. Nutritional and prenatal influences, while they have some effect, are perhaps less important. The black–white difference is already present in first grade at school (by the age of five or six) and does not change much thereafter, suggesting that the relevant factors act before six years of age.

6. Shockley's argument about the apparent correlation between IQ and white ancestry is also very weak, again because of the complete confounding of genetic and culturally transmissible environmental factors. Blacks in the Northern states are probably less sharply segregated than those in the South, and may have undergone a greater acculturation than those in Southern states. Even the estimate derived by Shockley—that a 1 percent increase in Caucasian ancestry carries a one-point increase in IQ—shows immediately that the claimed effect is unlikely to be genetic. Only an enormous heterotic effect, which is very unlikely, could achieve such a remarkable increase.

One observation that indicates that the cultural environment may be responsible for the black–white difference in IQ is the comparison between children from marriages between a black woman and a white man, and those from marriages between a white woman and a black man. It has been reported that, on average, the IQ of children is higher if the mother is white and the father black. Such a reciprocal difference (if it is confirmed in further studies) has no sensible explanation in terms of genetic factors. On the other hand, this observation is quite consistent with the supposition that the higher average IQ of whites is largely due to cultural traits transmitted to the child during infancy and early childhood, when the mother is normally the major influence on the child's experiences.

It has been argued that—because the gene pools of whites and blacks are known to differ with respect to all sorts of polymorphic markers, as well as for genes determining outward features such as skin color and face shape—there is no reason to suppose that there would not be similar differences with respect to IQ.

But this would mean that the genes affecting IQ would differ in such a way that, on average, whites had higher frequencies for the IQ-increasing genes than blacks. Of course, if there were a single gene with a major effect, this could differ in frequency between populations, giving rise to a genetically determined mean-IQ difference. It is, however, hard to believe on present evidence that one gene (or even a very small number of genes) could be responsible for the 15-point IQ difference. When there are many genes affecting a quantitative character, such as IQ, the only sensible explanation for a mean difference between populations (that is in part genetically determined) is natural selection in favor of an increase or a decrease in the character in one or the other population. This would ensure that, say, all genes that increased IQ would increase in frequency in the population in which an increased IQ had a selective advantage. Drift or other factors would cause gene-frequency changes that were independent of a gene's effect on IQ, and so could not lead to a systematic difference between two populations—unless a very small number of genes were involved, when this could happen by chance. It clearly is very difficult, if not impossible, to assess the role that natural selection might have played in accentuating IQ differences between the races. But it at least seems clear that it is most unlikely that a 15-point IQ difference could have been generated by selection in the comparatively short time (not more than 10 generations) since slavery was started. Certainly any hypothesis one chooses to put forward for the action of natural selection on IQ is in the realm of unsubstantiated speculation and cannot be claimed as even suggestive evidence for a substantial genetic component to the racial IQ difference.

The complete answer has yet to come for the question of whether there is a significant genetic component to the black–white IQ difference.

In any case, why is it important to answer the question? The problem does not seem to be of any great scientific or theoretical interest, and it is not clear just what practical actions would follow a demonstration of a given level of genetic determination for the IQ difference. Because the distributions of IQ scores in the two races largely overlap, a significant proportion of the black population has IQ scores higher than the mean of the white population. Therefore, any differing approaches to education (such as have been proposed by Jensen) would still have to be based upon individual testing and screening programs, not upon racial discrimination. Shockley has suggested a eugenics program that would involve payment of a bonus to low-IQ individuals who agree to be sterilized before having any children; the bonus would be proportional to the extent to which the individuals' IQ is below 100. Shockley offers this plan to counteract the supposed dysgenic effect due to the fact that groups with lower mean IQ in the present Western populations may have larger families on the average than those groups with higher mean IQs. In the first place, more detailed and carefully controlled studies (allowing, for example, for the fact that fewer low-IQ women tend to have families at all) have failed to confirm the existence of such a dysgenic effect (as discussed later). In the second place, such a program would run into all the difficulties and problems outlined in Section 21.4.

We ask again, why devote all this attention to a question whose answer seems to have neither theoretical nor practical applications? One reason may be simply that the question has been raised, that it has aroused a great deal of understandable passion, and that unwise actions might be taken by various authorities on the basis of partial or erroneous answers to the question.

One important source of information could come from children adopted "transracially." There are not many such, but one analysis at least has recently been published of black children adopted into white English families, or brought up in good nurseries together with children of other racial origins. These children had an average IQ slightly above that of white controls, thus apparently denying the existence of genetic differences. Very similar results have emerged from an even more recent and wider (as yet unpublished) investigation which involved 176 children of various ethnic groups, but mostly black, adopted by white families in Minnesota. Adoption studies are always subject to the difficulty that human observations of this kind do not use "randomized" material, and so may always leave open some uncertainty as to their interpretation. These results, however, constitute evidence against Jensen's hypothesis that is of a more direct nature than any of the facts used by him to support his contention.

Whatever the final answer about the genetic component, is the difference in IQ itself important?

If the difference is important, it surely does not matter whether the cause is genetic or environmental, for the difference is with us now, and efforts should be directed at eliminating it and its adverse effects. The question of whether or not there is some genetic basis is mainly relevant to the prediction of future changes. Environmental modification in one form or another would have in any case to be used to modify the phenotype, if this were deemed desirable. "Eugenic measures" directed against a clearly identified fraction of the population can only, and quite rightly, lead to accusations of genocide. We believe that there may well be enough scope for the manipulation of environmental effects to compensate for the observed differences. It is, therefore, on the environment, as well as on the genotype–environment interaction, that research efforts should be concentrated, independently of the problem of the existence of a genetic component.

The question as to whether IQ is really an important trait is difficult to answer. IQ is just one facet, probably a narrow one, of our very complex behavioral phenotype. Many psychologists disagree with the validity of the concept of "general intelligence" on which the significance of IQ as a measure of innate ability rests. The importance of IQ is emphasized mainly by those who consider that it is a good predictor of academic success. However, what about other types of success? Economic success, for example, is found to be correlated with IQ, but is more highly correlated with scholastic achievement and with the family socioeconomic background. The difference between the scholastic achievement of blacks and whites is less than that for IQ, while the most striking difference of all is in socioeconomic level. At present, on average, black Americans earn about $\frac{2}{3}$ as much as whites and, until recently, this ratio was nearer to $\frac{1}{2}$. The relative proportion of blacks entering college, however, is much larger than it used to be—18 percent now versus 3 percent thirty years ago, as compared to 24 and 8 percent, respectively, for whites. The effects of an improved education in increasing income are only now beginning to show in the younger age groups. It will take fifty years and perhaps more before the process of equalization is anywhere near stable values.

The second major question we raised at the beginning of this section is the relation between IQ and social class.

Several bodies of data point to such a correlation, and one well-known example is shown in Figure 21.11. Social class is generally defined by an index that combines aspects of occupation, income, and education. The data given in Figure 21.11 show an enormous difference of over 50 IQ points between the highest and the lowest class. However, in spite of this, the variation in IQ among parents within social classes is almost three-fifths of the total for the entire group, and that for children is even higher. The fact that the mean IQ of the offspring lies just about halfway between the class and overall-population means, is just another example of "regression to the mean," which can be explained on a genetic and/or a cultural transmission basis.

Wealth is inherited (as well as genes!), and the family background certainly influences the success of a child. Is IQ a direct determinant of economic success—that is, may the effect of scholastic achievement be through its influence on IQ? Again, this is a difficult question to answer, because of the confounding of many factors, all of which are highly correlated with one another.

There are a number of features that distinguish the problems of social class differences in IQ from the problems of the black–white difference.

1. The differences in IQ between upper and lower socioeconomic groups are much larger than the IQ differences between blacks and whites.

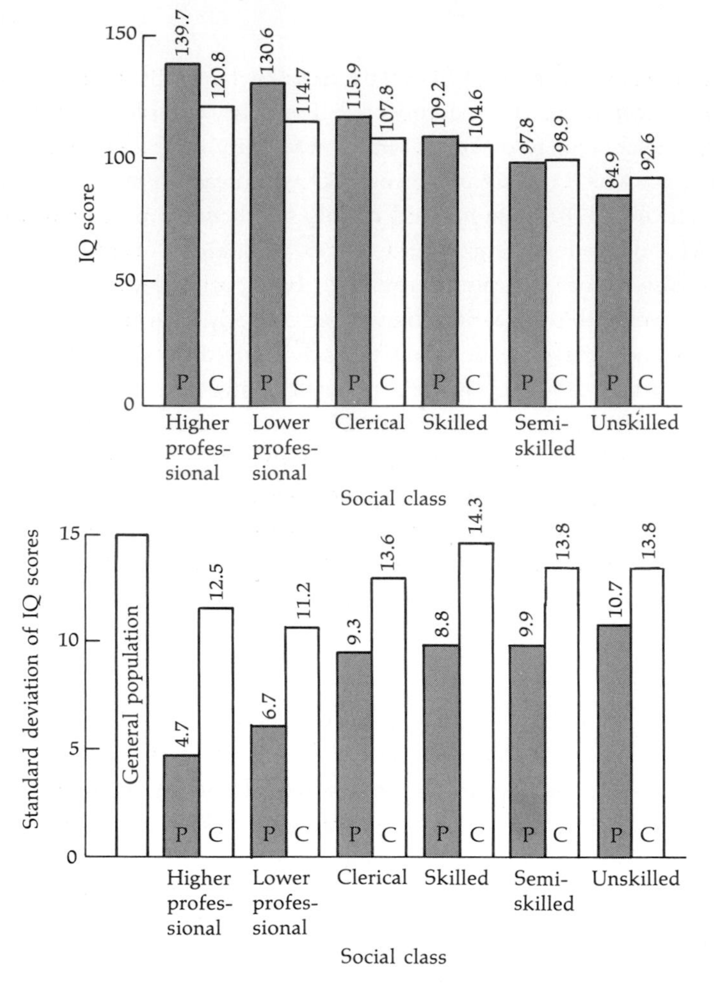

Figure 21.11
IQ and social class. (*Above*) The close relationship between social class and intelligence (as measured by IQ tests) is indicated in these data on London schoolchildren, collected by Cyril Burt (P = parents, C = children). In all classes, the IQ of the children shows a regression toward the mean. Such regression occurs for both genetically transmitted and culturally transmitted traits. (*Below*) The standard deviation of the children's IQ scores is closer to that of the general population than is that of the parental scores. (After W. F. Bodmer and L. L. Cavalli-Sforza, "Intelligence and race," *Scientific American*, October 1970. Copyright © 1970 by Scientific American, Inc. All rights reserved.)

2. These differences are found within a racial group in which mating does occur between different social strata (though not with complete randomness), so that heritability measurements are not irrelevant.
3. Unlike the black-white differences, the class differences might logically be predicted, given our present meritocratic society. It is not possible to change one's race, even after many generations, let alone within a lifetime. Yet there

can be movement from one class to another within a lifetime. If meritocracy implies selective social migration, favoring the movement of people with higher IQs to the upper classes, and if IQ has an inherited component, then we must expect that some of the differences between social classes are genetic. But the problem remains complex, because it is difficult to separate biological from nonbiological transmission of IQ, especially when studying the stratification of IQ by social class. Only adoption studies (as discussed in Part III) can separate the effects of cultural and biological transmission. The adoption data reported by Skodak and Skeel (Section 17.4) are particularly interesting in this respect, because they show an IQ increase of one standard deviation for white children from the lower socioeconomic strata who are adopted into higher strata. Thus, the environment is certainly important, but it is difficult with present data to say whether environmental effects can account for all of the IQ difference between social classes. The answer, as is so often the case, probably is that both biological and cultural differences matter, but that their precise relative contributions are hard to estimate at present.

This otherwise quiet field of discussion was stirred up in 1971, when Harvard psychologist Richard Herrnstein published an article on IQ in The Atlantic Monthly. *Herrnstein prophesied that, if IQ is inherited and if it contributes to social success, meritocracy will eventually produce genetic castes that are stratified by IQ, leading perhaps to a new form of feudal society.*

Of course, this is not a very original prophecy, and it incorporates at least one important fallacy. The prediction does not take into account a simple property of the social system that makes it virtually impossible to produce substantial change in a class with a large population. Thus, it is very likely that, if a dominant class were to emerge and eventually to stop exchanging genes with the rest of the population—thereby becoming, by definition, a caste—it would be a numerically small elite.

Consider a simple model in which the upper and lower classes exchange migrants in such a way that those who have an allele A tend to migrate to the upper class, and those with allele a tend to migrate to the lower class. Let us assume that this process continues for a sufficient number of generations (which may be quite large) so that the upper class contains only A genes. What frequency of A genes will then still remain in the lower class? If the population of the lower class is much larger than that of the upper class, it can be shown that there will be only a slight reduction in the frequency of A genes in the lower class, even if the upper class ends up with a gene frequency of 100 percent for the A allele.

Suppose, for example, that the A gene frequency is initially 50 percent in both classes. Table 21.1 shows the A frequency that will exist in the lower class after

Table 21.1
Gene frequency in lower class after frequency reaches 100 percent in upper class

Ratio of upper to lower class populations	Final gene frequency of *A* in lower class
1:5	40%
1:10	45%
1:50	49%

NOTE: Initial gene frequencies of 50 percent are assumed in both classes; equal fitness is assumed for both alleles. Individuals with the *A* allele tend to move to the upper class; those with the *a* allele tend to move to the lower class. After the upper class reaches a frequency of 100 percent for *A*, all genetic exchange between the classes is halted. If the population of the lower class is several times larger than that of the upper class, this process does not greatly reduce the frequency of *A* in the lower class.

the frequency reaches 100 percent in the upper class, assuming equal fitnesses for the two alleles. Because of the equal-fitness assumption, the overall gene frequency in the population does not change; only the distribution of the two alleles among the two classes is affected. Therefore, if the upper class is considerably smaller than the lower class, selective migration of *A* genes to the upper class can have little effect on the frequency in the lower class. Even when all alleles in the upper class are *A*, relative few *A* alleles have been removed from the much larger pool in the lower class. If the lower class is at least a few times larger than the upper class, most of the *A* alleles in the population must remain in the lower class.

Thus, the reduction in the number of "good" genes—or, more generally, the decrease in genetic variability—in the lower class will, in general, be trivial. The effect of selective migration into the upper class might be much larger if the "good" gene were rare initially, or if the two classes were more nearly equal in size. However, the effect on genetic variability would still be trivial in the first case, because the rare gene would have contributed little to overall genetic variability anyhow.

The fact that there is likely to be no substantial genetic impoverishment of the lower classes—even if "good" genes are concentrated in the upper classes—is confirmed by Cyril Burt's data shown in Figure 21.11.

As already pointed out, the variability of IQ among children from the larger lower classes is essentially the same as that of the whole population. In fact, the marked stratification for IQ seems altogether to have remarkably little effect on the variability among offspring. This suggests that the stratification is maintained, as Burt himself pointed out, by a relatively high level of intergenerational mobility between occupations and classes. The single gene model is, of course, a much oversimplified model. In particular, it does not take into account the homeostatic selective forces expected in natural populations. These could well, for example, counterbalance the class-based selective migration, and work toward maintaining the status quo.

It seems to be far too early to worry about Herrnstein's prediction. Even if it were correct, it would take a very long time to show any effect—whereas nowadays, changes in our society occur frequently and rapidly. Any prediction must be based on the assumption that a given social system remains essentially unchanged for a long time. Even assuming that a system of equal opportunity would generate a closed caste system, and that such forces are actually operating now, how long would this be with us? Are all the other assumptions implicit in the model valid? Long-term predictions are obviously the most difficult ones to make and, so far at least, there does not seem to be any cause for alarm.

There are significant fertility differences between the social classes, generally in the direction of increasing numbers of children per couple with decreasing socioeconomic (more specifically, educational) level (Fig. 21.12). Does this lead to a prediction of a decline in IQ?

The combination of an increasing fertility with decreasing socioeconomic status, and so with decreasing IQ, has often led to dire predictions of an overall decline in the mean IQ of our Western industrialized society. Certainly, the theory of quantitative genetics (discussed in Part III) and the experience of plant and animal breeders would predict that, if IQ has a narrow heritability greater than zero and if IQ is negatively correlated with reproductive ability, then IQ should decline from generation to generation. Equally, of course, an increased fitness associated with higher IQ should lead to the eugenist's goal of a positive improvement in overall IQ.

It is very difficult to obtain direct data on IQ changes in the human population over a sufficiently long period of time, but available studies do not suggest a decline.

There was a major study, the Scottish mental survey, that was directed specifically at this question. In 1932, some 87,000 eleven-year-old school-children in Scotland (about 90 percent of all Scottish eleven-year-olds in that year) were given group verbal intelligence tests. The *same* test was given to a similar group of about 70,000 children in 1947, with the results shown in Figure 21.13. The mean score changed from 34.5 in 1932 to 36.7 in 1947, an *increase* of about 6 percent. Other studies have given similar results with, on the whole, increases of a comparable magnitude.

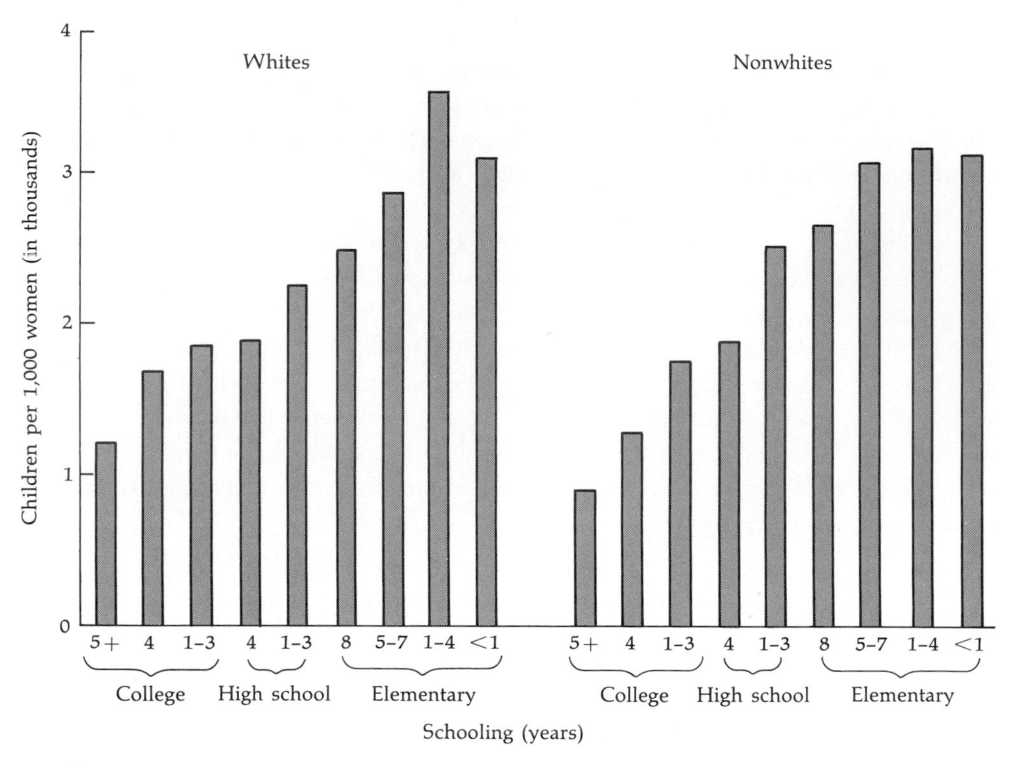

Figure 21.12
Average number of children born as a function of mother's education. The graph shows the average number of children born per 1,000 women as a function of the number of years of school completed by the women. The data for white and nonwhite populations are based upon samples of women 45 to 59 years of age in the United States during 1960. These women are assumed to be past the period of reproduction, so that their families may be assumed to be complete. (From C. Stern, *Principles of Human Genetics,* 3rd ed., W. H. Freeman and Company. Copyright © 1973.)

What is the explanation for this disagreement between the observations and the simple predictions of a decline in IQ?

One obvious possibility is that improvements in schooling over the years 1932 to 1947 were really responsible for the increase in IQ. Certainly, it is impossible to conceive that such a marked change over so short a period of time could have a genetic basis. Height offers an interesting parallel with IQ. It also is negatively correlated with increased family size and also has, nevertheless, increased substantially this century (as mentioned earlier) by an average 1 to 2 cm per decade. The increase in height must be attributed to as-yet-unidentified environmental factors, possibly either dietary or hygienic.

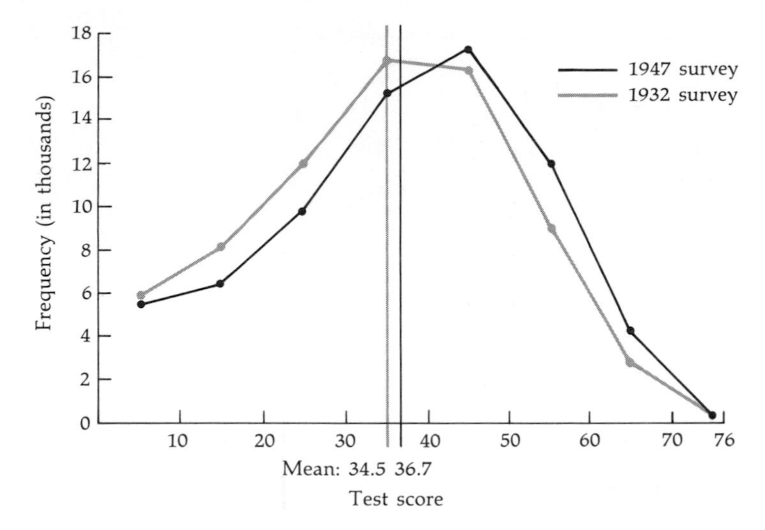

Figure 21.13
Distribution of group-test scores for all pupils in the 1932 and 1947 Scottish surveys. The same test was given in both years and was administered in the same manner as the SAT or National Merit exams in the U.S.A. This test is regarded as a general intelligence test, although scores have not been standardized to the usual IQ form. The results show an increase—not a decline—in the intelligence of the later generation. (From J. Maxwell, *The Trend of Scottish Intelligence,* Univ. of London Press, 1949.)

Let us now examine a little more closely the premises on which the predicted decline in IQ is based.

There can certainly be no doubt of the marked negative correlation between IQ and average family size, as indicated by the summary of a series of studies shown in Figure 21.14, which give an average correlation coefficient of about −0.3, and an approximately threefold variation in average family size. It is not enough, however, simply to examine the relation of IQ with family size. Any measure of overall fitness (see Chapter 7) must take into account other factors—such as mortality and also, of course, the probability of getting married. Data such as those shown in Figure 21.14 are based on families with at least one child, and do not, therefore, take into account possible fertility effects of IQ associated with the chance of not getting married or of having no children. The data illustrated in Figure 21.15, for example, (based on a study in Minnesota) show that more than 30 percent of persons whose IQ is 70 or below have no children, as compared to about 10 percent for the IQ range 101 to 110, and 3 or 4 percent for persons with IQ greater than 131.

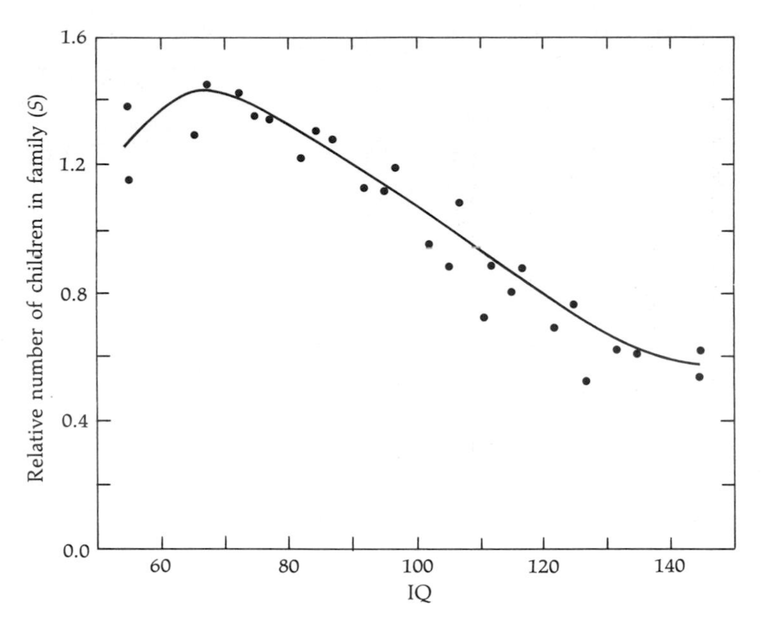

Figure 21.14
Relation of family size to average IQ of children in family. The family size (*S*) is given as a relative value, obtained by dividing the number of children in a family by the mean number of children per family for the whole sample. (From K. Mather, *Human Diversity,* Free Press, 1964.)

Further doubts come from a consideration of the relationship between birth order and IQ.

As we saw in Chapter 17, the later the birth order of children (almost independently of family size), the lower their average IQ. This phenomenon can itself hardly be genetic. It will inevitably build into the relationship between family size and IQ a component that is entirely environmental. If the relationship between family size and IQ were entirely environmental, it would not lead to predictions of genetic decline. Finally, the patterns of correlation between family size and socioeconomic or educational class are likely to be an expression of the demographic transition, a highly transient phenomenon, so that predictions (if any can be made) are likely to be valid only over a very short term, in view of their transiency.

There have been at least two studies that measured overall fitness as a function of IQ.

Fitness was measured in each case by the intrinsic rate of population increase which, as discussed in Chapter 7, takes account of both age-specific birth and death rates, and so necessarily includes the contribution of zero-child families.

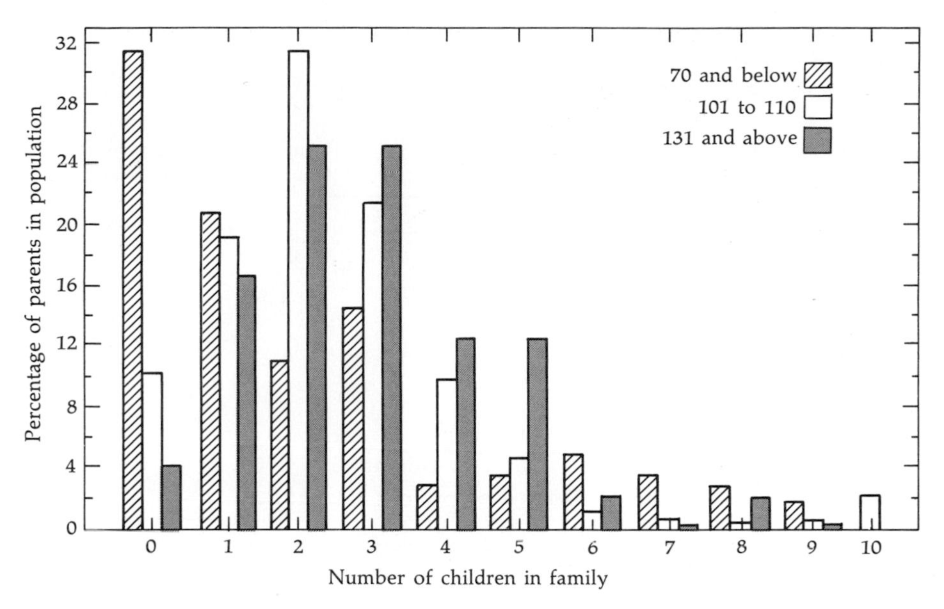

Figure 21.15
Distributions of family size for three different IQ groups of parents. The striped bars represent a group in which the average IQ of the parents was 70 or below. The white bars represent a group in which the average IQ of the parents was from 101 to 110. The gray bars represent a group in which the average IQ of the parents was 131 or above. (From J. V. Higgins, E. W. Reed, and S. C. Reed, *Eugenics Quarterly*, vol. 9, p. 84, 1962.)

The results of one study, based on a followup of a sample of about 1,000 children from Kalamazoo, Michigan, whose IQ was measured at school, are shown in Table 21.2. The relative fitness, given in the fourth column of the table, does not vary linearly with IQ, but seems to be highest for high and relatively low IQ, and low at intermediate and very low IQ values. The predicted change in IQ from those data is actually an *increase* of just over one IQ point. The second study, based on a sample from Michigan, also leads to a predicted increase, but of only about half an IQ point. Certainly, neither of those studies provides a case for predicting a decline in IQ.

Nevertheless, the fact remains that all direct studies on actual IQ changes and predicted changes came up with increases rather than decreases.

The measurements are all surely subject to a substantial margin of error. They inevitably refer to phenotypic measurements of IQ, making predictions on genotype changes even more uncertain. Furthermore, rapidly changing conditions of modern industrial society—especially economic pressures and changing patterns

Table 21.2
Changes in mean IQ due to natural selection

1 IQ range	2 Mean value	3 Proportion in range	4 Relative fitness	5 Proportion after selection
≥120	126	0.092	1	0.106
105–119	112.5	0.279	0.87	0.284
95–104	100	0.258	0.78	0.235
80–94	87.5	0.279	0.96	0.313
<80	74	0.092	0.58	0.062
Total		1.000		1.000

Mean IQ before selection (Σ column 2 $\times$ column 3) = 100
Mean IQ after selection (Σ column 2 $\times$ column 5) = 101.01

NOTE: Column 2 gives approximate mean IQs for the ranges included in each group (see column 1). Column 3 gives the proportion of the sample found in each IQ range, as calculated from a normal distribution with mean of 100 and standard error of 15. Column 4 gives relative fitnesses according to C. J. Bajema, *Eugenics Quarterly*, vol. 10, pp. 175–187. Column 5 gives the relative proportion of the sample in each IQ range after selection, computed as (column 3 $\times$ column 4)/Σ(column 3 $\times$ column 4).
SOURCE: L. L. Cavalli-Sforza and W. F. Bodmer, *The Genetics of Human Populations*, San Francisco: W. H. Freeman and Company, 1971, p. 620.

of birth control—must create varying patterns of selection that may indeed be quite transient. The predictions of changing IQ, in either direction, are almost as uncertain as Herrnstein's predictions of genetically stratified IQ castes. Present demographic trends point towards a reduction in the variation in fitness and a corresponding evening-out of the socioeconomic correlations with fitness. This emphasizes once again the need to concentrate on environmental improvements, which can achieve comparatively rapid results, rather than on altering hypothetical long-range genetic selection patterns.

21.6 The Changing Environment

The word environment means different things to different people. The geneticist's usage is perhaps the widest of all, for it includes everything other than the individual's genes as part of his environment. The environment, to the geneticist, is the total set of conditions in which the development of a fertilized egg takes place. The maternal uterus is, of course, an important part of this environment. After birth, the environment around each of us increases greatly in complexity, including physical, social, and cultural components that would require many dimensions for their complete quantitative description.

Our natural physical environment has, on the whole, changed slowly, while the social and cultural environments have changed very rapidly, and their rate of change still appears to be increasing

Technological innovation is also clearly having a major effect on the physical environment. As we all know, world population pressure is now putting a great strain on the physical environment. This strain has been created in large part by technological advances. To what extent are these pressures changing the direction and rate of our evolution? During much of the period of human evolution (discussed in Chapters 18 and 19) and at least until the beginnings of agriculture, conditions for evolution were determined largely by the natural physical environment and cannot on the whole have changed very rapidly. The major environmental adaptations must have occurred as man spread to different parts of the world and adapted to different climatic conditions. Subsequently, however, it has been man's own cultural development, leading to agriculture and other much more recent technologies, that has created major new challenges for adaptation. Some of the changes that may have followed the introduction of agricultural practices were discussed in Chapter 19. These may have included adaptations to the high carbohydrate diet of the cereal eater, and to the use of lactose-containing cow's milk.

The transition to agriculture, the Neolithic "revolution," ushered in an important further transition: that from the rural to the urban civilization. With food becoming easier to store and transport (though perhaps not easier to obtain), only a fraction of the population had to be occupied in food production. As a result, new occupations developed and trade appeared.

The earliest cities developed in Neolithic times, but the major transition from rural to urban civilization really came much later. Even 200 years ago, more than 80 percent of the population of Europe was rural and engaged in agricultural work. It is only quite recently, in areas such as the U.S.A. and Europe, that agricultural work and the truly rural population declined dramatically in size.

We do not know what genetic adaptations, if any, may have been caused by the transition to an urban society, but there is one peculiarity of the urban environment, which until comparatively recently may have precluded any significant evolutionary change.

The rate of population growth in cities was, until recently, never sufficient for replacement. In spite of this, cities have been growing almost all the time, because of the continual influx of immigrants from the rural areas. A major reason for the slow growth rate of urban populations must have been epidemics, which originated in the high population densities of cities. People of all times knew that a fairly safe protection against an epidemic was to go to the countryside—if they could afford it. But there are other ways in which cities have reduced net growth rates. Birth control, for example, probably started in the cities because, while the rural population needs large families, the urban one does not. Thus, for some time the gene pool of the cities may have been dominated by that from the country. Genes, with the people carrying them, originated in the country and died in the cities. So it may be that there was not much selection due to the urban transition except for those few populations, like the Jews, who have constantly and sometimes exclusively lived in an urban environment for at least the last 2,000 years.

The demographic transition to low birth and death rates certainly has had major effects on the pattern of natural selection.

This transition, as discussed in Chapters 7 and 19, started around the middle of the last century and has not yet quite spread throughout the world. Prereproductive mortality (which is still about 50 percent and even higher in populations that have not yet benefited from the impact of modern hygiene) is now very close to zero in those populations that have undergone the full transition. Thus, the probability of surviving to age 30, which is close to the mean age at reproduction, was in 1967 about 95 percent for males and 97 percent for females in the United Kingdom, most of Europe, North America, and Australia. Prereproductive mortality is now practically nonexistent in these countries, and cannot therefore any longer contribute significantly to natural selection. These figures, it should be noted, do not include abortion and stillbirth, which still occur at a level of at least 15 to 20 percent of all births. Many of these are, as discussed in the last chapter, the result of natural selection against lethals, including many chromosomal aberrations. Postreproductive mortality is probably of almost no selective importance, as the chance for survival of an orphan in modern society is probably not markedly lower than that of a child both of whose parents are alive. Of course, not all prereproductive mortality that was in existence was necessarily selective, and only mortality having differential effects on different genotypes would have any selective significance.

Differential fertility, in contrast to differential mortality, has not disappeared with the demographic transition.

The variance of the number of children born per woman provides an upper limit to the opportunity for selection by differential fertility, and it has not become smaller, even after the demographic transition. Birth control, however, may reduce the variation in family size, as most families achieve their desired size. If each couple simply had two children, the variance of children born per couple would of course shrink to zero, and there would be no room at all left for natural selection. Whatever evolution then took place would be due to mutation, which is very slow, and to drift, which with the present size of human populations, is minimal, so that human evolution would essentially be frozen. While we may be approaching this state gradually, it seems likely that there will always be a little room left for natural selection, however slow its effects may be in comparison with cultural changes.

Mutation rates deserve some further comment, as "the atomic age" is likely to increase the background of radioactivity to which we are exposed (see Section 4.3).

There is a certain "normal" background radiation that is unavoidable, deriving in part from cosmic rays and in part from radioactive elements—including tritium, carbon 14 (written ^{14}C) and, above all, potassium (^{40}K), which is present in our bodies (Fig. 21.16). Only the radiation that reaches the gonads can cause mutation, and this is estimated to be about 0.1 rad per year (the rad is a unit that measures the amount of radiation absorbed per gram of tissue; see Section 4.3). In a few areas that are near uranium-rich ore deposits—or deposits of other radioactive elements, such as are found in Brazil and India—the background levels may be higher.

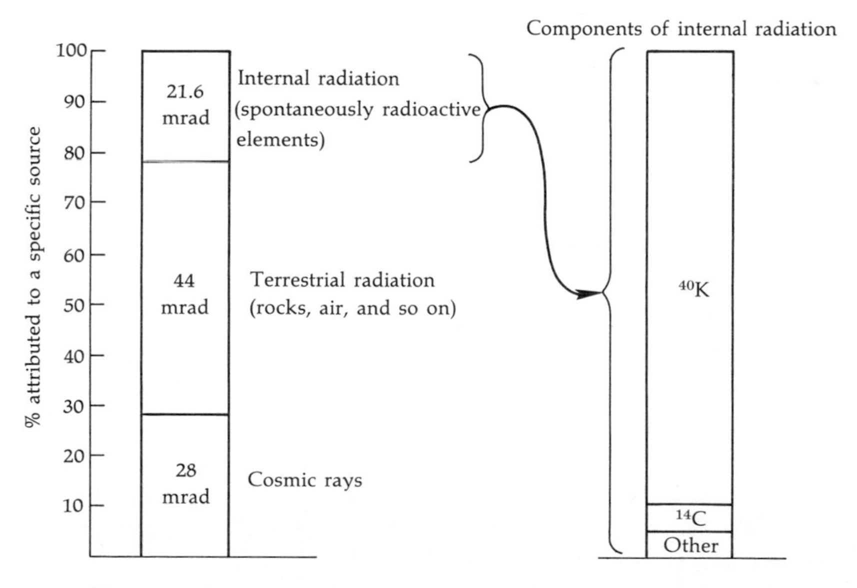

Figure 21.16
Sources of natural radiation. The graph shows the average number of mrads reaching the gonads per year from various sources.

This unavoidable radiation seems to be responsible for only a small part of the spontaneous mutation rate. Very approximate estimates indicate that only some 10 percent of the spontaneous mutation rate is due to natural radiation—the remaining 90 percent probably being due to as yet unknown chemical mutagens or unavoidable errors of DNA metabolism.

Most mutations (perhaps 95 percent or more) are, at least to some extent, deleterious—and so it is desirable to keep mutation rates as low as possible. Almost all the dysgenic effects due to selection relaxation following medical and hygienic advances will be the result of the accumulation of new mutations, not counterbalanced by natural selection. Thus, it is actually desirable to decrease mutation rates. This should eventually be within our technical competence, and would be a significant way to reduce, in the long run, the overall load of genetic disease. As discussed in Section 4.3, there is really no threshold below which radiation is ineffective. Therefore, any increase in environmental radiation is certain to have detrimental effects, however slight.

What about man-made radiation?

The early work of geneticists, which proved the deleterious effects of increased mutation rates, has been fundamental in making political authorities aware of the danger of continued experimental atomic explosions. The total genetically significant radiation accumulated to date from atmospheric tests of atom bombs is 120 millirems. This is about the equivalent of one extra year of spontaneous radiation, and thus is very small. Radiation from cratering experiments and from present atomic power is, so far, negligible. The same is not true, however, of the medical use of X rays, which contributes a dose that is far from trivial (Fig. 21.17). Averaging the data from 16 countries gives a genetically significant dose of radiation (to the gonads) of 10 to 50 millirems per year per person, or up to 50 percent of the natural background level. The medical advantages of giving an X-ray examination must, of course, be weighed against the disadvantages due to induced mutations and other effects. The genetic danger to the progeny is not the only risk, as X rays can induce cancers and leukemias, due presumably to somatic mutation. Because these effects will show only after a long time (5 to 20 years), and because the increase in incidence over the spontaneous level due to irradiation is not dramatic, the role of radiation cannot be determined in individual cases. Still, it is worth recalling that almost the only clearly recognizable effect on survivors of the Hiroshima and Nagasaki atomic bombs was not mutations in their progeny, but their increased incidence of leukemia.

What about consequences of the increasing dependence on nuclear power plants?

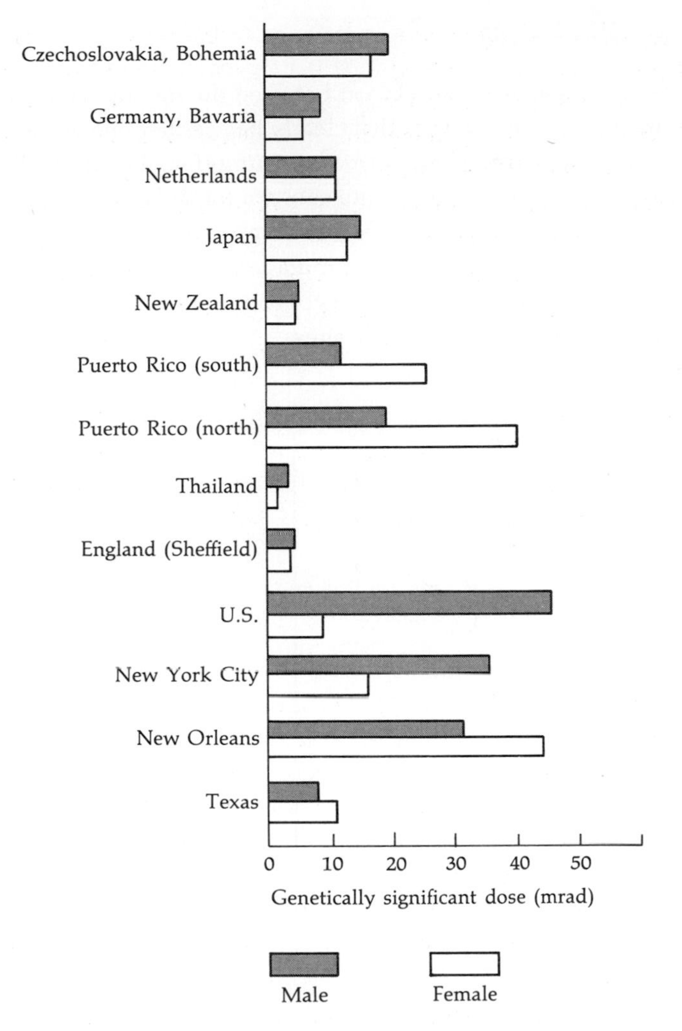

Figure 21.17
Genetically significant doses (GSD) of irradiation from medical uses of X rays, in millirads. The variations among countries are striking, as are the variations among cities within the United States.

The inevitable increase in nuclear power plants raises the problem of a potential radiation increase that may have both mutagenic and carcinogenic effects. Some increase is inevitable. For instance, the discharge of water used for cooling reactors will cause the radioactive contamination of fish; the gaseous wastes of the plant will cause contamination of the air. But the radiation increase can be contained in limits that do not seem to cause concern. Naturally, it is impossible to predict the consequence of contaminations resulting from accidents.

What about chemical mutagens?

There is a very significant correlation between the mutagenicity of a substance and its carcinogenicity. Substances that clearly have either one or the other activity obviously should be barred from normal consumption. The difficulties of testing for these activities, and also, inadequate concern for their effects, have meant that some carcinogens have until recently been very widely used (Fig. 21.18). Undoubtedly, many other carcinogens and mutagens still go undetected in our food. Atmospheric pollutants almost certainly contribute to lung-cancer rates and perhaps also to those of cancers at other sites.

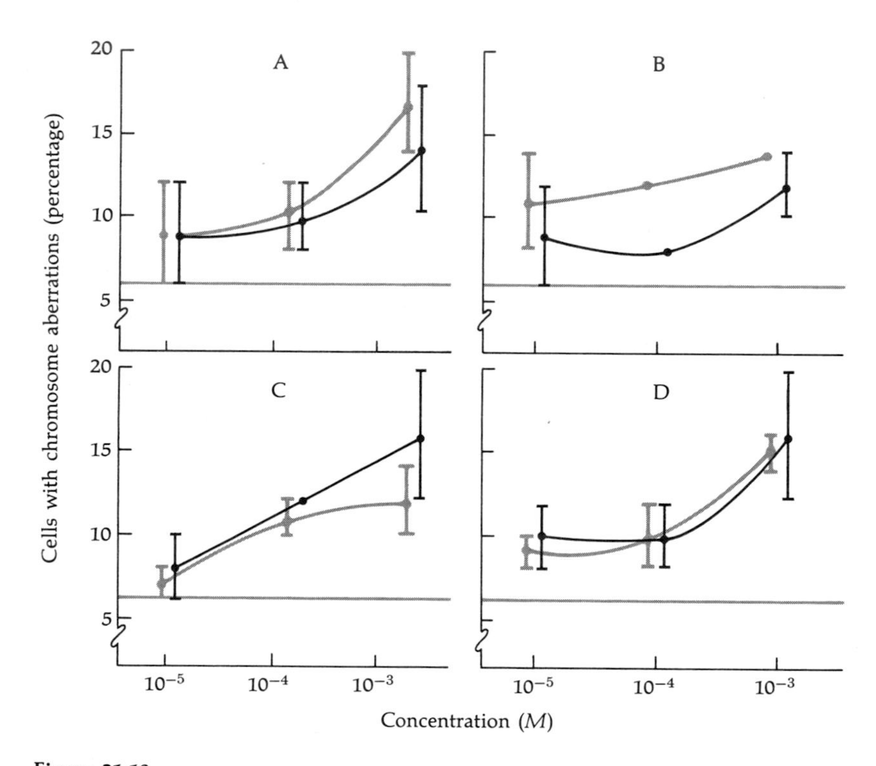

Figure 21.18
Chemical mutagens. The graphs represent frequencies of chromosomal abnormalities as a function of dose; each point represents analysis of 100 human lymphocytes (in culture). (A) Cyclamate, an artificial sweetener. (B) Cyclohexylamine, a degradation product of cyclamate. (C) N-hydroxycyclohexylamine, a metabolite of cyclohexylamine. (D) Dicyclohexylamine, an occasional contaminant of cyclamate. In each graph, the dark line represents a 25-hour treatment, and the lighter line represents a 5-hour treatment with the mutagen; the light horizontal line is the control (no mutagen). (From D. R. Stolz et al., *Science,* vol. 167, p. 1501, 1970. Copyright 1970 by the American Association for the Advancement of Science.)

Many substances are potentially mutagenic or carcinogenic, but research is still needed to improve existing methods of detecting and predicting the mutagenicity of a compound. The task of screening all substances that we may eat, breathe, or somehow contact is indeed a formidable one.

Another difficult problem is created by the introduction of new drugs and food additives—or, in general, new chemical components in the environment—that may be harmful, or even lethal, to people with previously undetected genetic idiosyncrasies.

One such example, discussed in the previous chapter, involves the muscle relaxants of the suxamethonium type employed in general anesthesia. There is only one individual in every thousand who may show abnormal behavior toward this drug because of a genetic defect in the enzyme pseudocholinesterase. The rarity of such idiosyncrasies makes it difficult to anticipate them until they are revealed by an accident. Similarly, individuals who carry rare mutant genes for α-antitrypsin, a protein normally found in blood, appear to have an inordinate lung sensitivity to toxic atmospheric pollutants and to smoking. Yet others may carry mutations that lead to the conversion of normally harmless substances into carcinogens. Thus, there are a number of genetic variants whose detrimental effects are only revealed by exposure to substances that may, for example, be found only or mainly in atmospheric pollution.

Economic and food-producing technological developments are changing the environment around us at an ever increasing rate. Europe was once covered with forests that have almost totally been destroyed over the last few thousand years to give way to the cultivation of crops. Residual forests in the tropics are now being destroyed at a rate that is one hundred times faster. The climate of major regions of the world has been, and will be, further modified by these interventions.

Plant and animal breeding have been a major force in the continuous quest for more and better food.

Whatever were the forces that catalyzed the agricultural revolution in the Near East over ten thousand years ago, increased yields were made possible by true genetic discoveries for which archeology and paleobotany have provided ample evidence. These included the cultivation of wheat mutants whose spikes do not scatter seeds around them and can therefore be harvested with a sickle. The common wild wheat used earlier had to be plucked by hand—which was, of course, a less efficient procedure. Similarly, the development of cultivated corn was accompanied by extensive genetic changes, and it is difficult now to recognize the original wild progenitor.

Scientific plant and animal breeding has had a short history, but a very considerable record of accomplishments. Perhaps the most significant is the production of hybrid corn. The American geneticist G. M. Shull, who chose corn as an experimental organism for the study of quantitative inheritance, developed inbred lines to fix their characters and then studied crosses between the lines. It was only later that he realized that the yield of corn could be improved by systematically using the heterosis or hybrid vigor he observed in hybrids between inbred lines. A number of other geneticists later contributed to this program, whose economic impact has been enormous, having increased the overall yield of corn by about 50 percent. The transition from the use of traditional techniques to the use of hybrid corn took place in just about two decades in the United States, starting shortly after 1930, and it has since been exported to a number of other countries (Fig. 21.19).

Another interesting advance in recent years has been the development of new lines of corn that have been modified in their proteins to produce an altered amino-acid composition. One such significant achievement is the discovery of hi-

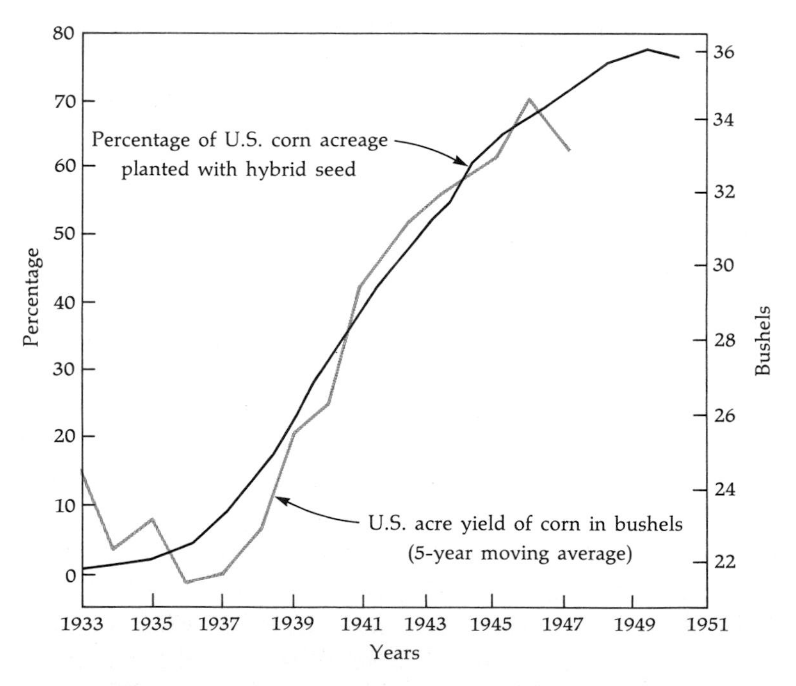

Figure 21.19
The transition to the use of hybrid corn. As the proportion of total acreage planted to hybrid corn in the U.S.A. has increased, the average yield has gone up correspondingly. (From P. C. Manglesdorf, in *Genetics in the Twentieth Century,* ed. L. C. Dunn, Macmillan, 1951.)

lysine strains, which have an increased amount of the essential amino acid lysine and so provide a more satisfactory source of protein from a nutritional point of view.

The yield of wheat has been increased, in part by the creation of genetically improved varieties, since the beginning of this century by an amount that is comparable in magnitude to that for corn, but the change has taken place somewhat more slowly. In recent years an important component of the "green revolution" has been the introduction of the semi-dwarf varieties of wheat, especially the new Mexican varieties, which have considerably increased yields. These new strains have almost doubled overall yields in some parts of Asia within just a few years.

It is very unfortunate that the quality of human "social engineering" is still so much poorer than that of the technology of food production or other technologies.

Any technological change that is introduced on a large scale cannot fail to have a profound social impact. The physical as well as the sociocultural environment will be changed, and there is often a considerable lag in the social response to these changes. During this lag, pressures of all sorts may arise.

Consider, for example, the long-range effects of the introduction of hygienic measures that greatly reduced the human death rate. Prophets such as Thomas Malthus had correctly anticipated the effects of a declining death rate half a century before the beginning of the population explosion. Yet it took a very long time before birth rates began to decrease—even though the need for a decline in birth rates had been clearly recognized. In fact, birth rates still remain high in many parts of the world.

The increase of population size has not always necessarily been disadvantageous (or appeared so, even if it was in the long range), and leaders of many different groups have advocated population growth for economic or political reasons. But now that the world population is reaching saturation, concern is mounting. Of course, even if we had clearly foreseen the present situation, we would not have kept in check the development of hygiene and thus the consequent decrease of mortality. The present mobilization of forces advocating family planning—though perhaps not yet sufficiently developed—indicates that the crisis of population growth is now fairly widely understood. Some believe that this recognition has come too late and that serious and irreparable damage has already been done to the future of humanity. However, such extreme pessimists may underestimate the power of new technologies and the capacity of man to adapt. But it certainly is discouraging that a danger, perceived almost two hundred years ago by Malthus, is only now becoming a common concern. Hopefully, our capacity to recognize impending social problems and to take prompt and effective action to counter them will improve in the future.

21.7 Equality

Genetic studies have shown that many of the differences we observe between people are inherited and that there is a large store of genetic variability in the population. Social progress, on the other hand, has in general moved in the direction of treating all men as being born equal. Humanitarian ideals, developed mostly in the last century, have destroyed many myths—for instance, that the blind and deaf are insane—and have helped to give handicapped people a chance to develop a quasinormal life.

Clearly, if by equal is meant the same, genetics has shown us that we are not at all equal.

One must therefore either say that equality means equality of opportunity, or else accept equal treatment for unequal people. Some opposition to genetics arises, in part because of this dilemma, from several quarters. Some opposition to genetics and evolution comes from fundamentalist religious beliefs. The fundamentalists' recent success in requiring that biblical creation be given "equal time" as evolutionary theory in California schools is a reminder that the dogmatic approach is far from dead. In other (more academically qualified) circles, opposition to genetic ideas probably arises from a fear of political abuse deriving from, or exploiting, scientific misconceptions. "Scientific" racism, as discussed in a previous section, is prominent among these abuses. But wrongly denying the validity of genetic ideas or their potential usefulness solves no problems.

Unquestionably, some fields of applied genetics pose legal and social problems—for example, in connection with genetic screening programs and selective abortion, as discussed earlier in this and the preceding chapter. Although the scientist is the instigator of these advances, he cannot act alone when it comes to applications. The scientist must explain the problems to the nonscientist so that both can act together. Obviously, scientists should become increasingly aware of and sensitive to the social problems that their own discoveries can both help solve and sometimes help aggravate. The idea, however, that scientists alone should be responsible for the effects of their discoveries is as unreasonable as that of making the inventor of the piston engine solely responsible for present-day air pollution.

The fact that all men are not born the same has been obvious for a long time. It does not require modern genetics to convince one that skin color and facial and body traits are to some extent inherited. The concept of genetic disease is also an ancient one, as exemplified by the fact that the X-linked pattern of inheritance of hemophilia was apparently recognized (as mentioned in Chapter 2) some 1,300 years ago by Jewish rabbis writing the Talmud. Modern genetics, however, has helped us to prove that genetic differences between individuals—even when they come from the same village, or the same family—are extremely numerous; so much so, in fact, that every individual is identified by his own genes much more than by his fingerprints. The chance of finding two individuals (who are not identical twins) that are identical on the basis of the genes we can now identify is extremely small, perhaps less than one in a billion. We can now recognize how many subtle differences exist between individuals, which by simple inspection of families can be seen to be to some extent inherited.

Modern genetics has also shown that the inherited differences between two individuals taken at random from a group are very many. It does not matter very much from which group these two individuals are taken. Even taking them from different races does not increase the differences between them very much, as discussed in Chapter 19. The basic variation is that between individuals and not that between groups. In this sense, genetic findings have reaffirmed the basic unity of man, irrespective of the group of origin. But they have also reaffirmed the uniqueness of the individual.

The range of genetic differences so far known is very considerable. Some, if not most of them, may be trivial—or at least relate to traits whose importance is far from clear, as is true of most protein differences detected electrophoretically or immunologically. But it does, for example, make a lot of difference to an individual whether or not he had the genes that determine normal development of vision or hearing. Genetic disease is clearly a major source of inequality and one to which we have already devoted a great deal of discussion. Genetically determined allergies and drug sensitivities, color blindness, and numerous other such variations all have obvious consequences in our society, which, however, we are on the whole learning to cope with in a relatively straightforward way. It is those genetic variations that affect the subtler aspects of human behavior, and that so far are very poorly understood, that raise the most difficult social questions.

What kind of decisions society should make may hardly seem to be an appropriate subject for a textbook on genetics and evolution, but it does become relevant as soon as it becomes clear that a decision might (even should) be influenced by existing genetic variation.

Sometimes the introduction of genetic arguments seems to detract from a rational decision. For example, a court case may be examined on the basis of whether a person charged with a crime is insane or not, and some genetic defect, such as a chromosomal aberration, may then be considered by the judge or jury to be a mitigating factor. It is, however, difficult to see why the fact that a predisposing condition is genetic, should affect the legal outcome of a case. Nongenetic predisposing factors—connected, for example, with an adverse childhood environment—may be just as much beyond the individual's overt control as genetic factors. Society rightly wants to defend itself against crime. The uncertainties connected with the XYY genotype are an example of the difficulties that are encountered in considering genetic predisposition to crime. On the other hand, there are clearly many environmental inducements to crime, and it is on these that society should try to act, by following a preventive rather than a punitive approach.

The legal equality of the rights of individuals, which is a most important aspect of equality, has nothing to do with genetics.

But there are other, more subtle rights of individuals that may be affected by genetic differences. For example, under conditions where physical strength was a major asset, the strongest could make the weakest miserable, and this difference surely has a genetic component. Job opportunities, in fact, depend on different abilities that undoubtedly have a genetic component, and it may be difficult (if not impossible) to compensate for the genetic differences in such a way as to make the opportunities really equal. No society can any longer afford to ignore the existence of innate, genetically determined differences between people in planning its social programs. The main question to be asked is, how "plastic" is each individual, and to what extent can he or she be educated to perform any given social role? The strict hereditarian believes that education will achieve little, while the extreme environmentalist asserts that education can give anything desired. The truth, of course, lies somewhere in between. The problem is our great ignorance about the best way to educate a person to fulfill a given task, added to which is the difficulty that the best approach is most likely to vary from one individual to another, depending on his genetic constitution.

The equal-opportunity system is generally considered a road to meritocracy, though it is not usually specified what is meant by "merit," nor what should be the compensation for merit.

Presumably, merit is in some sense directed toward the good of society, while compensation is usually thought of in terms of success: money, power, and prestige. Most modern societies try to provide at least a basic common level of educational opportunities up to a certain age. But no society, however egalitarian, assigns jobs to individuals at random. A major characteristic of our modern society is the *variety* of jobs and skills on which it depends. It is therefore most important to try to match the right job to the right person. Although the efficiency of this matching process may vary greatly between countries and between political systems, it always tends to be "meritocratic." The following two premises must surely be generally accepted: (1) it is more satisfactory for everyone in society if each job goes to the person who is best able to discharge it; (2) not everybody is equally suited to every kind of job. The cost of training a person for a given occupation, even if it were in principle possible to train everyone for any occupation, is likely to vary widely from one individual to another, and the bases for these differences are likely to be at least to some extent genetic (Fig. 21.20). In a world of limited resources, as ours is probably always going to be, one must try to fit people as best as possible to the right jobs.

The existence of individual differences in abilities, in addition to that of differences in the desirability of jobs, creates an important source of social tension.

Work necessarily occupies a large part of our lives. The importance of satisfaction in one's job is illustrated by an observation we have already cited. A series of monozygous twins who were discordant for heart disease—one of each pair having had an attack of angina pectoris, while the other had no history of heart attacks—were screened using a variety of physical and psychological tests. No differences were found between the members of each discordant pair for any of the physiological correlates that might have been suspected, such as cholesterol level, EEG, smoking, or obesity. The only clearcut difference was in their satisfaction with their jobs. Those twins who had had a heart attack were unsatisfied with their occupations.

Social systems favoring maximization of satisfaction with one's work and in general with one's life seem to be preferable *a priori* to all other systems, in that they try to maximize what other systems (which aim for example, at equal opportunities, incentives, or outcomes) hope to achieve indirectly. This is, however,

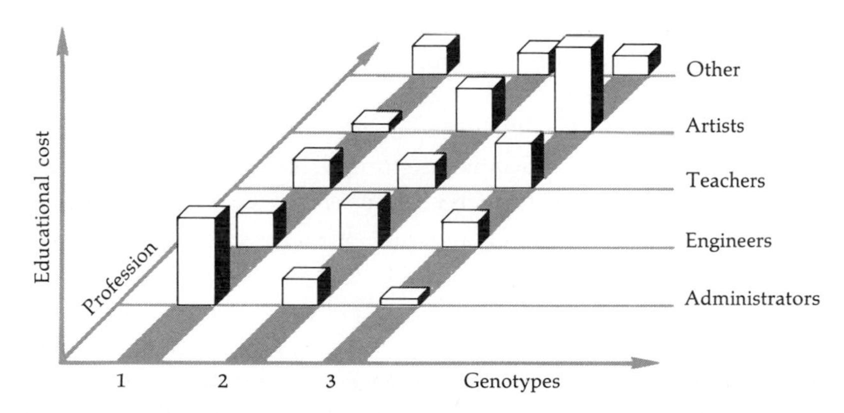

Figure 21.20
A hypothetical model of the effect of a diallelic locus on the average educational cost of training individuals of each genotype for different professions. Only one pair of alleles at a single locus is considered for simplicity, although the number of significant genotypes is likely to be very large. A great deal of research would be needed to obtain the data needed to construct such diagrams for actual genotypes. In each case, the height of the bar represents the cost to society per individual in order to achieve similar results with individuals that differ in their genetic constitutions.

not a goal that is either easily defined or easily achieved. Maximum *average* satisfaction might be achieved through exploitation of a minority of unhappy slaves, but this surely clashes with all our current moral precepts. Satisfaction must be maximized for the individual, but not at the expense of the overall level of satisfaction enjoyed by the rest of society. Thus, no one person should be in a position to increase his own satisfaction at the expense of the general average for the society in which he lives. We may call this a process of "equimaximation." Such a situation appears, for the present at least, to be (unfortunately) quite utopian. It would involve, among other things, the total disappearance of crime. Research, however, can and should be done on how a system approaching equimaximization may be achieved that is compatible with the existing individual variability.

Education aims to give everybody a general background, as well as a specific preparation for a range of occupations. There is clearly still a great deal of scope for improving the means by which an individual enters any given educational channel and emerges with a particular occupation. Furthermore, the nature of jobs may nowadays change rapidly, leading to a greater need for continuing adult education. At all levels and stages of education, it must surely be important to maintain a variety of approaches to cater for the wide range of individual differences. Differences in educability may be either genetically or environmentally determined; this fact in itself is not so important as the fact of the existence of the differences themselves. The need for a great variety of skills in modern society makes the existence of such variation an asset and not a liability if proper consideration is given to it. The aim of genetics is not just the bare study of mechanisms of biological inheritance, but more generally, the analysis of variation.

Education and the choice of occupations are perhaps the areas of social concern in which individual variation will have to be taken into account much more than has been the case so far. The difficulties are formidable, but surely no society will be successful in the future if it ignores the existence of individual variability.

Summary

1. Genetic diseases differ in a number of respects from other types of diseases and, as a result, their treatment raises a number of social, legal, and ethical problems.

2. Genetic screening strategies depend on (a) whether corrective treatment is available, and (b) if not, whether the condition can be detected by amniocentesis in time to permit abortion.

3. The cost-effectiveness of a screening program is very dependent on the incidence of the disease being screened for. PKU satisfies all the criteria needed to justify a screening program in Caucasians. In the case of prenatal detection of Down's syndrome through amniocentesis, a screening program probably is justified for older mothers, but it is not easy to decide what maternal age to choose as the threshold for a universal screening program.

4. Modern medicine may permit the survival of patients with previously incurable genetic diseases, and so may have a dysgenic effect. Though there are possible examples of such selection relaxation, it seems likely that dysgenic effects will be relatively small and will appear very slowly.

5. Eugenics aims at improving the future qualities of mankind through selective breeding. It can be applied positively (encouragement of desirable matings) or negatively (discouragement of undesirable matings). Elimination of some recessive diseases may have a beneficial effect, but the response will be very slow. One form of negative eugenics that might be reasonable is to discourage matings between individuals heterozygous for the *same* deleterious gene. Although this approach avoids the need to totally bar any individual from reproduction, it still encounters several practical difficulties.

6. Genetic counseling may have, on average, some modest eugenic value, but its primary purpose (as in nearly all medical work) is the benefit of the individual patient and the family.

7. Positive eugenics poses many problems, including the difficulty of deciding what traits are to be favored, the unacceptability of restrictions on the liberty of individuals to reproduce, and the dangers of reducing the stock of human genetic variability. Sex selection is a form of eugenics that is now technically possible, but it also poses serious difficulties.

8. It has been claimed that the difference in average IQ between blacks and whites in the U.S. may have a substantial genetic component. The arguments for this hereditarian point of view do not seem to hold up to careful scrutiny. The question of whether this difference has a substantial genetic component cannot be answered definitely now, and the question does not in itself seem to be of theoretical or practical importance.

9. There are significant IQ differences between social classes, and these may in part have a genetic component. There is, however, little evidence that this genetic factor could have any long-term selective effect, nor is there any evidence that the apparent negative correlation between IQ and family size is leading to a tendency for a decline in IQ.

10. Man-made radiation and some chemicals released in our environment are significant sources of increased mutagenicity and carcinogenicity.

11. Genetic contributions to plant and animal breeding have a considerable record of achievement in contributing to increased and improved supplies of food. The quality of "social engineering" is still much poorer than that of food production and other technologies.

12. If by "equal" is meant "the same," genetics has shown us that individuals are not equal. Legal equality of the rights of individuals, however, has nothing to do with genetics.

13. Some social decisions, especially in the educational and medical spheres, are bound to be influenced by the existence of genetic variation. No society can afford to ignore the existence of our much prized individual variability.

Exercises

1. In what ways do genetic diseases differ from other types of diseases?

2. Outline the criteria you would establish before starting up a genetic screening program.

3. What factors would you consider in setting the age above which all pregnant women would be screened by amniocentesis for the presence of Down's syndrome in a fetus?

4. Why are the dysgenic effects of medicine likely to act quite slowly? Under what conditions may medical treatment of a disease have a eugenic (rather than dysgenic) effect?

5. What are the problems involved in any attempt to remove deleterious recessive genes from the human gene pool?

6. What are the main factors that make it difficult, if not impossible, to establish whether an IQ difference between two groups (such as that between U.S. blacks and whites) has a significant genetic component?

7. Why is the IQ distribution among social classes apparently stable?

8. Under what circumstances, if any, would you expect average IQ to change from generation to generation? Would it be possible to detect such a change if it occurred? Give reasons for your answers.

9. What, in your view, are the major social implications of the existence of extensive genetic variability in human populations?

References

Baer, A.
1973. (Ed.) *Heredity and Society.* New York: Macmillan.

Bergsma, D.
1972. (Ed.) "Advances in human genetics and their impact on society," *Birth Defects: Original Article Series,* vol. 8, no. 4.

Bodmer, W. F., and L. L. Cavalli-Sforza
1970. "Intelligence and race," *Scientific American,* vol. 223, no. 4, pp. 19–29.

Burt, C.
1961. "Intelligence and social mobility," *British Journal of Statistical Psychology,* vol. 14, pp. 3–24.

Cattell, R. B.
1974. "Differential fertility and normal selection for IQ: Some required conditions in their investigation," *Social Biology,* vol. 21, pp. 168–177.

Cavalli-Sforza, L. L., and W. F. Bodmer
1971. *The Genetics of Human Populations.* San Francisco: W. H. Freeman and Company. [Chapter 12 includes discussion of the topics of this chapter.]

Committee for the Study of Inborn Errors of Metabolism
1975. *Genetic Screening.* Washington, D.C.: National Academy of Sciences.

Dobzhansky, Th.
1973. "Is genetic diversity compatible with human equality?" *Social Biology,* vol. 20, pp. 280–288.

Etzioni, A. M.
1968. "Sex control, science and society," *Science,* vol. 161, pp. 1107–1112.

Gajdusek, D. C.
1973. "Kuru in the New Guinea highlands," in *Tropical Neurology,* ed. J. D. Spillane, Oxford: Oxford Univ. Press.

Haller, M. H.
1963. *Eugenics.* New Brunswick, N.J.: Rutgers Univ. Press.

Halsey, A.
1958. "Genetics, social structure and intelligence," *British Journal of Statistical Psychology,* vol. 9, pp. 15–28.

Herrnstein, R. J.
1973. *IQ in the Meritocracy.* Boston: Little, Brown.

Higgins, J. V., E. W. Reed, and S. C. Reed
1962. "Intelligence and family size: A paradox resolved," *Eugenics Quarterly,* vol. 9, pp. 84–90.

Hollaender, A.
1971. (Ed.) *Chemical Mutagens: Principles and Methods for Their Detection.* 2 vols. New York: Plenum Press.

Jensen, A. R.
1973. *Educability and Group Difference.* New York: Harper and Row.

Jones, A., and W. F. Bodmer
1974. *Our Future Inheritance: Choice or Chance.* Oxford: Oxford Univ. Press.
Kaback, M. M., R. S. Zeiger, L. W. Reynolds, and M. Sonneborn
1974. "Approaches to the control and prevention of Tay–Sachs disease," *Progress in Medical Genetics,* vol. 10, pp. 103–134.
Kamin, L. J.
1975. *The Science and Politics of IQ.* New York: Wiley.
Loehlin, J. C., G. Lindzey, and J. N. Spuhler
1975. *Race Differences in Intelligence.* San Francisco: W. H. Freeman and Company.
Lorber, J.
1971. "Results of treatment of myelomeningocele," *Developmental Medicine and Child Neurology,* vol. 13, pp. 279–303.
Motulsky, A. G.
1971. "Human and medical genetics: A scientific discipline and expanding horizon," *American Journal of Human Genetics,* vol. 23, pp. 107–123.
Muller, H. J.
1966. "What genetic course will man steer?" in *Proceedings of the Third International Congress of Human Genetics,* Baltimore, Md.: Johns Hopkins Press, pp. 521–543.
Osborn, F.
1968. *The Future of Human Heredity.* New York: Weybright and Talley.
Osborn, F., and C. J. Bajema
1972. "The eugenic hypothesis," *Social Biology,* vol. 19, pp. 337–345.
Post, R.
1971. "Possible cases of relaxed selection in civilized populations," *Humangenetik,* vol. 13, pp. 253–284.
Tizard, B.
1974. "IQ and race," *Nature,* vol. 247, p. 316.
Tobias, P. V.
1970. "Brain-size, grey matter and race: Fact or fiction?" *American Journal of Physical Anthropology,* vol. 32, pp. 3–26.
United Nations Scientific Committee
1962. *Report on the Effects of Atomic Radiation* (17th Session of the General Assembly, Official Records, Supplement No. 16, A/5216). New York: United Nations.
Waller, J. H.
1971. "Differential reproduction: Its relation to IQ test score, education and occupation," *Social Biology,* vol. 18, pp. 122–131.

APPENDIX

Genetic Applications of Probability and Statistics

Each section of the Appendix expands somewhat on certain mathematical and statistical aspects of topics considered in the book; chapter references are given for each section. However, in order to follow a logical mathematical development of the subject, the sequence of topics here differs slightly from the sequence followed in the main body of the book.

A.1 Probability and Frequency (Chapter 2 and Later Chapters)

The concept of *probability* (and related concepts such as chance, odds, likelihood, and so on) can be defined in several ways. Here we consider only two major definitions.

Probability may be defined theoretically *as the ratio of the number of "favorable" outcomes of a situation to the total number of possible outcomes.*

For example, consider the probability (or chances) of getting heads when tossing an ideal coin. There are two possible outcomes of the toss: heads or tails. Only the first outcome is favorable, so the probability of tossing heads is $\frac{1}{2}$. Hidden in the stipulation of "an ideal coin" is the assumption that there is no reason for the coin to fall on one face more often than on the other—in other words, that the two possible outcomes are equally likely, or equally probable. The definition of probability thus makes implicit use of the concept that is being defined. In spite of this apparent logical fault, the theoretical definition is useful in practice, for it is intuitively clear what is meant by an ideal (or "fair") coin. To find out whether a *real* coin is fair, we must collect and analyze real data: if heads and tails turn up about equally in a long series of tosses, we conclude that the coin is fair.

As another example, consider the probability of throwing a six when rolling a cubic die. In this case there are six possible outcomes, of which only one is favorable, so the probability is $\frac{1}{6}$. What is the probability of rolling an even number? In this case, there are three favorable outcomes (two, four, or six) out of the six possible outcomes, so the probability is $\frac{3}{6}$, or $\frac{1}{2}$.

The theoretical probability of various outcomes for a Mendelian cross is evaluated in a similar fashion. We list the possible combinations of gametes, assume that each possible combination is as likely as any other, and count up favorable combinations to find the probability of the particular zygote genotype being discussed. To test whether this probability is valid, we must examine actual data—ratios of genotypes that appear among the offspring of actual crosses.

Probability may be defined empirically *as the relative frequency with which a particular outcome occurs.*

From actual data, we can estimate the frequency at which an event occurs. For example, we toss a coin several times, recording whether we obtain heads or tails for each throw. Adding up the total number of heads obtained and the total number of tails obtained, we have *absolute frequencies* of the two outcomes. Dividing each absolute frequency by the total number of tosses, we obtain *relative frequencies* for the two outcomes. As the total number of tosses is increased, the relative frequency (say, of heads) tends to stabilize closer and closer to a particular value (which is $\frac{1}{2}$ if the coin and the tossing procedure are fair; see Fig. A.1). When the theoretical probability is unknown, the observed relative frequency provides an estimate of it. The precision of this estimate increases with a larger number of individual events included in the observations.

A probability must lie between zero and one.

At most, the number of favorable outcomes is equal to the total number of possible outcomes, and then the probability is one. A probability of one corresponds to *certainty* that an event will occur. If the number of favorable outcomes is zero, then the probability is zero. A probability of zero corresponds to certainty that the event will *not* occur. If P_A is the probability that an event A will occur, the probability that it will not occur is $1 - P_A$. Note that $P_A + (1 - P_A) = 1$,

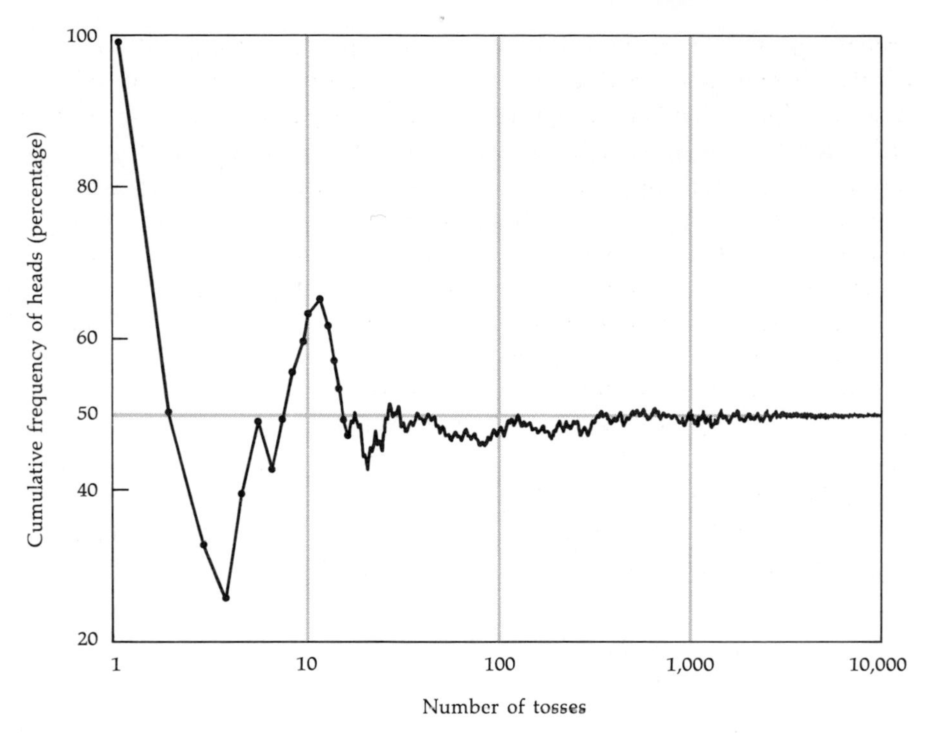

Figure A.1
Frequency of heads with increasing number of throws. As the number of observations increases, the relative frequency approaches a stable limiting value, which may be taken as an approximation of the probability of throwing heads on a single throw.

indicating a probability of one (a certainty) that the event either will or will not happen; there are no other possible outcomes. In every case, the sum of the probabilities for all possible outcomes should equal one.

A.2 Sum and Product of Probabilities (Chapter 2 and Later Chapters)

Suppose that an event can have several different outcomes: say, A, B, and C. The areas in Figure A.2 are sectors of a circle, drawn so that the area of each sector is proportional to the probability of the corresponding outcome. We assume that A, B, and C are mutually exclusive events—that is, if any one of them occurs, the others will not. We also assume that they are exhaustive events—that is, they exhaust the list of possible outcomes. One and only one of these three outcomes

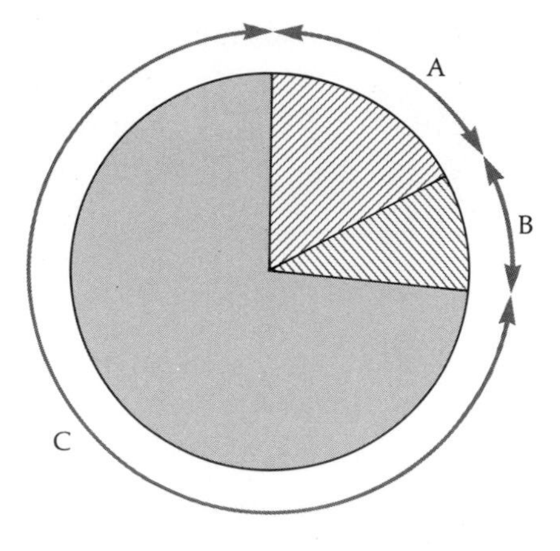

Figure A.2
Diagram representing probabilities of three mutually exclusive events. The area of each sector is proportional to the probability of the corresponding event. If the three events are exhaustive (there are no other possible outcomes), the sum of the areas equals one.

must be observed. In Figure A.2, the requirement of mutual exclusivity corresponds to a requirement that none of the areas overlap, and the requirement of exhaustiveness corresponds to a requirement that the sum of the areas equal one.

What is the probability that *either* A *or* B will be observed? In this case, the area corresponding to a favorable outcome is the sum of the areas for A and B. The probability of (A or B) is equal to the sum of the probabilities of A and B. What is the probability that *both* A *and* B will be observed? In this case, there are no favorable outcomes, because both events cannot occur together, so the probability is zero.

Suppose that some of the outcomes are not mutually exclusive. In this case, a diagram shows the corresponding areas overlapping (Fig. A.3). In the figure, some outcomes included in A are also included in B. P_C is the probability that *neither* A *nor* B will occur. Note that the sum of $P_A + P_B + P_C$ does not equal one in this case, because part of the total area (equal to one) is being counted twice in the summation.

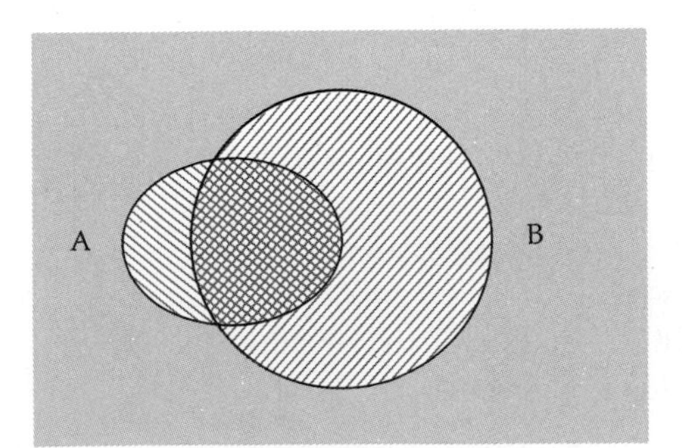

Figure A.3
Diagram representing probabilities of nonexclusive events. A and B are not mutually exclusive because they can occur together. Events A and B are not independent, because P_{AB} is not equal to $P_A \times P_B$.

We can usefully distinguish four possible outcomes in Figure A.3: A happens alone, B happens alone, both A and B happen, or C happens alone. These four outcomes are mutually exclusive, so the sum of their probabilities will equal one. By redefining the outcomes, we have converted this situation into one similar to that considered in Figure A.2. In Figure A.3, what is the probability that either A or B will occur? Any of the outcomes included in the striped areas are favorable, but the probability is *not* the sum of P_A and P_B because that sum counts part of the striped area twice. The easiest way to compute the desired probability is as $1 - P_C$, or the probability that the outcome will *not* be C, because all other possible outcomes are favorable. What is the probability that both A and B will occur? This probability corresponds to the area of overlap between A and B. Using Figure A.3, verify that this probability can be computed as $P_A + P_B + P_C - 1$.

If P_{AB} (the probability that both A and B will occur) is equal to $P_A \times P_B$, then the outcomes A and B are said to be *independent*. In this case, the probability of A occurring with B is the same as the probability of A occurring alone, and the probability of B occurring with A is the same as the probability of B occurring alone. This definition of statistical independence means that the phenomena A and B have no influences on each other. Knowing whether or not one event has occurred gives you no useful information to help decide whether or not the other event also occurred.

A.3 Binomial (or Bernoulli) Distribution (Chapter 2)

Let the probability of event A be p, and the probability of non-A be $q = 1 - p$. Say that A is the birth of a recessive homozygote in a mating between two heterozygotes. Then, $p = \frac{1}{4}$ and $q = \frac{3}{4}$. (Note that q in this case represents the probability of the offspring having the dominant phenotype.) The same probabilities are valid for each successive birth. The outcomes of a first and a second birth to the same parents are independent events. That is, the types of gametes that happen to join in the second mating are not influenced by the outcome of the first mating. Therefore, we can obtain probabilities for various combinations of phenotypes in a two-child family by multiplying various probabilities for the outcome of each birth, as shown in Table A.1 (where R represents a recessive phenotype and D a dominant phenotype).

In many cases, we are not interested in distinguishing between families of types 2 and 3. Either type is a family with one D and one R child, and we usually do not care which child was born first. The probability that a family will be *either* type 2 *or* type 3 is equal to the sum of the corresponding probabilities, or $2pq$. Thus, we obtain the following distribution for two-child families.

Table A.1
Distribution of various types of two-child families (see text)

	Phenotypes		
Type of family	First child	Second child	Probability
1	R	R	$p \times p = p^2$
2	R	D	$p \times q = pq$
3	D	R	$q \times p = pq$
4	D	D	$q \times q = q^2$

NOTE: The probability for each type of family is obtained as the product of the probabilities for the given phenotypes of the first child and the second child. The probability of a birth of a child with phenotype R is assumed to be p; the probability of a birth of a child with phenotype D is $q = 1 - p$. The phenotype of the second birth is assumed to be independent of the phenotype of the first birth.

Both children D	p^2
One D and one R	$2pq$
Both children R	q^2

The sum of these probabilities is $p^2 + 2pq + q^2 = (p + q)^2$. By definition, $p + q = 1$, so the probabilities of these three outcomes do sum to one, indicating that we have correctly defined exhaustive and mutually exclusive outcomes.

Table A.2 shows a generalization of this approach to families of n children. The term in the table needing explanation is written as $\binom{n}{k}$. This is the number of possible outcomes in which an n-child family has k children of type D and $n - k$ children of type R. The value of this expression can be shown to be

$$\binom{n}{k} = \frac{n!}{k!(n-k)!} = \frac{n(n-1)(n-2)\cdots(n-k+1)}{1 \times 2 \times 3 \times \cdots \times (k-1) \times k}, \tag{1}$$

where $a!$ (the "factorial" of the number a) is defined as

$$a! = 1 \times 2 \times 3 \times \cdots \times (a-1) \times a. \tag{2}$$

The expressions shown in the table for the case of k children of type D and $n - k$ of type R are actually general expressions that can be applied to all family types. When all children are type R, then $k = 0$, The number of outcomes fitting in this type is $\binom{n}{0}$, which is equal to one because $0!$ is defined as one. The probability that a family will be of this type is $\binom{n}{0}p^n q^0 = p^n$. Because $\binom{n}{n} = 1$, there is also only a single outcome giving an all-D family (when $k = n$). There are n possible outcomes yielding families with one D and all other children of type R; the only D birth can be either the first, the second, . . . , or the nth. This result can

Table A.2
Distribution of various types of n-child families (see text)

Family Type	Phenotype of Child						Probability	Number of Outcomes	Total Probability
	1	2	3	$\cdots$	$(n-1)$	n			
All R	R	R	R	$\cdots$	R	R	$p \times p \times p \times \cdots \times p \times p = p^n$	1	p^n
1 D	D	R	R	$\cdots$	R	R	$q \times p \times p \times \cdots \times p \times p = p^{n-1}q$		
and	R	D	R	$\cdots$	R	R	$p \times q \times p \times \cdots \times p \times p = p^{n-1}q$		
$(n-1)$ R	.	.	.	...	.	.	. .	n	$np^{n-1}q$
	.	.	.	...	.	.	. .		
	.	.	.	...	.	.	. .		
	R	R	R	$\cdots$	R	D	$p \times p \times p \times \cdots \times p \times q = p^{n-1}q$		
	D	D	D	$\cdots$	R	R	$p \times p \times p \times \cdots \times q \times q = p^{n-k}q^k$		
k D	.	.	.	...	.	.	. .		
and	.	.	.	...	.	.	. .	$\binom{n}{k}$	$\binom{n}{k}p^{n-k}q^k$
$(n-k)$ R	.	.	.	...	.	.	. .		
	R	R	R	$\cdots$	D	D	$q \times q \times q \times \cdots \times p \times p = p^{n-k}q^k$		
	D	D	D	$\cdots$	D	R	$q \times q \times q \times \cdots \times q \times p = pq^{n-1}$		
1 R	D	D	D	$\cdots$	R	D	$q \times q \times q \times \cdots \times p \times q = pq^{n-1}$		
and	.	.	.	...	.	.	. .	n	npq^{n-1}
$(n-1)$ D	.	.	.	...	.	.	. .		
	.	.	.	...	.	.	. .		
	R	D	D	$\cdots$	D	D	$p \times q \times q \times \cdots \times q \times q = pq^{n-1}$		
All D	D	D	D	$\cdots$	D	D	$q \times q \times q \times \cdots \times q \times q = q^n$	1	q^n

also be obtained by noting that $\binom{n}{1} = n$. The type of family with one R and all others D also includes n outcomes, and $\binom{n}{n-1} = n$.

The general result can be derived in the following way. Given n objects that are all different, they can be arranged in $n!$ different sequences. This follows from the fact that there are n different objects to choose from for the first position. Then, given the choice for the first, there remain $n-1$ objects from which to make a choice for the second position. Given these choices, $n-2$ objects remain from which to choose the third position. And so on, yielding a total of $n(n-1)(n-2) \cdots 1 = n!$ different sequences in which the objects can be arranged. If the objects are not all different, but k are of one type and $n-k$ are of another type, then there are $k!$ sequences that do not differ among themselves because k objects are all of a single type. (These k objects, in any given sequence, could be arranged in $k!$ ways if they were all different.) Similarly, there are $(n-k)!$ sequences that do not differ because $n-k$ of the objects are of the other type. The total number of distinguishable sequences, therefore, is $n!/(n-k)!k!$.

By analogy with the treatment of two-child families, we can write the distribution of types of n-child families in the following form:

$$p^n + np^{n-1}q + [n(n-1)/2]p^{n-2}q^2 + \cdots + \binom{n}{k}p^{n-k}q^k + \cdots + npq^{n-1} + q^n, \quad (3)$$

which is the binomial expansion of $(p + q)^n$—hence the name "binomial distribution."

The general term (giving the probability that an n-child family will have k children of a given type) is

$$P_{k,n} = \frac{n!}{k!(n-k)!} p^{n-k}q^k, \quad (4)$$

where p is the probability that any single birth will result in a child of the given type, and where $q = 1 - p$. Remember that $0! = 1! = 1$.

As an example, what is the probability that a seven-child family will contain one female and six males? If p is the probability of a male birth, then

$$P_{6,7} = \frac{7!}{6!\,1!} p^1q^6 = 7pq^6.$$

If $p = q = \frac{1}{2}$, then $P_{6,7} = 7 \times (\frac{1}{2})^7 = \frac{7}{128}$.

If a pair of heterozygotes has four children, what is the probability that none of them will be recessive homozygotes? Here we are asking for $P_{0,4}$ when $p = \frac{1}{4}$. The probability is $(\frac{3}{4})^4 = \frac{81}{256} = 0.32$. If the heterozygotes have five children, the probability that they will have no recessive-homozygote offspring becomes $(\frac{3}{4})^5 = \frac{273}{1,024} = 0.27$. If they have six children, the probability drops to $(\frac{3}{4})^6 = \frac{719}{4,096} = 0.18$.

If a couple has ten children, what is the probability that no more than one of them will be male? This is the sum of the probabilities of the mutually exclusive outcomes of no males (q^{10}) and of one male and nine females ($10pq^9$). If $p = q = \frac{1}{2}$, then $q^{10} = pq^9 = \frac{1}{1,024}$, and the sum of the two probabilities is $\frac{11}{1,024} = 0.011$.

A.4 The Hardy–Weinberg Law Derived for Random-Mating Pairs (Chapter 6)

In the text, the Hardy–Weinberg law is derived for random union of gametes. It can be demonstrated for random-mating pairs as follows. Suppose that the relative frequencies of genotypes are the same in males and females, and are u for genotype AA, v for genotype Aa, and w for genotype aa, where $u + v + w = 1$. Under conditions of random mating, the determinations of the genotypes of the two mates will be independent events, so we obtain the probability of each type of mating as the product of the probabilities of the two parental genotypes

Table A.3
Probabilities of various random matings

		Male		
		AA	Aa	aa
	AA	u^2	uv	uw
Female	Aa	uv	v^2	vw
	aa	uw	vw	w^2

NOTE: For both males and females, genotype frequencies are assumed to be u for AA, v for Aa, and w for aa.

(Table A.3). We are not interested in the distinction between matings involving the same genotypes in different sex combinations. For example, we can consider ♀ $AA \times$ ♂ Aa and ♀ $Aa \times$ ♂ AA as different cases of the same mating type, $AA \times Aa$. The probability that either of these mutually exclusive outcomes will occur is the sum of the individual probabilities shown in Table A.3, or $uv + uv = 2uv$. By similarly pooling other cases in Table A.3 that differ only in sexual arrangements, we obtain the six different matings listed in Table A.4, with the probabilities shown in the second column of that table.

The next three columns of the table show the expected Mendelian proportions for the progeny of each mating. Multiplying each of these proportions by the corresponding probability of the mating, we obtain the relative frequencies of progeny genotypes shown in the last three columns of the table. The sums of these three columns represent the probable relative frequencies of genotypes in the progeny generation. We have assumed that each mating is equally fertile, and that each genotype is equally viable—in other words, that there is no natural selection. We also assume that there is no mutation, and we neglect random

Table A.4
Relative frequencies of genotypes among progeny of random matings

Mating		Frequency	Mendelian proportions among progeny of each mating			Relative frequencies of genotypes among all progeny		
			AA	Aa	aa	AA	Aa	aa
AA	AA	u^2	all	—	—	u^2	—	—
AA	Aa	$2uv$	1/2	1/2	—	uv	uv	—
AA	aa	$2uw$	—	all	—	—	$2uw$	—
Aa	Aa	v^2	1/4	1/2	1/4	$v^2/4$	$v^2/2$	$v^2/4$
Aa	aa	$2vw$	—	1/2	1/2	—	vw	vw
aa	aa	w^2	—	—	all	—	—	w^2

fluctuations that would be caused by sampling effects in a real experiment. All of these assumptions are made in addition to the main assumption of random mating.

Among the parents, the gene frequency of A is $p = u + (v/2)$; that is, the frequency of AA plus one-half the frequency of Aa. The parental frequency of a is derived similarly as $q = (v/2) + w$.

Among the offspring, the frequency of AA is $u^2 + uv + (v^2/4)$, which can be rewritten as $[u + (v/2)]^2 = p^2$. Similarly, the frequency of aa is

$$(v^2/4) + vw + w^2 = [(v/2) + w]^2 = q^2.$$

Finally, the frequency of Aa is

$$2uw + uv + vw + (v^2/2) = 2[u + (v/2)][(v/2) + w] = 2pq.$$

Thus—whatever the genotype frequencies u, v, and w in the first generation—we find that the frequencies p^2, $2pq$, and q^2 are obtained in the second generation. The same computations apply to each following generation, so these frequencies will be maintained indefinitely. The gene frequencies remain constant through all generations. For example, the frequency of A is $p^2 + pq = p(p + q)$; because $p + q = 1$ by definition, $p(p + q) = p$.

A.5 The "Chi-Square" (χ^2) Goodness-of-Fit Test (Chapter 2 and Later Chapters)

The χ^2 test is used when one wishes to compare observed frequencies with those expected under some hypothesis (for example, a given Mendelian segregation, or expectations based on the Hardy–Weinberg law). Any real set of observed frequencies will almost certainly differ, at least slightly, from the frequencies predicted by theoretical considerations of probabilities. Are the deviations due to random statistical sampling errors, or do they indicate that the real results fail to match the theoretical model? The χ^2 test provides an estimate of the probability that results of a certain deviation from expectations could occur through random fluctuations, given the size of sample used.

The function χ^2 is defined by

$$\chi^2 = \Sigma[(E - O)^2/E], \tag{5}$$

where O is the observed absolute frequency for a class, E is the corresponding absolute frequency expected on the basis of the hypothesis, and the summation (Σ) is extended over all classes. Associated with the χ^2 value is a quantity called the "number of degrees of freedom" (d.f.). This quantity is a little hard to explain in words, but its computation is relatively simple. It is equal to the number of classes whose frequency must be known in order to determine the frequencies of all classes (given the total absolute frequency). The computation of d.f. is explained in the discussion of the examples that follow.

It can be shown that, on average, the χ^2 value should approximately equal d.f., if the hypothesis is correct. A χ^2 value much above the d.f. value is indicative of significant disagreement—in other words, indicative of a very small probability that such large deviations from expected frequencies would occur through random fluctuations. In practice, some χ^2 value is chosen arbitrarily for each d.f. value to define a threshold of acceptability. If the χ^2 value obtained for a given set of observations is below this threshold, the results are considered within tolerable agreement with the hypothesis. If the χ^2 value is larger than the threshold, the results are considered to be in significant disagreement with the hypothesis. The threshold normally used in genetic experiments is a χ^2 value that would be exceeded because of random fluctuations in only 5 percent of similar experiments. In other words, there is a probability of 5 percent that a χ^2 value this large or larger could be obtained if the hypothesis is correct. A very small value of χ^2 (far below the threshold) would seem superficially to indicate very good agreement between hypothesis and observations; in practice, such a result may also cast doubt upon the validity of the experiment, because if it tends to recur, it suggests that the sampling procedure is not correct. However, we will not deal further with the problem of very small χ^2 values here. Table A.5 gives threshold values of χ^2 (at the 5 percent significance level) for various values of d.f. The value of d.f. is often written in brackets as a subscript to the chi-square symbol—for example, $\chi^2_{[3]}$ indicates a chi-square value with three degrees of freedom.

Table A.5
Values of χ^2 corresponding to the 5 percent significance level

Degrees of freedom	χ^2
1	3.84
2	5.99
3	7.82
4	9.49
5	11.07
10	15.99
20	28.41
30	40.26

SOURCE: Sample of values from statistical tables compiled by R. A. Fisher and F. Yates, *Statistical Tables for Biological, Agricultural, and Medical Research,* Oliver and Boyd, 1953.

EXAMPLE 1 *Testing a 1:1 hypothesis.* Suppose that we find, among a sample of 90 patients with a certain disease, 51 females and 39 males. Is this observation significantly different from an expectation that males and females should appear in a 1:1 ratio among patients (in other words, an expectation that the frequencies of males and of females among patients should be each 50 percent)? Table A.6 shows the computation of the χ^2 value to test this hypothesis. Note that expected frequencies are computed by multiplying the expected proportions by the total number of individuals observed.

In this example, there are two classes (males and females). If we know the total number of individuals observed, we need only know the observed frequency for one class to completely specify the observed results. For example, if we know that 51 of the 90 observed individuals are female, we can easily compute that 90 − 51 = 39 are male. The d.f. value is the number of classes needed to specify the results, so d.f. = 1.

The threshold value of $\chi^2_{[1]}$ is 3.8 (from Table A.5). Because we obtain a value of χ^2 that is less than 3.8, we conclude that there is *not* significant disagreement between the hypothesis and the observations. This is the same as saying that our observations are in *agreement with the hypothesis* of a 1:1 expectation for the two sexes. Put another way, we have found that there is a probability of greater than 5 percent that the observed 51:39 ratio could be obtained through random statistical fluctuations, even if the underlying mechanism is one that would be expected to produce a 1:1 ratio over a very large sample population.

EXAMPLE 2 *A shortcut method for testing a 1:1 hypothesis.* In the case of a 1:1 hypothesis, the general formula for computing χ^2 is equivalent to

$$\chi^2_{[1]} = (a - b)^2/(a + b), \tag{6}$$

where a and b are the observed frequencies in the two classes. In the case of Example 1,

$$\chi^2_{[1]} = (51 - 39)^2/(51 + 39) = 144/90 = 1.6.$$

EXAMPLE 3 *Testing the hypothesis of Mendelian segregation.* Among the offspring of the cross $AaBb \times Aabb$, the Mendelian expectations for the phenotypes

Table A.6
Computation of χ^2 for Example 1

Class	O	E	$O - E$	$(O - E)^2$	$(O - E)^2/E$
Males	39	$\frac{1}{2} \times 90 = 45$	−6	36	$\frac{36}{45} = 0.8$
Females	51	$\frac{1}{2} \times 90 = 45$	+6	36	$\frac{36}{45} = 0.8$
Total	90	90	0		$1.6 = \chi^2_{[1]}$

AB:Ab:aB:ab are 3:3:1:1 (confirm this expectation for yourself). Converting these expectations to relative frequencies, we have $\frac{3}{8}:\frac{3}{8}:\frac{1}{8}:\frac{1}{8}$. Suppose that 64 offspring observed in an experimental cross show the phenotype frequencies 21:25:10:8. Are the observed results in agreement with the hypothesis of Mendelian segregation?

The χ^2 value is computed in the same way as for Example 1; the computations are shown in Table A.7. This time, however, we must specify three of the class frequencies (in addition to the total) in order to fully specify the observed results; the frequency for the fourth class can be obtained by subtraction from the total. Therefore, d.f. = 3 in this example. The computed χ^2 value of 0.916 is less than the threshold value of 7.82 for $\chi^2_{[3]}$, so we conclude that the observations are in agreement with the hypothesis of Mendelian segregation.

EXAMPLE 4 *Testing the hypothesis of Hardy–Weinberg equilibrium.* In a population of 150 individuals, the genotype frequencies *AA*:*Aa*:*aa* are found to be 112:32:6. Are these frequencies in agreement with the hypothesis that the population is in Hardy–Weinberg equilibrium for these alleles? First, we must compute the gene frequencies:

$$p = [112 + (32/2)]/150 = 0.853;$$
$$q = 1 - 0.853 = 0.147.$$

The expected frequencies, of course, are p^2, $2pq$, and q^2. Table A.8 shows the computation of χ^2. There are three classes in this distribution, so the value of d.f. would normally be 2. However, in this case we have used values of p and q computed from the observed frequencies to calculate the expected frequencies. Given these expected frequencies (and a knowledge of the way they were computed), all of the observed frequencies can be derived from knowledge of the observed frequency for a single class. For example, verify that

$$O_{Aa} = 2(Np - O_{AA}) \qquad \text{and} \qquad O_{aa} = N(1 - 2p) + O_{AA}$$

where O_{AA} and O_{Aa} and O_{aa} are the observed absolute frequencies of the three genotypes and N is the total number of individuals in the sample. Because the

Table A7
Computation of χ^2 for Example 3

Class	O	E	$O - E$	$(O - E)^2$	$(O - E)^2/E$
AB	21	$\frac{3}{8} \times 64 = 24$	-3	9	$\frac{9}{24} = 0.375$
Ab	25	$\frac{3}{8} \times 64 = 24$	$+1$	1	$\frac{1}{24} = 0.041$
aB	10	$\frac{1}{8} \times 64 = 8$	$+2$	4	$\frac{4}{8} = 0.500$
ab	8	$\frac{1}{8} \times 64 = 8$	0	0	$\frac{0}{8} = 0.000$
Total	64	64	0		$0.916 = \chi^2_{[3]}$

Table A.8
Computation of χ^2 for Example 4

Class	O	Expected relative frequency	Expected absolute frequency (E)	$O - E$	$(O - E)^2$	$(O - E)^2/E$
AA	112	$p^2 = (0.853)^2 = 0.7276$	$0.7276 \times 150 = 109.14$	2.86	8.1796	0.075
Aa	32	$2pq = 2 \times 0.853 \times 0.147 = 0.2508$	$0.2508 \times 150 = 37.62$	−5.62	31.5844	0.840
aa	6	$q^2 = (0.147)^2 = 0.0216$	$0.0216 \times 150 = 3.24$	2.76	7.6176	2.351
Total	150		150.00			$3.266 = \chi^2_{[1]}$

complete observation can be specified by a frequency for a single class (plus knowledge of the expected frequencies and total number of individuals), d.f. in this case is 1. The computed χ^2 value of 3.266 is less than the threshold value for $\chi^2_{[1]}$, so we conclude that the observed results are in agreement with the hypothesis of Hardy–Weinberg equilibrium.

It should be noted, however, that the validity of the χ^2 test is considered doubtful in any case where the value of E for any class is less than 5, if d.f. equals 1. Therefore, the χ^2 test should not really be used in this example—or, at any rate, its results should not be fully trusted. When d.f. is larger than 1, the results of a χ^2 test can usually be considered valid as long as the value of E for every class is at least 1.

EXAMPLE 5 *Testing a 2 × 2 table (the test of independence).* Suppose that we have observed frequencies of individuals classified according to two criteria, each of which distinguishes two or more classes. For example, we might classify twin pairs as MZ or DZ, and also as concordant or discordant for some trait. We wish to test the hypothesis that the MZ/DZ classification of a twin pair is independent of its concordant/discordant classification. The χ^2 test can be used for this purpose.

With two criteria of classification, each distinguishing two classes, there is a total of four classes. If a, b, c, and d are the numbers of observations in the four classes, then we can set out the data in a 2 × 2 table, as shown in Table A.9. The

Table A.9
A 2 × 2 table summarizing observations on twin pairs

	MZ	DZ	Total
Concordant	a	b	$a + b$
Discordant	c	d	$c + d$
Total	$a + c$	$b + d$	$N = a + b + c + d$

marginal totals are computed for each of the four criteria. The expected frequency for each class is computed as (row total) × (column total)/N. Thus, the expected frequency of concordant MZ twins is $(a + b)(a + c)/N$, and the other three expected frequencies are computed similarly. The general formula for χ^2 can be used. The following shortcut, derived from the general formula, is commonly used for a 2 × 2 table:

$$\chi^2_{[1]} = [(ad - bc)^2N]/[(a + c)(b + d)(a + b)(c + d)]. \quad (7)$$

There is only one degree of freedom for a 2 × 2 table, because all of the observed frequencies can be derived from a single class frequency, plus the marginal totals (confirm this for yourself). In general, d.f. for an $m \times n$ table is equal to $(m - 1) \times (n - 1)$.

Table A.10 shows observed frequencies of twin pairs, tested for concordance of liability to tuberculosis. Are these data in agreement with the hypothesis that concordance is independent of the MZ/DZ nature of the twin pair? Using equation 7, we compute

$$\chi^2_{[1]} = \{[(50 \times 435) - (78 \times 85)]^2 \times 648\}/(135 \times 513 \times 128 \times 520) = 32.14.$$

This value is very much larger than the threshold value for $\chi^2_{[1]}$, so we conclude that the observed data are *not* in agreement with the hypothesis of independence. In other words, we conclude that the concordance in MZ twins is significantly different from that in DZ twins. In this case, we note that the proportion of concordant twin pairs is much greater in MZ than in DZ twins. Because we have found this difference to be statistically significant, we can conclude that the observations support a theory that there is an inherited component to the trait of liability to tuberculosis. Of course, the χ^2 test tells us nothing about the validity of the design or interpretation of the experiment. It only tells us, in this case, that the observed higher frequency of concordance among MZ twins is very unlikely to be due to a sampling error caused by a small sample population. If the same experimental techniques were applied to a much larger population, we would still expect to find a higher frequency of concordance among MZ twins.

Table A.10
A 2 × 2 table summarizing observations for Example 5

	MZ	DZ	Total
Concordant	50	78	128
Discordant	85	435	520
Total	135	513	648

A.6 Segregation Analysis and Ascertainment (Chapter 2)

The analysis of pedigrees provides the basis for formulating hypotheses concerning the mode of inheritance of a trait. A single pedigree, however, rarely provides enough information to establish an inheritance pattern. Usually, it is necessary to collect data from many pedigrees segregating for the same trait. This raises the problem of genetic heterogeneity. The trait may not be determined by the same gene with the same mode of inheritance in the various pedigrees, in which case the results will be difficult to interpret. The statistical analysis of a collection of pedigrees can be very complicated. Another key problem is that the mode of *ascertainment* of the pedigrees plays a major role in their analysis.

The problem of biases due to ascertainment difficulties can be illustrated using as an example a rare recessive trait. Almost all the informative families are those in which one or more sibs have the trait. Their parents will nearly always both be heterozygotes and so both normal. Clearly, if only sibships containing at least one affected individual are ascertained, then a number of sibships from heterozygous parents will be missed because they happen to have all normal progeny. From the binomial distribution, the expected proportion of families with n children that are missed in this way is $(\frac{3}{4})^n$—that is, 75 percent of those with one child only, 56 percent of those with two children, 42 percent of those with three children, and so on. Knowing the distribution of the sizes of families studied, one can correct for this bias, but this correction is valid only if all affected homozygous recessives in the population are detected. This latter case (detection of all affected) is called *complete selection,* and it is difficult to achieve, particularly with rare traits. Unless the survey of a population is exhaustive or the trait is common, only a fraction of the affected individuals will actually be found. When this fraction is very small (a situation which is called *single selection*), the ascertainment bias can be corrected relatively simply. The method used is to discard the affected individual through which a sibship has been traced, namely the *proband* or the *propositus,* and to consider only his sibs for assessing segregation ratios. For example, 37 probands were found for alkaptonuria in 37 different families, with a total of $A = 66$ alkaptonurics and $T = 181$ offspring overall. The proportion of alkaptonurics computed directly from all the material is $\frac{66}{181} = 0.365$, which is significantly different from the expected 3:1 for an autosomal recessive trait ($\chi^2_{[1]} = 12.69$). Ignoring the $N = 37$ probands and considering only the remainder gives

$$\frac{A - N}{T - N} = \frac{66 - 37}{181 - 37} = \frac{29}{144} = 0.201 \tag{8}$$

for the proportion of homozygous recessives. This is less than $\frac{1}{4}$, but not significantly so ($\chi^2_{[1]} = 1.81$). If this difference were significant, one might suspect that the fitness of alkaptonurics was lower than that of normals, or that the method used to eliminate the ascertainment bias had overcorrected.

The case lying between the two extreme ascertainment situations of complete and single selection is called *multiple selection.* This, therefore, refers to the case where some of the sibships with more than one recessive homozygote have been ascertained more than once in the survey procedure. There can then be more than one proband per family, but not all affected individuals are probands. That is, some of the affected were not ascertained directly, but only through the fact that they were sibs of probands. There is, in other words, a probability of ascertaining or detecting an affected individual that is neither very small as in the case of single selection, nor close to one as in complete selection. A satisfactory estimate of the segregation probability is then given by

$$P = \Sigma[a(r - 1)]/\Sigma[a(s - 1)], \tag{9}$$

where a is the number of probands, r is the number of affected, and s is the total size for each family, and the sums (Σ) are taken over all families. This method of correcting for ascertainment bias is called Weinberg's proband method. The probability of ascertainment can be estimated as

$$\pi = \Sigma[a(a - 1)]/\Sigma[a(r - 1)]. \tag{10}$$

Other methods for dealing with more complex situations are quite involved from a numerical viewpoint, and they usually require the use of a computer. For a general introduction to this problem and a list of further references, see L. L. Cavalli-Sforza and W. F. Bodmer, *The Genetics of Human Populations* [San Francisco: W. H. Freeman and Company, 1971], pp. 850–890.

A.7 Change of Gene Frequencies Under Selection (Chapter 9)

The simplest description of gene-frequency changes under selection is that for one locus with two alleles, *A* and *a*, assuming fitnesses of w_1, w_2, and w_3 for the genotypes *AA*, *Aa*, and *aa*, respectively. As discussed in Chapter 7, fitness depends on both fertility and the probability of survival to reproductive age, though its interpretation in the usual models is in terms of viability only. More complex models are needed to take into account adequately the effects of both fertility and viability, but the basic conclusions are, for the most part, not changed by the more complete models.

EFFECT OF ONE GENERATION OF SELECTION The composition of the population after selection can be computed as shown in Table A.11. The last row of the table gives genotype frequencies (fw/T) that sum to one after selection. The gene frequency after selection can be computed by adding the frequency of *AA* homozygotes to one-half the frequency of heterozygotes:

$$p' = (w_1p^2 + w_2pq)/T, \tag{11}$$

where p' is the gene frequency of A after one generation of random mating and selection in the absence of mutation and drift. The change in gene frequency during the single generation can be computed as

$$\begin{aligned}\Delta p &= p' - p \\ &= [(w_1p^2 + w_2pq)/T] - p \\ &= (w_1p^2 + w_2pq - pT)/T. \end{aligned} \tag{12}$$

Substituting $T = w_1p^2 + 2w_2pq + w_3q^2$ (see Table A.11) in equation 12 yields

$$\Delta p = [w_1p^2(1 - p) + w_2pq(1 - 2p) - w_3pq^2]/(w_1p^2 + 2w_2pq + w_3q^2). \tag{13}$$

Because $1 - 2p = (1 - p) - p = q - p$, equation 13 can be reduced to the following form:

$$\Delta p = pq[(w_1 - w_2)p + (w_2 - w_3)q]/T. \tag{14}$$

EQUILIBRIUM GENE FREQUENCIES *Equilibrium* is reached when the gene frequency no longer changes—that is, when $p' = p$, and therefore $\Delta p = 0$. Assuming that fitnesses are the same in each generation (constant selection), we can find equilibrium gene frequencies by setting the expression in equation 14 equal to zero. This yields three solutions. The first is $p = 0$ (all genes are a); the second is $q = 0$ (all genes are A); and the third is

$$(w_1 - w_2)p + (w_2 - w_3)q = 0. \tag{15}$$

Solving equation 15 for p, we obtain

$$\begin{aligned}(w_1 - w_2)p + (w_2 - w_3)(1 - p) &= 0, \\ (w_1 - 2w_2 + w_3)p + (w_2 - w_3) &= 0,\end{aligned}$$

and

$$p = (w_3 - w_2)/(w_1 - 2w_2 + w_3). \tag{16}$$

Thus, we find three possible equilibrium values for p (zero, one, or the value given by equation 16). The first two values correspond to fixation of one allele or the other. There will be a genetically meaningful equilibrium (p_E), corresponding

Table A.11
Computation of genotype frequencies after one generation of selection

	AA	*Aa*	*aa*	Total
f = frequency at conception	p^2	$2pq$	q^2	$(p + q)^2 = 1$
w = fitness	w_1	w_2	w_3	
$f \times w$	w_1p^2	$2w_2pq$	w_3q^2	$T = w_1p^2 + 2w_2pq + w_3q^2$
fw/T = frequency after selection	w_1p^2/T	$2w_2pq/T$	w_3q^2/T	1

to the value of equation 16, if the value of that expression lies between zero and one. It can be seen that this will be true if either (1) w_2 is greater than either w_1 or w_3 (heterozygote advantage), or (2) w_2 is smaller than either w_1 or w_3 (heterozygote disadvantage). If the fitness of the heterozygote is intermediate between the fitnesses of the homozygotes, there will be no stable equilibrium other than fixation.

The equilibrium formula (equation 16) can be simplified if fitnesses are written in the following modified form:

$$\begin{aligned} &\text{fitness of } AA = w_1/w_2 = 1 - s, \text{ where } s = 1 - (w_1/w_2);\\ &\text{fitness of } Aa = w_2/w_2 = 1;\\ &\text{fitness of } aa = w_3/w_2 = 1 - t, \text{ where } t = 1 - (w_3/w_2). \end{aligned}$$

Dividing both numerator and denominator of the expression in Equation 16 by w_2, we obtain the following expression for the equilibrium gene frequency of A:

$$p_E = t/(s + t). \tag{17}$$

The equilibrium frequency of a is then given by

$$q_E = 1 - p_E = s/(s + t). \tag{18}$$

There will be stable equilibrium (other than fixation) either if s and t are both positive, or if s and t are both negative.

Using the new expressions for fitness, we can also simplify equation 14. Dividing the numerator of the expression for Δp by w_2, we obtain

$$\begin{aligned} pq[(w_1 - w_2)p + (w_2 - w_3)q] &= pq\{[(w_1/w_2) - 1]p + [1 - (w_3/w_2)]q\}\\ &= pq(tq - sp). \end{aligned}$$

Dividing the denominator of the expression by w_2, we obtain

$$\begin{aligned} T/w_2 &= (w_1/w_2)p^2 + 2pq + (w_3/w_2)q^2\\ &= (1 - s)p^2 + 2pq + (1 - t)q^2\\ &= p^2 + 2pq + q^2 - sp^2 - tq^2\\ &= (p + q)^2 - sp^2 - tq^2\\ &= 1 - sp^2 - tq^2. \end{aligned}$$

Using these new forms of the numerator and denominator, we can rewrite equation 14 as

$$\Delta p = pq(tq - sp)/(1 - sp^2 - tq^2). \tag{19}$$

Noting from equations 17 and 18 that $p_E/q_E = t/s$, we can rewrite equation 19 in another form that is sometimes useful:

$$\begin{aligned} \Delta p &= spq^2[(t/s) - (p/q)]/(1 - sp^2 - tq^2)\\ &= spq^2[(p_E/q_E) - (p/q)]/(1 - sp^2 - tq^2). \end{aligned} \tag{20}$$

Note that w_1 and w_3 cannot be less than zero, so s and t cannot be greater than one. Therefore, the denominator of equation 20 is never less than $1 - p^2 - q^2 = 2pq$, and hence can never be negative. The sign of Δp is therefore determined by the numerator of the expression in equation 20.

HETEROZYGOTE ADVANTAGE LEADS TO STABLE EQUILIBRIUM We are interested in the *stability* of an equilibrium—namely, the conditions under which the gene frequency, if displaced from its equilibrium value, will tend to return toward the equilibrium. We will show that, of the two possible types of equilibria given by equation 16, only the equilibrium involving heterozygote advantage (with s and t both positive, and w_1 and w_3 both less than w_2) is stable. In this case, the gene frequency of A tends to move toward the equilibrium value $p_E = t/(s + t)$ regardless of the initial conditions, so long as the initial value of p is at least slightly greater than zero or less than one.

In order to evaluate the stability condition, consider what happens to a gene frequency p_1 that is less than p_E. Under conditions of heterozygote advantage, s is positive; p_1 and q_1 are also positive. As we have seen, the denominator of equation 20 is always positive. Therefore, the sign of Δp_1 is the same as the sign of $(p_E/q_E) - (p_1/q_1)$. Given that p_1 is less than p_E, then p_1/q_1 must be less than p_E/q_E, so the sign of Δp_1 must be positive. But $\Delta p_1 = p' - p_1$, and so p' must be larger than p_1, and the gene frequency moves in the direction of p_E.

If, on the other hand, we consider an initial gene frequency p_2 that is larger than p_E, then $(p_E/q_E) - (p_2/q_2)$ is negative and so is Δp_2; the gene frequency decreases, again moving toward the equilibrium value. Thus, we have shown that the equilibrium is stable under conditions of heterozygote advantage. The equilibrium value will be approached from any initial frequency that is not actually either zero or one. After equilibrium is reached, any random fluctuations that move the frequency away from equilibrium will be counteracted by the tendency to move back toward equilibrium values.

HETEROZYGOTE DISADVANTAGE LEADS TO AN UNSTABLE EQUILIBRIUM When the heterozygote is at a disadvantage to both homozygotes, w_1 and w_3 are both greater than w_2, and s and t are both negative. The situation is now the reverse of that for heterozygote advantage. Becuase s is negative, Δp_1 is negative for $p_1 < p_E$; the gene frequency decreases, and so moves away from the equilibrium value. Similarly, Δp_2 is positive for $p_2 > p_E$, and again the frequency moves away from equilibrium. The equilibrium will be maintained only so long as the gene frequency remains exactly p_E. Even a very slight deviation from p_E (produced by random statistical fluctuations) will cause the gene frequency to begin its steady movement away from equilibrium toward fixation of one allele or the other. This equilibrium, therefore, is unstable.

DIRECTIONAL SELECTION The model we have been using does not include mutation or migration, which are the sources of new alleles. According to this simplified model, one allele or the other will remain permanently fixed if the gene frequency ever reaches zero or one. But what would happen in such a condition of fixation if the absent allele is introduced into the population by mutation or migration? From the analyses just given, we can see that the gene frequency would then tend to move toward the equilibrium value under conditions of heterozygote advantage. Under heterozygote disadvantage, the gene frequency would tend to return toward zero or one. What happens if the heterozygote fitness w_2 is intermediate between the fitnesses w_1 and w_3 of the homozygotes? An analysis similar to those already given shows that the gene frequency

1. will always move toward one (that is, toward fixation of allele A) if the order of fitnesses is $w_1 > w_2 > w_3$, or $w_1 = w_2 > w_3$, or $w_1 > w_2 = w_3$;
2. will always move toward zero (that is, toward fixation of allele a) if $w_1 < w_2 < w_3$, or $w_1 = w_2 < w_3$, or $w_1 < w_2 = w_3$; and
3. will tend to remain unchanged if $w_1 = w_2 = w_3$.

The third case (equal fitnesses for all three genotypes) is, of course, the set of conditions assumed for formulation of the Hardy–Weinberg law.

A.8 The Balance Between Mutation and Selection (Chapter 8)

Mutation from A to A', occurring at a rate μ (mutations per gamete per generation), changes gene frequencies according to the following formulas:

$$p' = p - p\mu; \qquad \Delta p = p' - p = -p\mu; \qquad \Delta q = -\Delta p = p\mu; \tag{21}$$

where p and q are the frequencies of A and A', respectively, and where p' is the gene frequency of A in the next generation. Equation 21 follows from the fact that there are p A genes that can mutate, and a proportion μ of them do so in each generation. Thus, in each generation, an overall proportion $p\mu$ of A genes are converted by mutation to A'.

We consider the two extreme cases of selection against deleterious mutations—namely, a deleterious dominant and a deleterious recessive (called A' in both cases). Following the same procedure used in Table A.11, we can compute the effects of selection on these mutant alleles (Table A.12). Note that a *lethal* is a gene for which either s (if recessive) or t (if dominant) is zero.

BALANCE FOR DELETERIOUS RECESSIVES Because of adverse selection, A' genes will be lost at each generation. For a recessive, the proportion of individuals lost

Table A.12
Computation of genotype frequencies for deleterious mutant alleles after one generation of selection

	AA	AA'	$A'A'$	T
Frequency before selection	p^2	$2pq$	q^2	1
A' is deleterious recessive:				
Fitness	1	1	$1 - s$	
Frequency after selection	p^2/T	$2pq/T$	$(1 - s)q^2/T$	$1 - sq^2$
A' is deleterious dominant:				
Fitness	1	$1 - t$	0	
Frequency after selection	p^2/T	$(1 - t)2pq/T$	0	$1 - 2tpq - q^2$

is sq^2 (note in Table A.12 that T, the total frequency after selection, is $1 - sq^2$). The individuals lost would produce only A' gametes, and hence the loss of A' genes in the next generation is also approximately sq^2. We say "approximately" because the loss of genes is actually $sq^2/(1 - sq^2)$, but the difference is very little if q is small.

When the loss of A' genes because of adverse selection ($\Delta q = -sq^2$) equals the gain due to $A \rightarrow A'$ mutation (which from equation 21 is $\Delta q = p\mu$)—that is, when $sq^2 = p\mu$—then the gene frequency will not change. Because q is very small, p is almost one, so that $p\mu$ is very nearly equal to μ. Therefore we can write the equilibrium condition as

$$\mu = sq^2. \tag{22}$$

The recessive gene frequency at equilibrium therefore is $q = \sqrt{\mu/s}$, and the frequency of homozygotes conceived is $q^2 = \mu/s$. The latter is also the incidence at birth if selection against A' takes place entirely after birth.

BALANCE FOR DELETERIOUS DOMINANTS If the deleterious mutant allele is dominant, the loss of affected individuals due to adverse selection is approximately $2tpq$. The loss of $A'A'$ homozygotes (q^2) can be ignored, because they are extremely rare in comparison with the heterozygotes when q is very small. Therefore the $2tpq$ loss of individuals can be regarded as a loss of AA' heterozygotes, half of whose genes are A'. Thus the loss of A' genes is approximately $\Delta q = -tpq$. Again there is an approximation involved here, because the gene loss would actually be $tpq/(1 - 2tpq - q^2)$. The increase due to mutation is, as before, $\Delta q = p\mu$. Equilibrium will be reached when $tpq = p\mu$, or

$$tq = \mu. \tag{23}$$

The gene frequency of A' at equilibrium is $q = \mu/t$. The frequency of heterozygotes at conception is $2pq$. However, note that

$$\begin{aligned} 2pq &= 2q(1 - q) \\ &= 2q - 2q^2. \end{aligned}$$

Because q is very small, q^2 is approximately zero, and the equilibrium incidence of affected at conception is approximately equal to $2q = 2\mu/t$.

A.9 Inbreeding Effects (Chapter 11)

The inbreeding coefficient of an individual is the probability that the individual is homozygous for a gene coming from an ancestor who was common to both his parents. A fully inbred individual ($F = 1$) is therefore always homozygous for such a gene. Consider as usual a locus with just two alleles A and a, whose relative gene frequencies are p and q. Fully inbred individuals then will be either AA with probability p, or aa with probability q. Randomly bred individuals will be AA, Aa, or aa with the usual Hardy–Weinberg frequencies. Individuals with an inbreeding coefficient F have a probability F of being like a fully inbred individual, and a probability $1 - F$ of being like a randomly bred individual. This observation leads to the computation shown in Table A.13. The frequency of heterozygotes is reduced in comparison with Hardy–Weinberg proportions by an amount $2Fpq$, while that of each homozygote is increased by Fpq.

Consider a large number of rare deleterious recessive genes which are such that the ith gene has a low gene frequency q_i and reduces the relative viability of homozygotes by an amount s_i. Among the progeny of consanguineous matings with inbreeding coefficient F, homozygotes occur with a frequency $q_i^2 + Fp_iq_i$ (see Table A.13), so the overall viability loss is $s_i(q_i^2 + Fp_iq_i)$. Because p_i is close to one, Fp_iq_i is approximately equal to Fq_i. Also, because F (at least for first cousins) is large compared to q_i, Fq_i is much greater than q_i^2. Therefore, the quantity $s_i(q_i^2 + Fp_iq_i)$ is approximately equal to s_iFq_i.

Table A.13
Computation of genotype frequencies among individuals with inbreeding coefficient F

Inbreeding	AA	Aa	aa
F	p	0	q
$1 - F$	p^2	$2pq$	q^2
Total	$pF + p^2(1 - F) = p^2 + Fpq$	$2pq(1 - F)$	$qF + q^2(1 - F) = q^2 + Fpq$

Because each of the recessives is rare, mutants at different loci will usually occur in different individuals, and so their separate occurrences can be summed to give the total number of deaths due to deleterious recessives as

$$\Sigma(s_i F q_i) = F\Sigma(s_i q_i), \tag{24}$$

where the sum is extended over all values of i (that is, over all loci).

Assuming that deaths due to other causes—say, a proportion A of all births—occur with the same frequency among inbreds and noninbreds, then the total mortality among inbreds is $A + BF$, where $B = \Sigma(s_i qi)$. Therefore, under the simplifying assumptions we have made, mortality should increase linearly with F. The slope (B) of this linear relationship gives the number of "lethal equivalents." For full lethals, $s_i = 1$ for all i, giving $B = \Sigma q_i$, the sum of the gene frequencies of the lethals.

A.10 Populations with different Gene frequencies

Two populations with different gene frequencies for allele A (versus allele a), say p_1 in one and p_2 in the other population, each in Hardy–Weinberg equilibrium, have the following genotype frequencies.

genotype	AA	Aa	aa
population 1	p_1^2	$2p_1\,(1-p_1)$	$(1-p_1)^2$
population 2	p_2^2	$2p_2\,(1-p_2)$	$(1-p_2)^2$

Let us compute the heterozygosity in three different situations.

1. Individuals mate only within their population (1 only with 1 and 2 only with 2). We call this heterozygosity *within* populations H_W, and obtain it by averaging the heterozygosities of H_1, H_2 of each of the two populations breeding separately:

$$H_W = \frac{H_1 + H_2}{2} = \frac{2p_1(1 - p_1) + 2p_2(1 - p_2)}{2} = p_1 + p_2 - p_1^2 - p_2^2.$$

Note that heterozygosity within a population is a measure of the genetic differences between individuals of that population (for that locus). It tells us, in fact, what is the probability that one gene taken at random from the two existing at an autosomal locus in a random individual is different from a gene similarily taken from another random individual of the same population.

2. We mix the two populations in equal proportions, and let this new mixed population interbreed freely. We have a new population with gene frequency $(p_1 + p_2)/2$, and heterozygosity (at Hardy–Weinberg equilibrium)

$$H_M = 2\left[\frac{p_1 + p_2}{2}\left(2 - \frac{p_1 + p_2}{2}\right)\right] = p_1 + p_2 - \frac{p_1^2}{2} - \frac{p_2^2}{2} - p_1p_2.$$

3. We form another hypothetical population, in which an individual of population 1 mates only with an individual of population 2. In their progeny heterozygotes will be

$$H_T = p_1(1 - p_2) + p_2(1 - p_1) = p_1 + p_2 - 2p_1p_2.$$

This last quantity is also the probability that one gene out of the two present at an autosomal locus of a random individual in population 1 is different from one gene at the same locus from a random individual of population 2. It thus measures the total differences between two populations for that locus; but it clearly contains also the difference between individuals of the *same* population, which is given by the heterozygosity within populations H_W. If we subtract H_W from H_T we have

$$H_T - H_W = p_1^2 + p_2^2 - 2p_1p_2 = (p_1 - p_2)^2 = d^2.$$

We thus have a measure of "genetic distance" between two populations, which is simply the square of d, the difference of the two gene frequencies.

Numerical values of differences between racial groups given in chapter 19 were d^2 values averaged over 25 polymorphic loci (from the 27 listed in L. Cavalli-Sforza and W. Bodmer, 1971).

Note that the heterozygosity in the mixed, random breeding population shown above in case 2, H_M, is exactly intermediate between H_W and H_T. In fact,

$$H_M - H_W = \frac{p_1^2}{2} + \frac{p_2^2}{2} - p_1p_2 = \frac{1}{2}d^2.$$

Thus the frequency of heterozygotes in a population separated in two subgroups which do not interbreed (as in case 1), is less than that in a random breeding population by the quantity $d^2/2$. Isolation in subpopulations has formally the same effect as inbreeding—it reduces heterozygotes. If instead of 2, we take k subpopulations, one can show by an extension of the above argument (verify this for the case of two populations) that k noninterbreeding populations, each with gene frequency p_i ($i = 1,2, \ldots, k$) would have average heterozygosity

$$2\bar{p}\bar{q} - 2\sigma^2,$$

where $\bar{p}$ = the average gene frequency, $\bar{q} = 1 - \bar{p}$, and σ^2 = the variance of gene frequencies (with k in the nominator instead of the usual $k - 1$). That is,

$$\bar{p} = \sum_{i=1}^{k} p_i; \qquad \sigma^2 = \sum_{i=1}^{k} (p_i^2 - \bar{p})^2/k.$$

Homozygotes have frequencies $\bar{p} + \sigma^2$ and $\bar{q} + \sigma^2$, respectively. Under inbreeding, heterozygotes have a frequency $2\bar{p}\bar{q}(1 - F)$ (see Table A.13) and thus F is replaced in the case of noninterbreeding populations by $f = \sigma^2/\overline{p}\,\overline{q}$, since

$$2\,\overline{p}\,\overline{q} - 2\sigma^2 = 2\,\overline{p}\,\overline{q}\left(1 - \frac{\sigma^2}{\overline{p}\,\overline{q}}\right) = 2\,\overline{p}\,\overline{q}\,(1 - f).$$

The modified variance $f = \sigma^2/\overline{p}\,\overline{q}$ was used in chapter 19 to measure the world variation of the frequency of an allele. It is expected to be the same for all genes (independently of $\overline{p}$) if drift is the only cause of variation between populations.

A.11 Population Growth (Chapter 7)

EXPONENTIAL GROWTH A population-growth rate of r per year means that, if there were N_t individuals at time t, then at time $t + 1$ (that is, after one year) the population has increased by rN_t individuals. The total number of individuals at time $t + 1$ therefore is

$$N_{t+1} = N_t + rN_t = (1 + r)N_t. \tag{25}$$

If there are N_0 individuals at time $t = 0$, and N_1 at $t = 1$, and so on, and if the growth rate r remains constant, then

$$\begin{aligned} N_1 &= N_0 + rN_0 &&= (1 + r)N_0;\\ N_2 &= N_1 + rN_1 &&= (1 + r)N_1 = (1 + r)^2N_0;\\ N_3 &= N_2 + rN_2 &&= (1 + r)^3N_0;\\ &\vdots && \quad\vdots \end{aligned}$$

$$N_t = (1 + r)N_{t-1} = (1 + r)^tN_0. \tag{26}$$

When r is small, it is convenient to make use of the approximation

$$(1 + r) \cong e^r, \tag{27}$$

where $e = 2.718\ldots$ is the base of natural logarithms. Under most circumstances, r is sufficiently small for the approximation to be valid in practice. Using the approximation (equation 27) in the growth equation (equation 26), we obtain

$$N_t = N_0e^{rt}. \tag{28}$$

Taking logarithms to the base e of both sides of the equation, we have

$$\log N_t = \log N_0 + rt. \tag{29}$$

Hence a plot of $\log N_t$ against time (that is, a plot of N_t on a semilogarithmic graph) should give a straight line with slope r.

Population growth that follows equation 26 or 28 is called geometric or exponential (sometimes also logarithmic) growth.

DOUBLING TIME UNDER EXPONENTIAL GROWTH The time needed for a population to double in size under exponential growth can be computed by letting $N_t = 2N_0$ in equation 29. Solving for t, we obtain the following expression for the doubling time T:

$$\begin{aligned} T &= (\log N_t - \log N_0)/r \\ &= \log (N_t/N_0)/r \\ &= (\log 2)/r = 0.69/r. \end{aligned} \tag{30}$$

Table A.14 gives representative values of doubling times for growth rates that are relevant to human populations.

GROWTH RATES As time goes from t_1 to t_2, the number of individuals in the population goes from N_1 to N_2. The change in population size is

$$\Delta N = N_2 - N_1, \tag{31}$$

and the time interval (say, a year) during which this change takes place is

$$\Delta t = t_2 - t_1. \tag{32}$$

The ratio $\Delta N/\Delta t$ is an absolute measure of growth rate over this time interval. The relative growth rate is equal to the absolute rate divided by the number of individuals present at time t_1, or $(1/N_1)(\Delta N/\Delta t)$. As the time interval Δt becomes very small, the ratio $\Delta N/\Delta t$ approaches the instantaneous growth rate (or differential coefficient) dN/dt, and the distinction between N_1 and N_2 disappears, so that the relative growth rate becomes $(1/N)(dN/dt)$. It is then a standard result of differential calculus that

$$\frac{1}{N}\frac{dN}{dt} = \frac{d(\log N)}{dt}.$$

Table A.14
Doubling times for some growth rates typical of human populations

Growth rate (r)	Doubling time (t)
0.04	17 years
0.03	23 years
0.02	34 years
0.01	69 years

Differentiation of equation 29 shows that

$$\frac{d(\log N)}{dt} = r,$$

so we find that the instantaneous relative growth rate is just the annual growth r:

$$\frac{1}{N}\frac{dN}{dt} = r. \tag{33}$$

LOGISTIC GROWTH It is clear that no real population can continue to grow indefinitely at a constant relative growth rate. For example, the present human population of the world averages more than 70 persons per square mile of land surface, and the population is doubling in slightly more than 30 years. If this growth rate were to remain constant, there would be one person for each square foot of land area within about six centuries. It would not take many more centuries after that for the mass of human beings to exceed the mass of the earth! Obviously, any real population must be limited by some maximum population size, N_{max}, that cannot be exceeded because of limits on food, space, and so on.

Let us assume that the growth rate r decreases linearly with population size as N approaches N_{max}—from its highest value when the population is very small, to zero when $N = N_{max}$. Then we can express r in the form

$$r = r_0 \, [1 - (N/N_{max})], \tag{34}$$

where r_0 is a constant, independent of t and N. From equation 33, we know that

$$r = \frac{1}{N}\frac{dN}{dt},$$

so we can change equation 34 to the following form:

$$\frac{dN}{dt} = r_0 N[1 - (N/N_{max})]. \tag{35}$$

This is the equation for logistic growth. If there were N_0 individuals at time $t = 0$, then we can solve for the number of individuals at any time t by integrating equation 35. The result can be shown to be

$$N_t = N_{max}/(1 + e^{-r_0(t-t_{1/2})}), \tag{36}$$

where $t_{1/2}$ is the time at which the population reaches half its maximum size. The curve obtained by plotting N_t against t has a sigmoid shape, as illustrated in Figure 7.4.

Although the logistic growth equation seems to be based upon a very simplified assumption about the way r will change with time, laboratory experiments with populations growing in limited environments do often yield growth curves quite similar to the logistic curve.

POPULATION PROJECTIONS As discussed in the text, population projections for the human population (or almost any animal population) must be based upon considerations of age distributions and age-specific birth and death rates if they are to be usefully accurate over periods of a few decades. Here we use a simple numerical example, based on data collected in 1960 about females in the United States, to illustrate the application of population projections to the determination of the intrinsic rate of increase and the stable equilibrium age distribution. The projection is limited to females. For simplicity, we measure time in 15-year intervals, although demographers usually use smaller age classes (such as 5-year or even 1-year intervals).

Let p_0 be the mean proportion of women in the age group of 0 to 15 years who survive for 15 years, p_1 be the mean proportion of the 15-to-30 age group who survive for 15 years, and so on. Similarly, let b_1 be the mean number of female children born per woman in the 15-to-30 age group, b_2 be the mean number born to the 30-to-45 age group, and so on. Table A.15 gives the values of these statistics for the female population of the United States in 1960.

Let $N_{g,c}$ represent the number of individuals in age group g at the end of cycle c, where each cycle consists of 15 years. Suppose that we begin with an initial population $N = 100$, with the populations of the age groups as $N_{0,0} = 40$, $N_{1,0} = 30$, $N_{2,0} = 20$, and $N_{3,0} - 10$. Table A.16 shows the computation of the projected population at the end of the first 15-year cycle.

The population of the 0-to-15 age group at the end of the first cycle ($N_{0,1}$) is made up of the female children that have been born during the cycle. The contribution of each age group g in the initial population is equal to $b_g N_{g,0}$.

The population of the 15-to-30 age group at the end of the first cycle ($N_{1,1}$) is made up only of the survivors from the original 0-to-15 age group. Similarly, $N_{2,1} = p_2 N_{2,0}$. The population of the over-45 age group at the end of the first cycle includes survivors from both the 30-to-45 and the over-45 groups of the initial population.

Subsequent changes in the population can be predicted by repeating the same series of calculations, using the populations at the end of the first cycle instead

Table A.15
Age-specific vital statistics for the female population of the United States in 1960

Age (years)	Age-specific birth rate	Age-specific survival probability
0–15	$b_0 = 0$	$p_0 = 0.992$
15–30	$b_1 = 1.37$	$p_1 = 0.988$
30–45	$b_2 = 0.465$	$p_2 = 0.964$
Over 45	$b_3 = 0$	$p_3 = 0.888$

Table A.16
Computation of age-group distribution for model population after one 15-year cycle

Age group	Initial population	Population after one cycle		Percentage distribution
0–15	$N_{0,0} = 40$	$N_{0,1} = \sum_{g=0}^{3} b_g N_{g,0} = (0 \times 40) + (1.37 \times 30) + (0.465 \times 20) + (0 \times 10)$	$= 50.4$	34.1
15–30	$N_{1,0} = 30$	$N_{1,1} = p_0 N_{0,0} = 0.992 \times 40$	$= 39.7$	26.8
30–45	$N_{2,0} = 20$	$N_{2,1} = p_1 N_{1,0} = 0.988 \times 30$	$= 29.6$	20.1
>45	$N_{3,0} = 10$	$N_{3,1} = p_2 N_{2,0} + p_3 N_{3,0} = (0.964 \times 20) + (0.888 \times 10)$	$= 28.1$	19.0
Total	100		147.8	100.0

of the initial populations as the base for the new computations. Table A.17 shows a sampling of the results obtained over subsequent cycles.

The first cycle results in a drastic change in the population structure. There is a considerable leveling of the age distribution and a 50 percent increase in population size. These effects arise because the age distribution of the initial population was very far from the equilibrium distribution, being heavily weighted toward the younger age groups, which of course are more fertile. This type of rapid temporary change in population size and age structure may be expected whenever there is a sudden change in patterns of mortality and/or fertility.

Table A.17 shows changes in the age structure for this model population at intervals of 60 years (four cycles), as calculated by computer. The proportions of the population in the various age groups stabilize very quickly. In fact, they are near the equilibrium distribution after only 60 years, or about two generations. The rate of growth of the total population also stabilized rapidly, as shown by the last column of the table.

Table A.17
Age-group distribution of model population through establishment of equilibrium

After cycle	After years	Percentage distribution by age group				Total population size	Population growth per cycle*
		0–15	15–30	30–45	over 45		
0	0	40.00	30.00	20.00	10.00	100.0	—
4	60	28.6	22.1	17.1	31.2	390.5	1.35
8	120	27.9	21.1	16.1	34.7	1,198.1	1.32
12	180	27.7	20.9	15.0	35.5	3,529.7	1.31
16	240	27.6	20.8	20.8	35.6	10,315.6	1.31
Equilibrium		27.55	20.92	15.81	35.72	—	1.307

*Population growth per cycle is computed as the ratio of population size in any given cycle to that in the previous cycle (one cycle = 15 years).

A.12 Frequency Distributions (Part III)

Suppose that the statures of ten men are measured in centimeters, yielding the following data:

168, 178, 171, 165, 183, 170, 177, 180, 173, 170.

This is a typical sample of observations on a continuous trait. With such a small sample size (number of observations), there is little point in constructing a frequency distribution. However, we use this example here for convenience in illustrating the principles in distributions.

A distribution is formed by combining the data into groups of *classes*—say, from 160 to 164, 165 to 169, 170 to 174 cm, and so on—and plotting the number in each class against the class values. With a larger sample size (say, 40 to 50 observations or more), such grouping can be extremely useful, especially for graphical presentation of the data. A simple approach to grouping is to rewrite all the observations in rows that correspond to the individual classes, as shown in Table A.18. The number of individuals in each class is the absolute frequency for that class, as given in the last column of the table. The sum of all the frequencies is equal to the size of the sample.

The number of classes and the intervals used for each are arbitrary—though the intervals are almost always chosen to be of equal length, as in this example. In later computations, each individual observation will be replaced by counting it as one occurrence with the value of the class midpoint. For example, the class including all observations between 165 and 169 cm has a midpoint of $\frac{(165 + 169)}{2} = 167$. Therefore, the individual observations of 165 and 168 cm will later be considered only as two occurrences with the value 167. Note that the midpoint value, which is taken to be representative of the class, depends on how the measurements were made. The value 167 is appropriate for the class 165–169 if measurements were made to the nearest centimeter, for then a value recorded as 165

Table A.18
A frequency distribution prepared from observations of statures of ten men

Class	Individual observations	Class midpoint (x)	Frequency (f)
165–169	168, 165	167	2
170–174	171, 170, 173, 170	172	4
175–179	178, 177	177	2
180–184	180, 183	182	2
Total			10

would fall somewhere between 164.5 and 165.5, and so on. If the class included actual values between 165.0 and 169.9, then 167.5 would be a more appropriate midpoint.

The size of the class interval is chosen to give the most useful set of grouped data for further computations. Fewer classes means less work (and a larger absolute frequency for each class, hence less uncertainty due to random fluctuations in sampling), but fewer classes also means a greater loss of precision because a wider range of observed values is replaced by a single midpoint value. At any rate, after the interval has been chosen and the observations grouped in classes, the frequency distribution of the sample is given by the corresponding values of x and f (see Table A.18).

The arithmetic mean of the sample is given by

$$m = \Sigma fx / \Sigma f, \tag{37}$$

where Σfx is the sum over all classes of the product $f \times x$ computed for each class. In the example of Table A.18,

$$\Sigma fx = (167 \times 2) + (172 \times 4) + (177 \times 2) + (182 \times 2) = 1{,}740.$$

The mean, therefore, is

$$m = 1{,}740/10 = 174.$$

The grouping does have an effect, as we can see if we compute the arithmetic mean from ungrouped observations. In this case, the mean is

$$\bar{x} = (\Sigma x_i)/n, \tag{38}$$

where the x_is are the original observed values, n is the sample size, and the sum is taken over all the observations. The sum of the ten original observations is 1,735, so the arithmetic mean is

$$\bar{x} = 1{,}735/10 = 173.5.$$

We note that grouping into classes does result in some distortion of the data, but the effect of the grouping is quite small.

BAR DIAGRAMS AND FREQUENCY CURVES A bar diagram is constructed by drawing a bar for each class, with the height of the bar proportional to the class frequency and with the width of the bar centered on the class midpoint. The bar may be drawn as a thin line or a rectangle whose width is arbitrary (Fig. A.4B), or as a rectangle whose base is equal to the class interval (Fig. A.4C). The bar diagrams of the type shown in Figure A.4C are sometimes called histograms; they are usually drawn so that the area of each rectangle is proportional to the frequency of the corresponding class, even if the size of intervals varies.

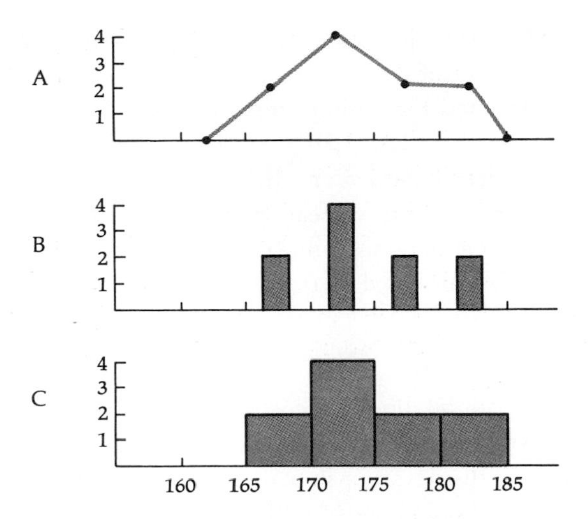

Figure A.4
Examples of bar graphs for frequency distributions. (A) Frequency diagram. (B) Bar diagram. (C) Histogram.

Where bars (Fig. A.4B) are drawn as thin lines, it is easy to see that the tops of the bars could be connected by straight lines to construct a curve that also would represent the distribution. Such a graph with the bars omitted is sometimes called a frequency diagram (Fig. A.4A). As the class interval becomes smaller, the frequency diagram approaches a continuous curve representing the actual distribution of the continuous trait.

A.13 Measure of Dispersion: Variance (Part III, Especially Chapters 14 and 15)

Average or representative values—such as the arithmetic mean, the median, and the mode—are measures of *location:* they indicate where, on the scale of a trait, a "typical" value can be found. They say nothing about the variation between individuals in the sample—that is, about the dispersion of individual values around the typical value. The usual measurement of dispersion or variation is the *variance,* which is defined by

$$V = [\Sigma(x_i - \overline{x})^2]/(n - 1), \qquad (39)$$

where the x_is are individual values (with i ranging from 1 through n), $\overline{x}$ is the mean computed from them (equation 38), and n is the sample size. The summation is taken over all values of i. This sum of the squares of the deviations from the mean is often called SS (for Sum of Squares). The denominator of equation 39 $(n - 1)$ is also the number of degrees of freedom of the variance, given that the mean $\bar{x}$ has been computed from the observations.

SHORTCUTS FOR COMPUTING VARIANCE It is often convenient to compute first the deviations from some arbitrary mean A (chosen to be not too far from the true mean), and then to square and sum these deviations. The sum of squares obtained from A can be corrected to obtain the true SS by making use of the identity

$$\Sigma(x_i - \bar{x})^2 = [\Sigma(x_i - A)^2] - n(\bar{x} - A)^2. \quad (40)$$

Table A.19 shows the computation of the variance for the ten stature observations from Table A.18, using both the full method (equation 39) and the shortcut (equation 40). Note that the deviations obtained with the shortcut are much easier to square and sum than those obtained with the full method, while the correction term $n(A - \bar{x})^2$ is relatively simple to compute. Both methods yield the same value for SS, and hence the same value for the variance.

VARIANCE OF A FREQUENCY DISTRIBUTION For a frequency distribution (as described in Section A.11), the formula for the variance is

$$V = \{\Sigma[f(x - \bar{x})^2]\}/[\Sigma(f) - 1], \quad (41)$$

Table A.19
Computation of variance for observations of statures of ten men

		Full method		Shortcut method ($A = 175$)	
i	x_i	$x_i - \bar{x}$	$(x_i - \bar{x})^2$	$x_i - A$	$(x_i - A)^2$
1	168	−5.5	30.25	−7	49
2	178	4.5	20.25	3	9
3	171	−2.5	6.25	−4	16
4	165	−8.5	72.25	−10	100
5	183	9.5	90.25	8	64
6	170	−3.5	12.25	−5	25
7	177	3.5	12.25	2	4
8	180	6.5	42.25	−5	25
9	173	−0.5	0.25	−2	4
10	170	−3.5	12.25	−5	25
$\Sigma x_i = 1{,}735$		$\Sigma(x_i - \bar{x})^2 = 298.50 = SS$		$\Sigma(x_i - A)^2 = 321$	
		$\bar{x} = 1{,}735/10 = 173.5$		$n(\bar{x} - A)^2 = 10(175 - 173.5)^2 = 22.5$	
		$V = 298.5/9 = 33.2$		$SS = 321 - 22.5 = 298.5$	
				$V = 298.5/9 = 33.2$	

where the sums are taken over all classes in the distribution. Equation 41 can be converted algebraically into the following more convenient form:

$$V = \frac{\Sigma(fx^2) - [\Sigma(fx)]^2/\Sigma(f)}{\Sigma(f) - 1}. \tag{42}$$

As in equation 40, one can simplify computations with these formulas by using a convenient quantity A instead of the original mean $\bar{x}$. No subsequent correction is needed in equation 42 because $\bar{x}$ does not appear in the formula; each class midpoint x is simply replaced by a more convenient "coded" variable $X = x - A$. Table A.20 shows the computation of the variance for the distribution of Table A.18, using equation 42 with x replaced by $X = x - 170$ in order to simplify the computations.

Note that the variance of 28.9 obtained for the frequency distribution differs from the variance of 33.2 obtained for the original set of ungrouped data (Table A.19). Again, this is an imprecision introduced by grouping the data and using a class midpoint to represent a number of individual observations. This imprecision is, in part at least, of random nature, but the deviation from the true variance is likely to be larger if larger class intervals are used.

THEORETICAL DISTRIBUTIONS Let $p_1, p_2, \ldots, p_n$ be the expected proportions of individuals that fall into each of n classes with midpoint values of $x_1, x_2, \ldots, x_n$. The mean of this distribution is

$$m = \Sigma(p_i x_i), \tag{43}$$

where $\Sigma p_i = 1$, and the sums are taken over all values of i from 1 through n. The variance is

$$V = [\Sigma(p_i x_i^2)] - m^2. \tag{44}$$

Table A.20
Computation of variance for frequency distribution of Table A.18

x	f	$X = x - 170$	fX	fX^2
167	2	−3	−6	18
172	4	2	8	16
177	2	7	14	98
182	2	12	24	288
	$\Sigma(f) = 10$		$\Sigma(fX) = 40$	$\Sigma(fX^2) = 420$

$[\Sigma(fX)]^2 = 1{,}600$

$[\Sigma(fX)]^2/\Sigma(f) = 160$

$SS = \Sigma(fX^2) - [\Sigma(fX)]^2/\Sigma(f) = 420 - 160 = 260$

$V = SS/[\Sigma(f) - 1] = 260/9 = 28.9$

As an example, consider a random-mating population with two alleles A and a having gene frequencies p and q, respectively. These alleles act on a trait whose average phenotypic value is a for AA homozygotes, $-d$ for heterozygotes, and $-a$ for aa homozygotes (Table A.21).

The phenotype scale used needs some comment. First, note that the symbol a used as a variable in measuring the phenotype is unrelated to the allele a. The a and d values are called the *additive* and *dominance* values, respectively. They are defined in the following way. First, obtain sample populations of each of the three genotypes AA, Aa, and aa. These samples should be chosen so that each sample shows the same gene frequencies for all other alleles at other loci that affect the trait being studied—in other words, so that the only relevant difference between the populations is the genotype at the A,a locus. Then measure the mean value of the trait for each of the three populations—call these mean values $\bar{x}_{AA}$, $\bar{x}_{Aa}$, and $\bar{x}_{aa}$. Now we can define a and d as

$$a = (\bar{x}_{AA} - \bar{x}_{aa})/2,$$
$$d = [(\bar{x}_{AA} + \bar{x}_{aa})/2] - \bar{x}_{Aa}$$

From these expressions, you can show that each of the phenotype values given in Table A.21 represents the mean value for the corresponding genotype minus the quantity $(\bar{x}_{AA} + \bar{x}_{aa})/2$. Thus the a and d values are simply coded variables for the mean phenotypic values of the genotypes (as discussed in connection with Table A.20).

Using formulas 43 and 44, we find that the mean of the distribution in Table A.21 is

$$m = p^2a - 2pqd - q^2a$$
$$= a(p^2 - q^2) - 2pqd.$$

However, $p^2 - q^2 = (p - q)(p + q) = p - q$, because $p + q = 1$. Therefore,

$$m = a(p - q) - 2pqd. \quad (45)$$

The variance of the distribution is

$$V = a^2p^2 + 2pqd^2 + a^2q^2 - m^2. \quad (46)$$

Table A.21
A theoretical distribution of phenotype values for genotypes determined by two alleles at a single locus

	AA	Aa	aa
Frequency (p_i)	p^2	$2pq$	q^2
Phenotype value (x_i)	a	$-d$	$-a$
p_ix_i	p^2a	$-2pqd$	$-q^2a$
$p_ix_i^2$	p^2a^2	$2pqd^2$	q^2a^2

First, consider the simple case of a population in which $p = q = \frac{1}{2}$. In this case (using equations 45 and 46),

$$m = -d/2; \tag{47}$$

$$V = (a^2/2) + (d^2/2) - m^2$$
$$= (a^2/2) + (d^2/4). \tag{48}$$

The variance is made up of two terms, one of which depends only on the additive value and the other only on the dominance value.

If there are many genes with alleles A and a, B and b, . . . , and with additive and dominance values a_a and d_a, a_b and d_b, . . . , respectively, and if their effects are independent so that they can be added up to produce the final phenotype, then

$$V = [(\Sigma a_i^2)/2] + [(\Sigma d_i^2)/4].$$

If we define $V_A = \frac{1}{2}\Sigma a_i^2$ and $V_D = \frac{1}{4}\Sigma d_i^2$, then

$$V = V_A + V_D. \tag{49}$$

The value V is the *genotypic* variance, because it is computed on the assumption that each individual with a given genotype has a phenotype x_i that is the *average* phenotypic value for that genotype. In other words, there is no allowance for "environmental" variation among individuals with the same genotype. When there is such variation, measured by V_E (which is the same for all genotypes)—and assuming further that there is no interaction (or correlation) between genotype and environment—then the total *phenotypic* variance is equal to V_E plus the genotypic variance V given by equation 49.

For a random-mating population in which p and q are not both $\frac{1}{2}$, the formulas are more complex. The genotypic variation V can be partitioned into $V_A + V_D$ by defining V_A and V_D as

$$V_A = \Sigma\{2pq[a + d(p - q)]^2\}, \tag{50}$$

and

$$V_D = \Sigma[(2pqd)^2], \tag{51}$$

where, as usual, the sums are taken over all genes (if there is more than one gene affecting the trait). Formulas 50 and 51 can be derived from equations 45 and 46. (For the full derivation, see L. L. Cavalli-Sforza and W. F. Bodmer, *The Genetics of Human Populations,* San Francisco: W. H. Freeman and Company, 1971, p. 531.)

A.14 Covariance, Regression, and Correlation (Part III)

Suppose that we have n pairs of observations, such as the observations of the stature of fathers and their sons summarized in Table A.22. The joint variation (or covariation) of the two variables is measured by the covariance:

Table A.22
A set of paired observations

Pair number (i)	1	2	3	4	5	6	7	8	9	10
Father's stature (x_i)	168	178	171	165	183	170	177	180	173	170
Son's stature (y_i)	169	172	175	162	180	180	173	179	178	172

$$W = \{\Sigma[(x_i - \bar{x})(y_i - \bar{y})]\}/(n - 1). \tag{52}$$

The numerator of this expression is sometimes called the "sum of products," although a more complete description would be "the sum of products of the deviations of x and y from their respective arithmetic means, $\bar{x}$ and $\bar{y}$." The denominator of the expression represents the degrees of freedom for the variance. The following formulas, which are algebraically equivalent to equation 52, are more convenient for computation:

$$W = [\Sigma(x_i y_i) - n\bar{x}\bar{y}]/(n - 1) \tag{53}$$
$$= \{\Sigma(x_i y_i) - [\Sigma(x_i)][\Sigma(y_i)]/n\}/(n - 1). \tag{54}$$

When carrying out computations, you can subtract arbitrary quantities A from all x_i and B from all y_i without affecting the result for the covariance; this procedure often helps simplify computations. In Table A.23, we have computed the covariance for the data of Table A.22, using $A = B = 175$. In other examples, it may be more convenient to use differing values for A and B.

The correlation coefficient is given by

$$r = W/\sqrt{V_x V_y}. \tag{55}$$

For this example, therefore,

$$r = 20.9/\sqrt{33.2 \times 32.4} = 0.637.$$

As a numerical exercise, recompute the covariance and correlation coefficient using values for A and B other than 175; you should obtain the same results.

The major use of the covariance is for the computation of correlation and regression coefficients.

LINEAR REGRESSION Given pairs of observations (x_i, y_i) with means $\bar{x}$ and $\bar{y}$, the problem of regression is that of finding the straight-line function,

$$Y = a + bx, \tag{56}$$

that best fits the observed data. This "regression line" is obtained by finding the values for a and b in equation 56 that minimize the quantity

$$SS_{res} = \Sigma[(y_i - Y_i)^2], \tag{57}$$

Table A.23
Computation of covariance for data from Table A.22

i	x_i	y_i	$X_i = x_i - 175$	$Y_i = y_i - 175$	X_iY_i	X_i^2	Y_i^2
1	168	169	−7	−6	42	49	36
2	178	172	3	−3	−9	9	9
3	171	175	−4	0	0	16	0
4	165	162	−10	−13	130	100	169
5	183	180	8	5	40	64	25
6	170	180	−5	5	−25	25	25
7	177	173	2	−2	−4	4	4
8	180	179	5	4	20	25	16
9	173	178	−2	3	−6	4	9
10	170	172	−5	−3	15	25	9
			$\Sigma X_i = -15$	$\Sigma Y_i = -10$	$\Sigma(X_iY_i) = 203$	$\Sigma X_i^2 = 321$	$\Sigma Y_i^2 = 302$

$$\text{Covariance} = W = \{\Sigma(X_iY_i) - [\Sigma(X_i)][\Sigma(Y_i)]/n)/(n-1)$$
$$= [203 - (-15)(-10)/10]/(10-1)$$
$$= 188/9 = 20.9.$$

$$\text{Variance of } x = V_x = \{\Sigma(X_i^2) - [(\Sigma X_i)^2/n]\}/(n-1)$$
$$= \{321 - [(-15)^2/10]\}/(10-1)$$
$$= 33.2.$$

$$\text{Variance of } y = V_y = \{\Sigma(Y_i^2) - [(\Sigma Y_i)^2/n]\}/(n-1)$$
$$= \{302 - [(-10)^2/10]\}/(10-1)$$
$$= 32.4.$$

where y_i is an observed value and Y_i is a value computed from the corresponding x_i using equation 56. The quantity being minimized (by finding the "best" values for a and b) is the sum of squares of the "residuals"—that is, the deviations from the predicted regression line.

Minimization of SS_{res} by the appropriate mathematical procedures (which we shall not give here) leads to the following estimated values for a and b:

$$a = \bar{y} - b\bar{x}; \quad (58)$$
$$b = W/V_x = \Sigma[(x_i - \bar{x})(y_i - \bar{y})]/\Sigma(x_i - \bar{x})^2. \quad (59)$$

Note that a is the intercept and b is the slope of the straight line defined by equation 56 (see Fig. A.5).

Using the data from Table A.23, we find that

$$b = 20.9/33.2 = 0.630;$$
$$a = 174.0 - (0.630 \times 173.5) = 64.7;$$
$$Y = 64.7 + 0.630\ x.$$

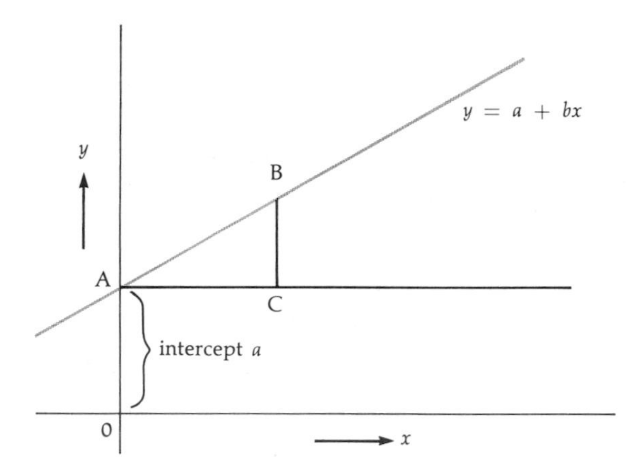

Figure A.5
Intercept and slope of a straight line. The slope, b, is equal to the ratio of line segments $\overline{BC}$ to $\overline{AC}$.

Figure A.6 shows the regression line, together with the original data. This regression line is that of y on x; it would be used to predict the stature of the child, given the stature of the father. The quantity b is called the regression coefficient of y on x. If one wished to predict the stature of the father, given that of the child, one should use the regression of x on y. To obtain this regression line, replace V_x by V_y in the denominator of equation 59. This will yield b', the regression coefficient of x on y. The intercept for this regression line is given by $a' = \bar{x} - b'\bar{y}$.

CORRELATION COEFFICIENTS A linear correlation coefficient r between two variables x and y is defined as in equation 55 by

$$\begin{aligned} r &= W/\sqrt{V_x V_y} \\ &= \Sigma[(x_i - \bar{x})(y_i - \bar{y})]/\sqrt{\Sigma(x_i - \bar{x})^2 \times \Sigma(y_i - \bar{y})^2}. \end{aligned} \quad (60)$$

This coefficient measures the fit of a straight line to a set of data—which is not the same thing as measuring the general association between two variables. It is possible to have a very exact nonlinear relationship between x and y that gives a linear correlation coefficient of zero. The square of the correlation coefficient, r^2, gives the proportion of the variance of one of the two variables (say, y) that is explained by the linear regression of it on the other variable x. The residual proportion of the variance, $1 - r^2$, is equal to the ratio between SS_{res} (see equation 57) and $\Sigma(y_i - \bar{y})^2$.

Correlation coefficients, like many other statistical indices, are subject to sampling errors, whose magnitude can be predicted if the sample size is known. In particular, Table A.24 gives the values of the correlation coefficient that are significantly different from zero at the 5 percent probability level for different sample

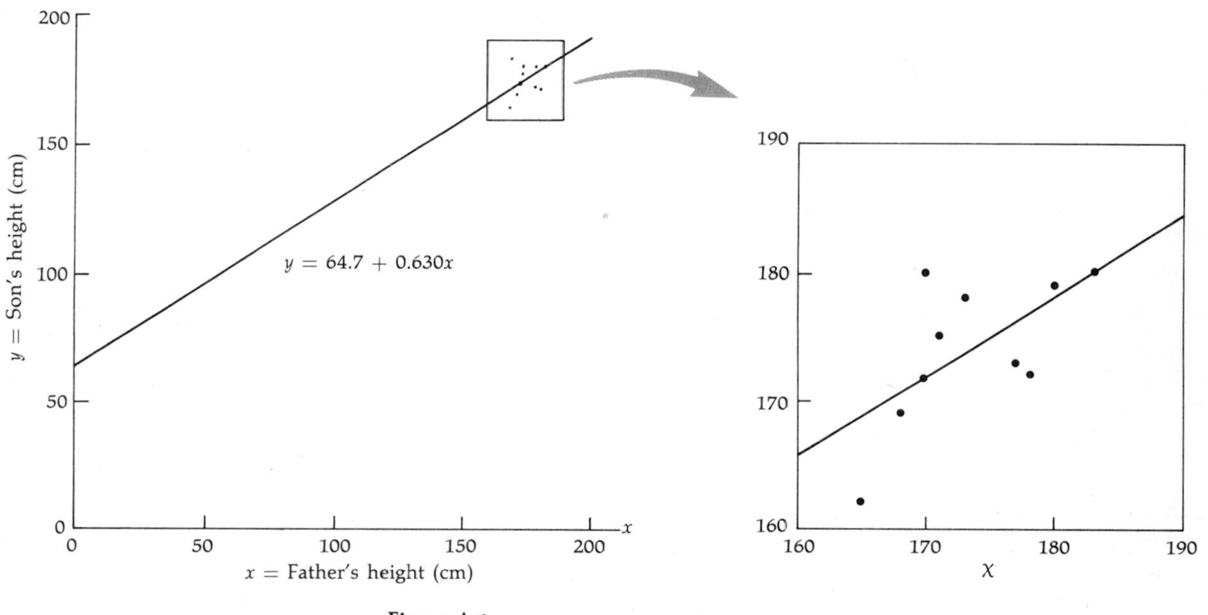

Figure A.6
Regression of stature of son versus stature of father.

Table A.24
Values of the correlation coefficient r that are significantly different from zero (at the 5 percent probability level), as a function of the number of pairs of observations on which the coefficient is based

Sample size (number of pairs)	Smallest significant value of r
5	0.89
10	0.63
20	0.40
30	0.34
40	0.30
50	0.27
100	0.19

SOURCE: Based on R. A. Fisher and F. Yates, *Statistical Tables for Biological, Agricultural, and Medical Research*, Oliver and Boyd, 1953.

sizes. These values correspond to those that would be exceeded in only 5 percent of random samples of the size given, for uncorrelated "normal" variables. (Standard errors of correlation coefficients are discussed in Section A.14.)

Note that genetic correlations are very frequently similar to regression coefficients; in theory, the correlation and regression coefficients are in fact expected to be identical. The correlation between parent (x) and child (y) is $W/\sqrt{V_x V_y}$, while the regression of child on parent is W/V_x. If the population is in equilibrium, the variance is expected to remain the same from generation to generation, and so V_y is expected to equal V_x. Therefore, $\sqrt{V_x V_y} = \sqrt{V_x^2} = V_x$, and the correlation and regression coefficients should be identical. In practice, however, there will usually be differences between the estimated values of V_x and V_y, because of sampling effects, and so there will also be differences between the regression and correlation coefficients computed from them. Usually, these differences are quite small.

A.15 Standard Errors and the Normal Distribution (Part III)

The variance is measured in units that are squares of the units used for the original variate. Thus, for stature measured in cm, the variance is in units of cm^2. Taking the square root of the variance yields a quantity called the standard deviation

(SD, or σ), which is measured in the same units as the original variate—in this case, cm.

The arithmetic mean m and the standard deviation σ (or the variance, σ^2) fully define the most commonly used theoretical distribution, which is the *normal* (or Gaussian) curve. The frequency of a "normal" variate x is given by the function

$$f(x) = e^{-(x-m)^2/2\sigma^2}/\sigma\sqrt{2\pi}, \tag{61}$$

where e is the base of natural logarithms (2.718 . . .) and $\pi = 3.14159$. . . . The value of $f(x)$ defines the height of the frequency curve for each value x. The area under this curve between any two values of x is proportional to the frequency of observations that fall within that range. The curve corresponding to the function $f(x)$ is shown in Table A.25. It has the well-known, single-peaked, bell-shaped, symmetrical form. The areas under the curve are given by the integral of the function $f(x)$, as it is defined by equation 61. Table A.25 gives a sample of numerical values.

Table A.25 is useful for predicting what fraction of individuals is expected within a given range of variation for any trait that is normally distributed. The table is useful for predicting the range within which an estimate of a quantity may be expected to lie.

The most typical application is to arithmetic means. The means of independent samples of size n taken from the same population will vary because of random sampling effects. The values obtained for the means of the samples form a distribution whose standard deviation, called the *standard error*, is a useful indicator of the confidence to be placed in an estimate of the population mean that has been obtained through sampling. The standard error (SE) can be shown to be related to the standard deviation (σ) of the total population distribution and to the sample size n as follows:

$$SE = \sigma/\sqrt{n}. \tag{62}$$

Table A.25
Area under a normal curve (expressed as a percentage of the total area under the curve) between the mean and a point that is $t\sigma$ distant from the mean

t	Area
0.0	0.0%
0.5	19.1%
1.0	34.1%
1.5	43.3%
2.0	47.7%
2.5	49.4%
3.0	49.86%

$f(x)$ Area x m $m + t\sigma$

The standard deviation of a single sample can be used as an estimate of σ for the total population.

If the original distribution is normal, then the means of the samples are also normally distributed. If the population distribution is not normal, the distribution of the sample means is almost always closer to a normal distribution than is the original distribution—the more so as the sample size becomes larger. On the assumption that the distribution is normal, one can compute an interval within which the unknown population mean that is being estimated may be expected to lie (with a particular level of probability). The interval of $\pm 2\text{SE}$ around the observed mean is often used to indicate the range in which the population mean (if it were known without sampling error—the "true" mean) probably lies. From Table A.25, we see that 47.7 percent of a population lies within two standard deviations of the mean in one direction, so 95.4 percent of the population lies within the range from $m - 2\text{SE}$ to $m + 2\text{SE}$. This means that 95 percent of the sample means fall in the interval of $\pm 2\text{SE}$ around the observed mean, and we conclude that there is a 95 percent probability that the true mean lies in this range. If the SD from which the SE is computed is obtained from a small sample (n less than 20), a slightly more elaborate procedure, using Student's t distribution, should be used to estimate this interval (called "fiducial"); for details, consult one of the statistical textbooks listed in the references for this Appendix.

In the following example, we compute the fiducial interval for the mean stature of the ten fathers of Table A.19. As the sample size from which the SD was estimated is less than 20, we should actually use, instead of twice the SE, a factor slightly greater than 2, obtained from Student's t table with degrees of freedom equal to the size of the sample from which the SD was obtained, minus one. We ignore here this complication, which would increase the size of the interval only slightly. The mean of the ten statures is 173.5, with variance of 33.2 (see Table A.19). The SD is $\sqrt{33.2} = 5.76$, and the SE of the mean is $5.76/\sqrt{10} = 1.82$. Thus we conclude that the "true" mean for the stature of the larger population of men from which the fathers were randomly selected has about a 95 percent probability of lying between $173.5 - (2 \times 1.82) = 169.9$ cm and $173.5 + (2 \times 1.82) = 177.1$ cm. This result is often expressed by reporting the observed mean stature as 173.5 ± 3.6 cm. At times, however, the quantity indicated after the $\pm$ sign is the SE (or some other measure of dispersion). To avoid ambiguity, it is important to state which quantity is appended after the $\pm$ sign.

The standard error of a difference between two means, m_A and m_B—or, more generally, between any two estimates whose standard errors SE_A and SE_B are known—is computed as

$$\text{SE}_{\text{diff}} = \sqrt{\text{SE}_A{}^2 + \text{SE}_B{}^2}. \qquad (63)$$

In words, the standard error of the difference is the square root of the sum of squares of the standard errors of the two estimates. Again, normality usually applies fairly well to the distribution of differences between means. When a difference is greater than twice its standard error, it is significantly different from zero at the 5 percent probability level. Again, the Student's t distribution should be used with small samples.

In Table A.22, the mean stature of the fathers ($\pm$SE) is 173.5 $\pm$ 1.8 cm, as computed previously, and the mean stature of the ten sons can be computed to be 174.0 $\pm$ 1.8 cm. The difference between the means of fathers and sons is 173.5 $-$ 174.0 $=$ $-$0.5 cm. The SE of this difference is $\sqrt{(1.8)^2 + (1.8)^2} = 2.6$. The difference between means (ignoring the sign, which is arbitrary) divided by its SE is 0.5/2.6 $=$ 0.2. Only if the difference were greater than 2(SE) could it be considered as significantly different from zero. Hence, there is no significant difference between the mean stature of fathers and the mean stature of their sons.

The standard error of a relative frequency p, obtained on a sample of n observations, is given by

$$SE_p = \sqrt{p(1 - p)/n}. \tag{64}$$

This SE is distributed nearly normally only if n is large, while values of p near zero or one will lead to a very poor approximation to normal distribution. In general, the difference between two relative frequencies is tested by χ^2 for a 2×2 table, as described in Section A.5. Consider the frequency of concordant MZ twin pairs reported in Table A.10. This frequency is $\frac{50}{135} = 0.370$, or 37 percent. The SE of this frequency is

$$\sqrt{0.37 \times (1 - 0.37)/135} = \pm 0.042.$$

Therefore, the frequency of concordant MZ twin pairs can be reported as 0.370 $\pm$ 0.042 (or 37.0 $\pm$ 4.2 percent) with about 68 percent probability, or as 0.370 $\pm$ 0.084 with about 95 percent probability.

From Table A.10 we can also obtain a frequency of 0.152 for concordant DZ twins, with an SE of 0.016. The difference between the two frequencies is 0.370 $-$ 0.152 $=$ 0.218. Using equation 63, we find that the SE of this difference is $\sqrt{(0.042)^2 + (0.016)^2} = 0.0449$. The ratio of the difference to its SE is $\frac{0.218}{0.0449} = 4.85$. Because the difference is much greater than twice its SE, the difference between the two means is highly significant. It is not surprising (rather, it is expected) that this test gives the same result as the χ^2 test used in the original discussion of Table A.10.

The standard error of a correlation coefficient r estimated on the basis of a sample of n pairs is known, but the distribution of the sample r values is far from normal, especially if the true r is different from zero. For comparison of correlation coefficients, it is best to use Fisher's transformation and convert each coefficient r to a corresponding z value, computed as

$$z = \frac{1}{2} \log_e[(1 + r)/(1 - r)].$$

The z values are very nearly normally distributed with SE $= 1/\sqrt{n - 3}$. A difference between z values can then be tested for significance using equation 63, as discussed for preceding examples.

References

TEXTBOOKS OF STATISTICS

Mather, K.
1972. *Statistical Analysis in Biology.* Gloucester, Mass.: Peter Smith.
Snedecor, G. W., and W. G. Cochran
1967. *Statistical Methods.* Iowa State Univ. Press.
Sokal, R. R., and F. J. Rohlf
1973. *Introduction to Biostatistics.* San Francisco: W. H. Freeman and Company.

HANDBOOK OF STATISTICAL TABLES

Fisher, R. A., and F. Yates
1964. *Statistical Tables for Biological, Agricultural, and Medical Research.* Edinburgh: Oliver and Boyd.

TEXTBOOKS OF POPULATION GENETICS

Cavalli-Sforza, L. L., and W. F. Bodmer
1971. *The Genetics of Human Populations.* San Francisco: W. H. Freeman and Company.
Crow, J., and M. Kimura
1970. *Introduction to Population Genetics: Theory.* New York: Harper and Row.
Li, C. C.
1955. *Population Genetics.* Univ. of Chicago Press.

Glossary

acentric: A fragment of a chromosome without a centromere.

acrocentric: A chromosome with centromere near one end, so that one arm is very short.

additive portion of genetic variance (V_A): Genetic variance due to the difference between homozygotes (for any locus). Heritability in the narrow sense is V_A divided by the total variance.

age-specific death (and birth) rates: Estimates of the probability that an individual of a given age dies (or gives birth) within the next year.

agglutination: The formation of clumps—usually of red cells of the blood, held together by antibodies attached to antigens on the cells' surfaces.

"all-or-none" trait: A trait that is either present or absent in a given individual.

alleles: Alternative forms of the same genetic locus.

amino acid: Small molecules that are the building blocks of proteins. There are 20 amino acids that commonly make up proteins. All amino acids have the same general structure with one acidic (carboxyl) and one alkaline (amino) end, but they differ in the side groups (R groups).

aminoacidurias: Diseases in which there is an abnormally high excretion in the urine of one or more amino acids.

amniocentesis: A clinical procedure by which a few milliliters of the amniotic fluid surrounding the fetus are withdrawn. The fluid and fetal cells contained in the fluid may then be subjected to tests for various genetic diseases.

anaphase: The period of cell division during which the chromosomes begin migration toward opposite poles of the cell.

aneuploidy: A karyotypic abnormality resulting from the presence of extra chromosomes or absence of chromosomes, such that the karyotype is neither haploid nor an exact multiple thereof.

antibody: A protein produced in the immune reaction with property of binding to specific foreign molecules.

anticodon: A triplet of nucleotides specific to each tRNA, corresponding to a particular amino acid and complementary to the codon (the triplet of nucleotides in mRNA).

antigen: A molecule that stimulates the production of specific antibodies.

artificial selection: The process of choosing reproducers in a given population on the basis of a given trait or traits.

assortative mating: Nonrandom mate selection with respect to a given trait or traits. Positive (negative) assortative mating occurs when individuals who are similar for a given characteristic mate more (less) frequently than would be predicted by chance.

Australopithecus: Genus of fossil Hominidae found in the old world. Bipedal, with a brain size intermediate between modern men and other modern primates. Lived between approximately 5 and 1 million years ago. There is no general agreement as to whether they are true ancestors of *Homo* or are on a collateral line.

auto-antibodies: Antibodies against components of the self.

autoimmune disease: A disease resulting from the production of antibodies against one's own tissues or organs (autoantibodies).

autosome: Any chromosome other than the sex chromosome.

backcross: A cross between a heterozygote (*Aa*) and a corresponding homozygote (*AA* or *aa*).

bacteriophage or **phage:** A virus parasitic to a bacterium.

balanced polymorphism: A polymorphism that is stable (tends to remain unchanged over time) and is probably maintained by advantage of the heterozygote over both homozygotes.

Barr body (sex chromatin): A mass of chromatin in the nucleus of resting cells, resulting from inactivation of an X chromosome. A cell ordinarily contains a number of Barr bodies that is equal to the number of X chromosomes minus one.

base: A substance with alkaline reaction. Bases forming nucleic acids are purines and pyrimidines.

base substitution: Substitution of one base for another in the DNA.

bimodal distribution: A distribution characterized by two modes (frequency peaks).

"blending" inheritance: The hypothesis used by Galton that the hereditary characteristics of the parents are irreversibly mixed in the progeny.

C region (constant region): The region of heavy and light chains of immunoglobulins that does not vary.

carcinogen: A chemical substance or physical agent (such as radiation) that causes cancer.

centriole: The cellular organelle that migrates to opposite poles of the cell during meiosis and mitosis, thus ensuring (through the spindle apparatus) the separation of each replicated chromosome and the equal partition of the chromosomes among the daughter cells.

centromere (also called **kinetochore** or **primary constriction):** The constricted portion of the chromosome by which the chromosome attaches to the spindle fibers at mitosis.

chiasma (plural **chiasmata):** An attachment between homologous chromosomes in diplotene, believed to correspond to (or be the consequence of) physical exchanges between homologs that are the basis of crossing-over.

chimera: An individual whose cells are not all of the same genotype. DZ twins are occasionally chimeras. See **mosaic,** which is distinct from a chimera.

chi-square (χ^2) test: A statistical test used to determine if a set of observed frequencies differs (to a degree that would be improbable by chance alone) from those expected on the basis of a specific hypothesis.

chromatid: One of the two visibly distinct longitudinal subunits of any reduplicated chromosome.

chromatin: The basic substance of the chromosome, including both proteins and DNA. This term essentially means "chromosome material."

chromosome: A threadlike body found in the nucleus of a cell and containing the genes. Under the light microscope, the chromosomes are not visible during interphase.

chromosomal aberrations: Karyotypic alterations involving whole chromosomes or portions of them, sufficiently large to be detectable through light microscopy.

cistron: A chromosome segment corresponding to the gene considered as a functional unit; defined operationally by complementation tests. Independent mutants of the same cistron should not complement or should do so less efficiently than mutants of different cistrons.

cline: A relatively regular change in a biological trait (a continuous unit or a gene frequency) over a stretch of territory.

clone: A line of cells derived from a single cell (usually presumed to contain the same genetic information). The term may be applied also to multicellular organisms that can propagate by vegetative (nonsexual) reproduction.

codominant: Said of two alleles that are both expressed in the heterozygote.

codon: A triplet of nucleotides in mRNA, coding for an amino acid.

complement: A series of serum proteins that under appropriate conditions cause antibody-coated cells to lyse.

complementation: The production of normal offspring from a mating between two homozygotes affected by a similar defect, showing that the parentals actually suffered from different genetic lesions.

complementation test: The introduction of two independently-occurred mutations into the same cell for the purpose of determining whether the mutations occurred in the same gene. This test can be accomplished either by mating two homozygous organisms or by somatic cell fusion.

concordance: Similarity between individuals for an "all-or-none" trait. Usually this term is applied in twin research; the comparison of concordance of MZ and DZ twins is a test for heritability.

congenital defect: A defect present at birth; it may be determined genetically or by external influences during intrauterine life.

consanguinity: Two or more individuals are said to be consanguineous if they have a common recent ancestor (usually not further back than three or four generations).

correlation coefficient (r): A coefficient measuring the degree to which points in a Cartesian diagram tend to fall near a straight line. The value of the coefficient is one if all points fall exactly on a straight line and if y increases with x (the value is minus one if y decreases with increasing x). The value of the coefficient is zero if the points are scattered randomly about the diagram, with no tendency to cluster around a straight line. Intermediate values are obtained in intermediate cases.

coupling: Dominant alleles at two different loci are said to be in coupling in a heterozygote if they are located on the same chromosome. Thus, the double heterozygote AB/ab is said to be in coupling, while Ab/aB is said to be in **repulsion.**

crossing-over: The process of exchange of genetic information between two homologous chromosomes, presumed to occur through breakage of both chromosomes at homologous sites followed by reunion after exchange.

Darwinian fitness: The fitness of a given genotype in a given environment is measured by its relative contribution to the ancestry of future generations—that is, by the change in the frequency of this genotype from one generation of parents to the next generation of parents. It depends on both fertility and survival.

deletion: At the molecular level, the removal of one or more bases from a DNA sequence. At the cytological level, the absence of a segment of a chromosome (also known as a deficiency).

demic diffusion (demic spread): The process of colonization of an area by a population. This term is used in contrast with cultural diffusion, which involves movement of ideas rather than individuals.

demic selection: Disproportionate growth of a subset of a population or species. Demic selection has an effect on the general genetic composition of the population if the subsets have different gene frequencies (as is usually the case). Demic selection is a type of intergroup selection, not necessarily involving direct competition.

demographic transition: A significant change in the pattern of birth and/or death rates. Usually used in reference to the recent transition that began with the industrial revolution, involving a decline in death rate followed after some delay by a decline in birth rate, causing a rapid population increase.

demography: The study of populations and how they survive, die, reproduce, and grow.

deoxyribonucleic acid (DNA): A polymer of nucleotides in which the sugar residue is deoxyribose. DNA is found primarily in the double-helical conformation.

dicentric: A chromosome with two centromeres; it can arise because of crossing-over in a heterozygote for a paracentric inversion.

diploid: A chromosome complement that contains two copies of each chromosome. Normal human somatic cells are diploid.

diplotene: Late meiotic prophase; chiasmata are visible at this stage.

discontinuous variation: Variation of a trait that occurs in such a way that the trait can be classified into two or more easily distinguishable, clearly separated categories.

dizygous (DZ) twins: Twins that arise from two different eggs fertilized by two different sperm. The genotypes of dizygous twins are thus no more related than those of any two sibs.

dominance portion of genetic variance (V_D): That portion of the genetic variance for a given trait that results from the fact that heterozygotes do not always score exactly midway between the homozygotes.

dominant: An allele manifesting its phenotypic effect also in the heterozygotes; a trait determined by a dominant allele.

dosage compensation: The mechanism by which genes on the X chromosome have the same physiological effect in females (who have two X chromosomes) as in males (who have only one).

double backcross: Mating of an individual who is heterozygous at two loci with an individual who is homozygous recessive at the same two loci.

Down's syndrome (trisomy 21, or **mongolism):** A syndrome characterized by mental, behavioral, and physiological defects, caused by the presence of an extra copy of the genetic material contained on chromosome 21. The third copy of this information may be present as an extra chromosome 21, or as a segment of it translocated to another chromosome.

drift: See **random genetic drift.**

duplication: Presence of two copies of a chromosome segment, usually in the same chromosome —sometimes in immediate sequence (tandem duplication), or elsewhere in the same or other chromosomes.

dysgenic: Due to, or determining, an increase in the frequency of deleterious genes; the opposite of *eugenic.* Dysgenic effects may be spontaneous, or they may result from medical or social interventions that improve the fitness of the handicapped.

egg (egg cell): The female **gamete.**

effective population size (N_e): The population size relevant for random genetic drift—that is, the portion of the total population that is reproducing. In human populations, this is roughly one-third the census size of the population.

electrophoresis: A technique for separating molecules, particularly proteins, according to the overall electric charge of the molecules.

epistasis: Nonadditive interaction between two or more different loci.

euchromatin: Differentiated from **heterochromatin** by its staining behavior during the cell cycle.

eugenic: Due to, or determining, a decrease in the frequency of deleterious genes.

eugenics: A program of decreasing the frequency of deleterious genes in a human population (negative eugenics) or of increasing that of advantageous genes (positive eugenics) through artificial selection against the genetically handicapped or in favor of the types considered especially desirable.

eutelegenesis: The use of sperm from selected donors for voluntary artificial insemination; suggested as a eugenic program.

enzyme: A protein that catalyzes a specific chemical reaction.

exponential growth (geometric growth, or logarithmic growth): A phase of growth of a culture or a population in which the time of doubling in size (or more generally, the population growth rate) is constant. Under exponential growth the logarithm of the number of individuals increases linearly with time.

F_1 generation: First filial generation. The progeny resulting from a parental cross between two pure lines or, more precisely, between homozygotes for different alleles.

F_2 generation: Second filial generation, resulting from the matings of two F_1 individuals (or selfing of an F_1 individual in species where this is possible).

feedback inhibition: Inhibition of a sequence of reactions by the end product of those reactions.

fitness: See **Darwinian fitness.**

founder effect: An expression emphasizing the drift effect (on gene frequencies) that results when a new population is founded by a small group of individuals selected from an old population.

frameshift mutation: A mutation resulting from the insertion or deletion in DNA of one or more bases (but not a multiple of three), thus changing the boundaries of the codons and altering the whole string of amino acids subsequent to the insertion or deletion.

frequency distribution: A synthetic presentation of a series of observations, obtained by specifying the number of observations falling in each "class"—that is, the frequencies of individuals with values of x between x_1 and x_2, between x_2 and x_3, and so on. The interval between x_1 and x_2, or x_2 and x_3, and so on, defines a class. May be graphically presented as frequency diagrams, histograms, or line graphs.

gamete: The haploid cell generated by meiosis that may fuse with another appropriate gamete to form a zygote. In a bisexual species, a gamete is either male (sperm) or female (egg).

gametic selection: Differential survival (and/or capacity to fertilize) of sperm or egg cells.

gamma (γ) globulins (immunoglobulins): A heterogeneous class of globular proteins of large molecular weight. Electrophoresis of blood plasma or serum separates four major protein components (in order of decreasing mobility, and approximately of increasing molecular weight): albumin, α globulins, β globulins, and γ globulins. Also see **immunoglobulins.**

gene: A segment of chromosome with a detectable function. Used as a synonym for **locus** or **cistron,** and sometimes for **allele.**

gene-dosage effect: The relation between the number of functional copies of a given gene present and the level of their activity (for instance, enzymatic activity).

gene flow: The exchange of genes at a low rate (in one or both directions) between two initially different groups. Because the rate of exchange is low, the groups may retain their identity.

gene frequency: The proportion (of all alleles at a locus) in which a given allele is found in the individuals forming a specified population.

genetic code: The code that relates nucleotide sequences in nucleic acids to amino-acid sequences. Each triplet of nucleotides designates a particular amino acid; thus, the genetic code allows the translation of information stored in DNA and the use of that information in protein synthesis.

genetic drift: See random genetic drift.

genetic equilibrium: A state reached by a population when gene and genotype frequencies do not change in successive generations.

genetic load: The decrease in the average fitness of the population due to the presence of genes that decrease survival relative to the maximum possible.

genetic marker: A gene mutation that has phenotypic effects useful for tracing the chromosome on which it is located.

genetic screening: A systematic testing of individuals to ascertain potential genetic handicaps in them or in their progeny—handicaps that may require treatment or prophylaxis.

genome: The ensemble of genetic material in a cell.

genotype: The genetic constitution of an individual at one or more loci.

genotype–environment correlation: The nonrandom assortment of particular genotypes among particular environments within a given population.

genotype–environment interaction: The interplay of a specific genotype and a specific environment, affecting the phenotype. The extent and nature of this interaction varies with each genotype and environment. The variance due to this interaction (V_I) is part of the total phenotypic variance of a trait.

genotype frequency: The relative proportion of a particular genotype in a population.

germ cells: Gametes, or their precursors.

germ line: The line of cells that produce gametes.

globulins: Proteins whose molecules have a globular shape. The globulins found in blood serum are classified as alpha, beta, and gamma globulins, in order of decreasing electrophoretic activity.

haploid: A cell is said to be haploid if it contains one copy of each chromosome.

haplotype: A combination of alleles from closely linked loci (usually having functional affinity, such as HLA loci) found in a single chromosome.

haptoglobins: Proteins in blood serum that bind hemoglobin and transport it from broken red cells to sites where it is further metabolized.

Hardy–Weinberg law: A rule for predicting genotype frequencies on the basis of gene frequencies, under the assumption of random mating in the absence of selection.

heavy (H) chain: One of the polypeptide chains present in every immunoglobulin molecule.

hemizygous: Having only one copy of a particular locus present in the genome. For example, males are hemizygous for genes on the X chromosome.

hemoglobin: A globin (globular) molecule found in the blood; it transports oxygen from the lungs to other tissues. The molecule is a tetramer (formed by four polypeptide chains); in the adult, it contains two α chains and two β chains.

heritability (broad): The fraction of the total variance remaining after exclusion of the fraction due to environmental effects. A measure of the degree of genetic determination of a trait.

heritability (narrow): The ratio of the additive genetic variance to the total variance. An estimate of the efficiency of selection.

heterochromatin: Chromatin that stains heavily in certain phases of the cell cycle; usually located near centromeres.

heterogametic: Referring to the sex whose gametes differ in their sex chromosomes. In humans, males are heterogametic because sperm carry either an X or a Y chromosome.

heterokaryon: A cell having two different types of nuclei, as a result of fusion of two cell types without nuclear fusion.

heterozygote: A cell or individual that is heterozygous.

heterozygous: Having different alleles at a given locus on homologous chromosomes.

histocompatibility: Capacity to accept a tissue or organ graft.

Homo: Genus including modern man (*Homo sapiens*) and fossils in the human line of descent (*Homo habilis, Homo erectus*).

homogametic: Referring to the sex whose gametes all carry the same sex chromosome. In humans, females are homogametic because all eggs carry an X chromosome.

homologous: Referring to homologs.

homologs: Chromosomes (or chromosome segments) that carry genes governing the same characteristics, that have similar morphology, and that pair during meiosis.

homozygote: A cell or individual that is homozygous.

homozygous: Having the same allele at a given locus on homologous chromosomes.

hybrid: The progeny resulting from a cross between different parental stocks.

immunoglobulin (Ig): Antibody molecule. There are five classes (IgG, IgM, IgA, IgD, and IgE), which serve different immunological functions.

in utero: (Latin) in the womb.

in vitro: (Latin, in glass); pertaining to experiments done on cells grown outside of the animal.

in vivo: (Latin, in the living); pertaining to experiments on a whole living animal.

inborn errors of metabolism: Inherited disorders that can be explained as genetic blocks in specific metabolic pathways, usually due to recessive alleles determining a decreased activity (or the absence) of a specific enzyme. Examples include albinism and alkaptonuria.

inbreeding: When consanguineous (closely related) individuals mate, their progeny is said to be inbred. Such mating is called inbreeding.

inbreeding coefficient (F)**:** The probability that an allele present in a common ancestor is homozygous in the inbred individuals.

incidence: The relative frequency of a trait or disease at birth.

incompatibility: The presence in a fetus (or graft, or donated blood) of antigens that can evoke an immune response in the mother (or graft recipient, or blood recipient).

inducer: A molecule that binds to a repressor, preventing it from blocking transcription.

interaction: See **genotype–environment interaction.**

intercross: A cross between two heterozygotes for the same alleles.

intergroup selection: An evolutionary process due to differences in growth rate or survival of different competing or noncompeting groups.

interphase: The phase between two successive mitoses.

inversion: A chromosomal aberration that arises when two breaks occur in the same chromosome and the region between the breaks is reinserted after a 180° rotation.

isochromosome: A chromosome with two morphologically identical arms.

isozymes: Enzymes performing similar functions (in the same or different organs). Isozymes may be present in the same individual; they are usually differentiated by electrophoresis.

karyotype: The chromosome complement of any organism, analyzed according to size and banding patterns of each chromosome.

kinetochore: See **centromere.**

Klinefelter's syndrome: A syndrome caused by the presence of an extra X chromosome in a male karyotype (XXY). Affected individuals are phenotypically male, but with underdeveloped gonads; other physical and behavioral problems are present in many cases.

leptotene: Early meiotic prophase.

lethal: An allele that kills all carriers (or only homozygotes in the case of a recessive lethal) before reproductive age and usually in the first years of life.

light (L) chain: One of the polypeptide chains present in every immunoglobulin molecule. There are two types of light chains: kappa and lambda.

linkage: The presence of two or more loci on a single chromosome, causing a tendency for alleles at the linked loci to be inherited together. Linkage is observed only when the loci are sufficiently close to one another; crossing-over can lead to random assortment of loci that are far apart on the same chromosome.

linkage group: A group of linked loci. A linkage group must correspond to the chromosome on which the genes are located.

locus (plural, **loci**)**:** Position of a gene on a chromosome. Used as a synonym for gene or cistron, but not for allele.

logistic growth: A pattern of population growth in which the growth rate decreases with increasing number of individuals until it becomes zero when the population reaches a maximum (saturation) size. When the number of individuals is small, the population increases almost exponentially, but as the growth rate decreases the growth curve (population size versus time) gradually levels out toward a constant value at the saturation size, thus forming a characteristic "S-shaped" curve.

lymphocytes: A subgroup of the white blood cells. B lymphocytes are the main producers of antibodies, while T lymphocytes control cell-mediated immune responses and antibody production.

Lyon hypothesis: A theory of dosage compensation that proposes the genetic inactivation of all X chromosomes in excess of one (selected randomly in each cell) at an early stage in embryogenesis.

major gene: A gene that may cause sufficiently large variation in the trait studied to be easily detected.

mean: An estimate or estimator of the central tendency of a variate. Most commonly used is the arithmetic mean, found by summing all values of the variate and dividing by the number of values.

median: The value of a variate chosen so that half the observations forming a sample have values exceeding the median, while the other half have values less than the median.

meiosis (reduction division): A series of two modified mitoses, generating haploid gametes from a diploid precursor cell.

meiotic drive: A disturbance in the expected one-to-one ratio of homologous chromosomes at meiosis.

messenger RNA (mRNA): RNA that is synthesized using one strand of DNA as a template, and that is then used to direct protein synthesis on ribosomes.

metacentric: A chromosome with the centromere near the middle.

metaphase: The phase of mitosis or meiosis in which the condensed chromosomes attached to the spindle fibers line up on an equatorial plane between the two poles of the cell.

minor gene: A gene whose effect on a given trait is so small that it is not easily detected.

missense mutation: The substitution of one amino acid for another, resulting from the substitution of one base for another in the DNA.

mitosis: The process of cell division in somatic cells, in which duplication and assortment of the chromosomes ensures the identity of genetic information in the parental and the two daughter cells.

mode: A peak in a frequency distribution. The modal value represents a trait value whose frequency is greater than those of adjacent higher or lower values. If the distribution is unimodal (has only one mode), the modal value corresponds to the trait value having the highest frequency in the distribution.

mongolism: An old term (now seldom used) for **Down's syndrome.**

monogenic: A monogenic (single-gene) model postulates that one gene (usually plus environmental factors) determines a trait.

monosomy: The presence of only one member of a chromosome pair.

monozygous (MZ) twins: Twins that develop from a single zygote, which divides to give rise to two complete embryos. Monozygous twins have identical genotypes.

morgan: A unit of recombination; a centimorgan indicates 1 percent of recombination; 1 morgan equals 100 centimorgans. One chiasma corresponds on the average to 50 centimorgans.

mosaicism: The presence in an individual of two or more cell genotypes arising by mutation or by chromosomal aberration (including nondisjunction).

mousterian: A tool-making tradition related to Neanderthal man.

multifactorial trait (polygenic trait): A trait whose phenotypic expression is influenced by the cumulative effects of many genes.

mutagen: A physical or chemical agent that increases the mutation rate.

mutation: A heritable change in the genetic material, or its detectable effects in the phenotype.

mutation rate: The frequency of mutations per generation at a given locus.

myeloma: A tumor consisting of a clone of antibody-forming cells.

natural selection: The process by which changes occur spontaneously in the proportions of genetic types within populations of a living organism, due to differences in the (Darwinian) fitnesses of these genetic types in the existing environment.

Neolithic (New Stone Age): The period of development of domestication of plants and animals, beginning about 5,000 to 10,000 years ago (the time varying in different areas). The name was originally derived from the appearance of new techniques used in making many stone tools. Some archeologists have identified the Neolithic with the appearance of pottery, which usually (but not always) coincides with the appearance of domesticated plants and animals.

nondisjunction: Failure of homologous chromosomes to separate during the first stage of meiosis (primary nondisjunction) or of chromatids to separate in the second division of meiosis (secondary nondisjunction). Nondisjunction can also occur in mitosis.

nonsense mutation: A mutation involving a base change in DNA that converts a triplet specifying an amino acid into one specifying chain termination, thus causing the premature termination of a polypeptide chain.

nucleic acids: Deoxyribonucleic acid (DNA) and ribonucleic acid (RNA).

nucleolus: A body in the nucleus involved in rRNA synthesis; it is usually associated with secondary constriction of certain chromosomes.

nucleoside: A purine or pyrimidine base attached to a sugar (deoxyribose or ribose); it is a nucleotide without the phosphate group.

nucleotide: A purine or pyrimidine base attached to a sugar (deoxyribose or ribose) and a phosphate group; nucleotides are the subunits of nucleic acid polymers.

nullisomy: The absence of a particular chromosome.

oncogenic: Having a tendency to "cause" cancer.

operator: A genetic region that, by interaction with a repressor, can prevent transcription of the DNA in a specific set of adjacent genes.

operon: A chromosome region containing a minimal functional unit, such as an operator, a promoter, and one or more structural genes under their control.

overdominance: A situation in which the heterozygote exhibits a more extreme manifestation of the trait under study than does either homozygote.

pachytene: The stage of meiotic prophase at which chromosomes are fully paired.

Paleolithic (Old Stone Age): The long period during which humans, living as hunter–gatherers, made stone tools. The end of the Paleolithic corresponds approximately with the first domestication of plants and animals (see **Neolithic**). The period is subdivided into Early, Middle, and Late Paleolithic; fossils or other finds and rock layers formed during these subperiods are called Lower, Middle, and Upper Paleolithic, respectively.

panmixia: Random mating.

paracentric inversion: An inversion in which both breaks occur on the same arm of the chromosome.

"particulate" inheritance: The theory of inheritance proposed by Mendel, in which trait-determining elements (now called genes) keep their individuality in the progeny of any mating. Progeny may show characteristics intermediate between those of the parents, but the parental characteristics may reappear in later generations. See by contrast **"blending" inheritance.**

pedigree: A diagram delineating the genetic relationships of members of a family over two or more generations.

penetrance: The relative incidence of a trait in a given genotype.

peptide: A string of a few amino acids joined by peptide bonds.

peptide bond: A covalent bond formed between the amino group of one amino acid and the carboxyl group of another.

phage: See **bacteriophage.**

pericentric inversion: An inversion in which the breaks occur on opposite sides of the centromere.

permanent cell line: A cell line that survives in culture through an unlimited number of generations.

pharmacogenetics: The study of the relation between an individual's genotype and his reaction to various pharmaceutical agents.

phenocopy: An individual who has a phenotype similar to that produced by a certain mutant genotype, even though the individual may not have that genotype.

phenotype: The observable characteristics of an organism, resulting from the interplay of the genotype and the environment in which development takes place.

phenylketonuria (PKU): A genetic disease, transmitted by simple Mendelian recessive inheritance. The gene alteration causes low activity of the enzyme phenylalanine hydroxylase, leading to a toxic accumulation of phenylalanine and of its metabolites.

plasma (blood plasma): The yellow fluid in which red blood cells are suspended. Pure samples of plasma are obtained by adding suitable agents to prevent coagulation and then sedimenting the red cells (usually by centrifugation). See **serum.**

pleiotropy: The capacity of a gene to influence a variety of phenotypic traits.

polar bodies: Products of meiosis in the female. Of the four daughter cells, three are polar bodies and one is an egg. The polar bodies are smaller than the egg and are unable to form a zygote.

polygenic trait: See **multifactorial trait.**

polymerase: An enzyme that catalyzes polymerization. In particular, DNA and RNA polymerases catalyze the formation of the nucleic acids from the nucleotide constituents on the basis of a single-stranded DNA template. Reverse transcriptase catalyzes the formation of DNA on the basis of an RNA template.

polymorphism: The occurrence of two or more alleles for a given locus in a population, where at least two alleles appear with frequencies of more than 1 percent.

polynucleotide: A polymer formed by the covalent linkage of nucleotides; the phosphate group of one nucleotide is bonded to the sugar of the next.

polypeptide: A polymer of amino acids joined linearly by peptide bonds.

polyploid: Having a simple multiple (greater than two) of the haploid number of chromosomes.

prevalence: The relative frequency of a trait or

disease in the individuals of a specified population.

primary response: The immunological response to the first challenge with a particular antigen.

proband (propositus, or index case): The affected individual who first brings the attention of the genetic researcher to a particular family.

propositus (feminine, **proposita;** plurals, **propositi** and **propositae).** See **proband.**

promoter: A region in the operon between the operator and the structural genes; RNA polymerase attaches to the promoter.

prophase: The first phase of mitosis or meiosis, in which the chromosomes condense and become visible as distinct entities under the light microscope.

proteolysis: The breakdown of proteins (usually into peptides and/or single amino acids, catalyzed by proteolytic enzymes).

"pure" line (inbred line): A population obtained by continuous inbreeding over many generations, such that each individual has essentially the same genome as every other member of the inbred line and that all (or most) loci are homozygous.

purines: A class of organic compounds to which the nucleic acid bases adenine and guanine belong.

pyrimidines: A class of organic compounds to which the nucleic acid bases cytosine, thymine, and uracil belong.

quantitative trait: A trait that varies continuously.

races: Subdivisions of a species, recognizably different from one another.

random genetic drift: Variation in gene frequency due to chance fluctuations.

random mating: A situation in which the genetic character being studied has no influence upon the choice of a mate.

random sample: A sample obtained in such a way that each individual in the population being studied has the same chance of being selected.

recessive: An allele that causes a phenotypic effect different from that of other alleles (dominant) only when present in the homozygous state. A trait is said to be recessive if it is due to a recessive allele.

reciprocal translocation: A translocation in which breaks occur in two different chromosomes and the resulting fragments are exchanged.

recombination: The genetic result of **crossing-over.**

recombination fraction: A measure of the frequency of crossing-over occurring between two specific loci. Recombination fractions of linked loci range from just over zero to just less than one-half. One-half is the value expected for unlinked loci.

reduction: See **meiosis.**

regression coefficient: The slope of the regression line (of variate y on variate x).

regression line: The line drawn between points corresponding to a plot of y data versus x data so as to minimize the (squared) deviations of the observed y values from those predicted by the line itself.

regression to the general mean: The phenomenon discovered by Galton that the trait value for offspring of a mating tends to be intermediate on average between the trait value in a parent and the mean trait value for the population.

regulatory gene: A gene responsible for "switching on or off" other genes, usually through production of a repressor that regulates the activity of the other genes.

repressor: A molecule that binds to the operator to prevent attachment of the RNA polymerase to the promoter.

repulsion: Two dominant alleles at two different loci are said to be in repulsion if they are located on homologous chromosomes. Thus, the double heterozygote Ab/aB is said to be in repulsion. See **coupling.**

restriction enzymes: Enzymes that break DNA is such a way that the ends remain "sticky" and can easily be joined to other similarly treated DNA fragments with a complementary sequence of bases.

ribonucleic acid (RNA): A polynucleotide similar to DNA, but with ribose as the sugar and with uracil present instead of thymine. See **messenger RNA, ribosomal RNA, transfer RNA.**

ribosomal RNA (rRNA): A form of RNA that is a major constituent of ribosomes; the largest portion of RNA in a cell consists of rRNA.

ribosomes: Small particles, made of rRNA and proteins, that are the site of protein synthesis.

ring chromosome: A chromosome formed by breaks at the two extremities and rejoining of the broken ends; acentric fragments are excluded in the process.

Robertsonian translocation: The joining together of two acrocentric chromosomes at the centromeres.

sample: A selection of individuals drawn from a population. See **random sample.**

sampling: The operation of collecting a sample from a population.

satellite region: A region of DNA that contains highly repetitive base sequences.

secondary response: The immunological response to a second challenge with a particular antigen. Secondary responses give rise to much higher levels of antibody in a much shorter time than do the primary responses.

segregation: The separation of homologous alleles at meiosis.

segregation ratio: The expected (Mendelian) or observed ratio between genotypes or phenotypes in the progeny of a cross.

selection coefficient: The difference between the fitness of a particular genotype and that of a "normal" genotype chosen as a standard of reference.

"selfing": Self-fertilization.

serum (blood serum): The yellow fluid remaining after blood is allowed to clot and the clot is removed. See **plasma.**

sex chromatin: See **Barr body.**

sex chromosomes: The chromosomes (X and Y) that differ in male and female karyotypes and thus can be said to be (among other things) normal genetic determinants of the sex of an individual.

sex-influenced trait: An autosomal trait that appears predominantly in the members of a specific sex.

sex-limited trait: A trait that is expressed *only* in members of a specific sex. This limitation is due to anatomical or physiological effects rather than to sex linkage.

sex-linked trait: A trait that is determined by genes located on the sex chromosomes.

sibs (siblings, or **sibship):** Offspring of the same parental combination. Brothers and sisters of a proband.

somatic cell hybrid: A uninuclear cell that results from the fusion of two somatic cells in culture.

somatic cells: Cells that are not part of the germ line.

species: A set of individuals who can interbreed and produce fertile progeny.

sperm (spermatozoon): The male gamete.

spindle apparatus: The bundle of fibers that joins the centrioles to the centromeres and ensures accurate partitioning of the genetic material during cell division.

standard deviation: A statistical measure of the variation of individual values about the mean, given in terms of the same units as the individual values.

structural gene: A gene coding for a polypeptide. Sometimes used to mean genes other than regulatory genes or genes coding for rRNA, tRNA, and mRNA; however, those regulatory genes that code for a repressor do code for a polypeptide.

sum of squares (SS): The sum of the squares of the deviations of individual values from the arithmetic mean; used in computing variance. SS_G refers to the sum of squares of deviations due to genetic factors; SS_E refers to the SS due to environmental differences; and SS_T is sometimes used for the total sum of squares ($SS_G + SS_E$).

suppressor: A mutation or allele that suppresses the phenotypic action of an allele at another locus.

Tay–Sachs disease: A degenerative brain disorder of infancy due to an autosomal recessive allele in the gene controlling the enzyme hexosaminidase A. The age of onset of the disease is 4 to 6 months. Affected children show progressive mental deterioration, paralysis, deafness, blindness, and convulsions, leading to death usually between the ages of 3 and 5 years.

telocentric: A chromosome with the centromere located at one end.

telophase: The final phase of mitosis, when nuclear envelopes form around the newly partitioned chromosomes.

teratogenic: Causing a serious deformity in fetal development.

tetraploid: Having four times the haploid complement of chromosomes.

thalassemias: A class of genetically transmitted anemias associated with reduced production (or absence) of β or α hemoglobin chains. Heterozygotes for the allele causing thalassemia have a selective advantage in malarial areas.

threshold model: A model proposing that there is a variation of individual liability for a certain trait; the trait appears only when this liability exceeds a threshold value.

tolerance (immunological): Antibodies are not formed during the fetal period of development, and the postnatal individual does not form antibodies against those antigens with which there has been contact during the fetal period. This tolerance for antigens encountered before birth ensures that individuals do not normally make antibodies against their own proteins.

transduction: The transfer of genetic information (DNA) from one bacterial strain to another, mediated by a phage that kills the DNA donor and carries some of its DNA to a recipient cell, which is not killed by the phage.

transfer RNA (tRNA): RNA with a partially double-stranded structure which functions as an agent of transfer of information between mRNA and the protein being synthesized. One end of the molecule is a triplet (anticodon) complementary to the mRNA triplet (codon). The other end of the tRNA holds the amino acid that is specified by the codon.

transformation: The incorporation into the chromosome of a recipient bacterial cell of a chromosome fragment (usually in the form of purified DNA) from another bacterium.

transformed cells: Cells that have become capable of sustained proliferation *in vitro.*

transcription: The formation of an mRNA strand complementary to one strand of DNA.

transient polymorphism: A polymorphism at a particular locus that exists because one allele is in the process of replacing another through selection or random genetic drift.

translation: The generation of a polypeptide chain on the basis of the information contained in an mRNA chain.

translocation: A chromosomal abnormality in which a chromosome (or portion thereof) becomes attached to another chromosome. A balanced translocation exists when an individual (or gamete) with a translocation carries the same number of copies of each locus as exist in the normal diploid (or haploid) genome.

triplet: A sequence of three nucleotides in a polynucleotide. Each triplet (except for nonsense or chain-end triplets) codes for a particular amino acid.

triploid: Having three times the number of chromosomes in the haploid complement.

trisomy: The presence of three chromosomes of one type in an individual. The most common in humans is trisomy 21 (Down's syndrome).

Turner's syndrome: A syndrome caused by monosomy for the X chromosome in the absence of a Y chromosome; the sex-chromosome complement is symbolized as XO. Affected individuals are phenotypically female, but in most cases have underdeveloped gonads. Other physical and behavioral abnormalities may be present.

unequal crossing-over (illegitimate crossing-over): A situation that results when breaks leading to crossing-over occur in nonhomologous regions of the two chromosomes that exchange.

variable (V) region: The region of light and heavy immunoglobulin chains that shows extreme variation from protein to protein; antibody specificity lies in the V region.

variance: A statistical measure of the variation of individual values about the mean, given in terms of the square of the units used to measure the values. In genetics, V_T refers to the total variance of a character, while V_G and V_E refer to the variances due to the genotype and environment, respectively. V_I refers to the variance resulting from genotype-environment interaction. See also **additive variance, dominance portion of genetic variance.**

wild-type allele: The allele that is most frequent in natural populations; usually indicated by the symbol +. This term cannot be applied for a locus that is recognizably polymorphic.

X inactivation: The genetic inactivation of all X chromosomes in excess of one, taking place on a random basis in each cell at an early stage in embryogenesis.

X-linked gene: A gene located on the X chromosome. A trait determined by such a gene is called an X-linked trait.

Y-linked gene: A gene located on the Y chromosome. A trait determined by such a gene is called a Y-linked trait.

zygote: The primordial cell of a new organism, formed by the fusion of an egg and a sperm.

zygotene: The stage of meiotic prophase in which chromosome pairing begins.

Index